NATURAL PHILOSOPHY

London

MACMILLAN AND CO.

PUBLISHERS TO THE UNIVERSITY OF

Oxford

TREATISE

ON

NATURAL PHILOSOPHY

BY

SIR WILLIAM THOMSON, LL.D., D.C.L., F.R.S.,

PROFESSOR OF NATURAL PHILOSOPHY IN THE UNIVERSITY OF GLASGOW,

AND

PETER GUTHRIE TAIT, M.A.,

PROFESSOR OF NATURAL PHILOSOPHY IN THE UNIVERSITY OF EDINBURGH,

FORMERLY FELLOWS OF ST. PETER'S COLLEGE, CAMBRIDGE.

VOL. I.

Oxford

AT THE CLARENDON PRESS

M.DCCC.LXVII

EDINBURGH: T. CONSTABLE,
PRINTER TO THE QUEEN, AND TO THE UNIVERSITY.

PREFACE.

Les causes primordiales ne nous sont point connues; mais elles sont assujetties à des lois simples et constantes, que l'on peut découvrir par l'observation, et dont l'étude est l'objet de la philosophie naturelle.—FOURIER.

THE term Natural Philosophy was used by NEWTON, and is ..l used in British Universities, to denote the investigation of ..ws in the material world, and the deduction of results not ...tly observed. Observation, classification, and description ..phenomena necessarily precede Natural Philosophy in every ..partment of natural science. The earlier stage is, in some ..anches, commonly called Natural History; and it might with ..ual propriety be so called in all others.

Our object is twofold: to give a tolerably complete account .f what is now known of Natural Philosophy, in language ..apted to the non-mathematical reader; and to furnish, to ..ose who have the privilege which high mathematical acquire-..nts confer, a connected outline of the analytical processes ..y which the greater part of that knowledge has been extended ..to regions as yet unexplored by experiment.

In the present volume, the mathematical development ..rinted in smaller type) necessarily occupies considerably ..re space than the experimental and descriptive portion.

We commence with a chapter on *Motion*, a subject totally ..dependent of the existence of *Matter* and *Force*. In this ..e are naturally led to the consideration of the curvature and ..rtuosity of curves, the curvature of surfaces, and various other ..urely geometrical subjects.

The *Laws of Motion*, the *Law of Gravitation and of Electric ..d Magnetic Attractions*, *Hooke's Law*, and other fundamental principles derived directly from experiment, lead by mathematical processes to interesting and useful results, for the testing of which our most delicate experimental methods are as yet totally insufficient. A large part of our first volume is devoted to these deductions; which, though not immediately proved by experiment, are as certainly true as the elementary laws from which mathematical analysis has evolved them.

The analytical processes which we have employed are, as a rule, such as lead most directly to the results aimed at, and are therefore in great part unsuited to the general reader. A smaller book, embodying much of the non-mathematical portion of the present one, and so much of the developments as can be easily obtained by the help of elementary geometry and algebra, will speedily appear, a portion of it being already in type.

We adopt the suggestion of AMPÈRE, and use the term *Kinematics* for the purely geometrical science of motion in the abstract. Keeping in view the proprieties of language, and following the example of the most logical writers, we employ the term *Dynamics* in its true sense as the science which treats of the action of *force*, whether it maintains relative rest, or produces acceleration of relative motion. The two corresponding divisions of Dynamics are thus conveniently entitled *Statics* and *Kinetics*.

One object which we have constantly kept in view is the grand principle of the *Conservation of Energy*. According to modern experimental results, especially those of JOULE, Energy is as real and as indestructible as Matter. It is satisfactory to find that NEWTON anticipated, so far as the state of experimental science in his time permitted him, this magnificent modern generalization.

We desire it to be remarked that in much of our work, where we may appear to have rashly and needlessly interfered with methods and systems of proof in the present day generally accepted, we take the position of Restorers, and not of Innovators.

In our introductory chapter on Kinematics, the consideration of Harmonic Motion naturally leads us to *Fourier's Theorem*, one of the most important of all analytical results as regards usefulness in physical science. In the Appendices to that chapter we have introduced an extension of *Green's Theorem*, and a short treatise on the remarkable functions known as *Laplace's Coefficients*. There can be but one opinion as to the beauty and utility of this analysis of Laplace; but the manner in which it has been hitherto presented has seemed repulsive to the ablest mathematicians, and difficult to ordinary mathematical students. In the simplified and symmetrical form in which we give it, it will be found quite within the reach of readers moderately familiar with modern mathematical methods.

In the second chapter we give NEWTON'S Laws of Motion in his own words, and with some of his own commentaries—every attempt that has yet been made to supersede them having ended in utter failure. Perhaps nothing so simple, and at the same time so comprehensive, has ever been given as the foundation of a system in any of the sciences. The dynamical use of the *Generalized Cöordinates* of LAGRANGE, and the *Varying Action* of HAMILTON, with kindred matter, complete the chapter.

The third chapter, "Experience," briefly treats of Observation and Experiment as the basis of Natural Philosophy.

The fourth chapter deals with the fundamental Units, and the chief Instruments used for the measurement of Time, Space, and Force.

Thus closes the First Division of the work, which is strictly preliminary.

The Second Division is devoted to Abstract Dynamics (commonly of late years, but not well, called Mechanics). Its object is briefly explained in the introductory (fifth) chapter, and the rest of the present volume is devoted to Statics.

Chapter VI., after a short notice of the Statics of a Particle, enters into considerable detail on the important subject of Attraction. In Chapter VII. the Statics of Solids and Fluids is treated with special detail in various important branches; such as the Deformation of Elastic Solids, the Statical Theory of the Tides, and the Figure and Rigidity of the Earth.

In the next volume, Division II. will be completed by chapters on the Kinetics of a Particle, and the Kinetics of Solids and Fluids. The Vibrations of Solids, and Wave-motion in general, will be fully treated. This volume will probably also contain Division III., which is to deal with Properties of Matter.

We believe that the mathematical reader will especially profit by a perusal of the large type portion of this volume; as he will thus be forced to think out for himself what he has been too often accustomed to reach by a mere mechanical application of analysis. Nothing can be more fatal to progress than a too confident reliance on mathematical symbols; for the student is only too apt to take the easier course, and consider the *formula* and not the *fact* as the physical reality.

A great deal of apparently purposeless matter has been introduced into the present volume, but it will be found to have a direct bearing on portions of the remaining three. The necessity of thus anticipating the wants of future volumes has been one of the main reasons for the delay in the publication of the present instalment, the printing having gone on at irregular intervals since November 1862.

The present volume has been printed by T. CONSTABLE, Esq., Printer to the Queen, and to the University of Edinburgh; and, as a specimen of mathematical printing, can scarcely be surpassed. The volume was not far from completion when we were informed that the Delegates of the Clarendon Press were desirous of publishing the work as one of their educational series. This gave us much pleasure, as it appeared likely to assist us in one of our main objects, the introduction into University study and examinations of something like a complete course of Natural Philosophy. The three remaining volumes will, of course, be printed at Oxford.

Those illustrations in which accuracy was considered essential have been photographed on the wood-block by E. W. DALLAS, Esq., F.R.S.E., from large drawings carefully executed by ourselves, or by Mr. D. MACFARLANE, Assistant to the Professor of Natural Philosophy in the University of Glasgow; and all have been engraved by Mr. J. ADAM with a skill which leaves nothing to be desired.

W. THOMSON.
P. G. TAIT.

July 1867.

ERRATA.

[The following are all we have noticed, but there can hardly fail to be a good many more. We shall be much obliged by being informed of any that may happen to be detected.]

PAGE

41, line 7, *after* tide, *insert* in the equilibrium theory (§ 811).
" line 8, *for* at open coast stations, *read* of this theory.
115, line 5 from foot, *for* Chap. II., *read* Vol. II.
131, upper marginal, *for* Freedon, *read* Freedom.
143, line 7 from foot, *for* equations, *read* equation.
" line 2 from foot, *for* degrees, *read* orders.
150, equation (40), in denominators, *after* $(s+1)$ and $(s+2)$, *insert* .1 and .1.2.
" line 7 from foot, *after* $\frac{1}{r}$, *insert* , if $i - s$ is odd.
152, line 5 from foot, *for* valuation, *read* evaluation.
159, end of line 2, *for* $d\eta'$, *read* $d\xi'$; and end of line 11, *for* r, *read* r'.
" equation (60), same correction as for page 150, equation (40).
" lines 10 and 11 from foot, *for* previous notation, *read* notation of (40).
" equation (61), in numerator and denominator of first factor, *for* $(i - \frac{1}{2})$ and i, *read* $(s - \frac{1}{2})$ and s.
160, equation (65), *before* $2s + i - s$, *insert* (.
212, line 3 from foot, *after* momentum, *insert* as.
" last line, *transpose* comma from motion *to* than.

PAGE

213, lines 10 and 25, *for* gravity, *read* inertia.

228, line 1, *for* (c), *read* (e).

236, equation (15), *for* $\zeta d\psi$, *read* $\zeta d\theta$.

" " (16), *for* $\zeta\dot{\psi}$, *read* $\zeta\dot{\theta}$.

237, in first member of equation (22), *delete* vincula.

238, line 23, *for* $\frac{d\dot{x}}{dx_1}$, *read* $\frac{d\dot{x}_1}{dx_1}$; and *for* $\frac{d\dot{x}}{dx_1}$, *read* $\frac{d\dot{x}_1}{dx_1}$.

" last line, *for* $\ddot{x}_1 dt$, *read* $m_1\ddot{x}_1 dt$.

239, line after (26), *for* ψ and ϕ, *read* $\dot{\psi}$ and $\dot{\phi}$.

240, in equation (31), *for* dT, *read* ∂T.

243, line 3 from foot, *after* expression, *insert* , as distinguished from those obtainable by differentiation of (34), which are now denoted simply by $\frac{dA}{d\psi}$, $\frac{dA}{d\phi}$, etc.

286, second line of § 352, *for* rectineal, *read* rectilineal.

304, 7th line of small type, *for* a, *read* the.

323, line 12, *after* 720 B.C., *insert* : but (§ 830) an error has been found in this calculation, and the corrected result renders it probable that the time of the earth's rotation is longer by $\frac{1}{2,700,000}$ now than at that date.

335, line 7 from foot, *for* $Q=\frac{Wa^2}{p}$, etc., *read* $Q=\frac{Wa^2}{l}$, etc.

357, line 3, *for* $+\frac{D^2-a^2}{ax}$, *read* $-\frac{D^2-a^2}{ax}$.

372, line 7 of case 1, *for* case, *read* cone.

377, line 7, *for* to, *read* to be.

401, line 8 from foot, *for* gravity, *read* inertia.

443, line 6, *for* axes, *read* axis.

582, line 2 from foot, *for* vanishes, *read* vanish.

CONTENTS.

DIVISION I.—PRELIMINARY NOTIONS.

CHAPTER I.—KINEMATICS.

CHAPTER III.—EXPERIENCE.

CHAPTER IV.—MEASURES AND INSTRUMENTS.

DIVISION II.—ABSTRACT DYNAMICS.

CHAPTER V.—INTRODUCTORY.

CHAPTER VI.—STATICS OF A PARTICLE—ATTRACTION.

DIVISION I.

PRELIMINARY NOTIONS.

CHAPTER I.—KINEMATICS.

1. There are many properties of motion, displacement, and deformation, which may be considered altogether independently of such physical ideas as force, mass, elasticity, temperature, magnetism, electricity. The preliminary consideration of such properties in the abstract is of very great use for Natural Philosophy, and we devote to it, accordingly, the whole of this our first chapter; which will form, as it were, the Geometry of our subject, embracing what can be observed or concluded with regard to actual motions, as long as the *cause* is not sought.

2. In this category we shall first take up the free motion of a point, then the motion of a point attached to an inextensible cord, then the motions and displacements of rigid systems—and finally, the deformations of surfaces and of solid or fluid masses. Incidentally, we shall be led to introduce a good deal of elementary geometrical matter connected with the curvature of lines and surfaces.

3. When a point moves from one position to another it must evidently describe a *continuous* line, which may be curved or straight, or even made up of portions of curved and straight lines meeting each other at any angles. If the motion be that of a *material particle*, however, there can be no such abrupt changes of direction, since (as we shall afterwards see) this would imply the action of an *infinite* force. It is useful to consider at the outset various theorems connected with the geometrical notion of the path, described by a moving point, Motion of a point.

Motion of a point.

and these we shall now take up, deferring the consideration of Velocity to a future section, as being more closely connected with physical ideas.

4. The *direction* of motion of a moving point is at each instant the tangent drawn to its path, if the path be a curve, or the path itself if a straight line.

Curvature of a plane curve.

5. If the path be not straight the direction of motion changes from point to point, and the *rate* of this change, per unit of length of the curve, is called the *curvature*. To exemplify this, suppose two tangents drawn to a circle, and radii to the points of contact. The angle between the tangents is the change of direction required, and the rate of change is evidently to be measured by the relation between this angle and the length of the circular arc. Now, if θ be the angle, s the arc, and ρ the radius, we see at once that (as the angle between the radii is equal to the angle between the tangents)

$$\rho\theta = s,$$

and therefore $\frac{\theta}{s} = \frac{1}{\rho}$ is the measure of the curvature. Hence the curvature of a circle is inversely as its radius, and is measured, in terms of the proper unit of curvature, simply by the reciprocal of the radius.

6. Any small portion of a curve may be approximately taken as a circular arc, the approximation being closer and closer to the truth, as the assumed arc is smaller. The curvature is then the reciprocal of the radius of this circle.

If $\delta\theta$ be the angle between two tangents at points of a curve distant by an arc δs, the definition of curvature gives us at once its measure, the limit of $\frac{\delta\theta}{\delta s}$ when δs is diminished without limit; or, according to the notation of the differential calculus, $\frac{d\theta}{ds}$. But we have

$$\tan\theta = \frac{dy}{dx},$$

if, the curve being a plane curve, we refer it to two rectangular axes OX, OY, according to the Cartesian method, and if θ denote the inclination of its tangent, at any point x, y, to OX. Hence

$$\theta = \tan^{-1}\frac{dy}{dx};$$

and, by differentiation with reference to any independent variable, t, we have Curvature of a plane curve.

$$d\theta = \frac{d\left(\frac{dy}{dx}\right)}{1+\left(\frac{dy}{dx}\right)^2} = \frac{dx\, d^2y - dy\, d^2x}{dx^2+dy^2}.$$

Also,
$$ds = (dx^2 + dy^2)^{\frac{1}{2}}.$$

Hence, if ρ denote the radius of curvature, so that $\frac{1}{\rho} = \frac{d\theta}{ds}$,

we conclude, $\frac{1}{\rho} = \frac{dx d^2y - dy d^2x}{(dx^2+dy^2)^{\frac{3}{2}}}$.

Although it is generally convenient, in kinematical and kinetic formulæ, to regard time as the independent variable, and all the changing geometrical elements as functions of it, there are cases in which it is useful to regard the length of the arc or path described by a point as the independent variable. On this supposition we have

$$d\,(ds^2) = d\,(dx^2 + dy^2) = 0.$$

Hence denoting by the suffix to the letter d, the independent variable understood in the differentiation, we have

$$\frac{d_s^2 y}{dx} = -\frac{d_s^2 x}{dy} = \frac{\{(d_s^2 y)^2 + (d_s^2 x)^2\}^{\frac{1}{2}}}{(dx^2 + dy^2)^{\frac{1}{2}}},$$

or
$$\frac{dx}{d_s^2 y} \cdot \frac{\{(d_s^2 x)^2 + (d_s^2 y)^2\}^{\frac{1}{2}}}{(dy^2+dx^2)^{\frac{1}{2}}} = \frac{-dy}{d_s^2 x} \cdot \frac{\{(d_s^2 x)^2 + (d_s^2 y)^2\}^{\frac{1}{2}}}{(dy^2+dx^2)^{\frac{1}{2}}} = 1$$

Multiplying the first and second terms of the numerator of the expression for $\frac{1}{\rho}$ just given, by the first and second terms of this equality, we have

$$\frac{1}{\rho} = \frac{\{(d_s^2 y)^2 + (d_s^2 x)^2\}^{\frac{1}{2}}}{ds^2};$$

or, according to the usual short, although not quite complete, notation,

$$\frac{1}{\rho} = \left\{\left(\frac{d^2y}{ds^2}\right)^2 + \left(\frac{d^2x}{ds^2}\right)^2\right\}^{\frac{1}{2}}.$$

7. If all points of the curve lie in one plane, it is called a *plane curve*, and in the same way we speak of a *plane* polygon or broken line. If various points of the line do not lie in one plane, we have in one case what is called a *curve of double curvature*, in the other a *gauche polygon*. The term "curve of double curvature" is a very bad one, and, though in very general use, is, we hope, not ineradicable. Tortuous curve.

Tortuous curve.

The fact is, that there are not two curvatures, but only a curvature (as above defined), of which the plane is continuously changing, or twisting, round the tangent line; thus exhibiting a torsion. The course of such a curve is, in common language, well called "tortuous;" and the measure of the corresponding property is conveniently called *Tortuosity*.

8. The nature of this will be best understood by considering the curve as a polygon whose sides are indefinitely small. Any two consecutive sides, of course, lie in a plane—and in that plane the curvature is measured as above, but in a curve which is not plane the third side of the polygon will not be in the same plane with the first two, and, therefore, the new plane in which the curvature is to be measured is different from the old one. The plane of the curvature on each side of any point of a tortuous curve is sometimes called the *Osculating Plane* of the curve at that point. As two successive positions of it contain the second side of the polygon above mentioned, it is evident that the osculating plane passes from one position to the next by revolving about the tangent to the curve.

Curvature and tortuosity.

9. Thus, as we proceed along such a curve, the curvature in general varies; and, at the same time, the plane in which the curvature lies is turning about the tangent to the curve. The rate of torsion, or the tortuosity, is therefore to be measured by the rate at which the osculating plane turns about the tangent, per unit length of the curve.

To express the radius of curvature, the direction cosines of the osculating plane, and the tortuosity, of a curve not in one plane, in terms of Cartesian triple co-ordinates, let, as before, $\delta\theta$ be the angle between the tangents at two points at a distance δs from one another along the curve, and let $\delta\phi$ be the angle between the osculating planes at these points. Thus, denoting by ρ the radius of curvature, and τ the tortuosity, we have

$$\frac{1}{\rho} = \frac{d\theta}{ds},$$

$$\tau = \frac{d\phi}{ds},$$

according to the regular notation for the limiting values of $\frac{\delta\theta}{\delta s}$, when δs is diminished without limit. Now, if l, m, n, l', m', n' be the direction cosines of two lines, $\delta\theta$ the angle between them, and λ, μ, ν, the direction cosines of their plane, we have by elementary formulæ, Curvature and tortuosity.

$$\sin\delta\theta=\{(mn'-nm')^2+(nl'-ln')^2+(lm'-ml')^2\}^{\frac{1}{2}}$$

$$\lambda=\frac{mn'-nm'}{\sin\delta\theta}, \text{ etc.}$$

Let $l'=l+\alpha\delta s, m'=m+\beta\delta s, n'=n+\gamma\delta s$, where δs is any quantity whatever, and we have $mn'-nm'=(m\gamma-n\beta)\delta s$, etc. Hence

$$\sin\delta\theta=\{(m\gamma-n\beta)^2+(n\alpha-l\gamma)^2+(l\beta-m\alpha)^2\}^{\frac{1}{2}}\delta s$$

and

$$\lambda=\frac{m\gamma-n\beta}{\{(m\gamma-n\beta)^2+(n\alpha-l\gamma)^2+(l\beta-m\alpha)^2\}^{\frac{1}{2}}}, \text{ etc.}$$

In the limit, we have $\frac{\sin\delta\theta}{\delta s}=\frac{\delta\theta}{\delta s}$, where δs is indefinitely small.

The formula may be applied first, to give the value of $\frac{1}{\rho}$, by taking

$$l=\frac{dx}{ds},\quad m=\frac{dy}{ds},\quad n=\frac{dz}{ds},$$

$$\alpha=\frac{d_t\left(\frac{dx}{ds}\right)}{ds},\quad \beta=\frac{d_t\left(\frac{dy}{ds}\right)}{ds},\quad \gamma=\frac{d_t\left(\frac{dz}{ds}\right)}{ds},$$

the suffix denoting, as before, the independent variable, in this case, t. Thus we find,

$$\frac{1}{\rho}=\frac{\left\{\left[dyd\left(\frac{dz}{ds}\right)-dzd\left(\frac{dy}{ds}\right)\right]^2+\left[dzd\left(\frac{dx}{ds}\right)-dxd\left(\frac{dz}{ds}\right)\right]^2+\left[dxd\left(\frac{dy}{ds}\right)-dyd\left(\frac{dx}{ds}\right)\right]^2\right\}^{\frac{1}{2}}}{ds^2},$$

and then we have

$$\lambda=\frac{dyd\left(\frac{dz}{ds}\right)-dzd\left(\frac{dy}{ds}\right)}{\rho^{-1}ds^2}, \text{ etc.}$$

We may simplify these expressions by actually differentiating the fractions $\frac{dx}{ds}$, etc. Thus we have

$$d\left(\frac{dx}{ds}\right)=\frac{dsd^2x-dxd^2s}{ds^2}, \text{ etc.},$$

and thence

$$dyd\left(\frac{dz}{ds}\right)-dzd\left(\frac{dy}{ds}\right)=\frac{dyd^2z-dzd^2y}{ds}.$$

Curvature and tortuosity.

We have therefore

$$\frac{1}{\rho}=\frac{\{(dyd^2z-dzd^2y)^2+(dzd^2x-dxd^2z)^2+(dxd^2y-dyd^2x)^2\}^{\frac{1}{2}}}{ds^3},$$

$$\lambda=\frac{dyd^2z-dzd^2y}{\rho^{-1}ds^3}, \text{ etc.}$$

The expression for $\frac{1}{\rho}$ may be modified algebraically to the following:—

$$\frac{1}{\rho}=\frac{\{ds^2[(d^2x)^2+(d^2y)^2+(d^2z)^2]-(dxd^2x+dyd^2y+dzd^2z)^2\}^{\frac{1}{2}}}{ds^3}$$

$$\text{or}\quad \frac{1}{\rho}=\frac{\{(d^2x)^2+(d^2y)^2+(d^2z)^2-(d^2s)^2\}^{\frac{1}{2}}}{ds^2},$$

which is useful as showing what the expression becomes when s is chosen as independent variable.

To find the tortuosity, $\frac{d\phi}{ds}$, we have only to apply the general equation above, with λ, μ, ν substituted for l, m, n, and $\frac{d_t\lambda}{ds}$, $\frac{d_t\mu}{ds}$, $\frac{d_t\nu}{ds}$ for α, β, γ. Thus we have

$$\tau=\{(\mu\frac{d_t\nu}{ds}-\nu\frac{d_t\mu}{ds})^2+(\nu\frac{d_t\lambda}{ds}-\lambda\frac{d_t\nu}{ds})^2+(\lambda\frac{d_t\mu}{ds}-\mu\frac{d_t\lambda}{ds})^2\}^{\frac{1}{2}},$$

where λ, μ, ν denote the direction cosines of the osculating plane, given by the preceding formulæ.

Integral curvature.

10. The *integral curvature*, or *whole change of direction* of an arc of a plane curve, is the angle through which the tangent has turned as we pass from one extremity to the other. The *average curvature* of any portion is its whole curvature divided by its length. Suppose a line, drawn from a fixed point, to move so as always to be parallel to the direction of motion of a point describing the curve, the angle through which this turns during the motion of the point exhibits what we have thus defined as the integral curvature. In estimating this, we must of course take the enlarged modern meaning of an angle, including angles greater than two right angles, and also negative angles. Thus the integral curvature of any closed curve, whether everywhere concave to the interior or not, is four right angles, provided it does not cut itself. That of a Lemniscate, or figure of 8, is *zero*. That of the Epicycloid  is eight right angles; and so on. Integral curvature.

11. The definition in last section may evidently be extended to a plane polygon, and the integral change of direction, or the angle between the first and last sides, is then the sum of its exterior angles, all the sides being produced each in the direction in which the moving point describes it while passing round the figure. This is true whether the polygon be closed or not. If closed, then, as long as it is not crossed, this sum is four right angles,—an extension of the result in Euclid, where all *re-entrant* polygons are excluded. In the case of the star-shaped figure , it is ten right angles, wanting the sum of the five acute angles of the figure; and so on.

12. The *integral curvature* and the *average curvature* of a curve which is not plane, may be defined as follows:—Let successive lines be drawn from a fixed point, parallel to tangents at successive points of the curve. These lines will form a conical surface. Suppose this to be cut by a sphere of unit radius having its centre at the fixed point. The *length* of the curve of intersection measures the *integral curvature* of the given curve. The *average curvature* is, as in the case of a plane curve, the integral curvature divided by the length of the curve.

13. Two consecutive tangents lie in the osculating plane. This plane is therefore parallel to the tangent plane to the cone described in the preceding section. Thus the tortuosity may be measured by the help of the same spherical curve we have just used for defining integral curvature. We cannot as yet complete the explanation, as it depends on the theory of curves on surfaces, which will be treated afterwards. But we shall see that if a plane roll on the sphere, so as always to touch it along the curve in question, and so that the instantaneous axis may always be at right angles to the curve, and tangential to the sphere, the integral curvature of the curve of contact or trace of the rolling on the plane, is a proper measure of the *whole torsion*, or integral of tortuosity. Farther, we shall see that the curvature of this plane curve at any point, or, which is

Curvature and tortuosity. the same, the projection of the curvature of the spherical curve on a tangent plane of the spherical surface, is equal to the tortuosity divided by the curvature of the given curve.

Let $\frac{1}{\rho}$ be the curvature and τ the tortuosity of the given curve, and ds an element of its length. Then $\int \frac{ds}{\rho}$ and $\int \tau ds$, each integral extended over any stated length, l, of the curve, are respectively the integral curvature and the integral tortuosity. The mean curvature and the mean tortuosity are respectively

$$\frac{1}{l}\int \frac{ds}{\rho} \text{ and } \frac{1}{l}\int \tau ds.$$

Infinite tortuosity will be easily understood, by considering a helix, of inclination α, described on a right circular cylinder of radius r. The curvature in a circular section being $\frac{1}{r}$, that of the helix is, of course, $\frac{\cos^2\alpha}{r}$. The tortuosity is $\frac{\sin\alpha\cos\alpha}{r}$, or $\tan\alpha \times$ curvature. Hence, if $\alpha = \frac{\pi}{4}$ the curvature and tortuosity are equal.

Let the curvature be denoted by $\frac{1}{\rho}$, so that $\cos^2\alpha = \frac{r}{\rho}$. Let ρ remain finite, and let r diminish without limit. The step of the helix being $2\pi r \tan\alpha = 2\pi\sqrt{\rho r}\,(1-\frac{r}{\rho})^{\frac{1}{2}}$, is, in the limit, $2\pi\sqrt{\rho r}$, which is infinitely small. Thus the motion of a point in the curve, though infinitely nearly in a straight line (the path being always at the infinitely small distance r from the fixed straight line, the axis of the cylinder), will have finite curvature $\frac{1}{\rho}$. The tortuosity, being $\frac{1}{\rho}\tan\alpha$ or $\frac{1}{\sqrt{\rho r}}(1-\frac{r}{\rho})^{\frac{1}{2}}$, will in the limit be a mean proportional between the curvature of the circular section of the cylinder and the finite curvature of the curve.

The acceleration (or force) required to produce such a motion of a point (or material particle) will be afterwards investigated.

Flexible line. **14.** A chain, cord, or fine wire, or a fine fibre, filament, or hair, may suggest what is not to be found among natural or artificial productions, a perfectly *flexible and inextensible line.* The elementary kinematics of this subject require no investigation. The mathematical condition to be expressed in any case of it is simply that the distance measured along the line from any one point to any other, remains constant, however the line be bent. Flexible line.

15. The use of a cord in mechanism presents us with many practical applications of this theory, which are in general extremely simple; although curious, and not always very easy, geometrical problems occur in connexion with it. We shall say nothing here about such cases as knots, knitting, weaving, etc., as being excessively difficult in their general development, and too simple in the ordinary cases to require explanation.

16. In the mechanical tracing of curves, a flexible and inextensible cord is often supposed. Thus, in drawing an ellipse, the focal property of the curve shows us that by fixing the ends of such a cord to the foci and keeping it stretched by a pencil, the latter will trace the curve.

By a ruler moveable about one focus, and a string attached to a point in the ruler and to the other focus, the hyperbola may be described by the help of its analogous focal property; and so on.

17. But the consideration of evolutes is of some importance in Natural Philosophy, especially in certain optical questions, and we shall therefore devote a section or two to this application of kinematics. Evolute.

Def. If a flexible and inextensible string be fixed at one point of a plane curve, and stretched along the curve, and be then unwound in the plane of the curve, its extremity will describe an *Involute* of the curve. The original curve, considered with reference to the other, is called the *Evolute.*

18. It will be observed that we speak of *an* involute, and of *the* evolute, of a curve. In fact, as will be easily seen, a curve can have but one evolute, but it has an infinite number of involutes. For all that we have to do to vary an involute, is to change the point of the curve from which the tracing point starts, or consider the involute described by each point of the string, and these will, in general, be different curves. But the following section shows that there is but one evolute.

Evolute. **19.** Let AB be any curve, PQ a portion of an involute, pP, qQ positions of the free part of the string. It will be seen at once that these must be tangents to the arc AB at p and q. Also (see a subsequent section), the string at any stage, as pP, ultimately revolves about p. Hence pP is *normal* to the curve PQ. And thus the evolute of PQ is a definite curve, viz., the envelope of the normals drawn at every point of PQ, or, which is the same thing, the locus of the centres of curvature of the curve PQ. And we may merely mention, as an obvious result of the mode of tracing, that the arc qp is equal to the difference of qQ and pP, or that the arc pA is equal to pP.

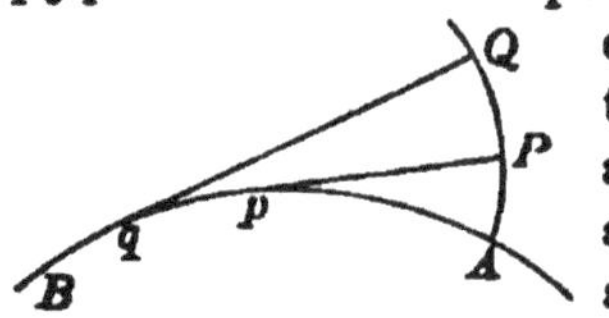

Velocity. **20.** The rate of motion of a point is called its *Velocity*. It will evidently be greater or less as the space passed over in a given time is greater or less: and it may be *uniform*, *i.e.*, the same at every instant; or it may be *variable*.

Uniform velocity is measured by the space passed over in unit of time, and is, in general, expressed in feet per second; if very great, as in the case of light, it may be measured in miles per second. It is to be observed, that time is here used in the abstract sense of a uniformly-increasing quantity—what in the differential calculus is called an independent variable. Its physical definition is given in the next chapter.

21. Thus, if v be the velocity of a point moving uniformly, it describes a space of v feet each second, and therefore vt feet in t seconds, t being any number whatever. Putting s for the space described, we have

$$s = vt.$$

The unit of velocity is thus that of a point which describes unit of space in unit of time.

22. It is well to observe here, that since, by our formula, we have generally

$$v = \frac{s}{t};$$

and since nothing has been said as to the magnitudes of s and t, we may take these as small as we choose. Thus *we get the same result whether we derive v from the space described in a million seconds, or from that described in a millionth of a second.* This idea is very useful, as it will give confidence in the result f a subsequent section, where a variable velocity has to be measured, and where we find ourselves obliged to approximate to its value by considering the space described in an interval so short, that during its lapse the velocity does not sensibly alter in value. Velocity.

23. When the point does not move uniformly, the velocity is variable, or different at different successive instants; but we define the *average* velocity during any time as the space described in that time, divided by the time. Or again, we define the exact velocity at any instant as the space which the point would have described in one second, if for such a period it kept its value unchanged. That there is at every instant a definite value of the velocity of any moving body, is evident to all, and is matter of everyday conversation. Thus, a railway train, after starting, gradually increases its speed, and we see no absurdity in saying that at a particular instant it moves at the rate of ten or of fifty miles an hour,—although, in the course of an hour, it may not have moved a mile altogether. Indeed, we may suppose that, at any instant during the motion, the steam is so adjusted as to keep the train running for some time at a perfectly uniform velocity. This would be the velocity which the train had at the instant in question. Without supposing any such definite adjustment of the driving power to be made, we can evidently obtain an approximation to this instantaneous velocity by considering the motion for so short a time, that during it the actual variation of speed may be small enough to be neglected.

24. In fact, if v be the velocity at either beginning or end, or at any instant of the interval, and s the space actually described in time t, the equation $v = \frac{s}{t}$ is more and more nearly true, as the velocity is more nearly uniform during the interval t; so that if we take t as $\frac{1}{10}$th of a second, and find s_1 to be the corresponding space described, $\frac{s_1}{\frac{1}{10}}$ or $10s_1$, is an approximation to the velocity at any instant of t; if we take, for t, $\frac{1}{100}$th of a second, s_2 being the corresponding space, $\frac{s_2}{\frac{1}{100}}$ or $100s_2$ is a closer approximation; and so on. Hence the set of values—

Velocity. Space described in one second,

Ten times the space described in the first tenth of a second.

A hundred „ „ „ hundredth „

and so on, give nearer and nearer approximations to the velocity at the beginning of the first second. The whole foundation of the differential calculus is, in fact, contained in this simple question, "What is the rate at which this space described increases?" *i.e.*, What is the velocity of the moving point?

Let a point which has described a space s in time t proceed to describe an additional space δs in time δt, and let v_1 be the greatest, and v_2 the least, velocity which it has during the interval δt. Then, evidently,

$$\delta s < v_1 \delta t, \quad \delta s > v_2 \delta t,$$

$$i.e., \quad \frac{\delta s}{\delta t} < v_1, \quad \frac{\delta s}{\delta t} > v_2.$$

But as δt diminishes, the values of v_1 and v_2 become more and more nearly equal, and in the limit, each is equal to the velocity at time t. Hence

$$v = \frac{ds}{dt}.$$

Resolution of velocity. **25.** The above definition of velocity is equally applicable whether the point move in a straight or curved line; but, since in the latter case the direction of motion continually changes the mere amount of the velocity is not sufficient completely to describe the motion, and we must have in every such case additional data to remove the uncertainty.

In such cases as this (we may explain once for all) the method commonly employed, whether we deal with velocities, or as we shall do farther on with accelerations and forces, consists mainly in studying, not the velocity, acceleration, or force, *directly*, but its resolved parts parallel to any three assumed directions at right angles to each other. Thus, for a train moving up an incline in a NE direction, we may have given the whole velocity and the steepness of the incline, or we may express the same ideas thus—the train is moving simultaneously northward, eastward, and upward—and the motion as to amount and direction will be completely known if we know separately the northward, eastward, and upward velocities—these being called the *components* of the whole velocity in the three mutually perpendicular directions N, E, and up.

In general the velocity of a point at x, y, z, is (as we have seen) $\frac{ds}{dt}$, or, which is the same, $\left\{\left(\frac{dx}{dt}\right)^2 + \left(\frac{dy}{dt}\right)^2 + \left(\frac{dz}{dt}\right)^2\right\}^{\frac{1}{2}}$. Resolution of velocity.

Now $\frac{dx}{dt} = v_x$ (suppose) is the rate at which x increases, or the velocity parallel to the axis of x, and so of the other two. Hence if α, β, γ be the angles which the direction of motion makes with the axes,

$$\cos\alpha = \frac{dx}{ds} = \frac{\frac{dx}{dt}}{\frac{ds}{dt}} = \frac{v_x}{v}.$$

Hence $v_x = v\cos\alpha$, and therefore

26. A velocity in any direction may be resolved in, and perpendicular to, any other direction. The first component is formed by multiplying the velocity by the cosine of the angle between the two directions—the second by using as factor the sine of the same angle. Or, it may be resolved into components in any three rectangular directions, each component being formed by multiplying the whole velocity by the cosine of the angle between its direction and that of the component.

It is useful to remark that if the axes of x, y, z are not rectangular, $\frac{dx}{dt}, \frac{dy}{dt}, \frac{dz}{dt}$ will still be the velocities parallel to the axes, but we shall no longer have

$$\left(\frac{ds}{dt}\right)^2 = \left(\frac{dx}{dt}\right)^2 + \left(\frac{dy}{dt}\right)^2 + \left(\frac{dz}{dt}\right)^2.$$

We leave as an exercise for the student the determination of the correct expression for the whole velocity in terms of its components.

If we resolve the velocity along a line whose inclinations to the axes are λ, μ, ν, and which makes an angle θ with the direction of motion, we find the two expressions below (which must of course be equal) according as we resolve v directly or by its components, v_x, v_y, v_z,

$$v\cos\theta = v_x\cos\lambda + v_y\cos\mu + v_z\cos\nu.$$

Substitute in this equation the values of v_x, v_y, v_z already given, § 25, and we have the well-known geometrical theorem for the angle between two straight lines which make given angles with the axes,

$$\cos\theta = \cos\alpha\cos\lambda + \cos\beta\cos\mu + \cos\gamma\cos\nu.$$

From the above expression we see at once that

[illegible]

[illegible]
[illegible]
[illegible]
and [illegible]
[illegible] the [illegible]
the [illegible]
[illegible]
[illegible]

27. [illegible]

...ocity of a point is said to be accelerated or re- Acceleration.
...g as it increases or diminishes, but the word
...nerally used in either sense, on the understand-
...y regard its quantity as either positive or nega-
...tion of velocity may of course be either uniform
It is said to be uniform when the point receives
...nts of velocity in equal times, and is then mea-
...ctual increase of velocity per unit of time. If we
... unit of acceleration that which adds a unit of
unit of time to the velocity of a point, an accelera-
...d by a will add a units of velocity in unit of
therefore, at units of velocity in t units of time.
... be the change in the velocity during the interval t,

$$V = at.$$

...leration is variable when the point's velocity does
... equal increments in successive equal periods of
...s then measured by the increment of velocity, which
...e been generated in a unit of time had the acceleration
... throughout that unit the same as at its commence-
The *average* acceleration during any time is the whole
gained during that time, divided by the time.

...et v be the velocity at time t, δv its change in the interval δt,
...nd a_2 the greatest and least values of the acceleration during
... interval δt. Then, evidently,

$$\delta v < a_1 \delta t,\ \delta v > a_2 \delta t,$$

$$\text{or}\quad \frac{\delta v}{\delta t} < a_1,\ \frac{\delta v}{\delta t} > a_2.$$

...s δt is taken smaller and smaller, the values of a_1 and a_2 approxi-
...ate infinitely to each other, and to that of a the required ac-
...celeration at time t. Hence

$$\frac{dv}{dt} = a.$$

It is useful to observe that we may also write (by changing the independent variable)

$$a = \frac{dv}{ds}\frac{ds}{dt} = v\frac{dv}{ds}.$$

Since $v = \frac{ds}{dt}$, we have $a = \frac{d^2s}{dt^2}$, and it is evident from similar
...oning that the accelerations parallel to the axes are

$$a_x = \frac{d^2x}{dt^2},\ \text{etc.}$$

Acceleration.

But it is to be carefully observed that $\frac{d^2s}{dt^2}$ is *not* the complete resultant of a_x, a_y, a_z; for we have not in general

$$\left(\frac{d^2s}{dt^2}\right)^2 = \left(\frac{d^2x}{dt^2}\right)^2 + \left(\frac{d^2y}{dt^2}\right)^2 + \left(\frac{d^2z}{dt^2}\right)^2.$$

The direction cosines of the tangent to the path at any point x, y, z are

$$\frac{1}{v}\frac{dx}{dt}, \frac{1}{v}\frac{dy}{dt}, \frac{1}{v}\frac{dz}{dt}.$$

Those of the line of resultant acceleration are

$$\frac{1}{f}\frac{d^2x}{dt^2}, \frac{1}{f}\frac{d^2y}{dt^2}, \frac{1}{f}\frac{d^2z}{dt^2},$$

where, for brevity, we denote by f the resultant acceleration Hence the direction cosines of the plane of these two lines are

$$\frac{dyd^2z-dzd^2y}{\{(dyd^2z-dzd^2y)^2+(dzd^2x-dxd^2z)+(dxd^2y-dyd^2x)\}^{\frac{1}{2}}}, \text{ etc.}$$

These (§ 9) show that this plane is the osculating plane of the curve. Again, if θ denote the angle between those two lines, we have

$$\sin\theta = \frac{\{(dyd^2z-dzd^2y)^2+(dzd^2x-dxd^2z)^2+(dxd^2y-dyd^2x)^2\}^{\frac{1}{2}}}{vfdt^3},$$

or, according to the expression for the curvature (§ 9),

$$\sin\theta = \frac{ds^3}{\rho vfdt^3} = \frac{v^2}{f\rho}.$$

Hence $$f\sin\theta = \frac{v^2}{\rho}.$$

Again, $$\cos\theta = \frac{1}{vf}\left(\frac{dx}{dt}\frac{d^2x}{dt^2} + \frac{dy}{dt}\frac{d^2y}{dt^2} + \frac{dz}{dt}\frac{d^2z}{dt^2}\right) = \frac{dsd^2s}{vfdt^3} = \frac{d^2s}{fdt^2}.$$

Hence $$f\cos\theta = \frac{d^2s}{dt^2},$$ and therefore,

Resolution and composition of accelerations.

30. The whole acceleration in any direction is the sum of the components (in that direction) of the accelerations parallel to any three rectangular axes—each component acceleration being found by the same rule as component velocities, that is, by multiplying by the cosine of the angle between the direction of the acceleration and the line along which it is to be resolved.

31. When a point moves in a curve the whole acceleration may be resolved into two parts, one in the direction of the motion and equal to the acceleration of the velocity—the other towards the centre of curvature (perpendicular therefore to the direction of motion), whose magnitude is proportional to the square of the velocity and also to the curvature of the path. The former of these changes the velocity, the other affects only the form of the path, or the direction of motion. Hence if a moving point be subject to an acceleration, constant or not, whose direction is continually perpendicular to the direction of motion, the velocity will not be altered—and the only effect of the acceleration will be to make the point move in a curve whose curvature is proportional to the acceleration at each instant. Resolution and composition of accelerations.

32. In other words, if a point move in a curve, whether with a uniform or a varying velocity, its change of direction is to be regarded as constituting an acceleration towards the centre of curvature, equal in amount to the square of the velocity divided by the radius of curvature. The whole acceleration will, in every case, be the resultant, of the acceleration thus measuring change of direction, and the acceleration of actual velocity along the curve.

We may take another mode of resolving acceleration for a plane curve, which is sometimes useful; along, and perpendicular to, the radius-vector. By a method similar to that employed in § 27, we easily find for the component along the radius-vector

$$\frac{d^2r}{dt^2} - r\left(\frac{d\theta}{dt}\right)^2,$$

and for that perpendicular to the radius-vector

$$\frac{1}{r}\frac{d}{dt}\left(r^2\frac{d\theta}{dt}\right).$$

33. If for any case of motion of a point we have given the whole velocity and its direction, or simply the components of the velocity in three rectangular directions, at any *time*, or, as is most commonly the case, for any *position*, the determination of the form of the path described, and other circumstances of the motion, is a question of pure mathematics, and in all cases is capable, if not of an exact solution, at all events of a solution to any degree of approximation that may be desired. Determination of the motion from given velocity or acceleration.

The same is true if the total acceleration and its direction at every instant, or simply its rectangular components, be given,

Determination of the motion from given velocity or acceleration.

provided the velocity and direction of motion, as well as the position, of the point at any one instant, be given.

For we have in the first case

$$\frac{dx}{dt}=v_x=v\cos a,\text{ etc.},$$

three simultaneous equations which can contain only x, y, z, and t, and which therefore suffice when integrated to determine x, y, and z in terms of t. By eliminating t among these equations, we obtain two equations among x, y, and z—each of which represents a surface on which lies the path described, and whose intersection therefore completely determines it.

In the second case we have

$$\frac{d^2x}{dt^2}=a_x,\text{ etc.},$$

to which equations the same remarks apply, except that here each has to be twice integrated.

The arbitrary constants introduced by integration are determined at once if we know the co-ordinates, and the components of the velocity, of the point at a given epoch.

Examples of velocity.

34. From the principles already laid down, a great many interesting results may be deduced, of which we enunciate a few of the most important.

a. If the velocity of a moving point be uniform, and if its direction revolve uniformly in a plane, the path described is a circle.

b. If a point moves in a plane, and if its component velocity parallel to each of two rectangular axes is proportional to its distance from that axis, the path is a conic section whose principal diameters coincide with those axes; and the acceleration is directed to or from the origin at every instant.

c. If the components of the velocity parallel to each axis be equimultiples of the distances from the other axis, the path is a straight line passing through the origin.

d. When the velocity is uniform, but in direction revolving uniformly in a right circular cone, the motion of the point is in a circular helix whose axis is parallel to that of the cone.

(*a*.) Let a be the velocity, and α the angle through which its direction turns in unit of time; then, by properly choosing the axes, we have Examples of velocity.

$$\frac{dx}{dt} = -a \sin \alpha t, \quad \frac{dy}{dt} = a \cos \alpha t,$$

whence
$$(x-A)^2+(y-B)^2 = \frac{a^2}{\alpha^2}.$$

(*b*.)
$$\frac{dx}{dt} = \mu y, \quad \frac{dy}{dt} = \nu x.$$

Hence $\frac{d^2x}{dt^2} = \mu\nu x$, $\frac{d^2y}{dt^2} = \mu\nu y$, and the whole acceleration is towards or from O.

Also $\frac{dy}{dx} = \frac{\nu}{\mu}\frac{x}{y}$, from which $\mu y^2 - \nu x^2 = C$, a conic section referred to its principal axes.

(*c*.)
$$\frac{dx}{dt} = \mu x, \quad \frac{dy}{dt} = \mu y,$$

$$\therefore \frac{dx}{x} = \frac{dy}{y},$$

or
$$y = Cx.$$

35. *a*. When a point moves uniformly in a circle of radius R, with velocity V, the whole acceleration is directed towards the centre, and has the constant value $\frac{V^2}{R}$. See § 31. Examples of acceleration.

b. With uniform acceleration in the direction of motion, a point describes spaces proportional to the squares of the times elapsed since the commencement of the motion.

In this case the space described in any interval is that which would be described in the same time by a point moving uniformly with a velocity equal to that at the middle of the interval. In other words, the average velocity (when the acceleration is uniform) is, during any interval, the arithmetical mean of the initial and final velocities. This is the case of a stone falling vertically.

For if the acceleration be parallel to x, we have

$$\frac{d^2x}{dt^2} = a, \text{ therefore } \frac{dx}{dt} = v = at, \text{ and } x = \tfrac{1}{2}at^2.$$

And we may write the equation (§ 29) $v\frac{dv}{dx} = a$, whence $\frac{v^2}{2} = ax$.

Examples of acceleration.

If at time $t=0$ the velocity was V, these equations become at once

$$v=V+at,\ x=Vt+\tfrac{1}{2}at^2,\ \text{and}\ \frac{v^2}{2}=\frac{V^2}{2}+ax.$$

$$\begin{aligned} \text{And initial velocity} &= V \\ \text{final}\quad ,, \quad &= V+at \\ \text{Arithmetical mean} &= V+\tfrac{1}{2}at \\ &= \frac{x}{t}, \end{aligned}$$

whence the second part of the above statement.

c. When there is uniform acceleration in a constant direction, the path described is a parabola, whose axis is parallel to that direction. This is the case of a projectile moving in vacuo.

For if the axis of y be parallel to the acceleration a, and if the plane of xy be that of motion at any time,

$$\frac{d^2z}{dt^2}=0,\ \frac{dz}{dt}=0, z=0,$$

and therefore the motion is wholly in the plane of xy.

Then $$\frac{d^2x}{dt^2}=0,\ \frac{d^2y}{dt^2}=a;$$

and if $$\frac{dx}{dt}=\text{const.}=U, \text{ we get at once}$$

$$\frac{d^2y}{dx^2}=\frac{a}{U^2},$$

$$\text{or } y=\frac{ax^2}{2U^2}+Cx+C'=\frac{ax^2}{2U^2}+\frac{Vx}{U}$$

if $y=0$, and $\frac{dy}{dt}=V$ when $x=0$. The equation may be written

$$\frac{2U^2}{a}\left(y+\frac{V^2}{2a}\right)=\left(x+\frac{UV}{a}\right)^2,$$

a parabola of which the axis is parallel to y, whose parameter is $\frac{U^2}{2a}$, and whose vertex is at the point whose co-ordinates are

$$x=-\frac{UV}{a},\ y=-\frac{V^2}{2a}.$$

d. As an illustration of acceleration in a tortuous curve, we take the case of § 13, or of § 34, *d.*

Let a point move in a circle of radius r with uniform angular velocity ω (about the centre), and let this circle move perpendicular to its plane with velocity V. The point describes a helix on a cylinder of radius r, and the inclination α is given by Examples of acceleration.

$$\tan\alpha=\frac{V}{r\omega}.$$

The curvature of the path is $\frac{1}{r}\frac{r^2\omega^2}{V^2+r^2\omega^2}$ or $\frac{r\omega^2}{V^2+r^2\omega^2}$, and the tortuosity $\frac{\omega}{V}\frac{V^2}{V^2+r^2\omega^2}=\frac{V\omega}{V^2+r^2\omega^2}$.

The acceleration is $r\omega^2$, directed perpendicularly towards the axis of the cylinder.—Call this A.

$$\text{Curvature}=\frac{A}{V^2+Ar}=\frac{A}{V^2+\frac{A^2}{\omega^2}}.$$

$$\text{Tortuosity}=\frac{V}{\sqrt{Ar}}\frac{A}{V^2+Ar}=\frac{V\omega}{V^2+\frac{A^2}{\omega^2}}.$$

Let A be finite, r indefinitely small, and therefore ω indefinitely great.

$$\text{Curvature (in the limit)}=\frac{A}{V^2}.$$

$$\text{Tortuosity (\quad ,, \quad)}=\frac{\omega}{V}.$$

Thus, if we have a material particle moving in the manner specified, and if we consider the force (see Chap. II.) required to produce the acceleration, we find that a finite force perpendicular to the line of motion, and whose direction revolves with an infinitely great angular velocity, maintains constant infinitely small deflection (in a direction opposite to its own) from the line of undisturbed motion, *finite* curvature, and infinite tortuosity.

e. When the acceleration is perpendicular to a given plane and proportional to the distance from it, the path is a plane curve, which is the harmonic curve if the acceleration be *towards* the plane, and a more or less fore-shortened catenary if *from* the plane.

As in case *c*, $\frac{d^2z}{dt^2}=\frac{dz}{dt}=z=0$, if the axis of z be perpendicular to the acceleration and to the direction of motion at any instant. Also, if we choose the origin *in* the plane

$$\frac{d^2x}{dt^2}=0,\ \frac{d^2y}{dt^2}=\mu y.$$

Examples of acceleration.

Hence $$\frac{dx}{dt}=\text{const}=a \text{ (suppose)}$$

and $$\frac{d^2y}{dx^2}=\frac{\mu}{a^2}y=\mp\frac{y}{b^2}$$

This gives, if μ is negative,

$$y=P\cos\left(\frac{x}{b}+Q\right),$$ the harmonic curve, or curve of sines.

If μ be positive,

$$y=P\epsilon^{\frac{x}{b}}+Q\epsilon^{-\frac{x}{b}};$$

and by shifting the origin along the axis of x this can be put in the form

$$y=R\left(\epsilon^{\frac{x}{b}}+\epsilon^{-\frac{x}{b}}\right),$$

which is the catenary if $2R=b$, otherwise it is the catenary stretched or fore-shortened in the direction of y.

Acceleration directed to a fixed centre.

36. *a.* When the acceleration is directed to a fixed point, the path is in a plane passing through that point; and in this plane the areas traced out by the radius-vector are proportional to the times employed. This includes the case of a satellite or planet revolving about its primary.

Evidently there is no acceleration perpendicular to the plane containing the fixed and moving points and the direction of motion of the second at any instant; and, there being no velocity perpendicular to this plane at starting, there is therefore none throughout the motion; thus the point moves in the plane. And had there been no acceleration, the point would have described a straight line with uniform velocity, so that in this case the areas described by the radius-vector would have been proportional to the time. Also, the area actually described in any instant depends on the length of the radius-vector and the velocity perpendicular to it, and is shown below to be unaffected by an acceleration parallel to the radius-vector. Hence the second part of the proposition.

We have $$\frac{d^2x}{dt^2}=P\frac{x}{r}, \quad \frac{d^2y}{dt^2}=P\frac{y}{r}, \quad \frac{d^2z}{dt^2}=P\frac{z}{r},$$

the fixed point being the origin, and P being some function of x, y, z; in *nature* a function of r only.

Hence $$x\frac{d^2y}{dt^2}-y\frac{d^2x}{dt^2}=0, \text{ etc.},$$ Acceleration directed to a fixed centre.

which give on integration

$$y\frac{dz}{dt}-z\frac{dy}{dt}=C_1,\ z\frac{dx}{dt}-x\frac{dz}{dt}=C_2,\ x\frac{dy}{dt}-y\frac{dx}{dt}=C_3.$$

Hence at once $C_1x+C_2y+C_3z=0$, or the motion is in a plane through the origin. Take this as the plane of xy, then we have only the one equation

$$x\frac{dy}{dt}-y\frac{dx}{dt}=C_3=h \text{ (suppose).}$$

In polar co-ordinates this is

$$h=r^2\frac{d\theta}{dt}=2\frac{dA}{dt}$$

if A be the area intercepted by the curve, a fixed radius-vector, and the radius-vector of the moving point. Hence the area increases uniformly with the time.

b. In the same case the velocity at any point is inversely as the perpendicular from the fixed point upon the tangent to the path, the momentary direction of motion.

For evidently the product of this perpendicular and the velocity gives double the area described in one second about the fixed point.

Or thus—if p be the perpendicular on the tangent

$$p=x\frac{dy}{ds}-y\frac{dx}{ds},$$

and therefore

$$p\frac{ds}{dt}=x\frac{dy}{dt}-y\frac{dx}{dt}=h.$$

If we refer the motion to co-ordinates in its own plane, we have only the equations

$$\frac{d^2x}{dt^2}=\frac{Px}{r},\ \frac{d^2y}{dt^2}=\frac{Py}{r},$$

whence, as before, $$r^2\frac{d\theta}{dt}=h.$$

If, by the help of this last equation we eliminate t from $\frac{d^2x}{dt^2}=\frac{Px}{r}$, substituting polar for rectangular co-ordinates, we arrive at the polar differential equation to the path.

Acceleration directed to a fixed centre.

For variety, we may derive it from the formulæ of § 32.

They give $$\frac{d^2r}{dt^2}-r\left(\frac{d\theta}{dt}\right)^2=P,\ r^2\frac{d\theta}{dt}=h.$$

Putting $\frac{1}{r}=u$, we have

$$\frac{d^2\left(\frac{1}{u}\right)}{dt^2}-\frac{1}{u}\left(\frac{d\theta}{dt}\right)^2=P, \text{ and } \frac{d\theta}{dt}=hu^2.$$

But $\frac{d\left(\frac{1}{u}\right)}{dt}=hu^2\frac{d\left(\frac{1}{u}\right)}{d\theta}=-h\frac{du}{d\theta}$, therefore $\frac{d^2\left(\frac{1}{u}\right)}{dt^2}=-h^2u^2\frac{d^2u}{d\theta^2}$.

Also $\frac{1}{u}\left(\frac{d\theta}{dt}\right)^2=h^2u^3$, the substitution of which values gives us $\frac{d^2u}{d\theta^2}+u=-\frac{P}{h^2u^2}$, the equation required. The integral of this equation involves *two* arbitrary constants besides h, and the remaining constant belonging to the two differential equations of the second order above is to be introduced on the farther integration of

$$\frac{d\theta}{dt}=hu^2$$

when the value of u in terms of θ is substituted from the equation to the path.

Other examples of these principles will be met with in the chapters on kinetics.

Hodograph.

37. If, from any fixed point, lines be drawn at every instant representing in magnitude and direction the velocity of a point describing any path in any manner, the extremities of these lines form a curve which is called the *Hodograph.* The invention of this construction is due to Sir W. R. Hamilton, and the most beautiful of the many remarkable theorems to which it leads is this : *The Hodograph for the motion of a planet or comet is always a circle, whatever be the form and dimensions of the orbit.*

Since the radius-vector of the hodograph represents the velocity at each instant, it is evident (§ 27) that an elementary arc represents the acceleration, and thus a finite arc represents the whole acceleration of the moving point during the corresponding time; and it is evident also that the tangent to the hodograph is parallel to the direction of the acceleration of the moving point in the corresponding position of its orbit.

If x, y, z be the co-ordinates of the moving point, ξ, η, ζ those of the corresponding point of the hodograph, then evidently Hodograph.

$$\xi=\frac{dx}{dt}, \quad \eta=\frac{dy}{dt}, \quad \zeta=\frac{dz}{dt},$$

and therefore

$$\frac{d\xi}{\frac{d^2x}{dt^2}}=\frac{d\eta}{\frac{d^2y}{dt^2}}=\frac{d\zeta}{\frac{d^2z}{dt^2}},$$

or the tangent to the hodograph is parallel to the acceleration in the orbit. Also if σ be the arc of the hodograph

$$\frac{d\sigma}{dt}=\sqrt{\left(\frac{d\xi}{dt}\right)^2+\left(\frac{d\eta}{dt}\right)^2+\left(\frac{d\zeta}{dt}\right)^2}$$

$$=\sqrt{\left(\frac{d^2x}{dt^2}\right)^2+\left(\frac{d^2y}{dt^2}\right)^2+\left(\frac{d^2z}{dt^2}\right)^2},$$

or the velocity in the hodograph is equal to the rate of acceleration in the orbit.

38. In the case of a planet or comet, the acceleration is directed towards the sun's centre. Hence (§ 36, *b.*) the velocity is inversely as the perpendicular from that point upon the tangent to the orbit. The orbit we assume to be a conic section, whose focus is the sun's centre. But we know that the intersection of the perpendicular with the tangent lies in the circle whose diameter is the major axis, if the orbit be an ellipse or hyperbola; in the tangent to the vertex if a parabola. Measure off on the perpendicular a third proportional to its own length and any constant line; this portion will thus represent the velocity in magnitude and in a direction perpendicular to its own—so that the locus of the new points in each perpendicular will be the hodograph turned through a right angle. But we see by geometry that the locus of these points is always a circle. Hence the proposition in § 37. The hodograph surrounds its origin if the orbit be an ellipse, passes through it if a parabola, and the origin is without the hodograph if the orbit is a hyperbola.

For a projectile unresisted by the air, it will be shown in Kinetics that we have the equations (assumed in § 35, *c*).

$$\frac{d^2x}{dt^2}=0, \quad \frac{d^2y}{dt^2}=-g,$$

if the axis of y be taken vertically upwards.

Hodograph.

Hence for the hodograph

$$\frac{d\xi}{dt}=0, \quad \frac{d\eta}{dt}=-g,$$

or $\xi=C$, $\eta=C'-gt$, and the hodograph is a vertical straight line along which the describing point moves uniformly.

Also it will be shown in Kinetics that the equations of motion of a planet or comet in the plane of its orbit are

$$\frac{d^2x}{dt^2}=\frac{\mu x}{r^3}, \quad \frac{d^2y}{dt^2}=\frac{\mu y}{r^3},$$

where $$r^2=x^2+y^2.$$

Hence, as in § 36,

$$x\frac{dy}{dt}-y\frac{dx}{dt}=h, \qquad \text{.............(1)}$$

and therefore

$$\frac{d^2x}{dt^2}=\frac{\mu x}{h}\frac{x\frac{dy}{dt}-y\frac{dx}{dt}}{r^3},$$

$$=\frac{\mu}{h}\frac{(x^2+y^2)\frac{dy}{dt}-y(x\frac{dx}{dt}+y\frac{dy}{dt})}{r^3},$$

$$=\frac{\mu}{h}\frac{r^2\frac{dy}{dt}-yr\frac{dr}{dt}}{r^3}.$$

Hence $$\frac{dx}{dt}+A=\frac{\mu}{h}\frac{y}{r} \qquad \text{.............(2)}$$

Similarly

$$\frac{dy}{dt}+B=-\frac{\mu}{h}\frac{x}{r} \qquad \text{..........(3)}$$

Hence for the hodograph

$$(\xi+A)^2+(\eta+B)^2=\frac{\mu^2}{h^2},$$

the circle as before stated.

We may merely mention that the equation to the orbit will be found at once by eliminating $\frac{dx}{dt}$ and $\frac{dy}{dt}$ among the three first integrals (1), (2), (3) above. We thus get

$$-h+Ay-Bx=\frac{\mu}{h}r,$$

a conic section of which the origin is a focus.

Applications of the hodograph.

39. The intensity of heat and light emanating from a point, or from an uniformly radiating spherical surface, diminishes with increasing distance according to the same law as gravitation. Hence the amount of heat and light, which a planet receives from the sun during any interval, is proportional to the whole acceleration during that interval, *i.e.*, to the corresponding arc of the hodograph. From this it is easy to see, for example, that if a comet move in a parabola, the amount of heat it receives from the sun in any interval is proportional to the angle through which its direction of motion turns during that interval. There is a corresponding theorem for a planet moving in an ellipse, but somewhat more complicated. **Applications of the hodograph.**

40. If two points move, each with a definite uniform velocity, one in a given curve, the other at every instant directing its course towards the first describes a path which is called a *Curve of Pursuit*. The idea is said to have been suggested by the old rule of steering a privateer always directly for the vessel pursued. (Bouguer, *Mém. de l'Acad.* 1732.) It is the curve described by a dog running to its master. **Curves of pursuit.**

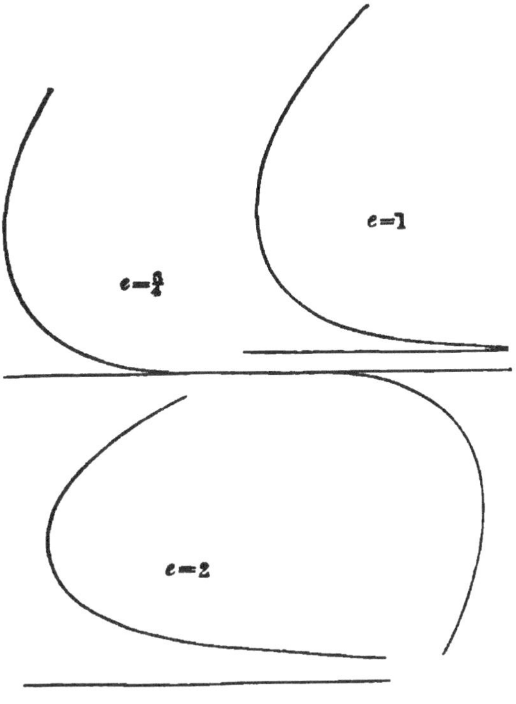

Curves of pursuit.

The simplest cases are of course those in which the first point moves in a straight line, and of these there are three, for the velocity of the first point may be greater than, equal to, or less than, that of the second. The figures in the preceding page represent the curves in these cases, the velocities of the pursuer being $\frac{4}{3}$, 1, and $\frac{1}{2}$ of those of the pursued, respectively. In the second and third cases the second point can never overtake the first, and consequently the line of motion of the first is an asymptote. In the first case the second point overtakes the first, and the curve at that point touches the line of motion of the first. The remainder of the curve satisfies a modified form of statement of the original question, and is called the *Curve of Flight*.

We will merely form the differential equation of the curve, and give its integrated form, leaving the work to the student.

Suppose Ox to be the line of motion of the first point, whose velocity is v, AP the curve of pursuit, in which the velocity is u.

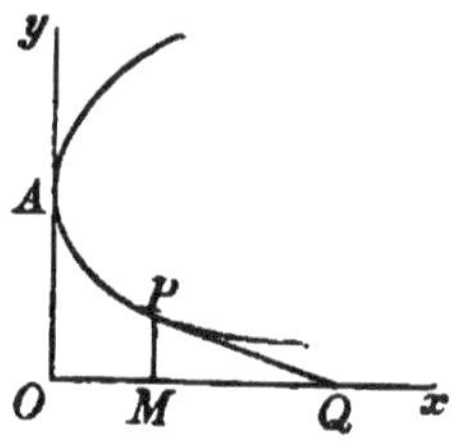

then the tangent at P always passes through Q, the instantaneous position of the first point. It will be evident, on a moment's consideration, that the curve AP must have a tangent perpendicular to Ox. Take this as the axis of y, and let $OA = a$. Then, if $OQ = \xi$, $AP = s$, and if x, y be the co-ordinates of P, we have

$$\frac{AP}{u} = \frac{OQ}{v},$$

because A, O and P, Q are pairs of simultaneous positions of the two points. This gives

$$\frac{v}{u}s = es = x - y\frac{dx}{dy}.$$

From this we find, unless $e = 1$,

$$2\left(x + \frac{ae}{e^2 - 1}\right) = \frac{y^{e+1}}{a^e(e+1)} + \frac{a^e}{y^{e-1}(e-1)};$$

and if $e = 1$,

$$2\left(x + \frac{a}{4}\right) = \frac{y^2}{2a} - a\log_e\frac{y}{a},$$

the only case in which we do not get an algebraic curve. The axis of x is easily seen to be an asymptote if $e \nless 1$.

Angular velocity.

41. When a point moves in any manner, the line joining it with a fixed point generally changes its direction. If, for simplicity, we consider the motion as confined to a plane passing through the fixed point, the angle which the joining line makes with a fixed line in the plane is continually altering, and its rate of alteration at any instant is called the *Angular Velocity* of the first point about the second. If uniform, it is of course measured by the angle described in unit of time; if variable, by the angle which would have been described in unit of time if the angular velocity at the instant in question were maintained constant for so long. In this respect, the process is precisely similar to that which we have already explained for the measurement of velocity and acceleration. Angular velocity.

Unit of angular velocity is that of a point which describes, or would describe, unit angle about a fixed point in unit of time. The usual unit angle is (as explained in treatises on plane trigonometry) that which subtends at the centre of a circle an arc whose length is equal to the radius; being an angle of $\frac{180°}{\pi} = 57°{\cdot}29578... = 57°\,17'\,44''{\cdot}8$ nearly.

42. The rate of increase or diminution of the angular velocity when variable is called the *angular acceleration*, and is measured in the same way and by the same unit. Angular acceleration.

By methods precisely similar to those employed for linear velocity and acceleration we see that if θ be the angle-vector of a point moving in a plane—the

$$\text{Angular velocity is } \omega = \frac{d\theta}{dt}\text{, and the}$$

$$\text{Angular acceleration is } \frac{d\omega}{dt} = \frac{d^2\theta}{dt^2} = \omega\frac{d\omega}{d\theta}.$$

Since (§ 27) $r\frac{d\theta}{dt}$ is the velocity perpendicular to the radius-vector, we see that

The angular velocity of a point in a plane is found by dividing the velocity perpendicular to the radius-vector by the length of the radius-vector.

43. When one point describes uniformly a circle about another, the time of describing a complete circumference being T, we have the angle 2π described uniformly in T; and, therefore, Angular velocity.

Angular velocity.

the angular velocity is $\frac{2\pi}{T}$. Even when the angular velocity is not uniform, as in a planet's motion, it is useful to introduce the quantity $\frac{2\pi}{T}$, which is then called the *mean* angular velocity.

When a point moves uniformly in a straight line its angular velocity evidently diminishes as it recedes from the point about which the angles are measured.

The polar equation of a straight line is

$$r = a\sec\theta.$$

But the length of the line between the limiting angles 0 and θ is $a\tan\theta$, and this increases with uniform velocity v. Hence

$$v = \frac{d}{dt}(a\tan\theta) = a\sec^2\theta\frac{d\theta}{dt} = \frac{r^2}{a}\frac{d\theta}{dt}.$$

Hence $\frac{d\theta}{dt} = \frac{av}{r^2}$, and is therefore inversely as the square of the radius-vector.

Similarly for the angular acceleration, we have by a second differentiation,

$$\frac{d^2\theta}{dt^2} + 2\tan\theta\left(\frac{d\theta}{dt}\right)^2 = 0,$$

i. e., $\frac{d^2\theta}{dt^2} = -\frac{2av^2}{r^3}\left(1 - \frac{a^2}{r^2}\right)^{\frac{1}{2}}$, and ultimately varies inversely as the third power of the radius-vector.

Angular velocity of a plane.

44. We may also talk of the angular velocity of a moving plane with respect to a fixed one, as the rate of increase of the angle contained by them—but unless their line of intersection remain fixed, or at all events parallel to itself, a somewhat more laboured statement is required to give definite information. This will be supplied in a subsequent section.

Relative motion.

45. All motion that we are, or can be, acquainted with, is *Relative* merely. We can calculate from astronomical data for any instant the direction in which, and the velocity with which, we are moving on account of the earth's diurnal rotation. We may compound this with the equally calculable velocity of the earth in its orbit. This resultant again we may compound with the (roughly known) velocity of the sun relatively to the so-called fixed stars; but, even if all these elements were accurately known, it could not be said that we had attained any idea of an *absolute* velocity; for it is only the sun's relative motion among the stars that we can observe; and, in all probability, sun and stars are moving on (it may be with inconceivable rapidity) relatively to other bodies in space. We must therefore consider how, from the actual motions of a set of points, we may find their relative motions with regard to any one of them; and how, having given the relative motions of all but one with regard to the latter, and the actual motion of the latter, we may find the actual motions of all. The question is very easily answered. Consider for a moment a number of passengers walking on the deck of a steamer. Their relative motions with regard to the deck are what we immediately observe, but if we compound with these the velocity of the steamer itself we get evidently their actual motion relatively to the earth. Again, in order to get the relative motion of all with regard to the deck, we *abstract from* the motion of the steamer altogether—that is, we alter the velocity of each by compounding it with the actual velocity of the vessel taken in a reversed direction. Relative motion.

Hence to find the relative motions of any set of points with regard to one of their number, imagine, impressed upon each in composition with its own velocity, a velocity equal and opposite to the velocity of that one; it will be reduced to rest, and the motions of the others will be the same with regard to it as before.

Thus, to take a very simple example, two trains are running in opposite directions, say north and south, one with a velocity of fifty, the other of thirty, miles an hour. The relative velocity of the second with regard to the first is to be found by impressing on both a southward velocity of fifty miles an hour; the effect of this being to bring the first to rest, and to give the second a southward velocity of eighty miles an hour, which is the required relative motion.

Or, given one train moving north at the rate of thirty miles an hour, and another relatively to it moving south at the rate of twenty-five miles an hour, the actual motion of the second is thirty miles north, and twenty-five south, per hour, *i.e.*, five miles north. It is needless to multiply such examples, as they must occur to every one.

Relative acceleration.

46. Exactly the same remarks apply to relative as compared with absolute acceleration, as indeed we may see at once, since accelerations are in all cases resolved and compounded by the same law as velocities.

If x, y, z, and x', y', z' be the co-ordinates of two points referred to axes regarded as fixed; and ξ, η, ζ their relative co-ordinates —we have

$$\xi = x' - x, \quad \eta = y' - y, \quad \zeta = z' - z,$$

and, differentiating,

$$\frac{d\xi}{dt} = \frac{dx'}{dt} - \frac{dx}{dt}, \text{ etc.,}$$

which give the relative, in terms of the absolute velocities, and

$$\frac{d^2\xi}{dt^2} = \frac{d^2x'}{dt^2} - \frac{d^2x}{dt^2}, \text{ etc.,}$$

proving our assertion about relative and absolute accelerations.

The corresponding expressions in polar co-ordinates in a plane are somewhat complicated, and by no means convenient. The student can easily write them down for himself.

Relative motion.

47. The following proposition in relative motion is of considerable importance :—

Any two moving points describe similar paths about each other, or relatively to any point which divides in a constant ratio the line joining them.

Let A and B be any simultaneous positions of the points. Take G or G' in AB such that the ratio $\frac{GA}{GB}$ or $\frac{G'A}{G'B}$ has a constant value.

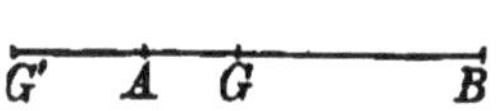

Then as the form of the relative path depends only upon the *length* and *direction* of the line joining the two points at any instant, it is obvious that these will be the same for A with regard to B, as for B with regard to A, saving only the inversion of the direction of the joining line. Hence B's path about A, is A's about B turned through two right angles. And with regard to G and G' it is evident that the directions remain the same, while the lengths are altered in a given ratio; but this is the definition of similar curves.

48. As a good example of relative motion, let us consider that of the two points involved in our definition of the curve of pursuit, § 40. Since, to find the relative position and motion of the pursuer with regard to the pursued, we must impress on both a velocity equal and opposite to that of the latter, we see at once that the problem becomes the same as the following. A boat crossing a stream is impelled by the oars with uniform velocity relatively to the water, and always towards a fixed point in the opposite bank; but it is also carried down stream at a uniform rate; determine the path described and the time of crossing. Here, as in the former problem, there are three cases, figured below. In the first, the boat, moving faster than the current, reaches the desired point; in the second, the velocities of boat and stream being equal, the boat gets across only after an infinite time—describing a parabola—but does not land at the desired point, which is indeed the focus of the parabola, while the landing point is the vertex. In the third case, its proper velocity being less than that of the water, it never reaches the other bank, and is carried indefinitely down stream. The comparison of the figures in § 40 with those in the present section cannot fail to be instructive. They are drawn to the same scale, and for the same relative velocities. The horizontal lines represent the farther bank of the river, and the vertical lines the path of the boat if there were no current. Relative motion.

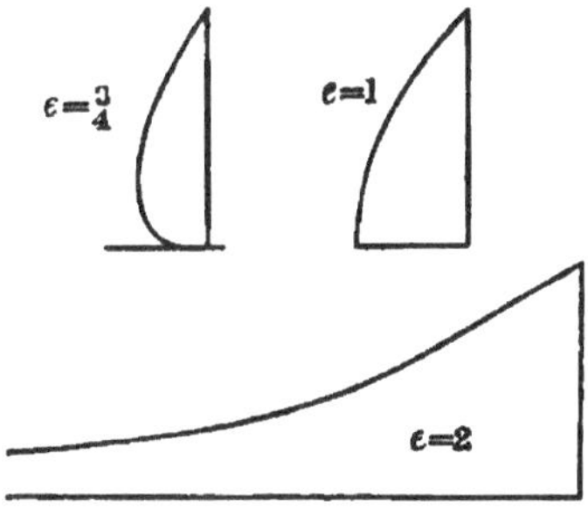

We leave the solution of this question as an exercise, merely noting that the equation of the curve is

$$\frac{y^{1+e}}{a^e}=\sqrt{x^2+y^2}-x,$$

in one or other of the three cases, according as e is $>$, $=$, or <1.

When $e=1$ this becomes

$$y^2=a^2-2ax, \text{ the parabola.}$$

Relative motion.

The time of crossing is $\dfrac{a}{u(1-e^2)}$,

which is finite only for $e < 1$, because of course a negative value is inadmissible.

49. Another excellent example of the transformation of relative into absolute motion is afforded by the family of cycloids. We shall in a future section consider their mechanical description, by the *rolling* of a circle on a fixed straight line or circle. In the meantime, we take a different form of enunciation, which, however, leads to precisely the same result.

Find the actual path of a point which revolves uniformly in a circle about another point—the latter moving uniformly in a straight line or circle in the same plane.

Take the former case first: let a be the radius of the relative circular orbit, and ω the angular velocity in it, v being the velocity of its centre along the straight line.

The relative co-ordinates of the point in the circle are $a\cos\omega t$ and $a\sin\omega t$, and the actual co-ordinates of the centre are vt and 0. Hence for the actual path

$$\xi = vt + a\cos\omega t, \quad \eta = a\sin\omega t.$$

Hence $\xi = \dfrac{v}{\omega}\sin^{-1}\dfrac{\eta}{a} + \sqrt{a^2-\eta^2}$, an equation which, by giving different values to v and ω, may be made to represent the cycloid itself, or either form of trochoid. See § 92.

For the epicycloids, let b be the radius of the circle which B describes about A, ω_1 the angular velocity; a the radius of A's path, ω the angular velocity.

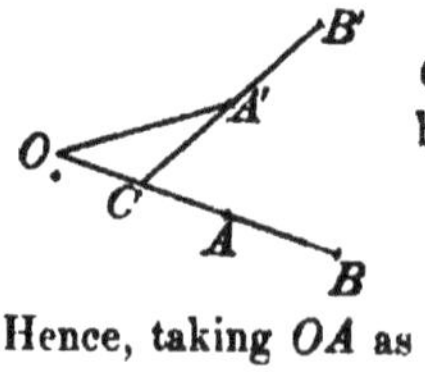

Also at time $t=0$, let B be in the radius OA of A's path. Then at time t, if A', B' be the positions, we see at once that

$$\overset{<}{AOA'} = \omega t, \quad \overset{<}{B'CA} = \omega_1 t.$$

Hence, taking OA as axis of x,

$$x = a\cos\omega t + b\cos\omega_1 t, \quad y = a\sin\omega t + b\sin\omega_1 t,$$

which, by the elimination of t, give an algebraic equation between x and y whenever ω and ω_1 are commensurable.

Thus, for $\omega_1 = 2\omega$, suppose $\omega t = \theta$, and we have

$$x = a\cos\theta + b\cos 2\theta, \quad y = a\sin\theta + b\sin 2\theta,$$

or, by an easy reduction, *Relative motion.*

$$(x^2+y^2-b^2)^2=a^2\{(x+b)^2+y^2\}.$$

Put $x-b$ for x, *i.e.*, change the origin to a distance AB to the left of O, the equation becomes

$$a^2(x^2+y^2)=(x^2+y^2-2bx)^2,$$

or, in polar co-ordinates,

$$a^2=(r-2b\cos\theta)^2,\quad r=a+2b\cos\theta,$$

and when $2b=a$, $r=a\,(1+\cos\theta)$, the cardioid. (See § 94.)

50. As an additional illustration of this part of our subject, we may define as follows:— *Resultant motion.*

If one point A executes any motion whatever with reference to a second point B; if B executes any other motion with reference to a third point C; and so on—the first is said to execute, with reference to the last, a movement which is the resultant of these several movements.

The relative position, velocity, and acceleration are in such a case the geometrical resultants of the various components combined according to preceding rules.

51. The following practical methods of effecting such a combination in the simple case of the movements of two points are useful in scientific illustrations and in certain mechanical arrangements. Let two moving points be joined by an elastic string; the middle point of this string will evidently execute a movement which is *half* the resultant of the motions of the two points. But for drawing, or engraving, or for other mechanical applications, the following method is preferable:—

CF and *ED* are rods of equal length moving freely round a pivot at *P*, which passes through the middle point of each—*CA*, *AD*, *EB*, and *BF*, are rods of half the length of the two former, and so pivoted to them as to form a pair of equal rhombi *CD*, *EF*, whose angles can be altered at will. Whatever motions, whether in a plane, or in space of three dimensions, be given to *A* and *B*, *P* will evidently be subjected to half their resultant.

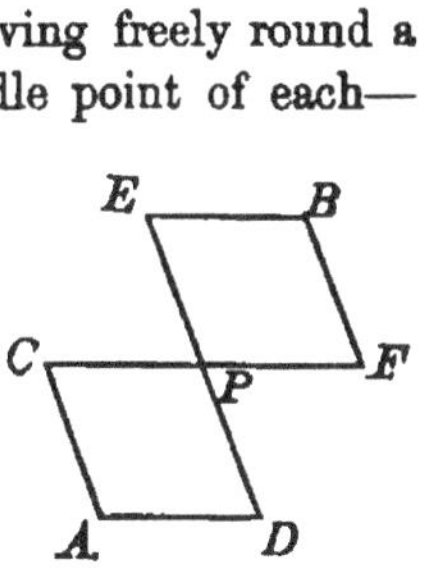

52. Amongst the most important classes of motions which we have to consider in Natural Philosophy, there is one, namely *Harmonic motion.*

Harmonic motion.

Harmonic Motion, which is of such immense use, not only in ordinary kinetics, but in the theories of sound, light, heat, etc., that we make no apology for entering here into considerable detail regarding it.

Simple harmonic motion.

53. *Def.* When a point Q moves uniformly in a circle, the perpendicular QP drawn from its position at any instant to a fixed diameter AA' of the circle, intersects the diameter in a point P, whose position changes by a *simple harmonic motion.*

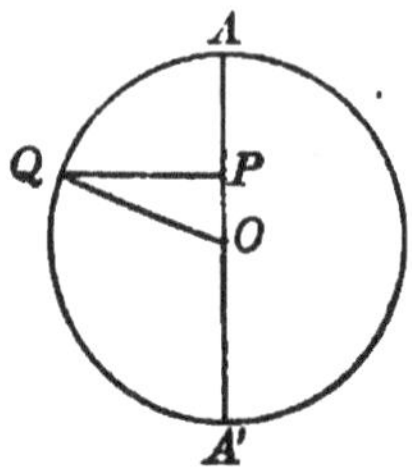

Thus, if a planet or satellite, or one of the constituents of a double star, be supposed to move uniformly in a circular orbit about its primary, and be viewed from a very distant position in the plane of its orbit, it will appear to move backwards and forwards in a straight line with a simple harmonic motion. This is nearly the case with such bodies as the satellites of Jupiter when seen from the earth.

Physically, the interest of such motions consists in the fact of their being approximately those of the simplest vibrations of sounding bodies, such as a tuning-fork or pianoforte wire; whence their name; and of the various media in which waves of sound, light, heat, etc., are propagated.

54. The *Amplitude* of a simple harmonic motion is the range on one side or the other of the middle point of the course, *i.e.*, OA or OA' in the figure.

An arc of the circle referred to, measured from any fixed point to the uniformly moving point Q, is the *Argument* of the harmonic motion.

The distance of a point, performing a simple harmonic motion, from the middle of its course or range, is a *simple harmonic function of the time.* The *argument* of this function is what we have defined as the argument of the motion.

The *Epoch* in a simple harmonic motion is the interval of time which elapses from the era of reckoning till the moving point first comes to its greatest elongation in the direction reckoned as positive, from its mean position or the middle of its range. Epoch in angular measure is the angle described on the circle of reference in the period of time defined as the epoch

The *Period* of a simple harmonic motion is the time which elapses from any instant until the moving point again moves in the same direction through the same position. Simple harmonic motion.

The *Phase* of a simple harmonic motion at any instant is the fraction of the whole period which has elapsed since the moving point last passed through its middle position in the positive direction.

55. Those common kinds of mechanism, for producing rectilineal from circular motion, or *vice versa*, in which a crank moving in a circle works in a straight slot belonging to a body which can only move in a straight line, fulfil strictly the definition of a simple harmonic motion in the part of which the motion is rectilineal, if the motion of the rotating part is uniform. Simple harmonic motion in mechanism.

The motion of the treadle in a spinning-wheel approximates to the same condition when the wheel moves uniformly; the approximation being the closer, the smaller is the angular motion of the treadle and of the connecting string. It is also approximated to more or less closely in the motion of the piston of a steam-engine connected, by any of the several methods in use, with the crank, provided always the rotatory motion of the crank be uniform.

56. The velocity of a point executing a simple harmonic motion is a simple harmonic function of the time, a quarter of a period earlier in phase than the displacement, and having its maximum value equal to the velocity in the circular motion by which the given function is defined. Velocity in S. H. motion.

For, in the fig. of § 53, if V be the velocity in the circle, it may be represented by OQ in a direction perpendicular to its own, and therefore by OP and PQ in directions perpendicular to those lines. That is, the velocity of P in the simple harmonic motion is $\frac{V}{OQ}PQ$; which, when P is at O, becomes V.

57. The acceleration of a point executing a simple harmonic motion is at any time simply proportional to the displacement from the middle point, but in opposite direction, or always towards the middle point. Its maximum value is that with which a velocity equal to that of the circular motion would be acquired in the time in which an arc equal to the radius is described. Acceleration in S. H. motion.

Acceleration in S. H. motion.

For, in the fig. of § 53, the acceleration of Q (by § 35, a) is $\frac{V^2}{QO}$ along QO. Supposing, for a moment, QO to represent the magnitude of this acceleration, we may resolve it in QP, PO. The acceleration of P is therefore represented on the same scale by PO. Its magnitude is therefore $\frac{V^2}{QO} \cdot \frac{PO}{QO} = \frac{V^2}{QO^2} PO$, which is proportional to PO, and has at A its maximum value, $\frac{V^2}{QO}$, an acceleration under which the velocity V would be acquired in the time $\frac{QO}{V}$ as stated.

Let a be the amplitude, ϵ the epoch, and T the period, of a simple harmonic motion. Then if s be the displacement from middle position at time t, we have

$$s = a \cos \left(\frac{2\pi t}{T} - \epsilon\right).$$

Hence, for velocity, we have $v = \frac{ds}{dt} = -\frac{2\pi a}{T} \sin \left(\frac{2\pi t}{T} - \epsilon\right)$.

Hence V, the maximum value, is $\frac{2\pi a}{T}$, as above stated (§ 56).

Again, for acceleration,

$$\frac{dv}{dt} = -\frac{4\pi^2 a}{T^2} \cos \left(\frac{2\pi t}{T} - \epsilon\right) = -\frac{4\pi^2}{T^2} s. \quad \text{(See § 57.)}$$

Lastly, for the maximum value of the acceleration,

$$\frac{4\pi^2 a}{T^2} = \frac{V}{\frac{T}{2\pi}},$$

where, it may be remarked, $\frac{T}{2\pi}$ is the time of describing an arc equal to radius in the relative circular motion.

Composition of S. H. M. in one line.

58. Any two simple harmonic motions in one line, and of one period, give, when compounded, a single simple harmonic motion; of the same period; of amplitude equal to the diagonal of a parallelogram described on lengths equal to their amplitudes measured on lines meeting at an angle equal to their difference of epochs; and of epoch differing from their

Composition of S. H. M. in one line.

epochs by angles equal to those which this diagonal makes with the two sides of the parallelogram. Let P and P' be two points executing simple harmonic motions of one period, and in one line $B'BCAA'$. Let Q and Q' be the uniformly moving points in the relative circles. On CQ and CQ' describe a parallelogram $SQCQ'$; and through S draw SR perpendicular to $B'A'$ produced. We have obviously $P'R = CP$ (being projections of the equal and parallel lines $Q'S$, CQ, on CR). Hence

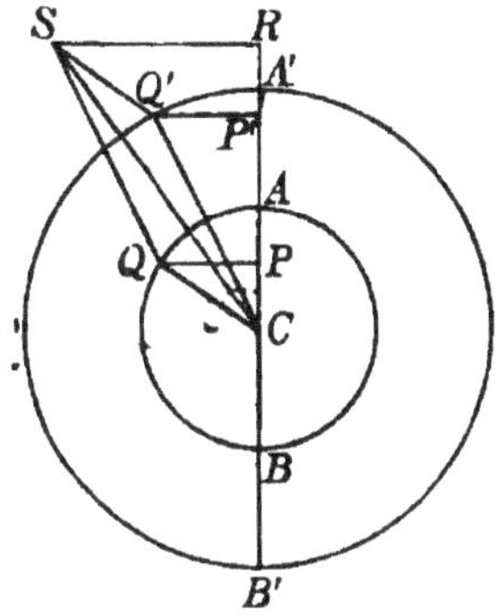

$CR = CP + CP'$; and therefore the point R executes the resultant of the motions of P and P'. But CS, the diagonal of the parallelogram, is constant, and therefore the resultant motion is simple harmonic, of amplitude CS, and of epoch exceeding that of the motion of P, and falling short of that of the motion of P', by the angles QCS and SCQ' respectively.

An analytical proof of the same proposition is useful, being as follows:—

$$a\cos\left(\frac{2\pi t}{T}-\epsilon\right)+a'\cos\left(\frac{2\pi t}{T}-\epsilon'\right)$$

$$=(a\cos\epsilon+a'\cos\epsilon')\cos\frac{2\pi t}{T}+(a\sin\epsilon+a'\sin\epsilon')\sin\frac{2\pi t}{T}=r\cos\left(\frac{2\pi t}{T}-\theta\right)$$

where $\quad r=\{(a\cos\epsilon+a'\cos\epsilon')^2+(a\sin\epsilon+a'\sin\epsilon')^2\}^{\frac{1}{2}}$

and $\quad \tan\theta=\dfrac{a\sin\epsilon+a'\sin\epsilon'}{a\cos\epsilon+a'\cos\epsilon'}$.

59. The construction described in the preceding section exhibits the resultant of two simple harmonic motions, whether of the same period or not. Only, if they are not of the same period, the diagonal of the parallelogram will not be constant, but will diminish from a maximum value, the sum of the component amplitudes, which it has at the instant when the phases of the component motions agree; to a minimum, the difference of those amplitudes, which is its value when the phases differ by half a period. Its direction, which always must be nearer to the greater than to the less of the two radii constituting the sides

Composition of S. H. M. in one line.

of the parallelogram, will oscillate on each side of the greater radius to a maximum deviation amounting on either side to the angle whose sine is the less radius divided by the greater, and reached when the less radius deviates more than this by a quarter circumference from the greater. The full period of this oscillation is the time in which either radius gains a full turn on the other. The resultant motion is therefore not simple harmonic, but is, as it were, simple harmonic with periodically increasing and diminishing amplitude, and with periodical acceleration and retardation of phase. This view is most appropriate for the case in which the periods of the two component motions are nearly equal, but the amplitude of one of them much greater than that of the other.

To express the resultant motion, let s be the displacement at time t; and let a be the greater of the two component half-amplitudes.

$$s = a\cos(nt-\epsilon) + a'\cos(n't-\epsilon')$$
$$= a\cos(nt-\epsilon) + a'\cos(nt-\epsilon+\phi)$$
$$= (a+a'\cos\phi)\cos(nt-\epsilon) - a'\sin\phi\sin(nt-\epsilon),$$

if $$\phi = (n't-\epsilon') - (nt-\epsilon);$$

or, finally, $$s = r\cos(nt-\epsilon+\theta),$$

if $$r = (a^2+2aa'\cos\phi+a'^2)^{\frac{1}{2}}$$

and $$\tan\theta = \frac{a'\sin\phi}{a+a'\cos\phi}.$$

The maximum value of $\tan\theta$ in the last of these equations is found by making $\phi = \frac{\pi}{2} + \sin^{-1}\frac{a'}{a}$, and is equal to $\frac{a'}{(a^2-a'^2)^{\frac{1}{2}}}$, and hence the maximum value of θ itself is $\sin^{-1}\frac{a'}{a}$. The geometrical methods indicated above (§ 58) lead to this conclusion by the following very simple construction.

To find the time and the amount of the maximum acceleration or retardation of phase, let CA be the greater half-amplitude. From A as centre, with AB the less half-amplitude as radius, describe a circle. CB touching this circle is the generating radius of the most deviated resultant. Hence CBA is a right angle; and

$$\sin BCA = \frac{AB}{CA}.$$

60. A most interesting application of this case of the composition of harmonic motions is to the lunar and solar tides; which, except in tidal rivers, or long channels, or deep bays, follow each very nearly the simple harmonic law, and produce, as the actual result, a variation of level equal to the sum of variations that would be produced by the two causes separately.

Examples of composition of S. H. M. in one line.

The amount of the lunar tide is about 2·1 times that of the solar. Hence the spring tides at open coast stations are 3·1, and the neap tides only 1·1, each reckoned in terms of the solar tide; and at spring and neap tides the hour of high water is that of the lunar tide alone. The greatest deviation of the actual tide from the phases (high, low, or mean water) of the lunar tide alone, is about ·95 of a lunar hour, that is, ·98 of a solar hour (being the same part of 12 lunar hours that 28° 26′, or the angle whose sine is $\frac{1}{2\cdot1}$, is of 360°). This maximum deviation will be in advance or in arrear according as the crown of the solar tide precedes or follows the crown of the lunar tide; and it will be exactly reached when the interval of phase between the two component tides is 3·95 lunar hours. That is to say, there will be maximum advance of the time of high water 4½ days after, and maximum retardation the same number of days before, spring tides.

61. We may consider next the case of equal amplitudes in the two given motions. If their periods are equal, their resultant is a simple harmonic motion, whose phase is at every instant the mean of their phases, and whose amplitude is equal to twice the amplitude of either multiplied by the cosine of half the difference of their phase. The resultant is of course nothing when their phases differ by half the period, and is a motion of double amplitude and of phase the same as theirs when they are of the same phase.

When their periods are very nearly, but not quite, equal (their amplitudes being still supposed equal), the motion passes very slowly from the former (zero, or no motion at all) to the latter, and back, in a time equal to that in which the faster has gone once oftener through its period than the slower has.

In practice we meet with many excellent examples of this case, which will, however, be more conveniently treated of

Examples of composition of S. H. M. in one line.

when we come to apply kinetic principles to various subjects in practical mechanics, acoustics, and physical optics; such as the marching of troops over a suspension bridge, the sympathy of pendulums or tuning-forks, etc.

Graphical representation of H. motions in one line.

62. We may exhibit, graphically, the various preceding cases of single or compound simple harmonic motions in one line by curves in which the abscissæ represent intervals of time, and the ordinates the corresponding distances of the moving point from its mean position. In the case of a single simple harmonic motion, the corresponding curve would be that described by the point P in § 53, if, while Q maintained its uniform circular motion, the circle were to move with uniform velocity in any direction perpendicular to OA. This construction gives the harmonic curve, or curve of sines, in which the ordinates are proportional to the sines of the abscissæ, the straight line in which O moves being the axis of abscissæ. It is the simplest possible form assumed by a vibrating string. When the harmonic motion is complex, but in one line, as is the case for any point in a violin-, harp-, or pianoforte-string (differing, as these do, from one another in their motions on account of the different modes of excitation used), a similar construction may be made. Investigation regarding complex harmonic functions has led to results of the highest importance, having their most general expression in *Fourier's Theorem*, to which we will presently devote several pages. We give below graphic representations of the composition of two simple harmonic motions in one line, of *equal* amplitudes and of periods which are as 1 : 2 and as 2 : 3, for differences of epoch corresponding to 0, 1, 2, etc., sixteenths of a circumference. In each case the epoch of the component of greater period is a quarter circumference. In the first, second, third, etc., of each series respectively, the epoch of the component of shorter period is less than a quarter period by 0, 1, 2, etc., sixteenths of a circumference. The successive horizontal lines are the axes of abscissæ of the successive curves; the vertical line to the left of each series being the common axis of ordinates. In each of the first set the slower motion goes through one complete period, in the second it goes through two periods.

Graphical representation of harmonic motions in one line.

1 : 2 (Octave) $y = \sin x + \sin (2x + \frac{n\pi}{8})$

2 : 3 (Fifth) $y = \sin 2x + \sin (3x + \frac{n\pi}{8})$

Both, from $x=0$ to $x=2\pi$; and for $n=0, 1, 2, \ldots\ldots 15$, in succession.

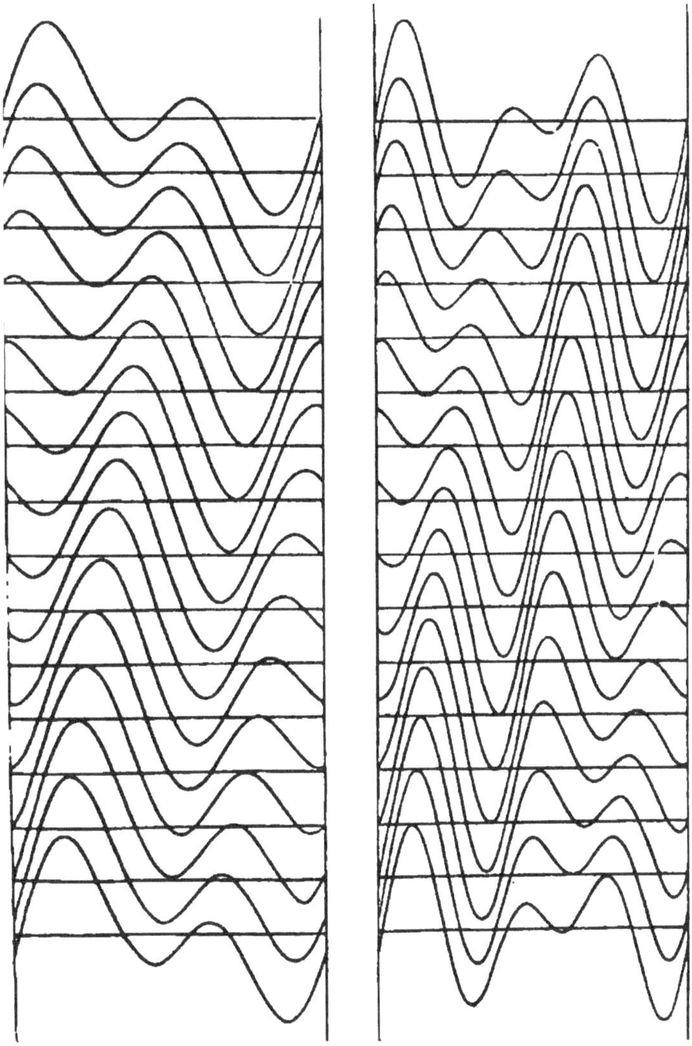

These, and similar cases, when the periodic times are not commensurable, will be again treated of under Acoustics.

S. H. motions in different directions.

63. We have next to consider the composition of simple harmonic motions in different directions. In the first place, we see that any number of simple harmonic motions of one period, and of the same phase, superimposed, produce a single simple harmonic motion of the same phase. For, the displacement at any instant being, according to the principle of the composition of motions, the geometrical resultant (see above, § 50) of the displacements due to the component motions separately, these component displacements in the case supposed, all vary in simple proportion to one another, and are in constant directions. Hence the resultant displacement will vary in simple proportion to each of them, and will be in a constant direction.

But if, while their periods are the same, the phases of the several component motions do not agree, the resultant motion will generally be elliptic, with equal areas described in equal times by the radius-vector from the centre; although in particular cases it may be uniform circular, or, on the other hand, rectilineal and simple harmonic.

64. To prove this, we may first consider the case, in which we have two equal simple harmonic motions given, and these in perpendicular lines, and differing in phase by a quarter period. Their resultant is a uniform circular motion.

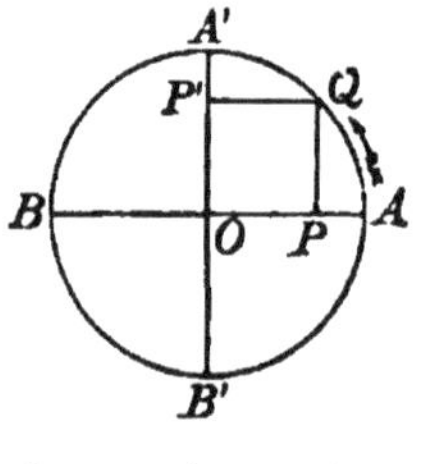

For, let BA, $B'A'$ be their ranges; and from O, their common middle point as centre describe a circle through $AA'BB'$. The given motion of P in BA will be (§ 53) defined by the motion of a point Q round the circumference of this circle; and the same point, if moving in the direction indicated by the arrow, will give a simple harmonic motion of P', in $B'A'$, a quarter of a period behind that of the motion of P in BA. But, since $A'OA$, QPO, and $QP'O$ are right angles, the figure $QP'OP$ is a parallelogram, and therefore Q is in the position of the displacement compounded of OP and OP'. Hence two equal simple harmonic motions in perpendicular lines, of phases ffering by a quarter period, are equivalent to a uniform cular motion of radius equal to the maximum displacement either singly, and in the direction from the positive end the range of the component in advance of the other towards e positive end of the range of this latter.

S. H. motions in different directions.

65. Now, orthogonal projections of simple harmonic motions e clearly simple harmonic with unchanged phase. Hence, if e project the case of § 64 on any plane, we get motion in an lipse, of which the projections of the two component ranges re conjugate diameters, and in which the radius-vector from the ntre describes equal areas (being the projections of the areas scribed by the radius of the circle) in equal times. But the lane and position of the circle of which this projection is aken may clearly be found so as to fulfil the condition of aving the projections of the ranges coincident with any two iven mutually bisecting lines. Hence any two given simple armonic motions, equal or unequal in range, and oblique or t right angles to one another in direction, provided only they iffer by a quarter period in phase, produce elliptic motion, aving their ranges for conjugate axes, and describing, by the adius-vector from the centre, equal areas in equal times.

66. Returning to the composition of any number of equal imple harmonic motions in lines in all directions and of all phases: each component simple harmonic motion may be determinately resolved into two in the same line, differing in phase by a quarter period, and one of them having any given epoch. We may therefore reduce the given motions to two sets, differing in phase by a quarter period, those of one set agreeing in phase with any one of the given, or with any other simple harmonic motion we please to choose (*i.e.*, having their epoch anything we please).

All of each set may (§ 58) be compounded into one simple harmonic motion of the same phase, of determinate amplitude, in a determinate line; and thus the whole system is reduced to two simple fully determined harmonic motions differing from one another in phase by a quarter period.

Now the resultant of two simple harmonic motions, one a quarter of a period in advance of the other, in different lines, has been proved to be motion in an ellipse of which the ranges

S. H. motions in different directions.

of the component motions are conjugate axes, and in which equal areas are described by the radius-vector from the centre in equal times. Hence the proposition of § 63.

Let

$$\left.\begin{aligned} x_1 &= l_1 a \cos(\omega t - \epsilon_1) \\ y_1 &= m_1 a \cos(\omega t - \epsilon_1) \\ z_1 &= n_1 a \cos(\omega t - \epsilon_1) \end{aligned}\right\} \quad (1)$$

be the Cartesian specification of the first of the given motions, and so with varied suffixes for the others;

l, m, n denoting the direction cosines,
a „ „ half amplitude,
ϵ „ „ epoch,

the proper suffix being attached to each letter to apply it to each case, and ω denoting the common relative angular velocity. The resultant motion, specified by x, y, z without suffixes, is

$$x = \Sigma l_1 a_1 \cos(\omega t - \epsilon_1) = \cos \omega t \Sigma l_1 a_1 \cos \epsilon_1 + \sin \omega t \Sigma l_1 a_1 \sin \epsilon_1,$$

$$y = \text{etc.};\ z = \text{etc.};$$

or, as we may write them for brevity,

$$\left.\begin{aligned} x &= P \cos \omega t + P' \sin \omega t \\ y &= Q \cos \omega t + Q' \sin \omega t \\ z &= R \cos \omega t + R' \sin \omega t \end{aligned}\right\} \quad (2)$$

where

$$\left.\begin{aligned} P &= \Sigma l_1 a_1 \cos \epsilon_1,\quad P' = \Sigma l_1 a_1 \sin \epsilon_1' \\ Q &= \Sigma m_1 a_1 \cos \epsilon_1,\quad Q' = \Sigma m_1 a_1 \sin \epsilon_1' \\ R &= \Sigma n_1 a_1 \cos \epsilon_1,\quad R' = \Sigma n_1 a_1 \sin \epsilon_1' \end{aligned}\right\} \quad (3)$$

The resultant motion thus specified, in terms of six component simple harmonic motions, may be reduced to two by compounding P, Q, R, and P', Q', R', in the elementary way. Thus if

$$\left.\begin{aligned} &\zeta = (P^2 + Q^2 + R^2)^{\frac{1}{2}} \\ &\lambda = \frac{P}{\zeta},\quad \mu = \frac{Q}{\zeta},\quad \nu = \frac{R}{\zeta} \\ &\zeta' = (P'^2 + Q'^2 + R'^2)^{\frac{1}{2}} \\ &\lambda' = \frac{P'}{\zeta'},\quad \mu' = \frac{Q'}{\zeta'},\quad \nu' = \frac{R'}{\zeta'} \end{aligned}\right\} \quad (4)$$

the required motion will be the resultant of $\zeta \cos \omega t$ in the line (λ, μ, ν), and $\zeta' \sin \omega t$ in the line (λ', μ', ν'). It is therefore motion in an ellipse, of which 2ζ and $2\zeta'$ in those directions are conjugate diameters; radius-vector from centre tracing equal areas in equal times; and period being $\frac{2\pi}{\omega}$.

H. motions of different kinds and in different lines.

67. We must next take the case of the composition of simple ıarmonic motions of *different* kinds and in different lines. In ;eneral, whether these lines be in one plane or not, the line of notion returns into itself if the periods are commensurable; and f not, not. This is evident without proof.

If a be the amplitude, ϵ the epoch, and n the angular velocity in the relative circular motion, for a component in a line whose direction cosines are λ, μ, ν—and if ξ, η, ζ be the co-ordinates in the resultant motion

$$\xi=\Sigma.\lambda_1 a_1 \cos(n_1 t-\epsilon_1),\quad \eta=\Sigma.\mu_1 a_1 \cos(n_1 t-\epsilon_1),$$
$$\zeta=\Sigma.\nu_1 a_1 \cos(n_1 t-\epsilon_1).$$

Now it is evident that at time $t+T$ the values of ξ, η, ζ will recur as soon as $n_1 T$, $n_2 T$, etc., are multiples of 2π, that is, when T is the least common multiple of $\frac{2\pi}{n_1}$, $\frac{2\pi}{n_2}$, etc.

If there be such a common multiple, the trigonometrical functions may be eliminated, and the equations (or equation, if the motion is in one plane) to the path are algebraic. If not, they are transcendental.

68. From the above we see generally that the composition f any number of simple harmonic motions in any directions :d of any periods, may be effected by compounding, according previously explained methods, their resolved parts in each f any three rectangular directions, and then compounding the :al resultants in these directions.

S. H. motions in two rectangular directions.

69. By far the most interesting case, and by far the simplest, that of *two* simple harmonic motions of any periods, whose :ections must of course be in one plane.

Mechanical methods of obtaining such combinations will be ·rwards described, as well as cases of their occurrence in ·tics and Acoustics.

We may suppose, for simplicity, the two component motions take place in perpendicular directions. Also, as we can only ave a re-entering curve when their periods are commensurable, : will be advisable to commence with such a case.

The following figures represent the paths produced by the ombination of simple harmonic motions of *equal* amplitude n two rectangular directions, the periods of the components

S. H. motions in two rectangular directions.

being as 1 : 2, and the epochs differing successively by 0, 1, 2 etc., sixteenths of a circumference.

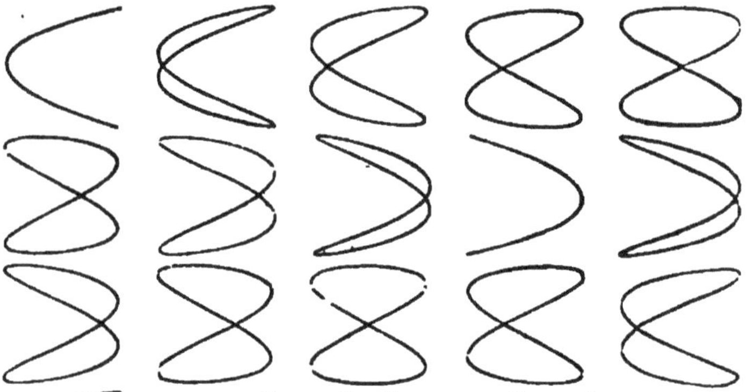

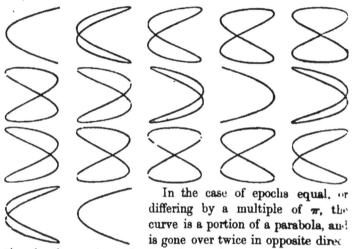

In the case of epochs equal, or differing by a multiple of π, the curve is a portion of a parabola, and is gone over twice in opposite directions by the moving point in each complete period.

For the case figured above,

$$x = a\cos(2nt - \epsilon), \quad y = a\cos nt.$$

Hence $$x = a\{\cos 2nt \cos\epsilon + \sin 2nt \sin\epsilon\}$$

$$= a\{(\frac{2y^2}{a^2} - 1)\cos\epsilon + 2\frac{y}{a}\sqrt{1 - \frac{y^2}{a^2}}\sin\epsilon\}$$

which for any given value of ϵ is the equation to the corresponding curve. Thus for $\epsilon = 0$,

$$\frac{x}{a} = \frac{2y^2}{a^2} - 1, \text{ or } y^2 = \frac{a}{2}(x + 2a), \text{ the parabola as above.}$$

$$\text{For } \epsilon = \frac{\pi}{4} \text{ we have } \frac{x}{a} = 2\frac{y}{a}\sqrt{1 - \frac{y^2}{a^2}}, \text{ or } a^2x^2 = 4y^2(a^2 - y^2),$$

the equation to the 5th and 13th of the above curves.

In general $x = a\cos(nt + \epsilon)$, $y = a\cos(n_1 t + \epsilon_1)$, from which t is to be eliminated if possible.

Composition of two uniform circular motions.

70. Another very important case is that of two groups of two simple harmonic motions in one plane, such that the resultant of each group is uniform circular motion.

If their periods are equal, we have a case belonging to those already treated (§ 63), and conclude that the resultant is, in general, motion in an ellipse, equal areas being described in equal times about the centre. As particular cases we may have simple harmonic, or uniform circular, motion. Composition of two uniform circular motions.

If the circular motions are in the *same* direction, the resultant is evidently circular motion in the same direction. This is the case of the motion of S in § 58, and requires no further comment, as its amplitude, epoch, etc., are seen at once from the figure.

71. If the periods of the two are very nearly equal, the resultant motion will be at any moment very nearly the circular motion given by the preceding construction. Or we may regard it as rigorously a motion in a circle with a varying radius decreasing from a maximum value, the sum of the radii of the two component motions, to a minimum, their difference, and increasing again, alternately; the direction of the resultant radius oscillating on each side of that of the greater component (as in corresponding case, § 59, above). Hence the angular velocity of the resultant motion is periodically variable. In the case of equal radii, next considered, it is constant.

72. When the radii of the two component motions are equal, we have the very interesting and important case figured below.

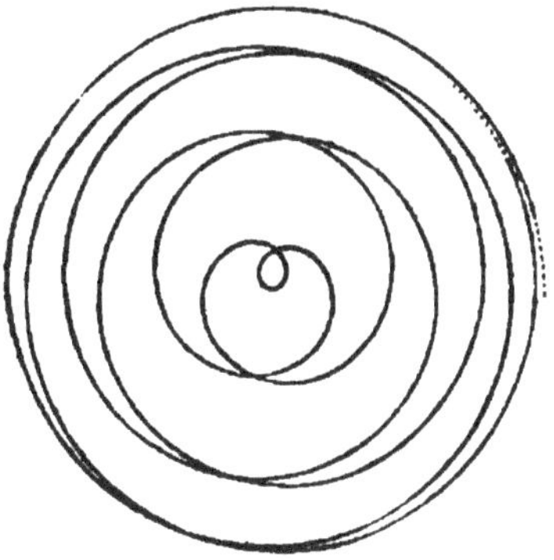

Here the resultant radius bisects the angle between the component radii. The resultant angular velocity is the arithmetical mean of its components. We will explain in a future section how this epitrochoid is traced by the rolling of one circle on another. (The particular case above delineated is that of a non-reëntrant curve.)

S. H. motions in two rectangular directions.

being a

etc., six

is along a telegraph wire, and the conduction of heat ... earth's crust, as subjects in their generality intractable ...ut it, is to give but a feeble idea of its importance. ...llowing seems to be the most intelligible form in which ... be presented to the general reader :— Fourier's Theorem.

...OREM.—*A complex harmonic function, with a constant added, is the proper expression, in mathematical language, ...y arbitrary periodic function; and consequently can ex- ...any function whatever between definite values of the variable.*

6. Any arbitrary periodic function whatever being given, ...mplitudes and epochs of the terms of a complex harmonic ...ion which shall be equal to it for every value of the in- ...ndent variable, may be investigated by the "method of ...erminate co-efficients." Such an investigation is suffi- ... as a solution of the problem,—to find a complex harmonic ...tion expressing a given arbitrary periodic function,—when ... we are assured that the problem is possible; and when ...have this assurance, it proves that the resolution is deter- ...nate; that is to say, that no other complex harmonic function ...n the one we have found can satisfy the conditions.

77. We might give the analytical proof of the theorem by process mentioned in last section, but it appears to us that ... nature of the expression will be made more intelligible by investigation from a different point of view.

Let $F(x)$ be any periodical function, of period p. That is to say, let $F(x)$ be any function fulfilling the condition

$$F(x+ip)=F(x) \qquad (1)$$

where i denotes any positive or negative integer. Consider the integral

$$\int_{c'}^{c} \frac{F(x)dx}{a^2+x^2},$$

where a, c, c' denote any three given quantities. Its value is less than $F(z)\int_{c'}^{c}\frac{dx}{a^2+x^2}$, and greater than $F(z')\int_{c'}^{c}\frac{dx}{a^2+x^2}$, if z and z' denote the values of x, either equal to or intermediate between the limits c and c', for which $F(x)$ is greatest and least respectively. But

$$\int_{c'}^{c}\frac{dx}{a^2+x^2}=\frac{1}{a}\left(\tan^{-1}\frac{c}{a}-\tan^{-1}\frac{c'}{a}\right) \qquad (2)$$

Fourier's Theorem.

and therefore

$$\left.\begin{aligned}&\int_{c'}^{c}\frac{F(x)a\,dx}{a^2+x^2} < F(z)\left(\tan^{-1}\frac{c}{a}-\tan^{-1}\frac{c'}{a}\right)\\ \text{and}\qquad&\qquad\text{,,}\qquad > F(z')\left(\tan^{-1}\frac{c}{a}-\tan^{-1}\frac{c'}{a}\right)\end{aligned}\right\}\quad(3)$$

Hence if A be the greatest of all the values of $F(x)$, and B the least,

$$\left.\begin{aligned}&\int_{c}^{\infty}\frac{F(x)a\,dx}{a^2+x^2} < A\left(\frac{\pi}{2}-\tan^{-1}\frac{c}{a}\right)\\ \text{and}\qquad&\qquad\text{,,}\qquad > B\left(\frac{\pi}{2}-\tan^{-1}\frac{c}{a}\right)\end{aligned}\right\}\quad(4)$$

Also, similarly,

$$\left.\begin{aligned}&\int_{-\infty}^{c'}\frac{F(x)a\,dx}{a^2+x^2} < A\left(\tan^{-1}\frac{c'}{a}+\frac{\pi}{2}\right)\\ \text{and}\qquad&\qquad\text{,,}\qquad > B\left(\tan^{-1}\frac{c'}{a}+\frac{\pi}{2}\right)\end{aligned}\right\}\quad(5).$$

Adding the first members of (3), (4), and (5), and comparing with the corresponding sums of the second members, we find

$$\left.\begin{aligned}&\int_{-\infty}^{\infty}\frac{F(x)a\,dx}{a^2+x^2} < F(z)\left(\tan^{-1}\frac{c}{a}-\tan^{-1}\frac{c'}{a}\right)+A\left(\pi-\tan^{-1}\frac{c}{a}+\tan^{-1}\frac{c'}{a}\right)\\ \text{and}\quad&\text{,,}\quad > F(z')\left(\tan^{-1}\frac{c}{a}-\tan^{-1}\frac{c'}{a}\right)+B\left(\pi-\tan^{-1}\frac{c}{a}+\tan^{-1}\frac{c'}{a}\right)\end{aligned}\right\}\quad(6)$$

But, by (1),

$$\int_{-\infty}^{\infty}\frac{F(x)dx}{a^2+x^2}=\int_0^p F(x)dx\left\{\sum_{i=-\infty}^{i=\infty}\left(\frac{1}{a^2+(x+ip)^2}\right)\right\}\quad(7)$$

Now if we denote $\sqrt{-1}$ by v,

$$\frac{1}{a^2+(x+ip)^2}=\frac{1}{2av}\left(\frac{1}{x+ip-av}-\frac{1}{x+ip+av}\right),$$

and therefore, taking the terms corresponding to positive and equal negative values of i together, and the terms for $i=0$ separately, we have

$$\sum_{i=-\infty}^{i=\infty}\left(\frac{1}{a^2+(x+ip)^2}\right)=\frac{1}{2av}\left\{\frac{1}{x-av}-2\sum_{i=1}^{i=\infty}\frac{x-av}{i^2p^2-(x-av)^2}\right.$$

$$\left.-\frac{1}{x+av}+2\sum_{i=1}^{i=\infty}\frac{x+av}{i^2p^2-(x+av)^2}\right\}$$

$$=\frac{\pi}{2apv}\left\{\cot\frac{\pi(x-av)}{p}-\cot\frac{\pi(x+av)}{p}\right\}$$

$$=\frac{\dfrac{\pi}{2apv}\sin\dfrac{2\pi av}{p}}{\cos^2\dfrac{\pi av}{p}-\cos^2\dfrac{\pi x}{p}}=\frac{\dfrac{\pi}{apv}\sin\dfrac{2\pi av}{p}}{\cos\dfrac{2\pi av}{p}-\cos\dfrac{2\pi x}{p}}$$

Fourier's Theorem.

$$=\frac{\pi}{ap}\frac{\varepsilon^{\frac{2\pi a}{p}}-\varepsilon^{-\frac{2\pi a}{p}}}{\varepsilon^{\frac{2\pi a}{p}}-2\cos\frac{2\pi x}{p}+\varepsilon^{-\frac{2\pi a}{p}}}.$$

Hence,

$$\int_{-\infty}^{\infty}\frac{F(x)dx}{a^2+x^2}=\frac{\pi}{ap}(\varepsilon^{\frac{2\pi a}{p}}-\varepsilon^{-\frac{2\pi a}{p}})\int_0^p\frac{F(x)dx}{\varepsilon^{\frac{2\pi a}{p}}-2\cos\frac{2\pi x}{p}+\varepsilon^{-\frac{2\pi a}{p}}} \quad (8).$$

Next, denoting temporarily, for brevity, $\varepsilon^{\frac{2\pi x\upsilon}{p}}$, by ζ, and putting

$$\varepsilon^{-\frac{2\pi a}{p}}=e \quad (9),$$

we have

$$\frac{1}{\varepsilon^{\frac{2\pi a}{p}}-2\cos\frac{2\pi x}{p}+\varepsilon^{-\frac{2\pi a}{p}}}=\frac{e}{1-e(\zeta+\zeta^{-1})+e^2}$$

$$=\frac{e}{1-e^2}\left(\frac{1}{1-e\zeta}+\frac{1}{1-e\zeta^{-1}}-1\right)$$

$$=\frac{e}{1-e^2}\{1+e(\zeta+\zeta^{-1})+e^2(\zeta^2+\zeta^{-2})+e^3(\zeta^3+\zeta^{-3})+\text{ etc.}\}$$

$$=\frac{e}{1-e^2}(1+2e\cos\frac{2\pi x}{p}+2e^2\cos\frac{4\pi x}{p}+2e^3\cos\frac{6\pi x}{p}+\text{ etc.})$$

Hence, according to (8) and (9),

$$\int_{-\infty}^{\infty}\frac{F(x)dx}{a^2+x^2}=\frac{\pi}{ap}\int_0^p F(x)\,dx\,(1+2e\cos\frac{2\pi x}{p}+2e^2\cos\frac{4\pi x}{p}+\text{ etc.}) \quad (10)$$

Hence, by (6), we infer that

$$F(z)(\tan^{-1}\frac{c}{a}-\tan^{-1}\frac{c'}{a})+A(\pi-\tan^{-1}\frac{c}{a}+\tan^{-1}\frac{c'}{a})>$$

and

$$F(z')(\tan^{-1}\frac{c}{a}-\tan^{-1}\frac{c'}{a})+B\,(\pi-\tan^{-1}\frac{c}{a}+\tan^{-1}\frac{c'}{a})<$$

$$\frac{\pi}{p}\int_0^p F(x)dx(1+2e\cos\frac{2\pi x}{p}+\text{ etc.}).$$

Now let $c'=-c$, and $x=\xi'-\xi$,

ξ' being a variable, and ξ constant, so far as the integration is concerned; and let $F(x)=\phi(x+\xi)=\phi(\xi')$,

and therefore $F(z)=\phi(\xi+z)$

$F(z')=\phi(\xi+z')$.

Fourier's Theorem.

The preceding pair of inequalities becomes

$$\left.\begin{aligned}&\phi(\xi+z).2\tan^{-1}\frac{c}{a}+A\left(\pi-2\tan^{-1}\frac{c}{a}\right)>\\ \text{and}\quad&\phi(\xi+z').2\tan^{-1}\frac{c}{a}+B\left(\pi-2\tan^{-1}\frac{c}{a}\right)<\\ \frac{\pi}{p}\Big\{\int_0^p&\phi(\xi')d\xi'+2\sum_{i=1}^{i=\infty}e^i\int_0^p\phi(\xi')d\xi'\cos\frac{2i\pi(\xi'-\xi)}{p}\Big\}\end{aligned}\right\}\quad(11)$$

where ϕ denotes any periodical function whatever, of period p.

Now let c be a very small fraction of p. In the limit, where c is infinitely small, the greatest and least values of $\phi(\xi')$ for values of ξ' between $\xi+c$ and $\xi-c$ will be infinitely nearly equal to one another and to $\phi(\xi)$; that is to say,

$$\phi(\xi+z)=\phi(\xi+z')=\phi(\xi).$$

Next, let a be an infinitely small fraction of c. In the limit

$$\tan^{-1}\frac{c}{a}=\frac{\pi}{2},$$

and

$$e=\varepsilon^{-\frac{2\pi a}{p}}=1.$$

Hence the comparison (11) becomes in the limit an equation which, if we divide both members by π, gives

$$\phi(\xi)=\frac{1}{p}\Big\{\int_0^p\phi(\xi')d\xi'+2\sum_{i=1}^{i=\infty}\int_0^p\phi(\xi')d\xi'\cos\frac{2i\pi(\xi'-\xi)}{p}\Big\}\quad(12)$$

This is the celebrated theorem discovered by Fourier[1] for the development of an arbitrary periodic function in a series of simple harmonic terms. A formula included in it as a particular case had been given previously by Lagrange.[2]

If, for $\cos\frac{2i\pi(\xi'-\xi)}{p}$, we take its value $\cos\frac{2i\pi\xi'}{p}\cos\frac{2i\pi\xi}{p}+\sin\frac{2i\pi\xi'}{p}\sin\frac{2i\pi\xi}{p}$ and introduce the following notation:—

$$\left.\begin{aligned}A_0&=\frac{1}{p}\int_0^p\phi(\xi)d\xi\\ A_i&=\frac{2}{p}\int_0^p\phi(\xi)\cos\frac{2i\pi\xi}{p}d\xi\\ B_i&=\frac{2}{p}\int_0^p\phi(\xi)\sin\frac{2i\pi\xi}{p}d\xi\end{aligned}\right\}\quad(13)$$

we reduce (12) to this form:—

$$\phi(\xi)=A_0+\sum_{i=1}^{i=\infty}A_i\cos\frac{2i\pi\xi}{p}+\sum_{i=1}^{i=\infty}B_i\sin\frac{2i\pi\xi}{p}\quad(14)$$

[1] *Théorie Analytique* de la Chaleur. Paris, 1822.

[2] *Anciens Mémoires de l'Académie de Turin.* Tome iii. p. 126.

which is the general expression of an arbitrary function in terms of a series of cosines and of sines. Or if we take Fourier's Theorem.

$$P_i=(A_i^2+B_i^2)^{\frac{1}{2}}, \text{ and } \tan\epsilon_i=\frac{B_i}{A_i} \qquad (15)$$

we have

$$\phi(\xi)=A_0+\sum_{i=1}^{i=\infty} P_i \cos\left(\frac{2i\pi\xi}{p}-\epsilon_i\right), \qquad (16)$$

which is the general expression in a series of single simple harmonic terms of the successive multiple periods.

To prevent misconception, it should be remarked that each of the equations and comparisons (2), (7), (8), (10), and (11) is a true arithmetical expression, and that it may be verified by actual calculation of the numbers, for any particular case; provided only that $F(x)$ has no infinite value in its period. Hence, with this exception, (12) or either of its equivalents, (14), (16), is a true arithmetical expression; and the series which it involves is therefore convergent. Hence we may with perfect rigour conclude that even the extreme case in which the arbitrary function experiences an abrupt finite change in its value when the independent variable, increasing continuously, passes through some particular value or values, is included in the general theorem. In such a case, if any value be given to the independent variable differing however little from one which corresponds to an abrupt change in the value of the function, the series must, as we may infer from the preceding investigation, converge and give a definite value for the function. But if exactly the critical value is assigned to the independent variable, the series cannot converge to any definite value. The consideration of the limiting values shown in the comparison (11), does away with all difficulty in understanding how the series (12) gives definite values having a finite difference, for two particular values of the independent variable on the two sides of a critical value, but differing infinitely little from one another. Convergency of Fourier's series.

If the differential co-efficient $\frac{d\phi(\xi)}{d\xi}$ is finite for every value of ξ within the period, it too is arithmetically expressible by a series of harmonic terms, which cannot be other than the series obtained by differentiating the series for $\phi(\xi)$. Hence

$$\frac{d\phi(\xi)}{d\xi}=-\frac{2\pi}{p}\sum_{i=1}^{i=\infty} i\,P_i \sin\left(\frac{2i\pi\xi}{p}-\epsilon_i\right), \qquad (17)$$

and this series is convergent; and we may therefore conclude that

Convergency of Fourier's series.

the series for $\phi(\xi)$ is more convergent than a harmonic series with

$$1, \ \tfrac{1}{2}, \ \tfrac{1}{3}, \ \tfrac{1}{4}, \ \text{etc.},$$

for its co-efficients. If $\dfrac{d^2\phi(\xi)}{d\xi^2}$ has no infinite values within the period, we may differentiate both members of (17) and still have an equation arithmetically true; and so on. We conclude that if the nth differential co-efficient of $\phi(\xi)$ has no infinite values, the harmonic series for $\phi(\xi)$ must converge more rapidly than a harmonic series with

$$1, \ \frac{1}{2^n}, \ \frac{1}{3^n}, \ \frac{1}{4^n}, \ \text{etc.},$$

for its co-efficients.

Displacement of a rigid body.

78. We now pass to the consideration of the displacement of a rigid body or group of points whose relative positions are unalterable. The simplest case we can consider is that of the motion of a plane figure in its own plane, and this, as far as kinematics is concerned, is entirely summed up in the result of the next section.

Displacements of a plane figure in its plane.

79. If a plane figure be displaced in any way in its own plane, there is always (with an exception treated in § 81) one point of it common to any two positions; that is, it may be moved from any one position to any other by rotation in its own plane about one point held fixed.

To prove this, let A, B be any two points of the plane figure in its first position, A', B' the positions of the same two after a displacement. The lines AA', BB' will not be parallel, except in one case to be presently considered. Hence the line equidistant from A and A' will meet that equidistant from B and B' in some point O. Join OA, OB, OA', OB'. Then, evidently, because $OA' = OA$, $OB' = OB$, and $A'B' = AB$, the triangles $OA'B'$ and OAB are equal and similar. Hence O is similarly situated with regard to $A'B'$ and AB, and is therefore one and the same point of the plane figure in its two positions. If, for the sake of illustration, we actually trace the triangle OAB upon the plane, it becomes $OA'B'$ in the second position of the figure.

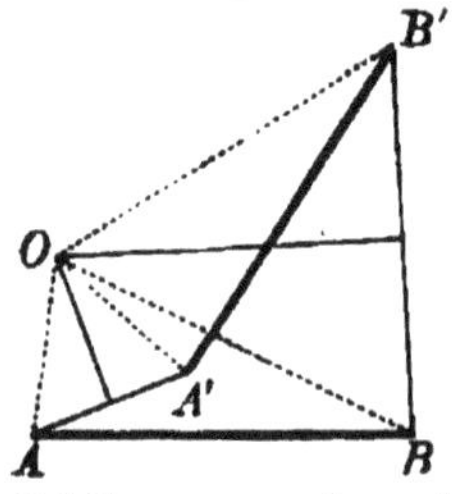

80. If from the equal angles $A'OB'$, AOB of these similar triangles we take the common part $A'OB$, we have the remaining angles AOA', BOB' equal, and each of them is clearly equal to the angle through which the figure must have turned round the point O to bring it from the first to the second position. Displacements of a plane figure in its plane.

The preceding simple construction therefore enables us not only to demonstrate the general proposition, § 79, but also to determine from the two positions of one line AB, $A'B'$ of the figure the common centre and the amount of the angle of rotation.

81. The lines equidistant from A and A', and from B and B', are parallel if AB is parallel to $A'B'$; and therefore the construction fails, the point O being infinitely distant, and the theorem becomes nugatory. In this case the motion is in fact a simple translation of the figure in its own plane without rotation—since, AB being parallel and equal to $A'B'$, we have AA' parallel and equal to BB'; and instead of there being one point of the figure common to both positions, the lines joining the successive positions of every point in the figure are equal and parallel.

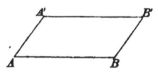

82. It is not necessary to suppose the figure to be a mere flat disc or plane—for the preceding statements apply to any one of a set of parallel planes in a rigid body, moving in any way subject to the condition that the points of any one plane in it remain always in a fixed plane in space.

83. There is yet a case in which the construction in § 79 is nugatory—that is when AA' is parallel to BB', but AB intersects $A'B'$. In this case, however, it is easy to see at once that this point of intersection is the point O required, although the former method would not have enabled us to find it.

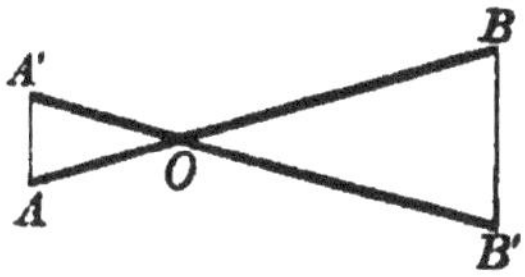

84. Very many interesting applications of this principle may be made, of which, however, few belong strictly to our subject, and we shall therefore give only an example or two. Thus we know that if a line of given length AB move with Examples of displacement in one plane.

Examples of displacement in one plane.

its extremities always in two fixed lines OA, OB, any point in it as P describes an ellipse. It is required to find the direction of motion of P at any instant, *i.e.*, to draw a tangent to the ellipse. BA will pass to its next position by rotating about the point Q; found by the method of § 79 by drawing perpendiculars to OA and OB at A and B. Hence P for the instant revolves about Q, and thus its direction of motion, or the tangent to the ellipse, is perpendicular to QP. Also AB in its motion always touches a curve (called in geometry its envelop); and the same principle enables us to find the point of the envelop which lies in AB, for the motion of that point must evidently be ultimately (that is for a very small displacement) along AB, and the only point which so moves is the intersection of AB, with the perpendicular to it from Q. Thus our construction would enable us to trace the envelop by points.

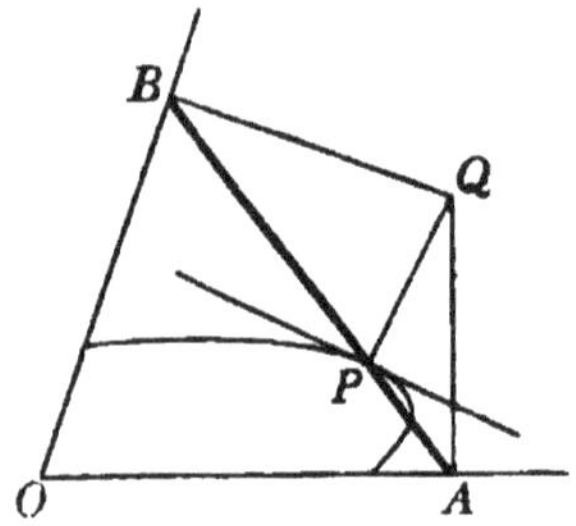

85. Again, suppose AB to be the beam of a stationary engine having a reciprocating motion about A, and by a link BD turning a crank CD about C. Determine the relation between the angular velocities of AB and CD in any position. Evidently the instantaneous direction of motion of B is transverse to AB, and of D transverse to CD—hence if AB, CD produced meet in O, the motion of BD is for an instant as if it turned about O. From this it may easily be seen that if the angular velocity of AB be ω, that of CD is $\frac{AB}{OB}\frac{OD}{CD}\omega$. A similar process is of course applicable to any combination of machinery, and we shall find it very convenient when we come to consider various dynamical problems connected with virtual velocities.

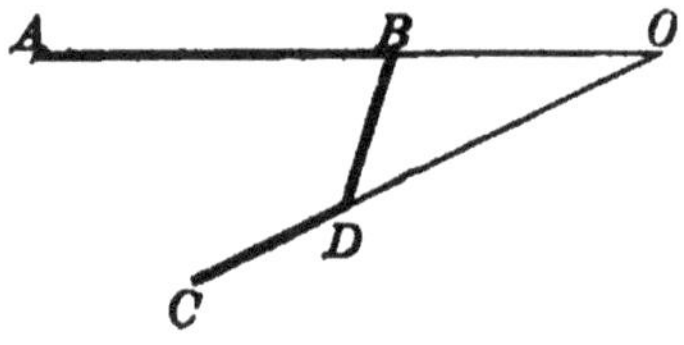

Composition of rotations about parallel axes.

86. Since in general any movement of a plane figure in its plane may be considered as a rotation about one point, it is evident that two such rotations may in general be compounded into one; and therefore, of course, the same may be done with any number of rotations. Thus let A and B be the points of the figure about which in succession the rotations are to take place. By rotation about A, B is brought say to B', and by a rotation about B', A is brought to A'. The construction of § 79 gives us at once the point O and the amount of rotation about it which singly gives the same effect as those about A and B in succession. But there is one case of exception, viz., when the rotations about A and B are of equal amount and in opposite directions. In this case $A'B'$ is evidently parallel to AB, and therefore the compound result is a *translation* only. That is, if a body revolve in succession through equal angles, but in opposite directions about two parallel axes, it finally takes a position to which it could have been brought by a simple translation perpendicular to the lines of the body in its initial or final position, which were successively made axes of rotation; and inclined to their plane at an angle equal to half the supplement of the common angle of rotation. **Composition of rotations about parallel axes**

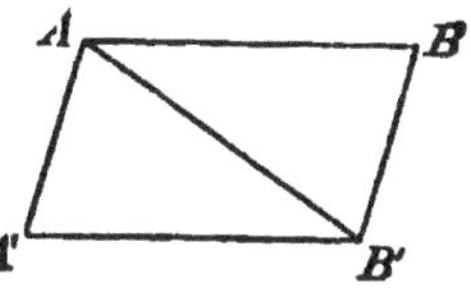

87. Hence to compound into an equivalent rotation a rotation and a translation, the latter being effected parallel to the plane of the former, we may decompose the translation into two rotations of equal amount and opposite direction, compound one of them with the given rotation by § 86, and then compound the other with the resultant rotation by the same process. Or we may adopt the following far simpler method. Let OA be the translation common to all points in the plane, and let BOC be the angle of rotation about O, BO being drawn so that OA bisects the exterior angle COB'. Evidently there is a point B' in BO produced, such that $B'C'$, the space through which the rotation carries it, is equal and opposite to OA. This point retains its former position after the performance of the compound operation; so that a rotation and a translation in one plane can be compounded into an equal rotation about a different axis. **Composition of rotations and translations in one plane.**

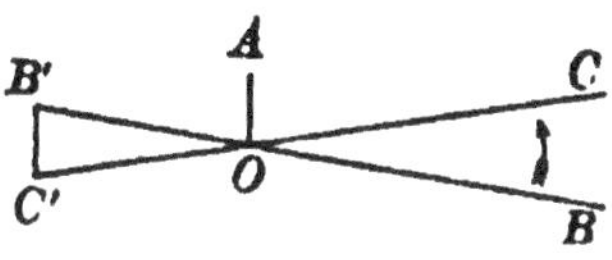

Composition of rotations and translations in one plane.

In general if the origin be taken as the point about which rotation takes place in the plane of xy, and if it be through an angle θ, a point whose co-ordinates were originally x, y, will have them changed to

$$\xi = x\cos\theta - y\sin\theta,\ \eta = x\sin\theta + y\cos\theta,$$

or, if the rotation be very small,

$$\xi = x - y\theta,\ \eta = y + x\theta.$$

Omission of the second and higher orders of small quantities.

88. In considering the composition of angular velocities about different axes, and other similar cases, we may deal with infinitely small displacements only; and it results at once from the principles of the differential calculus, that if these displacements be of the *first* order of small quantities, any point whose displacement is of the *second* order of small quantities is to be considered as rigorously at rest. Hence, for instance, if a body revolve through an angle of the first order of small quantities about an axis belonging to the body which during the revolution is displaced through an angle or space, also of the first order, the displacement of any point of the body is rigorously what it would have been had the axis been fixed during the rotation about it, and its own displacement made either before or after this rotation. Hence in any case of motion of a rigid system the angular velocities about a system of axes moving *with* the system are the same at any instant as those about a system fixed in space, provided only that the latter coincide at the instant in question with the moveable ones.

Superposition of small motions.

89. From similar considerations follows also the general principle of *Superposition of small motions*. It asserts that if several causes act *simultaneously* on the same particle or rigid body, and if the effect produced by each is of the first order of small quantities, the joint effect will be obtained if we consider the causes to act *successively*, each taking the point or system in the position in which the preceding one left it. It is evident at once that this is an immediate deduction from the fact that the second order of small quantities may be with rigorous accuracy neglected. This principle is of very great use, as we shall find in the sequel; its applications are of constant occurrence.

A plane figure has given angular velocities about given axes perpendicular to its plane, find the resultant. Superposition of small motions.

Let there be an angular velocity ω about an axis at the point a, b.

The consequent motion of the point x, y in the time δt is, as as we have just seen (§ 87),

$$-(y-b)\omega\delta t \text{ parallel to } x, \text{ and } (x-a)\omega\delta t \text{ parallel to } y.$$

Hence, by the superposition of small motions, the whole motion parallel to x is

$$-(y\Sigma\omega - \Sigma b\omega)\delta t,$$

and that parallel to y

$$(x\Sigma\omega - \Sigma a\omega)\delta t.$$

Hence the point whose co-ordinates are

$$x=\frac{\Sigma a\omega}{\Sigma\omega} \text{ and } y=\frac{\Sigma b\omega}{\Sigma\omega}$$

is at rest, and the resultant axis passes through it. Any other point ξ, η moves through spaces

$$-(\eta\Sigma\omega - \Sigma b\omega)\delta t, \quad (\xi\Sigma\omega - \Sigma a\omega)\delta t.$$

But if the whole had turned about x, y with velocity Ω, we should have had for the displacements of ξ, η

$$-(\eta - y)\Omega\delta t, \quad (\xi - x)\Omega\delta t.$$

Comparing, we find $\Omega=\Sigma\omega$.

Hence if the sum of the angular velocities be zero, there is no rotation, and indeed the above formulæ show that there is then merely translation

$$\Sigma(b\omega)\delta t \text{ parallel to } x, \text{ and } -\Sigma(a\omega)\delta t \text{ parallel to } y.$$

These formulæ suffice for the consideration of any problem on the subject.

90. Any motion whatever of a plane figure in its own plane might be produced by the rolling of a curve fixed to the figure upon a curve fixed in the plane. Rolling of curve on curve.

For we may consider the whole motion as made up of successive elementary displacements, each of which corresponds, as we have seen, to an elementary rotation about some point in the plane. Let O_1, O_2, O_3, etc., be the successive points of the *figure* about which the rotations take place, o_1, o_2, o_3, etc., the positions of these points on the *plane* when each is the instantaneous centre of rotation. Then the figure rotates about O_1 (or o_1, which coincides with it) till O_2 coincides with o_2, then

Rolling of curve on curve.

about the latter till O_3 coincides with o_3, and so on. Hence, if we join O_1, O_2, O_3, etc., in the plane of the figure, and o_1, o_2, o_3, etc., in the fixed plane, the motion will be the same as if the polygon $O_1O_2O_3$ etc., rolled upon the fixed polygon $o_1o_2o_3$ etc. By supposing the successive displacements small enough, the sides of these polygons gradually diminish, and the polygons finally become continuous curves. Hence the theorem.

From this it immediately follows, that any displacement of a rigid solid, which is in directions wholly perpendicular to a fixed line, may be produced by the rolling of a cylinder fixed in the solid on another cylinder fixed in space, the axes of the cylinders being parallel to the fixed line.

91. As an interesting example of this theorem, let us recur to the case of § 84 :—A circle may evidently be circumscribed about *OBQA* ; and it must be of invariable magnitude, since in it a chord of given length *AB* subtends a given angle *O* at the circumference. Also *OQ* is a diameter of this circle, and is therefore constant. Hence, as *Q* is momentarily at rest, the motion of the circle circumscribing *OBQA* is one of internal rolling on a circle of double its diameter. Hence if a circle roll internally on another of twice its diameter, any point in its circumference describes a diameter of the fixed circle, any other point in its plane an ellipse. This is precisely the same proposition as that of § 70, although the ways of arriving at it are very different. As it presents us with a particular case of the Hypocycloid, it warns us to return to the consideration of these and kindred curves which give good instances of kinematical theorems, but which besides are of great use in physics generally.

Cycloids and Trochoids.

92. When a circle rolls upon a straight line, a point in its circumference describes a Cycloid, an internal point describes a Prolate, an external one a Curtate, Cycloid. The two latter varieties are sometimes called Trochoids.

The general form of these curves will be seen in the annexed figures ; and in what follows we shall confine our remarks to the cycloid itself, as of immensely greater consequence than the others. The next section contains a simple investigation of those properties of the cycloid which are most useful in our subject.

Cycloids and Trochoids.

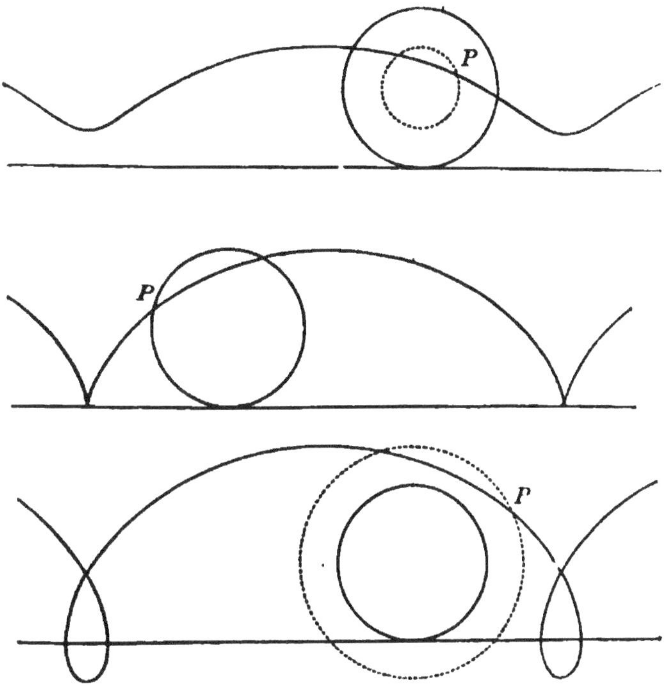

93. Let AB be a diameter of the generating (or rolling) circle, BC the line on which it rolls. The points A and B describe similar and equal cycloids, of which AQC and BS are portions. If PQR be any subsequent position of the generating circle, Q and S the new positions of A and B, $Q\hat{P}S$ is of course a right angle. If, therefore, QR be drawn parallel to PS, PR is a diameter of the rolling circle. Produce QR to T, making $RT = QR = PS$. Evidently the curve AT, which is the locus of T, is similar and equal to BS, and is therefore a cycloid similar and equal to AC. But QR is perpendicular to PQ, and is therefore the instantaneous direction of motion of Q, or is the tangent to the

Properties of the cycloid.

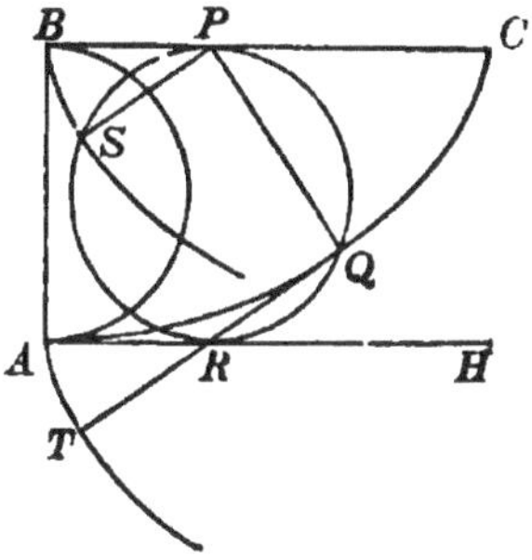

Properties of the cycloid. cycloid AQC. Similarly, PS is perpendicular to the cycloid BS at S, and so is therefore TQ to AT at T. Hence (§ 19 AQC is the evolute of AT, and arc $AQ = QT = 2\,QR$.

Epicycloids. Hypo-cycloids, etc. **94.** When the circle rolls upon another circle, the curve described by a point in its circumference is called an Epicycloid or a Hypocyloid, as the rolling circle is without or within the fixed circle; and when the tracing point is not in the circumference, we have Epitrochoids and Hypotrochoids. Of the latter we have already met with examples, § 70, 91, and others will be presently mentioned. Of the former, we have in the first of the appended figures the case of a circle rolling externally on another of equal size. The curve in this case is called the Cardioid (§ 49).

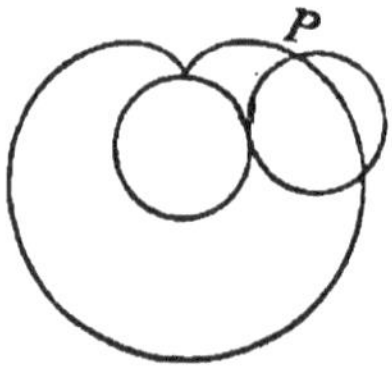

In the second, a circle rolls externally on another of twice its radius. The epicycloid so described is of importance in optics, and will, with others, be referred to when we consider the subject of Caustics by reflexion.

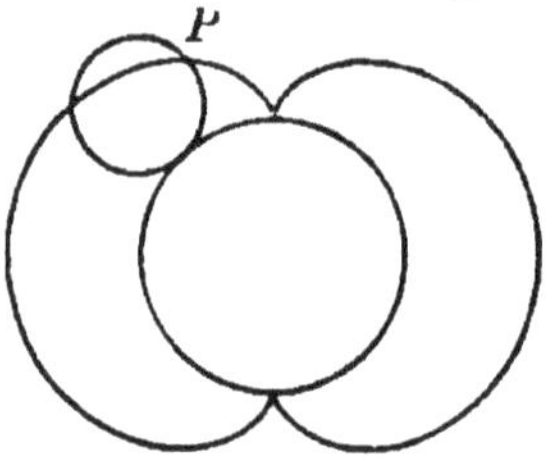

In the third, we have a hypocycloid traced by the rolling of one circle internally on another of four times its radius.

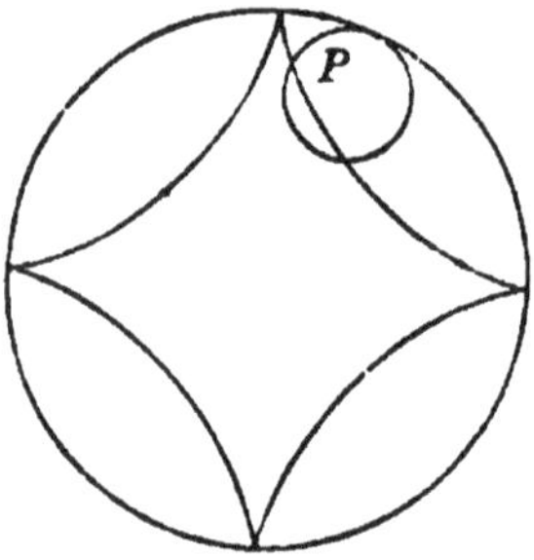

The curve figured in § 72 is an epitrochoid described by a point in the plane of a large circular disc which rolls upon a circular cylinder of small diameter, so that the point passes through the axis of the cylinder.

That of § 74 is a hypotrochoid described by a point in the plane of a circle which rolls internally on another of rather more than twice its diameter, the tracing point passing through the centre of the fixed circle. Had the diameters of the circles been exactly as 1 : 2, § 73 or § 91 shows us that this curve would have been reduced to a single straight line.

Epicycloids, Hypo-cycloids, etc.

The general equations of this class of curves are

$$x=(a+b)\cos\theta - eb\cos\frac{a+b}{b}\theta,$$

$$y=(a+b)\sin\theta - eb\sin\frac{a+b}{b}\theta,$$

where a is the radius of the fixed, b of the rolling circle; and eb is the distance of the tracing point from the centre of the latter.

Motion about a fixed point.

95. If a rigid solid body move in any way whatever, subject only to the condition that one of its points remains fixed, there is always (without exception) one line of it through this point common to the body in any two positions.

Consider a spherical surface within the body, with its centre at the fixed point C. All points of this sphere attached to the body will move on a sphere fixed in space. Hence the construction of § 79 may be made, only with great circles instead of straight lines; and the same reasoning will apply to prove that the point O thus obtained is common to the body in its two positions. Hence every point of the body in the line OC, joining O with the fixed point, must be common to it in the two positions. Hence the body may pass from any one position to any other by a definite amount of rotation about a definite axis. And hence, also, successive or simultaneous rotations about any number of axes through the fixed point may be compounded into one such rotation.

Composition of rotations.

Composition of angular velocities.

Let OA, OB, be two axes about which a body moves with angular velocities ω, ω_1.

Let the unit sphere be described with centre O, and cut these in A, B. Consider any other point P on the sphere. Then we may (§ 89) treat the displacements of P during an infinitely small interval δt as successive.

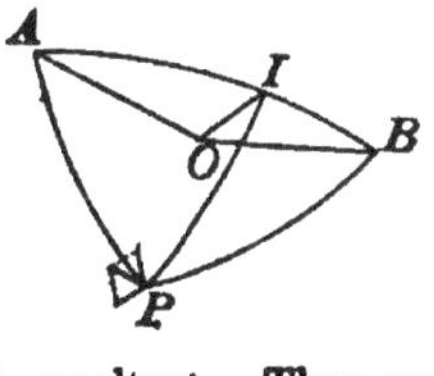

The displacements of P are in the tangent plane at P, and so is therefore their resultant. They are respectively perpendicular at P to the arcs AP, BP, their magnitudes are $\omega\sin AP\delta t$ and $\omega_1\sin BP\delta t$, and their directions contain an angle $=\overset{<}{APB}$.

Take a point I in AB, such that $\omega\sin AI=\omega_1\sin BI$. This point is at rest since its displacements are equal and opposite. Also, if Ω be the angular velocity about OI, the displacements

Motion about a fixed point.

of B must be equal, whether rotation take place about OI or OA. Hence $\Omega \sin IB = \omega \sin AB$.

As a verification, let us consider the motion of P. Join PI. Then

$$\frac{\sin API}{\sin BPI} = \frac{\sin PAI}{\sin PBI} \cdot \frac{\sin AI}{\sin BI} = \frac{\sin BP}{\sin AP} \cdot \frac{\omega_1}{\omega}$$

which is the ratio of the displacements of P. Hence evidently the whole displacement of P is perpendicular to PI, and is due to a single rotation about OI.

Its magnitude is $\omega \sin AP \cdot \frac{\sin APB}{\sin IPB}\delta t$

$$= \omega \sin AP \frac{\sin APB}{\sin PBI} \cdot \frac{\sin PBI}{\sin IPB}\delta t = \frac{\omega \sin AB}{\sin IB} \sin IP \delta t.$$

This is the result which would be given by a rotation during δt about OI, the angular velocity being

$$\Omega = \omega \frac{\sin AB}{\sin IB} = \omega_1 \frac{\sin AB}{\sin IA}, \text{ as before.}$$

Parallelogram of angular velocities.

The above formulæ show that if lengths proportional to the respective angular velocities about them be measured off on the component and resultant axes, the lines so determined will be the sides and diagonal of a parallelogram.

The following method of treating the subject is useful in connexion with the ordinary methods of co-ordinate geometry. It contains also, as will be seen, an independent demonstration of the parallelogram of angular velocities:—

Angular velocities ω_1, ω_2, ω_3 about the axes of x, y, and z respectively, produce in time δt displacements of the point at x, y, z (§§ 87, 89),

$$(\omega_2 z - \omega_3 y)\delta t \parallel x, \quad (\omega_3 x - \omega_1 z)\delta t \parallel y, \quad (\omega_1 y - \omega_2 x)\delta t \parallel z.$$

Hence points for which

$$\frac{x}{\omega_1} = \frac{y}{\omega_2} = \frac{z}{\omega_3}$$

are not displaced. These are therefore the equations of the axis.

Now the perpendicular from a point x, y, z to this line is, by co-ordinate geometry,

$$\left[x^2+y^2+z^2 - \frac{(\omega_1 x+\omega_2 y+\omega_3 z)^2}{\omega_1^2+\omega_2^2+\omega_3^2}\right]^{\frac{1}{2}}$$

$$= \frac{1}{\sqrt{\omega_1^2+\omega_2^2+\omega_3^2}} \sqrt{(\omega_2 z - \omega_3 y)^2 + (\omega_3 x - \omega_1 z)^2 + (\omega_1 y - \omega_2 x)^2}$$

$$= \frac{\text{whole displacement of } x, y, z}{\sqrt{\omega_1^2+\omega_2^2+\omega_3^2}\,\delta t}.$$

The actual displacement of x, y, z is therefore the same as would have been produced in time δt by a single angular velocity, $\Omega = \sqrt{\omega_1^2 + \omega_2^2 + \omega_3^2}$, about the axis determined by the above equations.

Motion about a fixed point.

Thus simultaneous rotations about any number of axes meeting in a point may be easily compounded. Let ω be the angular velocity about one of them whose direction-cosines are l, m, n; Ω, λ, μ, ν those of the resultant.

$$\lambda\Omega = \Sigma(l\omega), \quad \mu\Omega = \Sigma(m\omega), \quad \nu\Omega = \Sigma(n\omega)$$

and

$$\Omega^2 = \Sigma^2(l\omega) + \Sigma^2(m\omega) + \Sigma^2(n\omega),$$

whence the values of λ, μ, ν follow.

Composition of angular velocities about axes meeting in a point.

96. Hence the single angular velocity equivalent to three coexistent angular velocities about three mutually perpendicular axes, is determined in magnitude, and the direction of its axis is found, as follows:—The square of the resultant angular velocity is the sum of the squares of its components, and the ratios of the three components to the resultant are the direction-cosines of the axis.

Hence also, an angular velocity about any line may be resolved into three about any set of rectangular lines, the resolution in each case being (like that of simple velocities) effected by multiplying by the cosine of the angle between the directions.

Also, a rotation may be represented by a line in the direction of its axis, whose length is proportional to the angular velocity about it, and such axes are to be compounded as linear velocities are.

Hence, just as in § 31 a uniform acceleration, acting perpendicularly to the direction of motion of a point, produces a change in the *direction* of motion, but does not influence the *velocity*; so, if a body be rotating about an axis, and be subjected to an action tending to produce rotation about a perpendicular axis, the result will be a change of *direction* of the axis about which the body revolves, but no change in the *angular velocity*.

Composition of successive finite rotations.

97. We give next a few useful theorems relating to the composition of successive *finite* rotations.

If a pyramid or cone of any form roll on a similar pyramid (the image in a plane mirror of the first position of the first) all round, it clearly comes back to its primitive position. This (as all rolling of cones) is exhibited best by taking the intersection

Composition of successive finite rotations.

of each with a spherical surface. Thus we see that if a spherical polygon turns about its angular points in succession, always keeping on the spherical surface, and if the angle through which it turns about each point is twice the supplement of the angle of the polygon, or, which will come to the same thing, if it be in the other direction, but equal to twice the angle itself of the polygon, it will be brought to its original position.

The polar theorem (see § 134, below) to the above is, that a body, after successive rotations, represented by the doubles of the successive sides of a spherical polygon taken in order, is restored to its original position.

98. Another theorem is the following:—

If a pyramid rolls over all its sides on a plane, it leaves its track behind it as one plane angle, equal to the sum of the plane angles at its vertex.

Otherwise, in a spherical surface, a spherical polygon having rolled over all its sides along a great circle, is found in the same position as if the side first lying along that circle had been simply shifted along it through an arc equal to the polygon's periphery. The polar theorem is, if a body be made to take successive rotations, represented by the sides of a spherical polygon taken in order; it will finally remain as if it had revolved about the axis through the first angular point of the polygon through an angle equal to the spherical excess (§ 134) or area of the polygon.

Motion about a fixed point. Rolling cones.

99. The investigation of § 90 also applies to this case; and it is thus easy to show that the most general motion of a spherical figure on a fixed spherical surface is obtained by the rolling of a curve fixed in the figure on a curve fixed on the sphere. Hence as at each instant the line joining C and O contains a set of points of the body which are momentarily at rest, the most general motion of a rigid body of which one point is fixed consists in the rolling of a cone fixed in the body upon a cone fixed in space—the vertices of both being at the fixed point.

Position of the body due to given rotations.

100. To complete our kinematical investigation of the motion of a body of which one point is fixed, we require a solution of the following problem:—From the given angular velocities of the body about three rectangular axes attached to it, to determine the position of the body in space after a given time.

Refer the body to axes OX, OY, OZ passing through the fixed point O, and such as, at a given epoch, coincided with the axes OA, OB, OC, which move with the body. From the given angular velocities about OA, OB, OC, we know, § 95, the position of the instantaneous axis OI with reference to the body at every instant. Hence we know the conical surface in the body which rolls on the cone fixed in space. The data are sufficient also for the determination of this other cone; and these cones being known, and the portions of them which are in contact at any given instant, the motion is completely determined. Motion about a fixed point. Position due to given rotations.

If λ, μ, ν be the direction-cosines of OI referred to OA, OB, OC, ω_1, ω_2, ω_3 the angular velocities, and ϖ their resultant,

$$\frac{\lambda}{\omega_1}=\frac{\mu}{\omega_2}=\frac{\nu}{\omega_3}\left[=\frac{1}{\varpi}\right],$$

by § 95. These *two* equations in general contain t, by the elimination of which we get the equation of the cone fixed in the body. For the cone fixed in space: if σ be the radius of curvature of its trace on the unit sphere, ρ that of the rolling trace, we see, by § 95, or by § 105 below, that if s be the length of the arc of either trace from a common initial point,

$$\frac{ds}{\rho\sigma dt}=\frac{\varpi}{\sin(\sin^{-1}\rho+\sin^{-1}\sigma)}$$

$$=\frac{\varpi}{\rho\sqrt{1-\sigma^2}+\sigma\sqrt{1-\rho^2}}$$

which, as s, ρ and ϖ are known in terms of t, gives σ in terms of t, or of s, as we please—that is, the equation to the trace of the fixed cone.

A less symmetrical method is, however, often more convenient in applications to particular problems. If, for instance, the position of the body be determined at any instant by the angle $X\hat{Z}C$, by the supplement of $Z\hat{C}A$, and by the arc ZC, all on the unit sphere, the equations which connect the rate of change of these quantities with the angular velocities about the three axes are very easily obtained by the help of principles already explained.

To understand the meaning of these angular co-ordinates, suppose A, B, C initially to coincide with X, Y, Z respectively. Then let the body rotate about OZ through the angle XZC.

Motion about a fixed point.

Next let it rotate about the new position of OB through an angle equal to the arc ZC. And lastly, let it rotate about the new position of OC through an angle equal to the supplement of ZCA. It will be in the position specified by these three angles.

Let $\widehat{XZC}=\psi$, $\widehat{ZCA}=\pi-\phi$, and $ZC=\theta$, we have, by considering in succession the instantaneous motions of C along and perpendicular to ZC, and the motion of AB in its own plane,

$$\frac{d\theta}{dt}=\omega_1 \sin\phi+\omega_2\cos\phi, \qquad \sin\theta\frac{d\psi}{dt}=\omega_2\sin\phi-\omega_1\cos\phi,$$

and
$$\frac{d\psi}{dt}\cos\theta+\frac{d\phi}{dt}=\omega_3.$$

General motion of a rigid body.

101. We shall next consider the most general possible motion of a rigid body of which no point is fixed—and first we must prove the following theorem. There is one set of parallel planes in a rigid body which are parallel to each other in any two positions of the body. The parallel lines of the body perpendicular to these planes are of course parallel to each other in the two positions.

Let C and C' be any point of the body in its first and second positions. Move the body without rotation from its second position to a third in which the point at C' in the second position shall occupy its original position C. The preceding demonstration shows that there is a line CO common to the body in its first and third positions. Hence a line $C'O'$ of the body in its second position is parallel to the same line CO in the first position. This of course clearly applies to every line of the body parallel to CO, and the planes perpendicular to these lines also remain parallel.

102. Let S denote a plane of the body, the two positions of which are parallel. Move the body from its first position without rotation, in a direction perpendicular to S, till S comes into the plane of its second position. Then to get the body into its actual position, such a motion as is treated in § 79 is farther required. But by § 79 this may be effected by rotation about a certain axis perpendicular to the plane S, unless the motion required belongs to the exceptional case of pure translation. Hence [this case excepted] the body may be brought from the first position to the second by translation through a determinate distance perpendicular to a given plane, and rotation through a determinate angle about a determinate axis perpendicular to that plane. This is precisely the motion of a screw in its nut. General motion of a rigid body.

103. In the excepted case the whole motion consists of two translations, which can of course be compounded into a single one; and thus, in this case, there is no rotation at all, or every plane of it fulfils the specified condition for S of § 102.

104. Returning to the motion of a rigid body with one point fixed, let us consider the case in which the guiding cones, § 99, are both circular. The motion in this case may be called *Precessional Rotation.* Precessional Rotation.

The plane through the instantaneous axis and the axis of the fixed cone passes through the axis of the rolling cone. This plane turns round the axis of the fixed cone with an angular velocity Ω (see § 105 below), which must clearly bear a constant ratio to the angular velocity ω of the rigid body about its instantaneous axis.

105. The motion of the plane containing these axes is called the *precession* in any such case. What we have denoted by Ω is the angular velocity of the precession, or, as it is sometimes called, the rate of precession.

The angular motions ω, Ω are to one another inversely as the distances of a point in the axis of the rolling cone from the instantaneous axis and from the axis of the fixed cone.

For, let OA be the axis of the fixed cone, OB that of the rolling cone, and OI the instantaneous axis. From any point P in OB draw PN perpendicular to OI, and PQ perpendicular to OA. Then we perceive that P moves always in the circle whose centre is Q, radius PQ, and plane perpendicular to OA. Hence the actual velocity of the point P is ΩQP. But, by the principles explained above, § 99, the velocity of P is the same as that of a point moving in a circle whose centre is N, plane perpendicular to ON, and radius NP which, as this radius revolves with angular velocity ω, is ωNP. Hence

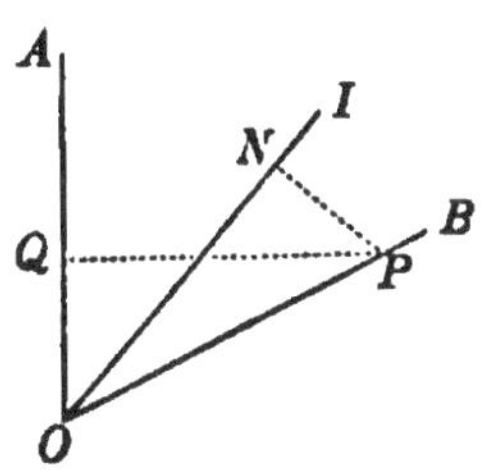

$\Omega QP = \omega NP$, or $\quad \omega : \Omega :: QP : NP$.

Precessional Rotation.

Let α be the semivertical angle of the fixed, β of the rolling, cone. Each of these may be supposed for simplicity to be acute, and their sum or difference less than a right angle—though, of course, the formulæ so obtained are (like all trigonometrical results) applicable to every possible case. We have the following three cases:—

I. Convex cone rolling on convex.

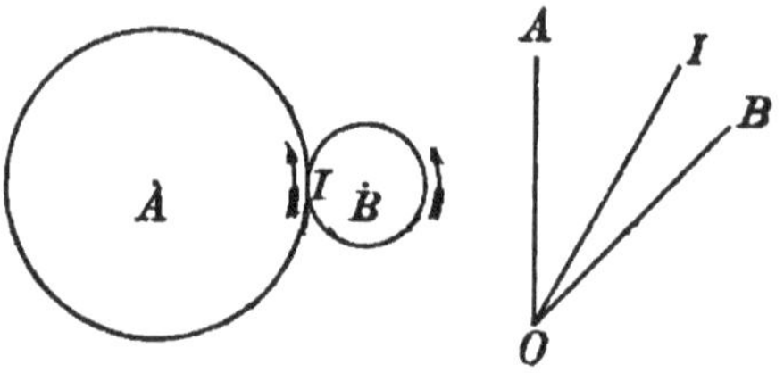

$\omega \sin \beta = \Omega \sin (\alpha + \beta)$.

II. Convex cone rolling inside concave.

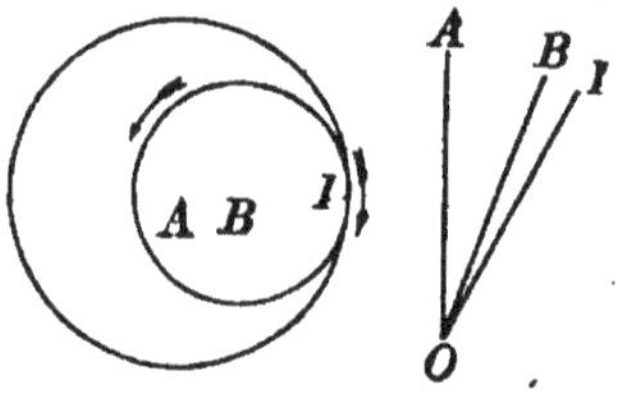

Let β be negative, and let $\beta' = -\beta$; then β' is positive, and we have

$$-\omega \sin \beta' = \Omega \sin (\alpha - \beta').$$

III. Concave cone rolling outside convex.

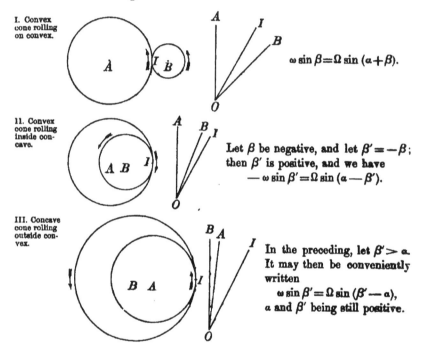

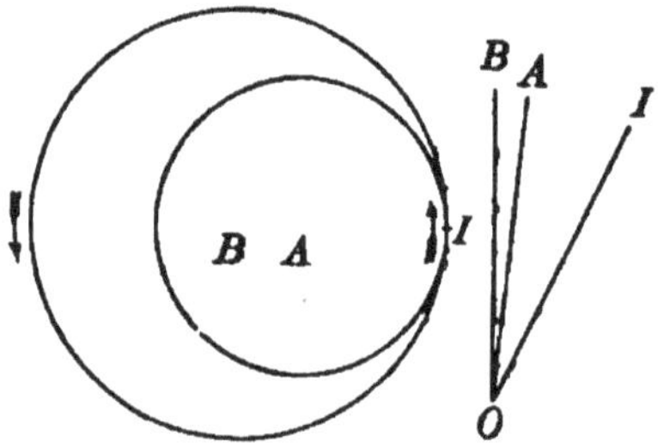

In the preceding, let $\beta' > \alpha$. It may then be conveniently written

$$\omega \sin \beta' = \Omega \sin (\beta' - \alpha),$$

α and β' being still positive.

Cases of precessional rotation.

106. If, as illustrated by the first of these diagrams, the case is one of a convex cone rolling on a convex cone, the precessional motion, viewed on an hemispherical surface having A for its pole and O for its centre, is in a similar direction to that of the angular rotation about the instantaneous axis. This we shall call *positive* precessional rotation. It is the case of a common spinning-top (peery), spinning on a very fine point which remains at rest in a hollow or hole bored by itself; not sleeping upright, nor nodding, but sweeping its axis round in a cone whose axis is vertical. In Case III. we have likewise *positive* precession. A good example of this occurs in the case of a coin spinning on a table when its plane is nearly horizontal. Cases of precessional rotation.

107. Case II., that of a convex cone rolling inside a concave one, gives an example of *negative* precession, as when viewed as before on the hemispherical surface the direction of angular rotation of the instantaneous axis is opposite to that of the rolling cone. This is the case of a symmetrical cup (or figure of revolution) supported on a point, and stable when balanced, *i.e.*, having its centre of gravity below the pivot; when inclined and set spinning non-nutationally. For instance, if a Troughton's top be placed on its pivot in any inclined position, and then spun off with very great angular velocity about its axis of figure, the nutation will be insensible; but there will be slow precession.

To this case also belongs the precessional motion of the earth's axis; for which the angle $\alpha = 23° 27' 28''$, while $\beta = 0''{\cdot}00867$. Or, if the second diagram represent a portion of the earth's surface round the pole, the arc $AI = 8{,}552{,}000$ feet, and therefore the circumference of the circle in which I moves $= 52{,}240{,}000$ feet; and $BI = 0{\cdot}88$ foot. The period of the rotation ω is the sidereal day; that of Ω is 25,868 years. Precession of the equinoxes.

108. Very interesting examples of Cases I. and III. are furnished by projectiles of different forms rotating about any axis. Thus the gyrations of an oval body or a rod or bar flung into the air, belong to Class I. (the body having one axis of less moment of inertia than the other two, equal); and the seemingly irregular evolutions of an ill-thrown quoit belong to Class III. (the quoit having one axis of greater moment of inertia than the other two, equal). Free rotation of a body kinetically symmetrical about an axis.

109. In various illustrations and arrangements of apparatus useful in Natural Philosophy, as well as in mechanics, it is required to connect two bodies, so that when either turns about a certain axis, the other shall turn with an equal angular velocity about another axis in the same plane with the former, but inclined to it at any angle. This is accomplished in mechanism by means of equal and similar bevelled wheels, or rolling cones; when the mutual inclination of two axes is given. It is approximately accomplished by means of Hooke's joint, when the two axes are nearly in the same line, but are Communication of angular velocity equally between inclined axes. Hooke's joint.

Flexible but untwistable cord.

required to be free to vary in their mutual inclination. A chain of an infinitely great number of Hooke's joints may be imagined as constituting a perfectly flexible, untwistable cord, which, if its end-links are rigidly attached to the two bodies, connects them so as to fulfil the condition rigorously without the restriction that the two axes remain in one plane. If we imagine an infinitely short length of such a chain (still, however, having an infinitely great number of links) to have its ends attached to two bodies, it will fulfil rigorously the condition stated, and at the same time keep a definite point of one body infinitely near a definite point of the other; that is to say, it will accomplish precisely for every angle of inclination what Hooke's joint does approximately for small inclinations.

Universal flexure joint.

Elastic universal flexure joint.

The same is dynamically accomplished with perfect accuracy for every angle, by a short, naturally straight, elastic wire of truly circular section, provided the forces giving rise to any resistance to equality of angular velocity between the two bodies are infinitely small. In many practical cases this mode of connexion is useful, and permits very little deviation from the conditions of a true universal flexure joint. It is used, for instance, in the suspension of the gyroscopic pendulum (§ 74) with perfect success.

Moving body attached to a fixed object by a universal flexure joint.

Of two bodies connected by a universal flexure joint, let one be held fixed. The motion of the other, as long as the angle of inclination of the axes remains constant, will be exactly that figured in Case I., § 105, above, when the angles α and β are equal. The supplement of the angle AOB is the mutual inclination of the axes; and the angle AOB itself is bisected by the instantaneous axis of the moving body. The annexed diagram shows a case of this motion,

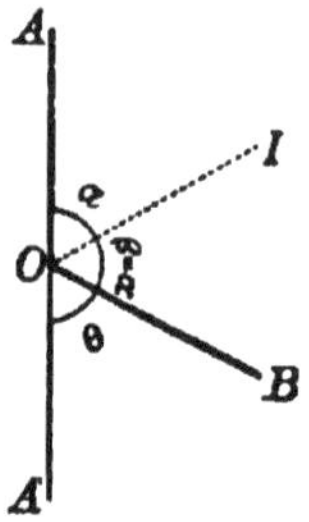

in which the mutual inclination, θ, of the axes is acute. According to the formulæ of Case I., § 105, we have

$$\omega \sin \alpha = \Omega \sin 2\alpha,$$

or

$$\omega = 2\Omega \cos \alpha = 2\Omega \sin \frac{\theta}{2},$$

where ω is the angular velocity of the moving body about its instantaneous axis, OI, and Ω is the angular velocity of its precession; that is to say, the angular velocity of the plane through the fixed axis AA', and the moving axis OB of the moving body.

Two degrees of freedom to move enjoyed by a body thus suspended.

Besides this motion, the moving body may clearly have any angular velocity whatever about an axis through O perpendicular to the plane AOB, which, compounded with ω round OI, gives the resultant angular velocity and instantaneous axis.

Two co-ordinates, $\theta = A'OB$, and ϕ measured in a plane perpendicular to AO, from a fixed plane of reference to the plane AOB, clearly specify the position of the moveable body in this case.

110. Suppose a rigid body bounded by any curved surface to be touched at any point by another such body. Any motion of one on the other must be of one or more of the forms *sliding*, *rolling*, or *spinning*. The consideration of the first is so simple as to require no comment.

General motion of one rigid body touching another.

Any motion in which the point of contact has no velocity must be rolling or spinning separately, or combined.

Let one of the bodies rotate about successive instantaneous axes, all lying in the common tangent plane at the point of instantaneous contact, and each passing through this point—the other body being fixed. This motion is what we call rolling, or simple rolling, of the moveable body on the fixed.

On the other hand, let the instantaneous axis of the moving body be the common normal at the point of contact. This is pure spinning, and does not change the point of contact.

Let it move, so that the instantaneous axis, still passing through the point of contact, is neither in, nor perpendicular to, the tangent plane. This motion is combined rolling and spinning.

Let ω be the angular velocity, α the inclination of the instantaneous axis to the normal, then $\omega \sin \alpha$ and $\omega \cos \alpha$ are the angular velocities of the rolling and spinning respectively.

111. When a body rolls and spins on another body, the *trace of either on the other* is the curved or straight line along which it has been successively touched. If the instantaneous axis is in a normal plane to the traces, the rolling is called *direct*. If not direct, the rolling may be resolved into a direct

Traces of rolling.

Direct rolling.

General motion of one rigid body touching another.

rolling, and a rotation round the tangent line to the traces. Imagine the traces constructed of rigid matter, and all the rest of each body removed. We may repeat the former motion with these curves alone. The difference of the circumstances now supposed will only be experienced if we vary the direction of the instantaneous axis. In the former case, if we do this we introduce more or less of spinning, and we *alter the trace* on each body; in the latter, we have always the same moveable curve rolling on the same fixed curve; with, in addition, any arbitrary velocity of twisting round the common tangent. The consideration of this case is very instructive, with regard to the general problem.

Curve rolling on curve.

112. It may be roughly imitated in practice by two stiff wires bent into the forms of the given curves, and prevented from crossing each other by a short piece of elastic tube clasping them together.

First let them be both plane curves, and kept in one plane. We have then pure *rolling*, as of one cylinder on another.

Let ρ be the radius of curvature of the rolling, σ of the fixed, cylinder; ω the angular velocity of the former, V the linear velocity of the point of contact. We have

$$\omega=\left(\frac{1}{\rho}+\frac{1}{\sigma}\right)V.$$

For, in the figure, suppose P to be at any time the point of contact, and Q and p the points which are to be in contact after an interval dt; O, O' the centres of curvature; $\angle POp=\theta$, $PO'Q=\phi$.

Then $PQ=Pp=$ space described by point of contact. In symbols $\rho\phi=\sigma\theta=Vdt$.

Also, before $O'Q$ and OP can coincide in direction, the former must evidently turn through an angle $\theta+\phi$.

Therefore $\omega dt=\theta+\phi$; and by eliminating θ and ϕ, and dividing by dt, we get the above result.

It is to be understood here, that as the radii of curvature have been considered positive here when both surfaces are convex, the negative sign must be introduced for either radius when the corresponding curve is concave.

Hence the angular velocity of the rolling curve is in this

case equal to the product of the linear velocity of the point of contact by the sum or difference of the curvatures, according as the curves are both convex, or one concave and the other convex. Angular velocity of rolling in a plane.

113. When the curves are both plane, but in different planes, the plane in which the rolling takes place divides the angle between the plane of one of the curves, and that of the other produced through the common tangent line, into parts whose sines are inversely as the curvatures in them respectively; and the angular velocity is equal to the linear velocity of the point of contact multiplied by the difference of the projections of the two curvatures on this plane. Plane curves not in same plane.

For, let PQ, Pp be equal arcs of the two curves as before, and let PR be taken in the common tangent (*i.e.*, the intersection of the planes of the curves) equal to either. Then QR, pR are ultimately perpendicular to PR.

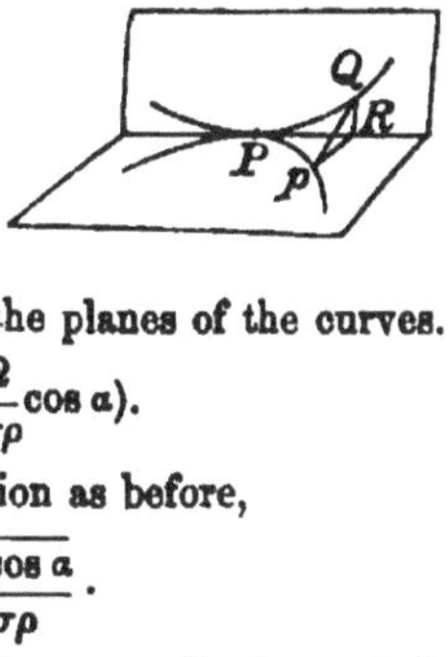

Hence
$$pR=\frac{PR^2}{2\sigma}$$
$$QR=\frac{PR^2}{2\rho}.$$

Also, $Q\hat{R}p=\alpha$, the angle between the planes of the curves.

We have
$$Qp^2=\frac{PR^4}{4}\left(\frac{1}{\sigma^2}+\frac{1}{\rho^2}-\frac{2}{\sigma\rho}\cos\alpha\right).$$

Therefore if ω be the velocity of rotation as before,
$$\omega=V\sqrt{\frac{1}{\sigma^2}+\frac{1}{\rho^2}-\frac{2\cos\alpha}{\sigma\rho}}.$$

Also the instantaneous axis is evidently perpendicular, and therefore the plane of rotation parallel, to Qp. Whence the above. In the case of $\alpha=\pi$, this agrees with the result of § 112.

A good example of this is the case of a coin spinning on a table (mixed rolling and spinning motion), as its plane becomes gradually horizontal. In this case the curvatures become more and more nearly equal, and the angle between the planes of the curves smaller and smaller. Thus the resultant angular velocity becomes exceedingly small, and the motion of the point of contact very great compared with it.

114. The preceding results are, of course, applicable to tortuous as well as to plane curves; it is merely requisite to

Curve rolling on curve: two degrees of freedom.

substitute the osculating plane of the former for the plane of the latter.

Curve rolling on surface: three degrees of freedom.

115. We come next to the case of a curve rolling, with or without spinning, on a surface.

It may, of course, roll on any curve traced on the surface. When this curve is given, the moving curve may, while rolling along it, revolve arbitrarily round the tangent. But the component instantaneous axis perpendicular to the common tangent, that is, the axis of the direct rolling of one curve on the other, is determinate, § 113. If this axis does not lie in the surface, there is spinning. Hence, when the trace on the surface is given, there are two independent variables in the motion; the space traversed by the point of contact, and the angular velocity about the tangent line at that point.

Trace prescribed and no spinning permitted:

116. If the trace is given, and it be prescribed as a condition that there shall be no spinning, the angular position of the rolling curve round the tangent at the point of contact is determinate. For in this case the instantaneous axis must be in the tangent plane to the surface. Hence, if we resolve the rotation into components round the tangent line, and round an axis perpendicular to it, the latter must be in the tangent plane. Thus the rolling, as of curve on curve, must be in a normal plane to the surface; and therefore at starting the given curve must be so applied to its trace on the surface that the projections of the two curves on the tangent plane may be of equal curvature.

Two degrees of freedom.

The curve, as it rolls on, must continually revolve about the tangent line to it at the point of contact with the surface, so as in every position to fulfil this condition.

Let a denote the inclination of the plane of curvature of the trace to the normal to the surface at any point, a' the same for the plane of the rolling curve; $\frac{1}{\rho}$, $\frac{1}{\rho'}$ their curvatures. We reckon a as obtuse, and a' acute, when the two curves lie on opposite sides of the tangent plane. Then

$$\frac{1}{\rho'}\sin a' = \frac{1}{\rho}\sin a,$$

which fixes a' or the position of the rolling curve when the point of contact is given.

Let ω be the angular velocity of rolling about an axis perpendicular to the tangent, ϖ that about the tangent, and let V be the linear velocity of the point of contact. Then, since $\frac{1}{\rho'}\cos\alpha'$ and $-\frac{1}{\rho}\cos\alpha$ (each positive when the curves lie on opposite sides of the tangent plane) are the projections of the two curvatures on a plane through the normal to the surface containing their common tangent, we have, by § 112, Angular velocity of direct rolling.

$$\omega = V\left(\frac{1}{\rho'}\cos\alpha' - \frac{1}{\rho}\cos\alpha\right),$$

α' being determined by the preceding equation. Let τ and τ' denote the tortuosities of the trace, and of the rolling curve, respectively. Then, first, if the curves were both plane, we see that one rolling on the other about an axis always perpendicular to their common tangent could never change the inclination of their planes. Hence, secondly, if they are both tortuous, such rolling will alter the inclination of their osculating planes by an indefinitely small amount $(\tau - \tau')\,ds$ during rolling which shifts the point of contact over an arc ds. Now α is a known function of s if the trace is given, and therefore so also is α'. But $\alpha - \alpha'$ is the inclination of the osculating planes, hence Angular velocity round tangent.

$$V\left\{\frac{d(\alpha-\alpha')}{ds} - (\tau - \tau')\right\} = \varpi.$$

117. Next, for one surface rolling and spinning on another. First, if the trace on each is given, we have the case of § 113 or § 115, one curve rolling on another, with this farther condition, that the former must *revolve* round the tangent to the two curves so as to keep the tangent planes of the two surfaces coincident. Surface on surface.

It is well to observe that when the points in contact, and the two traces, are given, the position of the moveable surface is quite determinate, being found thus:—Place it in contact with the fixed surface, the given points together, and *spin* it about the common normal till the tangent lines to the traces coincide. Both traces prescribed: one degree of freedom.

Hence when both the traces are given the condition of no spinning cannot be imposed. During the rolling there must in general be spinning, such as to keep the tangents to the two traces coincident. The instantaneous axis of the rolling must also so change as to give not only the rolling along the trace,

Surface on surface, both traces prescribed; one degree of freedom.

but also the necessary revolving about the tangent line to bring the tangent planes always to coincide.

In this case, then, there is but one independent variable—the space passed over by the point of contact, and when the velocity of the point of contact is given, the resultant angular velocity, and the direction of the instantaneous axis of the rolling body are determinate. We have thus a sufficiently clear view of the general character of the motion in question, but it is right that we consider it more closely, as it introduces us very naturally to an important question, the measurement of the *twist* of a rod, wire, or narrow plate, a quantity wholly distinct from the *tortuosity* of its axis (§ 7).

118. Suppose all of each surface cut away except an infinitely narrow strip, including the trace of the rolling. Then we have the rolling of one of these strips upon the other, each having at every point a definite curvature, tortuosity, and twist.

Twist.

119. Suppose a flat bar of small section to have been bent (the requisite amount of stretching and contraction of its edges being admissible) so that its axis assumes the form of any plane or tortuous curve. If it be unbent without twisting, *i.e.*, if the curvature of each element of the bar be removed by bending it through the requisite angle in the osculating plane, and it be found untwisted when thus rendered straight, it had no *twist* in its original form. This case is, of course, included in the general theory of *twist*, which is the subject of the following sections.

Axis and transverse.

120. A bent or straight rod of circular, or any other form of section being given, a line through the centres, or any other chosen points of its sections, may be called its *axis*. Mark a line on its side all along its length, such that it shall be a straight line parallel to the axis when the rod is unbent and untwisted. A line drawn from any point of the axis perpendicular to this side line of reference, is called the *transverse* of the rod at this point.

The whole twist of any length of a straight rod is the angle between the transverses of its ends. The average twist is the integral twist divided by the length. The twist at any point is the average twist in an infinitely short length through this point; in other words, it is the rate of rotation of its transverse per unit of length along it.

The twist of a curved, plane, or tortuous rod at any point is the rate of component rotation of its transverse round its tangent line, per unit of length along it. Twist.

If t be the twist at any point, $\int t ds$ over any length is the integral twist in this length.

121. Integral twist in a curved rod, although readily defined, as above, in the language of the integral calculus, cannot be exhibited as the angle between any two lines readily constructible. The following considerations show how it is to be reckoned, and lead to a geometrical construction exhibiting it in a spherical diagram, for a rod bent and twisted in any manner:—

122. If the axis of the rod forms a plane curve lying in one plane, the integral twist is clearly the difference between the inclinations of the transverses at its ends to its plane. For if it be simply unbent, without altering the twist in any part, the inclination of each transverse to the plane in which its curvature lay will remain unchanged; and as the axis of the rod now has become a straight line in this plane, the mutual inclination of the transverses at any two points of it has become equal to the difference of their inclinations to the plane. Estimation of integral twist: In a plane curve.

123. No simple application of this rule can be made to a tortuous curve, in consequence of the change of the plane of curvature from point to point along it; but, instead, we may proceed thus:—

First, Let us suppose the plane of curvature of the axis of the wire to remain constant through finite portions of the curve, and to change abruptly by finite angles from one such portion to the next (a supposition which involves no angular points, that is to say, no infinite curvature, in the curve). Let planes parallel to the planes of curvature of three successive portions, PQ, QR, RS (not shown in the diagram), cut a unit spherical surface in the great circles GAG', ACA', CE; and therefore, of course the radii of the sphere parallel to the tangents at the points Q and R of the curve where its curvature changes will cut its surface in A and C, the intersections of these circles. Let G be the In a curve consisting of plane portions in different planes.

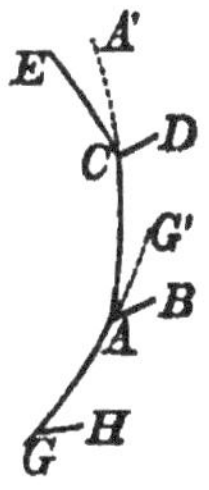

Estimation of integral twist: in a curve consisting of plane portions in different planes.

point in which the radius of the sphere parallel to the tangent at P cuts the surface; and let GH, AB, CD be parallels to the transverses of the bar drawn from the points P, Q, R of its axis. Then (§ 122) the twist from P to Q is equal to the difference of the angles HGA and BAG'; and the twist from Q to R is equal to the difference between BAC and DCA'. Hence the whole twist from P to R is equal to

$$HGA-BAG'+BAC-DCA',$$

or, which is the same thing,

$$A'CE+G'AC-(DCE-HGA).$$

Continuing thus through any length of rod, made up of portions curved in different planes, we infer that the integral twist between any two points of it is equal to the sum of the exterior angles in the spherical diagram, wanting the excess of the inclination of the transverse at the second point to the plane of curvature at the second point above the inclination at the first point to the plane of curvature at the first point. The sum of those exterior angles is what is defined below as the "change of direction in the spherical surface" from one side of the polygon of great circles to the other; and when the polygon is closed, and the sum includes all its exterior angles, it is (§ 134) equal to 2π wanting the area enclosed. The construction we have made obviously holds in the limiting case, when the lengths of the plane portions are infinitely small, and is therefore applicable to a wire forming a tortuous curve with continuously varying plane of curvature, for which it gives the following conclusion:—

In a continuously tortuous curve.

Parallel to the tangent to the axis of the bar, at a point moving along it, let a radius of a unit sphere be drawn, cutting the spherical surface in a curve (being the hodograph of a point moving with constant velocity along the bar). From points of this curve draw parallels to the transverses of the corresponding points of the bar. The excess of the change of direction (§ 135) from any point to another of the hodograph, above the increase of its inclination to the transverse, is equal to the twist in the corresponding part of the bar.

The annexed diagram, showing the hodograph and the parallels to the transverses, illustrates this rule. Thus, for instance, the excess of the change of direction in the spherical surface along the hodograph from A to C, above $DCS - BAT$ is equal to the twist in the bar between the points of it to which A and C correspond. Or, again, if we consider a portion of the bar from any point of it, to another point at which the tangent to its axis is parallel to the tangent at its first point, we shall have a closed curve as the spherical hodograph; and if A be the point of the hodograph corresponding to them, and AB and AB' the parallels to the transverses, the whole twist in the included part of the bar will be equal to the change of direction all round the hodograph, wanting the excess of the exterior angle $B'AT$ above the angle BAT; that is to say, the whole twist will be equal to the excess of the angle BAB' above the area enclosed by the hodograph. **Estimation of integral twist: in a continuously tortuous curve.**

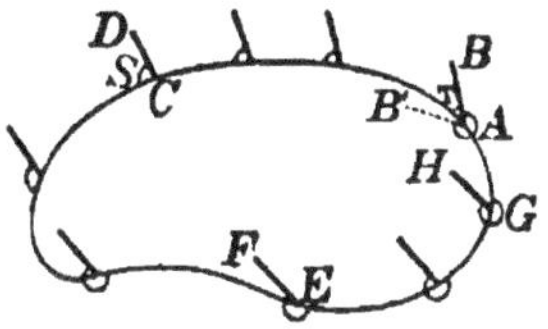

The principles of twist thus developed are of vital importance in the theory of rope-making, especially the construction and the dynamics of wire ropes and submarine cables, elastic bars, and spiral springs.

124. Returning to the motion of one surface rolling and spinning on another, the trace on each being given, we may consider that, of each, the curvature (§ 6), the tortuosity (§ 7), and the twist reckoned according to transverses in the tangent plane of the surface, are known; and the subject is fully specified in (§ 117) above. **Surface rolling on surface; both traces given.**

Let $\frac{1}{\rho'}$ and $\frac{1}{\rho}$ be the curvatures of the traces on the rolling and fixed surfaces respectively; α' and α the inclinations of their planes of curvature to the normal to the tangent plane, reckoned as in § 116; τ' and τ their tortuosities; t' and t their twists; and V the velocity of the point of contact. All these being known, it is required to find:—

ω the angular velocity of rotation about the transverse of the traces; that is to say, the line in the tangent plane perpendicular to their tangent line,

ϖ the angular velocity of rotation about the tangent line, and

σ ,, ,, of spinning.

Surface rolling on surface; both traces given.

We have

$$\omega = V\left(\frac{1}{\rho'}\cos\alpha' - \frac{1}{\rho}\cos\alpha\right)$$

$$\varpi = V(t-t') = V\left\{\frac{d(\alpha-\alpha')}{ds} - (\tau-\tau')\right\}$$

and $$\sigma = V\left(\frac{1}{\rho'}\sin\alpha' - \frac{1}{\rho}\sin\alpha\right)$$

Surface rolling on surface without spinning.

125. In the same case, suppose the trace on *one* only of the surfaces to be given. We may evidently impose the condition of no spinning, and then the trace on the other is determinate. This case of motion is thoroughly examined in § 137, below.

The condition is that the projections of the curvatures of the two traces on the common tangent plane must coincide.

If $\frac{1}{r'}$ and $\frac{1}{r}$ be the curvatures of the rolling and stationary surfaces in a normal section of each through the tangent line to the trace, and if α, α', ρ, ρ' have their late meanings,

$\rho' = r'\cos\alpha'$, $\rho = r\cos\alpha$ (Meunier's Theorem, below).

But $\frac{1}{\rho'}\sin\alpha' = \frac{1}{\rho}\sin\alpha$, hence $\tan\alpha' = \frac{r'}{r}\tan\alpha$, the condition required.

Examples of tortuosity and twist.

126. If a rod be bent along any curve on a spherical surface, so that a marked side line of reference on it lies all along in contact with the spherical surface, it acquires no twist in the operation. For if it is laid so along any finite arc of a small circle it will clearly have no twist. And no twist is produced in continuing from any point along another small circle having a common tangent with the first at this point.

If a rod be bent round a cylinder so that a line marked along one side of it may lie in contact with the cylinder, or if, what presents somewhat more readily the view now desired, we wind a straight ribband spirally on a cylinder, the axis of bending is parallel to that of the cylinder, and therefore oblique to the axis of the rod or ribband. We may therefore resolve the instantaneous rotation which constitutes the bending at any instant into two components, one round a line perpendicular to the axis of the bar, which is pure bending, and the other round the axis of the bar, which is pure twist.

The twist at any point in a rod or ribband, so wound on a circular cylinder, and constituting a uniform helix, is *Examples of tortuosity and twist.*

$$\frac{\cos a \sin a}{r}$$

if r be the radius of the cylinder and a the inclination of the spiral. For if V be the velocity at which the bend proceeds along the previously straight wire or ribband, $\frac{V \cos a}{r}$ will be the angular velocity of the instantaneous rotation round the line of bending (parallel to the axis), and therefore

$$\frac{V \cos a}{r} \sin a \text{ and } \frac{V \cos a}{r} \cos a$$

are the angular velocities of twisting and of pure bending respectively.

From the latter component we may infer that the curvature of the helix is

$$\frac{\cos^2 a}{r},$$

a known result, which agrees with the expression used above (§ 13).

127. The hodograph in this case is a small circle of the sphere. If the specified condition as to the mode of laying on of the rod on the cylinder is fulfilled, the transverses of the spiral rod will be parallel at points along it separated by one or more whole turns. Hence the integral twist in a single turn is equal to the excess of four right angles above the spherical area enclosed by the hodograph. If a be the inclination of the spiral, $\frac{1}{2}\pi - a$ will be the arc-radius of the hodograph, and therefore its area is $2\pi(1 - \sin a)$. Hence the integral twist in a turn of the spiral is $2\pi \sin a$, which agrees with the result previously obtained (§ 126).

128. As a preliminary to the further consideration of the rolling of one surface on another, and as useful in various parts of our subject, we may now take up a few points connected with the curvature of surfaces. *Curvature of surfaces.*

The tangent plane at any point of a surface may or may not cut it at that point. In the former case, the surface bends away from the tangent plane partly towards one side of it, and partly towards the other, and has thus, in some of its normal sections,

Curvature of surface.

curvatures oppositely directed to those in others. In the latter case, the surface on every side of the point bends away from the same side of its tangent plane, and the curvatures of all normal sections are similarly directed. Thus we may divide curved surfaces into *Anticlastic* and *Synclastic*. A saddle gives a good example of the former class; a ball of the latter. Curvatures in opposite directions, with reference to the tangent plane, have of course different signs. The outer portion of an anchor-ring is synclastic, the inner anticlastic.

Synclastic and anticlastic surfaces.

Curvature of oblique sections.

129. *Meunier's Theorem.*—The curvature of an oblique section of a surface is equal to that of the normal section through the same tangent line multiplied by the secant of the inclination of the planes of the sections. This is evident from the most elementary considerations regarding projections.

Principal curvatures.

130. *Euler's Theorem.*—There are at every point of a synclastic surface two normal sections, in one of which the curvature is a maximum, in the other a minimum; and these are at right angles to each other.

In an anticlastic surface there is maximum curvature (but in opposite directions) in the two normal sections whose planes bisect the angles between the lines in which the surface cuts its tangent plane. On account of the difference of sign, these may be considered as a maximum and a minimum.

Sum of curvatures in normal sections at right angles to each other.

Generally the sum of the curvatures at a point, in any two normal planes at right angles to each other, is independent of the position of these planes.

If the tangent plane be taken as that of x, y, and the origin at the point of contact, the equation to the surface is evidently (unless the origin be a singular point)

$$z = Ax^2 + 2Bxy + Cy^2 + \text{ etc.} \qquad (1)$$

The curvature of the normal section which passes through the point x, y, z is (in the limit)

$$\frac{1}{r} = \frac{2z}{x^2+y^2} = 2\frac{Ax^2+2Bxy+Cy^2}{x^2+y^2}.$$

If the section be inclined at an angle θ to the plane of XZ, this becomes

$$\frac{1}{r} = 2\{A\cos^2\theta + 2B\sin\theta\cos\theta + C\sin^2\theta\}. \qquad (2)$$

Hence, if $\frac{1}{r}$ and $\frac{1}{s}$ be curvatures in normal sections at right angles to each other, Sum of curvatures in normal sections at right angles to each other.

$$\frac{1}{r}+\frac{1}{s}=2(A+C)=\text{constant}.$$

(2) may be written

$$\frac{1}{r}=\{A(1+\cos 2\theta)+2B\sin 2\theta+C(1-\cos 2\theta)\}$$

$$=\{\overline{A+C}+\overline{A-C}\cos 2\theta+2B\sin 2\theta\}$$

or if $\qquad A-C=R\cos 2a,\ 2B=R\sin 2a,$

that is $\quad R=\sqrt{(A-C)^2+4B^2}$, and $\tan 2a=\frac{2B}{A-C}$,

we have

$$\frac{1}{r}=A+C+\sqrt{(A-C)^2+4B^2}\cos 2(\theta-a).$$

The maximum and minimum curvatures are therefore those in normal planes at right angles to each other for which $\theta=a$ and $\theta=a+\frac{\pi}{2}$, and are respectively Principal normal sections.

$$A+C\pm\sqrt{(A-C)^2+4B^2}.$$

Hence their product is $4(AC-B^2)$.

If this be positive we have a synclastic, if negative an anticlastic, surface. If it be zero we have one curvature only, and the surface is *cylindrical* at the point considered. It is demonstrated (§ 152, below) that if this condition is fulfilled at every point, the surface is "developable" (§ 139, below).

By (1) a plane parallel to the tangent plane and very near it cuts the surface in an ellipse, hyperbola, or two parallel straight lines (a variety of the parabola), in the three cases respectively. This section, whose nature informs us as to whether the curvature be synclastic, anticlastic, or cylindrical, at any point, was called by Dupin the *Indicatrix*.

131. Let P, p be two points of a surface indefinitely near to each other, and let r be the radius of curvature of a normal section passing through them. Then the radius of curvature of an oblique section through the same points, inclined to the former at an angle a, is (§ 129) $r\cos a$. Also the length along the normal section, from P to p, is less than that along the oblique section—since a given chord cuts off an arc from a circle, longer the less is the radius of that circle. Shortest line between two points on a surface.

Shortest line between two points on a surface.

If a be the length of the chord Pp, we have

$$\text{Distance } Pp \text{ along normal section} = 2r\sin^{-1}\frac{a}{2r} = a\left(1+\frac{a^2}{24r^2}\right),$$

$$\text{,, ,, oblique section} = a\left(1+\frac{a^2}{24r^2\cos^2\alpha}\right).$$

Geodetic Lines.

132. Hence, if the shortest possible line be drawn from one point of a surface to another, its plane of curvature is everywhere perpendicular to the surface.

Such a curve is called a *Geodetic* line. And it is easy to see that it is the line in which a flexible and inextensible string would touch the surface if stretched between those points, the surface being supposed smooth.

133. If an infinitely narrow ribband be laid on a surface along a geodetic line, its twist is equal to the tortuosity of its axis at each point. We have seen (§ 125) that when one body rolls on another without spinning, the projections of the trace on the common tangent plane agree in curvature at the point of contact. Hence, if one of the surfaces be a plane, and the trace on the other be a geodetic line, the trace on the plane is a straight line. Conversely, if the trace on the plane be a straight line, that on the surface is a geodetic line.

And, quite generally, if the given trace be a geodetic line, the other trace is also a geodetic line.

Spherical excess.

Area of spherical polygon.

134. The area of a spherical triangle is known to be proportional to the "spherical excess," *i.e.*, the excess of the sum of its angles over two right angles, or the excess of four right angles over the sum of its exterior angles. The area of a spherical polygon whose n sides are portions of great circles —*i.e.*, geodetic lines—is to that of the hemisphere as the excess of four right angles over the sum of its exterior angles is to four right angles. (We may call this the "spherical excess" of the polygon.)

For the area of a spherical triangle is known to be

$$(A+B+C-\pi)r^2.$$

Divide the polygon into n such triangles, with a common vertex, the angles about which, of course, amount to 2π.

$$\begin{aligned}\text{Area} &= (\text{sum of interior angles of triangles} - n\pi)r^2\\ &= (2\pi + \text{sum of interior angles of polygon} - n\pi)r^2\\ &= (2\pi - \text{sum of exterior angles of polygon})r^2.\end{aligned}$$

Given an open or closed spherical polygon, or line on the surface of a sphere composed of consecutive arcs of great circles. Take either pole of the first of these arcs, and the corresponding poles of all the others (all the poles to be on the right hand, or all on the left, of a traveller advancing along the given great circle arcs in order). Draw great circle arcs from the first of these poles to the second, the second to the third, and so on in order. Another closed or open polygon, constituting what is called the polar diagram to the given polygon, is thus obtained. The arcs of the second polygon are evidently equal to the exterior angles in the first; and the exterior angles of the second are equal to the sides of the first. Hence the relation between the two diagrams is reciprocal, or each is polar to the other. The polar figure to any continuous curve on a spherical surface is the locus of the ultimate intersections of great circles equatorial to points taken infinitely near each other along it.

Reciprocal polars on a sphere.

The area of a closed spherical figure is, consequently, according to what we have just seen, equal to the excess of 2π above the periphery of its polar.

135. If a point move on a surface along a figure whose sides are geodetic lines, the sum of the exterior angles of this polygon is defined to be the *integral change of the direction in the surface.*

Integral change of direction in a surface.

In great circle sailing, unless a vessel sail on the equator, or on a meridian, her course, as indicated by points of the compass (true, not magnetic, for the latter change even on a meridian), perpetually changes. Yet just as we say her direction does not change if she sail in a meridian, or in the equator, so we ought to say her direction does not change if she moves in any great circle. Now, the great circle is the geodetic line on the sphere, and by extending these remarks to other curved surfaces, we see the connexion of the above definition with that in the case of a plane polygon (§ 10).

Note.—We cannot define integral change of direction here by any angle directly constructible from the first and last tangents to the path, as was done (§ 10) in the case of a plane curve or polygon; but from §§ 125 and 133 we have the following statement:—The whole change of direction in a curved surface, from one end to another of any arc of a curve traced on it, is

Change of direction in a surface, of any arc traced on it.

equal to the change of direction from end to end of the trace of this arc on a plane by pure rolling.

Integral curvature. **136.** *Def.* The excess of four right angles above the integral change of direction from one side to the same side next time in going round a closed polygon of geodetic lines on a curved surface, is the *integral curvature* of the enclosed portion of surface. There is no such excess in the case of a polygon traced on a plane. We shall presently see that this corresponds exactly to what Gauss has called the *curvatura integra.*

Curvatura integra. *Def.* (Gauss.) The *curvatura integra* of any given portion of a curved surface, is the area enclosed on a spherical surface of unit radius by a straight line drawn from its centre, parallel to a normal to the surface, the normal being carried round the boundary of the given portion.

Horograph. The curve thus traced on the sphere is called the *Horograph* of the given portion of curved surface.

The *average curvature* of any portion of a curved surface is the integral curvature divided by the area. The *specific curvature* of a curved surface at any point is the average curvature of an infinitely small area of it round that point.

Change of direction round the boundary in the surface, together with area of the horograph, make up four right angles. **137.** The excess of 2π above the change of direction, in a surface, of a point moving round any closed curve on it, is equal to the area of the horograph of the enclosed portion of surface.

Let a tangent plane roll without spinning on the surface over every point of the bounding line. (Its instantaneous axis will always lie in it, and pass through the point of contact, but will not, as we have seen, be at right angles to the given bounding curve, except when the twist of a narrow ribband of the surface along this curve is nothing.) Considering the auxiliary sphere of unit radius, used in Gauss's definition, and the moving line through its centre, we perceive that the motion of this line is, at each instant, in a plane perpendicular to the instantaneous axis of the tangent plane to the given surface. The direction of motion of the point which cuts out the area on the spherical surface, is therefore perpendicular to this instantaneous axis. Hence, if we roll a tangent plane on the spherical surface also, making it keep time with the other, the trace on this tangent plane will be a curve always perpendicular to the instantaneous axis of each tangent plane. The change of direction, in the spherical surface, of the point moving round and cutting out the area, being equal to the change of direction in its own trace on its own tangent plane (§ 135), is therefore equal to the change of direction of the instantaneous axis in the tangent plane to the given surface reckoned from a line fixed relatively to this plane. But having rolled all round, and being in position to roll round again, the instantaneous axis of the fresh start must be inclined to the trace at the same angle as in the beginning. Hence the change of direction of the instantaneous axis in either tangent plane is equal to the change of direction, in the given surface, of a point going all round the boundary of the given portion of it; to which, therefore, the change of direction, in the spherical surface, of the point going all round the spherical area is equal. But, by the well-known theorem (§ 134) of the "spherical excess," this change of direction subtracted from 2π leaves the spherical area. Hence the spherical area, defined according to Gauss as the *curvatura integra*, is equal to 2π wanting the change of direction in going round the boundary. *Curvatura integra*, and horograph.

It will be perceived that when the two rollings we have considered are each complete, each tangent plane will have come back to be parallel to its original position, but any fixed line in it will have changed direction through an angle equal to the equal changes of direction just considered.

Note.—The two rolling tangent planes are at each instant parallel to one another, and a fixed line relatively to one drawn at any time parallel to a fixed line relatively to the other, remains parallel to the last-mentioned line.

If, instead of the closed curve, we have a closed polygon of geodetic lines on the given surface, the trace of the rolling of its tangent plane will be an unclosed rectilineal polygon. If each geodetic were a plane curve (which could only be if the given surface were spherical), the instantaneous axis would be always perpendicular to the particular side of this polygon which is rolled on at the instant; and, of course, the spherical area on the auxiliary sphere would be a similar polygon to the given one. But the given surface being other than spherical, there must be tortuosity in at least one geodetic of the closed polygon, and generally in all of them; or, which is the same thing, twist in the corresponding ribbands of the surface. Hence the portion of the whole trace on the second rolling tangent plane which corresponds to any one side of the given geodetic polygon, must in general be a curve; and as there will generally be finite angles in the second rolling corresponding to (but not equal to) those in the first, the trace of the second on

Curvatura integra.

its tangent plane will be an unclosed polygon of curves. The trace of the same rolling on the spherical surface in which it takes place will generally be a spherical polygon, not of great circle arcs, but of other curves. The sum of the exterior angles of this polygon, and of the changes of direction from one end to other of each of its sides, is the whole change of direction considered, and is, by the proper application of the theorem of § 134, equal to 2π wanting the spherical area enclosed.

Or again, if, instead of a geodetic polygon as the given curve, we have a polygon of curves, each fulfilling the condition that the normal to the surface through any point of it is parallel to a fixed plane; one plane for the first curve, another for the second, and so on; then the figure on the auxiliary spherical surface will be a polygon of arcs of great circles; its trace on its tangent plane will be an unclosed rectilineal polygon; and the trace of the given curve on the tangent plane of the first rolling will be an unclosed polygon of curves. The sum of changes of direction in these curves, and of exterior angles in passing from one to another of them, is of course equal to the change of direction in the given surface, in going round the given polygon of curves on it. The change of direction in the other will be simply the sum of the exterior angles of the spherical polygon, or of its rectilineal trace. Remark that in this case the instantaneous axis of the first rolling, being always perpendicular to that plane to which the normals are all parallel, remains parallel to one line, fixed with reference to the tangent plane, during rolling along each curved side, and also remains parallel to a fixed line in space.

Lastly, remark that although the whole change of direction of the trace in one tangent plane is equal to that in the trace on the other, when the rolling is completed round the given circuit; the changes of direction in the two are generally unequal in any part of the circuit. They may be equal for particular parts of the circuit, viz., between those points, if any, at which the instantaneous axis is equally inclined to the direction of the trace on the first tangent plane.

Analogy between lines and surfaces as regards curvature.

138. It appears from what precedes, that the same equality or identity subsists between "whole curvature" in a plane arc and the excess of π above the angle between the terminal tangents, as between "whole curvature" and excess of 2π above change of direction along the bounding line in the surface for any portion of a curved surface.

Or, according to Gauss, whereas the whole curvature in a plane arc is the angle between two lines parallel to the terminal normals, the whole curvature of a portion of curve surface is the solid angle of a cone formed by drawing lines from a point parallel to all normals through its boundary. Analogy between lines and surfaces as regards curvature.

Again, average curvature in a plane curve is $\frac{\text{change of direction}}{\text{length}}$; and specific curvature, or, as it is commonly called, curvature, at any point of it $= \frac{\text{change of direction in infinitely small length}}{\text{length}}$. Thus average curvature and specific curvature are for surfaces analogous to the corresponding terms for a plane curve.

Lastly, in a plane arc of uniform curvature, *i.e.*, in a circular arc, $\frac{\text{change of direction}}{\text{length}} = \frac{1}{\rho}$. And it is easily proved (as below) that, in a surface throughout which the specific curvature is uniform, $\frac{2\pi - \text{change of direction}}{\text{area}}$, or $\frac{\text{integral curvature}}{\text{area}} = \frac{1}{\rho\rho'}$, where ρ and ρ' are the principal radii of curvature. Hence in a surface, whether of uniform or non-uniform specific curvature, the specific curvature at any point is equal to $\frac{1}{\rho\rho'}$. In geometry of three dimensions, $\rho\rho'$ (an area) is clearly analogous to ρ in a curve and plane.

Consider a portion S, of a surface of any curvature, bounded by a given closed curve. Let there be a spherical surface, radius r, and C its centre. Through C draw a radius to the spherical surface CQ, parallel to the normal at any point P of S. If this be done for every point of the boundary also, the points so obtained enclose the spherical area used in Gauss's definition. Now let there be an infinitely small rectangle on S, at P, having for its sides arcs of angles ζ and ζ', on the normal sections of greatest and least curvature, and let their radii of curvature be denoted by ρ and ρ'. The lengths of these sides will be $\rho\zeta$ and $\rho'\zeta'$ respectively. Its area will therefore be $\rho\rho'\zeta\zeta'$. The corresponding figure at Q on the spherical surface will be bounded by arcs of equal angles, and, therefore, of lengths $r\zeta$ and $r\zeta'$ respectively, and its area will be $r^2\zeta\zeta'$. Hence if $d\sigma$ denote this area, the area of the infinitely small portion of the given surface will be $\frac{\rho\rho' d\sigma}{r^2}$. Area of the horograph.

Area of the horograph.

In a surface for which $\rho\rho'$ is constant, the area is therefore $= \frac{\rho\rho'}{r^2}\iint d\sigma = \rho\rho' \times$ integral curvature.

Flexible and inextensible surface.

139. A perfectly flexible but inextensible surface is suggested, although not realized, by paper, thin sheet metal, or cloth, when the surface is plane; and by sheaths of pods, seed vessels, or the like, when it is not capable of being stretched flat without tearing. The process of changing the form of a surface by bending is called "*developing.*" But the term "*Developable Surface*" is commonly restricted to such inextensible surfaces as can be developed into a plane, or, in common language, "smoothed flat."

140. The geometry or kinematics of this subject is a great contrast to that of the flexible line (§ 14), and, in its merest elements, presents ideas not very easily apprehended, and subjects of investigation that have exercised, and perhaps even overtasked, the powers of some of the greatest mathematicians.

141. Some care is required to form a correct conception of what is a perfectly flexible inextensible surface. First let us consider a plane sheet of paper. It is very flexible, and we can easily form the conception from it of a sheet of ideal matter perfectly flexible. It is very inextensible; that is to say, it yields very little to any application of force tending to pull or stretch it in any direction, up to the strongest it can bear without tearing. It does, of course, stretch a little. It is easy to test that it stretches when under the influence of force, and that it contracts again when the force is removed, although not always to its original dimensions, as it may and generally does remain to some sensible extent permanently stretched. Also, flexure stretches one side and condenses the other temporarily; and, to a less extent, permanently. Under elasticity we may return to this. In the meantime, in considering illustrations of our kinematical propositions, it is necessary to anticipate such physical circumstances.

Surface inextensible in two directions.

142. Cloth woven in the simple, common way, very fine muslin for instance, illustrates a surface perfectly inextensible in two directions (those of the warp and the woof), but susceptible of any amount of extension from 1 up to $\sqrt{2}$ along one diagonal, with contraction from 1 to 0 (each degree of

Surface inextensible in two directions.

xtension along one diagonal having a corresponding deter-inate degree of contraction along the other, the relation being $+ e'^2) = 2$, where $1 : e$ and $1 : e'$ are the ratios of elongation, hich will be contraction in the case in which e or e' is < 1) the other. What precedes supposes that the weaving is uare, in which case the diagonals remain at right angles one another. Oblong weaving gives a less simple relation, ough easily determinable. Cloth will hang very differently, cording as its rectangles are square, or oblong to any degree f inequality.

143. The flexure of a surface fulfilling any case of the ometrical condition just stated, presents an interesting sub-t for investigation, which we are reluctantly obliged to rego. The moist paper drapery that Albert Durer used on s little lay figures must hang very differently from cloth. erhaps the stiffness of the drapery in his pictures may be to me extent owing to the fact that he used the moist paper in eference to cloth on account of its superior flexibility, while naware of the great distinction between them as regards xtensibility. Fine muslin, prepared with starch or gum, is, luring the process of drying, kept moving by a machine, which, y producing a to-and-fro relative angular motion of warp and f, stretches and contracts the diagonals of its structure alter-ately, and thus prevents the parallelograms from becoming tiffened into rectangles.

Flexure of inextensible developable.

144. The flexure of an inextensible surface which can be lane, is a subject which has been well worked by geometrical nvestigators and writers, and, in its elements at least, presents ittle difficulty. The first elementary conception to be formed s, that such a surface (if perfectly flexible), taken plane in e first place, may be bent about any straight line ruled on t, so that the two plane parts may make any angle with one another.

Such a line is called a "generating line" of the surface to be formed.

Next, we may bend one of these plane parts about any other line which does not (within the limits of the sheet) intersect the former; and so on. If these lines are infinite in number, and the angles of bending infinitely small, but such that their

Flexure of inextensible developable. sum may be finite, we have our plane surface bent into a curved surface, which is of course "developable" (§ 139).

145. Lift a square of paper, free from folds, creases, or ragged edges, gently by one corner, or otherwise, without crushing or forcing it, or very gently by two points. It will hang in a form which is very rigorously a developable surface; for although it is not absolutely inextensible, yet the forces which tend to stretch or tear it, when it is treated as above described, are small enough to produce no sensible stretching. Indeed the greatest stretching it can experience without tearing, in any direction, is not such as can affect the form of the surface much when sharp flexures, singular points, etc., are kept clear of.

146. Prisms and cylinders (when the lines of bending, § 144, are parallel, and finite in number with finite angles, or infinite in number with infinitely small angles), and pyramids and cones (the lines of bending meeting in a point if produced), are clearly included.

147. If the generating lines, or line-edges of the angles of bending, are not parallel, they must meet, since they are in a plane when the surface is plane. If they do not meet all in one point, they must meet in several points: in general, let each one meet its predecessor and its successor in different points.

148. There is still no difficulty in understanding the form of, say a square, or circle, of the plane surface when bent as explained above, provided it does not include any of these points of intersection. When the number is infinite, and the surface finitely curved, the developable lines will in general be tangents to a curve (the locus of the points of intersection when the number is infinite). This curve is called the *edge of regression*. The surface must clearly, when *complete* (according to mathematical ideas), consist of two sheets meeting in this edge of regression (just as a cone consists of two sheets meeting in the vertex), because each tangent may be produced beyond the point of contact, instead of stopping at it, as in the annexed diagram.

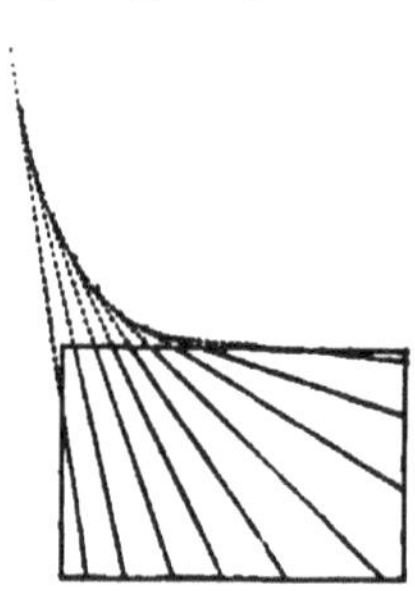

Edge of regression.

149. To construct a complete developable surface in two sheets from its edge of regression—

Practical construction of a developable from its edge.

Lay one piece of perfectly flat, unwrinkled, smooth-cut paper on the top of another. Trace any curve on the upper, and let it have no point of inflection, but everywhere finite curvature. Cut the paper quite away on the concave side. If the curve traced is closed, it must be cut open (see second diagram).

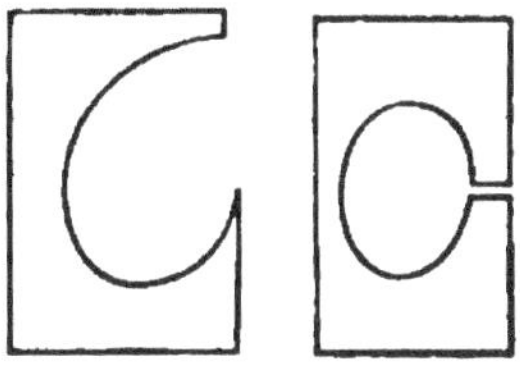

The limits to the extent that may be left uncut away, are the tangents drawn outwards from the two ends, so that, in short, no portion of the paper through which a real tangent does not pass is to be left.

Attach the two sheets together by very slight paper or muslin clamps gummed to them along the common curved edge. These must be so slight as not to interfere sensibly with the flexure of the two sheets. Take hold of one corner of one sheet and lift the whole. The two will open out into the two sheets of a developable surface, of which the curve, bending into a curve of double curvature, is the edge of regression. The tangent to the curve drawn in one direction from the point of contact, will always lie in one of the sheets, and its continuation on the other side in the other sheet. Of course a double-sheeted developable polyhedron can be constructed by this process, by starting from a polygon instead of a curve.

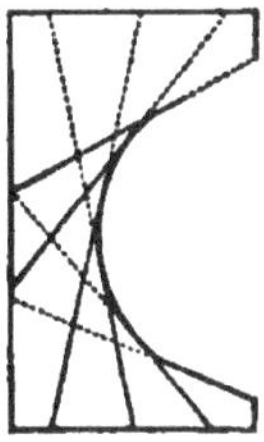

150. A flexible but perfectly inextensible surface, altered in form in any way possible for it, must keep any line traced on it unchanged in length; and hence any two intersecting lines unchanged in mutual inclination. Hence, also, geodetic lines must remain geodetic lines. Hence "the change of direction" in a surface, of a point going round any portion of it, must be the same, however this portion is bent. Hence (§ 136) the integral curvature remains the same in any and every portion however the surface is bent. Hence (§ 138, *Gauss's Theorem*) the product of the principal radii of curvature at each point remains unchanged.

General property of inextensible surface.

General property of inextensible surface.

151. The general statement of a converse proposition, expressing the condition that two given areas of curved surfaces may be bent one to fit the other, involves essentially some mode of specifying corresponding points on the two. A full investigation of the circumstances would be out of place here.

Surface of constant specific curvature.

152. In one case, however, a statement in the simplest possible terms is applicable. Any two surfaces, in each of which the specific curvature is the same at all points, and equal to that of the other, may be bent one to fit the other. Thus any surface of uniform positive specific curvature (*i.e.*, wholly convex one side, and concave the other) may be bent to fit a sphere whose radius is a mean proportional between its principal radii of curvature at any point. A surface of uniform negative, or anticlastic, curvature would fit an imaginary sphere, but the interpretation of this is not understood in the present condition of science. But practically, of any two surfaces of uniform anticlastic curvature, either may be bent to fit the other.

Geodetic triangles on such a surface

153. It is to be remarked, that geodetic trigonometry on any surface of uniform positive, or synclastic curvature, is identical with spherical trigonometry.

If $a=\frac{s}{\sqrt{\rho\rho'}}$, $b=\frac{t}{\sqrt{\rho\rho'}}$, $c=\frac{u}{\sqrt{\rho\rho'}}$, where s, t, u are the lengths of three geodetic lines joining three points on the surface, and if A, B, C denote the angles between the tangents to the geodetic lines at these points; we have six quantities which agree perfectly with the three sides and the three angles of a certain spherical triangle. A corresponding anticlastic trigonometry exists, although we are not aware that it has been worked out, for any surface of uniform anticlastic curvature. In a geodetic triangle on an anticlastic surface, the sum of the three angles is of course less than three right angles, and the difference, or "anticlastic defect" (like the "spherical excess"), is equal to the area divided by $\rho\times-\rho'$, when ρ and $-\rho'$ are positive.

Strain.

154. We have now to consider the very important kinematical conditions presented by the changes of volume or figure experienced by a solid or liquid mass, or by a group of points whose positions with regard to each other are subject to known conditions. Any such definite alteration of form or dimensions is called a *Strain*.

Thus a rod which becomes longer or shorter is strained. Water, when compressed, is strained. A stone, beam, or mass of metal, in a building or in a piece of framework, if condensed or dilated in any direction, or bent, twisted, or distorted in any way, is said to experience a strain. A ship is said to "strain" if, in launching, or when working in a heavy sea, the different parts of it experience relative motions. Strain.

155. If, when the matter occupying any space is strained in any way, all pairs of points of its substance which are initially at equal distances from one another in parallel lines remain equidistant, it may be at an altered distance; and in parallel lines, altered, it may be, from their initial direction; the strain is said to be homogeneous. Definition of homogeneous strain.

156. Hence if any straight line be drawn through the body in its initial state, the portion of the body cut by it will continue to be a straight line when the body is homogeneously strained. For, if ABC be any such line, AB and BC, being parallel to one line in the initial, remain parallel to one line in the altered, state; and therefore remain in the same straight line with one another. Thus it follows that a plane remains a plane, a parallelogram a parallelogram, and a parallelepiped a parallelepiped. Properties of homogeneous strain.

157. Hence, also, similar figures, whether constituted by actual portions of the substance, or mere geometrical surfaces, or straight or curved lines passing through or joining certain portions or points of the substance, similarly situated (*i.e.*, having corresponding parameters parallel) when altered according to the altered condition of the body, remain similar and similarly situated among one another.

158. The lengths of parallel lines of the body remain in the same proportion to one another, and hence all are altered in the same proportion. Hence, and from § 156, we infer that any plane figure becomes altered to another plane figure which is a diminished or magnified orthographic projection of the first on some plane. For example, if an ellipse be altered into a circle, its principal axes become radii at right angles to one another.

The elongation of the body along any line is the proportion which the addition to the distance between any two points in that line bears to their primitive distance.

Properties of homogeneous strain.

159. Every orthogonal projection of an ellipse is an ellipse (the case of a circle being included). Hence, and from § 158, we see that an ellipse remains an ellipse; and an ellipsoid remains a surface of which every plane section is an ellipse; that is, remains an ellipsoid.

A plane curve remains (§ 156) a plane curve. A system of two or of three straight lines of reference (Cartesian) remains a rectilineal system of lines of reference; but, in general, a rectangular system becomes oblique.

Let $$\frac{x^2}{a^2}+\frac{y^2}{b^2}=1$$

be the equation of an ellipse referred to any rectilineal conjugate axes, in the substance, of the body in its initial state. Let α and β be the proportions in which lines respectively parallel to OX and OY are altered. Thus, if we call ξ and η the altered values of x and y, we have

$$\xi=\alpha x,\ \eta=\beta y.$$

Hence $$\frac{\xi^2}{(\alpha a)^2}+\frac{\eta^2}{(\beta b)^2}=1,$$

which also is the equation of an ellipse, referred to oblique axes at, it may be, a different angle to one another from that of the given axes, in the initial condition of the body.

Or again, let $$\frac{x^2}{a^2}+\frac{y^2}{b^2}+\frac{z^2}{c^2}=1$$

be the equation of an ellipsoid referred to, three conjugate diametral planes, as oblique or rectangular planes of reference, in the initial condition of the body. Let α, β, γ be the proportion in which lines parallel to OX, OY, OZ are altered; so that if ξ, η, ζ be the altered values of x, y, z, we have

$$\xi=\alpha x,\ \eta=\beta y,\ \zeta=\gamma z.$$

Thus, $$\frac{\xi^2}{(\alpha a)^2}+\frac{\eta^2}{(\beta b)^2}+\frac{\zeta^2}{(\gamma c)^2}=1$$

which is the equation of an ellipsoid, referred to conjugate diametral planes, altered it may be in mutual inclination from those of the given planes of reference in the initial condition of the body.

Strain ellipsoid.

160. The ellipsoid which any surface of the body initially spherical becomes in the altered condition, may, to avoid circumlocutions, be called the strain ellipsoid.

161. In any absolutely unrestricted homogeneous strain there are three directions (the three principal axes of the strain ellipsoid), at right angles to one another, which remain at right angles to one another in the altered condition of the body (§ 158). Along one of these the elongation is greater, and along another less, than along any other direction in the body. Along the remaining one, the elongation is less than in any other line in the plane of itself and the first mentioned, and greater than along any other line in the plane of itself and the second. Strain ellipsoid.

Note.—Contraction is to be reckoned as a negative elongation: the maximum elongation of the preceding enunciation may be a minimum contraction: the minimum elongation may be a maximum contraction.

162. The ellipsoid into which a sphere becomes altered may be an ellipsoid of revolution, or, as it is called, a spheroid, prolate, or oblate. There is thus a maximum or minimum elongation along the axis, and equal minimum or maximum elongation along all lines perpendicular to the axis.

Or it may be a sphere; in which case the elongations are equal in all directions. The effect is, in this case, merely an alteration of dimensions without change of figure of any part.

The original volume (sphere) is to the new (ellipsoid) evidently as $1 : \alpha\beta\gamma$. Change of volume.

163. The principal axes of a strain are the principal axes of the ellipsoid into which it converts a sphere. The principal elongations of a strain are the elongations in the direction of its principal axes. Axes of a strain.

164. When the positions of the principal axes, and the magnitudes of the principal elongations of a strain are given, the elongation of any line of the body, and the alteration of angle between any two lines, may be obviously determined by a simple geometrical construction, Elongation and change of direction of any line of the body.

Analytically thus:—let $\alpha-1$, $\beta-1$, $\gamma-1$ denote the principal elongations, so that α, β, γ may be now the ratios of alteration along the three principal axes, as we used them formerly for the ratios for any three oblique or rectangular lines. Let l, m, n be the direction cosines of any line, with reference to the three principal axes. Thus,

$$lr, \; mr, \; nr$$

Elongation and change of direction of any line of the body.

being the three initial co-ordinates of a point P, at a distance $OP=r$, from the origin in the direction l, m, n; the co-ordinates of the same point of the body, with reference to the same rectangular axes, become, in the altered state,

$$\alpha lr,\ \beta mr,\ \gamma nr.$$

Hence the altered length of OP is

$$(\alpha^2 l^2+\beta^2 m^2+\gamma^2 n^2)^{\frac{1}{2}} r,$$

and therefore the "elongation" of the body in that direction is

$$(\alpha^2 l^2+\beta^2 m^2+\gamma^2 n^2)^{\frac{1}{2}}-1.$$

For brevity, let this be denoted by $\zeta-1$, *i.e.*

let $$\zeta=(\alpha^2 l^2+\beta^2 m^2+\gamma^2 n^2)^{\frac{1}{2}}.$$

The direction-cosines of OP in its altered position are

$$\frac{\alpha l}{\zeta},\ \frac{\beta m}{\zeta},\ \frac{\gamma n}{\zeta};$$

and therefore the angles XOP, YOP, ZOP are altered to having their cosines of these values respectively, from having them of the values l, m, n.

The cosine of the angle between any two lines OP and OP', specified in the initial condition of the body by the direction-cosines l', m', n', is

$$ll'+mm'+nn',$$

in the initial condition of the body, and becomes

$$\frac{\alpha^2 ll'+\beta^2 mm'+\gamma^2 nn'}{(\alpha^2 l^2+\beta^2 m^2+\gamma^2 n^2)^{\frac{1}{2}}(\alpha^2 l'^2+\beta^2 m'^2+\gamma^2 n'^2)^{\frac{1}{2}}}$$

in the altered condition.

Change of plane in the body.

165. With the same data the alteration of angle between any two planes of the body may also be easily determined, either geometrically or analytically.

Let l, m, n be the cosines of the angles which a plane makes with the planes YOZ, ZOX, XOY, respectively, in the initial condition of the body. The effects of the change being the same on all parallel planes, we may suppose the plane in question to pass through O; and therefore its equation will be

$$lx+my+nz=0.$$

In the altered condition of the body we shall have, as before,

$$\xi=\alpha x,\ \eta=\beta y,\ \zeta=\gamma z,$$

for the altered co-ordinates of any point initially x, y, z. Hence the equation of the altered plane is

$$\frac{l\xi}{\alpha}+\frac{m\eta}{\beta}+\frac{n\gamma}{\zeta}=0.$$

Change of plane in the body. But the planes of reference are still rectangular, according to our present supposition. Hence the cosines of the inclinations of the plane in question, to YOZ, ZOX, XOY, in the altered condition of the body, are altered from l, m, n to

$$\frac{l}{\alpha\vartheta},\ \frac{m}{\beta\vartheta},\ \frac{n}{\gamma\vartheta},$$

respectively, where for brevity

$$\vartheta=\left(\frac{l^2}{\alpha^2}+\frac{m^2}{\beta^2}+\frac{n^2}{\gamma^2}\right)^{\frac{1}{2}}.$$

If we have a second plane similarly specified by l', m', n', in the initial condition of the body, the cosine of the angle between the two planes, which is

$$ll'+mm'+nn'$$

in the initial condition, becomes altered to

$$\frac{\dfrac{ll'}{\alpha^2}+\dfrac{mm'}{\beta^2}+\dfrac{nn'}{\gamma^2}}{\left(\dfrac{l^2}{\alpha^2}+\dfrac{m^2}{\beta^2}+\dfrac{n^2}{\gamma^2}\right)^{\frac{1}{2}}\left(\dfrac{l'^2}{\alpha^2}+\dfrac{m'^2}{\beta^2}+\dfrac{n'^2}{\gamma^2}\right)^{\frac{1}{2}}}.$$

Conical surface of equal elongation. **166.** Returning to elongations, and considering that these are generally different in different directions, we perceive that all lines through any point, in which the elongations have any one value intermediate between the greatest and least, must lie on a determinate conical surface. This is easily proved to be in general a cone of the second degree.

For, in a direction denoted by direction cosines l, m, n, we have

$$\alpha^2l^2+\beta^2m^2+\gamma^2n^2=\zeta^2,$$

where ζ denotes the ratio of elongation, intermediate between α the greatest and γ the least. This is the equation of a cone of the second degree, l, m, n being the direction cosines of a generating line.

Two planes of no distortion. **167.** In one particular case this cone becomes two planes, the planes of the circular sections of the strain ellipsoid.

Let $\zeta=\beta$. The preceding equation becomes

$$\alpha^2l^2+\gamma^2n^2-\beta^2(1-m^2)=0$$

or since

$$1-m^2=l^2+n^2,$$

$$(\alpha^2-\beta^2)l^2-(\beta^2-\gamma^2)n^2=0.$$

The first member being the product of two factors, the equation

Two planes of no distortion,

is satisfied by putting either =0, and therefore the equation represents the two planes whose equations are

$$l(\alpha^2-\beta^2)^{\frac{1}{2}}+n(\beta^2-\gamma^2)^{\frac{1}{2}}=0,$$

and $$l(\alpha^2-\beta^2)^{\frac{1}{2}}-n(\beta^2-\gamma^2)^{\frac{1}{2}}=0,$$

respectively.

being the circular sections of the strain ellipsoid.

This is the case in which the given elongation is equal to that along the mean principal axis of the strain ellipsoid. The two planes are planes through the mean principal axis of the ellipsoid, equally inclined on the two sides of either of the other axes. The lines along which the elongation is equal to the mean principal elongation, all lie in, or parallel to, either of these two planes. This is easily proved as follows, without any analytical investigation.

168. Let the ellipse of the annexed diagram represent the section of the strain ellipsoid through the greatest and least principal axes. Let *SOS*, *TOT* be the two diameters of this ellipse, which are equal to the mean principal axis of the ellipsoid. Every plane through *O*, perpendicular to the plane of the diagram, cuts the ellipsoid in an ellipse of which one principal axis is the diameter in which it cuts the ellipse of the diagram, and the other, the mean principal diameter of the ellipsoid. Hence a plane through either *SS'*, or *TT'*, perpendicular to the plane of the diagram, cuts the ellipsoid in an ellipse of which the two principal axes are equal, that is to say, in a circle. Hence the elongations along all lines in either of these planes are equal to the elongation along the mean principal axis of the strain ellipsoid.

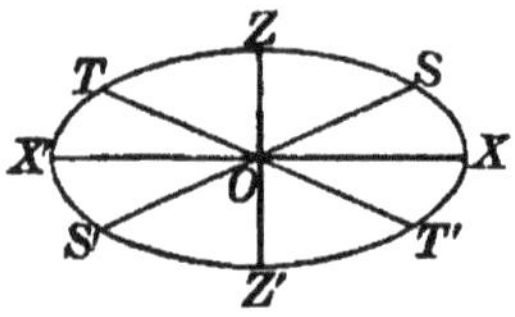

Distortion in parallel planes without change of volume.

169. The consideration of the circular sections of the strain ellipsoid is highly instructive, and leads to important views with reference to the analysis of the most general character of a strain. First let us suppose there to be no alteration of volume on the whole, and neither elongation nor contraction along the mean principal axis. That is to say, let $\beta=1$, and $\gamma=\frac{1}{\alpha}$ (§ 162).

Let OX and OZ be the directions of elongation $\alpha - 1$ and contraction $1 - \frac{1}{\alpha}$ respectively. Let A be any point of the body in its primitive condition, and $A_,$ the same point of the altered body, so that $OA_, = \alpha OA$. Now, if we take $OC = OA_,$, and if $C_,$ be the position of that point of the body which was in the position C initially, we shall have $OC_, = \frac{1}{\alpha} OC$, and therefore $OC_, = OA$. Hence the two triangles COA and $C_,OA_,$ are equal and similar. Distortion in parallel planes without change of volume.

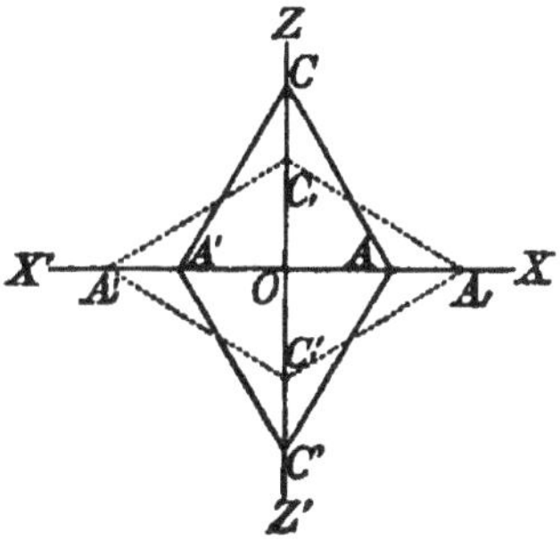

Hence CA experiences no alteration of length, but takes the altered position $C_,A_,$ in the altered position of the body. Similarly, if we measure on XO produced, OA' and $OA_,'$ equal respectively to OA and $OA_,$, we find that the line CA' experiences no alteration in length, but takes the altered position $C_,A_,'$. Initial and altered position of lines of no elongation.

Consider now a plane of the body initially through CA perpendicular to the plane of the diagram, which will be altered into a plane through $C_,A_,$, also perpendicular to the plane of the diagram. All lines initially perpendicular to the plane of the diagram remain so, and remain unaltered in length. AC has just been proved to remain unaltered in length. Hence (§ 158) all lines in the plane we have just drawn remain unaltered in length and in mutual inclination. Similarly we see that all lines in a plane through CA', perpendicular to the plane of the diagram, altering to a plane through $C_,A_,'$, perpendicular to the plane of the diagram, remain unaltered in length and in mutual inclination.

170. The precise character of the strain we have now under consideration will be elucidated by the following:—Produce CO, and take OC' and $OC_,'$ respectively equal to OC and $OC_,$. Join $C'A$, $C'A'$, $C_,'A_,$, and $C_,A_,'$, by plain and dotted lines as in the diagram. Then we see that the rhombus $CAC'A'$ (plain lines) of the body in its initial state becomes the rhombus $C_,A_,C_,'A_,'$ (dotted) in the altered condition. Now imagine

Initial and altered position of lines of no elongation.

the body thus strained to be moved as a rigid body (*i.e.*, with its state of strain kept unchanged) till $A_{\prime}$ coincides with A, and $C_{\prime}'$ with C', keeping all the lines of the diagram still in the same plane. $A_{\prime}'C$ will take a position in CA' produced, as shown in the new diagram, and the original and the altered parallelogram will be on the same base AC', and between the same parallels AC' and $CA_{\prime}'$, and their other sides will be equally inclined on the two sides of a perpendicular to them. Hence, irrespectively of any rotation, or other absolute motion of the body not involving change of form or dimensions, the strain under consideration may be produced by holding fast and unaltered the plane of the body through AC' perpendicular to the plane of the diagram, and making every plane parallel to it slide, keeping the same distance, through a space proportional to this distance (*i.e.*, different planes parallel to the fixed plane slide through spaces proportional to their distances).

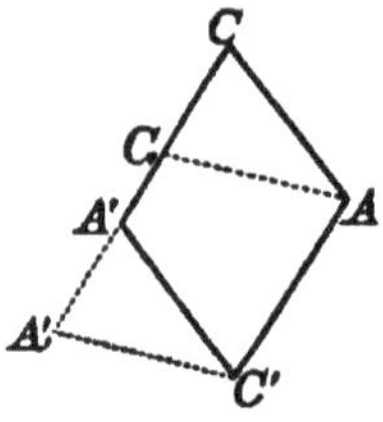

Simple shear.

171. This kind of strain is called a *simple shear*. The plane of a shear is a plane perpendicular to the undistorted planes, and parallel to the lines of their relative motion. It has (1.) the property that one set of parallel planes remain each unaltered in itself; (2.) that another set of parallel planes remain each unaltered in itself. This other set is found when the first set and the degree or amount of shear are given, thus:—Let $CC_{\prime}$ be the motion of one point of one plane, relative to a plane KL held fixed—the diagram being in a plane of the shear. Bisect $CC_{\prime}$ in N. Draw NA perpendicular to it. A plane perpendicular to the plane of the diagram, initially through AC, and finally through $AC_{\prime}$, remains unaltered in its dimensions.

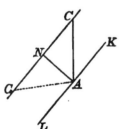

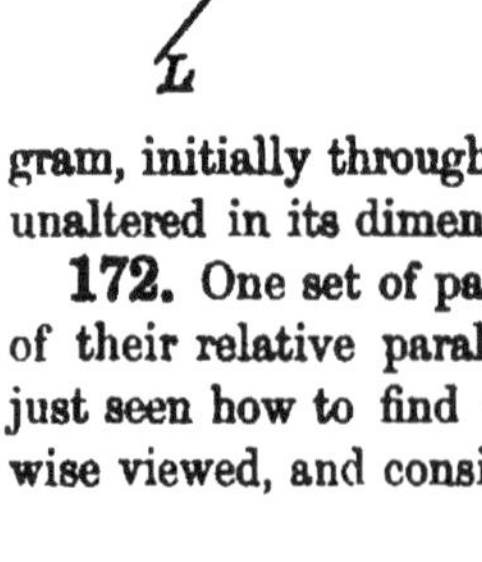

172. One set of parallel undistorted planes, and the amount of their relative parallel shifting having been given, we have just seen how to find the other set. The shear may be otherwise viewed, and considered as a shifting of this second set of arallel planes, relative to any one of them. The amount of his relative shifting is of course equal to that of the first set, elatively to one of them.

173. The principal axes of a shear are the lines of maximum elongation and of maximum contraction respectively. They may be found from the preceding construction (§ 171), hus:—In the plane of the shear bisect the obtuse and cute angles between the planes destined not to become deormed. The former bisecting line is the principal axis of longation, and the latter is the principal axis of contraction, n their initial positions. The former angle (obtuse) becomes qual to the latter, its supplement (acute), in the altered conlition of the body, and the lines bisecting the altered angles re the principal axes of the strain in the altered body. Axes of a shear.

Otherwise, taking a plane of shear for the plane of the liagram, let AB be a line in which it is cut by one of either set of parallel planes of no distortion. On any portion AB of this as diameter, lescribe a semicircle. Through C, its middle point, draw, by the preceding construction, CD the initial, and CE the final, position of an unstretched line. Join DA, DB, EA, EB. DA, DB are the initial, and EA, EB the final, positions of the principal axes.

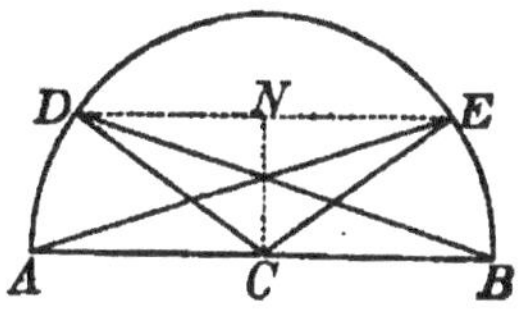

174. The ratio of a shear is the ratio of elongation and contraction of its principal axes. Thus if one principal axis is elongated in the ratio $1:a$, and the other therefore (§ 169) contracted in the ratio $a:1$, a is called the ratio of the shear. It will be convenient generally to reckon this as the ratio of elongation; that is to say, to make its numerical measure greater than unity. Measure of a shear.

In the diagram of § 173, the ratio of DB to EB, or of EA to DA, is the ratio of the shear.

175. The amount of a shear is the amount of relative motion per unit distance between planes of no distortion.

It is easily proved that this is equal to the excess of the ratio of the shear above its reciprocal.

Since $DCA = 2DBA$, and $\tan DBA = \frac{1}{a}$ we have $\tan DCA = \frac{2a}{a^2 - 1}$.

But $DE = 2CN \tan DCN = 2CN \cot DCA.$

Hence $$\frac{DE}{CN} = 2\frac{a^2-1}{2a} = a - \frac{1}{a}.$$

Ellipsoidal specification of a shear.

176. The planes of no distortion in a simple shear are clearly the circular sections of the strain ellipsoid. In the ellipsoid of this case, be it remembered, the mean axis remains unaltered, and is a mean proportional between the greatest and the least axis.

Shear, simple elongation, and expansion, combined.

177. If we now suppose all lines perpendicular to the plane of the shear to be elongated or contracted in any proportion, without altering lengths or angles in the plane of the shear, and if, lastly, we suppose every line in the body to be elongated or contracted in some other fixed ratio, we have clearly (§ 161) the most general possible kind of strain. Thus if s be the ratio of the simple shear, for which case s, 1, $\frac{1}{s}$ are the three principal ratios, and if we elongate lines perpendicular to its plane in the ratio $1 : m$, without any other change, we have a strain of which the principal ratios are

$$s,\ m,\ \frac{1}{s}.$$

If, lastly, we elongate all lines in the ratio $1 : n$, we have a strain in which the principal ratios are

$$ns,\ nm,\ \frac{n}{s},$$

where it is clear that ns, nm, and $\frac{n}{s}$ may have any values whatever. It is of course not necessary that nm be the mean principal ratio. Whatever they are, if we call them α, β, γ respectively, we have

$$s = \sqrt{\frac{\alpha}{\gamma}};\ n = \sqrt{\alpha\gamma};\ \text{and } m = \frac{\beta}{\sqrt{\alpha\gamma}}.$$

Analysis of a strain.

178. Hence any strain (α, β, γ) whatever may be viewed as compounded of a uniform dilatation in all directions, of linear ratio $\sqrt{\alpha\gamma}$, superimposed on a simple elongation $\frac{\beta}{\sqrt{\alpha\gamma}}$ in the direction of the principal axis to which β refers, superimposed on a simple shear, of ratio $\sqrt{\frac{\alpha}{\gamma}}$ (or of amount $\sqrt{\frac{\alpha}{\gamma}} - \sqrt{\frac{\gamma}{\alpha}}$) in the plane of the two other principal axes.

179. It is clear that these three elementary component strains may be applied in any other order as well as that stated. Thus, if the simple elongation is made first, the body thus altered must get just the same shear in planes perpendicular to the line of elongation, as the originally unaltered body gets when the order first stated is followed. Or the dilatation may be first, then the elongation, and finally the shear, and so on. Analysis of a strain.

180. In the preceding sections on strains, we have considered the alterations of lengths of lines of the body, and of angles between lines and planes of it; and we have, in particular cases, founded on particular suppositions (the principal axes of the strain remaining fixed in direction, § 169, or one of either set of undistorted planes in a simple shear remaining fixed, § 170), considered the actual displacements of parts of the body from their original positions. But to complete the kinematics of a non-rigid solid, it is necessary to take a more general view of the relation between displacements and strains. It will be sufficient for us to suppose one point of the body to remain fixed, as it is easy to see the effect of superimposing upon any motion with one point fixed, a motion of translation without strain or rotation. Displacement of a body, rigid or not, one point of which is held fixed.

181. Let us therefore suppose one point of a body to be held fixed, and any displacement whatever given to any point or points of it, subject to the condition that the whole substance if strained at all is homogeneously strained.

Let OX, OY, OZ be any three rectangular axes, fixed with reference to the initial position and condition of the body. Let x, y, z be the initial co-ordinates of any point of the body, and x_1, y_1, z_1 be the co-ordinates of the same point of the altered body, with reference to those axes unchanged. The condition that the strain is homogeneous throughout is expressed by the following equations :—

$$\left.\begin{aligned} x_1&=[Xx]x+[Xy]y+[Xz]z,\\ y_1&=[Yx]x+[Yy]y+[Yz]z,\\ z_1&=[Zx]x+[Zy]y+[Zz]z,\end{aligned}\right\}\qquad(1)$$

where $[Xx]$, $[Xy]$, etc., are nine quantities of absolutely arbitrary values, which are the same for all values of x, y, z.

$[Xx]$, $[Yx]$, $[Zx]$ denote the three final co-ordinates of a point

Displacement of a body, rigid or not, one point of which is held fixed.

originally at unit distance along OX, from O. They are, of course, proportional to the direction-cosines of the altered position of the line primitively coinciding with OX. Similarly for $[Xy]$, $[Yy]$, $[Zy]$, etc.

Let it be required to find, if possible, a line of the body which remains unaltered in direction, during the change specified by $[Xx]$, etc. Let x, y, z, and x_1, y_1, z_1, be the co-ordinates of the primitive and altered position of a point in such a line. We must have $$\frac{x_1}{x}=\frac{y_1}{y}=\frac{z_1}{z}=\epsilon,$$ where ϵ is the elongation of the line in question.

Thus we have $x_1=\epsilon x$, etc., and therefore

$$\left.\begin{aligned}\{[Xx]-\epsilon\}x \quad +[Xy]y \quad +[Xz]z=0,\\ [Yx]x+\{[Yy]-\epsilon\}y \quad +[Yz]z=0,\\ [Zx]x \quad +[Zy]y+\{[Zz]-\epsilon\}z=0.\end{aligned}\right\}\qquad(2)$$

From these equations, by eliminating the ratios $x:y:z$ according to the well-known algebraic process, we find

$$\big([Xx]-\epsilon\big)\big([Yy]-\epsilon\big)\big([Zz]-\epsilon\big)$$
$$-[Yz][Zy]\big([Xx]-\epsilon\big)-[Zx][Xz]\big([Yy]-\epsilon\big)-[Xy][Yx]\big([Zz]-\epsilon\big)$$
$$+[Xz][Yx][Zy]+[Xy][Yz][Zx]=0.$$

This cubic equation is necessarily satisfied by at least one real value of ϵ, and the two others are either both real or both imaginary. Each real value of ϵ gives a real solution of the problem, since any two of the preceding three equations with it, in place of ϵ, determine real values of the ratios $x:y:z$. If the body is rigid (*i.e.*, if the displacements are subject to the condition of producing no strain), we know (*ante*, § 95) that there is just one line common to the body in its two positions, the axis round which it must turn to pass from one to the other, except in the peculiar cases of *no* rotation, and of rotation through *two* right angles, which are treated below. Hence, in this case, the cubic equation has only one real, and therefore it has two imaginary, roots. The equations just formed solve the problem of finding the axis of rotation when the data are the actual displacements of the points primitively lying in three given fixed axes of reference, OX, OY, OZ; and it is worthy of remark, that the practical solution of this problem is founded on the one real root of a cubic which has two imaginary roots.

Again, on the other hand, let the given displacements be made so as to produce a strain of the body with no angular displacement of the principal axes of the strain. Thus three lines of the body remain unchanged. Hence there must be three real roots of the equation in ϵ, one for each such axis; and the three lines determined by them are necessarily at right angles to one another.

Displacement of a body, rigid or not, one point of which is held fixed.

But if neither of these conditions holds, we may have three real solutions and three oblique lines of directional identity; or we may have only one real root and only one line of directional identity.

An analytical proof of these conclusions may easily be given; thus we may write the cubic in the form—

$$\begin{vmatrix} [Xx], & [Xy], & [Xz] \\ [Yx], & [Yy], & [Yz] \\ [Zx], & [Zy], & [Zz] \end{vmatrix} - \epsilon \left\{ \begin{vmatrix} [Yy], & [Yz] \\ [Zy], & [Zz] \end{vmatrix} + \begin{vmatrix} [Zz], & [Zx] \\ [Xz], & [Xx] \end{vmatrix} + \begin{vmatrix} [Xx], & [Xy] \\ [Yx], & [Yy] \end{vmatrix} \right\} + \epsilon^2 \{ [Xx] + [Yy] + [Zz] \} - \epsilon^3 = 0. \qquad (3).$$

In the particular case of no strain, since $[Xx]$, etc., are then *equal*, not merely *proportional*, to the direction-cosines of three mutually perpendicular lines, we have by well-known geometrical theorems

$$\begin{vmatrix} [Xx], & [Xy], & [Xz] \\ [Yx], & [Yy], & [Yz] \\ [Zx], & [Zy], & [Zz] \end{vmatrix} = 1, \text{ and } \begin{vmatrix} [Yy], & [Yz] \\ [Zy], & [Zz] \end{vmatrix} = [Xx], \text{ etc.}$$

Hence the cubic becomes

$$1 - (\epsilon - \epsilon^2) \{ [Xx] + [Yy] + [Zz] \} - \epsilon^3 = 0,$$

of which one root is evidently $\epsilon = 1$. This leads to the above explained rotational solution, the line determined by the value 1 of ϵ being the axis of rotation. Dividing out the factor $1 - \epsilon$, we get for the two remaining roots the equation

$$1 - ([Xx] + [Yy] + [Zz] - 1)\epsilon + \epsilon^2 = 0,$$

whose roots are imaginary if the co-efficient of ϵ lies between $+2$ and -2. Now $+2$ is evidently its *greatest* value, and for that case the roots are real, each being unity. Here there is no rotation. Also -2 is its *least* value, and this gives us a pair of values each $= -1$, of which the interpretation is, that there is rotation through two right angles. In this case, as in general, one line (the axis of rotation) is determined by the equations (2) with the value $+1$ for ϵ; but with $\epsilon = -1$ these equations are satisfied by any line perpendicular to the former.

The limiting case of two equal roots, when there is strain, is an interesting subject which may be left as an exercise. It separates the cases in which there is only one axis of directional identity from those in which there are three.

Let it next be proposed to find those lines of the body whose

Displacement of a body, rigid or not, one point of which is held fixed.

elongations are greatest or least. For this purpose we must find the equations expressing that $x_1^2+y_1^2+z_1^2$ is a maximum, when $x^2+y^2+z^2=r^2$, a constant. First, we have

$$x_1^2+y_1^2+z_1^2=Ax^2+By^2+Cz^2+2(ayz+bzx+cxy) \qquad (4)$$

where

$$\left.\begin{aligned} A&=[Xx]^2+[Yx]^2+[Zx]^2\\ B&=[Xy]^2+[Yy]^2+[Zy]^2\\ C&=[Xz]^2+[Yz]^2+[Zz]^2\\ a&=[Xy][Xz]+[Yy][Yz]+[Zy][Zz]\\ b&=[Xz][Xx]+[Yz][Yx]+[Zz][Zx]\\ c&=[Xx][Xy]+[Yx][Yy]+[Zx][Zy]\end{aligned}\right\} \qquad (5)$$

The equation

$$Ax^2+By^2+Cz^2+2(ayz+bzx+cxy)=r_1^2 \qquad (6)$$

where r_1 is any constant, represents clearly the ellipsoid which a spherical surface, radius r_1, of the altered body, would become if if the body were restored to its primitive condition. The problem of making r_1 a maximum when r is a given constant, leads to the following equations:—

$$x^2+y^2+z^2=r^2, \qquad (7)$$

$$\left.\begin{aligned} xdx+ydy+zdz&=0,\\ (Ax+cy+bz)dx+(cx+By+az)dy+(bx+ay+Cz)dz&=0.\end{aligned}\right\} \qquad (8)$$

On the other hand, the problem of making r a maximum or minimum when r_1 is given, that is to say, the problem of finding maximum and minimum diameters, or principal axes, of the ellipsoid (6), leads to these same two differential equations (8), and only differs in having equation (6) instead of (7) to complete the determination of the absolute values of x, y, and z. Hence the ratios $x:y:z$ will be the same in one problem as in the other; and therefore the *directions* determined are those of the principal axes of the ellipsoid (6). We know, therefore, by the properties of the ellipsoid, that there are three real solutions, and that the directions of the three radii so determined are mutually rectangular. The ordinary method (Lagrange's) for dealing with the differential equations, being to multiply one of them by an arbitrary multiplier, then add, and equate the co-efficients of the separate differentials to zero, gives, if we take $-\epsilon$ as the arbitrary multiplier, and the first of the two equations the one multiplied by it,

$$\left.\begin{aligned} (A-\epsilon)x\quad &+cy &+bz&=0,\\ cx+(B-\epsilon)&y &+az&=0,\\ bx\quad &+ay+(C-\epsilon)&z&=0.\end{aligned}\right\} \qquad (9)$$

We may find what ϵ means if we multiply the first of these by x, the second by y, and the third by z, and add; because we thus obtain

$$Ax^2 + By^2 + Cz^2 + 2(ayz + bzx + cxy) - \epsilon(x^2 + y^2 + z^2) = 0,$$

or

$$r_1^2 - \epsilon r^2 = 0,$$

which gives

$$\epsilon = \left(\frac{r_1}{r}\right)^2. \tag{10}$$

Eliminating the ratios $x : y : z$ from (9), by the usual method, we have the well-known determinant cubic

$$-\epsilon)(B - \epsilon)(C - \epsilon) - a^2(A - \epsilon) - b^2(B - \epsilon) - c^2(C - \epsilon) + 2abc = 0, \tag{11}$$

of which the three roots are known to be all real. Any one of the three roots if used for ϵ, in (9), harmonizes these three equations for the true ratios $x : y : z$; and, making the co-efficients of x, y, z in them all known, allows us to determine the required ratios by any two of the equations, or symmetrically from the three, by the proper algebraic processes. Thus we have only to determine the absolute magnitudes of x, y, and z, which (7) enables us to do when their ratios are known.

It is to be remarked, that when $[Yz] = [Zy]$, $[Zx] = [Xz]$, and $[Xy] = [Yx]$, equation (3) becomes a cubic, the squares of whose roots are the roots of (11), and that the three lines determined by (2) in this case are identical with those determined by (9). The reader will find it a good analytical exercise to prove this directly from the equations. It is a necessary consequence of § 183, below.

We have precisely the same problem to solve when the question proposed is, to find what radii of a sphere remain perpendicular to the surface of the altered figure. This is obvious when viewed geometrically. The tangent plane is perpendicular to the radius when the radius is a maximum or minimum. Therefore, every plane of the body parallel to such tangent plane is perpendicular to the radius in the altered, as it was in the initial condition.

The analytical investigation of the problem, presented in the second way, is as follows:—

Let

$$l_1 x_1 + m_1 y_1 + n_1 z_1 = 0 \tag{12}$$

be the equation of any plane of the altered substance, through the origin of co-ordinates, the axes of co-ordinates being the same fixed axes, OX, OY, OZ, which we have used of late. The direction cosines of a perpendicular to it are, of course, proportional to l_1, m_1, n_1. If, now, for x_1, y_1, z_1, we substitute their values, as in (1), in terms of the co-ordinates which the same

Displacement of a body, rigid or not, one point of which is held fixed.

Displacement of a body, rigid or not, one point of which is held fixed.

point of the substance had initially, we find the equation of the same plane of the body in its initial position, which, when the terms are grouped properly, is this—

$$\{l_1[Xx]+m_1[Yx]+n_1[Zx]\}x+\{l_1[Xy]+m_1[Yy]+n_1[Zy]\}y$$
$$+\{l_1[Xz]+m_1[Yz]+n_1[Zz]\}z=0 \qquad (13)$$

The direction cosines of the perpendicular to the plane are proportional to the co-efficients of x, y, z. Now these are to be the direction cosines of the same line of the substance as was altered into the line $l_1 : m_1 : n_1$. Hence, if $l : m : n$ are quantities proportional to the direction cosines of this line in its initial position, we must have

$$\left.\begin{array}{l} l_1[Xx]+m_1[Yx]+n_1[Zx]=\epsilon l \\ l_1[Xy]+m_1[Yy]+n_1[Zy]=\epsilon m \\ l_1[Xz]+m_1[Zz]+n_1[Zz]=\epsilon n \end{array}\right\} \qquad (14)$$

where ϵ is arbitrary. Suppose, to fix the ideas, that l_1, m_1, n_1 are the co-ordinates of a certain point of the substance in its altered state, and that l, m, n are proportional to the initial co-ordinates of the same point of the substance. Then we shall have, by the fundamental equations, the expressions for l_1, m_1, n_1 in terms of l, m, n. Using these in the first members of (14), and taking advantage of the abbreviated notation (5), we have precisely the same equations for l, m, n as (9) for x, y, z above.

Analysis of a strain into distortion and rotation.

182. From the preceding analysis it follows that any homogeneous strain whatever applied to a body generally changes a sphere of the body into an ellipsoid, and causes the latter to rotate about a definite axis through a definite angle. In particular cases the sphere may remain a sphere. Also there may be no rotation. In the general case, when there is no rotation, there are three directions in the body (the axes of the ellipsoid) which remain fixed; when there *is* rotation, there are generally three such directions, but not rectangular. Sometimes, however, there is but one.

Pure strain.

183. When the axes of the ellipsoid are lines of the body whose direction does not change, the strain is said to be *pure*, or unaccompanied by rotation. The strains we have already considered were more general than this, being pure strains accompanied by rotation. We proceed to find the analytical conditions of the existence of a pure strain.

Let $O\Xi$, $O\Xi'$, $O\Xi''$ be the three principal axes of the strain, and let $l, m, n,\ l', m', n',\ l'', m'', n''$ be their direction cosines. Let $\alpha, \alpha', \alpha''$ be the principal elongations. Then, if ξ, ξ', ξ'' be the position of a point of the unaltered body, with reference to $O\Xi$, $O\Xi'$, $O\Xi''$, its position in the body when altered will be $\alpha\xi$, $\alpha'\xi'$, $\alpha''\xi''$. But if x, y, z be its initial, and x_1, y_1, z_1 its final, positions with reference to OX, OY, OZ, we have — Pure strain.

$$\xi = lx + my + nz,\quad \xi' = \text{etc.},\quad \xi'' = \text{etc.}, \tag{15}$$

and $x_1 = l\alpha\xi + l'\alpha'\xi' + l''\alpha''\xi''$, $y_1 = \text{etc.}$, $z_1 = \text{etc.}$

For ξ, ξ', ξ'' substitute their values (15), and we have x_1, y_1, z_1 in terms of x, y, z, expressed by the following equations:—

$$\left.\begin{array}{l} {}^2 + \alpha' l'^2 + \alpha'' l''^2)x + (\alpha lm + \alpha' l'm' + \alpha'' l''m'')y + (\alpha ln + \alpha' l'n' + \alpha'' l''n'')z \\ nl + \alpha' m'l' + \alpha'' m''l'')x + (\alpha m^2 + \alpha' m'^2 + \alpha'' m''^2)y + (\alpha mn + \alpha' m'n' + \alpha'' m''n'')z \\ l + \alpha' n'l' + \alpha'' n''l'')x + (\alpha nm + \alpha' n'm' + \alpha'' n''m'')y + (\alpha n^2 + \alpha' n'^2 + \alpha'' n''^2)z. \end{array}\right\} \tag{16}$$

Hence, comparing with (1) of § 181, we have

$$\left.\begin{array}{l} [Xx] = \alpha l^2 + \alpha' l'^2 + \alpha'' l''^2, \text{ etc.}; \\ [Zy] = [Yz] = \alpha mn + \alpha' m'n' + \alpha'' m''n'', \text{ etc.} \end{array}\right\} \tag{17}$$

In these equations, $l, l', l'',\ m, m', m'',\ n, n', n''$, are deducible from three independent elements, the three angular co-ordinates (§ 100, above) of a rigid body, of which one point is held fixed; and therefore, along with $\alpha, \alpha', \alpha''$, constituting in all six independent elements, may be determined so as to make the six members of these equations have any six prescribed values. Hence the conditions necessary and sufficient to insure no rotation are

$$[Zy] = [Yz],\ [Xz] = [Zx],\ [Xy] = [Yx] \tag{18}$$

184. If a body experience a succession of strains, each unaccompanied by rotation, its resulting condition will generally be producible by a strain and a rotation. From this follows the remarkable corollary that three pure strains produced one after another, in any piece of matter, each without rotation, may be so adjusted as to leave the body unstrained, but rotated through some angle about some axis. We shall have, later, most important and interesting applications to fluid motion, which (Chap. II.) will be proved to be instantaneously, or differentially, irrotational; but which may result in leaving a whole fluid mass merely turned round from its primitive position, as if it had been a rigid body. The following elementary geometrical investigation, though not bringing out a thoroughly — Composition of pure strains.

Analysis of strain into distortion and rotation

only, which is less than $P_{,}OP'$ by the excess of $P_{,}OP'$ above $P'OP$. Hence the resultant of the two shears, $PP_{,}$, $P_{,}P_{,,}$, each separately deprived of rotation, is a single shear $PP_{,,}$, and a rotation of its principal axes, in the direction of the hands of a watch through an angle equal to $P'OP_{,} - POP'$. Composition of pure strains.

185. Make the two partial shears each non-rotationally. Return from their resultant in a single non-rotational shear: we conclude with the body unstrained, but turned through an angle equal to $P'OP_{,} - POP'$, in the same direction as the hands of a watch.

$$x_1 = Ax + cy + bz$$
$$y_1 = cx + By + az$$
$$z_1 = bx + ay + Cz$$

is (§ 183) the most general possible expression for the displacement of any point of a body of which one point is held fixed, strained according to any three lines at right angles to one another, as principal axes, which are kept fixed in direction, relatively to the lines of reference OX, OY, OZ.

Similarly, if the body thus strained be again non-rotationally strained, the most general possible expressions for x_2, y_2, z_2, the co-ordinates of the position to which x_1, y_1, z_1, will be brought, are

$$x_2 = A_1x_1 + c_1y_1 + b_1z_1$$
$$y_2 = c_1x_1 + B_1y_1 + a_1z_1$$
$$z_2 = b_1x_1 + a_1y_1 + C_1z_1$$

Substituting in these, for x_1, y_1, z_1, their preceding expressions, in terms of the primitive co-ordinates, x, y, z, we have the following expressions for the co-ordinates of the position to which the point in question is brought by the two strains:—

$$x_2 = (A_1A + c_1c + b_1b)x + (A_1c + c_1B + b_1a)y + (A_1b + c_1a + b_1C)z$$
$$y_2 = (c_1A + B_1c + a_1b)x + (c_1c + B_1B + a_1a)y + (c_1b + B_1a + a_1C)z$$
$$z_2 = (b_1A + a_1c + C_1b)x + (b_1c + a_1B + C_1a)y + (b_1b + a_1a + C_1C)z$$

The resultant displacement thus represented is not generally of the non-rotational character, the conditions (18) of § 183 not being fulfilled, as we see immediately. Thus, for instance, we see that the co-efficient of y in the expression for x_2, is not necessarily equal to the co-efficient of x in the expression for y_2.

Cor.—If both strains are infinitely small, the resultant displacement is a pure strain without rotation. For A, B, C, A_1, B_1, C_1 are each infinitely nearly unity, and a, b, etc., each infinitely small. Hence, neglecting the products of these infinitely small quantities among one another, and of any of them with the differences between the former and unity, we have a resultant displacement

$$\begin{aligned} x_2 &= A_1Ax + (c+c_1)y + (b+b_1)z \\ y_2 &= (c_1+c)x + B_1By + (a+a_1)z \\ z_2 &= (b_1+b)x + (a_1+a)y + C_1Cz \end{aligned}$$

which represents a pure strain unaccompanied by rotation.

Displacement of a curve.

186. The measurement of rotation in a strained elastic solid, or in a moving fluid, is much facilitated by considering separately the displacement of any line of the substance. We are therefore led now to a short digression on the displacement of a curve, which may either belong to a continuous solid or fluid mass, or may be an elastic cord, given in any position. The propositions at which we shall arrive are, of course, applicable to a flexible but inextensible cord (§ 14, above) as a particular case.

It must be remarked, that the displacements to be considered do not depend merely on the curves occupied by the given line in its successive positions, but on the corresponding points of these curves.

Tangential displacement.

What we shall call tangential displacement is to be thus reckoned:—Divide the undisplaced curve into an infinite number of infinitely small equal parts. The product of the sums of the tangential components of the displacements from all the points of division, multiplied by the length of each of the infinitely small parts, is *the entire tangential displacement of the curve reckoned along the undisplaced curve.* The same reckoning carried out in the displaced curve is *the entire tangential displacement reckoned on the displaced curve.*

Two reckonings of tangential displacement compared.

187. The whole tangential displacement of a curve reckoned along the displaced curve, exceeds the whole tangential displacement reckoned along the undisplaced curve by half the rectangle under the sum and difference of the absolute terminal displacements, reckoned positive when the displacement of the end towards which the tangential components are reckoned positive exceeds that at the other. This theorem may be proved by a geometrical demonstration which the reader may easily supply.

Analytically thus :—Let x, y, z be the co-ordinates of any point, P, in the undisplaced curve; x_1, y_1, z_1, those of P_1 the point to which the same point of the curve is displaced. Let dx, dy, dz be the increments of the three co-ordinates corresponding to any infinitely small arc, ds of the first; so that

Two reckonings of tangential displacement compared.

$$ds=(dx^2+dy^2+dz^2)^{\frac{1}{2}},$$

and let corresponding notation apply to the corresponding element of the displaced curve. Let θ denote the angle between the line PP_1 and the tangent to the undisplaced curve through P; so that we have

$$\cos\theta=\frac{x_1-x}{D}\frac{dx}{ds}+\frac{y_1-y}{D}\frac{dy}{ds}+\frac{z_1-z}{D}\frac{dz}{ds}$$

where for brevity

$$D=\{(x_1-x)^2+(y_1-y)^2+(z_1-z)^2\}^{\frac{1}{2}}$$

being the absolute space of displacement. Hence

$$D\cos\theta ds=(x_1-x)dx+(y_1-y)dy+(z_1-z)dz.$$

Similarly we have

$$D\cos\theta_1 ds_1=(x_1-x)dx_1+(y_1-y)dy_1+(z_1-z)dz_1$$

and therefore

$$D\cos\theta_1 ds_1-D\cos\theta ds=(x_1-x)d(x_1-x)+(y_1-y)d(y_1-y)+(z_1-z)d(z_1-z)$$

or $$D\cos\theta_1 ds_1-D\cos\theta ds=\tfrac{1}{2}d(D^2).$$

To find the difference of the tangential displacements reckoned the two ways, we have only to integrate this expression. Thus we obtain

$$\int D\cos\theta_1 ds_1-\int D\cos\theta ds=\tfrac{1}{2}(D''^2-D'^2)=\tfrac{1}{2}(D''+D')(D''-D')$$

where D'' and D' denote the displacements of the two ends.

188. The entire tangential displacement of a closed curve is the same whether reckoned along the undisplaced or the displaced curve.

Tangential displacement of a closed curve.

189. The entire tangential displacement from one to another of two conterminous arcs, is the same reckoned along either as along the other.

190. The entire tangential displacement of a rigid closed curve when rotated through any angle about any axis, is equal to twice the area of its projection on a plane perpendicular to the axis, multiplied by the sine of the angle.

Rotation of a rigid closed curve.

(*a*) *Prop.*—The entire tangential displacement round a closed curve of a homogeneously strained solid, is equal to

Tangential displacement in a solid, in terms of components of strain.

$$2(P\varpi+Q\rho+R\sigma)$$

Tangential displacement in a solid, in terms of components of strain.

where P, Q, R denote, for its initial position, the areas of its projections on the planes YOZ, ZOX, XOY respectively, and ϖ, ρ, σ are as follows:—

$$\varpi=\tfrac{1}{2}\{[Zy]-[Yz]\}$$
$$\rho=\tfrac{1}{2}\{[Xz]-[Zx]\}$$
$$\sigma=\tfrac{1}{2}\{[Yx]-[Xy]\}$$

To prove this, let, farther,

$$a=\tfrac{1}{2}\{[Zy]+[Yz]\}$$
$$b=\tfrac{1}{2}\{[Xz]+[Zx]\}$$
$$c=\tfrac{1}{2}\{[Yx]+[Xy]\}$$

Thus we have

$$x_1=Ax+cy+bz+\sigma y-\rho z$$
$$y_1=cx+By+az+\varpi z-\sigma x$$
$$z_1=bx+ay+Cz+\rho x-\varpi y$$

Hence, according to the previously investigated expression, we have, for the tangential displacement, reckoned along the undisplaced curve,

$$\int\{(x_1-x)dx+(y_1-y)dy+(z_1-z)dz\}$$
$$=\int[\tfrac{1}{2}d\{(A-1)x^2+(B-1)y^2+(C-1)z^2+2(ayz+bzx+cxy)\}$$
$$+\varpi(ydz-zdy)+\rho(zdx-xdz)+\sigma(xdy-ydx)].$$

The first part, $\int\tfrac{1}{2}d\{\ \}$, vanishes for a closed curve.

The remainder of the expression is

$$\varpi\int(ydz-zdy)+\rho\int(zdx-xdz)+\sigma\int(xdy-ydx)$$

which, according to the formulæ for projection of areas, is equal to

$$2P\varpi+2Q\rho+2R\sigma.$$

For, as in § 36 (a), we have in the plane of xy

$$\int(xdy-ydx)=\int r^2d\theta$$

double the area of the orthogonal projection of the curve on that plane; and similarly for the other integrals.

(b) From this and § 190, it follows that if the body is rigid, and therefore only rotationally displaced, if at all, $[Zy]-[Yz]$ is equal to twice the sine of the angle of rotation multiplied by the cosine of the inclination of the axis of rotation to the line of reference OX.

(c) And in general $[Zy]-[Yz]$ measures the entire tangential displacement, divided by the area on ZOY, of any closed curve given, if a plane curve, in the plane YOZ, or, if a tortuous curve, given so as to have zero area projections on ZOX and XOY. The entire tangential displacement of any closed curve given in a plane, A, perpendicular to a line whose direction-cosines are proportional to ϖ, ρ, σ, is equal to twice its area multiplied by

$\sqrt{(\varpi^2+\rho^2+\sigma^2)}$. And the entire tangential displacement of any closed curve whatever is equal to twice the area of its projection on A, multiplied by $\sqrt{(\varpi^2+\rho^2+\sigma^2)}$.

Tangential displacement in a solid, in terms of components of strain.

In the transformation of co-ordinates, ϖ, ρ, σ transform by the elementary cosine law, and of course $\varpi^2+\rho^2+\sigma^2$ is an invariant; that is to say, its value is unchanged by transformation from one set of rectangular axes to another.

(*d*) In non-rotational homogeneous strain, the entire tangential displacement along any curve from the fixed point to (x, y, z), reckoned along the undisplaced curve, is equal to

$$\tfrac{1}{2}\{(A-1)x^2+(B-1)y^2+(C-1)z^2+2(ayz+bzx+cxy)\}.$$

Reckoned along displaced curve, it is, from this and § 187,

$$\tfrac{1}{2}\{(A-1)x^2+(B-1)y^2+(C-1)z^2+2(ayz+bzx+cxy)\}$$
$$+\tfrac{1}{2}\{[(A-1)x+cy+bz]^2+[cx+(B-1)y+az]^2$$
$$+[bx+ay+(C-1)z]^2\}.$$

And the entire tangential displacement from one point along any curve to another point, is independent of the curve, *i.e.*, is the same along any number of conterminous curves, this of course whether reckoned in each case along the undisplaced or along the displaced curve.

(*e*) Given the absolute displacement of every point to find the strain. Let α, β, γ be the components, relative to fixed axes, OX, OY, OZ, of the displacement of a particle, P, initially in the position x, y, z. That is to say, let $x+\alpha$, $y+\beta$, $z+\gamma$ be the co-ordinates, in the strained body, of the point of it which was initially at x, y, z.

Heterogeneous strain.

Consider the matter all round this point in its first and second positions. Taking this point P as moveable origin, let ξ, η, ζ be the initial co-ordinates of any other point near it, and ξ_1, η_1, ζ_1 the final co-ordinates of the same.

The initial and final co-ordinates of the last-mentioned point, with reference to the fixed axes OX, OY, OZ, will be

$$x+\xi,\ y+\eta,\ z+\zeta,$$

and $$x+\alpha+\xi_1,\ y+\beta+\eta_1,\ z+\gamma+\zeta_1,$$

respectively; that is to say,

$$\alpha+\xi_1-\xi,\ \beta+\eta_1-\eta,\ \gamma+\zeta_1-\zeta$$

are the components of the displacement of the point which had initially the co-ordinates $x+\xi$, $y+\eta$, $z+\zeta$, or, which is the same thing, are the values of α, β, γ, when x, y, z are changed into

$$x+\xi,\ y+\eta,\ z+\zeta.$$

Hetero-geneous strain.

Hence, by Taylor's theorem,

$$\xi_1-\xi=\frac{d\alpha}{dx}\xi+\frac{d\alpha}{dy}\eta+\frac{d\alpha}{dz}\zeta$$

$$\eta_1-\eta=\frac{d\beta}{dx}\xi+\frac{d\beta}{dy}\eta+\frac{d\beta}{dz}\zeta$$

$$\zeta_1-\zeta=\frac{d\gamma}{dx}\xi+\frac{d\gamma}{dy}\eta+\frac{d\gamma}{dz}\zeta,$$

the higher powers and products of ξ, η, ζ being neglected. Comparing these expressions with (1) of § 181, we see that they express the changes in the co-ordinates of any displaced point of a body relatively to three rectangular axes in fixed directions through one point of it, when all other points of it are displaced relatively to this one, in any manner subject only to the condition of giving a homogeneous strain. Hence we perceive that at distances all round any point, so small that the first terms only of the expressions by Taylor's theorem for the differences of displacement are sensible, the strain is sensibly homogeneous, and we conclude that the directions of the principal axes of the strain at any point (x, y, z), and the amounts of the elongations of the matter along them, and the tangential displacements in closed curves, are to be found according to the general methods described above, by taking

$$[Xx]=\frac{d\alpha}{dx}+1,\quad [Xy]=\frac{d\alpha}{dy},\quad [Xz]=\frac{d\alpha}{dz}$$

$$[Yx]=\frac{d\beta}{dx},\quad [Yy]=\frac{d\beta}{dy}+1,\quad [Yz]=\frac{d\beta}{dz}$$

$$[Zx]=\frac{d\gamma}{dx},\quad [Zy]=\frac{d\gamma}{dy},\quad [Zz]=\frac{d\gamma}{dz}+1.$$

Homo-geneous strain.

If each of these nine quantities is constant (*i.e.*, the same for all values of x, y, z), the strain is homogeneous: not unless.

Infinitely small strain

(*f*) The condition that the strain may be infinitely small is that

$$\frac{d\alpha}{dx},\quad \frac{d\alpha}{dy},\quad \frac{d\alpha}{dz},$$

$$\frac{d\beta}{dx},\quad \frac{d\beta}{dy},\quad \frac{d\beta}{dz},$$

$$\frac{d\gamma}{dx},\quad \frac{d\gamma}{dy},\quad \frac{d\gamma}{dz},$$

must be each infinitely small.

Most general motion of matter.

(*g*) These formulæ apply to the most general possible motion of any substance, and they may be considered as the fundamental equations of kinematics. If we introduce time as independent variable, we have for component velocities u, v, w, parallel to the fixed axes OX, OY, OZ, the following expressions; x, y, z, t being independent variables, and α, β, γ functions of them :— Most general motion of matter.

$$u=\frac{d\alpha}{dt},\quad v=\frac{d\beta}{dt},\quad w=\frac{d\gamma}{dt}.$$

(h) If we introduce the condition that no line of the body experiences any elongation, we have the general equations for the kinematics of a rigid body, of which, however, we have had enough already. The equations of condition to express this will be six in number, among the nine quantities $\frac{d\alpha}{dx}$, etc., which (g) are, in this case, each constant relatively to x, y, z. There are left three independent arbitrary elements to express any angular motion of a rigid body. Change of position of a rigid body

(i) If the disturbed condition is so related to the initial condition that every portion of the body can pass from its initial to its disturbed position and strain, by a translation and a strain without rotation; *i.e.*, if the three principal axes of the strain at any point are lines of the substance which retain their parallelism, we must have, § 183 (18), Non-rotational strain

$$\frac{d\beta}{dz}=\frac{d\gamma}{dy},\quad \frac{d\gamma}{dx}=\frac{d\alpha}{dz},\quad \frac{d\alpha}{dy}=\frac{d\beta}{dx};$$

and if these equations are fulfilled, the strain is non-rotational, as specified. But these three equations express neither more nor less than that

$$\alpha dx+\beta dy+\gamma dz$$

is the differential of a function of three independent variables. Hence we have the remarkable proposition, and its converse, that if $F(x, y, z)$ denote any function of the co-ordinates of any point of a body, and if every such point be displaced from its given position (x, y, z) to the point whose co-ordinates are

$$x_1=x+\frac{dF}{dx},\quad y_1=y+\frac{dF}{dy},\quad z_1=z+\frac{dF}{dz} \qquad (1),$$

the principal axes of the strain at every point are lines of the substance which have retained their parallelism. The displacement back from (x_1, y_1, z_1) to (x, y, z) fulfils the same condition, and therefore we must have

$$x=x_1+\frac{dF_1}{dx_1},\quad y=y_1+\frac{dF_1}{dy_1},\quad z=z_1+\frac{dF_1}{dz_1} \qquad (2),$$

Theorem of pure analysis.

where F_1 denotes a function of x_1, y_1, z_1, and $\frac{dF_1}{dx_1}$, etc., its partial differential co-efficients with reference to this system of variables. The relation between F and F_1 is clearly

$$F+F_1=-\tfrac{1}{2}D^2 \qquad (3),$$

Hetero-geneous strain.

where $D^2 = \frac{dF^2}{dx^2} + \frac{dF^2}{dy^2} + \frac{dF^2}{dz^2} = \frac{dF_1^2}{dx_1^2} + \frac{dF_1^2}{dy_1^2} + \frac{dF_1^2}{dz_1^2}$ (4).

This, of course, may be proved by ordinary analytical methods, applied to find x, y, z in terms of x_1, y_1, z_1, when the latter are given by (1) in terms of the former.

(*j*) Let α, β, γ be any three functions of x, y, z. Let dS be any element of a surface; l, m, n the direction-cosines of its normal.

$$\text{Then } \iint dS\{l(\frac{d\gamma}{dy}-\frac{d\beta}{dz})+m(\frac{d\alpha}{dz}-\frac{d\gamma}{dx})+n(\frac{d\beta}{dx}-\frac{d\alpha}{dy})\}$$
$$=\int(\alpha dx+\beta dy+\gamma dz)$$

the former integral being over any curvilinear area bounded by a closed curve; and the latter, which may be written

$$\int ds(\alpha\frac{dx}{ds}+\beta\frac{dy}{ds}+\gamma\frac{dz}{ds}),$$

being round the periphery of this curve line. To demonstrate this, it is only necessary to remark that

$$ldS=dydz,\ mdS=dzdx,\ ndS=dxdy,$$

and to prove that between the limits assigned

$$\iint\frac{d\alpha}{dz}dzdx-\iint\frac{d\alpha}{dy}dxdy=\int dx\int(\frac{d\alpha}{dz}dz+\frac{d\alpha}{dy}dy)=\int\alpha\frac{dx}{ds}ds;\ \text{etc.}$$

(*k*) It is remarkable that

$$\iint dS\{l(\frac{d\gamma}{dy}-\frac{d\beta}{dz})+m(\frac{d\alpha}{dz}-\frac{d\gamma}{dx})+n(\frac{d\beta}{dx}-\frac{d\alpha}{dy})\}$$

is the same for all surfaces having common curvilinear boundary; and when α, β, γ are the components of a displacement from x, y, z, it is the entire tangential displacement round the said curvilinear boundary, being a closed curve. It is therefore this that is nothing when the displacement of every part is non-rotational. And when it is not nothing, we see by the above propositions and corollaries precisely what the measure of the rotation is.

Displacement function.

(*l*) *Lastly*, We see what the meaning, for the case of no rotation, of $\int(\alpha dx+\beta dy+\gamma dz)$, or, as it has been called, "the displacement function," is. It is, the entire tangential displacement along any curve from the fixed point O, to the point P (x, y, z). And the entire tangential displacement, being the same along all different curves proceeding from one to another of any two points, is equal to the difference of the values of the displacement functions at those points.

"Equation of continuity."

191. As there can be neither annihilation nor generation of matter in any natural motion or action, the whole quantity of a fluid within any space at any time must be equal to the quantity originally in that space, increased by the whole quantity that has entered it and diminished by the whole quantity that has left it. This idea when expressed in a perfectly comprehensive manner for every portion of a fluid in motion constitutes what is called the "*equation of continuity*," a needlessly confusing expression. "Equation of continuity."

192. Two ways of proceeding to express this idea present themselves, each affording instructive views regarding the properties of fluids. In one we consider a definite portion of the fluid; follow it in its motions; and declare that the average density of the substance varies inversely as its volume. We thus obtain the equation of continuity in an integral form. Integral equation of continuity.

Let a, b, c be the co-ordinates of any point of a moving fluid, at a particular era of reckoning, and let x, y, z be the co-ordinates of the position it has reached at any time t from that era. To specify completely the motion, is to give each of these three varying co-ordinates as a function of a, b, c, t.

Let δa, δb, δc denote the edges, parallel to the axes of co-ordinates, of a very small rectangular parallelepiped of the fluid, when $t=0$. Any portion of the fluid, if only small enough in all its dimensions, must (§ 190, e), in the motion, approximately fulfil the condition of a body uniformly strained throughout its volume. Hence if δa, δb, δc are taken infinitely small, the corresponding portion of fluid must (§ 156) remain a parallelepiped during the motion.

If a, b, c be the initial co-ordinates of one angular point of this parallelepiped; and $a+\delta a$, b, c; a, $b+\delta b$, c; a, b, $c+\delta c$; those of the other extremities of the three edges that meet in it: the co-ordinates of the same points of the fluid at time t, will be

$$x, y, z;$$

$$x+\frac{dx}{da}\delta a,\ y+\frac{dy}{da}\delta a,\ z+\frac{dz}{da}\delta a;$$

$$x+\frac{dx}{db}\delta b,\ y+\frac{dy}{db}\delta b,\ z+\frac{dz}{db}\delta b;$$

$$x+\frac{dx}{dc}\delta c,\ y+\frac{dy}{dc}\delta c,\ z+\frac{dz}{dc}\delta c.$$

Hence the lengths and direction-cosines of the edges are respectively—

Integral equation of continuity.

$$\left(\frac{dx^2}{da^2}+\frac{dy^2}{da^2}+\frac{dz^2}{da^2}\right)^{\frac{1}{2}}\delta a,\quad \frac{\frac{dx}{da}}{\left(\frac{dx^2}{da^2}+\frac{dy^2}{db^2}+\frac{dz^2}{dc^2}\right)^{\frac{1}{2}}},\ \text{etc.}$$

$$\left(\frac{dx^2}{db^2}+\frac{dy^2}{db^2}+\frac{dz^2}{db^2}\right)^{\frac{1}{2}}\delta b,\quad \frac{\frac{dx}{db}}{\left(\frac{dx^2}{db^2}+\frac{dy^2}{db^2}+\frac{dz^2}{db^2}\right)^{\frac{1}{2}}},\ \text{etc.}$$

$$\left(\frac{dx^2}{dc^2}+\frac{dy^2}{dc^2}+\frac{dz^2}{dc^2}\right)^{\frac{1}{2}}\delta c,\quad \frac{\frac{dx}{dc}}{\left(\frac{dx^2}{dc^2}+\frac{dy^2}{dc^2}+\frac{dz^2}{dc^2}\right)^{\frac{1}{2}}},\ \text{etc.}$$

The volume of this parallelepiped is therefore

$$\left(\frac{dx}{da}\frac{dy}{db}\frac{dz}{dc}-\frac{dx}{da}\frac{dy}{dc}\frac{dz}{db}+\frac{dx}{db}\frac{dy}{dc}\frac{dz}{da}-\frac{dx}{db}\frac{dy}{da}\frac{dz}{dc}+\frac{dx}{dc}\frac{dy}{da}\frac{dz}{db}-\frac{dx}{dc}\frac{dy}{db}\frac{dz}{da}\right)\delta a\,\delta b\,\delta c,$$

or, as it is now usually written,

$$\begin{vmatrix} \frac{dx}{da}, & \frac{dy}{da}, & \frac{dz}{da} \\ \frac{dx}{db}, & \frac{dy}{db}, & \frac{dz}{db} \\ \frac{dx}{dc}, & \frac{dy}{dc}, & \frac{dz}{dc} \end{vmatrix} \delta a\delta b\delta c.$$

Now as there can be neither increase nor diminution of the quantity of matter in any portion of the fluid, the density, or the quantity of matter per unit of volume, in the infinitely small portion we have been considering, must vary inversely as its volume if this varies. Hence, if ρ denote the density of the fluid in the neighbourhood of (x, y, z) at time t, and ρ_0 the initial density, we have

$$\rho\begin{vmatrix} \frac{dx}{da}, & \frac{dy}{da}, & \frac{dz}{da} \\ \frac{dx}{db}, & \frac{dy}{db}, & \frac{dz}{db} \\ \frac{dx}{dc}, & \frac{dy}{dc}, & \frac{dz}{dc} \end{vmatrix} = \rho_0 \qquad (1)$$

which is the integral "equation of continuity."

Differential equation of continuity.

193. The form under which the equation of continuity is most commonly given, or the *differential equation of continuity*, as we may call it, expresses that the rate of diminution of the density bears to the density, at any instant, the same ratio as the rate of increase of the volume of an infinitely small portion bears to the volume of this portion at the same instant. Differential equation of continuity.

To find it, let a, b, c denote the co-ordinates, not when $t = 0$, but at any time $t - dt$, of the point of fluid whose co-ordinates are x, y, z at t; so that we have

$$x - a = \frac{dx}{dt}dt,\ y - b = \frac{dy}{dt}dt,\ z - c = \frac{dz}{dt}dt,$$

according to the ordinary notation for partial differential coefficients; or, if we denote by u, v, w, the components of the velocity of this point of the fluid, parallel to the axes of co-ordinates,

$$x - a = udt,\ y - b = vdt,\ z - c = wdt.$$

Hence

$$\frac{dx}{da} = 1 + \frac{du}{da}dt,\ \frac{dy}{da} = \frac{dv}{da}dt,\ \frac{dz}{da} = \frac{dw}{da}dt;$$

$$\frac{dx}{db} = \frac{du}{db}dt,\ \frac{dy}{db} = 1 + \frac{dv}{db}dt,\ \frac{dz}{db} = \frac{dw}{db}dt;$$

$$\frac{dx}{dc} = \frac{du}{dc}dt,\ \frac{dy}{dc} = \frac{dv}{dc}dt,\ \frac{dz}{dc} = 1 + \frac{dw}{dc}dt;$$

and, as we must reject all terms involving higher powers of dt than the first, the determinant becomes simply

$$1 + \left(\frac{du}{da} + \frac{dv}{db} + \frac{dw}{dc}\right)dt.$$

This therefore expresses the ratio in which the volume is augmented in time dt. The corresponding ratio of variation of density is

$$1 + \frac{D\rho}{\rho}$$

if $D\rho$ denote the differential of ρ, the density of one and the same portion of fluid as it moves from the position (a, b, c) to (x, y, z) in the interval of time from $t - dt$ to t. Hence

$$\frac{1}{\rho}\frac{D\rho}{dt} + \frac{du}{da} + \frac{dv}{db} + \frac{dw}{dc} = 0 \qquad (1).$$

Here ρ, u, v, w are regarded as functions of a, b, c, and t, and the variation of ρ implied in $\frac{D\rho}{dt}$ is the rate of the actual variation of the density of an indefinitely small portion of the fluid as it moves away from a fixed position (a, b, c). If we alter the principle of the notation, and consider ρ as the density of whatever portion of the fluid is at time t in the neighbourhood of the

Differential equation of continuity.

fixed point (a, b, c), and u, v, w the component velocities of the fluid passing the same point at the same time, we shall have

$$\frac{D\rho}{dt}=\frac{d_t\rho}{dt}+u\frac{d_a\rho}{da}+v\frac{d_b\rho}{db}+w\frac{d_c\rho}{dc} \quad . \qquad (2).$$

Omitting again the suffixes, according to the usual imperfect notation for partial differential co-efficients, which on our new understanding can cause no embarrassment, we thus have, in virtue of the preceding equation,

$$\frac{1}{\rho}\left(\frac{d\rho}{dt}+u\frac{d\rho}{da}+v\frac{d\rho}{db}+w\frac{d\rho}{dc}\right)+\frac{du}{da}+\frac{dv}{db}+\frac{dw}{dc}=0,$$

or,

$$\frac{d\rho}{dt}+\frac{d(\rho u)}{da}+\frac{d(\rho v)}{db}+\frac{d(\rho w)}{dc}=0 \qquad (3),$$

which is the differential equation of continuity, in the form in which it is most commonly given.

194. The other way referred to above (§ 192) leads immediately to the differential equation of continuity.

Imagine a space fixed in the interior of a fluid, and consider the fluid which flows into this space, and the fluid which flows out of it, across different parts of its bounding surface, in any time. If the fluid is of the same density and incompressible, the whole quantity of matter in the space in question must remain constant at all times, and therefore the quantity flowing in must be equal to the quantity flowing out in any time. If, on the contrary, during any period of motion, more fluid enters than leaves the fixed space, there will be condensation of matter in that space; or if more fluid leaves than enters, there will be dilatation. The rate of augmentation of the average density of the fluid, per unit of time, in the fixed space in question, bears to the actual density, at any instant, the same ratio that the rate of acquisition of matter into that space bears to the whole matter in that space.

Let the space S be an infinitely small parallelepiped, of which the edges α, β, γ are parallel to the axes of co-ordinates, and let x, y, z be the co-ordinates of its centre; so that $x \pm \frac{1}{2}\alpha$, $y \pm \frac{1}{2}\beta$, $z \pm \frac{1}{2}\gamma$ are the co-ordinates of its angular points. Let ρ be the density of the fluid at (x, y, z), or the mean density through the space S, at the time t. The density at the time $t + dt$ will be $\rho + \frac{d\rho}{dt}dt$; and hence the quantities of fluid contained in the space S, at the times t, and $t+dt$, are respectively $\rho\alpha\beta\gamma$ and $(\rho+\frac{d\rho}{dt}dt)\alpha\beta\gamma$. Hence the quantity of fluid lost (there will of course be an absolute gain if $\frac{d\rho}{dt}$ be positive) in the time dt is Differential equation of continuity.

$$-\frac{d\rho}{dt}\alpha\beta\gamma dt \qquad (a).$$

Now let u, v, w be the three components of the velocity of the fluid (or of a fluid particle) at P. These quantities will be functions of x, y, z (involving also t, except in the case of "steady motion"), and will in general vary gradually from point to point of the fluid; although the analysis which follows is not restricted by this consideration, but holds even in cases where in certain places of the fluid there are abrupt transitions in the velocity, as may be seen by considering them as limiting cases of motions in which there are very sudden continuous transitions of velocity. If ω be a small plane area, perpendicular to the axis of x, and having its centre of gravity at P, the volume of fluid which flows across it in the time dt will be equal to $u\omega dt$, and the mass or quantity will be $\rho u\omega dt$. If we substitute $\beta\gamma$ for ω, the quantity which flows across either of the sides $\beta\gamma$ of the parallelepiped S, will differ from this only on account of the variation in the value of ρu; and therefore the quantities which flow across the two sides $\beta\gamma$ are respectively "Steady motion" defined.

$$\{\rho u-\tfrac{1}{2}\alpha\frac{d(\rho u)}{dx}\}\beta\gamma dt,$$

and

$$\{\rho u+\tfrac{1}{2}\alpha\frac{d(\rho u)}{dx}\}\beta\gamma dt.$$

Hence $\alpha\frac{d(\rho u)}{dx}\beta\gamma dt$, or $\frac{d(\rho u)}{dx}\alpha\beta\gamma dt$ is the excess of the quantity of fluid which leaves the parallelepiped across one of the faces $\beta\gamma$ above that which enters it across the other. By considering in addition the effect of the motion across the other faces of the parallelepiped, we find for the total quantity of fluid lost from the space S, in the time dt,

$$\{\frac{d(\rho u)}{dx}+\frac{d(\rho v)}{dy}+\frac{d(\rho w)}{dz}\}\alpha\beta\gamma dt \qquad (b).$$

Equating this to the expression (a), previously found, we have

$$\{\frac{d(\rho u)}{dx}+\frac{d(\rho v)}{dy}+\frac{d(\rho w)}{dz}\}\alpha\beta\gamma dt=-\frac{d\rho}{dt}\alpha\beta\gamma dt;$$

Differential equation of continuity.

and we deduce

$$\frac{d(\rho u)}{dx}+\frac{d(\rho v)}{dy}+\frac{d(\rho w)}{dz}+\frac{d\rho}{dt}=0 \qquad (4),$$

which is the required equation.

Freedom and constraint.

195. Several references have been made in preceding sections to the number of independent variables in a displacement, or to the degrees of *freedom* or *constraint* under which the displacement takes place. It may be well, therefore, to take a general (but cursory) view of this part of the subject by itself.

Of a point.

196. A free point has *three* degrees of freedom, inasmuch as the most general displacement which it can take is resolvable into three, parallel respectively to any three directions, and independent of each other. It is generally convenient to choose these three directions of resolution at right angles to one another.

If the point be constrained to remain always on a given surface, *one* degree of constraint is introduced, or there are left but *two* degrees of freedom. For we may take the normal to the surface as one of three rectangular directions of resolution. No displacement can be effected parallel to it: and the other two displacements, at right angles to each other, in the tangent plane to the surface, are independent.

If the point be constrained to remain on *each* of two surfaces, it loses two degrees of freedom, and there is left but one. In fact, it is constrained to remain on the curve which is common to both surfaces, and along a curve there is at each point but one direction of displacement.

Of a rigid system.

197. Taking next the case of a free rigid system, we have evidently *six* degrees of freedom to consider—*three* independent displacements or translations in rectangular directions as a point has, and three independent rotations about three mutually rectangular axes.

If it have one point fixed, it loses *three* degrees of freedom: in fact, it has now only the rotations above mentioned.

This fixed point may be, and in general is, a point of a continuous surface of the body in contact with a continuous fixed surface. These surfaces must be supposed "perfectly rough," so that sliding may be impossible.

If a second point be fixed, the body loses *two* more degrees of freedom, and keeps only one freedom to rotate about the line joining the two fixed points. Freedom and constraint of a rigid system.

If a third point, not in a line with the other two, be fixed, the body is fixed.

198. If one point of the rigid system is forced to remain on a smooth surface, *one* degree of freedom is lost; there remain *five*, two displacements in the tangent plane to the surface, and three rotations. As an additional degree of freedom is lost by each successive limitation of a point in the body to a smooth surface, *six* such conditions completely determine the position of the body. Thus if six points properly chosen on the barrel and stock of a rifle be made to rest on six convex portions of the surface of a fixed rigid body, the rifle may be replaced any number of times in precisely the same position, for the purpose of testing its accuracy.

199. If one point be constrained to remain in a curve, there remain four degrees of freedom.

If two points be constrained to remain in given curves, there are four degrees of constraint, and we have left two degrees of freedom. One of these may be regarded as being a simple rotation about the line joining the constrained points, a motion which, it is clear, the body is free to receive. It may be shown that the other possible motion is of the most general character for one degree of freedom; that is to say, translation and rotation in any fixed proportions as of the nut of a screw.

If one line of a rigid system be constrained to remain parallel to itself, as, for instance, if the body be a three-legged stool standing on a perfectly smooth board fixed to a common window, sliding in its frame with perfect freedom, there remain *three* displacements and one rotation.

But we need not further pursue this subject, as the number of combinations that might be considered is almost endless; and those already given suffice to show how simple is the determination of the degrees of freedom or constraint in any case that may present itself.

200. One degree of constraint of the most general character, is not producible by constraining one point of the body to a curve surface; but it consists in stopping one line of the body One degree of constraint of the most general character.

One degree of constraint of the most general character.

from longitudinal motion, except accompanied by rotation round this line, in fixed proportion to the longitudinal motion, every other motion being left unimpeded; that is to say, free rotation about any axis perpendicular to this line (two degrees freedom); and translation in any direction perpendicular to the same line (two degrees freedom). These last four, with the one degree of freedom to screw, constitute the five degrees of freedom, which, with one degree of constraint, make up the six elements.

Mechanical illustration.

201. Let a screw be cut on one shaft of a Hooke's joint, and let the other shaft be joined to a fixed shaft by a second Hooke's joint. A nut turning on that screw shaft has the most general kind of motion admitted when there is one degree of constraint. Or it is subjected to just one degree of constraint of the most general character. It has five degrees of freedom; for it may move, 1*st*, by screwing on its shaft, the two Hooke's joints being at rest; 2*d*, it may rotate about either axis of the first Hooke's joint, or any axis in their plane (two more degrees of freedom: being freedom to rotate about two axes through one point); 3*d*, it may, by the two Hooke's joints, each bending, have translation without rotation in any direction perpendicular to the link, or shaft between the two Hooke's joints (two more degrees of freedom). But it cannot have a motion of translation parallel to the line of the link without a definite proportion of rotation round this line; nor can it have rotation round this line without a definite proportion of translation parallel to it.

No simpler mechanism can be easily imagined for producing one degree of constraint, of the most general kind.

Particular case (*a*).—Step of screw infinite (straight rifling), *i.e.*, the nut may slide freely, but cannot turn. Thus the one degree of constraint is, that there shall be no rotation about a certain axis. This is the kind and degree of freedom enjoyed by the outer ring of a gyroscope with its fly-wheel revolving infinitely fast. The outer ring cannot revolve about an axis perpendicular to the plane of the inner ring, but it may revolve freely about either of two axes at right angles to this, namely, the axis of the fly-wheel, and the axis of the outer, relative to the inner, ring; and it is of course perfectly free to translation in any direction.

Particular case (*b*).—Step of the screw = 0. In this case the nut may run round freely, but cannot move along the axis of the shaft. Hence the constraint is simply that the body can have no translation parallel to the line of shafts, but may have every other motion. This is the same as if any point of the body in this line were held to a fixed surface. This constraint may be produced less frictionally by not using a guiding surface, but the link and second Hooke's joint of the present arrangement, the first Hooke's joint being removed, and by pivoting one point of the body in a cup on the end of the link. Otherwise, let the end of the link be a continuous surface, and let a continuous surface of the body press on it, rolling or spinning when required, but not permitted to slide. Mechanical illustration.

A single degree of constraint is expressed by a single equation among the six co-ordinates specifying the position of one rigid body, relatively to another considered fixed. The effect of this on the body in any particular position is to prevent it from getting out of this position, except by means of component velocities (or infinitely small motions) fulfilling a certain linear equation among themselves. Constraint expressed analytically.

Thus if ϖ_1, ϖ_2, ϖ_3, ϖ_4, ϖ_5, ϖ_6 be the six co-ordinates, and $F(\varpi_1, \ldots\ldots\ldots) = 0$ the condition,

$$\frac{dF}{d\varpi_1}\delta\varpi_1 + \ldots\ldots\ldots\ldots = 0$$

is the linear equation which binds the motion through any particular position, the special values of ϖ_1, ϖ_2, ϖ_3, etc., for the particular position, being used in $\frac{dF}{d\varpi_1}$ and in each of the other partial differential co-efficients of F.

Now, whatever may be the co-ordinate system adopted, we may, if we please, reduce this equation to one between three velocities of translation u, v, w, and three angular velocities ω_1, ω_2, ω_3.

Let this equation be

$$Au + Bv + Cw + A'\omega_1 + B'\omega_2 + C'\omega_3 = 0.$$

This is equivalent to the following:—

$$q + a\omega = 0,$$

if q denote the component velocity along or parallel to the line whose direction-cosines are proportional to

$$A,\ B,\ C,$$

Constraint expressed analytically.

ω the component angular velocity round an axis through the origin and in the direction whose direction-cosines are proportional to

$$A', B', C',$$

and lastly,
$$a=\sqrt{\frac{A'^2+B'^2+C'^2}{A^2+B^2+C^2}}.$$

It might be supposed that by altering the origin of co-ordinates we could do away with the angular velocities, and leave only a linear equation among the components of velocity of translation. It is not so; for let the origin be shifted to a point whose co-ordinates are ξ, η, ζ. The angular velocities about the new axes, parallel to the old, will be unchanged; but the linear velocities which, in composition with these angular velocities about the new axes, give $\omega_1, \omega_2, \omega_3, u, v, w$, with reference to the old, are (§ 89)

$$u-\omega_2\eta+\omega_3\zeta=u',$$
$$v-\omega_1\zeta+\omega_3\xi=v',$$
$$w-\omega_2\xi+\omega_1\eta=w'.$$

Hence the equation of constraint becomes

$$Au'+Bv'+Cw'+(A'+B\zeta-C\eta)\omega_1+\text{etc.}=0.$$

Now we cannot generally determine ξ, η, ζ so as to make ω_1, etc., disappear, because this would require three conditions, whereas their co-efficients, as functions of ξ, η, ζ are not independent, since there exists the relation

$$A(B\zeta-C\eta)+B(C\xi-A\zeta)+C(A\eta-B\xi)=0.$$

The simplest form we can reduce to is

$$lu'+mv'+nw'+a(l\omega_1+m\omega_2+n\omega_3)=0,$$

that is to say, every longitudinal motion of a certain axis must be accompanied by a definite proportion of rotation about it.

Generalized co-ordinates.

202. These principles constitute in reality part of the general theory of "co-ordinates" in geometry. The three co-ordinates of either of the ordinary systems, rectangular or polar, required to specify the position of a point, correspond to the three degrees of freedom enjoyed by an unconstrained point. The most general system of co-ordinates of a point consists of three sets of surfaces, on one of each of which it lies. When one of these surfaces only is given, the point may be anywhere on it, or, in the language we have been using above, it enjoys two degrees of freedom. If a second and a third surface, on each of which also it must lie, it has, as we have seen, no freedom left: in other words, its position is completely

Of a point.

specified, being the point in which the three surfaces meet. The analytical ambiguities, and their interpretation, in cases in which the specifying surfaces meet in more than one point, need not occupy us here. Generalized co-ordinates.

To express this analytically, let $\psi=\alpha$, $\phi=\beta$, $\theta=\gamma$ where ψ, ϕ, θ are functions of the position of the point, and α, β, γ constants, be the equations of the three sets of surfaces, different values of each constant giving the different surfaces of the corresponding set. Any one value, for instance, of α, will determine one surface of the first set, and so for the others: and three particular values of the three constants, specify a particular point, P, being the intersection of the three surfaces which they determine. Thus α, β, γ are the "co-ordinates" of P; which may be referred to as "the point (α, β, γ)." The form of the co-ordinate surfaces of the (ψ, ϕ, θ) system is defined in terms of co-ordinates (x, y, z) on any other system, plane rectangular co-ordinates for instance, if ψ, ϕ, θ are given each as a function of (x, y, z). Of a point.

203. Component velocities of a moving point, parallel to the three axes of co-ordinates of the ordinary rectangular system, are, as we have seen, the rates of augmentation of the corresponding co-ordinates. These, according to the Newtonian fluxional notation, are written $\dot{x}, \dot{y}, \dot{z}$; or, according to Leibnitz's notation, which we have used above, $\frac{dx}{dt}, \frac{dy}{dt}, \frac{dz}{dt}$. Lagrange has combined the two notations with admirable skill and taste in his *Mécanique Analytique*, as we shall see in Chap. II. In specifying the motion of a point according to the generalized system of co-ordinates, ψ, ϕ, θ must be considered as varying with the time: $\dot{\psi}, \dot{\phi}, \dot{\theta}$, or $\frac{d\psi}{dt}, \frac{d\phi}{dt}, \frac{d\theta}{dt}$, will then be the generalized components of velocity: and $\ddot{\psi}, \ddot{\phi}, \ddot{\theta}$, or $\frac{d\dot{\psi}}{dt}, \frac{d\dot{\phi}}{dt}, \frac{d\dot{\theta}}{dt}$, or $\frac{d^2\psi}{dt^2}, \frac{d^2\phi}{dt^2}, \frac{d^2\theta}{dt^2}$ will be the generalized components of acceleration. Origin of the differential calculus.

204. On precisely the same principles we may arrange sets of co-ordinates for specifying the position and motion of a material system consisting of any finite number of rigid bodies, or Co-ordinates of any system.

Co-ordinates of any system. material points, connected together in any way. Thus if ψ, ϕ, θ, etc., denote any number of elements, independently variable, which, when all given, fully specify its position and configuration, being of course equal in number to the degrees of freedom to move enjoyed by the system, these elements are its *co-ordinates*. When it is actually moving, their rates of variation per unit of time, or $\dot{\psi}$, $\dot{\phi}$, etc., express what we shall call its generalized component velocities; and the rates at which $\dot{\psi}$, $\dot{\phi}$, etc., augment per unit of time, or $\ddot{\psi}$, $\ddot{\phi}$, etc., its component accelerations. Thus, for example, if the system consists of a single rigid body quite free, ψ, ϕ, etc., in number six, may be three common co-ordinates of one point of the body, and three angular co-ordinates (§ 100, above) fixing its position relatively to axes in a given direction through this point. Then $\dot{\psi}$, $\dot{\phi}$, etc., will be the three components of the velocity of this point, and the velocities of the three angular motions explained in § 100, as corresponding to variations in the angular co-ordinates. Or, again, the system may consist of one rigid body supported on a fixed axis; a second, on an axis fixed relatively to the first; a third, on an axis fixed relatively to the second, and so on. There will be in this case only as many co-ordinates as there are of rigid bodies. These co-ordinates might be, for instance, the angle between a plane of the first body and a fixed plane, through the first axis; the angle between planes through the second axis, fixed relatively to the first and second bodies, and so on; and the component velocities, $\dot{\psi}$, $\dot{\phi}$, etc. would then be the angular velocity of the first body relatively to directions fixed in space; the angular velocity of the second body relatively to the first; of the third relatively to the second, and so on. Or if the system be a set, i in number, of material points perfectly free, one of its $3i$ co-ordinates may be the sum of the squares of their distances from a certain point, either fixed or moving in any way relatively to the system, and the remaining $3i - 1$ may be angles, or may be mere ratios of distances between individual points of the system. But it is needless to multiply examples here. We shall have illustrations enough of the principle of generalized co-ordinates, by actual use of it in Chap. II., and other parts of this book.

Generalized components of velocity.

Examples.

APPENDIX TO CHAPTER I.

A.—Extension of Green's Theorem.

It is convenient that we should here give the demonstration of a few theorems of pure analysis, of which we shall have many and most important applications, not only in the subject of spherical harmonics, which follows immediately, but in the general theories of attraction, of fluid motion, and of the conduction of heat, and in the most practical investigations regarding electricity, and magnetic and electro-magnetic force.

(*a*) Let U and U' denote two functions of three independent variables, x, y, z, which we may conveniently regard as rectangular co-ordinates of a point P, and let α denote a quantity which may be either constant, or any arbitrary function of the variables. Let $\iiint dxdydz$ denote integration throughout a limited space bounded by a closed surface S; let $\iint dS$ denote integration over the whole surface S; and let ∂, prefixed to any function, denote its rate of variation at any point of S, per unit of length in the direction perpendicular to S outwards.

Then

$$\iiint \alpha^2\left(\frac{dU}{dx}\frac{dU'}{dx}+\frac{dU}{dy}\frac{dU'}{dy}+\frac{dU}{dz}\frac{dU'}{dz}\right)dxdydz$$

$$\left.\begin{aligned}&=\iint dS.U'\alpha^2\partial U-\iiint U'\left\{\frac{d(\alpha^2\frac{dU}{dx})}{dx}+\frac{d(\alpha^2\frac{dU}{dy})}{dy}+\frac{d(\alpha^2\frac{dU}{dz})}{dz}\right\}dxdydz\\&=\iint dS.U\alpha^2\partial U'-\iiint U\left\{\frac{d(\alpha^2\frac{dU'}{dx})}{dx}+\frac{d(\alpha^2\frac{dU'}{dy})}{dy}+\frac{d(\alpha^2\frac{dU'}{dz})}{dz}\right\}dxdydz\end{aligned}\right\}\quad(1)$$

α constant gives a theorem of Green's.

For, taking one term of the first member alone, and integrating "by parts," we have

$$\iiint \alpha^2\frac{dU}{dx}\frac{dU'}{dx}dxdydz=\iint U'\alpha^2\frac{dU}{dx}dydz-\iiint U'\frac{d(\alpha^2\frac{dU}{dx})}{dx}dxdydz,$$

the first integral being between limits corresponding to the surface S; that is to say, being from the negative to the positive end of the portion within S, or of each portion within S, of the line x through the point $(0, y, z)$. Now if A_2 and A_1 denote the inclination of the outward normal of the surface to this line, at points where it enters and emerges from S respectively, and if *a constant gives a theorem of Green's.*

dS_2 and dS_1 denote the elements of the surface in which it is cut at these points by the rectangular prism standing on $dydz$, we have

$$dydz = -\cos A_2 dS_2 = \cos A_1 dS_1.$$

Thus the first integral, between the proper limits, involves the elements $U'a^2\frac{dU}{dx}\cos A_1 dS_1$, and $-U'a^2\frac{dU}{dx}\cos A_2 dS_2$; the latter of which, as corresponding to the lower limit, is subtracted. Hence, there being in the whole of S an element dS_2 for each element dS_1, the first integral is simply

$$\iint U'a^2\frac{dU}{dx}\cos A\, dS,$$

for the whole surface. Adding the corresponding terms for y and z, and remarking that

$$\frac{dU}{dx}\cos A + \frac{dU}{dy}\cos B + \frac{dU}{dz}\cos C = \partial U$$

where B and C denote the inclinations of the outward normal through dS to lines drawn through dS in the positive directions parallel to y and z respectively, we perceive the truth of (1).

(*b*) Again, let U and U' denote two functions of x, y, z, which have equal values at every point of S, and of which the first fulfils the equation

Equation of the conduction of heat.

$$\frac{d(a^2\frac{dU}{dx})}{dx} + \frac{d(a^2\frac{dU}{dy})}{dy} + \frac{d(a^2\frac{dU}{dz})}{dz} = 0 \qquad (2),$$

for every point within S.

Then if $U' - U = u$, we have

$$\iiint\{(a\frac{dU'}{dx})^2 + (a\frac{dU'}{dy})^2 + (a\frac{dU'}{dz})^2\}dxdydz = \iiint\{(a\frac{dU}{dx})^2 + (a\frac{dU}{dy})^2 + (a\frac{dU}{dz})^2\}dxdydz$$

$$+\iiint\{(a\frac{du}{dx})^2 + (a\frac{du}{dy})^2 + (a\frac{du}{dz})^2\}dxdydz \qquad (3)$$

For the first number is equal identically to the second member with the addition of

$$2\iiint a^2(\frac{dU}{dx}\frac{du}{dx} + \frac{dU}{dy}\frac{du}{dy} + \frac{dU}{dz}\frac{du}{dz})dxdydz.$$

But, by (1) this is equal to

$$2\iint dS.ua^2\partial U - 2\iiint u\{\frac{d(a^2\frac{dU}{dx})}{dx} + \frac{d(a^2\frac{dU}{dy})}{dy} + \frac{d(a^2\frac{dU}{dz})}{dz}\}dxdydz,$$

of which each term vanishes; the first, or the double integral, because, by hypothesis, u is equal to nothing at every point of S, and the second, or the triple integral, because of (2).

(*c*) The second term of the second member of (3) is essentially positive, provided α has a real value, whether positive, zero, or negative, for every point (x, y, z) within S. Hence the first member of (3) necessarily exceeds the first term of the second member. But the sole characteristic of U is that it satisfies (2). Hence U' cannot also satisfy (2). That is to say, U being any one solution of (2), there can be no other solution agreeing with it at every point of S, but differing from it for some part of the space within S.

Property of solution with U given over S.

Solution proved to be determinate;

(*d*) One solution of (2) exists, satisfying the condition that U has an arbitrary value for every point of the surface S. For let U denote any function whatever, which has the given arbitrary value at each point of S; let u be any function whatever which is equal to nothing at each point of S, and which is of any real finite or infinitely small value, of the same sign as the value of

proved to be possible.

$$\frac{d(\alpha^2\frac{dU}{dx})}{dx}+\frac{d(\alpha^2\frac{dU}{dy})}{dy}+\frac{d(\alpha^2\frac{dU}{dz})}{dz}$$

at each internal point, and therefore, of course, equal to nothing at every internal point, if any, for which the value of this expression is nothing; and let $U'=U+\theta u$, where θ denotes any constant. Then, using the formulæ of (*b*), modified to suit the altered circumstances, and taking Q and Q' for brevity to denote

$$\iiint\{(\alpha\frac{dU}{dx})^2+(\alpha\frac{dU}{dy})^2+(\alpha\frac{dU}{dz})^2\}dxdydz,$$

and the corresponding integral for U', we have

$$Q'=Q-2\theta\iiint u\{\frac{d}{dx}(\alpha^2\frac{dU}{dx})+\frac{d}{dy}(\alpha^2\frac{dU}{dy})+\frac{d}{dz}(\alpha^2\frac{dU}{dz})\}dxdydz$$

$$+\theta^2\iiint\{(\alpha\frac{du}{dx})^2+(\alpha\frac{du}{dy})^2+(\alpha\frac{du}{dz})^2\}dxdydz.$$

The co-efficient of -2θ here is essentially positive, in consequence of the condition under which u is chosen, unless (2) is satisfied, in which case it is nothing; and the coefficient of θ^2 is essentially positive, if not zero, because all the quantities involved are real. Hence the equation may be written thus:—

$$Q'=Q-m\theta(n-\theta)$$

where m and n are each positive. This shows that if any positive value less than n is assigned to θ, Q' is made smaller than Q; that is to say, unless (2) is satisfied, a function, having the same value at S as U, may be found which shall make the Q integral smaller than for U. In other words, a function U, which, having

Solution proved possible.

any prescribed value over the surface S, makes the integral Q for the interior as small as possible, must satisfy equation (2). But the Q integral is essentially positive, and therefore there is a limit than which it cannot be made smaller. Hence there is a solution of (2) subject to the prescribed surface condition.

a constant gives "Green's theorem."

(e) We have seen (c) that there is, if one, only one, solution of (2) subject to the prescribed surface condition, and now we see that there is one. To recapitulate,—we conclude that if the value of U be given arbitrarily at every point of any closed surface, the equation

$$\frac{d}{dx}\left(a^2\frac{dU}{dx}\right)+\frac{d}{dy}\left(a^2\frac{dU}{dy}\right)+\frac{d}{dz}\left(a^2\frac{dU}{dz}\right)=0$$

determines its value without ambiguity for every point within that surface. That this important proposition holds also for the whole infinite space without the surface S, follows from the preceding demonstration, with only the precaution, that the different functions dealt with must be so taken as to render all the triple integrals convergent. S need not be merely a single closed surface, but it may be any number of surfaces enclosing isolated portions of space. The extreme case, too, of S, or any detached part of S, an open shell, that is a finite unclosed surface, is clearly included. Or lastly, S, or any detached part of S, may be an infinitely extended surface, provided the value of U arbitrarily assigned over it be so assigned as to render the triple and double integrals involved all convergent.

B.—Spherical Harmonic Analysis.

Object of spherical harmonic analysis.

The mathematical method which has been commonly referred to by English writers as that of "Laplace's Co-efficients," but which is here called *spherical harmonic analysis*, has for its object the expression of an arbitrary periodic function of two independent variables in the proper form for a large class of physical problems involving arbitrary data over a spherical surface, and the deduction of solutions for every point of space.

Definition of spherical harmonic functions.

(a) A *spherical harmonic function* is defined as a homogeneous function, V, of x, y, z, which satisfies the equation

$$\frac{d^2V}{dx^2}+\frac{d^2V}{dy^2}+\frac{d^2V}{dz^2}=0. \qquad (4)$$

Its degree may be any positive or negative integer; or it may be fractional; or it may be imaginary.

(*b*) A *spherical surface harmonic* is the function of two angular co-ordinates, or spherical surface co-ordinates, which a spherical harmonic becomes at any spherical surface described from O, the origin of co-ordinates, as centre. Sometimes a function which, according to the definition (a), is simply a spherical harmonic, will be called a *spherical solid harmonic*, when it is desired to call attention to its not being confined to a spherical surface. Definitions.

(*c*) A *complete spherical harmonic* is one which is finite and of single value for all finite values of the co-ordinates.

An *incomplete spherical harmonic* is a spherical harmonic which either does not satisfy the fundamental equation (4) for space completely surrounding the centre, or does not return to the same value in going once round every closed curve.

(*d*) It will be shown, later, that a complete spherical harmonic is necessarily either a rational integral function of the co-ordinates, or reducible to one by a factor of the form $(x^2+y^2+z^2)^m$. Algebraic quality of complete harmonics.

(*e*) The general problem of finding harmonic functions is most concisely stated thus :— Differential equations of a harmonic of degree n.

To find the most general integral of the equation

$$\frac{d^2u}{dx^2}+\frac{d^2u}{dy^2}+\frac{d^2u}{dz^2}=0 \qquad (4)$$

subject to the condition

$$x\frac{du}{dx}+y\frac{du}{dy}+z\frac{du}{dz}=nu \qquad (5)$$

the second of these equations being merely the analytical expression of the condition that u is a homogeneous function of x, y, z of the degree n, which may be any whole number positive or negative, any fraction, or any imaginary quantity.

(*f*) Analytical expressions in various forms for an absolutely general integration of these equations, may probably be found without much difficulty; but with us the only value or interest which any such investigation can have, depends on the availability of its results for solutions fulfilling the conditions at bounding surfaces presented by physical problems. In a very large and most important class of physical problems regarding space bounded by a complete spherical surface, or by two complete concentric spherical surfaces, or by closed surfaces differing very little from spherical surfaces, the case of n any positive or negative integer, integrated particularly under the restriction stated in (d) is of paramount importance. It Value of general symbolical expressions. Use of complete spherical harmonics in physical problems.

Uses of incomplete spherical harmonics in physical problems.

will be worked out thoroughly below. Again, in similar problems regarding sections cut out of spherical spaces by two diametral planes making any angle with one another ***not a sub-multiple of two right angles***, or regarding spaces bounded by two circular cones having a common vertex and axis, and by the included portion of two spherical surfaces described from their vertex as centre, solutions for cases of fractional and imaginary values of n are useful. Lastly, when the subject is a solid or fluid, shaped as a section cut from the last mentioned spaces by two planes through the axis of the cones, inclined to one another at any angle, whether a sub-multiple of π or not, we meet with the case of n either integral or not, but to be integrated under a restriction differing from that specified in (d). We shall accordingly, after investigating general expressions for complete spherical harmonics, give some indications as to the determination of the incomplete harmonics, whether of fractional, of imaginary, or of integral degrees, which are required for the solution of problems regarding such portions of spherical spaces as we have just described.

A few formulæ, which will be of constant use in what follows, are brought together in the first place.

Working formulæ.

(g) Calling O the origin of co-ordinates, and P the point x, y, z, let $OP=r$, so that $x^2+y^2+z^2=r^2$. Let ∂, prefixed to any function, denote its rate of variation per unit of space in the direction OP; so that

$$\partial=\frac{x}{r}\frac{d}{dx}+\frac{y}{r}\frac{d}{dy}+\frac{z}{r}\frac{d}{dz} \qquad (6).$$

If H_n denote any homogeneous function of x, y, z of order n, we have clearly

$$\partial H_n=\frac{n}{r}H_n \qquad (7);$$

whence
$$x\frac{dH_n}{dx}+y\frac{dH_n}{dy}+z\frac{dH_n}{dz}=nH_n \qquad (5) \text{ or } (8)$$

the well known differential equation of a homogeneous function, in which, of course, n may have any value, positive integral, negative, fractional, or imaginary. Again, denoting, for brevity, $\frac{d^2}{dx^2}+\frac{d^2}{dy^2}+\frac{d^2}{dz^2}$ by ∇^2, we have, by differentiation

$$\nabla^2(r^m)=m(m+1)r^{m-2} \qquad (9).$$

Also, if u, u' denote any two functions,

$$\nabla^2(uu')=u'\nabla^2u+2\left(\frac{du}{dx}\frac{du'}{dx}+\frac{du}{dy}\frac{du'}{dy}+\frac{du}{dz}\frac{du'}{dz}\right)+u\nabla^2u' \qquad (10);$$

whence, if u and u' are both solutions of (4), Working formulæ.

$$\nabla^2(uu') = 2\left(\frac{du}{dx}\frac{du'}{dx}+\frac{du}{dy}\frac{du'}{dy}+\frac{du}{dz}\frac{du'}{dz}\right) \qquad (11);$$

or, by taking $u=V_n$, a harmonic of degree n, and $u'=r^m$,

$$\nabla^2(r^m V_n) = 2mr^{m-2}\left(x\frac{dV_n}{dx}+y\frac{dV_n}{dy}+z\frac{dV_n}{dz}\right)+V_n\nabla^2(r^m),$$

or, by (8) and (9),

$$\nabla^2(r^m V_n) = m(2n+m+1)r^{m-2}V_n \qquad (12).$$

From this last it follows that $r^{-2n-1}V_n$ is a harmonic; which, being of degree $-n-1$, may be denoted by V_{-n-1}, so that we have

$$\left.\begin{array}{ll} & V_{-n-1}=r^{-2n-1}V_n \\ \text{or} & \dfrac{V_{n'}}{r^{n'}}=\dfrac{V_n}{r^n}, \\ \text{if} & n+n'=-1 \end{array}\right\} \qquad (13),$$

a formula showing a reciprocal relation between two solid harmonics which give the same form of surface harmonic at any spherical surface described from O as centre. Again, by taking $m=-1$, in (9), we have

$$\nabla^2\frac{1}{r}=0 \qquad (14).$$

Hence $\frac{1}{r}$ is a harmonic of degree -1. We shall see later that it is the only *complete harmonic*, of this degree.

If u be any solution of the equation $\nabla^2 u=0$, we have also

$$\nabla^2\frac{du}{dx}=0,$$

and so on for any number of differentiations. Hence if V_ι is a harmonic of any degree ι, $\frac{d^{j+k+l}V_\iota}{dx^j dy^k dz^l}$ is a harmonic of degree $\iota-j-k-l$; or, as we may write it,

$$\frac{d^{j+k+l}V_\iota}{dx^j dy^k dz^l}=V_{\iota-j-k-l} \qquad (15).$$

Again, we have a most important theorem expressed by the following equations:— Theorem due to Laplace.

$$\iint S_i S_{i'} d\varpi = 0 \qquad (16)$$

where $d\varpi$ denotes an element of a spherical surface, described from O as centre with radius unity; $\iint$ an integration over the whole of this surface; and S_i, $S_{i'}$ two complete surface harmonics, of which the degrees, i and i', are neither equal to one another nor such that $i+i'=-1$. For, denoting the solid harmonics

Uses of incomplete spherical harmonics in physical problems.

Working formulæ

...re $S_1, S_2, \ldots S_i$ denote the surface values of solid spherical ...monics of degrees 1, 2,...i, each a rational integral function ...every point within S. But

$$S_0+S_1\frac{r}{a}+S_2\frac{r^2}{a^2}+\ldots+S_i\frac{r^i}{a^i}+\text{ etc.} \qquad (19)$$

Harmonic solution of Green's problem for the space within a spherical surface.

... function fulfilling these conditions, and therefore, as was ...ved above, A (c), V cannot differ from it. Now, as a parti...ar case, let V be a harmonic function of positive degree ι, ...ch may be denoted by $S_\iota\frac{r^\iota}{a^\iota}$: we must have

$$S_\iota\frac{r^\iota}{a^\iota}=S_0+S_1\frac{r}{a}+S_2\frac{r^2}{a^2}+\ldots+S_i\frac{r^i}{a^i}+\text{ etc.}$$

This cannot be unless $\iota=i$, $S_\iota=S_i$, and all the other functions ... S_1, S_2, etc., vanish. Hence there can be no complete spheri...l harmonic of positive degree, which is not, as $S_i\frac{r^i}{a^i}$, of integral ...egree and an integral rational function of the co-ordinates.

Complete harmonics of positive degrees, proved rational and integral.

Again, let V be any function satisfying (17) for every point ...ithout the spherical surface S, and vanishing at an infinite dis...ance in any direction; and let, as before, (18) express its surface ...alue at S. We similarly prove that it cannot differ from

Harmonic solution of Green's problem for space external to a spherical surface.

$$\frac{aS_0}{r}+\frac{a^2S_1}{r^2}+\frac{a^3S_2}{r^3}+\ldots\ldots+\frac{a^{i+1}S_i}{r^{i+1}}+\text{etc.} \qquad (20).$$

Hence if, as a particular case, V be any complete harmonic $\frac{r^\kappa S_\kappa}{a^\kappa}$, of negative degree κ, we must have, for all points outside S,

$$\frac{r^\kappa S_\kappa}{a^\kappa}=\frac{aS_0}{r}+\frac{a^2S_1}{r^2}+\frac{a^3S_2}{r^3}+\ldots\ldots+\frac{a^{i+1}S_i}{r^{i+1}}+\text{etc.}$$

which requires that $\kappa=-(i+1)$, $S_\kappa=S_i$, and that all the other functions S_0, S_1, S_2, etc., vanish. Hence a complete spherical harmonic of negative degree cannot be other than $\frac{a^{i+1}S_i}{r^{i+1}}$, or $\frac{a^{i+1}}{r^{2i+1}}S_ir^i$, where S_ir^i is not only a rational integral function of the co-ordinates, as asserted in the enunciation, but is itself a spherical harmonic.

Complete harmonics of negative degree.

(*i*) Thus we have proved that a complete spherical harmonic, if of positive, is necessarily of integral, degree, and is, besides, a rational integral function of the co-ordinates, or if of negative degree, $-(i+1)$, is necessarily of the form $\frac{V_i}{r^{2i+1}}$, where V_i is a harmonic of positive degree, i. We shall therefore call the *order* of a complete spherical harmonic of negative degree, the *degree or order* of the complete harmonic of positive degree allied to it; and we shall call the *order* of a surface harmonic, the degree or order of the solid harmonic of positive degree, or the order of the solid harmonic of negative degree, which agrees with it at the spherical surface.

Orders and degrees of complete harmonics.

General expressions for complete harmonics.

(j) To obtain general expressions for complete spherical harmonics of all orders, we may first remark that, inasmuch as a constant is the only rational integral function of degree 0, a complete harmonic of degree 0 is necessarily constant. Hence, by what we have just seen, a complete harmonic of the degree -1 is necessarily of the form $\frac{A}{r}$. That this function is a harmonic we knew before, by (14).

By differentiation of harmonic of degree -1.

Hence, by (15), we see that

$$\left.\begin{array}{l} V_{-i-1}=\dfrac{d^{j+k+l}}{dx^j dy^k dz^l}\dfrac{1}{(x^2+y^2+z^2)^{\frac{1}{2}}} \\ \text{if} \quad j+k+l=i \end{array}\right\} \qquad (21),$$

where V_{-i-1} denotes a harmonic, which is clearly a complete harmonic, of degree $-(i+1)$. The differential coefficient here indicated, when worked out, is easily found to be a fraction, of which the numerator is a rational integral function of degree i, and the denominator is r^{2i+1}. By what we have just seen, the numerator must be a harmonic; and, denoting it by V_i, we thus have

$$V_i=r^{2i+1}\frac{d^{j+k+l}}{dx^j dy^k dz^l}\frac{1}{r} \qquad (22)$$

Number of independent harmonics of any order.

The number of independent harmonics of order i, which we can thus derive by differentiation from $\frac{1}{r}$, is $2i+1$. For, although there are $\frac{(i+2)(i+1)}{2}$ differential coefficients $\frac{d^{j+k+l}}{dx^j dy^k dz^l}$, for which $j+k+l=i$, only $2i+1$ of these are independent when $\frac{1}{r}$ is the subject of differentiation, inasmuch as

$$\left(\frac{d^2}{dx^2}+\frac{d^2}{dy^2}+\frac{d^2}{dz^2}\right)\frac{1}{r}=0 \qquad (14).$$

which gives $$\frac{d^{2n}}{dz^{2n}}\frac{1}{r}=(-)^n\left(\frac{d^2}{dx^2}+\frac{d^2}{dy^2}\right)^n\frac{1}{r} \quad (23),$$ Relation between differential coefficients of harmonics

n being any integer, and shows that,

$$\left.\begin{array}{ll}\dfrac{d^{j+k+l}}{dx^j dy^k dz^l}\dfrac{1}{r}=(-)^{\frac{l}{2}}\dfrac{d^{j+k}}{dx^j dy^k}\left(\dfrac{d^2}{dx^2}+\dfrac{d^2}{dy^2}\right)^{\frac{l}{2}}\dfrac{1}{r}, & \text{if } l \text{ is even,}\\ \text{or} \quad =(-)^{\frac{l-1}{2}}\dfrac{d^{j+k}}{dx^j dy^k}\left(\dfrac{d^2}{dx^2}+\dfrac{d^2}{dy^2}\right)^{\frac{l-1}{2}}\dfrac{d}{dz}\dfrac{1}{r}, & \text{if } l \text{ is odd.}\end{array}\right\} (24).$$

Hence, by taking $l=0$, and $j+k=i$, in the first place, we have $i+1$ differential coefficients $\frac{d^{j+k}}{dx^j dy^k}$; and by taking next $l=1$, and $j+k=i-1$, we have i varieties of $\frac{d^{j+k}}{dx^j dy^k}$; that is to say, we have in all $2i+1$ varieties, and no more, when $\frac{1}{r}$ is the subject. It is easily seen that these $2i+1$ varieties are in reality independent. We need not stop at present to show this, as it will be apparent in the actual expansions given below.

Now if $H_i(x, y, z)$ denote any rational integral function of x, y, z of degree i, $\nabla^2 H_i(x, y, z)$ is of degree $i-2$. Hence since in H_i there are $\frac{(i+2)(i+1)}{2}$ terms, in $\nabla^2 H_i$ there are $\frac{i(i-1)}{2}$. Hence if $\nabla^2 H_i=0$, we have $\frac{i(i-1)}{2}$ equations among the constant coefficients, and the number of independent constants remaining is $\frac{(i+2)(i+1)}{2}-\frac{i(i-1)}{2}$, or $2i+1$; that is to say, there are $2i+1$ constants in the general rational integral harmonic of degree i. But we have seen that there are $2i+1$ distinct varieties of differential coefficients of $\frac{1}{r}$ of order i, and that the numerator of each is a harmonic of degree i. Hence every complete harmonic of order i is expressible in terms of differential coefficients of $\frac{1}{r}$. It is impossible to form $2i+1$ functions symmetrically among three variables, except when $2i+1$ is divisible by 3; that is to say, when $i=3n+1$, n being any integer, a class of cases deserving special attention. But for every value of i the general harmonic may be exhibited as a function, with $2i+1$ constants, involving two out of the three variables symmetrically. This may be done in a variety of ways, of which we choose the two following, as being the most useful:—First,

Complete harmonic of any degree investigated algebraically.

General expression for complete harmonic of order i.

$$\frac{V_i}{r^{2i+1}}=\left\{A_0\left(\frac{d}{dx}\right)^i+A_1\left(\frac{d}{dx}\right)^{i-1}\frac{d}{dy}+A_2\left(\frac{d}{dx}\right)^{i-2}\left(\frac{d}{dy}\right)^2+\ldots+A_i\left(\frac{d}{dy}\right)^i\right\}\frac{1}{r}$$

$$+\left\{B_0\left(\frac{d}{dx}\right)^{i-1}+B_1\left(\frac{d}{dx}\right)^{i-2}\frac{d}{dy}+B_2\left(\frac{d}{dx}\right)^{i-3}\left(\frac{d}{dy}\right)^2+\ldots+B_{i-1}\left(\frac{d}{dy}\right)^{i-1}\right\}\frac{d}{dz}\frac{1}{r}$$

Secondly, let $x+yv=\xi,\ x-yv=\eta$ (26),

where, as formerly, v is taken to denote $\sqrt{-1}$.

This gives
$$x=\tfrac{1}{2}(\xi+\eta),\quad y=\frac{1}{2v}(\xi-\eta),\qquad \frac{1}{r}=\frac{1}{(\xi\eta+z^2)^{\frac{1}{2}}} \qquad (27);$$

Imaginary linear transformation.

$$\frac{d}{dx}[x,y]=\left(\frac{d}{d\xi}+\frac{d}{d\eta}\right)[\xi,\eta],\quad \frac{d}{dy}[x,y]=v\left(\frac{d}{d\xi}-\frac{d}{d\eta}\right)[\xi,\eta]$$
$$\frac{d}{d\xi}[\xi,\eta]=\tfrac{1}{2}\left(\frac{d}{dx}-v\frac{d}{dy}\right)[x,y],\quad \frac{d}{d\eta}[\xi,\eta]=\tfrac{1}{2}\left(\frac{d}{dx}+v\frac{d}{dy}\right)[x,y] \qquad (28),$$

where $[x, y]$ and $[\xi, \eta]$ denote the same quantity, expressed in terms of x, y, and of ξ, η respectively. From these we have, further,

$$\left(\frac{d^2}{dx^2}+\frac{d^2}{dy^2}+\frac{d^2}{dz^2}\right)[x,y,z]=\left(4\frac{d^2}{d\xi d\eta}+\frac{d^2}{dz^2}\right)[\xi,\eta,z],$$

or, according to our abbreviated notation, (29).

$$\nabla^2=4\frac{d^2}{d\xi d\eta}+\frac{d^2}{dz^2}.$$

Then, as before,

$$\frac{V_i}{r^{2i+1}}=\left\{\mathfrak{A}_0\left(\frac{d}{d\xi}\right)^i+\mathfrak{A}_1\left(\frac{d}{d\xi}\right)^{i-1}\frac{d}{d\eta}+\mathfrak{A}_2\left(\frac{d}{d\xi}\right)^{i-2}\left(\frac{d}{d\eta}\right)^2+\ldots+\mathfrak{A}_i\left(\frac{d}{d\eta}\right)^i\right\}\frac{1}{r}$$

$$+\left\{\mathfrak{B}_0\left(\frac{d}{d\xi}\right)^{i-1}+\mathfrak{B}_1\left(\frac{d}{d\xi}\right)^{i-2}\frac{d}{d\eta}+\mathfrak{B}_2\left(\frac{d}{d\xi}\right)^{i-3}\left(\frac{d}{d\eta}\right)^2+\ldots+\mathfrak{B}_{i-1}\left(\frac{d}{d\eta}\right)^{i-1}\right\}\frac{d}{dz}\frac{1}{r}$$

Expansion of elementary term.

The differentiations here are performed with great ease, by the aid of Leibnitz's theorem. Thus we have

$$r^{2(m+n)+1}\frac{d^{m+n}}{d\xi^m d\eta^n}\frac{1}{r}=(-)^{m+n}\tfrac{1}{2}.\tfrac{3}{2}.\tfrac{5}{2}\ldots(m+n-\tfrac{1}{2})\left[\eta^m\xi^n-\frac{mn}{1.(m+n-\frac{1}{2})}\eta^{m-1}\xi^{n-1}r^2\right.$$
$$\left.+\frac{m(m-1).n(n-1)}{1.2.(m+n-\frac{1}{2})(m+n-\frac{3}{2})}\eta^{m-2}\xi^{n-2}r^4-\text{etc.}\right]$$

and

$$r^{2(m+n)+3}\frac{d^{m+n+1}}{d\xi^m d\eta^n dz}\frac{1}{r}=(-)^{m+n+1}\tfrac{1}{2}.\tfrac{3}{2}.\tfrac{5}{2}\ldots(m+n+\tfrac{1}{2}).2z\left[\eta^m\xi^n-\frac{mn}{1.(m+n+\frac{1}{2})}\eta^{m-1}\xi^{n-1}r^2\right.$$
$$\left.+\frac{m(m-1).n(n-1)}{1.2.(m+n+\frac{1}{2})(m+n-\frac{1}{2})}\eta^{m-2}\xi^{n-2}r^4-\text{etc.}\right]$$

This expression leads at once to a real development, in terms of polar co-ordinates, thus :—Let

Polar transformation.

$$z=r\cos\theta,\ x=r\sin\theta\cos\phi,\ y=r\sin\theta\sin\phi \qquad (32);$$

so that $\quad \xi=r\sin\theta\epsilon^{\upsilon\phi},\ \eta=r\sin\theta\epsilon^{-\upsilon\phi} \qquad (33).$

Then, since $\quad \xi\eta=x^2+y^2=r^2\sin^2\theta,$

and

$$\eta^m=(\xi\eta)^m\xi^s=(\xi\eta)^m(r\sin\theta)^s(\cos\phi+\upsilon\sin\phi)^s=(r\sin\theta)^{m+n}(\cos s\phi+\upsilon\sin s\phi),$$

where $s=n-m$; and if, further, we take

$$\left.\begin{array}{l}\mathfrak{J}_n+\mathfrak{J}_m=\mathrm{A}_s,\ (\mathfrak{J}_n-\mathfrak{J}_m)\upsilon=\mathrm{A}_s'\\ \mathfrak{B}_n+\mathfrak{B}_m=\mathrm{B}_s,\ (\mathfrak{B}_n-\mathfrak{B}_m)\upsilon=\mathrm{B}_s'\end{array}\right\} \qquad (34)$$

we have

Trigonometrical expansions.

$$\left.\begin{array}{l}\ldots\frac{d^{n-m}}{\ldots d\eta^{n}}\frac{1}{r}+\mathfrak{J}_m\frac{d^{m+n}}{d\xi^n d\eta^m}\frac{1}{r}=(-)^{m+n}\tfrac{1}{2}\cdot\tfrac{3}{2}\cdot\tfrac{5}{2}\ldots(m+n-\tfrac{1}{2})r^{-(m+n+1)}(\mathrm{A}_s\cos s\phi+\mathrm{A}_s'\sin s\phi)\\ \quad[\sin^{m+n}\theta-\frac{mn}{1.(m+n-\frac{1}{2})}\sin^{m+n-2}\theta+\frac{m(m-1).n(n-1)}{1.2.(m+n-\frac{1}{2})(m+n-\frac{3}{2})}\sin^{m+n-4}\theta-\text{etc.}]\\ \ldots\frac{d^{n+1}}{\ldots\eta^n dz}\frac{1}{r}+\mathfrak{B}_m\frac{d^{m+n+1}}{d\xi^n d\eta^m dz}\frac{1}{r}=(-)^{m+n+1}\tfrac{3}{2}\cdot\tfrac{5}{2}\ldots(m+n+\tfrac{1}{2})r^{-(m+n+2)}(\mathrm{B}_s\cos s\phi+\mathrm{B}_s'\sin s\phi)\cos\theta\\ \quad[\sin^{m+n}\theta-\frac{mn}{1.(m+n+\frac{1}{2})}\sin^{m+n-2}\theta+\frac{m(m-1).n(n-1)}{1.2.(m+n+\frac{1}{2})(m+n-\frac{1}{2})}\sin^{m+n-4}\theta-\text{etc.}]\end{array}\right\}(35).$$

Setting aside now constant factors, which have been retained hitherto to show the relations of the expressions we have investigated, to differential co-efficients of $\frac{1}{r}$; and taking Σ to denote summation with respect to the arbitrary constants, A, A′, B, B′, we have the following perfectly general expression for a complete surface harmonic of order i:—

$$\left.\begin{array}{l}S_i=\sum^{m+n=i}(\mathrm{A}_s\cos s\phi+\mathrm{A}_s'\sin s\phi)\,\Theta_{(m,n)}+\sum^{m+n+1=i}(\mathrm{B}_s\cos s\phi+\mathrm{B}_s'\sin s\phi)\cos\theta Z_{(m,n)}\\ \text{where } s=m\sim n, \text{ and}\\ \Theta_{(m,n)}=\sin^{m+n}\theta-\frac{mn}{1.(m+n-\frac{1}{2})}\sin^{m+n-2}\theta+\frac{m(m-1).n(n-1)}{1.2.(m+n-\frac{1}{2})(m+n-\frac{3}{2})}\sin^{m+n-4}\theta-\text{etc.}\end{array}\right\}(36)$$

while $Z_{(m,n)}$ differs from $\Theta_{(m,n)}$ only in having $m+n+1$ in place of $m+n$, in the denominators.

The formula most commonly given for a spherical harmonic of order i (Laplace, *Mécanique Celeste*, livre III. chap. II., or Murphy's *Electricity*, Preliminary Prop. xi.) is somewhat simpler, being as follows :—

$$S_i=\sum_{s=0}^{s=i}(\mathrm{A}_s\cos s\phi+\mathrm{B}_s\sin s\phi)\,\Theta_i^{(s)} \qquad (37)$$

$$\Theta_i^{s}=\sin^s\theta[\cos^{i-s}\theta-\frac{(i-s)(i-s-1)}{2.(2i-1)}\cos^{i-s-2}\theta+\frac{(i-s)(i-s-1)(i-s-2)(i-s-3)}{2.4.(2i-1)(2i-3)}\cos^{i-s-4}\theta-\text{etc.}] \qquad (38).$$

Trigonometrical expansions.

where it may be remarked that $\Theta_i^{(s)}$ means the same as $\Theta_{(m,n)}$ if $m+n=i$, and $m \sim n=s$, or as $\cos\theta\, Z_{(m,n)}$ if $m+n+1=i$ and $m \sim n=s$. This may be derived from the preceding, by algebraic modification; or it may be obtained directly by the method of differentiation followed above, varied suitably. But it may also be obtained by assuming (with a_s and b_s as arbitrary constants)

$$V_i=S_i r^i=\Sigma(a_s\xi^s+b_s\eta^s)(z^{i-s}+pr^2z^{i-s-2}+qr^4z^{i-s-4}+\text{ etc.}),$$

which is obviously a proper form; and determining p, q, etc., by the differential equation $\nabla^2 V_i=0$, with (29).

Another form, perhaps the most useful, may be obtained with even greater ease, thus: Assuming

$$V_i=\Sigma(a_s\xi^s+b_s\eta^s)(z^{i-s}+p_1z^{i-s-2}\xi\eta+p_2z^{i-s-4}\xi^2\eta^2+\text{etc.})$$

and determining p_1, p_2, etc., by the differential equation, we have

$$V_i=\Sigma(a_s\xi^s+\beta_s\eta^s)\left[z^{i-s}-\frac{(i-s)(i-s-1)}{4.(s+1).1}z^{i-s-2}\xi\eta+\frac{(i-s)(i-s-1)(i-s-2)(i-s-3)}{4^2.(s+1)(s+2).1.2}z^{i-s-4}\xi^2\eta^2-\text{etc.}\right]$$

which might also have been found easily by the differentiation of $\frac{1}{r}$. Hence, eliminating imaginary symbols, and retaining the notation of (37) and (38), we have

$$\left.\begin{aligned}\Theta_i^{(s)}=C\sin^s\theta\,\Big[\cos^{i-s}\theta-\frac{(i-s)(i-s-1)}{4.(s+1)}\cos^{i-s-2}\theta\sin^2\theta\\ +\frac{(i-s)(i-s-1)(i-s-2)(i-s-3)}{4^2.(s+1)(s+2)}\cos^{i-s-4}\theta\sin^4\theta-\text{etc.}\Big]\\ \text{where}\qquad C=\frac{(2s+1)(2s+2)\dots(i+s)}{(2s+1)(2s+3)\dots(2i-1)}\end{aligned}\right\}\qquad(40)$$

This value of C is found either by comparing with (38), of which the present formula is an algebraic modification; or, perhaps more easily, thus:—We have, by (29),

$$\left.\begin{aligned}\frac{d^i}{dz^{i-s}d\eta^s}\frac{1}{r}&=(-)^{\frac{i-s}{2}}2^{i-s}\frac{d^i}{d\xi^{\frac{i-s}{2}}d\eta^{\frac{i+s}{2}}}\frac{1}{r},\text{ if } i-s \text{ is even,}\\ \text{or}\qquad&=(-)^{\frac{i-s-1}{2}}2^{i-s-1}\frac{d^i}{dz\,d\xi^{\frac{i-s-1}{2}}d\eta^{\frac{i+s-1}{2}}}\frac{1}{r}\end{aligned}\right\}\qquad(41)$$

Expanding the first member in terms of z, ξ, η, by successive differentiation, with reference first to η, s times, and then z, $i-s$ times, we find

$$(-)^i\tfrac{1}{2}.\tfrac{3}{2}\dots(s-\tfrac{1}{2})(2s+1)(2s+2)(2s+3)\dots(i+s)z^{i-s}\xi^s\qquad(42)$$

for a term in its numerator; whence by (41), (35), 39), (40) we find C.

(*k*) It is very important to remark, first, that

$$\iint U_i U_i' d\sigma = 0 \qquad (43)$$

Important properties of elementary terms and auxiliary functions.

where U_i and U_i' denote any two of the elements of which V is composed in one of the preceding expressions; and secondly, that

$$\int_0^{\pi} \Theta_i^{(s)} \Theta_{i'}^{(s)} \sin\theta d\theta = 0 \qquad (44)$$

the case of $i=i'$ being of course excluded. For, taking $r=a$, the radius of the spherical surface; and $d\sigma = a^2 d\varpi$, as above; we have $d\varpi = \sin\theta d\theta d\phi$, etc., the limits of θ and ϕ, in the integration for the whole spherical surface, being 0 to π, and 0 to 2π, respectively. Thus, since $\int_0^{2\pi} \cos s\phi \cos s'\phi \, d\phi = 0$, we see the truth of the first remark; and from (16) and (36) we infer the second, which the reader may verify algebraically, as an exercise.

(*l*) Each one of the preceding series may be taken by either end, and used with i, m, n, or s, one, any, or all of them fractional or imaginary. Whether finite or infinite in its number of terms, any series thus obtained expresses a harmonic of degree i; since it is of degree i, and satisfies $\nabla^2 V_i = 0$. In some of these cases it remains a finite series, and if it does, its application is obvious. If it does not, it may be used if it converges, and, except for particular limiting values of the variables, the infinite series obtained from any one of the preceding finite expressions taken by one of its two ends will converge. Thus each of the finite expressions always provides a series which converges, for any of the fractional or imaginary values now proposed. (It is easily proved, in fact, that when taken by one end it gives a divergent infinite series, and a convergent series when taken by the other end.) The determination of values of $i-s$, which shall make $\Theta_i^{(s)}$ or $\frac{d}{d\theta}\Theta_i^{(s)}$ vanish for each of two stated values of θ, is an analytical problem of high interest in connexion with these extensions of spherical harmonic analysis: and is essentially involved in the physical application referred to above where the spaces concerned are bounded partly by coaxal cones. When the boundary is completed by the intercepted portions of two concentric spherical surfaces, functions of the class described in (*o*) below also enter into the solution. When prepared to take advantage of physical applications we shall return to the subject; but it is necessary at present to restrict ourselves to these few observations.

Expansions of incomplete harmonics for cones and wedges.

Electric induction, motion of water, etc., in space between two coaxal cones.

(*m*) If, in physical problems such as those already referred to, the space considered is bounded by two planes meeting, at

Electric induction, motion of water, etc., in space between spherical surface and two planes meeting in a diameter.

any angle $\frac{\pi}{\iota}$, in a diameter, and the portion of spherical surface in the angle between them (the case of $\iota < 1$, that is to say, the case of angle exceeding two right angles, not being excluded) the harmonics required are all of fractional degrees, but each a finite algebraic function of the co-ordinates ξ, η, ζ if ι is any incommensurable number. Thus, for instance, if the problem be to find the internal temperature at any point of a solid of the shape in question, when each point of the curved portion of its surface is maintained permanently at any arbitrarily given temperature, and its plane sides at one constant temperature, the forms and the degrees of the harmonics referred to are as follows :—

Degree.	Harmonic.	Degree.	Harmonic.	Degree	Harmonic.
ι,	ξ^{ι}	2ι,	$\xi^{2\iota}$	3ι,	$\xi^{3\iota}$
$\iota+1$,	$r^{2\iota+3}\frac{d}{dz}\frac{\xi^{\iota}}{r^{2\iota+1}}$	$2\iota+1$,	$r^{4\iota+3}\frac{d}{dz}\frac{\xi^{2\iota}}{r^{4\iota+1}}$	$3\iota+1$,	$r^{6\iota+1}\frac{d}{dz}\frac{\xi^{3\iota}}{r^{6\iota+1}}$
$\iota+2$,	$r^{2\iota+5}\frac{d^2}{dz^2}\frac{\xi^{\iota}}{r^{2\iota+1}}$	$2\iota+3$,	$r^{4\iota+5}\frac{d^2}{dz^2}\frac{\xi^{2\iota}}{r^{4\iota+1}}$	......	....
$\iota+3$,	$r^{2\iota+7}\frac{d^3}{dz^3}\frac{\xi^{\iota}}{r^{2\iota+1}}$	$2\iota+5$,	$r^{4\iota+7}\frac{d^3}{dz^3}\frac{\xi^{2\iota}}{r^{4\iota+1}}$	......	....
......	....	......	....	......	....
......	....	......	....	......	....

These harmonics are expressed, by various formulæ (36)-(40), etc., in terms of real co-ordinates, in what precedes.

Harmonic functions of all degrees derived from that of degree -1 by generalized differentiation.

(*n*) It is worthy of remark that these, and every other spherical harmonic, of whatever degree, integral, real but fractional, or imaginary, are derivable by a general form of process, which includes differentiation as a particular case. Thus if $(\frac{d}{d\eta})^{\iota}$ denotes an operation which, when ι is an integer, constitutes taking the ιth differential coefficient, we have clearly

$$\xi^{\iota} = r^{2\iota+1}P_{\iota}(\frac{d}{d\eta})^{\iota}\frac{1}{(\xi\eta+z^2)^{\frac{1}{2}}},$$

where P_{ι} denotes a function of ι, which, when ι is a real integer, becomes $(-)^{\iota}\frac{1}{2}.\frac{3}{2}.\frac{5}{2}\ldots(\iota-\frac{1}{2})$.

The investigation of this generalized differentiation presents difficulties which are confined to the valuation of P_{ι}, and which have formed the subject of highly interesting mathematical investigations by Liouville, Gregory, Kelland, and others.

If we set aside the factor P_{ι}, and satisfy ourselves with determinations of *forms* of spherical harmonics, we have only to apply

Leibnitz's and other obvious formulae for differentiation with any fractional or imaginary number as index, to see that the equivalent expressions above given for a complete spherical harmonic of any degree, are derivable from $\frac{1}{r}$ by the process of generalized differentiation now indicated, so as to include every possible incomplete harmonic, of whatever degree, whether integral, or fractional and real, or imaginary. But, as stated above, those expressions may be used, in the manner explained, for incomplete harmonics, whether finite algebraic functions of ξ, η, ζ, or transcendents expressed by converging infinite series; quite irrespectively of the manner of derivation now remarked. Expansions of incomplete harmonics obtained by common formulae, with generalized indices.

(*o*) To illustrate the use of spherical harmonics of imaginary degrees, the problem regarding the conduction of heat specified above may be varied thus:—Let the solid be bounded by two concentric spherical surfaces, of radii a and a', and by the two cones or planes, and let every point of each of these flat or conical sides be maintained at any arbitrarily given temperature, and the whole spherical portion of the boundary at one constant temperature. Harmonics will enter into the solution, of degree Imaginary degrees useful when arbitrary functions of r are to be expressed.

$$-\frac{1}{2}+\frac{i\pi\sqrt{-1}}{\log\frac{a}{a'}},$$

where i denotes any integer. Converging series for these and the others required for the solution are included in our general formulae (36)-(40), etc.

(*p*) The method of finding complete spherical harmonics by the differentiation of $\frac{1}{r}$, investigated above, has this great advantage, that it shows immediately very important properties which they possess with reference to the values of the variables for which they vanish. Thus, inasmuch as $\frac{1}{r}$ and all its differential coefficients vanish for either $x=\pm\infty$, or $y=\pm\infty$, or $z=\pm\infty$, it follows that Derivation of any harmonic from that of degree -1 indicates the character and number of its nodes.

$$\frac{d^{j+k+l}}{dx^j dy^k dz^l}\frac{1}{r}$$

vanishes j times when x is increased from $-\infty$ to $+\infty$

" k " y " " " "

and " l " z " " " "

The reader who is not familiar with Fourier's theory of equations,

Expression of an arbitrary function in a series of surface harmonics.

will have no difficulty in verifying for himself the present application of the principles developed in that admirable work.

Thus it appears that spherical harmonics belong to the general class, to which Sir William R. Hamilton has applied the designation "Fluctuating Functions." This property is essentially involved in their capacity for expressing arbitrary functions, to the demonstration of which we now proceed, in conclusion.

Preliminary proposition.

(*r*) Let C be the centre and a be the radius of a spherical surface, which we shall denote by S. Let P be any external or internal point, and let f denote its distance from C. Let $d\sigma$ denote an element of S, at a point E, and let $EP=D$. Then, $\iint$ denoting an integration extended over S, it is easily proved that

$$\left.\begin{aligned} &\iint\frac{d\sigma}{D^3}=\frac{a}{f}\frac{4\pi a}{f^2-a^2} \text{ when } P \text{ is external to } S \\ \text{and}\quad &\iint\frac{d\sigma}{D^3}=\frac{4\pi a}{a^2-f^2} \text{ when } P \text{ is within } S \end{aligned}\right\} \quad (45).$$

This is merely a particular case of a very general theorem of Green's, included in that of A (a), above, as will be shown when we shall be particularly occupied, later, with the general theory of Attraction: a geometrical proof of a special theorem, of which it is a case, will occur in connexion with elementary investigations regarding the distribution of electricity on spherical conductors: and, in the meantime, the following direct evaluation of the integral itself is given, in order that no part of the important investigation with which we are now engaged may be even temporarily incomplete.

Choosing polar co-ordinates, $\theta=ECP$, and ϕ the angle between the plane of ECP and a fixed plane through CP, we have

$$d\sigma=a^2\sin\theta d\theta d\phi.$$

Hence, by integration from $\phi=0$ to $\phi=2\pi$,

$$\iint\frac{d\sigma}{D^3}=2\pi a^2\int_0^\pi\frac{\sin\theta d\theta}{D^3}.$$

But $$D^2=a^2-2af\cos\theta+f^2;$$

and therefore $$\sin\theta d\theta=\frac{DdD}{af};$$

the limiting values of D in the integral being

$$f-a,\ f+a, \text{ when } f>a$$

and $$a-f,\ a+f, \text{ when } f<a.$$

Hence we have

$$\iint\frac{d\sigma}{D^3}=\frac{2\pi a}{f}\left(\frac{1}{f-a}-\frac{1}{f+a}\right), \text{ or } =\frac{2\pi a}{f}\left(\frac{1}{a-f}-\frac{1}{a+f}\right),$$

in the two cases respectively, which proves (45).

(*s*) Let now $F(E)$ denote any arbitrary function of the position of E on S, and let

Solution of Green's problem for case of spherical surface, expressed by definite integral.

$$u=\iint\frac{(f^2\sim a^2)F(E)d\sigma}{D^3} \qquad (46).$$

When f is infinitely nearly equal to a, every element of this integral will vanish except those for which D is infinitely small. Hence the integral will have the same value as it would have if $F(E)$ had everywhere the same value as it has at the part of S nearest to P; and, therefore, denoting this value of the arbitrary function by $F(P)$, we have

$$u=F(P)\iint\frac{(f^2\sim a^2)d\sigma}{D^3},\text{ when } f \text{ differs infinitely little from } a;$$

or, by (45) $$u=4\pi aF(P) \qquad (46').$$

Now, if e denote any positive quantity less than unity, we have, by expansion in a convergent series,

Its expansion in harmonic series.

$$\frac{1}{(1-2e\cos\theta+e^2)^{\frac{1}{2}}}=1+Q_1e+Q_2e^2+\text{ etc.} \qquad (47),$$

Q_1, Q_2, etc., denoting functions of θ, for which expressions will be investigated below. Each of them is equal to $+1$, when $\theta=0$, and they are alternately equal to -1 and $+1$, when $\theta=\pi$. It is easily proved that each is >-1 and $<+1$, for all values of θ between 0 and π. Hence the series, which becomes the geometrical series $1\pm e+e^2\pm$ etc., in the extreme cases, converges more rapidly than the geometrical series, except in those extreme cases of $\theta=0$ and $\theta=\pi$.

$$\left.\begin{aligned}\text{Hence } \frac{1}{D}&=\frac{1}{f}\left(1+\frac{Q_1a}{f}+\frac{Q_2a^2}{f^2}+\text{ etc.}\right)\text{ when } f>a\\ \text{and } \frac{1}{D}&=\frac{1}{a}\left(1+\frac{Q_1f}{a}+\frac{Q_2f^2}{a^2}+\text{ etc.}\right)\text{ when } f<a\end{aligned}\right\} \qquad (48).$$

Now we have $$\frac{d\frac{1}{D}}{df}=\frac{a\cos\theta-f}{D^3},$$

and therefore $$\frac{f^2-a^2}{D^3}=-\left(2f\frac{d\frac{1}{D}}{df}+\frac{1}{D}\right).$$

Hence by (48), differentiated, etc.,

$$\left.\begin{aligned}\frac{f^2-a^2}{D^3}&=\frac{1}{f}\left(1+\frac{3Q_1a}{f}+\frac{5Q_2a^2}{f^2}+\ldots\ldots\ldots\right)\text{ when } f>a,\\ \text{and } \frac{a^2-f^2}{D^3}&=\frac{1}{a}\left(1+\frac{3Q_1f}{a}+\frac{5Q_2f^2}{a^2}+\ldots\ldots\ldots\right)\text{ when } f<a\end{aligned}\right\}(49).$$

Green's problem for case of spherical surface, solved explicitly in harmonic series.

Hence, for u (46), we have the following expansions :—

$$u=\frac{1}{f}\{\iint F(E)d\sigma+\frac{3a}{f}\iint Q_1 F(E)d\sigma+\frac{5a^2}{f^2}\iint Q_2 F(E)d\sigma+\ldots\},\text{ when } f>a.$$

and

$$u=\frac{1}{a}\{\iint F(E)d\sigma+\frac{3f}{a}\iint Q_1 F(E)d\sigma+\frac{5f^2}{a^2}\iint Q_2 F(E)d\sigma+\ldots\ldots\},\text{ when } f<a$$

These series being clearly convergent, except in the case of $f=a$, and, in this limiting case, the unexpanded value of u having been proved (46′) to be finite and equal to $4\pi a F(P)$, it follows that the sum of each series approaches more and more nearly to this value when f approaches to equality with a. Hence, in the limit,

Laplace's spherical harmonic expansion of an arbitrary function.

$$F(P)=\frac{1}{4\pi a^2}\{\iint F(E)d\sigma+3\iint Q_1 F(E)d\sigma+5\iint Q_2 F(E)d\sigma+\text{etc.},\} \quad (52).$$

which is the celebrated development of an arbitrary function in a series of "Laplace's coefficients," or, as we now call them, *spherical harmonics*.

(t) The preceding investigation shows that when there is one determinate value of the arbitrary function F for every point of S, the series (52) converges to the value of this function at P.

Convergence of series never lost except at abrupt changes in value of the function expressed.

The same reason shows that when there is an an abrupt transition in the value of F, across any line on S, the series cannot converge when P is *exactly on*, but must still converge, however near it may be to, this line.

Later we shall derive a rule for the degree of convergence of of the series (52) in any case according to the character of F.

(u) In the development (47) of $\frac{1}{(1-2e\cos\theta+e^2)^{\frac{1}{2}}}$, the coefficients of $e, e^2, \ldots e^i$, are clearly rational integral functions of $\cos\theta$, of degrees 1, 2...i, respectively. They are given explicitly below in (60) and (61), with $\theta'=0$. But, if x, y, z and x', y', z' denote rectangular co-ordinates of P and of E respectively, we have

Expansion of $\frac{1}{D}$ in symmetrical harmonic functions of the co-ordinates of the two points.

$$\cos\theta=\frac{xx'+yy'+zz'}{rr'},$$

where $r=(x^2+y^2+z^2)^{\frac{1}{2}}$, and $r'=(x'^2+y'^2+z'^2)^{\frac{1}{2}}$. Hence, denoting, as above, by Q_i the coefficient of e^i in the development, we have

$$Q_i=\frac{H_i[(x,y,z),(x',y',z')]}{r^i r'^i} \quad (53).$$

$H_i[(x,y,z),(x',y',z')]$ denoting a symmetrical function of (x, y, z) and (x', y', z'), which is homogeneous with reference to either set alone. An explicit expression for this function is of course found from the expression for Q_i in terms of $\cos\theta$.

Viewed as a function of (x, y, z), $Q_i r^i r'^i$ is symmetrical round OE; and as a function of (x', y', z') it is symmetrical round OP. We shall therefore call it the biaxal harmonic of $(x, y, z)(x', y', z')$ of degree i; and Q_i the biaxal surface harmonic of order i. Biaxal harmonic.

(v) But it is important to remark, that the coefficient of any term, such as $x'^j y'^k z'^l$, in it may be obtained alone, by means of Taylor's theorem, applied to a function of three variables, thus:— Expansion of $\frac{1}{D}$ by Taylor's theorem.

$$\frac{1}{(1-2e\cos\theta+e^2)^{\frac{1}{2}}}=\frac{r}{(r^2-2rr'\cos\theta+r'^2)^{\frac{1}{2}}}=\frac{r}{[(x-x')^2+(y-y')^2+(z-z')^2]^{\frac{1}{2}}}.$$

Now if $F(x, y, z)$ denote any function of x, y, and z, we have

$$F(x+f, y+g, z+h)=\sum_{j=0}^{j=\infty}\sum_{k=0}^{k=\infty}\sum_{l=0}^{l=\infty}\frac{f^j g^k h^l}{1.2\ldots j.1.2\ldots k.1.2\ldots l}\frac{d^{j+k+l}F(x, y, z)}{dx^j dy^k dz^l};$$

where it must be remarked that the interpretation of $1.2\ldots j$, when $j=0$, is unity, and so for k and l also. Hence, by taking

$$F(x, y, z)=\frac{1}{(x^2+y^2+z^2)^{\frac{1}{2}}},$$ we have

$$\frac{1}{[(x-x')^2+(y-y')^2+(z-z')^2]^{\frac{1}{2}}}=\Sigma\Sigma\Sigma\frac{(-1)^{j+k+l}x'^j y'^k z'^l}{1.2\ldots j.1.2\ldots k.1.2\ldots l}\frac{d^{j+k+l}}{dx^j dy^k dz^l}\frac{1}{(x^2+y^2+z^2)^{\frac{1}{2}}}$$

a development which, by comparing it with (48), above, we see to be convergent whenever

$$x'^2+y'^2+z'^2<x^2+y^2+z^2.$$

Hence

$$(rr')^i Q_i = r^{2i+1}\sum^{(j+k+l=i)}\sum\sum\frac{(-1)^{j+k+l}x'^j y'^k z'^l}{1.2\ldots j.1.2\ldots k.1.2\ldots l}\frac{d^{j+k+l}}{dx^j dy^k dz^l}\frac{1}{(x^2+y^2+z^2)^{\frac{1}{2}}} \quad (54),$$

Expression for biaxal harmonic deduced.

the summation including all terms which fulfil the indicated condition $(j+k+l=i)$. It is easy to verify that the second member is not only integral and homogeneous of the degree i, in x, y, z, as it is expressly in x', y', z'; but that it is symmetrical with reference to these two sets of variables. Arriving thus at the conclusion expressed above by (53), we have now, for the function there indicated, an explicit expression in terms of differential coefficients, which, further, may be immediately expanded into an algebraic form with ease.

(v') In the particular case of $x'=0$ and $y'=0$, (54) becomes reduced to a single term, a function of x, y, z symmetrical about the axis OZ; and, dividing each member by r'^i, or its equal, z'^i, we have

$$r^i Q_i = \frac{(-1)^i r^{2i+1}}{1.2.3\ldots i}\frac{d^i}{dz^i}\frac{1}{(x^2+y^2+z^2)^{\frac{1}{2}}} \quad (55).$$

Axial harmonic of order i.

By actual differentiation it is easy to find the law of successive derivation of the numerators; and thus we find, with about equal

Axial harmonic with its co-ordinates transformed becomes biaxal.

case, either of the expansions (31), (40), or (41), above, for the case $m=n$, or the trigonometrical formulæ, which are of course obtained by putting $z=r\cos\theta$ and $x^2+y^2=r^2\sin^2\theta$.

(*w*) If now, we put in these, $\cos\theta=\dfrac{xx'+yy'+zz'}{rr'}$, introducing again, as in (*u*) above, the notation (x, y, z), (x', y', z'), we arrive at expansions of Q_i in the terms indicated in (53).

Expansions of the biaxal harmonic, of order *i*.

(*x*) Some of the most useful expansions of Q_i are very readily obtained by introducing, as before, the imaginary co-ordinates (ξ, η) instead of (x, y), according to equations (26) of (*j*), and similarly, (ξ', η') instead of (x', y'). Thus we have

$$D^2=(\xi-\xi')(\eta-\eta')+(z-z')^2.$$

Hence, as above,

$$\frac{1}{[(\xi-\xi')(\eta-\eta')+(z-z')^2]^{\frac{1}{2}}}=\Sigma\Sigma\Sigma\frac{(-1)^{j+k+l}\xi'^j\eta'^k z'^l}{1.2\ldots j.1.2\ldots k.1.2\ldots l}\frac{d^{j+k+l}}{d\xi^j d\eta^k dz^l}\frac{1}{(\xi\eta+z^2)^{\frac{1}{2}}}$$

Hence

$$(rr')^i Q_i=r^{2i+1}\overset{(j+k+l=i)}{\Sigma\;\Sigma\;\Sigma}\frac{(-1)^{j+k+l}\xi'^j\eta'^k z'^l}{1.2\ldots j.1.2\ldots k.1.2\ldots l}\frac{d^{j+k+l}}{d\xi^j d\eta^k dz^l}\frac{1}{(\xi\eta+z^2)^{\frac{1}{2}}} \quad (56).$$

Of course we have in this case

$$r^2=\xi\eta+z^2,\ r'^2=\xi'\eta'+z'^2$$

and

$$\cos\theta=\frac{\xi\eta'+\xi'\eta+zz'}{rr'}.$$

And, just as above, we see that this expression, obviously a homogeneous function of ξ', η', z', of degree i, and also of η, ξ, z, involves these two systems of variables symmetrically.

Now, as we have seen above, all the ith differential coefficients of $\dfrac{1}{r}$ are reducible to the $2i+1$ independent forms

$$\left(\frac{d}{dz}\right)^i\frac{1}{r},\quad \begin{matrix}\left(\dfrac{d}{dz}\right)^{i-1}\dfrac{d}{d\eta}\dfrac{1}{r}, & \left(\dfrac{d}{dz}\right)^{i-2}\left(\dfrac{d}{d\eta}\right)^2\dfrac{1}{r}, & \ldots & \left(\dfrac{d}{d\eta}\right)^i\dfrac{1}{r} \\ \left(\dfrac{d}{dz}\right)^{i-1}\dfrac{d}{d\xi}\dfrac{1}{r}, & \left(\dfrac{d}{dz}\right)^{i-2}\left(\dfrac{d}{d\xi}\right)^2\dfrac{1}{r}, & \ldots & \left(\dfrac{d}{d\xi}\right)^i\dfrac{1}{r}.\end{matrix}$$

Hence $r^i Q_i$, viewed as a function of z, ξ, η, is expressed by these $2i+1$ terms, each with a coefficient involving z', ξ', η'. And because of the symmetry we see that this coefficient must be the same function of z', η', ξ', into some factor involving none of either set of variables (z, ξ, η), (z', η', ξ'). Also, by the symmetry with reference to ξ, η' and η, ξ', we see that the numerical factor must be the same for the terms similarly involving ξ, η' on the one hand, and η, ξ' on the other. Hence,

Biaxal harmonic expressed in symmetrical series of differential coefficients.

$$
\left.
\begin{aligned}
Q_i = (rr')^{i+1}\Big[E_0\Big(\frac{d}{dz'}\Big)^i\frac{1}{r'}\Big(\frac{d}{dz}\Big)^i\frac{1}{r} \\
+\sum_{s=1}^{s=i}E_i^{(s)}\Big\{\frac{d^i}{dz'^{i-s}d\xi'^s}\frac{1}{r'}\frac{d^i}{dz^{i-s}d\eta^s}\frac{1}{r}+\frac{d^i}{dz'^{i-s}d\eta'^s}\frac{1}{r'}\frac{d^i}{dz^{i-s}d\eta^s}\frac{1}{r}\Big\} \\
\text{where} \\
E_i^{(s)}=\frac{1}{1.2\ldots s\;1.2\ldots(i-s).\frac{1}{2}.\frac{3}{2}\ldots(s-\frac{1}{2}).(2s+1)(2s+2)\ldots(i+s)}
\end{aligned}
\right\}(57).
$$

The value of $E_i^{(s)}$ is obtained thus :—Comparing the coefficient of the term $(zz')^{i-s}(\xi\eta')^s$ in the numerator of the expression which (56) becomes when the differential coefficient is expanded, with the coefficient of the same term in (57), we have

$$\frac{(-)^i M}{1.2\ldots(i-s).1.2\ldots s}=E_i^{(s)}M^2 \qquad (58)$$

where M denotes the coefficient of $z^{i-s}\xi^s$ in $r^{2i+1}\dfrac{d_i}{dz^{i-s}d\eta^s}\dfrac{1}{r}$, or, which is the same, the coefficient of $z'^{i-s}\eta'^s$ in $r'^{2i+1}\dfrac{d^i}{dz'^{i-s}d\xi'^s}\dfrac{1}{r}$.

From this, with the value (42) for M, we find $E_i^{(s)}$ as above.

(y) We are now ready to reduce the expansion of Q_n to a real trigonometrical form. First, we have, by (33),

$$(\xi\eta')^s+(\xi'\eta)^s=2\,(rr'\sin\theta\sin\theta')^s\cos s(\phi-\phi') \qquad (59),$$

Let now

$$\vartheta_i^{(s)}=\sin^s\theta\Big[\cos^{i-s}\theta-\frac{(i-s)\,(i-s-1)}{4(s+1)}\cos^{i-s-2}\theta\sin^2\theta$$

$$+\frac{(i-s)(i-s-1)\,(i-s-2)\,(i-s-3)}{4^2(s+1)\,(s+2)}\cos^{i-s-4}\theta\sin^4\theta-\text{etc.}\Big] \;(60);$$

(that is to say, $C\vartheta^{(s)}=\Theta_i^{(s)}$, in accordance with the previous notation,) and let the corresponding notation with accents apply to θ'. Then, by the aid of (57), (58), and (59), we have

$$=2\sum_{s=0}^{s=i}\frac{\frac{1}{2}.\frac{3}{2}\ldots(i-\frac{1}{2})}{1.2\ldots i}.\frac{(2s+1)\,(2s+2)\ldots(2s+i-s)}{1.2\ldots(i-s)}\cos s(\phi-\phi')\vartheta_i^{(s)}\vartheta_i'^{(s)} \quad (61),$$

Trigonometrical expansion of biaxal surface harmonic.

of which, however, the first term (that for which $s=0$) must be halved.

(z) As a supplement to the fundamental proposition $\iint S_i S_{i'} d\varpi=0$, (17) of ($g$), and the corresponding propositions, (43) and (44), regarding elementary terms of harmonics, we are now prepared to evaluate $\iint S_i^2 d\varpi$.

First, using the general expression (37) investigated above for

Fundamental definite integral investigated.

S_i, and modifying the arbitrary constants to suit our present notation, we have

$$S_i=\sum_{s=0}^{s=i} A_s \cos(s\phi+a_s)\vartheta_i^{(s)} \qquad (62).$$

Hence

$$\iint S_i^2\, d\varpi=\pi\sum_0^i A_s^2 \int_0^\pi (\vartheta_i^{(s)})^2 \sin\theta d\theta \qquad (63).$$

To evaluate the definite integral in the second member, we have only to apply the general theorem (52) for expansion, in terms of surface harmonics, to the particular case in which the arbitrary function $F(E)$ is itself the harmonic, $\cos s\phi\, \vartheta_i^{(s)}$. Thus, remembering (16), we have

$$\cos s\phi\, \vartheta_i^{(s)}=\frac{2i+1}{4\pi}\int_0^\pi \sin\theta' d\theta' \int_0^{2\pi} d\phi' \cos s\phi' \vartheta_i'^{(s)} Q_i \qquad (64)$$

Using here for Q_i its trigonometrical expansion just investigated. and performing the integration for ϕ' between the stated limits, we find that $\cos s\phi\, \vartheta_i^{(s)}$ may be divided out, and (omitting the accents in the residual definite integral) we conclude,

$$\int_0^\pi \sin\theta (\vartheta_i^{(s)})^2 d\theta=\frac{2}{2i+1}\cdot\frac{1.2\ldots s}{\frac{1}{2}.\frac{3}{2}\ldots(s-\frac{1}{2})}\cdot\frac{1.2\ldots(i-s)}{(2s+1)(2s+2)\ldots 2s+i-s)} \qquad (65)$$

This holds without exception for the case $s=0$, in which the second member becomes $\frac{2}{2i+1}$. It is convenient here to recal equation (44), which, when expressed in terms of $\vartheta_i^{(s)}$ instead of $\Theta_{(m,n)}$ becomes

$$\int_0^\pi \sin\theta\, \vartheta_i^{(s)} \vartheta_{i'}^{(s)} d\theta=0 \qquad (66).$$

where i and i' must be different. The properties expressed by these two equations, (65) and (66), may be verified by direct integration, from the explicit expression (60) for $\vartheta_i^{(s)}$; and to do so will be a good, although possibly not a very easy, analytical exercise.

CHAPTER II.

DYNAMICAL LAWS AND PRINCIPLES.

205. In the preceding chapter we considered as a subject of pure geometry the motion of points, lines, surfaces, and volumes, whether taking place with or without change of dimensions and form; and the results we there arrived at are of course altogether independent of the idea of *matter*, and of the *forces* which matter exerts. We have heretofore assumed the *existence* merely of motion, distortion, etc.; we now come to the consideration, not of how we *might* consider such motion, etc., to be produced, but of the *actual* causes which in the material world *do* produce them. The results of the present chapter must therefore be considered to be due to actual experience, in the shape either of observation or experiment. How such experience is to be conducted will form the subject of a subsequent chapter. Ideas of matter and force introduced.

206. We cannot do better, at all events in commencing, than follow Newton somewhat closely. Indeed the introduction to the *Principia* contains in a most lucid form the general foundations of Dynamics. The *Definitiones* and *Axiomata sive Leges Motûs*, there laid down, require only a few amplifications and additional illustrations, suggested by subsequent developments, to suit them to the present state of science, and to make a much better introduction to dynamics than we find in even some of the best modern treatises.

207. We cannot, of course, give a definition of *Matter* which will satisfy the metaphysician, but the naturalist may be content to know matter as *that which can be perceived by the senses*, or as *that which can be acted upon by, or can exert, force.* The latter, and indeed the former also, of these definitions involves the idea of *Force*, which, in point of fact, is a direct object of Matter. Force.

Force. sense; probably of all our senses, and certainly of the "muscular sense." To our chapter on Properties of Matter we must refer for further discussion of the question, *What is matter?*

Mass. Density. **208.** The *Quantity of Matter* in a body, or, as we now call it, the *Mass* of a body, is proportional, according to Newton, to the *Volume* and the *Density* conjointly. In reality, the definition gives us the meaning of density rather than of mass; for it shows us that if twice the original quantity of matter, air for example, be forced into a vessel of given capacity, the density will be doubled, and so on. But it also shows us that, of matter of uniform density, the mass or quantity is proportional to the volume or space it occupies.

Let M be the mass, ρ the density, and V the volume, of a homogeneous body. Then

$$M = V\rho;$$

if we so take our units that unit of mass is that of unit volume of a body of unit density.

If the density vary from point to point of the body, we have evidently, by the above formula and the elementary notation of the integral calculus,

$$M = \iiint \rho \, dxdydz,$$

where ρ is supposed to be a known function of x, y, z, and the integration extends to the whole space occupied by the matter of the body whether this be continuous or not.

It is worthy of particular notice that, in this definition, Newton says, if there be anything which *freely* pervades the interstices of all bodies, this is *not* taken account of in estimating their Mass or Density.

Measurement of mass. **209.** Newton further states, that a practical measure of the mass of a body is its *Weight*. His experiments on pendulums, by which he establishes this most important remark, will be described later, in our chapter on Properties of Matter.

As will be presently explained, the unit mass most convenient for British measurements is an imperial pound of matter.

Momentum. **210.** The *Quantity of Motion*, or the *Momentum*, of a rigid body moving without rotation is proportional to its mass and velocity conjointly. The whole motion is the sum of the motions of its several parts. Thus a doubled mass, or a doubled velocity, would correspond to a double quantity of motion; and so on.

Hence, if we take as unit of momentum the momentum of a unit of matter moving with unit velocity, the momentum of a mass M moving with velocity v is Mv. Momentum.

211. *Change of Quantity of Motion*, or *Change of Momentum*, is proportional to the mass moving and the change of its velocity conjointly. Change of momentum

Change of velocity is to be understood in the general sense of § 27. Thus, in the figure of that section, if a velocity represented by OA be changed to another represented by OC, the change of velocity is represented in magnitude and direction by AC.

212. *Rate of Change of Momentum*, or *Acceleration of Momentum*, is proportional to the mass moving and the acceleration of its velocity conjointly. Thus (§ 35, *b*) the rate of change of momentum of a falling body is constant, and in the vertical direction. Again (§ 35, *a*) the rate of change of momentum of a mass M, describing a circle of radius R, with uniform velocity V, is $\frac{MV^2}{R}$, and is directed to the centre of the circle; that is to say, it is a change of direction, not a change of speed, of the motion. Acceleration of momentum.

Generally (§ 29), for a body of mass M moving anyhow in space there is change of momentum, at the rate, $M\frac{d^2s}{dt^2}$ in the direction of motion, and $M\frac{v^2}{\rho}$ towards the centre of curvature of the path; and, if we choose, we may exhibit the whole acceleration of momentum by its three rectangular components $M\frac{d^2x}{dt^2}$, $M\frac{d^2y}{dt^2}$, $M\frac{d^2z}{dt^2}$, or, according to the Newtonian notation, $M\ddot{x}$, $M\ddot{y}$, $M\ddot{z}$.

213. The *Vis Viva*, or *Kinetic Energy*, of a moving body is proportional to the mass and the square of the velocity, conjointly. If we adopt the same units of mass and velocity as before, there is particular advantage in defining kinetic energy as *half* the product of the mass and the square of its velocity. Kinetic energy.

214. *Rate of Change of Kinetic Energy* (when defined as above) is the product of the velocity into the component of acceleration of momentum in the direction of motion.

For $$\frac{d}{dt}\left(\frac{Mv^2}{2}\right) = vM\frac{dv}{dt}.$$

Particle and point.

215. It is to be observed that, in what precedes, with the exception of the definition of mass, we have taken no account of the dimensions of the moving body. This is of no consequence so long as it does not rotate, and so long as its parts preserve the same relative positions amongst one another. In this case we may suppose the whole of the matter in it to be condensed in one point or particle. We thus speak of a *material particle*, as distinguished from a *geometrical point*. If the body rotate, or if its parts change their relative positions, then we cannot choose any one point by whose motions alone we may determine those of the other points. In such cases the momentum and change of momentum of the whole body in any direction are, the sums of the momenta, and of the changes of momentum, of its parts, in these directions; while the kinetic energy of the whole, being non-directional, is simply the sum of the kinetic energies of the several parts or particles.

Inertia.

216. Matter has an innate power of resisting external influences, so that every body, as far as it can, remains at rest, or moves uniformly in a straight line.

This, the *Inertia* of matter, is proportional to the quantity of matter in the body. And it follows that some *cause* is requisite to disturb a body's uniformity of motion, or to change its direction from the natural rectilinear path.

Force

217. *Impressed Force*, or *Force*, simply, is any cause which tends to alter a body's natural state of rest, or of uniform motion in a straight line.

Force is wholly expended in the *Action* it produces; and the body, after the force ceases to act, retains by its inertia the direction of motion, and the velocity which were given to it. Force may be of divers kinds, as pressure, or gravity, or friction, or any of the attractive or repulsive actions of electricity, magnetism, etc.

Specification of a force.

218. The three elements specifying a force, or the three elements which must be known, before a clear notion of the force under consideration can be formed, are, its place of application, its direction, and its magnitude.

Place of application.

(*a*.) The place of application of a force. The first case to be considered is that in which the place of application is a point. It has been shown already in what sense the term "point" is to be taken, and, therefore, in what way a force may be imagined as acting at a point. In reality, however, the place of application of a force is always either a surface or a space of three dimensions occupied by matter. The point of the finest needle, or the edge of the sharpest knife, is still a surface, and acts as such on the bodies to which it may be applied. Even the most rigid substances, when brought together, do not touch at a point merely, but mould each other so as to produce a surface of application. On the other hand, gravity is a force of which the place of application is the whole matter of the body whose weight is considered; and the smallest particle of matter that has weight occupies some finite portion of space. Thus it is to be remarked, that there are two kinds of force, distinguishable by their place of application — force, whose place of application is a surface, and force, whose place of application is a solid. When a heavy body rests on the ground, or on a table, force of the second character, acting downwards, is balanced by force of the first character acting upwards. Place of application.

(*b*.) The second element in the specification of a force is its direction. The direction of a force is the line in which it acts. If the place of application of a force be regarded as a point, a line through that point, in the direction in which the force tends to move the body, is the direction of the force. In the case of a force distributed over a surface, it is frequently possible and convenient to assume a single point and a single line, such that a certain force acting at that point in that line would produce the same effect as is really produced. Direction.

(*c*.) The third element in the specification of a force is its magnitude. This involves a consideration of the method followed in dynamics for measuring forces. Before measuring anything, it is necessary to have a unit of measurement, or a standard to which to refer, and a principle of numerical specification, or a mode of referring to the standard. These will be supplied presently. See also § 258, below. Magnitude.

219. The *Accelerative Effect of a Force* is proportional to the velocity which it produces in a given time, and is measured by that which is, or would be, produced in unit of time; in other words, the rate of change of velocity which it produces. This is simply what we have already defined as acceleration, § 28. Accelerative effect.

Measure of force.

220. The *Measure of a Force* is the quantity of motion which it produces in unit of time.

The reader, who has been accustomed to speak of a force of so many pounds, or so many tons, may be reasonably startled when he finds that Newton gives no countenance to such expressions. The method is not correct unless it be specified at what part of the earth's surface the pound, or other definite quantity of matter named, is to be weighed; for the weight of a given quantity of matter differs in different latitudes. The clumsiness of this system is in great contrast to the clear and simple accuracy of the absolute method as stated above, to which we shall uniformly adhere, except when we wish, in describing results, to state forces in terms of the *vernacular* of engineers in any locality. Thus, let W be the mass of a body in pounds, g the velocity it would acquire in falling for a second under the influence of its weight, or the earth's attraction, and P the force of gravity upon it, measured in kinetic or absolute units. We thus have

$$P = Wg.$$

Inconvenient system of modern treatises.

221. According to the common system followed in modern mathematical treatises on dynamics, the unit of mass is g times the mass of the standard or unit weight. This definition, giving a varying and a very unnatural unit of mass, is exceedingly inconvenient. In reality, standards of weight are *masses*, not *forces*. They are employed primarily in commerce for the purpose of measuring out a definite *quantity* of matter; not an amount of matter which shall be attracted by the earth with a given force.

Standards of weight are masses, and not primarily intended for measurement of force.

Whereas a merchant, with a balance and a set of standard weights, would give his customers the same quantity of the same kind of matter however the earth's attraction might vary, depending as he does upon *masses* for his measurement; another, using a spring balance, would defraud his customers in high latitudes, and himself in low, if his instrument (which depends on forces and not on masses) were correctly adjusted in London.

It is a secondary application of our standards of weight to employ them for the measurement of *forces*, such as steam pressures, muscular power, etc. In all cases where great accuracy is required, the results obtained by such a method have to be reduced to what they would have been if the measurements of force had been made by means of a perfect spring-balance, graduated so as to indicate the forces of gravity on the standard weights in some conventional locality. **Standards of weight are *masses*, and not primarily intended for measurement of force.**

It is therefore very much simpler and better to take the imperial pound, or other national or international standard weight, as, for instance, the gramme (see the chapter on Measures and Instruments), as the unit of mass, and to derive from it, according to Newton's definition above, the unit of force. This is the method which Gauss has adopted in his great improvement of the system of measurement of forces; and by it we have, and by it only can we have, an *absolute unit of force*.

222. The formula, deduced by Clairault from observation, and a certain theory regarding the figure and density of the earth, may be employed to calculate the most probable value of the apparent force of gravity, being the resultant of true gravitation and centrifugal force, in any locality where no pendulum observation of sufficient accuracy has been made. This formula, with the two coefficients which it involves, corrected according to the best modern pendulum observations (*Airy, Encyc. Metr., Figure of the Earth*), is as follows:— **Clairault's formula for the amount of gravity.**

Let G be the apparent force of gravity on a unit mass at the equator, and g that in any latitude λ; then

$$g = G(1 + \cdot 005133 \sin^2 \lambda).$$

The value of G, in terms of the absolute unit, to be explained immediately, is

$$32 \cdot 088.$$

According to this formula, therefore, polar gravity will be

$$g = 32 \cdot 088 \times 1 \cdot 005133 = 32 \cdot 2527.$$

223. Gravity having failed to furnish a definite standard, independent of locality, recourse must be had to something else. The principle of measurement indicated as above by Newton, but first introduced practically by Gauss, furnishes us with what we want. According to this principle, the unit force is that force which, acting on a national standard unit of matter during the unit of time, generates the unit of velocity. **Gauss's absolute unit.**

Gauss's absolute unit.

This is known as Gauss's absolute unit; absolute, because it furnishes a standard force independent of the differing amounts of gravity at different localities.

224. The absolute unit depends on the unit of matter, the unit of time, and the unit of velocity; and as the unit of velocity depends on the unit of space and the unit of time, there is, in the definition, a single reference to mass and space, but a *double* reference to time; and this is a point that must be particularly attended to.

225. The unit of mass may be the British imperial pound; the unit of space the British standard foot; and the unit of time the mean solar second.

British absolute unit.

We accordingly define the British absolute unit force as "the force which, acting on one pound of matter for one second, generates a velocity of one foot per second."

Comparison with gravity.

226. To render this standard intelligible, all that has to be done is to find how many absolute units will produce, in any particular locality, the same effect as the force of gravity on a given mass. The way to do this is to measure the effect of gravity in producing acceleration on a body unresisted in any way. The most accurate method is indirect, by means of the pendulum. The result of pendulum experiments made at Leith Fort, by Captain Kater, is, that the velocity acquired by a body falling unresisted for one second is at that place 32·207 feet per second. The preceding formula gives exactly 32·2, for the latitude, 55° 33′, which is approximately that of Edinburgh. The variation in the force of gravity for one degree of difference of latitude about the latitude of Edinburgh is only ·0000832 of its own amount. It is nearly the same, though somewhat more, for every degree of latitude southwards, as far as the southern limits of the British Isles. On the other hand, the variation per degree would be sensibly less, as far north as the Orkney and Shetland Isles. Hence the augmentation of gravity per degree from south to north throughout the British Isles is at most about $\frac{1}{12000}$ of its whole amount in any locality. The average for the whole of Great Britain and Ireland differs certainly but little from 32·2. Our present application is, that the force of gravity at Edinburgh is 32·2 times the force which, acting on a pound for a second, would generate a velocity of one foot per second; in other words, 32·2 is the number of absolute units which measures the weight of a pound in this latitude. Thus, speaking very roughly, the British absolute unit of force is equal to the weight of about half an ounce. Comparison with gravity.

227. Forces (since they involve only direction and magnitude) may be represented, as velocities are, by straight lines in their directions, and of lengths proportional to their magnitudes, respectively.

Also the laws of composition and resolution of any number of forces acting at the same point, are, as we shall show later (§ 255), the same as those which we have already proved to hold for velocities; so that with the substitution of force for velocity, §§ 26, 27, are still true.

228. The *Component* of a force in any direction, sometimes called the *Effective Component* in that direction, is therefore found by multiplying the magnitude of the force by the cosine of the angle between the directions of the force and the component. The remaining component in this case is perpendicular to the other. Effective component of a force.

It is very generally convenient to resolve forces into components parallel to three lines at right angles to each other; each such resolution being effected by multiplying by the cosine of the angle concerned.

229. The point whose distances from three planes at right angles to one another are respectively equal to the mean distances of any group of points from these planes, is at a distance from any plane whatever, equal to the mean distance of the group from the same plane. Hence of course, if it is in motion, its velocity perpendicular to that plane is the mean of the velocities of the several points, in the same direction. Theorem.

Let (x_1, y_1, z_1), etc., be the points of the group in number i; and $\bar{x}, \bar{y}, \bar{z}$ be the co-ordinates of a point at distances respectively equal to their mean distances from the planes of reference; that is to say, let

$$\bar{x}=\frac{x_1+x_2+\text{etc.}}{i},\quad \bar{y}=\frac{y_1+y_2+\text{etc.}}{i},\quad \bar{z}=\frac{z_1+z_2+\text{etc.}}{i}.$$

Thus, if p_1, p_2, etc., and p, denote the distances of the points in question from any plane at a distance a from the origin of co-ordinates, perpendicular to the direction (l, m, n), the sum of a

Theorem.

and p_1 will make up the projection of the broken line x_1, y_1, z_1 on (l, m, n), and therefore

$$p_1 = lx_1 + my_1 + nz_1 - a, \text{ etc.};$$

and similarly, $p = l\bar{x} + m\bar{y} + n\bar{z} - a.$

Substituting in this last the expressions for $\bar{x}, \bar{y}, \bar{z}$, we find

$$p = \frac{p_1 + p_2 + \text{etc.}}{i},$$

which is the theorem to be proved. Hence, of course,

$$\frac{dp}{dt} = \frac{1}{i}\left(\frac{dp_1}{dt} + \frac{dp_2}{dt} + \text{etc.}\right)$$

Centre of Inertia.

230. The *Centre of Inertia* of a system of equal material points (whether connected with one another or not) is the point whose distance is equal to their average distance from any plane whatever (§ 229).

A group of material points of unequal masses may always be imagined as composed of a greater number of equal material points, because we may imagine the given material points divided into different numbers of very small parts. In any case in which the magnitudes of the given masses are incommensurable, we may approach as near as we please to a rigorous fulfilment of the preceding statement, by making the parts into which we divide them sufficiently small.

On this understanding the preceding definition may be applied to define the centre of inertia of a system of material points, whether given equal or not. The result is equivalent to this:—

The centre of inertia of any system of material points whatever (whether rigidly connected with one another, or connected in any way, or quite detached), is a point whose distance from any plane is equal to the sum of the products of each mass into its distance from the same plane divided by the sum of the masses.

We also see, from the proposition stated above, that a point whose distance from three rectangular planes fulfils this condition, must fulfil this condition also for every other plane.

The co-ordinates of the centre of inertia, of masses w_1, w_2, etc., at points (x_1, y_1, z_1), (x_2, y_2, z_2), etc., are given by the following formulæ:—

$$\bar{x} = \frac{w_1 x_1 + w_2 x_2 + \text{etc.}}{w_1 + w_2 + \text{etc.}} = \frac{\Sigma wx}{\Sigma w}, \quad \bar{y} = \frac{\Sigma wy}{\Sigma w}, \quad \bar{z} = \frac{\Sigma wz}{\Sigma w}.$$

These formulæ are perfectly general, and can easily be put into the particular shape required for any given case. Thus, suppose that, instead of a set of detached massive points, we have a continuous distribution of matter through certain definite portions of space; the density at x, y, z being ρ, the elementary principles of the integral calculus give us at once

Centre of Inertia.

$$\bar{x}=\frac{\iiint\rho x dx dy dz}{\iiint\rho dx dy dz}, \text{ etc.},$$

where the integrals extend through all the space occupied by the mass in question, in which ρ has a value different from zero.

The Centre of Inertia or Mass is thus a perfectly definite point in every body, or group of bodies. The term *Centre of Gravity* is often very inconveniently used for it. The theory of the resultant action of gravity which will be given under Abstract Dynamics shows that, except in a definite class of distributions of matter, there is no one fixed point which can properly be called the Centre of Gravity of a rigid body. In ordinary cases of terrestrial gravitation, however, an approximate solution is available, according to which, in common parlance, the term "*Centre of Gravity*" may be used as equivalent to *Centre of Inertia*; but it must be carefully remembered that the fundamental ideas involved in the two definitions are essentially different.

The second proposition in § 229 may now evidently be stated thus:—The sum of the momenta of the parts of the system in any direction is equal to the momentum in the same direction of a mass equal to the sum of the masses moving with a velocity equal to the velocity of the centre of inertia.

231. The *Moment* of any physical agency is the numerical measure of its importance. Thus, the moment of a force round a point or round a line, signifies the measure of its importance as regards producing or balancing rotation round that point or round that line.

Moment.

232. The *Moment* of a force about a point is defined as the product of the force into its perpendicular distance from the point. It is numerically double the area of the triangle whose vertex is the point, and whose base is a line representing the force in magnitude and direction. It is often convenient to represent it by a line numerically equal to it, drawn through the vertex of the triangle perpendicular to its plane, through the front of a watch held in the plane with its centre at the point, and facing

Moment of a force about a point.

Moment of a force about an axis. so that the force tends to turn round this point in a direction opposite to the hands. The moment of a force round any axis is the moment of its component in any plane perpendicular to the axis, round the point in which the plane is cut by the axis. Here we imagine the force resolved into two components, one parallel to the axis, which is ineffective so far as rotation round the axis is concerned; the other perpendicular to the axis (that is to say, having its line in any plane perpendicular to the axis). This latter component may be called the effective component of the force, with reference to rotation round the axis. And its moment round the axis may be defined as its moment round the nearest point of the axis, which is equivalent to the preceding definition. It is clear that the moment of a force round any axis, is equal to the area of the projection on any plane perpendicular to the axis, of the figure representing its moment round any point of the axis.

Digression on projection of areas. **233.** The projection of an area, plane or curved, on any plane, is the area included in the projection of its bounding line.

If we imagine an area divided into any number of parts, the projections of these parts on any plane make up the projection of the whole. But in this statement it must be understood that the areas of partial projections are to be reckoned as positive if particular sides, which, for brevity, we may call the outside of the projected area and the front of the plane of projection, face the same way, and negative if they face oppositely.

Of course if the projected surface, or any part of it, be a plane area at right angles to the plane of projection, the projection vanishes. The projections of any two shells having a common edge, on any plane, are equal. The projection of a closed surface (or a shell with evanescent edge), on any plane, is nothing.

Equal areas in one plane, or in parallel planes, have equal projections on any plane, whatever may be their figures.

Hence the projection of any plane figure, or of any shell, edged by a plane figure, on another plane, is equal to its area, multiplied by the cosine of the angle at which its plane is inclined to the plane of projection. This angle is acute or obtuse, according as the outside of the projected area, and the front of plane of projection, face on the whole towards the same parts, or oppositely. Hence lines representing, as above described, moments about a point in different planes, are to be compounded as forces are.—See an analogous theorem in § 96.

234. A *Couple* is a pair of equal forces acting in dissimilar directions in parallel lines. The *Moment* of a couple is the sum of the moments of its forces about any point in their plane, and is therefore equal to the product of either force into the shortest distance between their directions. This distance is called the *Arm* of the couple. Couple.

The *Axis of a Couple* is a line drawn from any chosen point of reference perpendicular to the plane of the couple, of such magnitude and in such direction as to represent the magnitude of the moment, and to indicate the direction in which the couple tends to turn. The most convenient rule for fulfilling the latter condition is this:—Hold a watch with its centre at the point of reference, and with its plane parallel to the plane of the couple. Then, according as the motion of the hands is contrary to, or along with the direction in which the couple tends to turn, draw the axis of the couple through the face or through the back of the watch. It will be found that a couple is completely represented by its axis, and that couples are to be resolved and compounded by the same geometrical constructions performed with reference to their axes as forces or velocities, with reference to the lines directly representing them.

235. If we substitute, for the force in § 232, a velocity, we have the moment of a velocity about a point; and by introducing the mass of the moving body as a factor, we have an important element of dynamical science, the *Moment of Momentum*. The laws of composition and resolution are the same as those already explained; but for the sake of some simple applications we give an elementary investigation. The moment of a rectilineal motion is the product of its length into the distance of its line from the point. Moment of velocity. Moment of momentum. Moment of a rectilineal displacement.

The moment of the resultant velocity of a particle about any point in the plane of the components is equal to the algebraic sum of the moments of the components, the proper sign of each moment being determined as above, § 233. The same is of course true of moments of displacements, of moments of forces, and of moments of momentum.

First, consider two component motions, AB and AC, and let

For two forces, motions, velocities, or moments, in one plane, the sum of their moments proved equal to the moment of their resultant round any point in that plane.

AD be their resultant (§ 27). Their half moments round the point O are respectively the areas OAB, OCA. Now OCA, together with half the area of the parallelogram $CABD$, is equal to OBD. Hence the sum of the two half moments, together with half the area of the parallelogram, is equal to AOB together with BOD, that is to say, to the area of the whole figure $OABD$. But ABD, a part of this figure, is equal to half the area of the parallelogram; and therefore the remainder, OAD, is equal to the sum of the two half moments. But OAD is half the moment of the resultant velocity round the point O. Hence the moment of the resultant is equal to the sum of the moments of the two components.

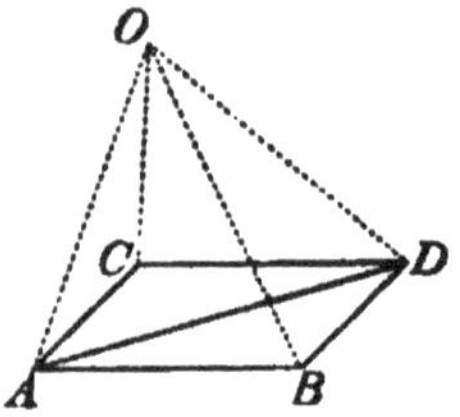

Any number of moments in one plane compounded by addition.

If there are any number of component rectilineal motions in one plane, we may compound them in order, any two taken together first, then a third, and so on; and it follows that the sum of their moments is equal to the moment of their resultant. It follows, of course, that the sum of the moments of any number of component velocities, all in one plane, into which the velocity of any point may be resolved, is equal to the moment of their resultant, round any point in their plane. It follows also, that if velocities, in different directions all in one plane, be successively given to a moving point, so that at any time its velocity is their resultant, the moment of its velocity at any time is the sum of the moments of all the velocities which have been successively given to it.

Cor.—If one of the components always passes through the point, its moment vanishes. This is the case of a motion in which the acceleration is directed to a fixed point, and we thus reproduce the theorem of § 36, *a*, that in this case the areas described by the radius-vector are proportional to the times; for, as we have seen, the moment of velocity is double the area traced out by the radius-vector in unit of time.

Moment round an axis.

236. The moment of the velocity of a point round any axis is the moment of the velocity of its projection on a plane perpendicular to the axis, round the point in which the plane is cut by the axis.

The moment of the whole motion of a point during any time, round any axis, is twice the area described in that time by the radius-vector of its projection on a plane perpendicular to that axis. Moment of a whole motion, round an axis.

If we consider the conical area traced by the radius-vector drawn from any fixed point to a moving point whose motion is not confined to one plane, we see that the projection of this area on any plane through the fixed point is half of what we have just defined as the moment of the whole motion round an axis perpendicular to it through the fixed point. Of all these planes, there is one on which the projection of the area is greater than on any other; and the projection of the conical area on any plane perpendicular to this plane, is equal to nothing, the proper interpretation of positive and negative projections being used.

If any number of moving points are given, we may similarly consider the conical surface described by the radius-vector of each drawn from one fixed point. The same statement applies to the projection of the many-sheeted conical surface, thus presented. The resultant axis of the whole motion in any finite time, round the fixed point of the motions of all the moving points, is a line through the fixed point perpendicular to the plane on which the area of the whole projection is greater than on any other plane; and the moment of the whole motion round the resultant axis, is twice the area of this projection. Resultant axis.

The resultant axis and moment of velocity, of any number of moving points, relatively to any fixed point, are respectively the resultant axis of the whole motion during an infinitely short time, and its moment, divided by the time.

The moment of the whole motion round any axis, of the the motion of any number of points during any time, is equal to the moment of the whole motion round the resultant axis through any point of the former axis, multiplied into the cosine of the angle between the two axes.

The resultant axis, relatively to any fixed point, of the whole motion of any number of moving points, and the moment of the whole motion round it, are deduced by the same elementary constructions from the resultant axes and moments of the individual points, or partial groups of points of the system, as the direction and magnitude of a resultant displacement, are

Moment of momentum. deduced from any given lines and magnitudes of component displacements.

Corresponding statements apply, of course, to the moments of velocity and of momentum.

Virtual velocity. **237.** If the point of application of a force be displaced through a small space, the resolved part of the displacement in the direction of the force has been called its *Virtual Velocity*. This is positive or negative according as the virtual velocity is in the same, or in the opposite, direction to that of the force.

The product of the force, into the virtual velocity of its point of application, has been called the *Virtual Moment* of the force. These terms we have introduced since they stand in the history and developments of the science; but, as we shall show further on, they are inferior substitutes for a far more useful set of ideas clearly laid down by Newton.

Work. **238.** A force is said to *do work* if its place of application has a positive component motion in its direction; and the work done by it is measured by the product of its amount into this component motion.

Thus, in lifting coals from a pit, the amount of work done is proportional to the weight of the coals lifted; that is, to the force overcome in raising them; and also to the height through which they are raised. The unit for the measurement of work adopted in practice by British engineers, is that required to overcome a force equal to the weight of a pound through the space of a foot; and is called a *Foot-Pound*.

Practical unit.

Scientific unit. In purely scientific measurements, the unit of work is not the foot-pound, but the kinetic unit force (§ 225) acting through unit of space. Thus, for example, as we shall show further on, this unit is adopted in measuring the work done by an electric current, the units for electric and magnetic measurements being founded upon the kinetic unit force.

If the weight be raised obliquely, as, for instance, along a smooth inclined plane, the space through which the force has to be overcome is increased in the ratio of the length to the height of the plane; but the force to be overcome is not the whole weight, but only the resolved part of the weight parallel to the plane; and this is less than the weight in the ratio of the height of the plane to its length. By multiplying these two expressions together, we find, as we might expect, that the amount of work required is unchanged by the substitution of the oblique for the vertical path. Work of a force.

239. Generally, for any force, the work done during an indefinitely small displacement of the point of application is the virtual moment of the force (§ 237), or is the product of the resolved part of the force in the direction of the displacement into the displacement.

From this it appears, that if the motion of the point of application be always perpendicular to the direction in which a force acts, such a force does no work. Thus the mutual normal pressure between a fixed and moving body, as the tension of the cord to which a pendulum bob is attached, or the attraction of the sun on a planet if the planet describe a circle with the sun in the centre, is a case in which no work is done by the force.

240. The work done by a force, or by a couple, upon a body turning about an axis, is the product of the moment of either into the angle (in circular measure) through which the body acted on turns, if the moment remains the same in all positions of the body. If the moment be variable, the above assertion is only true for indefinitely small displacements, but may be made accurate by employing the proper *average* moment of the force or of the couple. The proof is obvious. Work of a couple.

If Q be the moment of the force or couple for a position of the body given by the angle θ, $Q(\theta_1 - \theta_0)$ if Q is constant, or $\int_{\theta_0}^{\theta_1} Q d\theta = q(\theta_1 - \theta_0)$ where q is the proper average value of Q when variable, is the work done by the couple during the rotation from θ_0 to θ_1.

241. Work done on a body by a force is always shown by a corresponding increase of vis viva, or kinetic energy, if no other forces act on the body which can do work or have work done against them. If work be done against any forces, the increase of kinetic energy is less than in the former case by the amount of work so done. In virtue of this, however, the body possesses an equivalent in the form of *Potential Energy* (§ 273), if its physical conditions are such that these forces will act equally, and in the same directions, if the motion of the system is Transformation of work. Potential energy.

Potential energy.

reversed. Thus there may be no change of kinetic energy produced, and the work done may be wholly stored up as potential energy.

Thus a weight requires work to raise it to a height, a spring requires work to bend it, air requires work to compress it, etc.; but a raised weight, a bent spring, compressed air, etc., are *stores* of energy which can be made use of at pleasure.

Newton's Laws of Motion.

242. In what precedes we have given some of Newton's *Definitiones* nearly in his own words; others have been enunciated in a form more suitable to modern methods; and some terms have been introduced which were invented subsequent to the publication of the *Principia*. But the *Axiomata, sive Leges Motûs*, to which we now proceed, are given in Newton's own words; the two centuries which have nearly elapsed since he first gave them have not shown a necessity for any addition or modification. The first two, indeed, were discovered by Galileo, and the third, in some of its many forms, was known to Hooke, Huyghens, Wallis, Wren, and others; before the publication of the *Principia*. Of late there has been a tendency to split the second law into two, called respectively the second and third, and to ignore the third entirely, though using it *directly* in every dynamical problem; but all who have done so have been forced *indirectly* to acknowledge the completeness of Newton's system, by introducing as an axiom what is called D'Alembert's principle, which is really Newton's rejected third law in another form. Newton's own interpretation of his third law directly points out not only D'Alembert's principle, but also the modern principles of Work and Energy.

Axiom.

243. An Axiom is a proposition, the truth of which must be admitted as soon as the terms in which it is expressed are clearly understood. But, as we shall show in our chapter on "Experience," physical axioms are axiomatic to those only who have sufficient knowledge of the action of physical causes to enable them to see at once their necessary truth. Without further remark we shall give Newton's Three Laws; it being remembered that, as the properties of matter *might* have been such as to render a totally different set of laws axiomatic, these laws must be considered as resting on convictions drawn from observation and experiment, *not* on intuitive perception.

244. Lex I. *Corpus omne perseverare in statu suo quiescendi vel movendi uniformiter in directum, nisi quatenus illud à viribus impressis cogitur statum suum mutare.* Newton's first law.

Every body continues in its state of rest or of uniform motion in a straight line, except in so far as it may be compelled by impressed forces to change that state.

245. The meaning of the term *Rest*, in physical science, cannot be absolutely defined, inasmuch as absolute rest nowhere exists in nature. If the universe of matter were finite, its centre of inertia might fairly be considered as absolutely at rest; or it might be imagined to be moving with any uniform velocity in any direction whatever through infinite space. But it is remarkable that the first law of motion enables us (§ 249, below) to explain what may be called *directional* rest. Also, as will be seen farther on, a perfectly smooth spherical body, made up of concentric shells, each of uniform material and density throughout, if made to revolve about an axis, will, *in spite of impressed forces*, revolve with uniform angular velocity, and will maintain its axis of revolution in an absolutely fixed direction. Or, as will soon be shown, § 267, the plane in which the moment of momentum of the universe (if finite) round its centre of inertia is the greatest, which is clearly determinable from the actual motions at any instant, is fixed in direction in space. Rest.

246. We may logically convert the assertion of the first law of motion as to velocity into the following statements:—

The times during which any particular body, not compelled by force to alter the speed of its motion, passes through equal spaces, are equal. And, again—Every other body in the universe, not compelled by force to alter the speed of its motion, moves over equal spaces in successive intervals, during which the particular chosen body moves over equal spaces.

247. The first part merely expresses the convention universally adopted for the measurement of *Time*. The earth, in its rotation about its axis, presents us with a case of motion in which the condition, of not being compelled by force to alter its speed, is more nearly fulfilled than in any other which we can easily or accurately observe. And the numerical measurement of time practically rests on defining *equal intervals of time*, as *times during which the earth turns through equal* Time.

angles. This is, of course, a mere convention, and not a law of nature; and, as we now see it, is a part of Newton's first law.

Examples of the law.

248. The remainder of the law is not a convention, but a great truth of nature, which we may illustrate by referring to small and trivial cases as well as to the grandest phenomena we can conceive.

A curling-stone, projected along a horizontal surface of ice, travels equal distances, except in so far as it is retarded by friction and by the resistance of the air, in successive intervals of time during which the earth turns through equal angles. The sun moves through equal portions of interstellar space in times during which the earth turns through equal angles, except in so far as the resistance of interstellar matter, and the attraction of other bodies in the universe, alter his speed and that of the earth's rotation.

Directional fixings.

249. If two material points be projected from one position, A, at the same instant with any velocities in any directions, and each left to move uninfluenced by force, the line joining them will be always parallel to a fixed direction. For the law asserts, as we have seen, that $AP : AP' :: AQ : AQ'$, if P, Q, and again P', Q' are simultaneous positions; and therefore PQ is parallel to $P'Q'$. Hence if four material points O, P, Q, R are all projected at one instant from one position, OP, OQ, OR are fixed directions of reference ever after. But, practically, the determination of fixed directions in space, § 267, is made to depend upon the rotation of groups of particles exerting forces on each other, and thus involves the Third Law of Motion.

The "Invariable Plane" of the solar system.

250. The whole law is singularly at variance with the tenets of the ancient philosophers, who maintained that circular motion is perfect.

The last clause, "*nisi quatenus*," etc., admirably prepares for the introduction of the second law, by conveying the idea that *it is force alone which can produce a change of motion.* How, we naturally inquire, does the change of motion produced depend on the magnitude and direction of the force which produces it? And the answer is—

Newton's second law.

251. LEX II. *Mutationem motûs proportionalem esse vi motrici impressæ, et fieri secundum lineam rectam quâ vis illa imprimitur.*

Change of motion is proportional to the impressed force, and takes place in the direction of the straight line in which the force acts. Newton's second law.

252. If any force generates motion, a double force will generate double motion, and so on, whether simultaneously or successively, instantaneously, or gradually applied. And this motion, if the body was moving beforehand, is either added to the previous motion if directly conspiring with it; or is subtracted if directly opposed; or is geometrically compounded with it, according to the kinematical principles already explained, if the line of previous motion and the direction of the force are inclined to each other at any angle. (This is a paraphrase of Newton's own comments on the second law.)

253. In Chapter I. we have considered change of velocity, or acceleration, as a purely geometrical element, and have seen how it may be at once inferred from the given initial and final velocities of a body. By the definition of quantity of motion (§ 210), we see that, if we multiply the change of velocity, thus geometrically determined, by the mass of the body, we have the change of motion referred to in Newton's law as the measure of the force which produces it.

It is to be particularly noticed, that in this statement there is nothing said about the actual motion of the body before it was acted on by the force: it is only the *change* of motion that concerns us. Thus the same force will produce precisely the same change of motion in a body, whether the body be at rest, or in motion with any velocity whatever.

254. Again, it is to be noticed that nothing is said as to the body being under the action of *one* force only; so that we may logically put a part of the second law in the following (apparently) amplified form:—

When any forces whatever act on a body, then, whether the body be originally at rest or moving with any velocity and in any direction, each force produces in the body the exact change of motion which it would have produced if it had acted singly on the body originally at rest.

255. A remarkable consequence follows immediately from this view of the second law. Since forces are measured by the changes of motion they produce, and their directions assigned Composition of forces.

Composition of forces.

by the directions in which these changes are produced; and since the changes of motion of one and the same body are in the directions of, and proportional to, the changes of velocity—a single force, measured by the resultant change of velocity, and in its direction, will be the equivalent of any number of simultaneously acting forces. Hence

The resultant of any number of forces (applied at one point) is to be found by the same geometrical process as the resultant of any number of simultaneous velocities.

256. From this follows at once (§ 27) the construction of the *Parallelogram of Forces* for finding the resultant of two forces, and the *Polygon of Forces* for the resultant of any number of forces, in lines all through one point.

The case of the equilibrium of a number of forces acting at one point, is evidently deducible at once from this; for if we introduce one other force equal and opposite to their resultant, this will produce a change of motion equal and opposite to the resultant change of motion produced by the given forces; that is to say, will produce a condition in which the point experiences no change of motion, which, as we have already seen, is the only kind of rest of which we can ever be conscious.

257. Though Newton perceived that the Parallelogram of Forces, or the fundamental principle of Statics, is essentially involved in the second law of motion, and gave a proof which is virtually the same as the preceding, subsequent writers on Statics (especially in this country) have very generally ignored the fact; and the consequence has been the introduction of various unnecessary Dynamical Axioms, more or less obvious, but in reality included in or dependent upon Newton's laws of motion. We have retained Newton's method, not only on account of its admirable simplicity, but because we believe it contains the most philosophical foundation for the static as well as for the kinetic branch of the dynamic science.

Measurement of force and mass.

258. But the second law gives us the means of measuring force, and also of measuring the mass of a body.

For, if we consider the actions of various forces upon the same body for equal times, we evidently have changes of velocity produced which are *proportional to* the forces. The changes of velocity, then, give us in this case the means of comparing the magnitudes of different forces. Thus the velocities acquired in one second by the same mass (falling freely) at different parts of the earth's surface, give us the relative amounts of the earth's attraction at these places. *Measurement of force and mass.*

Again, if equal forces be exerted on different bodies, the changes of velocity produced in equal times must be *inversely* as the masses of the various bodies. This is approximately the case, for instance, with trains of various lengths started by the same locomotive : it is exactly realized in such cases as the action of an electrified body on a number of solid or hollow spheres of the same external diameter, and of different metals.

Again, if we find a case in which different bodies, each acted on by a force, acquire in the same time the same changes of velocity, the forces must be proportional to the masses of the bodies. This, when the resistance of the air is removed, is the case of falling bodies; and from it we conclude that the weight of a body in any given locality, or the force with which the earth attracts it, is proportional to its mass; a most important physical truth, which will be treated of more carefully in the chapter devoted to "Properties of Matter."

259. It appears, lastly, from this law, that every theorem of Kinematics connected with acceleration has its counterpart in Kinetics. *Translations from the kinematics of a point.*

For instance, suppose X, Y, Z to be the components, parallel to fixed axes of x, y, z respectively, of the whole force acting on a particle of mass M. We see by § 212 that

$$M\frac{d^2x}{dt^2}=X,\ M\frac{d^2y}{dt^2}=Y,\ M\frac{d^2z}{dt^2}=Z;$$

or

$$M\ddot x=X,\ M\ddot y=Y,\ M\ddot z=Z.$$

Also, from these, we may evidently write,

$$M\ddot s=X\frac{dx}{ds}+Y\frac{dy}{ds}+Z\frac{dz}{ds}=X\frac{\dot x}{\dot s}+Y\frac{\dot y}{\dot s}+Z\frac{\dot z}{\dot s},$$

$$0=X\frac{\dot y\ddot z-\dot z\ddot y}{\rho^{-1}\dot s^3}+Y\frac{\dot z\ddot x-\dot x\ddot z}{\rho^{-1}\dot s^3}+Z\frac{\dot x\ddot y-\dot y\ddot x}{\rho^{-1}\dot s^3}$$

$$\frac{M\dot s^2}{\rho}=X\frac{\dot s\ddot x-\dot x\ddot s}{\rho^{-1}\dot s^3}+Y\frac{\dot s\ddot y-\dot y\ddot s}{\rho^{-1}\dot s^3}+Z\frac{\dot s\ddot z-\dot z\ddot s}{\rho^{-1}\dot s^3}.$$

The second members of these equations are respectively the components of the impressed force, along the tangent (§ 9), perpendicular to the osculating plane (§ 9), and towards the centre of curvature, of the path described.

Measurement of force and mass.

260. We have, by means of the first two laws, arrived at a *definition* and a *measure* of force; and have also found how to compound, and therefore also how to resolve, forces; and also how to investigate the motion of a single particle subjected to given forces. But more is required before we can completely understand the more complex cases of motion, especially those in which we have mutual actions between or amongst two or more bodies; such as, for instance, attractions, or pressures, or transference of energy in any form. This is perfectly supplied by

Newton's third law.

261. LEX III. *Actioni contrariam semper et æqualem esse reactionem: sive corporum duorum actiones in se mutuò semper esse æquales et in partes contrarias dirigi.*

To every action there is always an equal and contrary reaction: or, the mutual actions of any two bodies are always equal and oppositely directed.

262. If one body presses or draws another, it is pressed or drawn by this other with an equal force in the opposite direction. If any one presses a stone with his finger, his finger is pressed with the same force in the opposite direction by the stone. A horse towing a boat on a canal is dragged backwards by a force equal to that which he impresses on the towing-rope forwards. By whatever amount, and in whatever direction, one body has its motion changed by impact upon another, this other body has its motion changed by the same amount in the opposite direction; for at each instant during the impact the force between them was equal and opposite on the two. When neither of the two bodies has any rotation, whether before or after impact, the changes of velocity which they experience are inversely as their masses.

When one body attracts another from a distance, this other attracts it with an equal and opposite force. This law holds not only for the attraction of gravitation, but also, as Newton himself remarked and verified by experiment, for magnetic attractions: also for electric forces, as tested by Otto-Guericke.

263. What precedes is founded upon Newton's own comments on the third law, and the actions and reactions contemplated are simple forces. In the scholium appended, he makes the following remarkable statement, introducing another specification of actions and reactions subject to his third law, the full meaning of which seems to have escaped the notice of commentators :— Newton's third law.

Si æstimetur agentis actio ex ejus vi et velocitate conjunctim ; et similiter resistentis reactio æstimetur conjunctim ex ejus partium singularum velocitatibus et viribus resistendi ab earum attritione, cohæsione, pondere, et acceleratione oriundis; erunt actio et reactio, in omni instrumentorum usu, sibi invicem semper æquales.

In a previous discussion Newton has shown what is to be understood by the velocity of a force or resistance; *i.e.*, that it is the velocity of the point of application of the force *resolved in the direction of the force*, in fact proportional to the virtual velocity. Bearing this in mind, we may read the above statement as follows :—

If the Action of an agent be measured by its amount and its velocity conjointly; and if, similarly, the Reaction of the resistance be measured by the velocities of its several parts and their several amounts conjointly, whether these arise from friction, cohesion weight, or acceleration ;—Action and Reaction, in all combinations of machines, will be equal and opposite.

Farther on we shall give a full development of the consequences of this most important remark.

264. Newton, in the passage just quoted, points out that forces of resistance against acceleration are to be reckoned as reactions equal and opposite to the actions by which the acceleration is produced. Thus, if we consider any one material point of a system, its reaction against acceleration must be equal and opposite to the resultant of the forces which that point experiences, whether by the actions of other parts of the system upon it, or by the influence of matter not belonging to the system. In other words, it must be in equilibrium with these forces. Hence Newton's view amounts to this, that all the forces of the system, with the reactions against acceleration of the material points composing it, form groups of equilibrating systems for these points considered individually. Hence, by the principle of superposition of forces in equilibrium, all the forces acting on points of the system form, with the reactions against acceleration, an equilibrating set of forces on the whole system. This is the celebrated principle first explicitly stated, D'Alembert's principle.

D'Alembert's principle.

and very usefully applied, by D'Alembert in 1742, and still known by his name. We have seen, however, that it is very distinctly implied in Newton's own interpretation of his third law of motion. As it is usual to investigate the general equations or conditions of equilibrium, in dynamical treatises, before entering in detail on the kinetic branch of the subject, this principle is found practically most useful in showing how we may write down at once the equations of motion for any system for which the equations of equilibrium have been investigated.

Mutual forces between particles of a rigid body.

265. Every rigid body may be imagined to be divided into indefinitely small parts. Now, in whatever form we may eventually find a *physical* explanation of the origin of the forces which act between these parts, it is certain that each such small part may be considered to be held in its position relatively to the others by mutual forces in lines joining them.

266. From this we have, as immediate consequences of the second and third laws, and of the preceding theorems relating to Centre of Inertia and Moment of Momentum, a number of important propositions such as the following:—

Motion of centre of inertia of a rigid body.

(*a*) The centre of inertia of a rigid body moving in any manner, but free from external forces, moves uniformly in a straight line.

(*b*) When any forces whatever act on the body, the motion of the centre of inertia is the same as it would have been had these forces been applied with their proper magnitudes and directions at that point itself.

Moment of momentum of a rigid body.

(*c*) Since the moment of a force acting on a particle is the same as the moment of momentum it produces in unit of time, the changes of moment of momentum in any two parts of a rigid body due to their mutual action are equal and opposite. Hence the moment of momentum of a rigid body, about any axis which is fixed in direction, and passes through a point which is either fixed in space or moves uniformly in a straight line, is unaltered by the mutual actions of the parts of the body.

(*d*) The rate of increase of moment of momentum, when the body is acted on by external forces, is the sum of the moments of these forces about the axis.

267. We shall for the present take for granted, that the mutual action between two rigid bodies may in every case be imagined as composed of pairs of equal and opposite forces in straight lines. From this it follows that the sum of the quantities of motion, parallel to any fixed direction, of two rigid bodies influencing one another in any possible way, remains unchanged by their mutual action; also that the sum of the moments of momentum of all the particles of the two bodies, round any line in a fixed direction in space, and passing through any point moving uniformly in a straight line in any direction, remains constant. From the first of these propositions we infer that the centre of inertia of any number of mutually influencing bodies, if in motion, continues moving uniformly in a straight line, unless in so far as the direction or velocity of its motion is changed by forces acting mutually between them and some other matter not belonging to them; also that the centre of inertia of any body or system of bodies moves just as all their matter, if concentrated in a point, would move under the influence of forces equal and parallel to the forces really acting on its different parts. From the second we infer that the axis of resultant rotation through the centre of inertia of any system of bodies, or through any point either at rest or moving uniformly in a straight line, remains unchanged in direction, and the sum of moments of momenta round it remains constant if the system experiences no force from without. This principle is sometimes called *Conservation of Areas*, a very misleading designation.

Conservation of momentum, and of moment of momentum.

The "Invariable Plane" is a plane through the centre of inertia, perpendicular to the resultant axis.

268. The foundation of the abstract theory of energy is laid by Newton in an admirably distinct and compact manner in the sentence of his scholium already quoted (§ 263), in which he points out its application to mechanics.[1] The *actio agentis*, as he defines it, which is evidently equivalent to the product of the effective component of the force, into the velocity of the point on which it acts, is simply, in modern English phraseology, the rate at which the agent works. The subject for measurement here is precisely the same as that for which Watt, a hundred years later, introduced the practical unit of a "*Horse-*

Rate of doing work.

Horse-power

[1] The reader will remember that we use the word "mechanics" in its true classical sense, the science of machines, the sense in which Newton himself used it, when he dismissed the further consideration of it by saying (in the scholium referred to), *Cæterum mechanicam tractare non est hujus instituti.*

Horse-power *power*," or the rate at which an agent works when overcoming 33,000 times the weight of a pound through the space of a foot in a minute; that is, producing 550 foot-pounds of work per second. The unit, however, which is most generally convenient is that which Newton's definition implies, namely, the rate of doing work in which the unit of energy is produced in the unit of time.

Energy in abstract dynamics. **269.** Looking at Newton's words (§ 263) in this light, we see that they may be logically converted into the following form:—

Work done on any system of bodies (in Newton's statement, the parts of any machine) has its equivalent in work done against friction, molecular forces, or gravity, if there be no acceleration; but if there be acceleration, part of the work is expended in overcoming the resistance to acceleration, and the additional kinetic energy developed is equivalent to the work so spent. This is evident from § 214.

When part of the work is done against molecular forces, as in bending a spring; or against gravity, as in raising a weight; the recoil of the spring, and the fall of the weight, are capable at any future time, of reproducing the work originally expended (§ 241). But in Newton's day, and long afterwards, it was supposed that work was *absolutely lost* by friction; and, indeed, this statement is still to be found even in recent authoritative treatises. But we must defer the examination of this point till we consider in its modern form the principle of *Conservation of Energy*.

270. If a system of bodies, given either at rest or in motion, be influenced by no forces from without, the sum of the kinetic energies of all its parts is augmented in any time by an amount equal to the whole work done in that time by the mutual forces, which we may imagine as acting between its points. When the lines in which these forces act remain all unchanged in length, the forces do no work, and the sum of the kinetic energies of the whole system remains constant. If, on the other hand, one of these lines varies in length during the motion, the mutual forces in it will do work, or will consume work, according as the distance varies with or against them.

Conservative system. **271.** A limited system of bodies is said to be *dynamically conservative* (or simply *conservative*, when force is understood to be the subject), if the mutual forces between its parts always perform, or always consume, the same amount of work during any motion whatever, by which it can pass from one particular configuration to another.

272. The whole theory of energy in physical science is founded on the following proposition :— Foundation of the theory of energy.

If the mutual forces between the parts of a material system are independent of their velocities, whether relative to one another, or relative to any external matter, the system must be dynamically conservative.

For if more work is done by the mutual forces on the different parts of the system in passing from one particular configuration to another, by one set of paths than by another set of paths, let the system be directed, by frictionless constraint, to pass from the first configuration to the second by one set of paths and return by the other, over and over again for ever. It will be a continual source of energy without any consumption of materials, which is impossible. Physical axiom that "the Perpetual Motion is impossible" introduced.

273. The *potential energy* of a conservative system, in the configuration which it has at any instant, is the amount of work that its mutual forces perform during the passage of the system from any one chosen configuration to the configuration at the time referred to. It is generally, but not always, convenient to fix the particular configuration chosen for the zero of reckoning of potential energy, so that the potential energy, in every other configuration practically considered, shall be positive. Potential energy of conservative system.

274. The potential energy of a conservative system, at any instant, depends solely on its configuration at that instant, being, according to definition, the same at all times when the system is brought again and again to the same configuration. It is therefore, in mathematical language, said to be a function of the co-ordinates by which the positions of the different parts of the system are specified. If, for example, we have a conservative system consisting of two material points; or two rigid bodies, acting upon one another with force dependent only on the relative position of a point belonging to one of them, and a point belonging to the other; the potential energy of the system depends upon the co-ordinates of one of these points relatively to lines of reference in fixed directions through the

Potential energy of conservative system.

other. It will therefore, in general, depend on three independent co-ordinates, which we may conveniently take as the distance between the two points, and two angles specifying the absolute direction of the line joining them. Thus, for example, let the bodies be two uniform metal globes, electrified with any given quantities of electricity, and placed in an insulating medium such as air, in a region of space under the influence of a vast distant electrified body. The mutual action between these two spheres will depend solely on the relative position of their centres. It will consist partly of gravitation, depending solely on the distance between their centres, and of electric force, which will depend on the distance between them, but also, in virtue of the inductive action of the distant body, will depend on the absolute direction of the line joining their centres. In our divisions devoted to gravitation and electricity respectively, we shall investigate the portions of the mutual potential energy of the two bodies depending on these two agencies separately. The former we shall find to be the product of their masses divided by the distance between their centres; the latter a somewhat complicated function of the distance between the centres and the angle which this line makes with the direction of the resultant electric force of the distant electrified body. Or again, if the system consist of two balls of soft iron, in any locality of the earth's surface, their mutual action will be partly gravitation, and partly due to the magnetism induced in them by terrestrial magnetic force. The portion of the mutual potential energy depending on the latter cause, will be a function of the distance between their centres and the inclination of this line to the direction of the terrestrial magnetic force. It will agree in mathematical expression with the potential energy of electric action in the preceding case, so far as the inclination is concerned, but the law of variation with the distance will be less easily determined.

Inevitable loss of energy of visible motions.

275. In nature the hypothetical condition of § 271 is *apparently violated* in all circumstances of motion. A material system can never be brought through any returning cycle of motion without spending more work against the mutual forces of its parts than is gained from these forces, because no relative motion can take place without meeting with frictional or other forms of resistance; among which are included (1.) mutual friction between solids sliding upon one another; (2.) resistances due to the viscosity of fluids, or imperfect elasticity of solids; (3.) resistances due to the induction of electric currents; (4.) resistances due to varying magnetization under the influence of imperfect magnetic retentiveness. No motion in nature can take place without meeting resistance due to some, if not to all, of these influences. It is matter of every day experience that friction and imperfect elasticity of solids impede the action of all artificial mechanisms; and that even when bodies are detached, and left to move freely in the air, as falling bodies, or as projectiles, they experience resistance owing to the viscosity of the air. Inevitable loss of energy of visible motions.

The greater masses, planets and comets, moving in a less resisting medium, show less indications of resistance.[1] Indeed it cannot be said that observation upon any one of these bodies, with the exception of Encke's comet, has demonstrated resistance. But the analogies of nature, and the ascertained facts of physical science, forbid us to doubt that every one of them, every star, and every body of any kind moving in any part of space, has its relative motion impeded by the air, gas, vapour, medium, or whatever we choose to call the substance occupying the space immediately round it; just as the motion of a rifle bullet is impeded by the resistance of the air.

276. There are also indirect resistances, owing to friction impeding the tidal motions, on all bodies which, like the earth, have portions of their free surfaces covered by liquid, which, as long as these bodies move relatively to neighbouring bodies, must keep drawing off energy from their relative motions. Thus, if we consider, in the first place, the action of the moon alone, on the earth with its oceans, lakes, and rivers, we perceive that it must tend to equalize the periods of the earth's rotation about its axis, and of the revolution of the two bodies about their centre of inertia; because as long as these periods differ, the tidal action of the earth's surface must keep subtracting energy from their motions. To view the subject more Effect of tidal friction.

[1] Newton, *Principia*. (Remarks on the first law of motion.) "Majora autem Planetarum et Cometarum corpora motus suos et progressivos et circulares, in spatiis minus resistentibus factos, conservant diutius."

Potential energy of conservative system.

other. It w[illegible] co-ordinates, wh[illegible] between the two p[illegible] direction of the l[illegible] Thus, for ex[illegible] bodies be two unif[illegible] quantities of electricity, and placed in an insulating [illegible] such as air, in a re[illegible] distant electrified body. The mutual action between t[illegible] spheres will depend solely on the relative position [illegible] centres. It will consist partly of gravitation, dep[illegible] on the distance between their centres, and of elect[illegible] which will depend on the distance between them, [illegible] virtue of the inductive action of the distant body, w[illegible] on the absolute direction of the line joining their c[illegible] our divisions devoted to gravitation and electricity [illegible] we shall investigate the portions of the mutual p[illegible] of the two bodies depending on these two agencies [illegible] The former we shall find to be the product of th[illegible] divided by the distance between their centres; the [illegible] somewhat complicated function of the distance between [illegible] centres and the angle which this line makes with the dir[illegible] of the resultant electric force of the distant electrifie[illegible] Or again, if the system consist of two balls of soft iron, [illegible] locality of the earth's surface, their mutual action will be [illegible] gravitation, and partly due to the magnetism induced in [illegible] by terrestrial magnetic force. The portion of the [illegible] potential energy depending on the latter cause, will be a [illegible] tion of the distance between their centres and the incli[illegible] of this line to the direction of the terrestrial magneti[illegible] It will agree in mathematical expression with the p[illegible] energy of electric action in the preceding case, so far as [illegible] inclination is concerned, but the law of variation with distance will be less easily determined.

275. In nature the hypothetical condition of § 271 [illegible] in all circumstances of motion. A [illegible] never be brought through any [illegible] without spending more work [illegible] than is gained from [illegible] can take place [illegible] or other forms of resistan... mutual friction between ... resistances due to the visc... of solids ; (3.) resistan... currents ; (4.) resistan... the influence of imp... in nature can take ... some, if not to all, o... day experience that ... impede the action o... when bodies are det... as falling bodies, or a... owing to the viscosity ...

The greater mass... resisting medium, show ... it cannot be said tha... with the exception of ... ance. But the anal... physical science, for ... every star, and every ... of space, has its rela... medium, or whatever ... the space immediat... bullet is impeded by ...

276. There are ... impeding the tidal ... have portions of the... as long as these bod... must keep drawing ... Thus, if we consider ... alone, on the earth ... ceive that it must ... rotation about its axis ... about their centre of ... ... tidal acti... ...ting energy from ...

...re
...in
...hat is
...e same
...of energy
...urse of the

Inevitable loss of energy of visible motions. Effect of tidal friction.

in detail, and, at the same time, to avoid unnecessary complications, let us suppose the moon to be a uniform spherical body. The mutual action and reaction of gravitation between her mass and the earth's, will be equivalent to a single force in some line through her centre; and must be such as to impede the earth's rotation as long as this is performed in a shorter period than the moon's motion round the earth. It must therefore lie in some such direction as the line *MQ* in the diagram, which represents, necessarily with enormous exaggeration, its deviation, *OQ*, from the earth's centre. Now the actual force on the moon in the line *MQ*, may be regarded as consisting of a force in the line *MO* towards the earth's centre, sensibly equal in amount to the whole force, and a comparatively very small force in the line *MT* perpendicular to *MO*. This latter is very nearly tangential to the moon's path, and is in the direction *with* her motion. Such a force, if suddenly commencing to act, would, in the first place, increase the moon's velocity; but after a certain time she would have moved so much farther from the earth, in virtue of this acceleration, as to have lost, by moving against the earth's attraction, as much velocity as she had gained by the tangential accelerating force. The effect of a continued tangential force, acting with the motion, but so small in amount as to make only a small deviation at any moment from the circular form of the orbit, is to gradually increase the distance from the central body, and to cause as much again as its own amount of work to be done against the attraction of the central mass, by the kinetic energy of motion lost. The circumstances will be readily understood, by considering this motion round the central body in a very gradual spiral path tending outwards. Provided the law of force is the inverse square of the distance, the tangential component of gravity against the motion will be twice as great as the disturbing tangential force in the direction with the motion; and therefore one-half of the amount of work done against the former, is done by the latter, and the other half by kinetic energy taken from the motion. The integral effect on

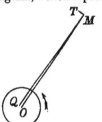

Inevitable loss of energy of visible motions. Tidal friction.

the moon's motion, of the particular disturbing cause now under consideration, is most easily found by using the principle of moments of momenta. Thus we see that as much moment of momentum is gained in any time by the motions of the centres of inertia of the moon and earth relatively to their common centre of inertia, as is lost by the earth's rotation about its axis. The sum of the moments of momentum of the centres of inertia of the moon and earth as moving at present, is about 4·45 times the present moment of momentum of the earth's rotation. The average plane of the former is the ecliptic; and therefore the axes of the two momenta are inclined to one another at the average angle of 23° 27½′, which, as we are neglecting the sun's influence on the plane of the moon's motion, may be taken as the actual inclination of the two axes at present. The resultant, or whole moment of momentum, is therefore 5·38 times that of the earth's present rotation, and its axis is inclined 19° 13′ to the axis of the earth. Hence the ultimate tendency of the tides is, to reduce the earth and moon to a simple uniform rotation with this resultant moment round this resultant axis, as if they were two parts of one rigid body: in which condition the moon's distance would be increased (approximately) in the ratio 1 : 1·46, being the ratio of the square of the present moment of momentum of the centres of inertia to the square of the whole moment of momentum; and the period of revolution in the ratio 1 : 1·77, being that of the cubes of the same quantities. The distance would therefore be increased to 347,100 miles, and the period lengthened to 48·36 days. Were there no other body in the universe but the earth and the moon, these two bodies might go on moving thus for ever, in circular orbits round their common centre of inertia, and the earth rotating about its axis in the same period, so as always to turn the same face to the moon, and therefore to have all the liquids at its surface at rest relatively to the solid. But the existence of the sun would prevent any such state of things from being permanent. There would be solar tides—twice high water and twice low water—in the period of the earth's revolution relatively to the sun (that is to say, twice in the solar day, or, which would be the same thing, the month). This could not go on without loss of energy by fluid friction. It is not easy to trace the whole course of the

Inevitable loss of energy of visible motions. Tidal friction.

disturbance in the earth's and moon's motions which this cause would produce, but its ultimate effect must be to bring the earth, moon, and sun to rotate round their common centre of inertia, like parts of one rigid body. It would carry us too far from our course to investigate at present which of all the configurations fulfilling this condition is the one that would be ultimately approximated to. We hope, however, to return to the subject later, and to consider the general problem of the motion of any number of rigid bodies or material points acting on one another with mutual forces, under any actual physical law, and therefore, as we shall see, necessarily subject to loss of energy as long as any of their mutual distances vary; that is to say, until all subside into a state of motion in circles round an axis passing through their centre of inertia, like parts of one rigid body. It is probable that the moon, in ancient times liquid or viscous in its outer layer if not throughout, was thus brought to turn always the same face to the earth.

Ultimate tendency of the solar system.

277. We have no data in the present state of science for estimating the relative importance of tidal friction, and of the resistance of the resisting medium through which the earth and moon move; but whatever it may be, there can be but one ultimate result for such a system as that of the sun and planets, if continuing long enough under existing laws, and not disturbed by meeting with other moving masses in space. That result is the falling together of all into one mass, which, although rotating for a time, must in the end come to rest relatively to the surrounding medium.

Conservation of energy.

278. The theory of energy cannot be completed until we are able to examine the physical influences which accompany loss of energy in each of the classes of resistance mentioned above, § 275. We shall then see that in every case in which energy is lost by resistance, heat is generated; and we shall learn from Joule's investigations that the quantity of heat so generated is a perfectly definite equivalent for the energy lost. Also that in no natural action is there ever a development of energy which cannot be accounted for by the disappearance of an equal amount elsewhere by means of some known physical agency. Thus we shall conclude, that if any limited portion of the material universe could be perfectly isolated, so as to be prevented from either giving energy to, or taking energy from, matter external to it, the sum of its potential and kinetic energies would be the same at all times: in other words, that every material system subject to no other forces than actions and reactions between its parts, is a dynamically conservative system, as defined above, § 271. But it is only when the inscrutably minute motions among small parts, possibly the ultimate molecules of matter, which constitute light, heat, and magnetism; and the intermolecular forces of chemical affinity; are taken into account, along with the palpable motions and measurable forces of which we become cognizant by direct observation, that we can recognise the universally conservative character of all natural dynamic action, and perceive the bearing of the principle of reversibility on the whole class of natural actions involving resistance, which seem to violate it. In the meantime, in our studies of abstract dynamics, it will be sufficient to introduce a special reckoning for energy lost in working against, or gained from work done by, forces not belonging palpably to the conservative class. Conservation of energy.

279. As of great importance in farther developments, we prove a few propositions intimately connected with energy.

280. The kinetic energy of any system is equal to the sum of the kinetic energies of a mass equal to the sum of the masses of the system, moving with a velocity equal to that of its centre of inertia, and of the motions of the separate parts relatively to the centre of inertia. Kinetic energy of a system.

For if x, y, z be the co-ordinates of any particle, m, of the system; ξ, η, ζ its co-ordinates relative to the centre of inertia; and $\bar{x}, \bar{y}, \bar{z}$, the co-ordinates of the centre of inertia itself; we have for the whole kinetic energy

$$\Sigma m\{(\frac{dx}{dt})^2+(\frac{dy}{dt})^2+(\frac{dz}{dt})^2\}=\tfrac{1}{2}\Sigma m\{(\frac{d(\bar{x}+\xi)}{dt})^2+(\frac{d(\bar{y}+\eta)}{dt})^2+(\frac{d(\bar{z}+\zeta)}{dt})^2\}$$

But by the properties of the centre of inertia, we have

$$\Sigma m\frac{d\bar{x}}{dt}\frac{d\xi}{dt}=\frac{d\bar{x}}{dt}\Sigma m\frac{d\xi}{dt}=0, \text{ etc. etc.}$$

Hence the preceding is equal to

$$\tfrac{1}{2}\Sigma m\{(\frac{d\bar{x}}{dt})^2+(\frac{d\bar{y}}{dt})^2+(\frac{d\bar{z}}{dt})^2\}+\tfrac{1}{2}\Sigma m\{(\frac{d\xi}{dt})^2+(\frac{d\eta}{dt})^2+(\frac{d\zeta}{dt})^2\}.$$

which proves the proposition.

Kinetic energy of a system.

281. The kinetic energy of rotation of a rigid system about any axis is (§ 95) expressed by $\frac{1}{2}\Sigma mr^2\omega^2$, where m is the mass of any part, r its distance from the axis, and ω the angular velocity of rotation. It may evidently be written in the form $\frac{1}{2}\omega^2\Sigma mr^2$. The factor Σmr^2 is of very great importance in kinetic investigations, and has been called the ***Moment of Inertia*** of the system about the axis in question. The moment of inertia about any axis is therefore found by summing the products of the masses of all the particles each into the square of its distance from the axis.

Moment of inertia.

It is worth while to notice that the moment of momentum of any rigid system about an axis, being $\Sigma mvr = \Sigma mr^2\omega$, is the product of the angular velocity into the moment of inertia.

If we take a quantity k, such that

$$k^2\Sigma m = \Sigma mr^2$$

Radius of gyration.

k is called the ***Radius of Gyration*** about the axis from which r is measured. The radius of gyration about any axis is therefore the distance from that axis at which, if the whole mass were placed, it would have the same moment of inertia as before. In a fly-wheel, where it is desirable to have as great a moment of inertia with as small a mass as possible, within certain limits of dimensions, the greater part of the mass is formed into a ring of the largest admissible diameter, and the radius of this ring is then approximately the radius of gyration of the whole.

Fly-wheel.

Moment of inertia about any axis.

A rigid body being referred to rectangular axes passing through any point, it is required to find the moment of inertia about an axis through the origin making given angles with the co-ordinate axes.

Let λ, μ, ν be its direction-cosines. Then the distance (r) of the point x, y, z from it is, by § 95,

$$r^2 = (\mu z - \nu y)^2 + (\nu x - \lambda z)^2 + (\lambda y - \mu x)^2,$$

and therefore

$$Mk^2 = \Sigma mr^2 = \Sigma m[\lambda^2(y^2+z^2) + \mu^2(z^2+x^2) + \nu^2(x^2+y^2) - 2\mu\nu yz - 2\nu\lambda zx - 2\lambda\mu xy]$$

which may be written

$$A\lambda^2 + B\mu^2 + C\nu^2 - 2\alpha\mu\nu - 2\beta\nu\lambda - 2\gamma\lambda\mu,$$

where A, B, C are the moments of inertia about the axes, and $\alpha = \Sigma myz$, $\beta = \Sigma mzx$, $\gamma = \Sigma mxy$. From its derivation we see that this quantity is *essentially positive*. Hence when, by a proper linear transformation, it is deprived of the terms containing the products of λ, μ, ν, it will be brought to the form Moment of inertia.

$$Mk^2 = A\lambda^2 + B\mu^2 + C\nu^2 = Q,$$

where A, B, C are essentially positive. They are evidently the moments of inertia about the new rectangular axes of co-ordinates, and λ, μ, ν the corresponding direction-cosines of the axis round which the moment of inertia is to be found.

Let $A > B > C$, if they are unequal. Then

$$A\lambda^2 + B\mu^2 + C\nu^2 = Q(\lambda^2 + \mu^2 + \nu^2)$$

shows than Q cannot be greater than A, nor less than C. Also, if A, B, C be equal, Q is equal to each.

If a, b, c be the radii of gyration about the new axes of x, y, z,

$$A = Ma^2, \; B = Mb^2, \; C = Mc^2,$$

and the above equation gives

$$k^2 = a^2\lambda^2 + b^2\mu^2 + c^2\nu^2.$$

But if x, y, z be any point in the line whose direction-cosines are λ, μ, ν, and r its distance from the origin, we have

$$\frac{x}{\lambda} = \frac{y}{\mu} = \frac{z}{\nu} = r, \text{ and therefore}$$

$$k^2r^2 = a^2x^2 + b^2y^2 + c^2z^2.$$

If, therefore, we consider the ellipsoid whose equation is

$$a^2x^2 + b^2y^2 + c^2z^2 = \epsilon^4,$$

we see that it intercepts on the line whose direction-cosines are λ, μ, ν—and about which the radius of gyration is k, a length r which is given by the equation

$$k^2r^2 = \epsilon^4;$$

or the rectangle under any radius-vector of this ellipsoid and the radius of gyration about it is constant. Its semi-axes are evidently $\frac{\epsilon^2}{a}$, $\frac{\epsilon^2}{b}$, $\frac{\epsilon^2}{c}$ where ϵ may have any value we may assign. Thus it is evident that

282. For every rigid body there may be described about ny point as centre, an ellipsoid (called *Poinsot's Momental llipsoid*) which is such that the length of any radius-vector is nversely proportional to the radius of gyration of the body bout that radius-vector as axis. Momental ellipsoid.

The axes of this ellipsoid are the *Principal Axes* of inertia f the body at the point in question. Principal axes.

283. The proposition of § 280 shows that the moment of ertia of a rigid body about any axis is equal to that which

Principal axes.

the mass, if collected at the centre of inertia, would have about this axis, together with that of the body about a parallel axis through its centre of inertia.

Let the origin, O, be the centre of inertia, and the axes the principal axes at that point. Then, by §§ 280, 281, we have for the moment of inertia about a line through the point P (ξ, η, ζ), whose direction-cosines are λ, μ, ν;

$$Q = A\lambda^2 + B\mu^2 + C\nu^2 + M(\overline{\mu\zeta - \nu\eta}^2 + \overline{\nu\xi - \lambda\zeta}^2 + \overline{\lambda\eta - \mu\xi}^2)$$
$$= \{A + M(\eta^2 + \zeta^2)\}\lambda^2 + \{B + M(\zeta^2 + \xi^2)\}\mu^2 + \{C + M(\xi^2 + \eta^2)\}\nu^2$$
$$- 2M(\mu\nu\eta\zeta + \nu\lambda\zeta\xi + \lambda\mu\xi\eta).$$

Substituting for Q, A, B, C their values, and dividing by M, we have

$$k^2 = (a^2 + \eta^2 + \zeta^2)\lambda^2 + (b^2 + \zeta^2 + \xi^2)\mu^2 + (c^2 + \xi^2 + \eta^2)\nu^2$$
$$- 2(\eta\zeta\mu\nu + \zeta\xi\nu\lambda + \xi\eta\lambda\mu).$$

Let it be required to find λ, μ, ν so that the direction specified by them may be a principal axis. Let $s = \lambda\xi + \mu\eta + \nu\zeta$, *i.e.*, let s represent the projection of OP on the axis sought.

The axes of the ellipsoid

$$(a^2 + \eta^2 + \zeta^2)x^2 + \ldots\ldots - 2(\eta\zeta yz + \ldots\ldots) = H \qquad (a).$$

are found by means of the equations

$$\left.\begin{array}{l} (a^2 + \eta^2 + \zeta^2 - p)\lambda - \xi\eta\mu - \zeta\xi\nu = 0 \\ -\xi\eta\lambda + (b^2 + \zeta^2 + \xi^2 - p)\mu - \eta\zeta\nu = 0 \\ -\zeta\xi\lambda - \eta\zeta\mu + (c^2 + \xi^2 + \eta^2 - p)\nu = 0 \end{array}\right\} \qquad (b)$$

If, now, we take f to denote OP, or $(\xi^2 + \eta^2 + \zeta^2)^{\frac{1}{2}}$, these equations, where p is clearly the square of the radius of gyration about the axis to be found, may be written

$$(a^2 + f^2 - p)\lambda - \xi(\xi\lambda + \eta\mu + \zeta\nu) = 0,$$
$$\text{etc.} = \text{etc.},$$

or

$$(a^2 + f^2 - p)\lambda - \xi s = 0,$$
$$\text{etc.} = \text{etc.},$$

or

$$\left.\begin{array}{l} (a^2 - K)\lambda - \xi s = 0 \\ (b^2 - K)\mu - \eta s = 0 \\ (c^2 - K)\nu - \zeta s = 0 \end{array}\right\} \qquad (c),$$

where $K = p - f^2$. Hence

$$\lambda = \frac{\xi s}{a^2 - K}, \text{ etc.}$$

Multiply, in order, by ξ, η, ζ, add, and divide by s, and we get

$$\frac{\xi^2}{a^2 - K} + \frac{\eta^2}{b^2 - K} + \frac{\zeta^2}{c^2 - K} = 1 \qquad (d).$$

By (c) we see that (λ, μ, ν) is the direction of the normal through the point P, (ξ, η, ζ) of the surface represented by the equation

$$\frac{x^2}{a^2-K}+\frac{y^2}{b^2-K}+\frac{z^2}{c^2-K}=1 \qquad (e),$$ Principal axes.

which is obviously a surface of the second degree confocal with the ellipsoid

$$\frac{x^2}{a^2}+\frac{y^2}{b^2}+\frac{z^2}{c^2}=1 \qquad (f),$$

and passing through P in virtue of (d), which determines K accordingly. The three roots of this cubic are clearly all real; one of them is less than the least of a^2, b^2, c^2, and positive or negative according as P is within or without the ellipsoid (f). And if $a>b>c$, the two others are between c^2 and b^2, and between b^2 and a^2, respectively. The addition of f^2 to each gives the square of the radius of gyration round the corresponding principal axis. Hence

284. The principal axes at any point of a rigid body are normals to the three surfaces of the second order which pass through that point, and are confocal with an ellipsoid, having its centre at the centre of inertia, and its three principal diameters coincident with the three principal axes through these points, and equal respectively to the doubles of the radii of gyration round them. This ellipsoid is called the *Central Ellipsoid*. Central ellipsoid.

285. A rigid body is said to be kinetically symmetrical about its centre of inertia when its moments of inertia about three principal axes through that point are equal; and therefore necessarily the moments of inertia about *all* axes through that point equal, § 281, and all these axes principal axes. About it uniform spheres, cubes, and in general any complete crystalline solid of the first system (see chapter on Properties of Matter) are kinetically symmetrical. Kinetic symmetry round a point;

A rigid body is kinetically symmetrical about an *axis* when this axis is one of the principal axes through the centre of inertia, and the moments of inertia about the other two, and therefore about any line in their plane, are equal. A spheroid, a square or equilateral triangular prism or plate, a circular ring, disc, or cylinder, or any complete crystal of the second or fourth system, is kinetically symmetrical about its axis. round an axis.

286. The only actions and reactions between the parts of a system, not belonging palpably to the conservative class, which we shall consider in abstract dynamics, are those of friction between solids sliding on solids, except in a few instances in which we shall consider the general character and ultimate Energy in abstract dynamics.

Energy in abstract dynamics.

results of effects produced by viscosity of fluids, imperfect elasticity of solids, imperfect electric conduction, or imperfect magnetic retentiveness. We shall also, in abstract dynamics, consider forces as applied to parts of a limited system arbitrarily from without. These we shall call, for brevity, the applied forces.

287. The law of energy may then, in abstract dynamics, be expressed as follows:—

The whole work done in any time, on any limited material system, by applied forces, is equal to the whole effect in the forms of potential and kinetic energy produced in the system, together with the work lost in friction.

288. This principle may be regarded as comprehending the whole of abstract dynamics, because, as we now proceed to show, the conditions of equilibrium and of motion, in every possible case, may be immediately derived from it.

Equilibrium.

289. A material system, whose relative motions are unresisted by friction, is in equilibrium in any particular configuration if, and is not in equilibrium unless, the work done by the applied forces is equal to the potential energy gained, in any possible infinitely small displacement from that configuration. This is the celebrated principle of virtual velocities which Lagrange made the basis of his *Mécanique Analytique*.

Principle of virtual velocities.

290. To prove it, we have first to remark that the system cannot possibly move away from any particular configuration except by work being done upon it by the forces to which it is subject: it is therefore in equilibrium if the stated condition is fulfilled. To ascertain that nothing less than this condition can secure its equilibrium, let us first consider a system having only one degree of freedom to move. Whatever forces act on the whole system, we may always hold it in equilibrium by a single force applied to any one point of the system in its line of motion, opposite to the direction in which it tends to move and of such magnitude that, in any infinitely small motion in either direction, it shall resist, or shall do, as much work as the other forces, whether applied or internal, altogether do or resist. Now, by the principle of superposition of forces in equilibrium we might, without altering their effect, apply to any one point of the system such a force as we have just seen would hold the system in equilibrium, and another force equal and opposite to it. All the other forces being balanced by one of these two, they and it might again, by the principle of superposition of forces in equilibrium, be removed; and therefore the whole set of given forces would produce the same effect, whether for equilibrium or for motion, as the single force which is left acting alone. This single force, since it is in a line in which the point of its application is free to move, must move the system. Hence the given forces, to which this single force has been proved equivalent, cannot possibly be in equilibrium unless their whole work for an infinitely small motion is nothing, in which case the single equivalent force is reduced to nothing. But whatever amount of freedom to move the whole system may have, we may always, by the application of frictionless constraint, limit it to one degree of freedom only; —and this may be freedom to execute any particular motion whatever, possible under the given conditions of the system. If, therefore, in any such infinitely small motion, there is variation of potential energy uncompensated by work of the applied forces, constraint limiting the freedom of the system to only this motion will bring us to the case in which we have just demonstrated there cannot be equilibrium. But the application of constraints limiting motion cannot possibly disturb equilibrium, and therefore the given system under the actual conditions cannot be in equilibrium in any particular configuration if there is more work done than resisted in any possible infinitely small motion from that configuration by all the forces to which it is subject.

Principle of virtual velocities.

291. If a material system, under the influence of internal and applied forces, varying according to some definite law, is balanced by them in any position in which it may be placed, its equilibrium is said to be neutral. This is the case with any spherical body of uniform material resting on a horizontal plane. A right cylinder or cone, bounded by plane ends perpendicular to the axis, is also in neutral equilibrium on a horizontal plane. Practically, any mass of moderate dimensions is in neutral equilibrium when its centre of inertia only is fixed, since, when its longest dimension is small in comparison with the earth's radius, gravity is, as we shall see, approximately equivalent to a single force through this point.

Neutral equilibrium.

Stable equilibrium. But if, when displaced infinitely little in any direction from a particular position of equilibrium, and left to itself, it commences and continues vibrating, without ever experiencing more than infinitely small deviation in any of its parts, from the position of equilibrium, the equilibrium in this position is said to be stable. A weight suspended by a string, a uniform sphere in a hollow bowl, a loaded sphere resting on a horizontal plane with the loaded side lowest, an oblate body resting with one end of its shortest diameter on a horizontal plane, a plank, whose thickness is small compared with its length and breadth, floating on water, etc. etc., are all cases of stable equilibrium; if we neglect the motions of rotation about a vertical axis in the second, third, and fourth cases, and horizontal motion in general, in the fifth, for all of which the equilibrium is neutral.

Unstable equilibrium. If, on the other hand, the system can be displaced in any way from a position of equilibrium, so that when left to itself it will not vibrate within infinitely small limits about the position of equilibrium, but will move farther and farther away from it, the equilibrium in this position is said to be unstable. Thus a loaded sphere resting on a horizontal plane with its load as high as possible, an egg-shaped body standing on one end, a board floating edgeways in water, etc. etc., would present, if they could be realized in practice, cases of unstable equilibrium.

When, as in many cases, the nature of the equilibrium varies with the direction of displacement, if unstable for any possible displacement it is practically unstable on the whole. Thus a coin standing on its edge, though in neutral equilibrium for displacements in its plane, yet being in unstable equilibrium for those perpendicular to its plane, is practically unstable. A sphere resting in equilibrium on a saddle presents a case in which there is stable, neutral, or unstable equilibrium, according to the direction in which it may be displaced by rolling, but, practically, it would be unstable.

Test of the nature of equilibrium. **292.** The theory of energy shows a very clear and simple test for discriminating these characters, or determining whether the equilibrium is neutral, stable, or unstable, in any case. If there is just as much work resisted as performed by the applied and internal forces in any possible displacement the equilibrium is neutral, but not unless. If in every possible infinitely small displacement from a position of equilibrium they do less work among them than they resist, the equilibrium is thoroughly stable, and not unless. If in any or in every infinitely small displacement from a position of equilibrium they do more work than they resist, the equilibrium is unstable. It follows that if the system is influenced only by internal forces, or if the applied forces follow the law of doing always the same amount of work upon the system passing from one configuration to another by all possible paths, the whole potential energy must be constant, in all positions, for neutral equilibrium; must be a minimum for positions of thoroughly stable equilibrium; must be either an absolute maximum, or a maximum for some displacements and a minimum for others when there is unstable equilibrium.

Test of the nature of equilibrium.

293. We have seen that, according to D'Alembert's principle, as explained above (§ 264), forces acting on the different points of a material system, and their reactions against the accelerations which they actually experience in any case of motion, are in equilibrium with one another. Hence in any actual case of motion, not only is the actual work done by the forces equal to the kinetic energy produced in any infinitely small time, in virtue of the actual accelerations; but so also is the work which would be done by the forces, in any infinitely small time, if the velocities of the points constituting the system, were at any instant changed to any possible infinitely small velocities, and the accelerations unchanged. This statement, when put in the concise language of mathematical analysis, constitutes Lagrange's application of the "principle of virtual velocities" to express the conditions of D'Alembert's equilibrium between the forces acting, and the resistances of the masses to acceleration. It comprehends, as we have seen, every possible condition of every case of motion. The "equations of motion" in any particular case are, as Lagrange has shown, deduced from it with great ease.

Deduction of the equations of motion of any system.

Let m be the mass of any one of the material points of the system; x, y, z its rectangular co-ordinates at time t, relatively to axes fixed in direction (§ 249) through a point reckoned as fixed (§ 245); and X, Y, Z the components, parallel to the same

Indeterminate equation of motion of any system.

axes, of the whole force acting on it. Thus $-m\frac{d^2x}{dt^2}$, $-m\frac{d^2y}{dt^2}$, $-m\frac{d^2z}{dt^2}$ are the components of the reaction against acceleration. And these, with X, Y, Z, for the whole system, must fulfil the conditions of equilibrium. Hence if δx, δy, δz denote any arbitrary variations of x, y, z consistent with the conditions of the system, we have

$$\Sigma\{(X-m\frac{d^2x}{dt^2})\delta x+(Y-m\frac{d^2y}{dt^2})\delta y+(Z-m\frac{d^2z}{dt^2})\delta z\}=0 \quad (1),$$

where Σ denotes summation to include all the particles of the system. This may be called the indeterminate, or the variational, equation of motion. Lagrange used it as the foundation of his whole kinetic system, deriving from it all the common equations of motion, and his own remarkable equations in generalized co-ordinates (presently to be given). We may write it otherwise as follows:

$$\Sigma m(\ddot{x}\delta x+\ddot{y}\delta y+\ddot{z}\delta z)=\Sigma(X\delta x+Y\delta y+Z\delta z) \quad (2),$$

where the first member denotes the work done by forces equal to those required to produce the real accelerations, acting through the spaces of the arbitrary displacements; and the second member the work done by the actual forces through these imagined spaces.

If the moving bodies constitute a conservative system, and if V denote its potential energy in the configuration specified by $(x, y, z,$ etc.), we have of course (§§ 241, 273)

$$\delta V=-\Sigma(X\delta x+Y\delta y+Z\delta z), \quad (3),$$

and therefore the indeterminate equation of motion becomes

Of conservative system.

$$\Sigma m(\ddot{x}\delta x+\ddot{y}\delta y+\ddot{z}\delta z)=-\delta V \quad (4).$$

where δV denotes the excess of the potential energy in the configuration $(x+\delta x, y+\delta y, z+\delta z,$ etc.) above that in the configuration $(x, y, z,$ etc.)

One immediate particular result must of course be the common equation of energy, which must be obtained by supposing δx, δy, δz, etc., to be the actual variations of the co-ordinates in an infinitely small time δt. Thus if we take $\delta x=\dot{x}\delta t$, etc., and divide both members by δt, we have

Equation of energy.

$$\Sigma(X\dot{x}+Y\dot{y}+Z\dot{z})=\Sigma m(\ddot{x}\dot{x}+\ddot{y}\dot{y}+\ddot{z}\dot{z}) \quad (5).$$

Here the first member is composed of Newton's *Actiones Agentium*; with his *Reactiones Resistentium* so far as friction, gravity, and molecular forces are concerned, subtracted: and the second consists of the portion of the *Reactiones* due to acceleration. As we have seen above (§ 214), the second member is the rate of increase of $\Sigma\frac{1}{2}m(\dot{x}^2+\dot{y}^2+\dot{z}^2)$ per unit of time. Hence, denoting by v the velocity of one of the particles, and by W the integral of the first member multiplied by dt, that is to say, the integral work done by the working and resisting forces in any time, we have

Equation of energy.

$$\Sigma\tfrac{1}{2}mv^2 = W + E_0 \qquad (6),$$

E_0 being the initial kinetic energy. This is the integral equation of energy. In the particular case of a conservative system, W is a function of the co-ordinates, irrespectively of the time, or of the paths which have been followed. According to the previous notation, with besides V_0 to denote the potential energy of the system in its initial configuration, we have $W = V_0 - V$, and the integral equation of energy becomes,

$$\Sigma\tfrac{1}{2}mv^2 = V_0 - V + E_0.$$

or, if E denote the sum of the potential and kinetic energies, a constant,

$$\Sigma\tfrac{1}{2}mv^2 = E - V \qquad (7).$$

The general indeterminate equation gives immediately, for the motion of a system of free particles,

$$m_1\ddot{x}_1 = X_1,\quad m_1\ddot{y}_1 = Y_1,\quad m_1\ddot{z}_1 = Z_1,\quad m_2\ddot{x}_2 = X_2,\ \text{etc.}$$

Of these equations the three for each particle may of course be treated separately if there is no mutual influence between the particles: but when they exert force on one another, X_1, Y_1, etc., will each in general be a function of all the co-ordinates.

From the indeterminate equation (1) Lagrange, by his method of multipliers, deduces the requisite number of equations for determining the motion of a rigid body, or of any system of connected particles or rigid bodies, thus:—Let the number of the particles be i, and let the connexions between them be expressed by n equations,

Constraint introduced into the indeterminate equation.

$$\left.\begin{array}{l} F_{,}(x_1, y_1, z_1, x_2, \ldots) = 0 \\ F_{,,}(x_1, y_1, z_1, x_2, \ldots) = 0 \\ \quad \text{etc.} \qquad \text{etc.} \end{array}\right\} \qquad (8),$$

being the *kinematical equations* of the system. By taking the variations of these we find that every possible infinitely small displacement $\delta x_1, \delta y_1, \delta z_1, \delta x_2, \ldots$ must satisfy the n linear equations

$$\frac{dF_{,}}{dx_1}\delta x_1 + \frac{dF_{,}}{dy_1}\delta y_1 + \text{etc.} = 0,\quad \frac{dF_{,,}}{dx_1}\delta x_1 + \frac{dF_{,,}}{dy_1}\delta y_1 + \text{etc.} = 0,\ \text{etc.} \quad (9).$$

Multiplying the first of these by $\lambda_{,}$, the second by $\lambda_{,,}$, etc., adding to the indeterminate equation, and then equating the coefficients of δx_1, δy_1, etc., each to zero, we have

Determinate equations of motion deduced.

$$\left.\begin{aligned}\lambda_{,}\frac{dF_{,}}{dx_1}+\lambda_{,,}\frac{dF_{,,}}{dx_1}+\ldots+X_1-m_1\frac{d^2x_1}{dt^2}&=0\\ \lambda_{,}\frac{dF_{,}}{dy_1}+\lambda_{,,}\frac{dF_{,,}}{dy_1}+\ldots+Y_1-m_1\frac{d^2y_1}{dt^2}&=0\\ \text{etc.}\qquad\qquad \text{etc.}\end{aligned}\right\}\qquad(10).$$

These are in all $3i$ equations to determine the n unknown quantities $\lambda_{,}$, $\lambda_{,,}$, ..., and the $3i-n$ independent variables to which x_1, y_1, ... are reduced by the kinematical equations (8).

The problem of finding the motion of a system subject to any *unvarying* kinematical conditions whatever, under the action of any given forces, is thus reduced to a question of pure analysis. In the still more general problem of determining the motion when certain parts of the system are constrained to move in a specified manner, the equations of condition (8) involve not only the co-ordinates, but also, t, the time. It is easily seen, however, that the equations (10) still hold.

Gauss's principle of least constraint.

When there are connexions between any parts of a system, the motion is in general not the same as if all were free. If we consider any particle during any infinitely small time of the motion, and call the product of its mass into the square of the distance between its positions at the end of this time, on the two suppositions, the *constraint:* the sum of the constraints is a minimum. This follows easily from (1).

Impact.

294. When two bodies, in relative motion, come into contact, pressure begins to act between them to prevent any parts of them from jointly occupying the same space. This force commences from nothing at the first point of collision, and gradually increases per unit of area on a gradually increasing surface of contact. If, as is always the case in nature, each body possesses some degree of elasticity, and if they are not kept together after the impact by cohesion, or by some artificial appliance, the mutual pressure between them will reach a maximum, will begin to diminish, and in the end will come to nothing, by gradually diminishing in amount per unit of area on a gradually diminishing surface of contact. The whole process would occupy not greatly more or less than an hour if the bodies were of such dimensions as the earth, and such degrees of rigidity as copper, steel, or glass. It is finished, probably, within a thousandth of a second if they are globes of any of these substances not exceeding a yard in diameter.

295. The whole amount, and the direction, of the "*Impact*" experienced by either body in any such case, are reckoned according to the "change of momentum" which it experiences. The amount of the impact is measured by the amount, and its direction by the direction of the change of momentum, which is produced. The component of an impact in a direction parallel to any fixed line is similarly reckoned according to the component change of momentum in that direction. Impact.

296. If we imagine the whole time of an impact divided into a very great number of equal intervals, each so short that the force does not vary sensibly during it, the component change of momentum in any direction during any one of these intervals will (§ 220) be equal to the force multiplied by the measure of the interval. Hence the component of the impact is equal to the sum of the forces in all the intervals, multiplied by the length of each interval.

Let P be the component force in any direction at any instant, τ, of the interval, and let I be the amount of the corresponding component of the whole impact. Then

$$I=\int P d\tau.$$

297. Any force in a constant direction acting in any circumstances, for any time great or small, may be reckoned on the same principle; so that what we may call its whole amount during any time, or its "*time-integral*," will measure, or be measured by, the whole momentum which it generates in the time in question. But this reckoning is not often convenient or useful except when the whole operation considered is over before the position of the body, or configuration of the system of bodies, involved, has altered to such a degree as to bring any other forces into play, or alter forces previously acting, to such an extent as to produce any sensible effect on the momentum measured. Thus if a person presses gently with his hand, during a few seconds, upon a mass suspended by a cord or chain, he produces an effect which, if we know the degree of the force at each instant, may be thoroughly calculated on elementary principles. No approximation to a full determination of the motion, or to answering such a partial question as "how great will be the whole deflection produced?" can be founded on a knowledge of the "*time-integral*" alone. If, for Time-integral.

Time-integral. instance, the force be at first very great and gradually diminish, the effect will be very different from what it would be if the force were to increase very gradually and to cease suddenly, even although the time-integral were the same in the two cases. But if the same body is "struck a blow," in a horizontal direction, either by the hand, or by a mallet or other somewhat hard mass, the action of the force is finished before the suspending cord has experienced any sensible deflection from the vertical. Neither gravity nor any other force sensibly alters the effect of the blow. And therefore the whole momentum at the end of the blow is sensibly equal to the "amount of the impact," which is, in this case, simply the time-integral.

Ballistic pendulum. **298.** Such is the case of Robins' *Ballistic Pendulum*, a massive block of wood movable about a horizontal axis at a considerable distance above it—employed to measure the velocity of a cannon or musket-shot. The shot is fired into the block in a horizontal direction perpendicular to the axis. The impulsive penetration is so nearly instantaneous, and the inertia of the block so large compared with the momentum of the shot, that the ball and pendulum are moving on as one mass *before the pendulum has been sensibly deflected from the vertical.* This is the essential peculiarity of the apparatus. A sufficiently great force might move it far from the vertical in a small fraction of its time of vibration. But in order that the time-integral may have its simplest application to such a case, the direction of the force would have continually to change so as to be always the same as that of the motion of the block.

299. Other illustrations of the cases in which the time-integral gives us the complete solution of the problem may be given without limit. They include all cases in which the direction of the force is always coincident with the direction of motion of the moving body, and those special cases in which the time of action of the force is so short that the body's motion does not, during its lapse, sensibly alter its relation to the direction of the force, or the action of any other forces to which it may be subject. Thus, in the vertical fall of a body, the time-integral gives us at once the change of momentum; and the same rule applies in most cases of forces of brief duration, as in a "drive" in cricket or golf.

300. The simplest case which we can consider, and the one usually treated as an introduction to the subject, is that of the collision of two smooth spherical bodies whose centres before collision were moving in the same straight line. The force between them at each instant must be in this line, because of the symmetry of circumstances round it; and by the third law it must be equal in amount on the two bodies. Hence (LEX II.) they must experience changes of motion at equal rates in contrary directions; and at any instant of the impact the integral amounts of these changes of motion must be equal. Let us suppose, to fix the ideas, the two bodies to be moving both before and after impact in the same direction in one line: one of them gaining on the other before impact, and either following it at a less speed, or moving along with it, as the case may be, after the impact is completed. Cases in which the former is driven backwards by the force of the collision, or in which the two moving in opposite directions meet in collision, are easily reduced to dependence on the same formula by the ordinary algebraic convention with regard to positive and negative signs. Direct impact of spheres.

In the standard case, then, the quantity of motion lost, up to any instant of the impact, by one of the bodies, is equal to that gained by the other. Hence at the instant when their velocities are equalized they move as one mass with a momentum equal to the sum of the momenta of the two before impact. That is to say, if v denote the common velocity at this instant, we have

$$(M+M')v=MV+M'V',$$

or

$$v=\frac{MV+M'V'}{M+M'},$$

if M, M' denote the masses of the two bodies, and V, V' their velocities before impact.

During this first period of the impact the bodies have been, on the whole, coming into closer contact with one another, through a compression or deformation experienced by each, and resulting, as remarked above, in a fitting together of the two surfaces over a finite area. No body in nature is perfectly inelastic; and hence, at the instant of closest approximation, the mutual force called into action between the two

Direct impact of spheres.

bodies continues, and tends to separate them. Unless prevented by natural surface cohesion or welding (such as is always found, as we shall see later in our chapter on Properties of Matter, however hard and well polished the surfaces may be), or by artificial appliances (such as a coating of wax, applied in one of the common illustrative experiments; or the coupling applied between two railway carriages when run together so as to push in the springs, according to the usual practice at railway stations), the two bodies are actually separated by this force, and move away from one another. Newton found that, *provided the impact is not so violent as to make any sensible permanent indentation in either body*, the relative velocity of separation after the impact bears a proportion to their previous relative velocity of approach, which is constant for the same two bodies. This proportion, always less than unity, approaches more and more nearly to it the harder the bodies are.

Effect of elasticity.

Newton's experiments.

Thus with balls of compressed wool he found it $\frac{5}{9}$, iron nearly the same, glass $\frac{15}{16}$. The results of more recent experiments on the same subject have confirmed Newton's law. These will be described later. In any case of the collision of two balls, let e denote this proportion, to which we give the name *Coefficient of Restitution*;[1] and, with previous notation, let in addition U, U' denote the velocities of the two bodies after the conclusion of the impact; in the standard case each being positive, but $U' > U$. Then we have

$$U' - U = e(V - V')$$

and, as before, since one has lost as much momentum as the other has gained,

$$MU + M'U' = MV + M'V'.$$

From these equations we find

$$(M+M')U = MV + M'V' - eM'(V - V'),$$

with a similar expression for U'.

Also we have, as above,

$$(M+M')v = MV + M'V'.$$

Hence, by subtraction,

$$(M+M')(v - U) = eM'(V - V') = e\{M'V - (M+M')v + MV\}$$

[1] In most modern treatises this is called a "coefficient of elasticity," which is clearly a mistake; suggested, it may be, by Newton's words, but inconsistent with his facts, and utterly at variance with modern language and modern knowledge regarding elasticity.

and therefore

Direct impact of spheres.

$$v-U=e(V-v).$$

Of course we have also

$$U'-v=e(v-V').$$

These results may be put in words thus :—The *relative* velocity of either of the bodies with regard to the centre of inertia of the two is, after the completion of the impact, reversed in direction, and diminished in the ratio $e:1$.

301. Hence the loss of kinetic energy, being, according to §§ 267, 280, due only to change of kinetic energy relative to the centre of inertia, is to this part of the whole as $1-e^2:1$.

Thus

Initial kinetic energy $=\frac{1}{2}(M+M')v^2+\frac{1}{2}M(V-v)^2+\frac{1}{2}M'(v-V')^2$.

Final ,, ,, $=\frac{1}{2}(M+M')v^2+\frac{1}{2}M(v-U)^2+\frac{1}{2}M'(U'-v)^2$.

Loss $=\frac{1}{2}(1-e^2)\{M(V-v)^2+M'(v-V')^2\}$.

302. When two elastic bodies, the two balls supposed above for instance, impinge, some portion of their previous kinetic energy will always remain in them as vibrations. A *portion* of the loss of energy (miscalled the effect of imperfect elasticity) is necessarily due to this cause in every real case.

Distribution of energy after impact.

Later, in our chapter on Properties of Matter, it will be shown as a result of experiment, that forces of elasticity are, to a very close degree of accuracy, simply proportional to the strains (§ 154), within the limits of elasticity, in elastic solids which, like metals, glass, etc., bear but small deformations without permanent change. Hence when two such bodies come into collision, sometimes with greater and sometimes with less mutual velocity, but with all other circumstances similar, the velocities of all particles of either body, at corresponding times of the impacts, will be always in the same proportion. Hence the velocity of separation of the centres of inertia after impact will bear a constant proportion to the previous velocity of approach; which agrees with the Newtonian Law. It is therefore probable that a very sensible portion, if not the whole, of the loss of energy in the visible motions of two elastic bodies, after impact, experimented on by Newton, may have been due to vibrations; but unless some other cause also was largely operative, it is difficult to see how the loss was so much greater with iron balls than with glass.

Newton's experimental law consistent with *perfect* elasticity.

Distribution of energy after impact.

303. In certain definite extreme cases, imaginable although not realizable, no energy will be spent in vibrations, and the two bodies will separate, each moving simply as a rigid body, and having in this simple motion the whole energy of work done on it by elastic force during the collision. For instance, let the two bodies be cylinders, or prismatic bars with flat ends, of the same kind of substance, and of equal and similar transverse sections; and let this substance have the property of compressibility with perfect elasticity, in the direction of the length of the bar, and of absolute resistance to change in every transverse dimension. Before impact, let the two bodies be placed with their lengths in one line, and their transverse sections (if not circular) similarly situated, and let one or both be set in motion in this line. The result, as regards the motions of the two bodies after the collision, will be sensibly the same if they are of any real ordinary elastic solid material, provided the greatest transverse diameter of each is very small in comparison with its length. Then, if the lengths of the two be equal, they will separate after impact with the same relative velocity as that with which they approached, and neither will retain any vibratory motion after the end of the collision.

304. If the two bars are of unequal length, the shorter will, after the impact, be exactly in the same state as if it had struck another of its own length, and it therefore will move as a rigid body after the collision. But the other will, along with a motion of its centre of gravity, calculable from the principle that its whole momentum must (§ 267) be changed by an amount equal exactly to the momentum gained or lost by the first, have also a vibratory motion, of which the whole kinetic and potential energy will make up the deficiency of energy which we shall presently calculate in the motions of the centres of inertia. For simplicity, let the longer body be supposed to be at rest before the collision. Then the shorter on striking it will be left at rest; this being clearly the result in the case of $e=1$ in the preceding formulæ (§ 300) applied to the impact of one body striking another of equal mass previously at rest. The longer bar will move away with the same momentum, and therefore with less velocity of its centre of inertia, and less kinetic energy of this motion, than the other body had before impact, in the ratio of the smaller to the greater mass. It will also have a very remarkable vibratory motion, which, when its length is more than double of that of the other, will consist of a wave running backwards and forwards through its length, and causing the motion of its ends, and, in fact, of every particle of it, to take place by "fits and starts," not continuously. The full analysis of these circumstances, though very simple, must be reserved until we are especially occupied with waves, and the kinetics of elastic solids. It is sufficient at present to remark, that the motions of the centres of gravity of the two bodies after impact, whatever they may have been previously, are given by the preceding formulæ with for e the value $\frac{M'}{M}$, where M' and M are the smaller and the larger mass respectively. Distribution of energy after impact.

305. The mathematical theory of the vibrations of solid elastic spheres has not yet been worked out; and its application to the case of the vibrations produced by impact presents considerable difficulty. Experiment, however, renders it certain, that but a small part of the whole kinetic energy of the previous motions can remain in the form of vibrations after the impact of two equal spheres of glass or of ivory. This is proved, for instance, by the common observation, that one of them remains nearly motionless after striking the other previously at rest; since, the velocity of the common centre of gravity of the two being necessarily unchanged by the impact, we infer that the second ball acquires a velocity nearly equal to that which the first had before striking it. But it is to be expected that unequal balls of the same substance coming into collision will, by impact, convert a very sensible proportion of the kinetic energy of their previous motions into energy of vibrations; and generally, that the same will be the case when equal or unequal masses of different substances come into collision; although for one particular proportion of their diameters, depending on their densities and elastic qualities, this effect will be a minimum, and possibly not much more sensible than it is when the substances are the same and the diameters equal.

306. It need scarcely be said that in such cases of impact as that of the tongue of a bell, or of a clock-hammer striking its bell (or spiral spring as in the American clocks), or of pianoforte hammers striking the strings, or of the drum struck with the proper implement, a large part of the kinetic energy of the blow is spent in generating vibrations.

Moment of an impact about an axis.

307. The *Moment of an Impact* about any axis is derived from the line and amount of the impact in the same way as the moment of a velocity or force is determined from the line and amount of the velocity or force, §§ 235, 236. If a body is struck, the change of its moment of momentum about any axis is equal to the moment of the impact round that axis. But, without considering the measure of the impact, we see (§ 267) that the moment of momentum round any axis, lost by one body in striking another, is, as in every case of mutual action, equal to that gained by the other.

Ballistic pendulum.

Thus, to recur to the ballistic pendulum—the line of motion of the bullet at impact may be in any direction whatever, but the only part which is effective is the component in a plane perpendicular to the axis. We may therefore, for simplicity, consider the motion to be in a line perpendicular to the axis, though not necessarily horizontal. Let m be the mass of the bullet, v its velocity, and p the distance of its line of motion from the axis. Let M be the mass of the pendulum with the bullet lodged in it, and k its radius of gyration. Then if ω be the angular velocity of the pendulum when the impact is complete,

$$mvp = Mk^2\omega,$$

from which the solution of the question is easily determined.

For the kinetic energy after impact is changed (§ 241) into its equivalent in potential energy when the pendulum reaches its position of greatest deflection. Let this be given by the angle θ: then the height to which the centre of inertia is raised is $h(1-\cos\theta)$ if h be its distance from the axis. Thus

$$Mgh(1-\cos\theta) = \tfrac{1}{2}Mk^2\omega^2 = \tfrac{1}{2}\frac{m^2v^2p^2}{Mk^2},$$

or

$$2\sin\frac{\theta}{2} = \frac{mvp}{Mk\sqrt{gh}},$$

an expression for the chord of half the angle of deflection. In practice the chord of the angle θ is measured by means of a light tape or cord attached to a point of the pendulum, and slipping with small friction through a clip fixed close to the position occupied by that point when the pendulum hangs at rest.

308. *Work done by an impact* is, in general, the product of the impact into half the sum of the initial and final velocities of the point at which it is applied, resolved in the direction of the impact. In the case of direct impact, such as that treated in § 300, the initial kinetic energy of the body is $\frac{1}{2}MV^2$, the final $\frac{1}{2}MU^2$, and therefore the gain, by the impact, is Work done by impact.

$$\tfrac{1}{2}M(U^2-V^2),$$

or, which is the same,

$$M(U-V).\ \tfrac{1}{2}(U+V).$$

But $M(U-V)$ is (§ 295) equal to the amount of the impact. Hence the proposition: the extension of which to the most general circumstances is easily seen.

Let ι be the amount of the impulse up to time τ, and I the whole amount, up to the end, T. That is to say, let

$$\iota=\int_0^\tau Pd\tau,\ I=\int_0^T Pd\tau;\ \text{also}\ P=\frac{d\iota}{d\tau}\cdot$$

Whatever may be the conditions to which the body struck is subjected, the change of velocity in the point struck is proportional to the amount of the impulse up to any part of its whole time, so that, if $\mathfrak{M}$ be a constant depending on the masses and conditions of constraint involved, and if U, v, V denote the component velocities of the point struck, in the direction of the impulse, at the beginning, at the time τ, and at the end, respectively, we have

$$v=U+\frac{\iota}{\mathfrak{M}},\ V=U+\frac{I}{\mathfrak{M}}\cdot$$

Hence, for the rate of the doing of work by the force P, at the instant t, we have

$$Pv=PU+\frac{\iota P}{\mathfrak{M}}\cdot$$

Hence for the whole work (W) done by it,

$$\begin{aligned}W&=\int_0^T\left(PU+\frac{\iota P}{\mathfrak{M}}\right)d\tau\\&=UI+\frac{1}{\mathfrak{M}}\int_0^I \iota d\iota=UI+\tfrac{1}{2}\frac{I^2}{\mathfrak{M}}\\&=UI+\tfrac{1}{2}I(V-U)=I.\tfrac{1}{2}(U+V).\end{aligned}$$

309. It is worthy of remark, that if any number of impacts be applied to a body, their whole effect will be the same whether they be applied together or successively (provided that the

Work done by impact. whole time occupied by them be infinitely short), although the work done by each particular impact is in general different according to the order in which the several impacts are applied. The whole amount of work is the sum of the products obtained by multiplying each impact by half the sum of the components of the initial and final velocities of the point to which it is applied.

Equations of impulsive motion. **310.** The effect of any stated impulses, applied to a rigid body, or to a system of material points or rigid bodies connected in any way, is to be found most readily by the aid of D'Alembert's principle; according to which the given impulses, and the impulsive reaction against the generation of motion, measured in amount by the momenta generated, are in equilibrium; and are therefore to be dealt with mathematically by applying to them the equations of equilibrium of the system.

Let P_1, Q_1, R_1 be the component impulses on the first particle, m_1, and let $\dot{x}_1$, $\dot{y}_1$, $\dot{z}_1$ be the components of the velocity instantaneously acquired by this particle. Component forces equal to $(P_1 - m_1\dot{x}_1)$, $(Q_1 - m_1\dot{y}_1)$, ... must equilibrate the system, and therefore we have (§ 290)

$$\Sigma\{(P - m\dot{x})\delta x + (Q - m\dot{y})\delta y + (R - m\dot{z})\delta z\} = 0 \qquad (a)$$

where δx_1, δy_1, ... denote the components of any infinitely small displacements of the particles possible under the conditions of the system. Or, which amounts to the same thing, since any possible infinitely small displacements are simply proportional to any possible velocities in the same directions,

$$\Sigma\{(P - m\dot{x})u + (Q - m\dot{y})v + (Q - m\dot{z})w\} = 0 \qquad (b)$$

where u_1, v_1, w_1 denote any possible component velocities of the first particle, etc.

One particular case of this equation is of course had by supposing u_1, v_1, ... to be equal to the velocities $\dot{x}_1$, $\dot{y}_1$, ... actually acquired; and, by halving, etc., we find

$$\Sigma(P.\tfrac{1}{2}\dot{x} + Q.\tfrac{1}{2}\dot{y} + R.\tfrac{1}{2}\dot{z}) = T \qquad (c)$$

where T denotes the energy of the motion generated. This agrees with § 308 above.

Theorem of Euler, extended by Lagrange. **311.** Euler discovered that the kinetic energy acquired from rest by a rigid body in virtue of an impulse fulfils a maximum-minimum condition. Lagrange[1] extended this proposition to

[1] *Mécanique Analytique*, 2nde partie, 3me section, § 37.

a system of bodies connected by any invariable kinematic relations, and struck with any impulses. Delaunay found that it is really always a maximum *when the impulses are given, and when different motions possible under the conditions of the system, and fulfilling the law of energy* [§ 308 above, or § 310 (c)], *are considered.* Farther, Bertrand shows that the energy actually acquired is not merely a "maximum," but exceeds the energy of any other motion fulfilling these conditions; and that the amount of the excess is equal to the energy of the motion which must be compounded with either to produce the other. Equations of impulsive motion.

Let $\dot{x}_1'$, $\dot{y}_1'$ ... be the component velocities of any motion whatever fulfilling the equation (c), which becomes

$$\tfrac{1}{2}\Sigma(P\dot{x}'+Q\dot{y}'+R\dot{z}')=T'=\tfrac{1}{2}\Sigma m(\dot{x}'^2+\dot{y}'^2+\dot{z}'^2) \qquad (d).$$

If, then, we take $\dot{x}_1-\dot{x}_1'=u_1$, $\dot{y}_1-\dot{y}_1'=v_1$, etc., we have

$$T-T'=\tfrac{1}{2}\Sigma m\{(2\dot{x}-u)u+\ldots\}$$
$$=\Sigma m(\dot{x}u+\dot{y}v+\dot{z}w)-\tfrac{1}{2}\Sigma m(u^2+v^2+w^2) \qquad (e).$$

But, by (b),

$$\Sigma m(\dot{x}u+\dot{y}v+\dot{z}w)=\Sigma(Pu+Qv+Rw);$$

and, by (c) and (d),

$$\Sigma(Pu+Qv+Rw)=2T-2T'.$$

Hence (e) gives

$$T-T'=\tfrac{1}{2}\Sigma m(u^2+v^2+w^2) \qquad (f),$$

which is Bertrand's result.

312. The energy of the motion generated suddenly in a mass of incompressible liquid given at rest completely filling a vessel of any shape, when the vessel is suddenly set in motion, or when it is suddenly bent out of shape in any way whatever, subject to the condition of not changing its volume, *is less than the energy of any other motion it can have with the same motion of its bounding surface.* The consideration of this theorem, which, so far as we know, was first published in the *Cambridge and Dublin Mathematical Journal* [Feb. 1849], has led us to the general *minimum* property proved below regarding motion acquired by any system when *any given velocities* are generated suddenly in any of its parts. Liquid set in motion impulsively.

313. The method of generalized co-ordinates explained above (§ 204) is extremely useful in its application to the dynamics of a system; whether for expressing and working out the details of any particular case in which there is any Impulsive motion referred to generalized co-ordinates.

Impulsive generation of motion referred to generalized co-ordinates.

finite number of degrees of freedom, or for proving general principles applicable even to cases, such as that of a liquid, as described in the preceding section, in which there may be an infinite number of degrees of freedom. It leads us to generalize the measure of inertia, and the resolution and composition of forces, impulses, and momenta, on dynamical principles corresponding with the kinematical principles explained in § 204, which gave us generalized component velocities: and, as we shall see later, the generalized equations of continuous motion are not only very convenient for the solution of problems, but most *instructive* as to the nature of relations, however complicated, between the motions of different parts of a system. In the meantime we shall consider the generalized expressions for the impulsive generation of motion. We have seen above (§ 308) that the kinetic energy acquired by a system given at rest and struck with any given impulses, is equal to half the sum of the products of the component forces multiplied each into the corresponding component of the velocity acquired by its point of application, when the ordinary system of rectangular co-ordinates is used. Precisely the same statement holds on the generalized system, and if stated as the convention agreed upon, it suffices to define the generalized components of impulse, those of velocity having been fixed on kinematical principles (§ 204). Generalized components of momentum of any specified motion are, of course, equal to the generalized components of the impulse by which it could be generated from rest.

Generalized components of impulse or momentum.

(*a*) Let ψ, ϕ, θ, ... be the generalized co-ordinates of a material system at any time; and let $\dot{\psi}$, $\dot{\phi}$, $\dot{\theta}$, ... be the corresponding generalized velocity-components, that is to say, the rates at which ψ, ϕ, θ, ... increase per unit of time, at any instant, in the actual motion. If x_1, y_1, z_1 denote the common rectangular co-ordinates of one particle of the system, and $\dot{x}_1$, $\dot{y}_1$, $\dot{z}_1$ its component velocities, we have

$$\left.\begin{aligned}\dot{x}_1 &= \frac{dx_1}{d\psi}\dot{\psi} + \frac{dx_1}{d\phi}\dot{\phi} + \text{etc.}\\ \dot{y}_1 &= \frac{dy_1}{d\psi}\dot{\psi} + \frac{dy_1}{d\phi}\dot{\phi} + \text{etc.}\\ &\text{etc.} \qquad\quad \text{etc.}\end{aligned}\right\} \quad (1).$$

Hence the kinetic energy, which is $\Sigma\frac{1}{2}m(\dot{x}^2+\dot{y}^2+\dot{z}^2)$, in terms of rectangular co-ordinates, becomes a quadratic function of $\dot{\psi}$, $\dot{\phi}$, etc., when expressed in terms of generalized co-ordinates, so that if we denote it by T we have

$$T=\tfrac{1}{2}\{(\psi,\psi)\dot{\psi}^2+(\phi,\phi)\dot{\phi}^2+\ldots+2(\psi,\phi)\dot{\psi}\dot{\phi}+\ldots\} \qquad (2),$$

Generalized expression for kinetic energy.

where (ψ,ψ), (ϕ,ϕ), (ψ,ϕ), etc., denote various functions of the co-ordinates, determinable according to the conditions of the system. The only condition essentially fulfilled by these coefficients is, that they must give a finite positive value to T for all values of the variables.

(*b*) Again let (X_1, Y_1, Z_1), (X_2, Y_2, Z_2), etc., denote component forces on the particles (x_1, y_1, z_1), (x_2, y_2, z_2), etc., respectively; and let $(\delta x_1, \delta y_1, \delta z_1)$, etc., denote the components of any infinitely small motions possible without breaking the conditions of the system. The work done by those forces, upon the system when so displaced, will be

$$\Sigma(X\delta x+Y\delta y+Z\delta z) \qquad (3).$$

To transform this into an expression in terms of generalized co-ordinates, we have

$$\left.\begin{aligned} \delta x_1&=\frac{dx_1}{d\psi}\delta\psi+\frac{dx_1}{d\phi}\delta\phi+\text{etc.}\\ \delta y_1&=\frac{dy_1}{d\psi}\delta\psi+\frac{dy_1}{d\phi}\delta\phi+\text{etc.}\\ &\text{etc.} \qquad \text{etc.}\end{aligned}\right\} \qquad (4),$$

and it becomes

$$\Psi\delta\psi+\Phi\delta\phi+\text{etc.} \qquad (5),$$

where

$$\left.\begin{aligned} \Psi&=\Sigma\left(X\frac{dx}{d\psi}+Y\frac{dy}{d\psi}+Z\frac{dz}{d\psi}\right)\\ \Phi&=\Sigma\left(X\frac{dx}{d\phi}+Y\frac{dy}{d\phi}+Z\frac{dz}{d\phi}\right)\\ &\text{etc.} \qquad \text{etc.}\end{aligned}\right\} \qquad (6).$$

Generalized components of force,

These quantities, Ψ, Φ, etc., are clearly *the generalized components of the force on the system.*

Let $\underset{\cdot}{\Psi}$, $\underset{\cdot}{\Phi}$, etc., denote component impulses, generalized on the same principle; that is to say, let

$$\underset{\cdot}{\Psi}=\int_0^\tau \Psi dt,\ \underset{\cdot}{\Phi}=\int_0^\tau \Phi dt,\ \text{etc.,}$$

of impulse.

where Ψ, Φ, ... denote generalized components of the continuous force acting at any instant of the infinitely short time τ, within which the impulse is completed.

If this impulse is applied to the system, previously in motion

Impulsive generation of motion referred to generalized co-ordinates.

in the manner specified above, and if $\delta\dot{\psi}$, $\delta\dot{\phi}$, ... denote the resulting augmentations of the components of velocity, the means of the component velocities before and after the impulse will be

$$\dot{\psi}+\tfrac{1}{2}\delta\dot{\psi},\ \dot{\phi}+\tfrac{1}{2}\delta\dot{\phi},\ \ldots .$$

Hence, according to the general principle explained above for calculating the work done by an impulse, the whole work done in this case is

$$\Psi(\dot{\psi}+\tfrac{1}{2}\delta\dot{\psi})+\Phi(\dot{\phi}+\tfrac{1}{2}\delta\dot{\phi})+\text{etc.}$$

To avoid unnecessary complications, let us suppose $\delta\dot{\psi}$, $\delta\dot{\phi}$, etc., to be each infinitely small. The preceding expression for the work done becomes

$$\Psi\dot{\psi}+\Phi\dot{\phi}+\text{etc.};$$

and, as the effect produced by this work is augmentation of kinetic energy from T to $T+\delta T$, we must have

$$\delta T=\Psi\dot{\psi}+\Phi\dot{\phi}+\text{etc.}$$

Now let the impulses be such as to augment $\dot{\psi}$ to $\dot{\psi}+\delta\dot{\psi}$, and to leave the other component velocities unchanged. We shall have

$$\Psi\dot{\psi}+\Phi\dot{\phi}+\text{etc.}=\frac{dT}{d\dot{\psi}}\delta\dot{\psi}.$$

Dividing both members by $\delta\dot{\psi}$, and observing that $\frac{dT}{d\dot{\psi}}$ is a linear function of $\dot{\psi}$, $\dot{\phi}$, etc., we see that $\frac{\Psi}{\delta\dot{\psi}}$, $\frac{\Phi}{\delta\dot{\psi}}$, etc., must be equal to the coefficients of $\dot{\psi}$, $\dot{\phi}$, ... respectively in $\frac{dT}{d\dot{\psi}}$.

(*c*) From this we see, further, that the impulse required to produce the component velocity $\dot{\psi}$ from rest, or to generate it in the system moving with any other possible velocity, has for its components

$$(\psi,\psi)\dot{\psi},\ (\psi,\phi)\dot{\psi},\ (\psi,\theta)\dot{\psi},\ \text{etc.}$$

Hence we conclude that to generate the whole resultant velocity $(\dot{\psi}, \dot{\phi}, \ldots)$ from rest, requires an impulse, of which the components, if denoted by ξ, η, ζ, ..., are expressed as follows:—

Momenta in terms of velocities.

$$\left.\begin{aligned}\xi&=(\psi,\psi)\dot{\psi}+(\phi,\psi)\dot{\phi}+(\theta,\psi)\dot{\theta}+\ldots\\ \eta&=(\psi,\phi)\dot{\psi}+(\phi,\phi)\dot{\phi}+(\theta,\phi)\dot{\theta}+\ldots\\ \zeta&=(\psi,\theta)\dot{\psi}+(\phi,\theta)\dot{\phi}+(\theta,\theta)\dot{\theta}+\ldots\\ &\text{etc.}\end{aligned}\right\} \quad (7),$$

where it must be remembered that, as seen in the original expression for T, from which they are derived, (ϕ, ψ) means the same thing as (ψ, ϕ), and so on. The preceding expressions are the differential coefficients of T with reference to the velocities; that is to say, Momenta in terms of velocities.

$$\xi=\frac{dT}{d\dot{\psi}},\ \eta=\frac{dT}{d\dot{\phi}},\ \zeta=\frac{dT}{d\dot{\theta}},\ \ldots \qquad (8).$$

(*d*) The second members of these equations, being linear functions of $\dot{\psi}$, $\dot{\phi}$, ..., we may, by ordinary elimination, find $\dot{\psi}$, $\dot{\phi}$, etc., in terms of ξ, η, etc., and the expressions so obtained are of course linear functions of the last-named elements. And, since T is a quadratic function of $\dot{\psi}$, $\dot{\phi}$, etc., we have

$$2T=\xi\dot{\psi}+\eta\dot{\phi}+\zeta\dot{\theta}+\ldots \qquad (9).$$

Kinetic energy in terms of momenta and velocities.

From this, on the supposition that T, $\dot{\psi}$, $\dot{\phi}$, ... are expressed in terms of ξ, η, ..., we have by differentiation

$$2\frac{dT}{d\xi}=\dot{\psi}+\xi\frac{d\dot{\psi}}{d\xi}+\eta\frac{d\dot{\phi}}{d\xi}+\zeta\frac{d\dot{\theta}}{d\xi}+\ldots$$

Now the algebraic process by which $\dot{\psi}$, $\dot{\phi}$, etc., are obtained in terms of ξ, η, etc., shows that, inasmuch as the coefficient of $\dot{\phi}$ in the expression, (7), for ξ, is equal to the coefficient of $\dot{\psi}$, in the expression for η, and so on; the coefficient of η in the expression for $\dot{\psi}$ must be equal to the coefficient of ξ in the expression for $\dot{\phi}$, and so on; that is to say,

$$\frac{d\dot{\psi}}{d\eta}=\frac{d\dot{\phi}}{d\xi},\ \frac{d\dot{\psi}}{d\zeta}=\frac{d\dot{\theta}}{d\xi},\ \text{etc.}$$

Hence the preceding expression becomes

$$2\frac{dT}{d\xi}=\dot{\psi}+\xi\frac{d\dot{\psi}}{d\xi}+\eta\frac{d\dot{\psi}}{d\eta}+\zeta\frac{d\dot{\psi}}{d\zeta}+\ldots=2\dot{\psi},$$

and therefore

$$\left.\begin{aligned} &\dot{\psi}=\frac{dT}{d\xi}, \\ \text{Similarly}\qquad &\dot{\phi}=\frac{dT}{d\eta},\ \text{etc.} \end{aligned}\right\} \qquad (10)$$

Velocities in terms of momenta.

These expressions solve the direct problem,—to find the velocity produced by a given impulse (ξ, η,...), when we have the kinetic energy, T, expressed as a quadratic function of the components of the impulse.

(*e*) If we consider the motion simply, without reference to the impulse required to generate it from rest, or to stop it, the quantities ξ, η, ... are clearly to be regarded as the components of the momentum of the motion, according to the system of generalized co-ordinates.

Reciprocal relation between momenta and velocities in two motions.

(*f*) The following algebraic relation will be useful:—

$$\xi_{,}\dot{\psi}+\eta_{,}\dot{\phi}+\zeta_{,}\dot{\theta}+\text{ etc.}=\xi\dot{\psi}_{,}+\eta\dot{\phi}_{,}+\zeta\dot{\theta}_{,}+\text{ etc.} \qquad (11).$$

where, ξ, η, $\dot{\psi}$, $\dot{\phi}$, etc., having the same signification as before, $\xi_{,}$, $\eta_{,}$, $\zeta_{,}$, etc., denote the impulse-components corresponding to any other values, $\dot{\psi}_{,}$, $\dot{\phi}_{,}$, $\dot{\theta}_{,}$, etc., of the velocity-components. It is proved by observing that each member of the equation becomes a symmetrical function of $\dot{\psi}$, $\dot{\psi}_{,}$; $\dot{\phi}$, $\dot{\phi}_{,}$; etc.; when for $\xi_{,}$, $\eta_{,}$, etc., their values in terms of $\dot{\psi}_{,}$, $\dot{\phi}_{,}$, etc., and for ξ, η, etc., their values in terms of $\dot{\psi}$, $\dot{\phi}$, etc., are substituted.

Application of generalized coordinates to theorems of § 311.

314. A material system of any kind, given at rest, and subjected to an impulse in any specified direction, and of any given magnitude, moves off so as to take the greatest amount of kinetic energy which the specified impulse can give it.

Let ξ, η,... be the components of the given impulse, and $\dot{\psi}$, $\dot{\phi}$,... the components of the actual motion produced by it, which are determined by the equations (10) above. Now let us suppose the system be guided by means of merely directive constraint, to take, from rest, under the influence of the given impulse, some motion ($\dot{\psi}_{,}$, $\dot{\phi}_{,}$,...) different from the actual motion; and let $\xi_{,}$, $\eta_{,}$,... be the impulse which, with this constraint removed, would produce the motion ($\dot{\psi}_{,}$, $\dot{\phi}_{,}$,...). We shall have, for this case, as above,

$$T_{,}=\tfrac{1}{2}(\xi_{,}\dot{\psi}_{,}+\eta_{,}\dot{\phi}_{,}+\ldots)$$

But $\xi_{,}-\xi$, $\eta_{,}-\eta$.... are the components of the impulse experienced in virtue of the constraint we have supposed introduced. They can neither perform nor consume work on the system when moving as directed by this constraint; that is to say,

$$(\xi_{,}-\xi)\dot{\psi}_{,}+(\eta_{,}-\eta)\dot{\phi}_{,}+(\zeta_{,}-\zeta)\dot{\theta}_{,}+\text{etc.}=0 \qquad (12);$$

and therefore

$$2T_{,}=\xi\dot{\psi}_{,}+\eta\dot{\phi}_{,}+\zeta\dot{\theta}_{,}+\text{etc.}$$

Hence we have

$$\begin{aligned}2(T-T_{,})&=\xi(\dot{\psi}-\dot{\psi}_{,})+\eta(\dot{\phi}-\dot{\phi}_{,})+\text{etc.}\\&=(\xi-\xi_{,})(\dot{\psi}-\dot{\psi}_{,})+(\eta-\eta_{,})(\dot{\phi}-\dot{\phi}_{,})+\text{etc.}\\&\quad+\xi_{,}(\dot{\psi}-\dot{\psi}_{,})+\eta_{,}(\dot{\phi}-\dot{\phi}_{,})+\text{etc.}\end{aligned}$$

But, by (11) above, we have

$$\xi_{,}(\dot{\psi}-\dot{\psi}_{,})+\eta_{,}(\dot{\phi}-\dot{\phi}_{,})+\text{etc.}=(\xi-\xi_{,})\dot{\psi}_{,}+(\eta-\eta_{,})\dot{\phi}_{,}+\text{etc.},$$

and therefore, by (12), each vanishes: and we have

$$2(T-T_{,})=(\xi-\xi_{,})(\dot{\psi}-\dot{\psi}_{,})+(\eta-\eta_{,})(\dot{\phi}-\dot{\phi}_{,})+\text{etc.} \qquad (13).$$

that is to say, T exceeds $T_{,}$ by the value of the kinetic energy that would be generated by an impulse ($\xi-\xi_{\prime}$, $\eta-\eta_{\prime}$, $\zeta-\zeta_{\prime}$, etc.) applied simply to the system, which, of course, is essentially positive. In other words, Theorems of § 311 in terms of generalized co-ordinates.

315. If the system is guided to take, under the action of a given impulse, any motion ($\dot{\psi}_{\prime}$, $\dot{\phi}_{\prime}$, ...) different from the natural motion ($\dot{\psi}$, $\dot{\phi}$, ...), it will have less kinetic energy than that of the natural motion, by a difference equal to the kinetic energy of the motion ($\dot{\psi}-\dot{\psi}_{\prime}$, $\dot{\phi}-\dot{\phi}_{\prime}$, ...).

COR. If a set of material points are struck independently by impulses each given in amount, more kinetic energy is generated if the points are perfectly free to move each independently of all the others, than if they are connected in any way. And the deficiency of energy in the latter case is equal to the amount of the kinetic energy of the motion which geometrically compounded with the motion of either case would give that of the other.

(*a*) Hitherto we have either supposed the motion to be fully given, and the impulses required to produce them, to be to be found; or the impulses to be given and the motions produced by them to be to be found. A not less important class of problems is presented by supposing as many linear equations of condition between the impulses and components of motion to be given as there are degrees of freedom of the system to move (or independent co-ordinates). These equations, and as many more supplied by (8) or their equivalents (10), suffice for the complete solution of the problem, to determine the impulses and the motion. Problems whose data involve impulses and velocities.

(*b*) A very important case of this class is presented by prescribing, among the velocities alone, a number of linear equations with constant terms, and supposing the impulses to be so directed and related as to do no work on any velocities satisfying another prescribed set of linear equations with no constant terms; the whole number of equations of course being equal to the number of independent co-ordinates of the system. The equations for solving this problem need not be written down, as they are obvious; but the following reduction is useful, as affording the easiest proof of the *minimum* property stated below.

(*c*) The given equations among the velocities may be reduced to a set, each homogeneous, except one equation with a constant term. Those homogeneous equations diminish the number of degrees of freedom; and we may transform the co-ordinates so as

Problems whose data involve impulses and velocities.

to have the number of independent co-ordinates diminished accordingly. Farther, we may choose the new co-ordinates, so that the linear function of the velocities in the single equation with a constant term may be one of the new velocity-components; and the linear functions of the velocities appearing in the equation connected with the prescribed conditions as to the impulses may be the remaining velocity-components. Thus the impulse will fulfil the condition of doing no work on any other component velocity than the one which is given, and the general problem—

General problem.

316. Given any material system at rest. Let any parts of it be set in motion suddenly with any specified velocities possible, according to the conditions of the system; and let its other parts be influenced only by its connexions with those. It is required to find the motion:

takes the following very simple form:—An impulse of the character specified as a particular component, according to the generalized method of co-ordinates, acts on a material system; its amount being such as to produce a given velocity-component of the corresponding type. It is required to find the motion.

The solution of course is to be found from the equations

$$\dot\psi = A, \qquad \eta = 0, \qquad \zeta = 0, \text{ etc.}, \tag{15}$$

which are the special equations of condition of the problem; and the general kinetic equations (7), or (10). Choosing the latter, and denoting by $[\xi,\xi]$, $[\xi,\eta]$, etc., the coefficients of $\frac{1}{2}\xi^2$, $\xi\eta$, etc., in T, we have

$$\xi = \frac{A}{[\xi,\xi]}, \quad \dot\phi = \frac{[\xi,\eta]}{[\xi,\xi]}A, \quad \dot\theta = \frac{[\xi,\zeta]}{[\xi,\xi]}A, \text{ etc.} \tag{16}$$

for the result.

This result possesses the remarkable property, that the kinetic energy of the motion expressed by it is less than that of any other motion which fulfils the prescribed condition as to velocity. For, if $\xi_,$, $\eta_,$, $\zeta_,$, etc., denote the impulses required to produce any other motion, $\dot\psi_,$, $\dot\phi_,$, $\dot\theta_,$, etc., and $T_,$ the corresponding kinetic energy, we have, by (9),

$$2T_, = \xi_,\dot\psi_, + \eta_,\dot\phi_, + \zeta_,\dot\theta_, + \text{ etc.}$$

But by (11),

$$\xi_,\dot\psi + \eta_,\dot\phi + \zeta_,\dot\theta + \text{ etc.} = \xi\dot\psi_,,$$

since, by (15), we have $\eta = 0$, $\zeta = 0$, etc. Hence

$$2T_, = \xi\dot\psi_, + \xi_,(\dot\psi_, - \dot\psi) + \eta_,(\dot\phi_, - \dot\phi) + \zeta_,(\dot\theta_, - \dot\theta) + \ldots$$

Now let also this second case $(\dot{\psi}_{,}, \dot{\phi}_{,},\ldots)$ of motion fulfil the prescribed velocity-condition $\dot{\psi}_{,}=A$. We shall have

General problem.

$$\xi_{,}(\dot{\psi}_{,}-\dot{\psi})+\eta_{,}(\dot{\phi}_{,}-\dot{\phi})+\zeta_{,}(\dot{\theta}_{,}-\dot{\theta})+\ldots=(\xi_{,}-\xi)(\dot{\psi}_{,}-\dot{\psi})$$
$$+(\eta_{,}-\eta)(\dot{\phi}_{,}-\dot{\phi})+(\zeta-\zeta)(\dot{\theta}-\dot{\theta})+\ldots,$$

since $\dot{\psi}_{,}-\dot{\psi}=0$, $\eta=0$, $\zeta=0,\ldots$. Hence if $\mathfrak{T}$ denote the kinetic energy of the motion generated from rest, by the impulse $(\xi_{,}-\xi, \eta_{,}-\zeta_{,},\ldots)$, we have

$$2T_{,}=2T+2\mathfrak{T} \quad . \qquad (17)$$

But $\mathfrak{T}$ is essentially positive and therefore $T_{,}$, the kinetic energy of any motion fulfilling the prescribed velocity-condition, but differing from the actual motion, is greater than T the kinetic energy of the actual motion; and the amount, $\mathfrak{T}$, of the difference is given by the equation

$$2\mathfrak{T}=\eta_{,}(\dot{\phi}_{,}-\dot{\phi})+\zeta_{,}(\dot{\theta}_{,}-\dot{\theta})+\ldots \qquad (18).$$

In other words,

317. The solution of the problem is, that The motion actually taken by the system is that which has less kinetic energy than any other motion fulfilling the prescribed velocity conditions. And the excess of the energy of any other such motion, above that of the actual motion, is equal to the energy of the motion that would be generated by the action alone of the impulse which, if compounded with the impulse producing the actual motion, would produce this other supposed motion.

Kinetic energy a minimum in this case.

In dealing with cases it would rarely happen that the use of the particular co-ordinate system required for the application of the solution (16) could be convenient; but in all cases, even in such as in examples (2) and (3) below, which involve an infinite number of degrees of freedom, the minimum property now proved affords an easy solution.

Example (1). Let a smooth plane, constrained to keep moving with a given normal velocity, q, come in contact with a free rigid body at rest: to find the motion produced. The velocity condition here is, that the motion shall consist of any motion whatever giving to the point of the body which is struck a stated velocity, q, perpendicular to the impinging plane, compounded with any motion whatever giving to the same point any velocity parallel to this plane. To express this condition, let u, v, w be rectangular component linear velocities of the centre of gravity, and let ϖ, ρ, σ be component angular velocities round axes through the centre of gravity parallel to the lines of reference.

Impact of a smooth rigid plane of infinite mass on a free rigid body at rest.

Impact of a smooth rigid plane of infinite mass on a free rigid body at rest.

Thus, if x, y, z denote the co-ordinates of the point struck, relatively to these axes through the centre of gravity, and if l, m, n be the direction-cosines of the normal to the impinging plane, the prescribed velocity-condition becomes

$$(u+\rho z-\sigma y)l+(v+\sigma x-\varpi z)m+(w+\varpi y-\rho x)n=-q \quad (a),$$

the negative sign being placed before q on the understanding that the motion of the impinging plane is obliquely, if not directly, *towards* the centre of gravity, when l, m, n are each positive. If, now, we suppose the rectangular axes through the centre of gravity to be principal axes of the body, and denote by Mf^2, Mg^2, Mh^2 the moments of inertia round them, we have

$$T=\tfrac{1}{2}M(u^2+v^2+w^2+f^2\varpi^2+g^2\rho^2+h^2\sigma^2) \quad (b).$$

This must be made a minimum subject to the equation of condition (a). Hence, by the ordinary method of indeterminate multipliers,

$$\left.\begin{array}{c} Mu+\lambda l=0,\ Mv+\lambda m=0,\ Mw+\lambda n=0 \\ Mf^2\varpi+\lambda(ny-mz)=0,\ Mg^2\rho+\lambda(lz-nx)=0,\ Mh^2\sigma+\lambda(mx-ly)=0 \end{array}\right\} (c)$$

These six equations give each of them the value of one of the six unknown quantities u, v, w, ϖ, ρ, σ, in terms of λ and data. Using the values thus found in (a), we have an equation to determine λ; and thus the solution is completed. The first three of equations (c) show that λ, which has entered as an indeterminate multiplier, is to be interpreted as the measure of the amount of the impulse.

Generation of motion by impulse in an inextensible cord or chain.

Example (2). A stated velocity in a stated direction is communicated impulsively to each end of a flexible inextensible cord forming any curvilineal arc: it is required to find the initial motion of the whole cord.

Let x, y, z be the co-ordinates of any point P in it, and $\dot{x}$, $\dot{y}$, $\dot{z}$ the components of the required initial velocity. Let also s be the length from one end to the point P.

If the cord were extensible, the rate per unit of time of the stretching per unit of length which it would experience at P, in virtue of the motion $\dot{x}$, $\dot{y}$, $\dot{z}$, would be

$$\frac{dx}{ds}\frac{d\dot{x}}{ds}+\frac{dy}{ds}\frac{d\dot{y}}{ds}+\frac{dz}{ds}\frac{d\dot{z}}{ds}.$$

Hence, as the cord is inextensible, by hypothesis,

$$\frac{dx}{ds}\frac{d\dot{x}}{ds}+\frac{dy}{ds}\frac{d\dot{y}}{ds}+\frac{dz}{ds}\frac{d\dot{z}}{ds}=0 \quad (a)$$

Generation of motion by impulse in an inextensible cord or chain.

Subject to this, the kinematical condition of the system, and

$$\left.\begin{matrix}\dot{x}=u\\ \dot{y}=v\\ \dot{z}=w\end{matrix}\right\}\text{when } s=0 \qquad \left.\begin{matrix}\dot{x}=u'\\ \dot{y}=v'\\ \dot{z}=w'\end{matrix}\right\}\text{when } s=l,$$

l denoting the length of the cord, and (u, v, w), (u', v', w), the components of the given velocities at its two ends: it is required to find $\dot{x}$, $\dot{y}$, $\dot{z}$ at every point, so as to make

$$\int_0^l \tfrac{1}{2}\mu(\dot{x}^2+\dot{y}^2+\dot{z}^2)ds \qquad (b)$$

a minimum, μ denoting the mass of the string per unit of length, at the point P, which need not be uniform from point to point; and of course

$$ds=(dx^2+dy^2+dz^2)^{\frac{1}{2}} \qquad (c)$$

Multiplying (a) by λ, an indeterminate multiplier, and proceeding as usual according to the method of variations, we have

$$\int_0^l \{\mu(\dot{x}\delta\dot{x}+\dot{y}\delta\dot{y}+\dot{z}\delta\dot{z})+\lambda(\frac{dx}{ds}\frac{d\delta\dot{x}}{ds}+\frac{dy}{ds}\frac{d\delta\dot{y}}{ds}+\frac{dz}{ds}\frac{d\delta\dot{z}}{ds})\}ds=0,$$

in which we may regard x, y, z as known functions of s, and this it is convenient we should make independent variable. Integrating "by parts" the portion of the first member which contains λ, and attending to the terminal conditions, we find, according to the regular process, for the equations containing the solution

$$\mu\dot{x}=\frac{d}{ds}(\lambda\frac{dx}{ds}),\ \mu\dot{y}=\frac{d}{ds}(\lambda\frac{dy}{ds}),\ \mu\dot{z}=\frac{d}{ds}(\lambda\frac{dz}{ds}) \qquad (d)$$

These three equations suffice to determine the four unknown quantities, $\dot{x}$, $\dot{y}$, $\dot{z}$, and λ. Using them to eliminate $\dot{x}$, $\dot{y}$, $\dot{z}$ from (a), we have

$$0=\frac{d\frac{1}{\mu}}{ds}\{\frac{dx}{ds}\frac{d}{ds}(\lambda\frac{dx}{ds})+\ldots\}+\frac{1}{\mu}\{\frac{dx}{ds}\frac{d^2}{ds^2}(\lambda\frac{dx}{ds})+\ldots\}.$$

Supposing now, for simplicity, that s is independent variable, and performing the differentiation here indicated, with attention to the following relations:—

$$\frac{dx^2}{ds^2}+\ldots=1,\ \frac{dx}{ds}\frac{d^2x}{ds^2}+\ldots=0$$

$$\frac{dx}{ds}\frac{d^3x}{ds^3}+\ldots+(\frac{d^2x}{ds^2})^2+\ldots=0,$$

and the expression (§ 9) for ρ, the radius of curvature, we find

$$\frac{1}{\mu}\frac{d^2\lambda}{ds^2}+\frac{d(\frac{1}{\mu})}{ds}\frac{d\lambda}{ds}-\frac{\lambda}{\mu\rho^2}=0 \qquad (e)$$

Generation of motion by impulse in an inextensible cord or chain.

The interpretation of (c) is very obvious. It shows that λ is the impulsive tension at the point P of the string; and that the velocity which this point acquires instantaneously is the resultant of $\frac{1}{\mu}\frac{d\lambda}{ds}$ tangential, and $\frac{\lambda}{\rho\mu}$ towards the centre of curvature. The differential equation (e) therefore shows the law of transmission of the instantaneous tension along the string, and proves that it depends solely on the mass of the cord per unit of length in each part, and the curvature from point to point, but not at all on the plane of curvature, of the initial form. Thus, for instance, it will be the same along a helix as along a circle of the same curvature.

With reference to the fulfilling of the six terminal equations, a difficulty occurs inasmuch as $\dot{x}$, $\dot{y}$, $\dot{z}$ are expressed by (d) immediately, without the introduction of fresh arbitrary constants, in terms of λ, which, as the solution of a differential equation of the second degree, involves only two arbitrary constants. The explanation is, that at any point of the cord, at any instant, any velocity in any direction perpendicular to the tangent may be generated without at all altering the condition of the cord even at points infinitely near it. This, which seems clear enough without proof, may be demonstrated analytically by transforming the kinematical equation (a) thus. Let f be the component tangential velocity, q the component velocity towards the centre of curvature, and p the component velocity perpendicular to the osculating plane. Using the elementary formulæ for the direction cosines of these lines (§ 9), and remembering that s is now independent variable, we have

$$\dot{x}=f\frac{dx}{ds}+q\frac{\rho d^2x}{ds^2}+p\frac{\rho(dzd^2y-dyd^2z)}{ds^2},\ \dot{y}=\text{etc.}$$

Substituting these in (a) and reducing, we find

$$\frac{df}{ds}=\frac{q}{\rho} \qquad (f)$$

a form of the kinematical equation of a flexible line which will be of much use to us later.

We see, therefore, that if the tangential components of the impressed terminal velocities have any prescribed values, we may give besides, to the ends, any velocities whatever perpendicular to the tangents, without altering the motion acquired by any part of the cord. From this it is clear also, that the directions of the terminal impulses are necessarily tangential; or, in other words, that an impulse inclined to the tangent at either end, would generate an infinite transverse velocity.

Generation of motion by impulse in an inextensible cord or chain.

To express, then, the terminal conditions, let F and F' be the tangential velocities produced at the ends, which we suppose known. We have, for any point, P, as we have seen above,

$$f=\frac{1}{\mu}\frac{d\lambda}{ds} \qquad (g)$$

and hence when

$$\left.\begin{aligned} s=0,\ \frac{1}{\mu}\frac{d\lambda}{ds}&=F \\ \text{and when}\qquad s=l,\ \frac{1}{\mu}\frac{d\lambda}{ds}&=F' \end{aligned}\right\} \qquad (h)$$

which suffice to determine the constants of integration of (d). Or if the data are the tangential impulses, I, I', required at the ends to produce the motion, we have

$$\left.\begin{aligned} \text{when}\qquad s=0,\ \lambda&=I, \\ \text{and when}\qquad s=l,\ \lambda&=I' \end{aligned}\right\} \qquad (i)$$

Or if either end be free, we have $\lambda=0$ at it, and any prescribed condition as to impulse applied, or velocity generated, at the other end.

The solution of this problem is very interesting, as showing how rapidly the propagation of the impulse falls off with "change of direction" along the cord. The reader will have no difficulty in illustrating this by working it out in detail for the case of a cord either uniform or such that $\mu\frac{d\frac{1}{\mu}}{ds}$ is constant, and given in the form of a circle or helix. The results have curious, and dynamically most interesting, bearings on the motions of a whip lash, and of the rope in harpooning a whale.

Impulsive motion of incompressible liquid

Example (3). Let a mass of incompressible liquid be given at rest completely filling a closed vessel of any shape; and let, by suddenly commencing to change the shape of this vessel, any arbitrarily prescribed normal velocities be suddenly produced in the liquid at all points of its bounding surface, subject to the condition of not altering the volume: It is required to find the instantaneous velocity of any interior point of the fluid.

Let x, y, z be the co-ordinates of any point P of the space occupied by the fluid, and let u, v, w be the components of the generated velocity of the fluid at this point. Then ρ being the density of the fluid, and $\iiint$ denoting integration throughout the space occupied by the fluid, we have

$$T=\iiint\tfrac{1}{2}\rho(u^2+v^2+w^2)dxdydz \qquad (a)$$

Impulsive motion of incompressible liquid.

which, subject to the kinematical condition (§ 193),

$$\frac{du}{dx}+\frac{dv}{dy}+\frac{dw}{dz}=0 \qquad (b)$$

must be the least possible, with the given surface values of the normal component velocity. By the method of variation we have

$$\iiint\{\rho(u\delta u+v\delta v+w\delta w)+\lambda(\frac{d\delta u}{dx}+\frac{d\delta v}{dy}+\frac{d\delta w}{dz})\}dxdydz=0. \qquad (c)$$

But integrating by parts we have

$$\iiint\lambda(\frac{d\delta u}{dx}+\frac{d\delta v}{dy}+\frac{d\delta w}{dz})dxdydz=\iint\lambda(\delta udydz+\delta vdzdx+\delta wdxdy)$$

$$-\iiint(\delta u\frac{d\lambda}{dx}+\delta v\frac{d\lambda}{dy}+\delta w\frac{d\lambda}{dz})dxdydz, \qquad (d)$$

and if l, m, n denote the direction cosines of the normal at any point of the surface, dS an element of the surface, and $\iint$ integration over the whole surface, we have

$$\iint\lambda(\delta udydz+\delta vdzdx+\delta wdxdy)=\iint\lambda(l\delta u+m\delta v+n\delta w)dS=0,$$

since the normal component of the velocity is given, which requires that $l\delta u+m\delta v+n\delta w=0$. Using this in (c), (d), and equating the coefficients of δu, δv, δw, we have

$$\rho u=\frac{d\lambda}{dx},\quad \rho v=\frac{d\lambda}{dy},\quad \rho w=\frac{d\lambda}{dz} \qquad (e).$$

These, used to eliminate u, v, w from (b), give

$$\frac{d}{dx}(\frac{1}{\rho}\frac{d\lambda}{dx})+\frac{d}{dy}(\frac{1}{\rho}\frac{d\lambda}{dy})+\frac{d}{dz}(\frac{1}{\rho}\frac{d\lambda}{dz})=0 \qquad (f),$$

an equation for the determination of λ, whence by (e) the solution is completed.

The condition to be fulfilled, besides the kinematical equation (b), amounts to this merely,—that $\rho(udx+vdy+wdz)$ must be a complete differential. If the fluid is homogeneous, ρ is constant, and $udx+vdy+wdz$ must be a complete differential; in other words, the motion suddenly generated must be of the "non-rotational" character [§ 190, (i)] throughout the fluid mass. The equation to determine λ becomes, in this case,

$$\frac{d^2\lambda}{dx^2}+\frac{d^2\lambda}{dy^2}+\frac{d^2\lambda}{dz^2}=0 \qquad (g).$$

From the hydrodynamical principles explained later it will appear that λ, the function of which $\rho(udx+vdy+wdz)$ is the differential, is the impulsive pressure at the point (x, y, z) of the fluid. Hence we may infer that the equation (f), with the condition that λ shall have a given value at every point of a certain closed surface, has a possible and a determinate solution for every point within that surface. This is precisely the same problem as the determination of the permanent temperature at any point within a heterogeneous solid of which the surface is kept permanently with any non-uniform distribution of temperature over it, (f) being Fourier's equation for the uniform conduction of heat through a solid of which the conducting power at the point (x, y, z) is $\frac{1}{\rho}$. The possibility and the determinateness of this problem were both proved above [Chap. I. App. A, (e)] by a demonstration, the comparison of which with the present is instructive. The other case of superficial condition—that with which we have commenced here—shows that the equation (f), with $l\frac{d\lambda}{dx} + m\frac{d\lambda}{dy} + n\frac{d\lambda}{dz}$ given arbitrarily for every point of the surface, has also a possible and single solution for the whole interior space. This, as we shall see in examining the mathematical theory of magnetic induction, may also be inferred from the general theorem (e) of App. A above, by supposing a to be zero for all points without the given surface, and to have the value $\frac{1}{\rho}$ for any internal point (x, y, z). Impulsive motion of incompressible liquid.

318. Maupertuis' celebrated principle of *Least Action* has been, even up to the present time, regarded rather as a curious and somewhat perplexing property of motion, than as a useful guide in kinetic investigations. We are strongly impressed with the conviction that a much more profound importance will be attached to it, not only in abstract dynamics, but in the theory of the several branches of physical science now beginning to receive dynamic explanations. As an extension of it, Sir W. R. Hamilton[1] has evolved his method of *Varying Action*, which undoubtedly must become a most valuable aid in future generalizations. Least action.

What is meant by "Action" in these expressions is, unfortunately, something very different from the *Actio Agentis* defined by Newton, and, it must be admitted, is a much less judiciously chosen word. Taking it, however, as we find it, now universally used by writers on dynamics, we define the *Action of a Moving System* as proportional to the average kinetic energy, which it has possessed for the time from any convenient epoch of reckoning, multiplied by the time. Ac- Action. Time average of energy.

[1] *Phil. Trans.* 1834-1835.

Time average of energy doubled.

cording to the unit generally adopted, the action of a system which has not varied in its kinetic energy, is twice the amount of the energy multiplied by the time from the epoch. Or if the energy has been sometimes greater and sometimes less, the action at time t, is the double of what we may call the *time-integral* of the energy, that is to say, is denoted in the integral calculus by

$$2\int_0^t T d\tau$$

where T denotes the kinetic energy at any time τ, between the epoch and t.

Let m be the mass, and v the velocity at time τ, of any one of the material points of which the system is composed. We have

$$T = \Sigma \tfrac{1}{2} m v^2 \qquad (1),$$

and therefore, if A denote the action at time t,

$$A = \int_0^t \Sigma m v^2 d\tau \qquad (2).$$

This may be put otherwise by taking ds to denote the space described by a particle in time $d\tau$, so that $v d\tau = ds$, and therefore

$$A = \int \Sigma m v ds \qquad (3).$$

or, if x, y, z be the rectangular co-ordinates of m at any time,

$$A = \int \Sigma m (\dot{x} dx + \dot{y} dy + \dot{z} dz) \qquad (4).$$

Hence we might, as many writers in fact have virtually done, define action thus:—

Space average of momenta.

The action of a system is equal to the sum of the *average momenta for the spaces* described by the particles from any era each multiplied by the length of its path.

Least action.

319. The principle of Least Action is this:—Of all the different sets of paths along which a conservative system may be guided to move from one configuration to another, with the sum of its potential and kinetic energies equal to a given constant, that one for which the action is the least is such that the system will require only to be started with the proper velocities, to move along it unguided.

Let x, y, z be the co-ordinates of a particle, m, of the system, at time τ, and V the potential energy of the system in its particular configuration at this instant; and let it be required to find the way to pass from one given configuration to another with velocities at each instant satisfying the condition

$$\Sigma \tfrac{1}{2} m (\dot{x}^2 + \dot{y}^2 + \dot{z}^2) + V = E, \text{ a constant} \qquad (5).$$

so that A, or

$$\int \Sigma m(\dot{x}dx+\dot{y}dy+\dot{z}dz)$$

may be the least possible. Least action.

By the method of variations we must have $\delta A=0$, where

$$\delta A=\int \Sigma m(\dot{x}d\delta x+\dot{y}d\delta y+\dot{z}d\delta z+\delta\dot{x}dx+\delta\dot{y}dy+\delta\dot{z}dz) \qquad (6).$$

Taking in this $dx=\dot{x}d\tau$, $dy=\dot{y}d\tau$, $dz=\dot{z}d\tau$, and remarking that

$$\Sigma m(\dot{x}\delta\dot{x}+\dot{y}\delta\dot{y}+\dot{z}\delta\dot{z})=\delta T \qquad (7),$$

we have

$$\int \Sigma m(\delta\dot{x}dx+\delta\dot{y}dy+\delta\dot{z}dz)=\int_0^t \delta T d\tau \qquad (8).$$

Also, by integration by parts,

$$\int \Sigma m(\dot{x}d\delta x+\ldots)=\{\Sigma m(\dot{x}\delta x+\ldots)\}-[\Sigma m(\dot{x}\delta x+\ldots)]-\int \Sigma m(\ddot{x}\delta x+\ldots)d\tau,$$

where $[\ldots]$ and $\{\ldots\}$ denote the values of the quantities enclosed, at the beginning and end of the motion considered, and where, further, it must be remembered that $d\dot{x}=\ddot{x}d\tau$, etc. Hence, from above,

$$\delta A=\{\Sigma m(\dot{x}\delta x+\dot{y}\delta y+\dot{z}\delta z)\}-[\Sigma m(\dot{x}\delta x+\dot{y}\delta y+\dot{z}\delta z)]$$
$$+\int_0^t d\tau[\delta T-\Sigma m(\ddot{x}\delta x+\ddot{y}\delta y+\ddot{z}\delta z)] \qquad (9).$$

This, it may be observed, is a perfectly general kinematical expression, unrestricted by any terminal or kinetic conditions. Now in the present problem we suppose the initial and final positions to be invariable. Hence the terminal variations, δx, etc., must all vanish, and therefore the integrated expressions $\{\ldots\}$, $[\ldots]$ disappear. Also, in the present problem $\delta T=-\delta V$, by the equation of energy (5). Hence, to make $\delta A=0$, since the intermediate variations, δx, etc., are quite arbitrary, subject only to the conditions of the system, we must have

$$\Sigma m(\ddot{x}\delta x+\ddot{y}\delta y+\ddot{z}\delta z)+\delta V=0 \qquad (10),$$

which [(4), § 293 above] is the general variational equation of motion of a conservative system. This proves the proposition. Hence also it follows that

320. In any unguided motion whatever, of a conservative system, the Action from any one stated position to any other, though not necessarily a minimum, fulfils the *stationary condition*, that is to say, the condition that the variation vanishes, which secures either a minimum or maximum, or maximum-minimum. Why called "stationary action" by Hamilton.

This can scarcely be made intelligible without mathematical language. Let (x_1, y_1, z_1), (x_2, y_2, z_2), etc., be the co-ordinates

Stationary action.

of particles, m_1, m_2, etc., composing the system; at any time τ of the actual motion. Let V be the potential energy of the system, in this configuration; and let E denote the given value of the sum of the potential and kinetic energies. The equation of energy is—

$$\tfrac{1}{2}\{m_1(\dot{x}_1^2+\dot{y}_1^2+\dot{z}_1^2)+m_2(\dot{x}_2^2+\dot{y}_2^2+\dot{z}_2^2)+\text{etc.}\}+V=E \quad (1).$$

Choosing any part of the motion, for instance that from time 0 to time t, we have, for the action during it,

$$A=\int_0^t (E-V)d\tau=Et-\int_0^t Vd\tau \quad (11).$$

Let now the system be guided to move in any other way possible for it, with any other velocities, from the same initial to the same final configuration as in the given motion, subject only to the condition, that the sum of the kinetic and potential energies shall still be E. Let $(x_1', y_1' z_1')$, etc., be the co-ordinates, and V' the corresponding potential energy; and let $(\dot{x}_1', \dot{y}_1', z_1)$, etc., be the component velocities, at time τ in this arbitrary motion; equation (2) still holding, for the accented letters, with only E unchanged. For the action we shall have

$$A'=Et'-\int_0^{t'} V'd\tau \quad (12),$$

where t' is the time occupied by this supposed motion. Let now θ denote a small numerical quantity, and let ξ_1, η_1, etc., be finite lines such that

$$\frac{x_1'-x_1}{\xi_1}=\frac{y_1'-y_1}{\eta_1}=\frac{z_1-z_1}{\zeta_1}=\frac{x_2'-x_2}{\xi_2}=\text{etc.}=\theta.$$

The "principle of stationary action" is, that $\frac{V'-V}{\theta}$ vanishes when θ is made infinitely small, for every possible deviation $(\xi_1\theta, \eta_1\theta$, etc.) from the natural way and velocities, subject only to the equation of energy and to the condition of passing through the stated initial and final configurations: and conversely, that if $\frac{V'-V}{\theta}$ vanishes with θ for every possible such deviation, from a certain way and velocities, specified by (x_1, y_1, z_1), etc., as the co-ordinates at t, *this* way and *these* velocities are such that the system unguided will move accordingly if only started with proper velocities from the initial configuration.

Varying action.

321. From this principle of stationary action, founded, as we have seen, on a comparison between a natural motion, and any other motion, arbitrarily guided and subject only to the law of energy, the initial and final configurations of the system being the same in each case; Hamilton passes to the consideration of the variation of the action in a natural or unguided motion of the system produced by varying the initial and final configurations, and the sum of the potential and kinetic energies. The result is, that

Varying action.

322. The rate of *decrease* of the action per unit of increase of any one of the free (generalized) co-ordinates (§ 204) specifying the initial configuration, is equal to the corresponding (generalized) component momentum [§ 313, (*c*)] of the actual motion from that configuration: the rate of *increase* of the action per unit increase of any one of the free co-ordinates specifying the final configuration, is equal to the corresponding component momentum of the actual motion towards this second configuration: and the rate of increase of the action per unit increase of the constant sum of the potential and kinetic energies, is equal to the time occupied by the motion of which the action is reckoned.

To prove this we must, in our previous expression (9) for δA, now suppose the terminal co-ordinates to vary; δT to become $\delta E - \delta V$, in which δE is a constant during the motion; and each set of paths and velocities to belong to an unguided motion of the system, which requires (10) to hold. Hence

$$\delta A = \{\Sigma m(\dot{x}\delta x + \dot{y}\delta y + \dot{z}\delta z)\} - [\Sigma m(\dot{x}\delta x + \dot{y}\delta y + \dot{z}\delta z)] + t\delta E \quad (13).$$

If, now, in the first place, we suppose the particles constituting the system to be all free from constraint, and therefore (x, y, z) for each to be three independent variables, and if, for distinctness, we denote by (x_1', y_1', z_1') and (x_1, y_1, z_1) the co-ordinates of m_1 in its initial and final positions, and by $(\dot{x}_1', \dot{y}_1', \dot{z}_1')$, $(\dot{x}_1, \dot{y}_1, \dot{z}_1)$ the components of the velocity it has at those points, we have, from the preceding, according to the ordinary notation of partial differential coefficients,

$$\left.\begin{aligned} &\frac{dA}{dx_1'} = -m_1\dot{x}_1', \quad \frac{dA}{dy_1'} = -m_1\dot{y}_1', \quad \frac{dA}{dz_1'} = -m_1\dot{z}_1', \text{ etc.} \\ &\frac{dA}{dx_1} = m_1\dot{x}_1, \quad \frac{dA}{dy_1} = m_1\dot{y}_1, \quad \frac{dA}{dz_1} = m_1\dot{z}_1, \text{ etc.} \\ \text{and} \quad &\frac{dA}{dE} = t. \end{aligned}\right\} \quad (14).$$

In these equations we must suppose A to be expressed as a function of the initial and final co-ordinates, in all six times as many

Varying action.

independent variables as there are of particles; and E, one more variable, the sum of the potential and kinetic energies.

If the system consists not of free particles, but of particles connected in any way forming either one rigid body or any number of rigid bodies connected with one another or not, we might, it is true, be contented to regard it still as a system of free particles, by taking into account, among the impressed forces, the forces necessary to compel the satisfaction of the conditions of connexion. But although this method of dealing with a system of connected particles is very simple, so far as the law of energy merely is concerned, Lagrange's methods, whether that of "equations of condition," or, what for our present purposes is much more convenient, his "generalized co-ordinates," relieve us from very troublesome interpretations when we have to consider the displacements of particles due to arbitrary variations in the configuration of a system.

Let us suppose then, for any particular configuration (x_1, y_1, z_1) $(x_2, y_2, z_2)\ldots$, the expression

$m_1(\dot{x}_1\delta x_1+\dot{y}_1\delta y_1+\dot{z}_1\delta z_1)+$etc., to become $\xi\delta\psi+\eta\delta\phi+\zeta\delta\psi+$etc. (15),

when transformed into terms of ψ, ϕ, $\theta\ldots$, generalized co-ordinates, as many in number as there are of degrees of freedom for the system to move [§ 313, (c)].

The same transformation applied to the kinetic energy of the system would obviously give

$$\tfrac{1}{2}m_1(\dot{x}_1^2+\dot{y}_1^2+\dot{z}_1^2)+\text{etc.}=\tfrac{1}{2}(\xi\dot{\psi}+\eta\dot{\phi}+\zeta\dot{\psi}+\text{etc.}) \qquad (16).$$

and hence ξ, η, ζ, etc., are those linear functions of the generalized velocities which we have designated as "generalized components of momentum;" and which, when T, the kinetic energy, is expressed as a quadratic function of the velocities (of course with, in general, functions of the co-ordinates ψ, ϕ, θ, etc., for the coefficients) are derivable from it thus:

$$\xi=\frac{dT}{d\dot{\psi}},\quad \eta=\frac{dT}{d\dot{\phi}},\quad \zeta=\frac{dT}{d\dot{\theta}},\ \text{etc.} \qquad (17).$$

Hence, taking as before non-accented letters for the second, and accented letters for the initial, configurations of the system respectively, we have

$$\left.\begin{aligned}&\frac{dA}{d\psi'}=-\xi', \quad \frac{dA}{d\phi'}=-\eta', \quad \frac{dA}{d\theta'}=-\zeta', \text{ etc.}\\ &\frac{dA}{d\psi}=\xi, \quad \frac{dA}{d\phi}=\eta, \quad \frac{dA}{d\theta}=\zeta, \text{ etc.}\\ &\text{and, as before,}\quad \frac{dA}{dE}=t.\end{aligned}\right\} \qquad (18)$$

These equations (18), including of course (14) as a particular case, express in mathematical terms the proposition stated in words above, as the *Principle of Varying Action.* Varying action.

The values of the momenta, thus (14) and (18) expressed in terms of differential coefficients of A, must of course satisfy the equation of energy. Hence, for the case of free particles,

$$\Sigma\frac{1}{m}\left(\frac{dA^2}{dx^2}+\frac{dA^2}{dy^2}+\frac{dA^2}{dz^2}\right)=2(E-V) \qquad (19)$$

$$\Sigma\frac{1}{m}\left(\frac{dA^2}{dx'^2}+\frac{dA^2}{dy'^2}+\frac{dA^2}{dz'^2}\right)=2(E-V') \qquad (20).$$

Or, in general, for a system of particles or rigid bodies connected in any way, we have (16) and (18),

$$\dot\psi\frac{dA}{d\psi}+\dot\phi\frac{dA}{d\phi}+\dot\theta\frac{dA}{d\theta}+\text{etc.}=2(E-V) \qquad (21)$$

$$\left(\dot\psi'\frac{dA}{d\psi'}+\dot\phi'\frac{dA}{d\phi'}+\dot\theta'\frac{dA}{d\theta'}+\text{etc.}\right)=2(E-V') \qquad (22),$$

where $\dot\psi$, $\dot\phi$, etc., are expressible as linear functions of $\frac{dA}{d\psi}$, $\frac{dA}{d\phi}$, etc., by the solution of the equations

$$\left.\begin{aligned}(\psi,\psi)\dot\psi+(\psi,\phi)\dot\phi+(\psi,\theta)\dot\theta+\text{etc.}&=\xi=\frac{dA}{d\psi}\\ (\phi,\psi)\dot\psi+(\phi,\phi)\dot\phi+(\phi,\theta)\dot\theta+\text{etc.}&=\eta=\frac{dA}{d\phi}\\ \text{etc.}\qquad\qquad\qquad & \text{etc.}\end{aligned}\right\} \qquad (23);$$

and $\dot\psi'$, $\dot\phi'$, etc., as similar functions of $-\frac{dA}{d\psi'}$, $-\frac{dA}{d\phi'}$, etc., by

$$\left.\begin{aligned}(\psi',\psi')\dot\psi'+(\psi',\phi')\dot\phi'+(\psi',\theta')\dot\theta'+\text{etc.}&=\xi'=-\frac{dA}{d\psi'}\\ (\phi',\psi')\dot\psi'+(\phi',\phi')\dot\phi'+(\phi',\theta')\dot\theta'+\text{etc.}&=\eta'=-\frac{dA}{d\phi'}\\ \text{etc.}\qquad\qquad\qquad & \text{etc.}\end{aligned}\right\} \qquad (24),$$

where it must be remembered that (ψ,ψ), (ψ,ϕ), etc., are functions of the specifying elements, ψ, ϕ, θ, etc., depending on the kinematical nature of the co-ordinate system alone, and quite independent of the dynamical problem with which we are now concerned; being the coefficients of the half squares and the products of the generalized velocities in the expression for the kinetic energy of any motion of the system; and that (ψ',ψ'), (ψ',ϕ'), etc., are the same functions with ψ', ϕ', etc., written for ψ, ϕ, θ, etc.; but, on the other hand, that A is a function of all the

Varying action.

elements ψ, ϕ, etc., ψ', ϕ', etc. Thus the first member of (21) is a quadratic function of $\frac{dA}{d\psi}$, $\frac{dA}{d\phi}$, etc., with coefficients, known functions of ψ, ϕ, etc., depending merely on the kinematical relations of the system, and the masses of its parts, but not at all on the actual forces or motions; while the second member is a function of the co-ordinates ψ, ϕ, etc., depending on the forces in the dynamical problem, and a constant expressing the particular value given to the sum of the potential and kinetic energies in the actual motion; and so for (22), and ψ', ϕ', etc.

It is remarkable that the single linear partial differential equation (19) of the first order and second degree, for the case of free particles, or its equivalent (21), is sufficient to determine a function A, such that the equations (14) or (18) express the momenta in an actual motion of the system, subject to the given forces. For, taking the case of free particles first, and differentiating (19) still on the Hamiltonian understanding that A is expressed merely as a function of initial and final co-ordinates, and of E, the sum of the potential and kinetic energies, we have

$$2\Sigma\frac{1}{m}\left(\frac{dA}{dx}\frac{d^2A}{dx_1dx}+\frac{dA}{dy}\frac{d^2A}{dx_1dy}+\frac{dA}{dz}\frac{d^2A}{dx_1dz}\right)=-2\frac{dV}{dx}\,.$$

But, by (14)

$$\frac{1}{m_1}\frac{dA}{dx_1}=\dot{x}_1,\quad \frac{1}{m_1}\frac{dA}{dy_1}=\dot{y}_1,\ \text{etc.},$$

and therefore

$$\frac{d^2A}{dx_1^2}=m_1\frac{d\dot{x}}{dx_1}\,,\quad \frac{d^2A}{dx_1dy_1}=m_1\frac{d\dot{y}_1}{dx_1}=m_1\frac{d\dot{x}_1}{dy_1}\,,\quad \frac{d^2A}{dx_1dz_1}=m_1\frac{d\dot{z}}{dx_1}=m_1\frac{d\dot{x}_1}{dz_1}\,,$$

$$\frac{d^2A}{dx_1dx_2}=m_2\frac{d\dot{x}_2}{dx_1}=m_1\frac{d\dot{x}_1}{dx_2}\,,\ \text{etc.}$$

Using these properly in the preceding and taking half; and writing out for two particles to avoid confusion as to the meaning of Σ, we have

$$m_1\left(\dot{x}_1\frac{d\dot{x}_1}{dx_1}+\dot{y}_1\frac{d\dot{x}_1}{dy_1}+\dot{z}_1\frac{d\dot{x}_1}{dz_1}+\dot{x}_2\frac{d\dot{x}_1}{dx_2}+\dot{y}_2\frac{d\dot{x}_1}{dy_2}+\dot{z}_2\frac{d\dot{x}_1}{dz_2}+\text{etc.}\right)=-\frac{dV}{dx_1}\quad (25).$$

Now if we multiply the first member by dt, we have clearly the change of the value of $m_1\dot{x}_1$ due to varying, still on the Hamiltonian supposition, the co-ordinates of all the points, that is to say, the configuration of the system, from what it is at any moment to what it becomes at a time dt later; and it is therefore the actual change in the value of $m\dot{x}_1$, in the natural motion, from the time, t, when the configuration is $(x_1, y_1, z_1, x_2, \ldots, E)$, to the time $t+dt$. It is therefore equal to $\ddot{x}_1dt$, and hence (25) becomes simply $m_1\ddot{x}_1 = -\frac{dV}{dx_1}$. Similarly we find

Varying action.

$$m_1\ddot{y}_1 = -\frac{dV}{dy_1},\ m_1\ddot{z}_1 = -\frac{dV}{dz_1},\ m_2\ddot{x}_2 = -\frac{dV}{dx_2},\ \text{etc.}$$

But these are [§ 293, (4)] the elementary differential equations of the motions of a conservative system composed of free mutually influencing particles.

If next we regard x_1, y_1, z_1, x_2, etc., as constant, and go through precisely the same process with reference to x_1', y_1', z_1', x_2', etc., we have exactly the same equations among the accented letters, with only the difference that $-A$ appears in place of A; and end with $m_1\ddot{x}_1' = \frac{dV'}{dx_1'}$, from which we infer that, if (20) is satisfied, the motion represented by (14) is a natural motion through the configuration (x_1', y_1' z_1', x_2', etc.)

Hence if both (19) and (20) are satisfied, and if when $x_1 = x_1'$, $y_1 = y_1'$, $z_1 = z_1'$, $x_2 = x_2'$, etc., we have $\frac{dA}{dx_1} = -\frac{dA}{dx_1'}$, etc., the motion represented by (14) is a natural motion through the two configurations (x_1', y_1', z_1', x_2', etc.), and (x_1, y_1, z_1, x_2, etc.) Although the signs in the preceding expressions have been fixed on the supposition that the motion is *from* the former, to the latter configuration, it may clearly be from either towards the other, since whichever way it is, the reverse is also a natural motion (§ 271), according to the general property of a conservative system.

To prove the same thing for a conservative system of particles or rigid bodies connected in any way, we have, in the first place, from (18)

$$\frac{d\eta}{d\psi} = \frac{d\xi}{d\phi},\ \frac{d\zeta}{d\psi} = \frac{d\xi}{d\theta},\ \text{etc.} \qquad (26),$$

where, on the Hamiltonian principle, we suppose ψ, ϕ, etc., and ξ, η, etc., to be expressed as functions of ψ, ϕ, etc., ψ', ϕ', etc., and the sum of the potential and kinetic energies. On the same supposition, differentiating (21), we have,

$$\dot\psi\frac{d\xi}{d\psi} + \dot\phi\frac{d\eta}{d\psi} + \dot\theta\frac{d\zeta}{d\psi} + \text{etc.} + \xi\frac{d\dot\psi}{d\psi} + \eta\frac{d\dot\phi}{d\psi} + \zeta\frac{d\dot\theta}{d\psi} + \text{etc.} = -2\frac{dV}{d\psi} \qquad (27),$$

But, by (26), and by the considerations above, we have

$$\dot\psi\frac{d\xi}{d\psi} + \dot\phi\frac{d\eta}{d\psi} + \dot\theta\frac{d\zeta}{d\psi} + \text{etc.} = \dot\psi\frac{d\xi}{d\psi} + \dot\phi\frac{d\xi}{d\phi} + \dot\theta\frac{d\xi}{d\theta} + \text{etc.} = \dot\xi \qquad (28),$$

where $\dot\xi$ denotes the rate of variation of ξ per unit of time in the actual motion.

Varying action.

Again, we have

$$\frac{d\dot\psi}{d\psi}=\frac{\partial\dot\psi}{d\xi}\frac{d\xi}{d\psi}+\frac{\partial\dot\psi}{d\eta}\frac{d\eta}{d\psi}+\text{ etc.}+\frac{\partial\dot\psi}{d\psi},$$

$$\frac{d\dot\phi}{d\psi}=\frac{\partial\dot\phi}{d\xi}\frac{d\xi}{d\psi}+\frac{\partial\dot\phi}{d\eta}\frac{d\eta}{d\psi}+\text{ etc.}+\frac{\partial\dot\psi}{d\psi},\text{ etc.} \qquad (29).$$

if, as in Hamilton's system of canonical equations of motion, we suppose $\dot\psi$, $\dot\phi$, etc., to be expressed as linear functions of ξ, η, etc., with coefficients involving ψ, ϕ, θ, etc., and if we take ∂ to denote the partial differentiation of these functions with reference to the system ξ, η,...ψ, ϕ,..., regarded as independent variables. Let the coefficients be denoted by $[\psi, \psi]$, etc., according to the plan followed above; so that, if the formula for the kinetic energy be

$$T=\tfrac{1}{2}\{[\psi, \psi]\xi^2+[\phi, \phi]\eta^2+\ldots+2[\psi, \phi]\xi\eta+\text{ etc.}\} \qquad (30),$$

we have

$$\left.\begin{aligned}\dot\psi=\frac{dT}{d\xi}&=[\psi, \psi]\xi+[\psi, \phi]\eta+[\psi, \theta]\zeta+\text{ etc.}\\ \dot\phi=\frac{dT}{d\eta}&=[\phi, \psi]\xi+[\phi, \phi]\eta+[\phi, \phi]\zeta+\text{ etc.}\\ \text{etc.}&\qquad\qquad\text{etc.}\end{aligned}\right\} \qquad (31)$$

where of course $[\psi, \phi]$, and $[\phi, \psi]$, mean the same.

Hence $$\frac{\partial\dot\psi}{d\xi}=[\psi, \psi],\ \frac{\partial\dot\phi}{d\xi}=[\phi, \psi], \ldots;$$

$$\frac{\partial\dot\psi}{d\psi}=\frac{d[\psi, \psi]}{d\psi}\xi+\frac{d[\psi, \phi]}{d\psi}\eta+\text{ etc.};\ \frac{\partial\dot\phi}{d\psi}=\frac{d[\phi, \psi]}{d\psi}\xi+\text{ etc. etc.},$$

and therefore, by (29),

$$\xi\frac{d\dot\psi}{d\psi}+\eta\frac{d\dot\phi}{d\psi}+\zeta\frac{d\dot\theta}{d\psi}+\text{etc.}=\{[\psi,\psi]\xi+[\phi,\psi]\eta+\text{etc.}\}\frac{d\xi}{d\psi}+\{[\psi,\phi]\xi+[\phi,\phi]\eta+\text{etc.}$$

$$+\text{ etc.}+\frac{d[\psi, \psi]}{d\psi}\xi^2+\frac{d[\phi,\phi]}{d\psi}\eta^2+\text{ etc.}+2\frac{d[\psi,\phi]}{d\psi}\xi\eta+\text{etc.}$$

$$=\dot\psi\frac{d\xi}{d\psi}+\dot\phi\frac{d\eta}{d\psi}+\text{ etc.}+2\frac{\partial T}{d\psi};$$

whence, by (28), we see that

$$\xi\frac{d\dot\psi}{d\psi}+\eta\frac{d\dot\phi}{d\psi}+\zeta\frac{d\dot\theta}{d\psi}+\text{ etc.}=\dot\xi+2\frac{\partial T}{d\psi} \qquad (32).$$

This, and (28) reduce the first member of (27) to $2\dot\xi+2\frac{\partial T}{d\psi}$, and therefore, halving, we conclude

$$\dot\xi+\frac{\partial T}{d\psi}=-\frac{dV}{d\psi},\text{ and similarly, }\dot\eta+\frac{\partial T}{d\phi}=-\frac{dV}{d\phi},\text{ etc.,} \qquad (33)$$

These, in all as many differential equations as there are of vari ables, ψ, ϕ, etc., suffice for determining them in terms of t and twice as many arbitrary constants. But every solution of the dynamical problem, as has been demonstrated above, satisfies (21) and (23); and therefore it must satisfy these (33), which we have derived from them. These (33) are therefore *the* equations of motion, of the system referred to generalized co-ordinates, as many in number as it has of degrees of freedom. They are the Hamiltonian explicit equations of motion, of which a direct demonstration will be given below. Just as above, it appears therefore, that if (21) and (22) are satisfied, (18) expresses a natural motion of the system from one to another of the two configurations $(\psi, \phi, \theta, \ldots)$ $(\psi', \phi', \theta', \ldots)$. Hence Varying action.

323. The determination of the motion of any conservative system from one to another of any two configurations, when the sum of its potential and kinetic energies is given, depends on the determination of a single function of the co-ordinates specifying those configurations by means of two quadratic partial differential equations of the first order, with reference to those two sets of co-ordinates respectively, with the condition that the corresponding terms of the two differential equations become separately equal when the values of the two sets of co-ordinates agree. The function thus determined and employed to express the solution of the kinetic problem was called the *Characteristic Function* by Sir W. R. Hamilton, to whom the method is due. It is, as we have seen, the "action" from one of the configurations to the other; but its peculiarity in Hamilton's system is, that it is to be expressed as a function of the co-ordinates and a constant, the whole energy, as explained above. It is evidently symmetrical with respect to the two configurations, changing only in sign if their co-ordinates are interchanged. Characteristic function.

Since not only the complete solution of the problem of motion gives a solution, A, of the partial differential equation (19) or (21), but, as we have just seen [§ 322 (33), etc.], every solution of this equation corresponds to an actual problem relative to the motion, it becomes an object of mathematical analysis, which could not be satisfactorily avoided, to find what character of completeness a solution or integral of the differential equation must have in order that a complete integral of the dynamical equations may be derivable from it—a question which seems to have been first noticed by Jacobi. What Characteristic equation of motion.

Complete integral of characteristic equation.

is called a "complete integral" of the differential equation; that is to say, an expression,

$$A = A_0 + F(\psi, \phi, \theta, \ldots \alpha, \beta, \ldots) \qquad (34),$$

for A satisfying it and involving the same number, i, let us suppose, of independent arbitrary constants, A_0, α, β,... as there are of the independent variables, ψ, ϕ, etc.; leads, as he found, to a complete final integral of the equations of motion, expressed as follows:—

$$\frac{dF}{d\alpha} = \mathfrak{A}, \quad \frac{dF}{d\beta} = \mathfrak{B} \ldots, \qquad (35),$$

and, as above,

$$\frac{dF}{dE} = t + \epsilon \qquad (36),$$

where ϵ is the constant depending on the epoch, or era of reckoning, chosen, and $\mathfrak{A}$, $\mathfrak{B}$,... are $i-1$ other arbitrary constants, constituting in all, with E, α, β,..., the proper number, $2i$, of arbitrary constants. This is proved by remarking that (35) are the equations of the "course" (or *paths* in the case of a system of free particles), which is obvious. For they give

$$\left.\begin{aligned} 0 &= \frac{d}{d\psi}\frac{dF}{d\alpha}d\psi + \frac{d}{d\phi}\frac{dF}{d\alpha}d\phi + \frac{d}{d\theta}\frac{dF}{d\alpha}d\theta + \ldots \\ 0 &= \frac{d}{d\psi}\frac{dF}{d\beta}d\psi + \frac{d}{d\phi}\frac{dF}{d\beta}d\phi + \frac{d}{d\theta}\frac{dF}{d\beta}d\theta + \ldots \\ &\quad \text{etc.} \qquad\qquad\qquad \text{etc.} \end{aligned}\right\} \qquad (37),$$

in all $i-1$ equations to determine the ratios $d\psi : d\phi : d\theta : \ldots$ From these, and (21), we find

$$\frac{d\psi}{\dot\psi} = \frac{d\phi}{\dot\phi} = \frac{d\theta}{\dot\theta} \ldots \qquad (38),$$

[since (37) are the same as the equations which we obtain by differentiating (21) and (23) with reference to α, β,... successively, only that they have $d\psi$, $d\phi$, $d\theta$,... in place of $\dot\psi$, $\dot\phi$, $\dot\theta$,...].

General solution of characteristic equation.

A perfectly general solution of the partial differential equation, that is to say, an expression for A including every function of ψ, ϕ, θ,...which can satisfy (21) may of course be found, by the regular process, from the complete integral (34), by eliminating A_0, α, β,... from it by means of an arbitrary equation

$$f(A_0, \alpha, \beta, \ldots) = 0,$$

and the $(i-1)$ equations

$$\frac{1}{\frac{df}{dA_0}} = \frac{\frac{dF}{d\alpha}}{\frac{df}{d\alpha}} = \frac{\frac{dF}{d\beta}}{\frac{df}{d\beta}} = \ldots$$

General solution of characteristic equation.

where f denotes an arbitrary function of the i element, A_0, α, β,... now made to be variables depending on ψ, ϕ,... But the full meaning of the general solution of (21) will be better understood in connexion with the physical problem if we first go back to the Hamiltonian solution, and then from it to the general. Thus, first, let the equations (35) of the course be assumed to be satisfied for each of two sets ψ, ϕ, θ, ..., and ψ', ϕ', θ', ..., of the co-ordinates. They will give $2(i-1)$ equations for determining the $2(i-1)$ constants α, β, ..., $\mathfrak{A}$, $\mathfrak{B}$, ..., in terms of ψ, ϕ, ..., ψ', ϕ', ..., to fulfil these conditions. Using the values of α, β, ..., so found, and assigning A_0 so that A shall vanish when $\psi=\psi'$, $\phi=\phi'$, etc., we have the Hamiltonian expression for A in terms of ψ, ϕ, ..., ψ', ϕ', ..., and E, which is therefore equivalent to a "complete integral" of the partial differential equation (21). Now let ψ', ϕ', ..., be connected by any single arbitrary equation

$$\mathfrak{f}(\psi', \phi', \ldots)=0 \qquad (39),$$

and by means of this equation and the following $(i-1)$ equations, let their values be determined in terms of ψ, ϕ, ..., and E:—

$$\frac{\frac{dA}{d\psi'}}{\frac{d\mathfrak{f}}{d\psi'}}=\frac{\frac{dA}{d\phi'}}{\frac{d\mathfrak{f}}{d\phi'}}=\frac{\frac{dA}{d\theta'}}{\frac{d\mathfrak{f}}{d\theta'}}=\text{etc.} \qquad (40).$$

Substituting the values thus found for ψ', ϕ', θ', etc., in the Hamiltonian A, we have an expression for A, which is the general solution of (21). For we see immediately that (40) expresses that the values of A are equal for all configurations satisfying (39), that is to say, we have

$$\frac{dA}{d\psi'}d\psi'+\frac{dA}{d\phi'}d\phi'+\ldots=0$$

when ψ', ϕ', etc., satisfy (39) and (40). Hence when, by means of these equations, ψ', ϕ', ..., are eliminated from the Hamiltonian expression for A, the complete Hamiltonian differential

$$dA=\left(\frac{dA}{d\psi}\right)d\psi+\left(\frac{dA}{d\phi}\right)d\phi+\ldots+\frac{dA}{d\psi'}d\psi'+\frac{dA}{d\phi'}d\phi'+\ldots \qquad (41)$$

becomes merely

$$dA=\left(\frac{dA}{d\psi}\right)d\psi+\left(\frac{dA}{d\phi}\right)d\phi+\ldots \qquad (42)$$

where $\left(\frac{dA}{d\psi}\right)$, etc., denote the differential coefficients in the Hamiltonian expression. Hence, A being now a function of ψ, ϕ, etc., both as these appear in the Hamiltonian expression and as they are introduced by the elimination of ψ', ϕ', etc., we have

General solution of characteristic equation.

$$\frac{dA}{d\psi}=\left(\frac{dA}{d\psi}\right),\ \frac{dA}{d\phi}=\left(\frac{dA}{d\phi}\right),\ \text{etc.} \qquad (43):$$

and therefore the new expression satisfies the partial differential equation (21). That it is a completely general solution we see, because it satisfies the condition that the action is equal for all configurations fulfilling an absolutely arbitrary equation (39).

For the case of a single free particle, the interpretation of (39) is that the point (x', y', z') is on an arbitrary surface, and of (40) that each line of motion cuts this surface at right angles. Hence

324. The most general possible solution of the quadratic, partial, differential equation of the first order, which Hamilton showed to be satisfied by his Characteristic Function (either terminal configuration alone varying), when interpreted for the case of a single free particle, expresses the action up to any point (x, y, z), from some point of a certain arbitrarily given surface, from which the particle has been projected, in the direction of the normal, and with the proper velocity to make the sum of the potential and actual energies have a given value. In other words, the physical problem solved by the most general solution of that partial differential equation, is this :—

Properties of surfaces of equal action.

Let free particles, not mutually influencing one another, be projected normally from all points of a certain arbitrarily given surface, each with the proper velocity to make the sum of its potential and kinetic energies have a given value. To find, for the particle which passes through a given point (x, y, z), the "action" in its course from the surface of projection to this point. The Hamiltonian principles stated above, show that the surfaces of equal action cut the paths of the particles at right angles ; and give also the following remarkable properties of the motion :—

If, from all points of an arbitrary surface, particles not mutually influencing one another be projected with the proper velocities in the directions of the normals ; points which they reach with equal actions lie on a surface cutting the paths at right angles. The infinitely small thickness of the space between any two such surfaces corresponding to amounts of action differing by any infinitely small quantity, is inversely proportional to the velocity of the particle traversing it ; being equal to the infinitely small difference of action divided by the whole momentum of the particle.

Properties of surfaces of equal action.

Let λ, μ, ν be the direction-cosines of the normal to the surface of equal action through (x, y, z). We have

$$\lambda=\frac{\frac{dA}{dx}}{\left(\frac{dA^2}{dx^2}+\frac{dA^2}{dy^2}+\frac{dA^2}{dz^2}\right)^{\frac{1}{2}}}, \text{ etc.} \tag{1}$$

But $\frac{dA}{dx}=m\dot{x}$, etc., and, if q denote the resultant velocity,

$$mq=\left(\frac{dA^2}{dx^2}+\frac{dA^2}{dy^2}+\frac{dA^2}{dz^2}\right)^{\frac{1}{2}} \tag{2}$$

Hence $$\lambda=\frac{\dot{x}}{q},\ \mu=\frac{\dot{y}}{q},\ \nu=\frac{\dot{z}}{q},$$

which proves the first proposition. Again, if δA denote the infinitely small difference of action from (x, y, z) to any other point $(x+\delta x, y+\delta y, z+dz)$, we have

$$\delta A=\frac{dA}{dx}\delta x+\frac{dA}{dy}\delta y+\frac{dA}{dz}dz.$$

Let the second point be at an infinitely small distance, e, from the first, in the direction of the normal to the surface of equal action; that is to say, let

$$\delta x=e\lambda,\ \delta y=e\mu,\ \delta z=e\nu,$$

Hence, by (1), $$\delta A=e\left(\frac{dA^2}{dx^2}+\frac{dA^2}{dy^2}+\frac{dA}{dz^2}\right)^{\frac{1}{2}} \tag{3};$$

whence, by (2) $$e=\frac{\delta A}{mq} \tag{4},$$

which is the second proposition.

Examples of varying action.

325. Irrespectively of methods for finding the "characterstic function" in kinetic problems, the fact that any case of notion whatever can be represented by means of a single unction in the manner explained in § 323, is most remarkable, and, when geometrically interpreted, leads to highly important and interesting properties of motion, which have valuable applications in various branches of Natural Philosophy. One of the many applications of the general principle made by Hamilton[1] led to a general theory of optical instruments, comprehending the whole in one expression.

Some of its most direct applications; to the motions of planets, comets, etc., considered as free points, and to the celebrated problem of perturbations, known as the Problem of Three Bodies, are worked out in considerable detail by Hamilton

[1] *On the Theory of Systems of Rays.* Trans. R.I.A., 1824, 1830, 1832.

Examples of varying action. (*Phil. Trans.*, 1834-35), and in various memoirs by Jacobi, Liouville, Bour, Donkin, Cayley, Boole, etc. The now abandoned, but still interesting, corpuscular theory of light furnishes a host of good ones. In this theory light is supposed to consist of material particles not mutually influencing one another, but subject to molecular forces from the particles of bodies—not sensible at sensible distances, and therefore not causing any deviation from uniform rectilinear motion in a homogeneous medium, except within an indefinitely small distance from its boundary. The laws of reflection and of single refraction follow correctly from this hypothesis, which therefore suffices for what is called geometrical optics.

Application to common optics, We hope to return to this subject, with sufficient detail, in treating of Optics. At present we limit ourselves to state a theorem comprehending the known rule for measuring the magnifying power of a telescope or microscope (by comparing the diameter of the object-glass with the diameter of pencil of parallel rays emerging from the eye-piece, when a point of light is placed at a great distance in front of the object-glass), as a particular case.

or kinetics of a single particle. **326.** Let any number of attracting or repelling masses, or perfectly smooth elastic objects, be fixed in space. Let two stations, O and O' be chosen. Let a shot be fired with a stated velocity, V, from O, in such a direction as to pass through O'. There may clearly be more than one natural path by which this may be done; but, generally speaking, when one such path is chosen, no other, not sensibly diverging from it, can be found; and any infinitely small deviation in the line of fire from O, will cause the bullet to pass infinitely near to, but not through, O'. Now let a circle, with infinitely small radius r, be described round O as centre, in a plane perpendicular to the line of fire from this point, and let—all with infinitely nearly the same velocity, but fulfilling the condition that the sum of the potential and kinetic energies is the same as that of the shot from O—bullets be fired from all points of this circle, all directed infinitely nearly parallel to the line of fire from O, but each precisely so as to pass through O'. Let a target be held at an infinitely small distance, a', beyond O', in a plane perpendicular to the line of the shot reaching it from O. The bullets fired from the circumference of the circle round O, will, after passing through

O', strike this target in the circumference of an exceedingly small ellipse, each with a velocity (corresponding of course to its position, under the law of energy), differing infinitely little from V', the common velocity with which they pass through O'. Let now a circle, equal to the former, be described round O', in the plane perpendicular to the central path through O', and let bullets be fired from points in its circumference, each with the proper velocity, and in such a direction infinitely nearly parallel to the central path as to make it pass through O. These bullets, if a target is held to receive them perpendicularly at a distance $a = a'\frac{V}{V'}$, beyond O, will strike it along the circumference of an ellipse equal to the former and placed in a corresponding position; and the points struck by the individual bullets will correspond in the manner explained below. Let P and P' be points of the first and second circles, and Q and Q' the points on the first and second targets which bullets from them strike; then if P' be in a plane containing the central path through O', and the position which Q would take if its ellipse were made circular by a pure strain (§ 183); Q and Q' are similarly situated on the two ellipses.

Application to common optics, or kinetics of a single particle.

For, if XOY be a plane perpendicular to the central path through O; and $X'O'Y'$ the corresponding plane through O'. Let A be the "action" from O to O', and ϕ the action from a point $P(x, y, z)$, in the neighbourhood of O, specified with reference to the former axes of co-ordinates, to a point $P'(x', y', z')$, in the neighbourhood of O', specified with reference to the latter.

The function $\phi - A$ vanishes, of course, when $x = 0$, $y = 0$, $z = 0$, $x' = 0$, $y' = 0$, $z' = 0$. Also, for the same values of the co-ordinates, its differential coefficients $\frac{d\phi}{dx}$, $\frac{d\phi}{dy}$, and $\frac{d\phi}{dx'}$, $\frac{d\phi}{dy'}$, must vanish, and $\frac{d\phi}{dz}$, $-\frac{d\phi}{dz'}$ must be respectively equal to V and V', since, for any values whatever of the co-ordinates, $\frac{d\phi}{dx}$ and $\frac{d\phi}{dy}$ are the component velocities parallel to the two lines OX, OY, of the particle passing through P', when it comes from P, and $-\frac{d\phi}{dx'}$ and $-\frac{d\phi}{dy'}$, are the components parallel to OX', OY', of the velocity through P' directed so as to reach P. Hence by Taylor's (or Maclaurin's) theorem we have

Application to common optics, or kinetics of a single particle.

$$\phi - A = -V'z' + Vz$$
$$+\tfrac{1}{2}\{(X,X)x^2+(Y,Y)y^2+\ldots+(X',X')x'^2+\ldots$$
$$+2(Y,Z)yz+\ldots+2(Y',Z')y'z'+\ldots$$
$$+2(X,X')xx'+2(Y,Y')yy'+2(Z,Z')zz'$$
$$+2(X,Y')xy'+2(X,Z')xz'+\ldots+2(Z,Y')zy'\}+R, \quad (1),$$

where (X,X), (X,Y), etc., denote constants, viz., the values of the differential coefficients $\frac{d^2\phi}{dx^2}$, $\frac{d^2\phi}{dxdy}$, etc., when each of the six co-ordinates x, y, z, x', y', z' vanishes; and R denotes the remainder after the terms of the second degree. According to Cauchy's principles regarding the convergence of Taylor's theorem, we have a rigorous expression for $\phi - A$ in the same form, without R, if the coefficients (X,X), etc., denote the values of the differential coefficients with some variable values intermediate between 0 and the actual values of x, y, etc., substituted for these elements. Hence, provided the values of the differential coefficients are infinitely nearly the same for any infinitely small values of the co-ordinates as for the vanishing values, R becomes infinitely smaller than the terms preceding it, when x, y, etc., are each infinitely small. Hence when each of the variables x, y, z, x', y', z' is infinitely small, we may omit R in the expression (1) for $\phi - A$. Now, as in the proposition to be proved, let us suppose z and z' each to be rigorously zero: and we have

$$\frac{d\phi}{dx}=(X,X)x+(X,Y)y+(X,X')x'+(X,Y')y';$$
$$\frac{d\phi}{dy}=(Y,Y)y+(X,Y)x+(Y,X')x'+(Y,Y')y'.$$

These expressions, if in them we make $x=0$, and $y=0$, become the component velocities parallel to OX, OY, of a particle passing through O having been projected from P'. Hence, if ξ, η, ζ denote its co-ordinates, an infinitely small time, $\frac{a}{V}$, after it passes through O, we have $\zeta = a$, and

$$\xi=\{(X,X')x'+(X,Y')y'\}\frac{a}{V}, \quad \eta=\{(Y,X')x'+(Y,Y')y'\}\frac{a}{V}, \quad (2).$$

Here ξ and η are the rectangular co-ordinates of the point Q' in which, in the second case, the supposed target is struck. And by hypothesis $\qquad x'^2+y'^2=r^2 \qquad (3).$

If we eliminate x', y' between these three equations, we have clearly an ellipse; and the former two express the relation of the "corresponding" points. Corresponding equations with x and y for x' and y'; with ξ', η' for ξ, η; and with $-(X,X')$, $-(Y,X')$, $-(X,Y')$, $-(Y,Y')$, in place of (X,X'), (X,Y'),

(Y, X'), (Y, Y'), express the first case. Hence the proposition, as is most easily seen by choosing OX and $O'X'$ so that (X, Y') and (Y, X') may each be zero.

327. The most obvious optical application of this remarkable result is, that in the use of any optical apparatus whatever, if the eye and the object be interchanged without altering the position of the instrument, the magnifying power is unaltered. This is easily understood when, as in an ordinary telescope, microscope, or opera-glass (Galilean telescope), the instrument is symmetrical about an axis, and is curiously contradictory of the common idea that a telescope "diminishes" when looked through the wrong way, which no doubt is true if the telescope is simply reversed about the middle of its length, eye and object remaining fixed. But if the telescope be removed from the eye till its eye-piece is close to the object, the part of the object seen will be seen enlarged to the same extent as when viewed with the telescope held in the usual manner. This is easily verified by looking from a distance of a few yards, in through the object-glass of an opera-glass, at the eye of another person holding it to his eye in the usual way.

Application to common optics.

The more general application may be illustrated thus:—Let the points, O, O' (the centres of the two circles described in the preceding enunciation), be the optic centres of the eyes of two persons looking at one another through any set of lenses, prisms, or transparent media arranged in any way between them. If their pupils are of equal sizes in reality, they will be seen as similar ellipses of equal apparent dimensions by the two observers. Here the imagined particles of light, projected from the circumference of the pupil of either eye, are substituted for the projectiles from the circumference of either circle, and the retina of the other eye takes the place of the target receiving them, in the general kinetic statement.

328. If instead of one free particle we have a conservative system of any number of mutually influencing free particles, the same statement may be applied with reference to the initial position of one of the particles and the final position of another, or with reference to the initial positions, or to the final positions of two of the particles. It serves to show how the influence of an infinitely small change in one of those positions, on the direction of the other particle passing through the other position, is re-

Application to system of free mutually influencing particles.

Application to system of free mutually influencing particles, and to generalized system.

lated to the influence on the direction of the former particle passing through the former position produced by an infinitely small change in the latter position. A corresponding statement, in terms of generalized co-ordinates, may of course be adapted to a system of rigid bodies or particles connected in any way. All such statements are included in the following very general proposition:—

The rate of increase of any one component momentum, corresponding to any one of the co-ordinates, per unit of increase of any other co-ordinate, is equal to the rate of increase of the component momentum corresponding to the latter per unit increase or diminution of the former co-ordinate, according as the two co-ordinates chosen belong to one configuration of the system, or one of them belongs to the initial configuration and the other to the final.

Let ψ and χ be two out of the whole number of co-ordinates constituting the argument of the Hamiltonian characteristic function A; and ξ, η the corresponding momenta. We have [§ 322 (18)]

$$\frac{dA}{d\psi} = \pm\xi, \quad \frac{dA}{d\chi} = \pm\eta,$$

the upper or lower sign being used according as it is a final or an initial co-ordinate that is concerned. Hence

$$\frac{d^2A}{d\psi d\chi} = \pm\frac{d\xi}{d\chi} = \pm\frac{d\eta}{d\psi},$$

and therefore

$$\frac{d\xi}{d\chi} = \frac{d\eta}{d\psi},$$

if both co-ordinates belong to one configuration, or

$$\frac{d\xi}{d\chi} = -\frac{d\eta}{d\psi},$$

if one belongs to the initial configuration, and the other to the final, which is the second proposition. The geometrical interpretation of this statement for the case of a free particle, and two co-ordinates both belonging to one position, its final position, for instance, gives merely the proposition of § 324 above, for the case of particles projected from one point, with equal velocities in all directions; or, in other words, the case of the arbitrary surface of that enunciation, being reduced to a point. To complete the set of variational equations derived from § 322 we have $\frac{dt}{d\chi} = \pm\frac{d\eta}{dE}$ which expresses another remarkable property of conservative motion.

329. Lagrange's generalized form of the equations of motion has been already alluded to (§ 293). We shall now give it, and a few examples of its application. Lagrange's generalized form of the equations of motion.

As above, § 293, we have generally

$$\Sigma[(X-m\ddot{x})\delta x+\ldots]=0 \qquad (1).$$

Now suppose the equations of condition of the system (§ 293) to be

$$\left.\begin{array}{l} F_{,}(x_1, y_1, z_1, x_2, \ldots, t)=0, \\ F_{,,}(x_1, y_1, z_1, x_2, \ldots, t)=0, \\ \text{etc.} \qquad \text{etc.} \end{array}\right\} \qquad (2).$$

It is required to transform (1) into an equation involving only the time and the generalized co-ordinates ψ, ϕ, θ, etc., with their variations $\delta\psi$, $\delta\phi$, $\delta\theta$, etc.

We have, since each of the co-ordinates x_1, y_1, z_1, x_2, etc., is in general a function of t, ψ, ϕ, θ, etc.,

$$\dot{x}=(\frac{dx}{dt})+\frac{dx}{d\psi}\dot{\psi}+\frac{dx}{d\phi}\dot{\phi}+\ldots \qquad (3)$$

where $(\frac{dx}{dt})$ vanishes if t be not explicitly involved in (2), *i.e.*, if the kinematical relations do not vary with the time.

The kinetic energy is therefore

$$T=\tfrac{1}{2}\Sigma m\{[(\frac{dx}{dt})+\frac{dx}{d\psi}\dot{\psi}+\frac{dx}{d\phi}\dot{\phi}+\ldots]^2+[(\frac{dy}{dt})+\text{etc.}]^2+[(\frac{dz}{dt})+\text{etc.}]^2\} \quad (4),$$

which is a quadratic function of $\dot{\psi}$, $\dot{\phi}$, etc., the generalized components of velocity—homogeneous only when (as in § 313, a) the kinematical relations are invariable.

This gives for the partial differential coefficient with respect to $\dot{\psi}$, so far as it is *explicitly* contained in T,

$$(\frac{dT}{d\dot{\psi}})=\Sigma m\{[(\frac{dx}{dt})+\frac{dx}{d\psi}\dot{\psi}+\frac{dx}{d\phi}\dot{\phi}+\ldots]\frac{dx}{d\psi}+[\ldots]\frac{dy}{d\psi}+\text{etc.}\} \quad (5),$$

with similar expressions for the others.

To effect the required transformation of the general equation (1), it is sufficient to operate on one term, as $\ddot{x}\delta x$. We have, since $\frac{d\delta x}{dt}=\delta\dot{x}$,

$$\ddot{x}\delta x=\frac{d}{dt}(\dot{x}\delta x)-\dot{x}\delta\dot{x} \qquad (6).$$

Now $$\delta x=\frac{dx}{d\psi}\delta\psi+\frac{dx}{d\phi}\delta\phi+\text{etc.}$$

Also $$\delta\dot{x}=\frac{d\dot{x}}{d\psi}\delta\psi+\frac{d\dot{x}}{d\dot{\psi}}\delta\dot{\psi}+\text{etc},$$

Lagrange's generalized form of the equations of motion.

since these belong to the configuration at some *definite* instant. [This remark explains the assertion of § 293, that the equations (9) hold even if (8) involve t explicitly.]

But, by (8),

$$\frac{d\dot{x}}{d\dot{\psi}}=\frac{dx}{d\psi}, \text{ etc.}$$

Hence
$$\delta\dot{x}=\frac{d\dot{x}}{d\psi}\delta\psi+\frac{dx}{d\psi}\delta\dot{\psi}+\text{etc.}$$

With these values we find, exhibiting only the terms in $\delta\psi$ and $\delta\dot{\psi}$,

$$\frac{d}{dt}(\dot{x}\delta x)=\frac{d}{dt}\{[(\frac{dx}{dt})+\frac{dx}{d\psi}\dot{\psi}+\frac{dx}{d\phi}\dot{\phi}+\ldots]\frac{dx}{d\psi}\delta\psi+\ldots\}$$
$$=\delta\psi\frac{d}{dt}\{[(\frac{dx}{dt})+\frac{dx}{d\psi}\dot{\psi}+\frac{dx}{d\phi}\dot{\phi}+\ldots]\frac{dx}{d\psi}\}+\ldots$$
$$+\delta\dot{\psi}[(\frac{dx}{dt})+\frac{dx}{d\psi}\dot{\psi}+\frac{dx}{d\phi}\dot{\phi}+\ldots]\frac{dx}{d\psi}+\ldots \qquad (7).$$

Also

$$\dot{x}\delta\dot{x}=\delta\psi[(\frac{dx}{dt})+\frac{dx}{d\psi}\dot{\psi}+\frac{dx}{d\phi}\dot{\phi}+\ldots]\frac{d}{d\psi}[(\frac{dx}{dt})+\frac{dx}{d\psi}\dot{\psi}+\frac{dx}{d\phi}\dot{\phi}+\ldots]$$
$$+\delta\dot{\psi}[(\frac{dx}{dt})+\frac{dx}{d\psi}\dot{\psi}+\frac{dx}{d\phi}\dot{\phi}+\ldots]\frac{dx}{d\psi}+\ldots \qquad (8).$$

In (7) and (8) the terms in $\delta\dot{\psi}$ are equal.

By (6), (7), and (8) we now see that

$$\Sigma m\ddot{x}\delta x=\Sigma m\{\frac{d}{dt}(\dot{x}\delta x)-\dot{x}\delta\dot{x}\}$$
$$=\delta\psi\Big\{\frac{d}{dt}\Sigma m[(\frac{dx}{dt})+\frac{dx}{d\psi}\dot{\psi}+\frac{dx}{d\phi}\dot{\phi}+\ldots]\frac{dx}{d\psi}$$
$$-\Sigma m[(\frac{dx}{dt})+\frac{dx}{d\psi}\dot{\psi}+\frac{dx}{d\phi}\dot{\phi}+\ldots]\frac{d}{d\psi}[(\frac{dx}{dt})+\frac{dx}{d\psi}\dot{\psi}+\frac{dx}{d\phi}\dot{\phi}+]\Big\}$$
$$+\delta\phi\{\ldots\}+\text{ etc.}$$

Hence by inspection of (4) and (5),

$$\Sigma m(\ddot{x}\delta x+\ddot{y}\delta y+\ddot{z}\delta z)=\delta\psi\{\frac{d}{dt}(\frac{dT}{d\dot{\psi}})-\frac{dT}{d\psi}\}+\delta\phi\{\ldots\}+\text{etc.} \qquad (9).$$

But the first member is, by (1), equal to

$$\Sigma\{X\delta x+Y\delta y+Z\delta z\},$$

or in generalized co-ordinates [§ 315 (5)],

$$\Psi\delta\psi+\Phi\delta\phi+\ldots.$$

Equating this to the second number of (9), we have the required transformation of (1).

Hence, as $\delta\psi$, $\delta\phi$, etc., are independent, we have, for the equations of motion

$$\left.\begin{array}{l}\dfrac{d}{dt}\left(\dfrac{dT}{d\dot\psi}\right)-\dfrac{dT}{d\psi}=\Psi\\ \quad\text{etc.}\qquad\text{etc.}\end{array}\right\}\qquad(10),$$

Lagrange's generalized form of the equations of motion.

of which the number is, of course, the same as that of the degrees of freedom. If the system be a conservative one, and if V be the potential energy, we have

Conservative system.

$$-\delta V=\Psi\delta\psi+\ldots\ [\S\ 293\ (3)],$$

and the generalized equations become

$$\left.\begin{array}{l}\dfrac{d}{dt}\left(\dfrac{dT}{d\dot\psi}\right)-\dfrac{dT}{d\psi}=-\dfrac{dV}{d\psi}\\ \quad\text{etc.}\qquad\text{etc.}\end{array}\right\}\qquad(11).$$

Hamilton's form.

When the kinematical relations are invariable, that is to say, when t does not appear in the equations of condition (2), the equations of motion may be put under a slightly different form first given by Hamilton, which is often convenient; thus:—Let T, $\dot\psi$, $\dot\phi$, ..., be expressed in terms of ξ, η, ..., the impulses required to produce the motion from rest at any instant [§ 313 (d)]; so that T will now be a homogeneous quadratic function, and $\dot\psi$, $\dot\phi$, ... each a linear function, of these elements, with coefficients—functions of ψ, ϕ, etc., depending on the kinematical conditions of the system, but not on the particular motion. Thus, denoting, as in § 322 (29), by ∂, partial differentiation with reference to ξ, η, ..., ψ, ϕ, ..., considered as independent variables, we have [§ 313 (10)]

$$\dot\psi=\frac{\partial T}{d\xi},\quad \dot\phi=\frac{\partial T}{d\eta},\ \ldots\qquad(12),$$

and, allowing d to denote, as in what precedes, the partial differentiations with reference to the system $\dot\psi$, $\dot\phi$, ..., ψ, ϕ, ..., we have [§ 313, (8)]

$$\xi=\frac{dT}{d\dot\psi},\quad \eta=\frac{dT}{d\dot\phi},\ \ldots\qquad(13).$$

The two expressions for T being, as above, § 313,

$$=\tfrac{1}{2}\{(\psi,\psi)\dot\psi^2+\ldots+2(\psi,\phi)\dot\psi\dot\phi+\ldots\}=\tfrac{1}{2}\{[\psi,\psi]\xi^2+\ldots+2[\psi,\phi]\xi\eta+\ldots\}\ (14),$$

the second of these is to be obtained from the first by substituting for $\dot\psi$, $\dot\phi$, ..., their expressions in terms of ξ, η, ... Hence

$$\frac{\partial T}{d\psi}=\frac{dT}{d\psi}+\frac{dT}{d\dot\psi}\frac{\partial\dot\psi}{d\psi}+\frac{dT}{d\dot\phi}\frac{\partial\dot\phi}{d\psi}+\ldots=\frac{dT}{d\psi}+\xi\frac{\partial}{d\psi}\frac{\partial T}{d\xi}+\eta\frac{\partial}{d\phi}\frac{\partial T}{d\eta}+\ldots$$

$$=\frac{dT}{d\psi}+\frac{\partial}{d\psi}\left(\xi\frac{\partial T}{d\xi}+\eta\frac{\partial T}{d\eta}+\ldots\right)=\frac{dT}{d\psi}+2\frac{\partial T}{d\psi}\,.$$

Hamilton's form of Lagrange's generalized equations of motion.

From this we conclude

$$\frac{\partial T}{d\psi}=-\frac{dT}{d\psi}\text{; and, similarly, }\frac{\partial T}{d\phi}=-\frac{dT}{d\phi}\text{, etc.} \quad (15).$$

Hence Lagrange's equations become

$$\frac{d\xi}{dt}+\frac{\partial T}{d\psi}=\Psi\text{, etc.} \quad (16),$$

the same as those (33) which we derived, in § 322, from Hamilton's partial differential equation for his characteristic function and his expression of the motion by means of it.

If we put $\Psi=\frac{\partial T}{d\psi}+\Psi_{,}$, we have $\frac{d\xi}{dt}=\Psi_{,}$, and therefore

330. Hamilton's form of Lagrange's equations of motion in terms of generalized co-ordinates, expresses that what is required to prevent any one of the components of momentum from varying is a corresponding component force equal in amount to the rate of change of the kinetic energy per unit increase of the corresponding co-ordinate, with all components of momentum constant: and that whatever is the amount of the component force, its excess above this value measures the rate of increase of the component momentum.

In the case of a conservative system, the same statement takes the following form:—The rate at which any component momentum increases per unit of time is equal to the rate, per unit increase of the corresponding co-ordinate, at which the sum of the potential energy, and the kinetic energy for constant momenta, diminishes. This is the celebrated "canonical form" of the equations of motion of a system, though why it has been so called it would be hard to say.

"Canonical form" of Hamilton's general equations of motion of a conservative system.

Let U denote the algebraic expression for the sum of the potential energy, in terms of the co-ordinates, ψ, ϕ, ..., and the kinetic energy in terms of the co-ordinates and the components of momentum, ξ, η, ... Then

$$\left.\begin{aligned}&\frac{d\xi}{dt}=-\frac{\partial U}{d\psi}\text{, etc.}\\ \text{also}\quad &\frac{d\psi}{dt}=\frac{\partial U}{d\xi}\text{, etc.}\end{aligned}\right\} \quad (17),$$

the latter being equivalent to (12), since the potential energy does not contain ξ, η, etc.

In the following examples we shall adhere to Lagrange's form (10), as the most convenient for such applications.

Example (A.)—Motion of a single point (m) referred to polar co-ordinates (r, θ, ϕ). From the well-known geometry of this case we see that δr, $r\delta\theta$, and $r\sin\theta\delta\phi$ are the amounts of linear displacement corresponding to infinitely small increments, δr, $\delta\theta$, $\delta\phi$, of the co-ordinates: also that these displacements are respectively in the direction of r, of the arc $r\delta\theta$ (of a great circle) in the plane of r and the pole, and of the arc $r\sin\theta\delta\phi$ (of a small circle in a plane perpendicular to the axis); and that they are therefore at right angles to one another. Hence if F, G, H denote the components of the force experienced by the point, in these three rectangular directions, we have

Examples of the use of Lagrange's generalized equations of motion.

$$F=R,\ Gr=\Theta, \text{ and } Hr\sin\theta=\Phi;$$

R, Θ, Φ being what the generalized components of force (§ 318) become for this particular system of co-ordinates. We also see that $\dot r$, $r\dot\theta$, and $r\sin\theta\dot\phi$ are three components of the velocity, along the same rectangular directions. Hence

$$T=\tfrac{1}{2}m(\dot r^2+r^2\dot\theta^2+r^2\sin^2\theta\dot\phi^2).$$

From this we have

$$\frac{dT}{d\dot r}=m\dot r,\quad \frac{dT}{d\dot\theta}=mr^2\dot\theta,\quad \frac{dT}{d\dot\phi}=mr^2\sin^2\theta\dot\phi;$$

$$\frac{dT}{dr}=mr(\dot\theta^2+\sin^2\theta\dot\phi^2),\quad \frac{dT}{d\theta}=mr^2\sin\theta\cos\theta\dot\phi^2,\quad \frac{dT}{d\phi}=0.$$

Hence the equations of motion become

$$m\{\frac{d\dot r}{dt}-r(\dot\theta^2+\sin^2\theta\dot\phi^2)\}\quad =F,$$

$$m\{\frac{d(r^2\dot\theta)}{dt}-r^2\sin\theta\cos\theta\dot\phi^2\}=Gr,$$

$$m\frac{d(r^2\sin^2\theta\dot\phi)}{dt}\quad =Hr\sin\theta;$$

or, according to the ordinary notation of the differential calculus,

$$m\{\frac{d^2r}{dt^2}-r(\frac{d\theta^2}{dt^2}+\sin^2\theta\frac{d\phi^2}{dt^2})\}\quad =F,$$

$$m\{\frac{d}{dt}(r^2\frac{d\theta}{dt})-r^2\sin\theta\cos\theta\frac{d\phi^2}{dt^2}\}=Gr,$$

$$m\frac{d}{dt}(r^2\sin^2\theta\frac{d\phi}{dt})\quad =Hr\sin\theta.$$

If the motion is confined to one plane, that of r, θ, we have $\frac{d\phi}{dt}=0$, and therefore $H=0$, and the two equations of motion which remain are

$$m(\frac{d^2r}{dt^2}-r\frac{d\theta^2}{dt^2})=F,\quad m\frac{d}{dt}(r^2\frac{d\theta}{dt})=Gr.$$

Examples of the use of Lagrange's generalized equations of motion.

These equations might have been written down at once in terms of the second law of motion from the kinematical investigation of § 32, in which it was shown that $\frac{d^2r}{dt^2} - r\frac{d\theta^2}{dt^2}$, and $\frac{1}{r}\frac{d}{dt}(r^2\frac{d\theta}{dt})$ are the components of acceleration along and perpendicular to the radius-vector, when the motion of a point in a plane is expressed according to polar co-ordinates, r, θ.

The same equations, with ϕ instead of θ, are obtained from the polar equations in three dimensions by putting $\theta = \frac{1}{2}\pi$, which implies that $G = 0$, and confines the motion to the plane (r, ϕ).

Example (B.)—Two particles are connected by a string; one of them, m, moves in any way on a smooth horizontal plane, and the string, passing through a smooth infinitely small aperture in this plane, bears the other particle m', hanging vertically downwards, and only moving in this vertical line: (the string remaining always stretched in any practical illustration, but, in the problem, being of course supposed capable of transmitting negative tension with its two parts straight.) Let l be the whole length of the string, r that of the part of it from m to the aperture in the plane, and let θ be the angle between the direction of r and a fixed line in the plane. We have

$$T = \tfrac{1}{2}\{m(\dot{r}^2 + r^2\dot{\theta}^2) + m'\dot{r}^2\}$$

$$\frac{dT}{d\dot{r}} = (m+m')\dot{r}, \quad \frac{dT}{d\dot{\theta}} = mr^2\dot{\theta}$$

$$\frac{dT}{dr} = mr\dot{\theta}^2, \quad \frac{dT}{d\theta} = 0.$$

Also, there being no other external force than gm', the weight of the second particle,

$$R = -gm', \quad \Theta = 0.$$

Hence the equations of motion are

$$(m+m')\ddot{r} - mr\dot{\theta}^2 = -m'g, \quad m\frac{d(r^2\dot{\theta})}{dt} = 0.$$

The motion of m' is of course that of a particle influenced only by a force towards a fixed centre; but the law of this force, P (the tension of the string), is remarkable. To find it we have (§ 32), $P = m(-\ddot{r} + r\dot{\theta}^2)$. But, by the equations of the motion,

$$\ddot{r} - r\dot{\theta}^2 = -\frac{m'}{m+m'}(g + r\dot{\theta}^2), \text{ and } \dot{\theta} = \frac{h}{mr^2},$$

where h (according to the usual notation) denotes the moment of momentum of the motion, being an arbitrary constant of integration. Hence

$$P=\frac{mm'}{m+m'}\left(g+\frac{h^2}{m^2}r^{-3}\right).$$

The particular case of projection which gives m a circular motion and leaves m' at rest is interesting, inasmuch as (§ 350, below) the motion of m is stable, and therefore m' is in stable equilibrium. Case of stable equilibrium due to motion.

Example (C.)—A rigid body is supported on a fixed axis, and another rigid body is supported on the first, by another axis; the motion round each axis being perfectly free. Examples continued.

Case (*a*).—*The second axis parallel to the first.* At any time, t, let ϕ and ψ be the inclinations of a fixed plane through the first axis to the plane of it and the second axis, and to a plane through the second axis and the centre of inertia of the second body. These two co-ordinates, ϕ, ψ, it is clear, completely specify the position of the body. Now let a be the distance of the second axis from the first, and b that of the centre of inertia of the second body from the second axis. The velocity of the second axis will be $a\dot\phi$; and the velocity of the centre of inertia of the second body will be the resultant of two velocities

$$a\dot\phi, \text{ and } b\dot\psi,$$

in lines inclined to one another at an angle equal to $\psi-\phi$, and its square will therefore be equal to

$$a^2\dot\phi^2+2ab\dot\phi\dot\psi\cos(\psi-\phi)+b^2\dot\psi^2.$$

Hence, if m and m' denote the masses, j the radius of gyration of the first body about the fixed axis, and k that of the second body about a parallel axis through its centre of inertia; we have, according to §§ 280, 281,

$$T=\tfrac{1}{2}\{mj^2\dot\phi^2+m'[a^2\dot\phi^2+2ab\dot\phi\dot\psi\cos(\psi-\phi)+b^2\dot\psi^2+k^2\dot\psi^2]\}.$$

Hence we have,

$$=mj^2\dot\phi+m'a^2\dot\phi+m'ab\cos(\psi-\phi)\dot\psi;\ \frac{dT}{d\dot\psi}=m'ab\cos(\psi-\phi)\dot\phi+m(b^2+k^2)\dot\psi;$$

$$\frac{dT}{d\phi}=-\frac{dT}{d\psi}=m'ab\sin(\psi-\phi)\dot\phi\dot\psi.$$

The most general supposition we can make as to the applied forces, is equivalent to assuming a couple, Φ, to act on the first body, and a couple, Ψ, on the second, each in a plane perpendicular to the axes; and these are obviously what the generalized components of stress become in this particular co-ordinate system, ϕ, ψ. Hence the equations of motion are

Examples of the use of Lagrange's generalized equations of motion

$$(mj^2+m'a^2)\ddot{\phi}+m'ab\frac{d[\dot{\psi}\cos(\psi-\phi)]}{dt}-m'ab\sin(\psi-\phi)\dot{\phi}\dot{\psi}=\Phi,$$

$$m'ab\frac{d[\dot{\phi}\cos(\psi-\phi)]}{dt}+m'(b^2+k^2)\ddot{\psi}+m'ab\sin(\psi-\phi)\dot{\phi}\dot{\psi}=\Psi.$$

If there is no other applied force than gravity, and if, as we may suppose without losing generality, the two axes are horizontal, the potential energy of the system will be

$$gmh(1-\cos\phi)+gm'\{a[1-\cos(\phi+A)]+b[1-\cos(\psi+A)]\},$$

the distance of the centre of inertia of the first body from the fixed axis being denoted by h, the inclination of the plane through the fixed axis and the centre of inertia of the first body, to the plane of the two axes, being denoted by A, and the fixed plane being so taken that $\phi=0$ when the former plane is vertical. By differentiating this, with reference to ϕ and ψ, we therefore have

$$-\Phi=gmh\sin\phi+gm'a\sin(\phi+A),\quad -\Psi=gm'b\sin(\psi+A).$$

We shall examine this case in some detail later, in connexion with the interference of vibrations, a subject of much importance in physical science.

When there are no applied, or intrinsic working, forces, we have $\Phi=0$ and $\Psi=0$: or, if there are mutual forces between the two bodies, but no forces applied from without, $\Phi+\Psi=0$. In either of these cases we have the following first integral:—

$$(mj^2+m'a^2)\dot{\phi}+m'ab\cos(\psi-\phi)(\dot{\phi}+\dot{\psi})+m'(b^2+k^2)\dot{\psi}=C;$$

obtained by adding the two equations of motion and integrating. This, which clearly expresses the constancy of the whole moment of momentum, gives $\dot{\phi}$ and $\dot{\psi}$ in terms of $(\dot{\psi}-\dot{\phi})$ and $(\psi-\phi.)$ Using these in the integral equation of energy, provided the mutual forces are of the conservative class, we have a single equation between $\frac{d(\psi-\phi)}{dt}$, $(\psi-\phi)$, and constants, and thus the full solution of the problem is reduced to quadratures.

Case (*b*).—*The second axis perpendicular to the first.* This case we suggest as an exercise for the student. We may return to it later, as its application to the theory of centrifugal chronometric regulators is very important.

Example (D.)—*Gyroscopic pendulum.*—A rigid body, P, is attached to one axis of a universal flexure joint (§ 109), of which the other is held fixed, and a second body, Q, is supported on P by a fixed axis, in line with, or parallel to, the first-mentioned arm of the joint. For simplicity, we shall suppose Q to be kinetically symmetrical about its bearing axis, and OB to be a principal axis of a supposed rigid body, PQ, composed of P and a mass equal to Q collected the centre of inertia of Q and attached to P. Let AO be the fixed arm, O the joint, OB the movable arm bearing the body P, and coinciding with, or parallel to, the axis of Q. Let $BOA' = \theta$; let ϕ be the angle which the plane AOB makes with a fixed plane of reference, through OA, chosen so as to contain the principal axis of moment $\mathfrak{C}$ of the imagined rigid body, PQ, when OB is placed in line with AO; and let ψ be the angle between a plane of reference in Q through its axis of symmetry and the plane AOB. These three co-ordinates (θ, ϕ, ψ) clearly specify the position of the system at any time, t. Let the moments of inertia of the imagined rigid body PQ, round its principal axis OB, and its other two, be denoted by $\mathfrak{A}$, $\mathfrak{B}$, $\mathfrak{C}$; and let $\mathfrak{A}'$, $\mathfrak{B}'$ be the moments of inertia of Q round its bearing axis, and any perpendicular axis through its centre of inertia, respectively. Gyroscopic pendulum.

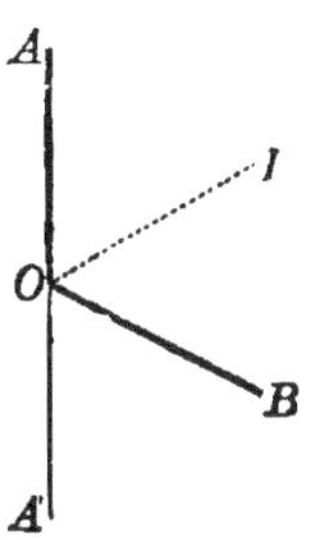

We have seen (§ 109) that, with the kind of joint we have supposed at O, every possible motion of a body rigidly connected with OB, is resolvable into a rotation round OI, the line bisecting the angle AOB, and a rotation round the line through O perpendicular to the plane AOB. The angular velocity of the latter is $\dot\theta$, according to our present notation. The former would give to any point in OB the same absolute velocity by rotation round OI, that it has by rotation with angular velocity $\dot\phi$ round AA'; and is therefore equal to

$$\frac{\sin A'OB}{\sin IOB}\dot\phi = \frac{\sin\theta}{\cos\frac{1}{2}\theta}\dot\phi = 2\dot\phi\sin\tfrac{1}{2}\theta.$$

This may be resolved into $2\dot\phi\sin^2\frac{1}{2}\theta = \dot\phi(1-\cos\theta)$ round OB, and $2\dot\phi\sin\frac{1}{2}\theta\cos\frac{1}{2}\theta = \dot\phi\sin\theta$ round the perpendicular to OB, in plane AOB. Again, in virtue of the symmetrical character of the joint with reference to the line OI, the angle ϕ, as defined above, will be equal to the angle between the plane of the body P, which coincided with the fixed plane of reference when $\phi = 0$, and the plane AOB. Hence the axis of the angular velocity $\dot\phi\sin\theta$, is inclined to the principal axis of moment $\mathfrak{B}$ at an angle equal to ϕ. Resolving therefore this angular velocity, and $\dot\theta$, into components round the axes of $\mathfrak{B}$ and $\mathfrak{C}$, we find, for the whole component angular velocities of the imagined rigid body

Gyroscopic pendulum.

PQ, round these axes, $\dot\phi\sin\theta\cos\phi+\dot\theta\sin\phi$, and $-\dot\phi\sin\theta\sin\phi+\dot\theta\cos\phi$, respectively. The whole kinetic energy, T, being composed of that of the imagined rigid body PQ, and that of Q about axes through its centre of inertia, we therefore have

$$2T=\mathfrak{A}(1-\cos\theta)^2\dot\phi^2+\mathfrak{B}(\dot\phi\sin\theta\cos\phi+\dot\theta\sin\phi)^2+\mathfrak{C}(\dot\phi\sin\theta\sin\phi-\dot\theta\cos\phi)^2$$
$$+\mathfrak{A}'\{\dot\psi-\dot\phi(1-\cos\theta)\}^2+\mathfrak{B}'(\dot\phi^2\sin^2\theta+\dot\theta^2)$$

Hence $\dfrac{dT}{d\dot\psi}=\mathfrak{A}'\{\dot\psi-\dot\phi(1-\cos\theta)\}$, $\dfrac{dT}{d\psi}=0$,

$$\frac{dT}{d\dot\phi}=\mathfrak{A}(1-\cos\theta)^2\dot\phi+\mathfrak{B}(\dot\phi\sin\theta\cos\phi+\dot\theta\sin\phi)\sin\theta\cos\phi+\mathfrak{C}(\dot\phi\sin\theta\sin\phi-\dot\theta\cos\phi)\sin\theta\ldots$$
$$-\mathfrak{A}'\{\dot\psi-\dot\phi(1-\cos\theta)\}(1-\cos\theta)+\mathfrak{B}'\dot\phi\ldots$$

$$\frac{dT}{d\phi}=-\mathfrak{B}(\dot\phi\sin\theta\cos\phi+\dot\theta\sin\phi)(\dot\phi\sin\theta\sin\phi-\dot\theta\cos\phi)+\mathfrak{C}(\dot\phi\sin\theta\sin\phi-\dot\theta\cos\phi)(\dot\phi\sin\ldots$$
$$+\dot\theta\ldots$$

$$\frac{dT}{d\dot\theta}=\mathfrak{B}(\dot\phi\sin\theta\cos\phi+\dot\theta\sin\phi)\sin\phi-\mathfrak{C}(\dot\phi\sin\theta\sin\phi-\dot\theta\cos\phi)\cos\phi+\mathfrak{B}'\dot\theta,$$

and $\dfrac{dT}{d\theta}=\mathfrak{A}(1-\cos\theta)\sin\theta\dot\phi^2+\mathfrak{B}\cos\theta\cos\phi\dot\phi(\dot\phi\sin\theta\cos\phi+\dot\theta\sin\phi)$

$$+\mathfrak{C}\cos\theta\sin\phi\dot\phi(\dot\phi\sin\theta\sin\phi-\dot\theta\cos\phi)+\mathfrak{A}'\sin\theta\dot\phi\{\dot\psi-(1-\cos\theta)\dot\phi\}+\mathfrak{B}'\sin\theta\ldots$$

Now let a couple, G, act on the body Q, in a plane perpendicular to its axis, and let L, M, N act on P, in the plane perpendicular to OB, in the plane $A'OB$, and in the plane through OB perpendicular to the diagram. If ψ is kept constant, and ϕ varied, the couple G will do or resist work in simple addition with L. Hence, resolving $L+G$ and N into components round OI, and perpendicular to it, rejecting the latter, and remembering that $2\sin\frac{1}{2}\theta\dot\phi$ is the angular velocity round OI, we have

$$\Phi=2\sin\tfrac{1}{2}\theta\{-(L+G)\sin\tfrac{1}{2}\theta+N\cos\tfrac{1}{2}\theta\}=\{-(L+G)(1-\cos\theta)+N\sin\theta\ldots$$

Also, obviously

$$\Psi=G,\quad \Theta=M.$$

Using these several expressions in Lagrange's general forms, we have the equations of motion of the system. They will be of great use to us later, when we shall consider several particular cases of remarkable interest and of very great importance.

Examples continued.

Example (E).—*Motion of a particle referred to rotating axes.* Let x, y, z be the co-ordinates of a moving particle referred to axes rotating with a constant or varying angular velocity round the axis OZ. Let x_1, y_1, z, be its co-ordinates referred to the same axis, OZ, and two axes OX_1, OY_1, fixed in the plane perpendicular to it. We have

$$x_1=x\cos\theta-y\sin\theta,\ y_1=x\sin\theta+y\cos\theta\,;$$
$$\dot x_1=\dot x\cos\theta-\dot y\sin\theta-(x\sin\theta+y\cos\theta)\dot\theta,\ \dot y_1=\text{etc.}$$

where θ, the angle X_1OX, must be considered as a given function of t. Hence

Examples of the use of Lagrange's generalized equations of motion.

$$T=\tfrac{1}{2}m\{\dot{x}^2+\dot{y}^2+\dot{z}^2+2(x\dot{y}-y\dot{x})\dot{\theta}+(x^2+y^2)\dot{\theta}^2\}$$

$$\frac{dT}{d\dot{x}}=m(\dot{x}-y\dot{\theta}),\ \frac{dT}{d\dot{y}}=m(\dot{y}+x\dot{\theta}),\ \frac{dT}{d\dot{z}}=m\dot{z}.$$

$$\frac{dT}{dx}=m(\dot{y}\dot{\theta}+x\dot{\theta}^2),\ \frac{dT}{dy}=m(-\dot{x}\dot{\theta}+y\dot{\theta}^2),\ \frac{dT}{dz}=0.$$

Also,

$$\frac{d}{dt}\frac{dT}{d\dot{x}}=m(\ddot{x}-\dot{y}\dot{\theta}-y\ddot{\theta}),\ \frac{d}{dt}\frac{dT}{d\dot{y}}=m(\ddot{y}+\dot{x}\dot{\theta}+x\ddot{\theta}),$$

and hence the equations of motion are

$$m(\ddot{x}-y\ddot{\theta}-2\dot{y}\dot{\theta}-x\dot{\theta}^2)=X,\ m(\ddot{y}+x\ddot{\theta}+2\dot{x}\dot{\theta}-y\dot{\theta}^2)=Y,\ m\ddot{z}=Z,$$

X, Y, Z denoting simply the components of the force on the particle, parallel to the moving axes at any instant. In this example t enters into the relation between fixed rectangular axes and the co-ordinate system to which the motion is referred; but there is no constraint. The next is given as an example of varying, or kinetic, constraint.

Example (F).—*A particle, influenced by any forces, and attached to one end of a string of which the other is moved with any constant or varying velocity in a straight line.* Let θ be the inclination of the string at time t, to the given straight line, and ϕ the angle between two planes through this line, one containing the string at any instant, and the other fixed. These two co-ordinates (θ, ϕ) specify the position, P, of the particle at any instant, the length of the string being a given constant, a, and the distance OE, of its other end E, from a fixed point, O, of the line in which it is moved, being a given function of t, which we shall denote by u. Let x, y, z be the co-ordinates of the particle referred to three fixed rectangular axes. Choosing OX as the given straight line, and YOX the fixed plane from which ϕ is measured, we have

$$x=u+a\cos\theta,\ y=a\sin\theta\cos\phi,\ z=a\sin\theta\sin\phi,$$
$$\dot{x}=\dot{u}-a\sin\theta\dot{\theta};$$

and for $\dot{y}$, $\dot{z}$ we have the same expression as in Example A. Hence

$$T=\tfrac{1}{2}m(\dot{u}^2-2\dot{u}\dot{\theta}a\sin\theta)+\mathfrak{T}$$

where $\mathfrak{T}$ denotes the same as T with $r=0$, and $r=a$, in that example. Hence, denoting as there, by G and H the two components of the force on the particle, perpendicular to EP, respectively in the plane of θ and perpendicular to it, we find, for the two required equations of motion,

Examples of the use of Lagrange's generalized equations of motion.

$$m\{a(\ddot{\theta}-\sin\theta\cos\theta\dot{\phi}^2)-\sin\theta\ddot{u}\}=G, \text{ and } ma\frac{d(\sin^2\theta\dot{\phi})}{dt}=H.$$

These show that the motion is the same as if E were fixed, and a force equal to $-m\ddot{u}$ were applied to the particle in a direction parallel to EX; a result that might have been arrived at at once by superimposing on the whole system an acceleration equal and opposite to that of E, to effect which on P the force $-m\ddot{u}$ is required.

Kinetics of a perfect liquid.

331. Problems in fluid motion of remarkable interest and importance, not hitherto attacked, are very readily solved by the aid of Lagrange's generalized equations of motion. For brevity we shall designate a mass which is absolutely incompressible, and absolutely devoid of resistance to change of shape, by the simple appellation of a *liquid*. We need scarcely say that matter perfectly satisfying this definition does not exist in nature: but we shall see (under properties of matter) how nearly it is approached by water and other common real liquids. And we shall find that much practical and interesting information regarding their true motions is obtained by deductions from the principles of abstract dynamics applied to the ideal perfect liquid of our definition. We shall see later, under hydrodynamics, that the motion of a homogeneous liquid, whether of infinite extent, or contained in a finite closed vessel of any form, with any rigid or flexible bodies moving through it, if it has ever been at rest, is the same at each instant as that determinate motion (fulfilling, § 312, the condition of having the least possible kinetic energy) which would be impulsively produced from rest by giving instantaneously to every part of the bounding surface, and of the surface of each of the solids within it, its actual velocity at that instant. So that, for example, however long it may have been moving, if all these surfaces were suddenly or gradually brought to rest, the whole fluid mass would come to rest at the same time. Hence, if none of the surfaces is flexible, but we have one or more rigid bodies moving in any way under the influence of any forces, through the liquid, the kinetic energy of the whole motion at any instant will depend solely on the finite number of co-ordinates and component velocities, specifying the position and motion of those bodies, whatever may be the positions reached by particles of the fluid (expressible only by an infinite number of co-ordinates). And an expression for the whole kinetic energy in terms of such elements, finite in number, is precisely what is wanted, as we have seen, as the foundation of Lagrange's equations in any particular case. Hydrodynamical examples of the application of Lagrange's equations of motion.

It will clearly, in the hydrodynamical, as in all other cases, be a homogeneous quadratic function of the components of velocity, if referred to an invariable co-ordinate system; and the coefficients of the several terms will in general be functions of the co-ordinates, the determination of which follows immediately from the solution of the minimum problem of Example (3), § 317, in each particular case.

Example (1.)—*A ball set in motion through a mass of incompressible fluid extending infinitely in all directions one side of an infinite plane, and originally at rest.* Let x, y, z be the co-ordinates of the centre of the ball at time t, with reference to rectangular axes through a fixed point O of the bounding plane, with OX perpendicular to this plane. If at any instant either component $\dot{y}$ or $\dot{z}$ of the velocity be reversed, the kinetic energy will clearly be unchanged, and hence no terms $\dot{y}\dot{z}$, $\dot{z}\dot{x}$, or $\dot{x}\dot{y}$ can appear in the expression for the kinetic energy: which, on this account, and because of the symmetry of circumstances with reference to y and z, is

$$T=\tfrac{1}{2}\{P\dot{x}^2+Q(\dot{y}^2+\dot{z}^2)\}.$$

Also, we see that P and Q are functions of x simply, since the circumstances are similar for all values of y and z. Hence, by differentiation,

$$\frac{dT}{d\dot{x}}=P\dot{x},\quad \frac{dT}{d\dot{y}}=Q\dot{y},\quad \frac{dT}{d\dot{z}}=Q\dot{z}.$$

$$\frac{d}{dt}\left(\frac{dT}{d\dot{x}}\right)=P\ddot{x}+\frac{dP}{dx}\dot{x}^2,\quad \frac{d}{dt}\left(\frac{dT}{d\dot{y}}\right)=Q\ddot{y}+\frac{dQ}{dx}\dot{y}\dot{x},\ \text{etc.}$$

$$\frac{dT}{dx}=\tfrac{1}{2}\left\{\frac{dP}{dx}\dot{x}^2+\frac{dQ}{dx}(\dot{y}^2+\dot{z}^2)\right\},\quad \frac{dT}{dy}=0,\ \text{etc.},$$

and the equations of motion are

$$P\ddot{x}+\tfrac{1}{2}\left\{\frac{dP}{dx}\dot{x}^2-\frac{dQ}{dx}(\dot{y}^2+\dot{z}^2)\right\}=X,$$

$$Q\ddot{y}+\frac{dQ}{dx}\dot{y}\dot{x}=Y,\quad Q\ddot{z}+\frac{dQ}{dx}\dot{z}\dot{x}=Z.$$

Principles sufficient for a practical solution of the problem of determining P and Q, will be given later. In the meantime, it is obvious that each decreases as x increases. Hence the equations of motion show that

Effect of a rigid plane on the motion of a ball through a liquid.

332. A ball projected through a liquid perpendicularly *from* an infinite plane boundary, and influenced by no other forces than those of fluid pressure, experiences a gradual acceleration, quickly approximating to a limiting velocity which it sensibly reaches when its distance from the plane is many times its diameter. But if projected *parallel* to the plane, it experiences, as the resultant of fluid pressure, a resultant attraction towards the plane. The former of these results is easily proved by first considering projection *towards* the plane (in which case the motion of the ball will obviously be retarded), and by taking into account the general principle of reversibility (§ 272) which has perfect application in the ideal case of a perfect liquid. The second result is less easily foreseen without the aid of Lagrange's analysis; but it is an obvious consequence of the Hamiltonian form of his equations, as stated in general terms in § 330 above. In the precisely equivalent case, of a liquid extending infinitely in all directions, and given at rest; and two equal balls projected through it with equal velocities perpendicular to the line joining their centres—the result that the two balls will seem to attract one another is most remarkable, and very suggestive.

Hydrodynamical examples continued.

Example (2.)—*A solid of revolution moving through a liquid so as to keep its axis always in one plane.* Let ω be the angular velocity of the body at any instant about any axis perpendicular to the fixed plane, and let u and q be the component velocities along and perpendicular to the axis of figure, of any chosen point, C, of the body in this line. By the general principle stated in § 331 (since changing the sign of u cannot alter the kinetic energy), we have

$$T=\tfrac{1}{2}(Au^2+Bq^2+\mu'\omega^2+E\omega q) \qquad (a).$$

where A, B, μ', and E are constants depending on the figure of the body, its mass, and the density of the liquid. Now let v denote the velocity, perpendicular to the axis, of a point which we shall call the *centre of reaction*, being a point in the axis and at a distance $\frac{E}{2B}$ from C, so that (§ 87) $q=v-\frac{E}{2B}\omega$. Then, denoting $\mu'-\frac{E^2}{2B}$ by μ, we have $T=\tfrac{1}{2}(Au^2+Bv^2+\mu\omega^2)$ $\qquad$ (a').

Let x and y be the co-ordinates of the centre of reaction relatively to any fixed rectangular axes in the plane of motion of the axis of figure, and let θ be the angle between this line and OX, at any instant, so that

Hydrodynamics examples of the application of Lagrange's equations of motion.

$$\omega = \dot\theta,\ u = \dot x\cos\theta + \dot y\sin\theta,\ v = -\dot x\sin\theta + \dot y\cos\theta \qquad (b).$$

Substituting in T, differentiating, and retaining the notation u, v where convenient for brevity, we have

$$\left.\begin{array}{l} \frac{dT}{d\dot\theta} = \mu\dot\theta,\ \frac{dT}{d\dot x} = Au\cos\theta - Bv\sin\theta,\ \frac{dT}{d\dot y} = Au\sin\theta + Bv\cos\theta, \\ \frac{dT}{d\theta} = (A-B)uv,\ \frac{dT}{dx} = 0,\ \frac{dT}{dy} = 0. \end{array}\right\} \qquad (c).$$

Hence the equations of motion are

$$\left.\begin{array}{l} \mu\ddot\theta - (A-B)uv = L, \\ \frac{d(Au\cos\theta - Bv\sin\theta)}{dt} = X,\quad \frac{d(Au\sin\theta + Bv\cos\theta)}{dt} = Y \end{array}\right\} \qquad (d),$$

where X, Y are the component forces in lines through C parallel to OX and OY, and L the couple, applied to the body.

Denoting by λ, ξ, η the impulsive couple, and the components of impulsive force through C required to produce the motion at any instant, we have of course [§ 313 (c)],

$$\lambda = \frac{dT}{d\dot\theta} = \mu\dot\theta,\ \xi = \frac{dT}{d\dot x},\ \eta = \frac{dT}{d\dot y} \qquad (e),$$

and therefore, by (c), and (b),

$$u = \frac{1}{A}(\xi\cos\theta + \eta\sin\theta),\quad v = \frac{1}{B}(-\xi\sin\theta + \eta\cos\theta),\quad \dot\theta = \frac{\lambda}{\mu} \qquad (f)$$

$$\left.\begin{array}{l} \dot x = (\frac{\cos^2\theta}{A} + \frac{\sin^2\theta}{B})\xi + (\frac{1}{A} - \frac{1}{B})\sin\theta\cos\theta\eta, \\ \dot y = (\frac{1}{A} - \frac{1}{B})\sin\theta\cos\theta\xi + (\frac{\sin^2\theta}{A} + \frac{\cos^2\theta}{B})\eta \end{array}\right\} \qquad (g);$$

and the equations of motion become

$$\mu\frac{d^2\theta}{dt^2} - \frac{A-B}{2AB}\{(-\xi^2+\eta^2)\sin 2\theta + 2\xi\eta\cos 2\theta\} = L,\ \frac{d\xi}{dt} = X,\ \frac{d\eta}{dt} = Y, \ (h).$$

The simple case of $X=0$, $Y=0$, $L=0$, is particularly interesting. In it ξ and η are each constant, and we may therefore choose the axes OX, OY, so that η shall vanish. Thus we have by (g) two first integrals of the equations of motion,

$$\dot x = \xi(\frac{\cos^2\theta}{A} + \frac{\sin^2\theta}{B}),\quad \dot y = -\frac{A-B}{2AB}\xi\sin 2\theta \qquad (k);$$

and the first of equations (h) becomes

$$\mu\frac{d^2\theta}{dt^2} + \frac{A-B}{2AB}\xi^2\sin 2\theta = 0 \qquad (l).$$

Motion of a solid of revolution through a liquid.

In this let, for a moment, $2\theta = \phi$, and $\frac{A-B}{AB} \cdot \xi^2 = gW$. It becomes

$$\mu \frac{d^2\phi}{dt^2} + gW \sin\phi = 0,$$

which is the equation of motion of a common pendulum, of mass W, and moment of inertia μ round its fixed axis, if ϕ be the angle from the position of equilibrium to the position at time t. As we shall see, under kinetics, the final integral of this equation expresses ϕ in terms of t by means of an elliptic function. By using the value thus found for θ or $\frac{1}{2}\phi$, in (k), we have equations giving x and y in terms of t by common integration; and thus the full solution of our present problem is reduced to quadratures. The detailed working out to exhibit both the actual curve described by the centre of reaction, and the position of the axis of the body at any instant, is highly interesting. It is very easily done approximately for the case of very small angular vibrations; that is to say, when either $A-B$ is positive, and ϕ always very small, or $A-B$ negative, and ϕ very nearly equal to $\frac{1}{2}\pi$. But without attending at present to the final integrals, rigorous or approximate, we see from (k) and (l) that

333. If a solid of revolution in an infinite liquid, be set in motion round any axis perpendicular to its axis of figure, or simply projected in any direction without rotation, it will move with its axis always in one plane, and every point of it moving only parallel to this plane; and the strange evolutions which it will, in general, perform, are perfectly defined by comparison with the common pendulum thus. First, for brevity, we shall call by the name of *quadrantal pendulum* (which will be further exemplified in various cases described later, under electricity and magnetism; for instance, an elongated mass of soft iron pivoted on a vertical axis, in a "uniform field of magnetic force"), a body moving about an axis, according to the same law with reference to a quadrant on each side of its position of equilibrium, as the common pendulum with reference to a half circle on each side.

Quadrantal pendulum defined.

Let now the body in question be set in motion by an impulse, ξ, in any line through the centre of reaction, and an impulsive couple in the plane of that line and the axis. This will (as will be proved later in the theory of statical couples) have the same effect as a simple impulse ξ (applied to a point,

Motion of a solid of revolution through a liquid.

if not of the real body, connected with it by an imaginary infinitely light framework) in a certain line, which we shall call the line of resultant impulse, or of resultant momenta, being parallel to the former line, and at a distance from it equal to $\frac{\lambda}{\xi}$. The whole momentum of the motion generated is of course (§ 295) equal to ξ. The body will move ever afterwards according to the following conditions:—(1.) The angular velocity round the centre of reaction follows the law of the quadrantal pendulum. (2.) The distance of the centre of reaction from the line of resultant impulse varies simply as the angular velocity. (3.) The velocity of the centre of reaction parallel to the line of impulse is found by dividing the excess of the whole constant energy of the motion above the part of it due to the angular velocity round the centre of reaction by half the momentum. (4.) If A, B, and μ denote constants, depending on the mass of the solid and its distribution, the density of the liquid, and the form and dimensions of the solid, such that $\frac{\xi^2}{A}$, $\frac{\xi^2}{B}$, $\frac{\lambda^2}{\mu}$ are the linear velocities, and the angular velocity, respectively produced by an impulse ξ along the axis, an impulse ξ in a line through the centre of reaction perpendicular to the axis, and an impulsive couple λ in a plane through the axis; the length of the simple gravitation pendulum, whose motion would keep time with the periodic motion in question, is $\frac{g\mu AB}{\xi^2(A-B)}$, and, when the angular motion is vibratory, the vibrations will, according as $A > B$, or $A < B$, consist of the axis, or of a line perpendicular to the axis, vibrating on each side of the line of impulse. The angular motion will in fact be vibratory if the distance of the line of resultant impulse from the centre of reaction is anything less than $\sqrt{\frac{(A\sim B)\mu}{AB}}\cos\alpha$ where α denotes either the inclination of the impulse to the initial position of the axis, or its complement, as A or B is the greater. In this case the path of the centre of reaction will be a sinuous curve symmetrical on the two sides of the line of impulse; every time it cuts this line, the angular motion will reverse, and the maximum inclination will be attained;

Motion of a solid of revolution through a liquid.

and every time the centre of reaction is at its greatest distance on either side, the angular velocity will be at its greatest, positive or negative, value, and the linear velocity of the centre of reaction will be at its least. If, on the other hand, the line of the resultant impulse be at a greater distance than $\sqrt{\frac{(A\sim B)\mu}{AB}}\cos\alpha$ from the centre of reaction, the angular motion will be always in one direction, but will increase and diminish periodically, and the centre of reaction will describe a sinuous curve on one side of that line; being at its greatest and least deviations when the angular velocity is greatest and least. At the same points the curvature of the path will be greatest and least respectively, and the linear velocity of the describing point will be least and greatest.

334. At any instant the component linear velocities along and perpendicular to the axis of the solid will be $\frac{\xi\cos\theta}{A}$ and $\frac{\xi\sin\theta}{B}$ respectively, if θ be its inclination to the line of resultant impulse; and the angular velocity will be $\frac{\xi y}{\mu}$ if y be the distance of the centre of reaction from that line. The whole kinetic energy of the motion will be

$$\frac{\xi^2\cos^2\theta}{2A}+\frac{\xi^2\sin^2\theta}{2B}+\frac{\xi^2y^2}{2\mu},$$

and the last term is what we have referred to above as the part due to rotation round the centre of reaction. To stop the whole motion at any instant, a simple impulse equal and opposite to ξ in the fixed "line of resultant impulse" will suffice (or, of course, an equal and parallel impulse in any line through the body, with the proper impulsive couple, according to the principle already referred to).

335. From Lagrange's equations applied as above to the case of a solid of revolution moving through a liquid, the couple which must be kept applied to it to prevent it from turning is immediately found to be

$$uv(A-B),$$

if u and v be the component velocities along and perpendicular to the axis, or [§ 332 (*f*)]

$$\xi^2 \frac{(A-B)\sin 2\theta}{2AB},$$

Motion of a solid of revolution through a liquid.

if, as before, ξ be the generating impulse, and θ the angle between its line and the axis. The direction of this couple must be such as to prevent θ from diminishing or from increasing, according as A or B is the greater. The former will clearly be the case of a flat disc, or oblate spheroid; the latter that of an elongated, or oval-shaped body. The actual values of A and B we shall learn how to calculate (hydrodynamics) for several cases, including a body bounded by two spherical surfaces cutting one another at any angle a submultiple of two right angles; two complete spheres rigidly connected; and an oblate or a prolate spheroid.

Observed phenomena.

336. The tendency of a body to turn its flat side, or its length (as the case may be) across the direction of its motion through a liquid, to which the accelerations and retardations of rotatory motion described in § 333 are due, and of which we have now obtained the statical measure, is a remarkable illustration of the statement of § 330; and is closely connected with the dynamical explanation of many curious observations well known in practical mechanics, among which may be mentioned; that the towing rope of a canal boat, when the rudder is left straight, takes a position in a vertical plane cutting the axis *before* its middle point; that a boat sculled rapidly across the direction of the wind, always (unless it is extraordinarily unsymmetrical in its draught of water, and in the amounts of surface exposed to the wind, towards its two ends) requires the weather oar to be worked hardest to prevent it from running up on the wind, and that a sailing vessel generally "carries a weather helm" for the same reason; that in a heavy gale it is exceedingly difficult, and often found impossible, to get a ship out of "the trough of the sea," and that it cannot be done at all without rapid motion ahead, whether by steam or sails; that an elongated rifle-bullet requires rapid rotation about its axis to keep its point foremost. The curious motions of a flat disc, oyster-shell, or the like, when dropped obliquely into water, resemble, no doubt, to some extent those described in § 333. But it must be remembered that the real circumstances differ greatly, because of fluid friction, from those of the abstract problem, of which we take leave for the present.

Slightly disturbed equilibrium.

337. By the help of Lagrange's form of the equations of motion, § 329, we may now, as a preliminary to the consideration of stability of motion, investigate the motion of a system infinitely little disturbed from a position of equilibrium, and left free to move, the velocities of its parts being initially infinitely small. The resulting equations give the values of the independent co-ordinates at any future time, provided the displacements *continue* infinitely small; and the mathematical expressions for their values must of course show the nature of the equilibrium, giving at the same time an interesting example of the *coexistence of small motions*, § 89. The method consists simply in finding what the equations of motion, and their integrals, become for co-ordinates which differ infinitely little from values corresponding to a configuration of equilibrium—and for an infinitely small initial kinetic energy. The solution of these differential equations is always easy, as they are linear and have constant coefficients. If the solution indicates that these differences *remain infinitely small*, the position is one of stable equilibrium; if it shows that one or more of them may *increase indefinitely*, the result of an infinitely small displacement from the position of equilibrium may be a finite departure from it—and thus the equilibrium is unstable.

Since there is a position of equilibrium, the kinematic relations must be invariable. As before,

$$T=\tfrac{1}{2}\{(\psi,\psi)\dot{\psi}^2+(\phi,\phi)\dot{\phi}^2+2(\psi,\phi)\dot{\psi}\dot{\phi}+\text{etc. } \ldots\} \qquad (1)$$

which cannot be negative for any values of the co-ordinates. Now, though the values of the coefficients in this expression are not generally constant, they are to be taken as constant in the approximate investigation, since their variations, depending on the infinitely small variations of ψ, ϕ, etc., can only give rise to terms of the third or higher orders of small quantities. Hence Lagrange's equations become simply

$$\frac{d}{dt}\left(\frac{dT}{d\dot{\psi}}\right)=\Psi,\quad \frac{d}{dt}\left(\frac{dT}{d\dot{\phi}}\right)=\Phi,\text{ etc.} \qquad (2),$$

and the first member of each of these equations is a linear function of $\ddot{\psi}$, $\ddot{\phi}$, etc, with constant coefficients.

Now, since we may take what origin we please for the generalised co-ordinates, it will be convenient to assume that ψ, ϕ, θ etc., are measured from the position of equilibrium considered; and that their values are therefore always infinitely small.

Hence infinitely small quantities of higher orders being neglected, and the forces being supposed to be independent of the velocities, we shall have linear expressions for Ψ, Φ, etc., in terms of ψ, ϕ, etc., which we may write as follows:— Slightly disturbed equilibrium.

$$\left.\begin{aligned}\Psi &= a\,\psi + b\,\phi + c\,\theta + \ldots \\ \Phi &= a'\psi + b'\phi + c'\theta + \ldots \\ &\text{etc.} \qquad \text{etc.}\end{aligned}\right\} \quad (3).$$

Equations (2) consequently become linear differential equations of the second order, with constant coefficients; as many in number as there are variables ψ, ϕ, etc., to be determined.

The regular processes explained in elementary treatises on differential equations, lead of course, independently of any particular relation between the coefficients, to a general form of solution (§ 343 below). But this form has very remarkable characteristics in the case of a conservative system; which we therefore examine particularly in the first place. In this case we have

$$\Psi = -\frac{dV}{d\psi},\quad \Phi = -\frac{dV}{d\phi},\ \ldots$$

where V is, in our approximation, a homogeneous quadratic function of ψ, ϕ,... if we take the origin, or configuration of equilibrium, as the configuration from which (§ 273) the potential energy is reckoned. Now, it is obvious,[1] from the common theory of the transformation of quadratic functions, that we may,

---

[1] For in the first place any such assumption as

$$\begin{aligned}\psi &= A\psi_{,} + B\phi_{,} + \ldots \\ \phi &= A'\psi_{,} + B'\phi_{,} + \ldots \\ &\text{etc., etc.}\end{aligned}$$

gives equations for $\dot\psi$, $\dot\phi$, etc., in terms of $\dot\psi_{,}$, $\dot\phi_{,}$, etc., with the same coefficients, A, B, etc., if these are independent of t. Hence (the co-ordinates being i in number) let the quadratic expression for $2T$ in terms of $\dot\psi^2$, $\dot\phi^2$, $\dot\psi\dot\phi$, etc., be reduced to the form $\dot\psi_{,}^2 + \dot\phi_{,}^2 + \ldots$ by proper assignment of values to A, B, etc. This may be done arbitrarily, in an infinite number of ways, without the solution of any algebraic equation of degree higher than the first; as we may easily see by working out a synthetical process algebraically according to the analogy of finding first the conjugate diametral plane to any chosen diameter of an ellipsoid, and then the diameter of its elliptic section, conjugate to any chosen diameter of this ellipse. Now let

$$\begin{aligned}\psi_{,} &= l\psi_{,,} + m\phi_{,,} + \ldots \\ \phi_{,} &= l'\psi_{,,} + m'\phi_{,,} + \ldots \\ &\text{etc., etc.,}\end{aligned}$$

where l, m, ..., l', m', ... satisfy the equations

Simplified expressions for the kinetic and potential energies.

by a determinate linear transformation of the co-ordinates, reduce the expression for $2T$, which is essentially positive, to a sum of squares, of generalized component velocities, and at the same time V to a sum of the squares of the corresponding co-ordinates, each multiplied by a constant, which may be either positive or negative, but is essentially real. Hence ψ, ϕ,... may be so chosen that

$$\left.\begin{aligned} T&=\tfrac{1}{2}(\dot{\psi}^2+\dot{\phi}^2+\text{etc.}),\\ \text{and}\quad V&=\tfrac{1}{2}(\alpha\psi^2+\beta\phi^2+\text{etc.}\end{aligned}\right\} \qquad (4),$$

α, β, etc., being real positive or negative constants. Hence Lagrange's equations become

$$\ddot{\psi}=-\alpha\psi,\quad \ddot{\phi}=-\beta\phi,\ \text{etc.}, \qquad (5).$$

Integrated equations of motion.

The solutions of these equations are

$$\psi=A\sin(t\sqrt{\alpha}+\epsilon),\quad \phi=A'\sin(t\sqrt{\beta}+\epsilon'),\ \text{etc.} \qquad (6),$$

A, ϵ, A', ϵ', etc., being the arbitrary constants of integration. Hence we conclude the motion consists of a simple harmonic variation of each co-ordinate, provided that α, β, etc., are all positive. This condition is satisfied when V is a true minimum at the configuration of equilibrium; which, as we have seen (§ 292), is necessarily the case when the equilibrium is stable. If any one or more of α, β, ... vanishes, the equilibrium might be either stable or unstable, or neutral; but terms of higher orders in the expansion of V in ascending powers and products of the co-ordinates would have to be examined to test it; and if it were stable, the period of an infinitely small oscillation in the value of the corresponding co-ordinate or co-ordinates would be infinitely great. If any or all of α, β, γ, ... are negative, V is not a minimum, and the equilibrium is (§ 292) essentially unstable. The

---

$$ll'+mm'+\ldots=0,\ l'l''+m'm''+\ldots=0,\ \text{etc.},$$
$$\text{and}\quad l^2+m^2+\ldots=1,\ l'^2+m'^2+\ldots=1,\ \text{etc.}$$

We shall still have, obviously, the same form for $2T$, that is.—

$$2T=\dot{\psi}^2_{,,}+\dot{\phi}^2_{,,}+\ldots$$

And, according to the known theory of the transformation of quadratic functions, we may find l, m, ..., l', m', ... so as to make the products of the co-ordinates disappear from the expression for V, and give

$$2V=\alpha\psi^2_{,,}+\beta\phi^2_{,,}+\ldots,$$

where α, β, γ, etc., are the roots, necessarily real, of an equation of the ith degree of which the coefficients depend on the coefficients of the squares and products in the expression for V in terms of $\psi_{,}$, $\phi_{,}$, etc. Later [(7) and (8) of § 343], a *single process* for carrying out this investigation, when T and V are given as any two homogeneous quadratic functions, will be indicated.

form (6) for the solution, for each co-ordinate for which this is the case, becomes imaginary, and is to be changed into the exponential form, thus; for instance, let $a = -a'$, a' being positive. Thus

Infinitely small disturbance from unstable equilibrium.

$$\psi = A\epsilon^{+t\sqrt{a'}} + B\epsilon^{-t\sqrt{a'}} \qquad (7)$$

which (unless the disturbance is so adjusted as to make the arbitrary constant A vanish) indicates an unlimited increase in the deviation. This form of solution expresses the approximate law of falling away from a configuration of unstable equilibrium. In general, of course, the approximation becomes less and less accurate as the deviation increases.

Example of normal displacements.

One example for the present will suffice. Let a solid, immersed in an infinite *liquid* (§ 331), be prevented from any motion of rotation, and left only freedom to move parallel to a certain fixed plane, and let it be influenced by forces subject to the conservative law, which vanish in a particular position of equilibrium Taking any point of reference in the body, choosing its position when the body is in equilibrium, as origin of rectangular coordinates OX, OY, and reckoning the potential energy from it, we shall have, as in general,

$$2T = A\dot{x}^2 + B\dot{y}^2 + 2C\dot{x}\dot{y}\,;\ 2V = ax^2 + by^2 + 2cxy,$$

the principles stated in § 331 above, allowing us to regard the co-ordinates x and y as fully specifying the system, provided always, as is understood, that if the body is given at rest, or is brought to rest, the whole liquid is at rest at the same time. By solving the obviously determinate problem of finding that pair of conjugate diameters which are in the same directions for the ellipse

$$Ax^2 + By^2 + 2Cxy = \text{const.},$$

and the ellipse or hyperbola,

$$ax^2 + by^2 + 2cxy = \text{const.},$$

and choosing these as oblique axes of co-ordinates (x_1, y_1), we shall have

$$2T = A_1\dot{x}_1^2 + B_1\dot{y}_1^2, \text{ and } 2V = a_1x_1^2 + b_1y_1^2.$$

And, as A_1, B_1 are essentially positive, we may, merely to shorten our expressions, take $x_1\sqrt{A_1} = \psi$, $y_1\sqrt{B_1} = \phi$; so that we shall have

$$2T = \dot{\psi}^2 + \dot{\phi}^2,\ 2V = \alpha\psi^2 + \beta\phi^2,$$

the normal expressions, according to the general form shown above.

The interpretation of the general solution is as follows:—

338. If a conservative system is infinitely little displaced from a configuration of stable equilibrium, it will ever after vibrate about this configuration, remaining infinitely near it;

General theorem regarding infinitely small motion about a configuration of equilibrium.

each particle of the system performing a motion which is composed of simple harmonic vibrations. If there are i degrees of freedom to move, and we consider any system (§ 202) of generalized co-ordinates specifying its position at any time, the deviation of any one of these co-ordinates from its value for the configuration of equilibrium will vary according to a complex harmonic function (§ 68), composed in general of i simple harmonics of incommensurable periods, and therefore (§ 67) the whole motion of the system will not recur periodically through the same series of configurations. There are in general, however, i distinct determinate displacements, which we shall call *the normal displacements*, fulfilling the condition, that if any one of them be produced alone, and the system then left to itself for an instant at rest, this displacement will diminish and increase periodically according to a simple harmonic function of the time, and consequently every particle of the system will execute a simple harmonic movement in the same period. This result, we shall see later, includes cases in which there are an infinite number of degrees of freedom; as for instance a stretched cord; a mass of air in a closed vessel; waves in water, or oscillations in a vessel of water of limited extent, or an elastic solid; and in these applications it gives the theory of the so-called "fundamental vibration," and successive "harmonics" of the cord, and of all the different possible simple modes of vibration in the other cases.

Normal displacements from equilibrium.

Case of equality among periods.

339. If, as may be in particular cases, the periods of the vibrations for two or more of the normal displacements are equal, any displacement compounded of them will also fulfil the condition of a normal displacement. And if the system be displaced according to any one such normal displacement, and projected with velocity corresponding to another, it will execute a movement, the resultant of two simple harmonic movements in equal periods. The graphic representation of the variation of the corresponding co-ordinates of the system, laid down as two rectangular co-ordinates in a plane diagram, will consequently (§ 65) be a circle or an ellipse; which will therefore, of course, be the form of the orbit of any particle of the system which has a distinct direction of motion, for two of the displacements in question. But it must be remembered that some of the principal parts (as for instance the body supported on the fixed axis, in the illustration of § 330, *Example* C) may have only one degree of freedom; or even that each part of the system may have only one degree of freedom, as for instance if the system is composed of a set of particles each constrained to remain on a given line, or of rigid bodies on fixed axes, mutually influencing one another by elastic cords or otherwise. In such a case as the last, no particle of the system can move otherwise than in one line; and the ellipse, circle, or other graphical representation of the composition of the harmonic motions of the system, is merely an aid to comprehension, and not a representation of any motion actually taking place in any part of the system. Graphic representation.

340. In nature, as has been said above (§ 278), every system uninfluenced by matter external to it is conservative, when the ultimate molecular motions constituting heat, light, and magnetism, and the potential energy of chemical affinities, are taken into account along with the palpable motions and measurable forces. But (§ 275) practically we are obliged to admit forces of friction, and resistances of the other classes there enumerated, as causing losses of energy to be reckoned, in abstract dynamics, without regard to the equivalents of heat or other molecular actions which they generate. Hence when such resistances are to be taken into account, forces opposed to the motions of various parts of a system must be introduced into the equations. According to the approximate knowledge which we have from experiment, these forces are independent of the velocities when due to the friction of solids; and are simply proportional to the velocities when due to fluid viscosity directly, or to electric or magnetic influences, with corrections depending on varying temperature, and on the varying configuration of the system. In consequence of the last-mentioned cause, the resistance of a *real liquid* (which is always more or less viscous) against a body moving very rapidly through it, and leaving a great deal of irregular motion, such as "eddies," in its wake, seems to be nearly in proportion to the square of the velocity; although, as Stokes has shown, at the lowest speeds the resistance is probably in simple proportion to the velocity, and for all speeds may, it is probable, be approxi- Dissipative systems. Views of Stokes on resistance to a solid moving through a fluid.

Stokes' probable law. mately expressed as the sum of two terms, one simply as the velocity, and the other as the square of the velocity.

Friction of solids. **341.** The effect of friction of solids rubbing against one another is simply to render impossible the *infinitely* small vibrations with which we are now particularly concerned; and to allow any system in which it is present, to rest balanced when displaced within certain finite limits, from a configuration of frictionless equilibrium. In mechanics it is easy to estimate its effects with sufficient accuracy when any practical case of finite oscillations is in question. But the other classes of dissipative agencies give rise to resistances simply as the velocities, without the corrections referred to, when the motions are infinitely small; and can never balance the system in a configuration deviating to any extent, however small, from a configuration of equilibrium without friction. In the theory of infinitely small vibrations, they are to be taken into account by adding to the expressions for the generalized components of force, terms consisting of the generalized velocities each multiplied by a constant, which gives us equations still remarkably amenable to rigorous mathematical treatment. The result of the integration for the case of a single degree of freedom is very simple; and it is of extreme importance, both for the explanation of many natural phenomena, and for use in a large variety of experimental investigations in Natural Philosophy. Partial conclusions from it, in the first place, stated in general terms, are as follows:—

Resistances varying as velocities.

342. If the resistance is less than a certain limit, in any particular case, the motion is a simple harmonic oscillation, with amplitude decreasing by equal proportions in equal successive intervals of time. But if the resistance exceeds this limit, the system when displaced from its position of equilibrium, and left to itself, returns gradually towards its position of equilibrium, never oscillating through it to the other side, and only reaching it after an infinite time.

In the unresisted motion, let n^2 be the rate of acceleration, when the displacement is unity; so that (§ 57) we have $T=\frac{2\pi}{n}$: and let the rate of retardation due to the resistance corresponding to unit velocity be k. Then the motion is of the oscillatory or non-oscillatory class according as $k < 2n$ or $k > 2n$. In the first case, the period of the oscillation is increased by the resistance from T to $T\frac{n}{(\frac{1}{4}k^2-n^2)^{\frac{1}{2}}}$, and the rate at which the Napierian logarithm of the amplitude diminishes per unit of time is $\frac{1}{2}k$.

Effect of resistance varying as velocity in a simple motion.

343. The general solution of the problem, to find the motion of a system having any number, i, of degrees of freedom, when infinitely little disturbed from a position of equilibrium, and left to move subject to resistances proportional to velocities, shows that the whole motion may be resolved, in general determinately, into $2i$ different motions each either simple harmonic with amplitude diminishing according to the law stated above (§ 342), or non-oscillatory, and consisting of equi-proportionate diminutions of the components of displacement in equal successive intervals of time.

Infinitely small motion of a dissipative system.

For the case of one degree of freedom, the differential equation of motion is

$$\ddot{\psi}+k\dot{\psi}+n^2\psi=0,$$

of which the complete integral is

$$\psi=\{A\sin n't+B\cos n't\}\epsilon^{-\frac{1}{2}kt},\text{ where } n'=\sqrt{(n^2-\tfrac{1}{4}k^2)},$$

or, which is the same,

$$\psi=(C\epsilon^{-n_,t}+C'\epsilon^{n_,t})\epsilon^{-\frac{1}{2}kt},\text{ where } n_,=\sqrt{(\tfrac{1}{4}k^2-n^2)}.$$

A and B in one case, or C and C' in the other, being the arbitrary constants of integration. Hence the propositions stated in § 342 for this case.

The most general suppositions we can make regarding the infinitely small motions of a system give, as the differential equations,

$$\left.\begin{array}{l}\frac{d}{dt}\left(\frac{dT}{d\dot{\psi}}\right)+\frac{d}{dt}(\mathfrak{A}\psi+\mathfrak{B}\phi+\ldots)+a\psi+b\phi+\ldots=0\\ \frac{d}{dt}\left(\frac{dT}{d\dot{\phi}}\right)+\frac{d}{dt}(\mathfrak{A}'\psi+\mathfrak{B}'\phi+\ldots)+a'\psi+b'\phi+\ldots=0\\ \qquad\qquad\text{etc.}\qquad\qquad\qquad\text{etc.}\end{array}\right\}\quad(1)$$

(forces of the non-conservative class, dependent on position not on motion, being not excluded unless the relations $b=a'$, $c=a''$, etc., hold).

The theory of simultaneous linear differential equations with constant coefficients shows that the general solution for each

Infinitely small motion of a dissipative system.

element is the sum of particular solutions, and that a particular solution is of the form

$$\psi = l\epsilon^{\lambda t},\ \phi = m\epsilon^{\lambda t},\ \text{etc.} \quad (2).$$

Assuming, then, this to be a solution, and substituting in the differential equations, we have

$$\left.\begin{array}{l} \lambda^2 \dfrac{d\mathfrak{T}}{dl} + \lambda(\mathfrak{A}l + \mathfrak{B}m + \ldots) + al + bm + \ldots = 0 \\ \lambda^2 \dfrac{d\mathfrak{T}}{dm} + \lambda(\mathfrak{A}'l + \mathfrak{B}'m + \ldots) + a'l + b'm + \ldots = 0 \\ \qquad\quad \text{etc.} \qquad\qquad\qquad \text{etc.} \end{array}\right\} \quad (3)$$

where $\mathfrak{T}$ denotes the same homogeneous quadratic function of $l, m, \ldots$, that T is of $\psi, \phi, \ldots$ These equations, i in number, determine λ by the determinantal equation

$$\begin{vmatrix} (\lambda^2 A + \lambda\mathfrak{A} + a), & (\lambda^2 B + \lambda\mathfrak{B} + b), & \ldots \\ (\lambda^2 A' + \lambda\mathfrak{A}' + a'), & (\lambda^2 B' + \lambda\mathfrak{B}' + b'), & \ldots \\ \ldots & \ldots & \ldots \\ \ldots & \ldots & \ldots \end{vmatrix} = 0 \quad (4)$$

if A, B, C, A', B', C', etc., denote the coefficients of l, m, n, etc., in $\dfrac{d\mathfrak{T}}{dl}, \dfrac{d\mathfrak{T}}{dm}$, etc., which are of course subject to the relations

$$B = A',\ C = A'',\ C' = B'',\ \text{etc.} \quad (5).$$

The equation (4) is of the degree $2i$, in λ; and if any one of its roots be used for λ in the i linear equations (3), these become harmonized and give the $i-1$ ratios $l:m$, $l:n$, etc.; and we have then, in (2) a particular solution with one arbitrary constant, l. Thus, from the $2i$ roots, when unequal, we have $2i$ distinct particular solutions, each with an arbitrary constant; and the addition of these solutions, as explained above, gives the general solution. Cases in which there are equal roots leave a corresponding number of degrees of indeterminateness in the ratios $l:m$, $l:n$, etc., and so allow the requisite number of arbitrary constants to be made up.

When the forces not due to motion are of the conservative class, we have

$$b = a',\ c = a'',\ c' = b'' \quad (6)$$

a, b, etc., being such that

$$V = \tfrac{1}{2}(a\psi^2 + 2b\psi\phi + 2c\psi\theta + \ldots + b'\phi^2 + 2c'\phi\theta + \ldots).$$

When $\mathfrak{A}, \mathfrak{B}, \mathfrak{A}', \mathfrak{B}'$, etc., are either all positive, or when those of them which are negative are limited to such magnitudes as they could have in nature, roots of the equation for λ, if real, must be negative, or if imaginary must have their real parts negative;

so that every particular solution may be composed of terms of either of the forms *Infinitely small motion of a dissipative system.*

$$C\epsilon^{-pt}\sin qt, \text{ or } C\epsilon^{-pt},$$

where p is essentially positive. This we see because terms such as $C\epsilon^{pt}\sin qt$, would represent a motion returning again and again with continually increasing energy through the configuration of equilibrium. The mathematical analysis of these conditions, which has not, so far as we are aware, been worked out, deserves attention from mathematicians.

We fall back on the case of no resistance, by taking *Case of no resistance.*

$$\mathfrak{A}=0, \ \mathfrak{B}=0, \ldots \ \mathfrak{A}'=0, \text{ etc.},$$

and the determinantal equation becomes

$$\begin{vmatrix} (\lambda^2A+a), & (\lambda^2B+b), & \ldots \\ (\lambda^2A'+a'), & (\lambda^2B'+b'), & \ldots \\ \ldots & \ldots & \ldots \end{vmatrix} = 0 \qquad (7).$$

This is of the degree i in λ^2. Its i roots are of course, for the case of a conservative system, the values of α, β, ..., of our first investigation (§ 337); and we infer that, for this case, they are all real from what was proved there. The equations (3) to determine l, m, ..., become

$$\left.\begin{array}{l} (\lambda^2A+a)l+(\lambda^2B+b)m+\ldots=0 \\ (\lambda^2A'+a')l+(\lambda^2B'+b')m+\ldots=0 \\ \quad\text{etc.} \qquad\qquad \text{etc.} \end{array}\right\} \qquad (8),$$

and thus, in (7) and (8), we have the promised solution in one completely expressed process. The property of the determinantal equation (7), that its roots are all real when the relations (5) and (6) are satisfied, is very remarkable. It seems to have escaped the notice of modern algebraists. When these relations are not satisfied [as with the well-known *wry cubic*, § 181 (3)], the values of λ^2 may be all real, or some of them, if not all, may be imaginary. When they are not all real, let $\rho\pm\sigma\sqrt{-1}$ be a pair of imaginary roots. The corresponding values of λ, or the square roots of those, may be denoted by $\pm(p\pm q\sqrt{-1})$. Hence in the general solution terms of the form

$$C\epsilon^{pt}\sin qt$$

will occur. That is to say, there are infinitely small displacements from a position of equilibrium which would give rise to harmonic oscillations with amplitude increasing according to the logarithmic law as long as the displacement remains small enough to allow our approximation to hold. This is of course a way of diverging from a position of unstable equilibrium which is impossible except with artificial arrangements giving not a conservative, but an *accumulative* system of force.

Artificial or imaginary accumulative system.

344. When the forces of a system depending on configuration, and not on motion, or, as we may call them for brevity, the forces of position, violate the law of conservatism, we have seen (§ 272) that energy without limit may be drawn from it by guiding it perpetually through a returning cycle of configurations, and we have inferred that in every real system, not supplied with energy from without, the forces of position fulfil the conservative law. But it is easy to arrange a system artificially, in connexion with a source of energy, so that its forces of position shall be non-conservative; and the consideration of the kinetic effects of such an arrangement, especially of its oscillations about or motions round a configuration of equilibrium, is most instructive, by the contrasts which it presents to the phenomena of a natural system. The preceding investigation gives the general solution of the problem—to find the infinitely small motion of a system infinitely near a position of equilibrium, when there is deviation from conservatism both by resistances, and by the character of the forces of position. In the case of no resistance, with which alone we need occupy ourselves at present, the character of the equilibrium as to stability or instability is discriminated according to the character of the roots of an algebraic equation of degree equal to the number of degrees of freedom of the system.

If the roots (λ^2) of the determinantal equation § 343 (7) are all real and negative, the equilibrium is stable: in every other case it is unstable.

345. But although, when the equilibrium is stable, no possible infinitely small displacement and velocity given to the system can cause it, when left to itself, to go on moving either farther and farther away till a finite displacement is reached, or till a finite velocity is acquired; it is very remarkable that stability should be possible, considering that even in the case of stability an endless increase of velocity may, as is easily seen from § 272, be obtained merely by *constraining* the system to a particular closed course, or circuit of configurations, nowhere deviating by more than an infinitely small amount from the configuration of equilibrium, and leaving it at rest anywhere in a certain part of this circuit. This result, and the distinct peculiarities of the cases of stability and instability, will be sufficiently illustrated by the simplest possible example,—that of a material particle moving in a plane.

Artificial or imaginary accumulative system.

Let the mass be unity, and the components of force parallel to two rectangular axes be $ax+by$, and $a'x+b'y$, when the position of the particle is (x, y). The equations of motion will be

$$\ddot{x}=ax+by,\ \ddot{y}=a'x+b'y \qquad (1).$$

Let $\tfrac{1}{2}(a'+b)=c$, and $\tfrac{1}{2}(a'-b)=e$:

the components of the force become

$$ax+cy-ey, \text{ and } cx+b'y+ex,$$

or

$$-\frac{dV}{dx}-ey, \text{ and} -\frac{dV}{dy}+ex,$$

where $V=-\tfrac{1}{2}(ax^2+b'y^2+2cxy)$.

The terms $-ey$ and $+ex$ are clearly the components of a force $e(x^2+y^2)^{\frac{1}{2}}$, perpendicular to the radius-vector of the particle. Hence if we turn the axes of co-ordinates through any angle, the corresponding terms in the transformed components are still $-ey$ and $+ex$. If, therefore, we choose the axes so that

$$V=\tfrac{1}{2}(\alpha x^2+\beta y^2) \qquad (2),$$

the equations of motion become, without loss of generality,

$$\ddot{x}=-\alpha x-ey,\ \ddot{y}=-\beta y+ex.$$

To integrate these, assume, as in general [§ 343 (2)],

$$x=l\epsilon^{\lambda t},\ y=m\epsilon^{\lambda t}.$$

Then, as before [§ 343 (8)],

$$(\lambda^2+\alpha)l+em=0, \text{ and } -el+(\lambda^2+\beta)m=0.$$

Whence $$(\lambda^2+\alpha)(\lambda^2+\beta)=-e^2 \qquad (3),$$

which gives

$$\lambda^2=-\tfrac{1}{2}(\alpha+\beta)\pm\{\tfrac{1}{4}(\alpha-\beta)^2-e^2\}^{\frac{1}{2}}$$

This shows that the equilibrium is stable if both $\alpha\beta+e^2$, and $\alpha+\beta$ are positive and $e^2<\tfrac{1}{4}(\alpha-\beta)^2$ [that is, if e is between the values $\tfrac{1}{2}(\beta-\alpha)$ and $-\tfrac{1}{2}(\beta-\alpha)$] but unstable in every other case.

But let the particle be constrained to remain on a circle, of radius r. Denoting by θ its angle-vector from OX, and, transforming (§ 27) the equations of motion, we have

$$\ddot{\theta}=-(\beta-\alpha)\sin\theta\cos\theta+e=-\tfrac{1}{2}(\beta-\alpha)\sin 2\theta+e \qquad (4).$$

If we had $e=0$ (a conservative system of force) the positions of equilibrium would be at $\theta=0$, $\theta=\tfrac{1}{2}\pi$, $\theta=\pi$, and $\theta=\tfrac{3}{2}\pi$; and the motion would be that of the quadrantal pendulum. But when e has any finite value less than $\tfrac{1}{2}(\beta-\alpha)$ which, for convenience, we may suppose positive, there are positions of equilibrium at

Artificial or imaginary accumulative system.

$$\theta=\vartheta,\ \theta=\frac{\pi}{2}-\vartheta,\ \theta=\pi+\vartheta,\ \text{and}\ \theta=\frac{3\pi}{2}-\vartheta,$$

where ϑ is half the acute angle whose sine is $\frac{2e}{\beta-\alpha}$: the first and third being positions of stable, and the second and fourth of unstable, equilibrium. Thus it appears that the effect of the constant tangential force is to displace the positions of stable and unstable equilibrium forwards and backwards on the circle through angles each equal to ϑ. And, by multiplying (4) by $2\dot{\theta}dt$ and integrating, we have as the integral equation of energy

$$\dot{\theta}^2=C+\tfrac{1}{2}(\beta-\alpha)\cos 2\theta+2e\theta. \qquad (5)$$

From this we see that the value of C, to make the particle just reach the position of unstable equilibrium, is

$$C=-\tfrac{1}{2}(\beta-\alpha)\cos(\pi-2\vartheta)-e(\pi-2\vartheta),$$
$$=\sqrt{\frac{(\beta-\alpha)^2}{4}-e^2}+e\left(\pi-\sin^{-1}\frac{2e}{\beta-\alpha}\right),$$

and by equating to zero the expression (5) for $\dot{\theta}^2$, with this value of C substituted, we have a transcendental equation in θ, of which the least negative root, $\theta_{,}$, gives the limit of vibrations on the side reckoned backwards from a position of stable equilibrium. If the particle be placed at rest on the circle at any distance less than $\frac{\pi}{2}-2\vartheta$ *before* a position of stable equilibrium, or less than $\vartheta-\theta_{,}$ *behind* it, it will vibrate. But if placed anywhere beyond those limits and left either at rest or moving with any velocity in either direction, it will end by flying round and round forwards with a periodically increasing and diminishing velocity, but increasing every half turn by equal additions to its squares.

If on the other hand $e>\tfrac{1}{2}(\beta-\alpha)$, the positions both of stable and unstable equilibrium are imaginary; the tangential force predominating in every position. If the particle be left at rest in any part of the circle it will fly round with continually increasing velocity, but periodically increasing and diminishing acceleration.

Kinetic stability.

346. There is scarcely any question in dynamics more important for Natural Philosophy than the stability or instability of motion. We therefore, before concluding this chapter, propose to give some general explanations and leading principles regarding it.

A "conservative disturbance of motion" is a disturbance in the motion or configuration of a conservative system, not altering the sum of the potential and kinetic energies. A conservative disturbance of the motion through any particular configuration is a change in velocities, or component velocities, not altering the whole kinetic energy. Thus, for example, a conservative disturbance of the motion of a particle through any point, is a change in the direction of its motion, unaccompanied by change of speed. Conservative disturbance.

347. The actual motion of a system, from any particular configuration, is said to be *stable* if every possible infinitely small conservative disturbance of its motion through that configuration may be compounded of conservative disturbances, any one of which would give rise to an alteration of motion which would bring the system again to some configuration belonging to the undisturbed path, in a finite time, and without more than an infinitely small digression. If this condition is not fulfilled, the motion is said to be *unstable*. Kinetic stability and instability discriminated.

348. For example, if a body, A, be supported on a fixed vertical axis; if a second, B, be supported on a parallel axis belonging to the first; a third, C, similarly supported on B, and so on; and if B, C, etc., be so placed as to have each its centre of inertia as far as possible from the fixed axis, and the whole set in motion with a common angular velocity about this axis, the motion will be stable, from every configuration, as is evident from the principles regarding the resultant centrifugal force on a rigid body, to be proved later. If, for instance, each of the bodies is a flat rectangular board hinged on one edge, it is obvious that the whole system will be kept stable by centrifugal force, when all are in one plane and as far out from the axis as possible. But if A consist partly of a shaft and crank, as a common spinning-wheel, or the fly-wheel and crank of a steam-engine, and if B be supported on the crank-pin as axis, and turned inwards (towards the fixed axis, or across the fixed axis), then, even although the centres of inertia of C, D, etc., are placed as far from the fixed axis as possible, consistent with this position of B, the motion of the system will be unstable. Examples.

349. The rectilinear motion of an elongated body lengthwise, or of a flat disc edgewise, through a fluid is unstable. But the motion of either body, with its length or its broadside perpendicular to the direction of motion, is stable. This is demon-

Kinetic stability. Hydrodynamic example.

strated for the ideal case of a perfect liquid (§ 331), in § 332, *Example* (2); and the results explained in § 333 show, for a solid of revolution, the precise character of the motion consequent upon an infinitely small disturbance in the direction of the motion from being exactly along or exactly perpendicular to the axis of figure; whether the infinitely small oscillation, in a definite period of time, when the rectilineal motion is stable, or the swing round to an infinitely nearly-inverted position when the rectilineal motion is unstable. Observation proves the assertion we have just made, for real fluids, air and water, and for a great variety of circumstances affecting the motion. Several illustrations have been referred to in § 336; and it is probable we shall return to the subject later, as being not only of great practical importance, but profoundly interesting although very difficult in theory.

Circular simple pendulum.

350. The motion of a single particle affords simpler and not less instructive illustrations of stability and instability. Thus if a weight, hung from a fixed point by a light inextensible cord, be set in motion so as to describe a circle about a vertical line through its position of equilibrium, its motion is stable. For, as we shall see later, if disturbed infinitely little in direction without gain or loss of energy, it will describe a sinuous path, cutting the undisturbed circle at points successively distant from one another by definite fractions of the circumference, depending upon the angle of inclination of the string to the vertical. When this angle is very small, the motion is sensibly the same as that of a particle confined to one plane and moving under the influence of an attractive force towards a fixed point, simply proportional to the distance; and the disturbed path cuts the undisturbed circle four times in a revolution. Or if a particle confined to one plane, move under the influence of a centre in this plane, attracting with a force inversely as the square of the distance, a path infinitely little disturbed from a circle will cut the circle twice in a revolution. Or if the law of central force be the nth power of the distance, and if $n+3$ be positive, the disturbed path will cut the undisturbed circular orbit at successive angular intervals, each equal to $\frac{\pi}{\sqrt{n+3}}$. But the motion will be unstable if n be negative, and $-n > 3$.

Circular orbit.

The criterion of stability is easily investigated for circular motion round a centre of force from the differential equation of the general orbit (§ 36), Kinetic stability in circular orbit.

$$\frac{d^2u}{d\theta^2}+u=\frac{P}{h^2u^2}.$$

Let the value of h be such that motion in a circle of radius a^{-1} satisfies this equation. That is to say, let $\frac{P}{h^2u^2}=u$, when $u=a$.

Let now $u=a+\rho$, ρ being infinitely small. We shall have

$$u-\frac{P}{h^2u^2}=\alpha\rho,$$

if α denotes the value of $\frac{d}{du}(u-\frac{P}{h^2u^2})$ when $u=a$: and therefore the differential equation for motion infinitely nearly circular is

$$\frac{d^2\rho}{d\theta^2}+\alpha\rho=0.$$

The integral of this is most conveniently written

$$\rho=A\sin(\theta\sqrt{\alpha}+\beta) \text{ when } \alpha \text{ is positive,}$$

and $\rho=C\epsilon^{\theta\sqrt{-\alpha}}+C'\epsilon^{-\theta\sqrt{-\alpha}}$ when α is negative.

Hence we see that the circular motion is stable in the former case, and unstable in the latter.

For instance, if $P=\mu r^n=\mu u^{-n}$, we have

$$\frac{d}{du}(u-\frac{P}{h^2u^2})=1+(n+2)\frac{P}{h^2u^3};$$

and putting $\frac{P}{h^2u^2}=u=a$, in this we find $\alpha=n+3$; whence the result stated above.

Or, taking example (B) of § 330, and putting mP for P, and mh for h,

$$\frac{P}{h^2u^2}=\frac{m'}{m+m'}(\frac{g}{h^2}u^{-2}+u),$$

$$\frac{d}{du}(u-\frac{P}{h^2u^2})=\frac{m+\frac{2m'g}{h^2u^3}}{m+m'}.$$

Hence, putting $u=a$, and making $h^2=\frac{gm'}{ma^3}$ so that motion in a circle of radius a^{-1} may be possible, we find

$$\alpha=\frac{3m}{m+m'}.$$

Hence the circular motion is always stable; and the period of the variation produced by an infinitely small disturbance from it is $2\pi\sqrt{\frac{m+m'}{3m}}$.

Kinetic stability of a particle moving on a smooth surface.

351. The case of a particle moving on a smooth fixed surface under the influence of no other force than that of the constraint, and therefore always moving along a geodetic line of the surface, affords extremely simple illustrations of stability and instability. For instance, a particle placed on the inner circle of the surface of an anchor ring, and projected in the plane of the ring, would move perpetually in that circle, but unstably, as the smallest disturbance would clearly send it away from this path, never to return until after a digression round the outer edge. (We suppose of course that the particle is held to the surface, as if it were placed in the infinitely narrow space between a solid ring and a hollow one enclosing it.) But if a particle is placed on the outermost, or greatest, circle of the ring, and projected in its plane, an infinitely small disturbance will cause it to describe a sinuous path cutting the circle at points round it successively distant by angles each equal to $\pi\sqrt{\frac{b}{a}}$, or intervals of time, $\frac{\pi}{\omega}\sqrt{\frac{b}{a}}$, where a denotes the radius of that circle, ω the angular velocity in it, and b the radius of the circular cross section of the ring. This is proved by remarking that an infinitely narrow band from the outermost part of the ring has, at each point, a and b for its principal radii of curvature, and therefore (§ 150) has for its geodetic lines the great circles of a sphere of radius $\sqrt{ab}$, upon which (§ 152) it may be bent.

352. In all these cases the undisturbed motion has been circular or rectineal, and, when the motion has been stable, the effect of a disturbance has been *periodic*, or recurring with the same phases in equal successive intervals of time. An illustration of thoroughly stable motion in which the effect of a disturbance is not "periodic," is presented by a particle sliding down an inclined groove under the action of gravity. To take the simplest case, we may consider a particle sliding down along the lowest straight line of an inclined hollow cylinder. If slightly disturbed from this straight line, it will oscillate on each side of it perpetually in its descent, but not with a uniform periodic motion, though the durations of its excursions to each side of the straight line are all equal.

353. A very curious case of stable motion is presented by a particle constrained to remain on the surface of an anchor-ring fixed in a vertical plane, and projected along the great circle from any point of it, with any velocity. An infinitely small disturbance will give rise to a disturbed motion of which the path will cut the vertical circle over and over again for ever, at unequal intervals of time, and unequal angles of the circle; and obviously not recurring periodically in any cycle, except with definite particular values for the whole energy, some of which are less and an infinite number are greater than that which just suffices to bring the particle to the highest point of the ring. The full mathematical investigation of these circumstances would afford an excellent exercise in the theory of differential equations, but it is not necessary for our present illustrations. Kinetic stability. Incommensurable oscillations.

354. In this case, as in all of stable motion with only two degrees of freedom, which we have just considered, there has been stability throughout the motion; and an infinitely small disturbance from any point of the motion has given a disturbed path which intersects the undisturbed path over and over again at finite intervals of time. But, for the sake of simplicity at present confining our attention to two degrees of freedom, we have a *limited* stability in the motion of an unresisted projectile, which satisfies the criterion of stability only at points of its upward, not of its downward, path. Thus if *MOPQ* be Oscillatory kinetic stability. Limited kinetic stability.

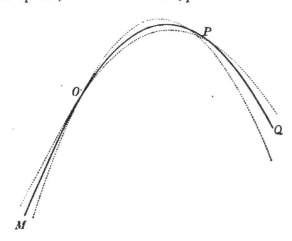

Kinetic stability of a projectile.

the path of a projectile, and if at O it be disturbed by an infinitely small force either way perpendicular to its instantaneous direction of motion, the disturbed path will cut the undisturbed infinitely near the point P where the direction of motion is perpendicular to that at O: as we easily see by considering that the line joining two particles projected from one point at the same instant with equal velocities in the directions of any two lines, will always remain perpendicular to the line bisecting the angle between these two.

General criterion.

355. The principle of varying action gives a mathematical criterion for stability or instability in every case of motion. Thus in the first place it is obvious, and it will be proved below (§§ 358, 361), that if the action is a true minimum in the motion of a system from any one configuration to the configuration reached at any other time, however much later, the motion is thoroughly unstable. For instance, in the motion of a particle constrained to remain on a smooth fixed surface, and uninfluenced by gravity, the action is simply the length of the path, multiplied by the constant velocity. Hence in the particular case of a particle uninfluenced by gravity, moving round the inner circle in the plane of an anchor-ring considered above, the action, or length of path, is clearly a minimum from any one point to the point reached at any subsequent time. (The action is not merely a minimum, but is the least possible, from any point of the circular path to any other, through less than half a circumference of the circle.) On the other hand, although the path from any point in the greatest circle of the ring to any other at a distance from it along the circle, less than $\pi\sqrt{ab}$, is clearly least possible if along the circumference; the path of absolutely least length is not along the circumference between two points at a greater circular distance than $\pi\sqrt{ab}$ from one another, nor is the path along the circumference between them a minimum at all in this latter case.

Examples.

Motion on an anticlastic surface proved unstable.

On any surface whatever which is everywhere anticlastic, or along a geodetic of any surface which passes altogether through an anticlastic region, the motion is thoroughly unstable. For if it were stable from any point O, we should have the given undisturbed path, and the disturbed path from O cutting it at some point Q;—two different geodetic lines joining two points; which is impossible on an anticlastic surface, inasmuch as the sum of the exterior angles of any closed figure of geodetic lines exceeds four right angles (§ 136) when the integral curvature of the enclosed area is negative, which (§§ 138, 128) is the case for every portion of surface thoroughly anticlastic. But, on the other hand, it is easily proved that if we have an endless rigid band of curved surface everywhere synclastic, with a geodetic line running through its middle, the motion of a particle projected along this line will be stable throughout, and an infinitely slight disturbance will give a disturbed path cutting the given undisturbed path again and again for ever at successive distances differing according to the different specific curvatures of the intermediate portions of the surface. If from any point, N, of the undisturbed path, a perpendicular be drawn to cut the infinitely near disturbed path in E, the angles OEN and NOE must (§ 138) be together greater than a right angle by an amount equal to the integral curvature of the area EON. From this the differential equation of the disturbed path may be obtained immediately.

Motion of a particle on an anticlastic surface, unstable; on a synclastic surface, stable.

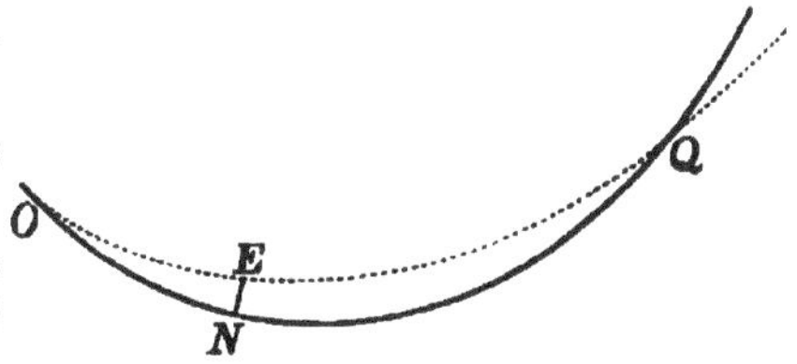

Differential equation of disturbed path.

Let $<EON=a$, $ON=s$, and $NE=u$; and let ϑ, a known function of s, be the specific curvature (§ 136) of the surface in the neighbourhood of N. Let also, for a moment, ϕ denote the complement of the angle OEN. We have

$$a-\phi=\int_0^s \vartheta u ds.$$

Hence $$\frac{d\phi}{ds}=-\vartheta u.$$

But, obviously, $$\phi=\frac{du}{ds};$$

hence $$\frac{d^2u}{ds^2}+\vartheta u=0.$$

When ϑ is constant (as in the case of the equator of a surface of revolution considered above, § 351), this gives

$$u=A\cos(s\sqrt{\vartheta}+E),$$

Differential equation of disturbed path.

agreeing with the result (§ 351) which we obtained by development into a spherical surface.

The case of two or more bodies supported on parallel axes in the manner explained above, and rotating with the centre of inertia of the whole at the least possible distance from the axis, affords a very good illustration also of this proposition which may be safely left as an exercise to the student.

General investigation of disturbed path.

356. To investigate the effect of an infinitely small conservative disturbance produced at any instant in the motion of any conservative system, may be reduced to a practicable problem (however complicated the required work may be) of mathematical analysis, provided the undisturbed motion is thoroughly known.

General equation of motion free in two degrees.

(*a*) First, for a system having but two degrees of freedom to move, let

$$2T = P\dot{\psi}^2 + Q\dot{\phi}^2 + 2R\dot{\psi}\dot{\phi} \qquad (1),$$

where P, Q, R are functions of the co-ordinates not depending on the actual motion. Then

$$\left.\begin{aligned} &\frac{dT}{d\dot{\psi}} = P\dot{\psi} + R\dot{\phi}, \quad \frac{dT}{d\dot{\phi}} = Q\dot{\phi} + R\dot{\psi} \\ &\frac{d}{dt}\frac{dT}{d\dot{\psi}} = P\ddot{\psi} + R\ddot{\phi} + \frac{dP}{d\psi}\dot{\psi}^2 + \left(\frac{dP}{d\phi} + \frac{dR}{d\psi}\right)\dot{\psi}\dot{\phi} + \frac{dR}{d\phi}\dot{\phi}^2, \end{aligned}\right\} \qquad (2);$$

and the Lagrangian equations of motion [§ 329 (10)] are

$$\left.\begin{aligned} &P\ddot{\psi} + R\ddot{\phi} + \tfrac{1}{2}\left\{\frac{dP}{d\psi}\dot{\psi}^2 + 2\frac{dP}{d\phi}\dot{\psi}\dot{\phi} + \left(2\frac{dR}{d\phi} - \frac{dQ}{d\psi}\right)\dot{\phi}^2\right\} = \Psi \\ &R\ddot{\psi} + Q\ddot{\phi} + \tfrac{1}{2}\left\{\left(2\frac{dR}{d\psi} - \frac{dP}{d\phi}\right)\dot{\psi}^2 + 2\frac{dQ}{d\psi}\dot{\psi}\dot{\phi} + \frac{dQ}{d\phi}\dot{\phi}^2\right\} = \Phi \end{aligned}\right\} \qquad (3)$$

We shall suppose the system of co-ordinates so chosen that none of the functions P, Q, R, nor their differential coefficients $\frac{dP}{d\phi}$, etc., can ever become infinite.

(*b*) To investigate the effects of an infinitely small disturbance, we may consider a motion in which, at any time t, the co-ordinates are $\psi + p$ and $\phi + q$, p and q being infinitely small; and, by simply taking the variations of equations (3) in the usual manner, we arrive at two simultaneous differential equations of the second degree, linear with respect to

$$p, \; q, \; \dot{p}, \; \dot{q}, \; \ddot{p}, \; \ddot{q},$$

but having variable coefficients which, when the undisturbed motion ψ, ϕ is fully known, may be supposed to be known functions of t. In these equations obviously none of the coefficients can at any time become infinite if the data correspond to a real dynamical problem, provided the system of co-ordinates is properly chosen (a); and the coefficients of $\ddot{p}$ and $\ddot{q}$ are the values, at the time t, of P, R, and R, Q, respectively, in the order in which they appear in (3), P, Q, R being the coefficients of a homogeneous quadratic function (1) which is essentially positive. These properties being taken into account, it may be shown that in no case can an infinitely small interval of time be the solution of the problem presented (§ 347) by the question of kinetic stability or instability, which is as follows:— General investigation of disturbed path.

(c) The component velocities $\dot{\psi}$, $\dot{\phi}$ are at any instant changed to $\dot{\psi}+a$, $\dot{\phi}+\beta$, subject to the condition of not changing the value of T. Then, a and β being infinitely small, it is required to find the interval of time until $\frac{q}{p}$ first becomes equal to $\frac{\dot{\phi}}{\dot{\psi}}$.

(d) The differential equations in p and q reduce this problem, and in fact the full problem of finding the disturbance in the motion when the undisturbed motion is given, to a practicable form. But, merely to prove the proposition that the disturbed course cannot meet the undisturbed course until after some finite time, and to estimate a limit which this time must exceed in any particular case, it may be simpler to proceed thus:—

(e) To eliminate t from the general equations (3), let them first be transformed so as not to have t independent variable. We must put

$$\ddot{\psi}=\frac{dtd^2\psi-d\psi d^2t}{dt^3},\quad \ddot{\phi}=\frac{dtd^2\phi-d\phi d^2t}{dt^3} \qquad (4).$$

And by the equation of energy we have

$$dt=\frac{(Pd\psi^2+Qd\phi^2+2Rd\psi d\phi)^{\frac{1}{2}}}{\{2(E-V)\}^{\frac{1}{2}}} \qquad (5),$$

it being assumed that the system is conservative. Eliminating dt and d^2t between this and the two equations (3), we find a differential equation of the second degree between ψ and ϕ, which is the differential equation of the course. For simplicity, let us suppose one of the co-ordinates, ϕ for instance, to be independent variable; that is, let $d^2\phi=0$. We have, by (4)

$$d^2t=-\ddot{\phi}\frac{dt^3}{d\phi},\ \text{and therefore}\ \ddot{\psi}dt^2=d^2\psi+\frac{d\psi}{d\phi}\ddot{\phi}dt^2$$

and the result of the elimination becomes

General investigation of disturbed path.

$$(PQ-R^2)\frac{d^2\psi}{d\phi^2}+F\left(\frac{d\psi}{d\phi}\right)=\frac{\left(P\frac{d\psi^2}{d\phi^2}+2R\frac{d\psi}{d\phi}+Q\right)\left[\left(Q+R\frac{d\psi}{d\phi}\right)\Psi-\left(R+P\frac{d\psi}{d\phi}\right)\Phi\right]}{2(E-V)} \quad (6),$$

$F(\frac{d\psi}{d\phi})$ denoting a function of $\frac{d\psi}{d\phi}$ of the third degree, with variable coefficients, none of which can become infinite as long as $E-V$, the kinetic energy, is finite.

(*f*) Taking the variation of this equation on the supposition that ψ becomes $\psi+p$, where p is infinitely small, we have

$$(PQ-R^2)\frac{d^2p}{d\phi^2}+L\frac{dp}{d\phi}+Mp=0 \quad (7),$$

where L and M denote known functions of ϕ, neither of which has any infinitely great value. This determines the deviation, p, of the course. Inasmuch as the quadratic (1) is essentially always positive, $PQ-R^2$ must be always positive. Hence, if for a particular value of ϕ, p vanishes, and $\frac{dp}{d\phi}$ has a given value which defines the disturbance we suppose made at any instant, ϕ must increase by a finite amount (and therefore a finite time must elapse) before the value of p can be again zero; that is to say, before the disturbed course can again cut the undisturbed course.

(*g*) The same proposition consequently holds for a system having any number of degrees of freedom. For the preceding proof shows it to hold for the system subjected to any frictionless constraint, leaving it only two degrees of freedom; including that particular frictionless constraint which would not alter either the undisturbed or the disturbed course. The full general investigation of the disturbed motion, with more than two degrees of freedom, takes a necessarily complicated form, but the principles on which it is to be carried out are sufficiently indicated by what we have done.

(*h*) If for $\frac{L}{PQ-R^2}$ we substitute a constant $2a$, less than its least value, irrespectively of sign, and for $\frac{M}{PQ-R^2}$, a constant β greater algebraically than its greatest value, we have an equation

$$\frac{d^2p}{d\phi^2}+2a\frac{dp}{d\phi}+\beta=0 \quad (8).$$

Here the value of p vanishes for values of ϕ successively ex-

General investigation of disturbed path.

ceeding one another by $\frac{\pi}{\sqrt{(\beta-\alpha^2)}}$, which is clearly less than the increase that ϕ must have in the actual problem before p vanishes a second time. Also, we see from this, that if $\alpha^2 > \beta$ the actual motion is unstable. It might of course be unstable even if $\alpha^2 < \beta$; and the proper analytical methods for finding either the rigorous solution of (7), or a sufficiently near practical solution, would have to be used to close the criterion of stability or instability, and to thoroughly determine the disturbance of the course.

Differential equation of disturbed path of single particle in a plane.

(i) When the system is only a single particle, confined to a plane, the differential equation of the deviation may be put under a remarkably simple form, useful for many practical problems. Let N be the normal component of the force, per unit of the mass, at any instant, v the velocity, and ρ the radius of curvature of the path. We have (§ 259),

$$N = \frac{v^2}{\rho}.$$

Let, in the diagram, ON be the undisturbed, and OE the disturbed path. Let EN, cutting ON at right angles, be denoted by u, and ON by s. If further we denote by ρ' the radius of curvature in the disturbed path,

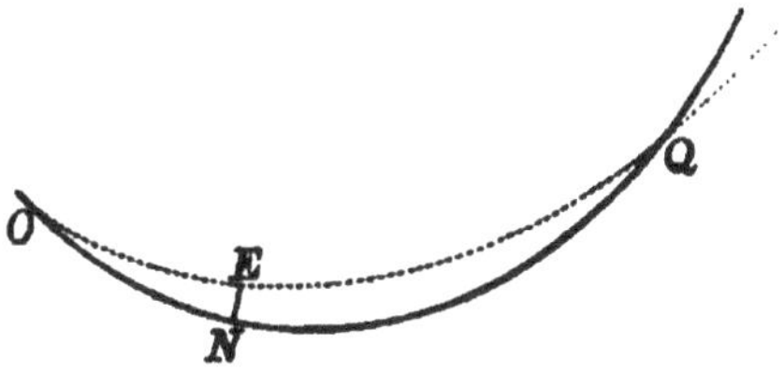

remembering that u is infinitely small, we easily find

$$\frac{1}{\rho'} = \frac{1}{\rho} + \frac{d^2u}{ds^2} + \frac{u}{\rho^2} \qquad (9).$$

Hence, using δ to denote variations from N to E, we have

$$\delta N = \delta\frac{v^2}{\rho} = \frac{\delta(v^2)}{\rho} + v^2\left(\frac{d^2u}{ds^2} + \frac{u}{\rho^2}\right) \qquad (10).$$

But, by the equation of energy,

$$v^2 = 2(E - V),$$

and therefore

$$\delta(v^2) = -2\delta V = 2Nu = \frac{2v^2}{\rho}u.$$

Hence (10) becomes

$$\frac{d^2u}{d\rho^2} + \frac{3u}{\rho^2} - \frac{\delta N}{v^2} = 0 \qquad (11),$$

Differential equation of disturbed path of single particle in a plane.

or, if we denote by ζ the rate of variation of N, per unit of distance from the point N in the normal direction, so that $\delta N = \zeta u$,

$$\frac{d^2u}{d\rho^2} + \left(\frac{3}{\rho^2} - \zeta\right)u = 0 \qquad (12).$$

This includes, as a particular case, the equation of deviation from a circular orbit, investigated above (§ 350).

Kinetic foci.

357. If, from any one configuration, two courses differing infinitely little from one another, have again a configuration in common, this second configuration will be called a kinetic focus relatively to the first: or (because of the reversibility of the motion) these two configurations will be called conjugate kinetic foci. Optic foci, if for a moment we adopt the corpuscular theory of light, are included as a particular case of kinetic foci in general. By § 356 (g) we see that there must be finite intervals of space and time between two conjugate foci in every motion of every kind of system, only provided the kinetic energy does not vanish.

Theorem of minimum action.

Action never a minimum in a course including kinetic foci.

358. Now it is obvious that, provided only a sufficiently short course is considered, the *action*, in any natural motion of a system, is less than for any other course between its terminal configurations. It will be proved presently (§ 361) that the first configuration up to which the action, reckoned from a given initial configuration, ceases to be a minimum, is the first kinetic focus; and conversely, that when the first kinetic focus is passed, the action, reckoned from the initial configuration, ceases to be a minimum; and therefore of course can never again be a minimum, because a course of shorter action, deviating infinitely little from it, can be found for a part, without altering the remainder of the whole, natural course.

Notation for configurations, courses, and action.

359. In such statements as this it will frequently be convenient to indicate particular configurations of the system by single letters, as O, P, Q, R; and any particular course, in which it moves through configurations thus indicated, will be called the course $O \ldots P \ldots Q \ldots R$. The *action* in any natural course will be denoted simply by the terminal letters, taken in the order of the motion. Thus OR will denote the action from O to R; and therefore $OR = -RO$. When there are more real natural courses from O to R than one, the analytical expression for OR will have more than one real value; and it may be necessary to specify for which of these courses the action is reckoned. Thus we may have

OR for $O\ldots E\ldots R$,
OR for $O\ldots E'\ldots R$,
OR for $O\ldots E''\ldots R$,

three different values of one algebraic irrational expression.

360. In terms of this notation the preceding statement (§ 358) may be expressed thus:—If, for a conservative system, moving on a certain course $O\ldots P\ldots O'\ldots P'$, the first kinetic focus conjugate to O be O', the action OP, in this course, will be less than the action along any other course deviating infinitely little from it: but, on the other hand, OP' is greater than the actions in some courses from O to P' deviating infinitely little from the specified natural course $O\ldots P\ldots O'\ldots P'$. Theorem of minimum action.

361. It must not be supposed that the action along OP is necessarily *the least possible* from O to P. There are, in fact, cases in which the action ceases to be *least of all possible*, before a kinetic focus is reached. Thus if $OEAPO'E'A'$ be a sinuous geodetic line cutting the outer circle of an anchor ring, or the equator of an oblate spheroid, in successive points O, A, A', it is easily seen that O', the first kinetic focus conjugate to O, must lie somewhat beyond A. But the length $OEAP$, although a *minimum* (a stable position for a Two or more courses of minimum action possible.

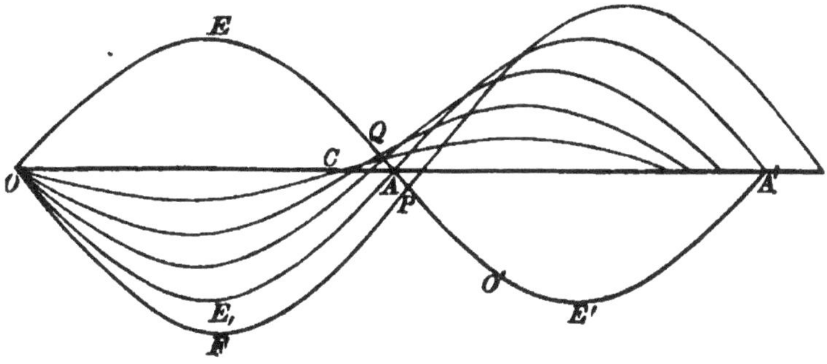

stretched string), is not the shortest distance on the surface from O to P, as *this* must obviously be a line lying entirely on one side of the great circle. From O, to any point, Q, short of A, the distance along the geodetic $OEQA$ is clearly the least possible: but if Q be near enough to A (that is to say, between A and the point in which the envelope of the geodetics drawn Case of two minimum, and one not minimum, geodetic lines between two points.

Case of two minimum, and one not minimum, geodetic lines between two points.

from O, cuts OEA), there will also be two other geodetics from O to Q. The length of one of these will be a minimum, and that of the other not a minimum. If Q is moved forward to A, the former becomes $OE_{,}A$, equal and similar to OEA, but on the other side of the great circle: and the latter becomes the great circle from O to A. If now Q be moved on, to P, beyond A, the minimum geodetic $OEAP$ ceases to be the less of the two minima, and the geodetic OFP lying altogether on the other side of the great circle becomes the least possible line from O to P. But until P is advanced beyond the point, O', in which it is cut by another geodetic from O lying infinitely nearly along it, the length $OEAP$ remains a minimum, according to the general proposition of § 358, which we now proceed to prove.

Difference between two sides and the third of a kinetic triangle

(*a*) Referring to the notation of § 360, let $P_{,}$ be any configuration differing infinitely little from P, but not on the course $O \ldots P \ldots O' \ldots P'$; and let S be a configuration on this course, reached at some finite time after P is passed. Let ψ, ϕ, ... be the co-ordinates of P, and $\psi_{,}$, $\phi_{,}$, ... those of $P_{,}$, and let

$$\psi_{,}-\psi=\delta\psi,\quad \phi_{,}-\phi=\delta\phi,\ \ldots$$

Thus, by Taylor's theorem,

$$OP_{,}+P_{,}S=OS+\{\frac{d(OP+PS)}{d\psi}\delta\psi+\frac{d(OP+PS)}{d\phi}\delta\phi+\ldots\}$$
$$+\tfrac{1}{2}\{\frac{d^2(OP+PS)}{d\psi^2}(\delta\psi)^2+2\frac{d^2(OP+PS)}{d\psi d\phi}\delta\psi\delta\phi+\ldots\}$$
$$+\text{ etc.}$$

But if ξ, η, ... denote the components of momentum at P in the course $O \ldots P$, which are the same as those at P in the continuation, $P \ldots S$, of this course, we have [§ 322 (18)]

$$\xi=\frac{dOP}{d\psi}=-\frac{dPS}{d\psi},\quad \eta=\frac{dOP}{d\phi}=-\frac{dPS}{d\phi},\ \ldots$$

Hence the coefficients of the terms of the first degree of $\delta\psi$, $\delta\phi$, in the preceding expression vanish, and we have

$$\left.\begin{aligned}OP_{,}+P_{,}S-OS=\tfrac{1}{2}\{\frac{d^2(OP+PS)}{d\psi^2}\delta\psi^2+2\frac{d^2(OP+PS)}{d\psi d\phi}\delta\psi\delta\phi+\ldots\}\\ +\text{ etc.}\end{aligned}\right\}\quad (1).$$

(*b*) Now, assuming
$$\left.\begin{aligned}x_1&=a_1\delta\psi+\beta_1\delta\phi+\ldots\\ x_2&=a_2\delta\psi+\beta_2\delta\phi+\ldots\\ &\text{etc.}\quad\text{etc.}\end{aligned}\right\}\quad (2),$$

according to the known method of linear transformations, let a_1, β_1, ... a_2, β_2, ... be so chosen that the preceding quadratic function be reduced to the form

$$A_1x_1^2+A_2x_2^2+\ldots+A_ix_i^2,$$

the whole number of degrees of freedom being i.

Difference between two sides and the third of a kinetic triangle.

This may be done in an infinite variety of ways; and, towards fixing upon one particular way, we may take $\alpha_i=\dot\psi$, $\beta_i=\dot\phi$, etc.; and subject the others to the conditions

$$\dot\psi\alpha_1+\dot\phi\beta_1+\ldots=0,\ \dot\psi\alpha_2+\dot\phi\beta_2+\ldots=0, \text{ etc.}$$

This will make $A_i=0$: for if for a moment we suppose $P_,$ to be on the course $O\ldots P\ldots O'$, we have

$$\frac{\delta\psi}{\dot\psi}=\frac{\delta\phi}{\dot\phi}=\ldots,$$

and therefore

$$x_i=\frac{\dot\psi}{\delta\psi}(\delta\psi^2+\delta\phi^2+\ldots),\ x_{i-1}=0,\ \ldots\ x_2=0,\ x_1=0.$$

But in this case $OP_,+P_,S=OS$; and therefore the value of the quadratic must be zero; that is to say, we must have $A_i=0$. Hence we have

$$\left.\begin{array}{l} OP_,+P_,S-OS=\frac{1}{2}(A_1x_1^2+A_2x_2^2+\ldots+A_{i-1}x^2_{i-1}) \\ \qquad +R \end{array}\right\} \quad (3)$$

where R denotes a remainder consisting of terms of the third and higher degrees in $\delta\psi$, $\delta\phi$, etc., or in x_1, x_2, etc.

(c) Another form, which will be used below, may be given to the same expression thus:—Let $(\xi_,, \eta_,, \zeta_,, \ldots)$ and $(\xi_,', \eta_,', \zeta_,', \ldots)$ be the components of momentum at $P_,$, in the courses $OP_,$ and $P_,S$ respectively. By § 322 (18) we have

$$\xi_,=\frac{dOP'}{d\psi_,},$$

and therefore by Taylor's theorem

$$\xi_,=\frac{dOP}{d\psi}+\frac{d^2OP}{d\psi^2}\delta\psi+\frac{d^2OP}{d\psi d\phi}\delta\phi+\ldots+\text{etc.}$$

Similarly,

$$-\xi_,'=\frac{dPS}{d\psi}+\frac{d^2PS}{d\psi^2}\delta\psi+\frac{d^2PS}{d\psi d\phi}\delta\phi+\ldots+\text{etc.};$$

and therefore, as $\dfrac{dOP}{d\psi}=-\dfrac{dPS}{d\psi}$,

$$\xi_,'-\xi_,=-\left\{\frac{d^2(OP+PS)}{d\psi^2}\delta\psi+\frac{d^2(OP+PS)}{d\psi d\phi}\delta\phi+\ldots\right\}+\text{etc.} \quad (4),$$

and so for $\eta_,'-\eta_,$, etc. Hence (1) is the same as

$$\left.\begin{array}{l} OP_,+P_,S-OS=-\frac{1}{2}\{(\xi_,'-\xi')\delta\psi+(\eta_,'-\eta_,)\delta\phi+\ldots\} \\ \qquad +R \end{array}\right\} \quad (5),$$

where R denotes a remainder consisting of terms of the third and higher degrees. Also the transformation from $\delta\psi$, $\delta\phi$, ... to x_1, x_2, ..., gives clearly

Difference between two sides and the third of a kinetic triangle.

$$\left.\begin{aligned}\xi_{,}'-\xi_{,}&=-(A_1a_1x_1+A_2a_2x_2+\ldots+A_{i-1}a_{i-1}x_{i-1})\\ \eta_{,}'-\eta_{,}&=-(B_1\beta_1x_1+B_2\beta_2x_2+\ldots+B_{i-1}\beta_{i-1}x_{i-1})\\ &\text{etc.}\qquad\qquad\text{etc.}\end{aligned}\right\}\quad(6).$$

(*d*) Now for any infinitely small time the velocities remain sensibly constant; as also do the coefficients (ψ,ψ), (ψ,ϕ), etc., in the expression [§ 313 (2)] for T: and therefore for the action we have
$$\int 2Tdt=\sqrt{2T}\int\sqrt{2T}dt$$
$$=\sqrt{2T}\{(\psi,\psi)(\psi-\psi_0)^2+2(\psi,\phi)(\psi-\psi_0)(\phi-\phi_0)+\text{etc.}\}^{\frac{1}{2}}$$
where $(\psi_0, \phi_0, \ldots)$ are the co-ordinates of the configuration from which the action is reckoned. Hence, if P, P', P'' be any three configurations infinitely near one another, and if Q, with the proper differences of co-ordinates written after it, be used to denote square roots of quadratic functions such as that in the preceding expression, we have

$$\left.\begin{aligned}PP'&=\sqrt{2T}.Q\{(\psi-\psi'),(\phi-\phi'),\ldots\}\\ P'P''&=\sqrt{2T}.Q\{(\psi'-\psi''),(\phi'-\phi''),\ldots\}\\ P''P&=\sqrt{2T}.Q\{(\psi''-\psi'),(\phi''-\phi'),\ldots\}\end{aligned}\right\}\quad(7).$$

In the particular case of a single free particle, these expressions become simply proportional to the distances PP', $P'P''$, $P''P$; and by Euclid we have
$$P'P+PP''<P'P''$$
unless P is in the straight line $P'P''$.

The verification of this proposition by the preceding expressions (7) is merely its proof by co-ordinate geometry with an oblique rectilineal system of co-ordinates, and is necessarily somewhat complicated. If $(\psi,\phi)=(\phi,\theta)=(\theta,\psi)=0$, the co-ordinates become rectangular and the algebraic proof is easy. There is no difficulty, by following the analogies of these known processes, to prove that, for any number of co-ordinates, ψ, ϕ, etc., we have
$$P'P+PP''>P'P'',$$
unless
$$\frac{\psi-\psi'}{\psi''-\psi}=\frac{\phi-\phi'}{\phi''-\phi}=\frac{\theta-\theta'}{\theta''-\theta}=\ldots,$$
(expressing that P is on the course from P to P'') in which case
$$P'P+PP''=P'P'',$$
$P'P$, etc., being given by (7). And further, by the aid of (1), it is easy to find the proper expression for $P'P+PP''-P'P''$, when P is infinitely little off the course from P' to P'': but it is quite unnecessary for us here to enter on such purely algebraic investigations.

(*e*) It is obvious indeed, as has been already said (§ 358), that the action along any natural course is *the least possible between its terminal configurations* if only a sufficiently short course is included. Hence for all cases in which the time from O to S is less than some particular amount, the quadratic term in the expression (3) for $OP_{,}+P_{,}S-OS$ is necessarily positive, for all values of x_1, x_2, etc.; and therefore A_1, A_2,...A_{i-1} must each be positive.

Difference between two sides and the third of a kinetic triangle.

(*f*) Let now S be removed further and further from O, along the definite course $O...P...O'$, until it becomes O'. When it is O', let $P_{,}$ be taken on a natural course through O and O', deviating infinitely little from the course OPO'. Then, as $OP_{,}O'$ is a natural course,

Actions on different courses infinitely near one another between two conjugate kinetic foci, proved ultimately equal.

$$\xi_{,}'-\xi_{,}=\eta_{,}'-\eta_{,},...=0;$$

and therefore (5) becomes

$$OP_{,}+P_{,}O'-OO'=R,$$

which proves that the chief, or quadratic, term in the other expression (3) for the same, vanishes. Hence one at least of the coefficients A_1, A_2,... must vanish, and if one only, $A_{i-1}=0$ for instance, we must have

$$x_1=0,\ x_2=0,...x_{i-2}=0.$$

These equations express the condition that $P_{,}$ lies on a natural course from O to O'.

(*g*) Conversely if one or more of the coefficients A_1, A_2, etc., vanishes, if for instance $A_{i-1}=0$, S must be a kinetic focus. For if we take $P_{,}$ so that

If two sides, deviating infinitely little from the third, are together equal to it, they constitute an unbroken natural course.

$$x_1=0,\ x_2=0,...x_{i-2}=0,$$

we have, by (6),

$$\xi_{,}'-\xi_{,}=\eta_{,}'-\eta_{,}=...=0.$$

(*h*) Thus we have proved that at a kinetic focus conjugate to O the action from O is not a minimum of the first order,[1] and that the last configuration, up to which the action from O *is* a minimum of the first order, is a kinetic focus conjugate to O.

(*i*) It remains to be proved that the action from O ceases to be a minimum when the first kinetic focus conjugate to O is passed. Let, as above (§ 360), $O...P...O'...P'$ be a natural course extending beyond O', the first kinetic focus conjugate to O. Let P and P' be so near one another that there is no focus conjugate to either, between them; and let $O...P_{,}...O'$ be a natural course from O to O' deviating infinitely little from $O...P...O'$. By what

[1] A maximum or minimum "of the first order" of any function of one or more variables, is one in which the differential of the first degree vanishes, but not that of the second degree.

Natural course proved not a course of minimum action, beyond a kinetic focus.

we have just proved (e), the action OO' along $O\ldots P_{,}\ldots O'$ differs only by R, an infinitely small quantity of the third order from the action OO' along $O\ldots P\ldots O'$, and therefore

$$Ac.(O\ldots P\ldots O'\ldots P') = Ac.(O\ldots P_{,}\ldots O') + O'P' + R.$$
$$= OP_{,} + P_{,}O' + O'P' + R.$$

But, by a proper application of (e) we see that

$$P_{,}O' + O'P' = P_{,}P' + Q$$

where Q denotes an infinitely small quantity of the second order, which is essentially positive. Hence

$$Ac(O\ldots P\ldots O'\ldots P') = OP_{,} + P_{,}P' + Q + R,$$

and therefore, as R is infinitely small in comparison with Q,

$$Ac(O\ldots P\ldots O'\ldots P') > OP_{,} + P_{,}P'.$$

Hence the broken course $O\ldots P_{,}$, $P_{,}\ldots P'$ has less action than the natural course $O\ldots P\ldots O'\ldots P'$, and therefore, as the two are infinitely near one another, the latter is not a minimum.

A course which includes no focus conjugate to either extremity includes no pair of conjugate foci.

362. As it has been proved that the action from any configuration ceases to be a minimum at the first conjugate kinetic focus, we see immediately that if O' be the first kinetic focus conjugate to O, reached after passing O, no two configurations on this course from O to O' can be kinetic foci to one another. For, the action from O just ceasing to be a minimum when O' is reached, the action between any two intermediate configurations of the same course is necessarily a minimum.

How many kinetic foci in any case.

363. When there are i degrees of freedom to move there are in general, on any natural course from any particular configuration, O, at least $i-1$ kinetic foci conjugate to O. Thus, for example, on the course of a ray of light emanating from a luminous point O, and passing through the centre of a convex lens held obliquely to its path, there are two kinetic foci conjugate to O, as defined above, being the points in which the line of the central ray is cut by the so-called "focal lines"[1] of a pencil of rays diverging from O and made convergent after passing through the lens. But some or all of these kinetic foci may be on the course previous to O; as for instance in the case of a common projectile when its course passes obliquely downwards through O. Or some or all may be lost; as when, in the optical illustration just referred to, the lens is only strong enough to produce convergence in one of the principal

[1] In our second volume we hope to give all necessary elementary explanations on this subject.

planes, or too weak to produce convergence in either. Thus also in the case of the undisturbed rectilineal motion of a point, or in the motion of a point uninfluenced by force, on an anticlastic surface (§ 355), there are no real kinetic foci. In the motion of a projectile (not confined to one vertical plane) there can only be one kinetic focus on each path, conjugate to one given point; though there are three degrees of freedom. Again, there may be any number more than $i-1$, of foci in one course, all conjugate to one configuration, as for instance on the course of a particle, uninfluenced by force moving round the surface of an anchor ring, along either the outer great circle, or along a sinuous geodetic such as we have considered in § 361, in which clearly there are an infinite number of foci each conjugate to any one point of the path, at equal successive distances from one another.

How many kinetic foci in any case.

Referring to the notation of § 361 (*f*), let S be gradually moved on until first one of the coefficients, A_{i-1} for instance, vanishes; then another, A_{i-2}, etc.; and so on. We have seen that each of these positions of S is a kinetic focus: and thus by the successive vanishing of the $i-1$ coefficients we have $i-1$ foci. If none of the coefficients can ever vanish, there are no kinetic foci. If one or more of them, after vanishing, comes to a minimum, and again vanishes, as S is moved on, there may be any number more than $i-1$ of foci each conjugate to the same configuration, O.

364. If $i-1$ distinct[1] courses from a configuration O, each differing infinitely little from a certain natural course $O\,..\,E\,.\,.\,O_1\,..\,O_2\,.....\,O_{i-1}\,..\,Q$, cut it in configurations $O_1, O_2, O_3, \ldots O_{i-1}$, and if, besides these, there are not on it any other kinetic foci conjugate to O, between O and Q, and no focus at all, conjugate to E, between E and Q, the action in this natural course from O to Q is the maximum for all courses $O\ldots P_{\prime}, P_{\prime}\ldots Q$; $P_{\prime}$ being a configuration infinitely nearly agreeing with some configuration between E and O_1 of the standard course $O\,..\,E\,..\,O_1\,..\,O_2\,....\,O_{i-1}\,..\,Q$, and $O\ldots P_{\prime}$, $P_{\prime}\ldots Q$ denoting the natural courses between O and $P_{\prime}$, and $P_{\prime}$ and Q, which deviate infinitely little from this standard course.

Theorem of maximum action.

[1] Two courses are not called distinct if they differ from one another only in the absolute magnitude, not in the proportions of the components, of the deviations by which they differ from the standard course.

Theorem of maximum action.

In § 361 (i), let O' be any one, O_1, of the foci, O_1, O_2,...O_{i-1}, and let $P_{,}$ be called P_1 in this case. The demonstration there given shows that $OQ > OP_1 + P_1Q$.

Hence there are $i-1$ different broken courses

$$O...P_1, P_1...Q;\ O...P_2, P_2...Q;\ \text{etc.},$$

in each of which the action is less than in the standard course from O to Q. But whatever be the deviation of $P_{,}$, it may clearly be compounded of deviations P to P_1, P to P_2, P to P_3, ..., P to P_{i-1}, corresponding to these $i-1$ cases respectively; and it is easily seen from the analysis that

$$OP_{,}+P_{,}Q-OQ=(OP_1+P_1Q-OQ)+(OP_2+P_2Q-OQ)+....$$

Hence $OP_{,}+P_{,}Q < OQ$, which was to be proved.

Applications to two degrees of freedom.

365. Considering now, for simplicity, only cases in which there are but two degrees (§§ 195, 204) of freedom to move, we see that after any infinitely small conservative disturbance of a system in passing through a certain configuration, the system will first again pass through a configuration of the undisturbed course, at the first configuration of the latter at which the action in the undisturbed motion ceases to be a minimum. For instance, in the case of a particle, confined to a surface, and subject to any conservative system of force, an infinitely small conservative disturbance of its motion through any point, O, produces a disturbed path, which cuts the undisturbed path at the first point, O', at which the action in the undisturbed path from O ceases to be a minimum. Or, if projectiles, under the influence of gravity alone, be thrown from one point, O, in all directions with equal velocities, in one vertical plane, their paths, as is easily proved, intersect one another consecutively in a parabola, of which the focus is O, and the vertex the point reached by the particle projected directly upwards. The actual course of each particle from O is the course of least possible action to any point, P, reached before the enveloping parabola, but is not a course of minimum action to any point, Q, in its path after the envelope is passed.

366. Or again, if a particle slides round along the greatest circle of the smooth inner surface of a hollow anchor ring, the "action," or simply the length of path, from point to point, will be least possible for lengths (§ 351) less than $\pi\sqrt{ab}$. Thus if a string be tied round outside on the greatest circle of a perfectly smooth anchor ring, it will slip off unless held in position by staples, or checks of some kind, at distances of not less than this amount, $\pi\sqrt{ab}$, from one another in succession round the circle. With reference to this example, see also § 361, above.

Applications to two degrees of freedom.

Or, of a particle sliding down an inclined cylindrical groove, the action from any point will be the least possible along the straight path to any other point reached in a time less than that of the vibration one way of a simple pendulum of length equal to the radius of the groove, and influenced by a force equal $g \cos i$, instead of g the whole force of gravity. But the action will not be a minimum from any point, along the straight path, to any other point reached in a longer time than this. The case in which the groove is horizontal ($i=0$) and the particle is projected along it, is particularly simple and instructive, and may be worked out in detail with great ease, without assuming any of the general theorems regarding action.

Hamilton's second form.

367. In the preceding account of the Hamiltonian principle, and of developments and applications which it has received, we have adhered to the system (§§ 321, 323) in which the initial and final co-ordinates and the constant sum of potential and kinetic energies are the elements of which the action is supposed to be a function. Another system was also given by Hamilton, according to which the action is expressed in terms of the initial and final co-ordinates and the *time prescribed for the motion*; and a set of expressions quite analogous to those with which we have worked, are established. For practical applications this method is generally less convenient than the other; and the analytical relations between the two are so obvious that we need not devote any space to them here.

Liouville's kinetic theorem.

368. We conclude by calling attention to a very novel analytical investigation of the motion of a conservative system, by Liouville (*Comptes Rendus*, June 16, 1856), which leads immediately to the principle of least action, and the Hamiltonian principle with the developments by Jacobi and others; but which also establishes a very remarkable and absolutely new theorem regarding the amount of the action along any constrained course. For brevity we shall content ourselves with

Liouville's kinetic theorem.

giving it for a single free particle, referring the reader to the original article for Liouville's complete investigation in terms of generalized co-ordinates, applicable to any conservative system whatever.

Let (x, y, z) be the co-ordinates of any point through which the particle may move: V its potential energy in this position: E the sum of the potential and kinetic energies of the motion in question: A the action, from any position (x_0, y_0, z_0) to (x, y, z) along any course arbitrarily chosen (supposing, for instance, the particle to be guided along it by a frictionless guiding tube). Then (§ 318), the mass of a particle being taken as unity,

$$A=\int v\,ds=\int\sqrt{2(E-V)}\sqrt{(dx^2+dy^2+dz^2)}.$$

Now let ϑ be a function of x, y, z, which satisfies the partial differential equation

$$\frac{d\vartheta^2}{dx^2}+\frac{d\vartheta^2}{dy^2}+\frac{d\vartheta^2}{dz^2}=2(E-V).$$

Then

$$A=\int\sqrt{\left(\frac{d\vartheta^2}{dx^2}+\frac{d\vartheta^2}{dy^2}+\frac{d\vartheta^2}{dz^2}\right)(dx^2+dy^2+dz^2)}$$

$$=\int\sqrt{\left[\left(\frac{d\vartheta}{dx}dx+\frac{d\vartheta}{dy}dy+\frac{d\vartheta}{dz}dz\right)^2+\left(\frac{d\vartheta}{dz}dy-\frac{d\vartheta}{dy}dz\right)^2+\left(\frac{d\vartheta}{dx}dz-\frac{d\vartheta}{dz}dx\right)^2+\left(\frac{d\vartheta}{dy}dx-\frac{d\vartheta}{dx}dy\right)^2\right]}.$$

But

$$\frac{d\vartheta}{dx}dx+\frac{d\vartheta}{dy}dy+\frac{d\vartheta}{dz}dz=d\vartheta,$$

and, if $\dot{x}$, $\dot{y}$, $\dot{z}$ denote the actual component velocities along the arbitrary path, and $\dot{\vartheta}$ the rate at which ϑ increases per unit of time in this motion,

$$dx=\dot{x}dt,\quad dy=\dot{y}dt,\quad dz=\dot{z}dt,\quad d\vartheta=\dot{\vartheta}dt.$$

Hence the preceding becomes

$$A=\int d\vartheta\sqrt{\left\{1+\frac{\left(\dot{y}\frac{d\vartheta}{dz}-\dot{z}\frac{d\vartheta}{dy}\right)^2+\left(\dot{z}\frac{d\vartheta}{dx}-\dot{x}\frac{d\vartheta}{dz}\right)^2+\left(\dot{x}\frac{d\vartheta}{dy}-\dot{y}\frac{d\vartheta}{dx}\right)^2}{\dot{\vartheta}^2}\right\}}.$$

CHAPTER III.

EXPERIENCE.

369. By the term Experience, in physical science, we desig- *Observation and experiment.*
ate, according to a suggestion of Herschel's, our means of ecoming acquainted with the material universe and the laws vhich regulate it. In general the actions which we see ever aking place around us are *complex*, or due to the simultaneous ction of many causes. When, as in astronomy, we endeavour o ascertain these causes by simply watching their effects, we *bserve;* when, as in our laboratories, we interfere arbitrarily vith the causes or circumstances of a phenomenon, we are said o *experiment.*

370. For instance, supposing that we are possessed of instru- *Observation*
nental means of measuring time and angles, we may trace out y successive observations the relative position of the sun and arth at different instants; and (the method is not susceptible f any accuracy, but is alluded to here only for the sake f illustration) from the variations in the apparent diameter f the former we may calculate the ratios of our distances from t at those instants. We have thus a set of observations in- olving time, angular position with reference to the sun, and atios of distances from it; sufficient (if numerous enough) to nable us to discover the laws which connect the variations f these co-ordinates.

Similar methods may be imagined as applicable to the notion of any planet about the sun, of a satellite about its rimary, or of one star about another in a binary group.

371. In general all the data of Astronomy are determined n this way, and the same may be said of such subjects as Tides and Meteorology. Isothermal Lines, Lines of Equal Dip

Observation. or Intensity, Lines of No Declination, the Connexion of Solar Spots with Terrestrial Magnetism, and a host of other data and phenomena, to be explained under the proper heads in the course of the work, are thus deducible from *Observation* merely. In these cases the apparatus for the gigantic experiments is found ready arranged in Nature, and all that the philosopher has to do is to watch and measure their progress to its last details.

372. Even in the instance we have chosen above, that of the planetary motions, the observed effects are complex; because, unless possibly in the case of a double star, we have no instance of the *undisturbed* action of one heavenly body on another; but to a first approximation the motion of a planet about the sun is found to be the same as if no other bodies than these two existed; and the approximation is sufficient to indicate the probable law of mutual action, whose full confirmation is obtained when, *its* truth being assumed, the disturbing effects thus calculated are allowed for, and found to account completely for the observed deviations from the consequences of the first supposition. This may serve to give an idea of the mode of obtaining the laws of phenomena, which can only be observed in a complex form—and the method can always be directly applied when one cause is known to be pre-eminent.

Experiment. **373.** Let us take a case of the other kind—that in which the effects are so complex that we cannot deduce the causes from the observation of combinations arranged in Nature, but must endeavour to form for ourselves other combinations which may enable us to study the effects of each cause separately, or at least with only slight modification from the interference of other causes.

A stone, when dropped, falls to the ground; a brick and a boulder, if dropped from the top of a cliff at the same moment, fall side by side, and reach the ground together. But a brick and a slate do not; and while the former falls in a nearly vertical direction, the latter describes a most complex path. A sheet of paper or a fragment of gold leaf presents even greater irregularities than the slate. But by a slight modification of the circumstances, we gain a considerable insight into the nature of the question. The paper and gold leaf, if rolled into balls, fall nearly in a vertical line. Here, then, there are evidently at least two causes at work, one which tends to make all bodies fall, and that vertically; and another which depends on the form and substance of the body, and tends to retard its fall and alter its vertical direction. How can we study the effects of the former on all bodies without sensible complication from the latter? The effects of Wind, etc., at once point out *what* the latter cause is, the air (whose existence we may indeed suppose to have been discovered by such effects); and to study the nature of the action of the former it is necessary to get rid of the complications arising from the presence of air. Hence the necessity for *Experiment*. By means of an apparatus to be afterwards described, we remove the greater part of the air from the interior of a vessel, and in *that* we try again our experiments on the fall of bodies; and now a general law, simple in the extreme, though most important in its consequences, is at once apparent—viz., that *all* bodies, of whatever size, shape, or material, if dropped side by side at the same instant, fall side by side in a space void of air. Before experiment had thus separated the phenomena, hasty philosophers had rushed to the conclusion that some bodies possess the quality of *heaviness*, others that of *lightness*, etc. Had this state of things remained, the law of gravitation, vigorous though its action be throughout the universe, could never have been recognised as a general principle by the human mind. Experiment.

Mere observation of lightning and its effects could never have led to the discovery of their relation to the phenomena presented by rubbed amber. A modification of the course of nature, such as the bringing down of atmospheric electricity into our laboratories, was necessary. Without experiment we could never even have learned the existence of terrestrial magnetism

374. In all cases when a particular agent or cause is to be studied, experiments should be arranged in such a way as to lead if possible to results depending on it alone; or, if this cannot be done, they should be arranged so as to increase the effects due to the cause to be studied till these so far exceed the unavoidable concomitants, that the latter may be considered as only disturbing, not essentially modifying, the effects of the principal agent. Rules for the conduct of experiments.

Rules for the conduct of experiments.

Thus, in order to find the nature of the action of a galvanic current upon a magnetized needle, we may adopt either of these methods. For instance, we may neutralize the disturbing effects of the earth's magnetism on the needle by properly placing a magnetized bar in its neighbourhood. This is an instance of the first method.

Or we may, by increasing the strength of the current, or by coiling the wire many times about the needle (as will be explained when we describe the galvanometer), multiply the effects of the current so that those of the earth's magnetism may be negligible in comparison.

375. In some cases, however, the latter mode of procedure is utterly deceptive—as, for instance, in the use of multiplying condensers for the detection of very small electro-motive forces. In this case the friction between the parts of the condenser often produces more electricity than that which is to be measured, so that the true results cannot be deduced: a feeble positive charge, for instance, may be trebled, neutralized, or even changed to a negative one, by variations of manipulation so delicate as to be undiscoverable, and therefore unavoidable.

376. We thus see that it is uncertain which of these methods may be preferable in any particular case; and indeed, in discovery, he is the most likely to succeed who, not allowing himself to be disheartened by the non-success of one form of experiment, carefully varies his methods, and thus interrogates in every conceivable manner the subject of his investigations.

Residual phenomena.

377. A most important remark, due to Herschel, regards what are called *residual* phenomena. When, in an experiment, all known causes being allowed for, there remain certain unexplained effects (excessively slight it may be), these must be carefully investigated, and every conceivable variation of arrangement of apparatus, etc., tried; until, if possible, we manage so to exaggerate the residual phenomenon as to be able to detect its cause. It is here, perhaps, that in the present state of science we may most reasonably look for extensions of our knowledge; at all events we are warranted by the recent history of Natural Philosophy in so doing. Thus, to take only a very few instances, and to say nothing of the discovery of electricity and magnetism by the ancients, the peculiar smell observed in a room in which an electrical machine is kept in action, was long ago observed, but called the "smell of electricity," and thus left unexplained. The sagacity of Schönbein led to the discovery that this is due to the formation of Ozone, a most extraordinary body, of enormous chemical energies; whose nature is still uncertain, though the attention of chemists has for years been directed to it. Residual phenomena.

378. Slight anomalies in the motion of Uranus led Adams and Le Verrier to the discovery of a new planet; and the fact that a magnetized needle comes to rest sooner when vibrating above a copper plate than when the latter is removed, led Arago to what was once called magnetism of rotation, but has since been explained, immensely extended, and applied to most important purposes. In fact, this accidental remark about the oscillation of a needle led to facts from which, in Faraday's hands, was evolved the grand discovery of the Induction of Electrical Currents by magnets or by other currents. We need not enlarge upon this point, as in the following pages the proofs of the truth and usefulness of the principle will continually recur. Our object has been not so much to give applications as methods, and to show if possible how to attack a new combination, with the view of separating and studying in detail the various causes which generally conspire to produce observed phenomena, even those which are apparently the simplest.

379. If on repetition several times, an experiment continually gives different results, it must either have been very carelessly performed, or there must be some disturbing cause not taken account of. And, on the other hand, in cases where no very great coincidence is likely on repeated trials, an unexpected degree of agreement between the results of various trials should be regarded with the utmost suspicion, as probably due to some unnoticed peculiarity of the apparatus employed. In either of these cases, however, careful observation cannot fail to detect the cause of the discrepancies or of the unexpected agreement, and may possibly lead to discoveries in a totally unthought-of quarter. Instances of this kind may be given without limit; one or two must suffice. Unexpected agreement or discordance of results of different trials.

380. Thus, with a *very* good achromatic telescope a star

Unexpected agreement or discordance of results of different trials.

appears to have a sensible disc. But, as it is observed that the discs of all stars appear to be of equal angular diameter, we of course suspect some common error. Limiting the aperture of the object-glass *increases* the appearance in question, which, on full investigation, is found to have nothing to do with discs at all. It is, in fact, a diffraction phenomenon, and will be explained in our chapters on Light.

Again, in measuring the velocity of Sound by experiments conducted at night with cannon, the results at one station were never found to agree exactly with those at the other; sometimes, indeed, the differences were very considerable. But a little consideration led to the remark, that on those nights in which the discordance was greatest a strong wind was blowing nearly from one station to the other. Allowing for the obvious effect of this, or rather eliminating it altogether, the mean velocities on different evenings were found to agree very closely.

Hypotheses.

381. It may perhaps be advisable to say a few words here about the use of hypotheses, and especially those of very different gradations of value which are promulgated in the form of Mathematical Theories of different branches of Natural Philosophy.

382. Where, as in the case of the planetary motions and disturbances, the forces concerned are thoroughly known, the mathematical theory is absolutely true, and requires only analysis to work out its remotest details. It is thus, in general, far ahead of observation, and is competent to predict effects not yet even observed—as, for instance, Lunar Inequalities due to the action of Venus upon the Earth, etc. etc., to which no amount of observation, unaided by theory, would ever have enabled us to assign the true cause. It may also, in such subjects as Geometrical Optics, be carried to developments far beyond the reach of experiment; but in this science the assumed bases of the theory are only approximate, and it fails to explain in all their peculiarities even such comparatively simple phenomena as Halos and Rainbows—though it is perfectly successful for the practical purposes of the maker of microscopes and telescopes—and has, in these cases, carried the construction of instruments to a degree of perfection which merely tentative processes never could have reached.

383. Another class of mathematical theories, based to a certain extent on experiment, is at present useful, and has even in certain cases pointed to new and important results, which experiment has subsequently verified. Such are the Dynamical Theory of Heat, the Undulatory Theory of Light, etc. etc. In the former, which is based upon the experimental fact that *heat is motion*, many formulæ are at present obscure and uninterpretable, because we do not know *what* is moving or *how* it moves. Results of the theory in which these are not involved, are of course experimentally verified. The same difficulties exist in the Theory of Light. But before this obscurity can be perfectly cleared up, we must know something of the ultimate, or *molecular*, constitution of the bodies, or groups of molecules, at present known to us only in the aggregate. Hypotheses.

384. A third class is well represented by the Mathematical Theories of Heat (Conduction), Electricity (Statical), and Magnetism (Permanent). Although we do not know *how* Heat is propagated in bodies, nor *what* Statical Electricity or Permanent Magnetism are—the laws of their forces are as certainly known as that of Gravitation, and can therefore like it be developed to their consequences, by the application of Mathematical Analysis. The works of Fourier,[1] Green,[2] and Poisson,[3] are remarkable instances of such development. Another good example is Ampère's Theory of Electro-dynamics. And this leads us to a fourth class, which, however ingenious, must be regarded as in reality pernicious rather than useful.

385. A good type of such a theory is that of Weber, which professes to supply a physical basis for Ampère's Theory of Electro-dynamics, just mentioned as one of the admirable and really useful third class. Ampère contents himself with experimental data as to the action of closed currents on each other, and from these he deduces mathematically the action which an element of one current ought to exert on an element of another—if such a case could be submitted to experiment. This cannot possibly lead to confusion. But Weber goes further, he assumes that an electric current consists in the

[1] *Théorie Analytique de la Chaleur.* Paris. 1822.

[2] *Essay on the Application of Mathematical Analysis to the Theories of Electricity and Magnetism.* Nottingham. 1828. Reprinted in Crelle's Journal.

[3] *Mémoires sur le Magnétisme.* Mém. de l'Acad. des Sciences. 1811.

Hypotheses. motion of particles of two kinds of electricity moving in opposite directions through the conducting wire; and that these particles exert forces on other such particles of electricity, when in relative motion, different from those they would exert if at relative rest. In the present state of science this is wholly unwarrantable, because it is impossible to conceive that the hypothesis of two electric fluids can be true, and besides, because the conclusions are inconsistent with the Conservation of Energy, which we have numberless experimental reasons for receiving as a general principle in nature. It only adds to the danger of such theories, when they happen to explain further phenomena, as those of induced currents are explained by that of Weber. Another of this class is the Corpuscular Theory of Light, which for a time did great mischief, and which could scarcely have been justifiable unless a luminous corpuscle had been actually seen and examined. As such speculations, though dangerous, are interesting, and often beautiful (as, for instance, that of Weber), we will refer to them again under the proper heads.

Different species of mathematical theories of physics.

386. Mathematical theories of physical forces are in general of one of two species. First, those in which the fundamental assumption is far more general than is necessary. Thus the equation of Laplace's Functions [Chap. I. App. B, (*a*)] contains the mathematical foundation of the theories of Gravitation, Statical Electricity, Permanent Magnetism, Permanent Flux of Heat, Motion of Incompressible Fluids, etc. etc., and has therefore to be accompanied by limiting considerations when applied to any one of these subjects.

Again, there are those which are built upon a few experiments, or simple but inexact hypotheses, only; and which require to be modified in the way of extension rather than limitation. As a notable example of such, we may give the whole subject of Abstract Dynamics, which requires extensive modifications (explained in DIVISION III.) before it can in general be applied to practical purposes.

Deduction of most probable result from a number of observations.

387. When the most probable result is required from a number of observations of the same quantity which do not exactly agree, we must appeal to the mathematical theory of probabilities to guide us to a method of combining the results of experience, so as to eliminate from them, as far as possible, the inaccuracies of observation. Of course it is to be understood that we do not here class as *inaccuracies of observation* any errors which may affect alike every one of a series of observations, such as the inexact determination of a zero point or of the essential units of time and space, the personal equation of the observer, etc. The process, whatever it may be, which is to be employed in the elimination of errors, is applicable even to these, but only when *several distinct series* of observations have been made, with a change of instrument, or of observer, or of both. **Deduction of most probable result from a number of observations.**

388. We understand as inaccuracies of observation the whole class of errors which are as likely to lie in one direction as another in successive trials, and which we may fairly presume would, on the average of an infinite number of repetitions, exactly balance each other in excess and defect. Moreover, we consider only errors of such a kind that their probability is the less the greater they are; so that such errors as an accidental reading of a wrong number of whole degrees on a divided circle (which, by the way, can in general be probably corrected by comparison with other observations) are not to be included.

389. Mathematically considered, the subject is by no means an easy one, and many high authorities have asserted that the reasoning employed by Laplace, Gauss, and others, is not well founded; although the results of their analysis have been generally accepted. As an excellent treatise on the subject has recently been published by Airy, it is not necessary for us to do more than sketch in the most cursory manner a simple and apparently satisfactory method of arriving at what is called the *Method of Least Squares.*

390. Supposing the zero-point and the graduation of an instrument (micrometer, mural circle, thermometer, electrometer, galvanometer, etc.) to be *absolutely* accurate, successive readings of the value of a quantity (linear distance, altitude of a star, temperature, potential, strength of an electric current, etc.) may, and in general do, continually differ. What is most probably the true value of the observed quantity?

The most probable value, in all such cases, if the observa-

Deduction of most probable result from a number of observations.

tions are all equally reliable, will evidently be the simple mean; or if they are not equally reliable, the mean found by attributing *weights* to the several observations in proportion to their presumed exactness. But if several such means have been taken, or several single observations, and if these several means or observations have been differently qualified for the determination of the sought quantity (some of them being likely to give a more exact value than others), we must assign *theoretically* the best method of combining them in practice.

391. Inaccuracies of observation are, in general, as likely to be in excess as in defect. They are also (as before observed) more likely to be small than great; and (practically) large errors are not to be expected at all, as such would come under the class of *avoidable mistakes*. It follows that in any one of a series of observations of the same quantity the probability of an error of magnitude x must depend upon x^2, and must be expressed by some function whose value diminishes very rapidly as x increases. The probability that the error lies between x and $x+\delta x$, where δx is very small, must also be proportional to δx.

Hence we may assume the probability of an error of any magnitude included in the range of x to $x+\delta x$ to be

$$\phi(x^2)\delta x.$$

Now the error must be included between $+\infty$ and $-\infty$. Hence, as a first condition,

$$\int_{-\infty}^{+\infty}\phi(x^2)dx=1 \qquad (1).$$

The consideration of a very simple case gives us the means of determining the form of the function ϕ involved in the preceding expression.[1]

Suppose a stone to be let fall with the object of hitting a mark on the ground. Let two perpendicular lines be drawn through the mark, and take them as axes of x and y respectively. The chance of the stone falling at a distance between x and $x+\delta x$ from the axis of y is $\qquad \phi(x^2)\delta x.$

Of its falling between y and $y+\delta y$ from the axis of x the chance is $\qquad \phi(y^2)\delta y.$

The chance of its falling on the elementary area $\delta x\delta y$, whose coordinates are x, y, is therefore (since these are independent events,

[1] Compare Boole, Trans. R.S.E., 1857.

and it is to be observed that this is the assumption on which the whole investigation depends) Deduction of most probable result from a number of observations.

$$\phi(x^2)\phi(y^2)\delta x\delta y, \text{ or } \alpha\phi(x^2)\phi(y^2),$$

if α denote the indefinitely small area about the point xy.

Had we taken any other set of rectangular axes with the same origin, we should have found for the same probability the expression

$$\alpha\phi(x'^2)\,\phi(y'^2),$$

x', y' being the new co-ordinates of α. Hence we must have

$$\phi(x^2)\phi(y^2)=\phi(x'^2)\phi(y'^2), \text{ if } x^2+y^2=x'^2+y'^2.$$

From this functional equation we have at once

$$\phi(x^2)=A\epsilon^{mx^2},$$

where A and m are constants. We see at once that m must be negative (as the chance of a large error is very small), and we may write for it $-\frac{1}{h^2}$, so that h will indicate the degree of delicacy or coarseness of the system of measurement employed.

Substituting in (1) we have

$$A\int_{-\infty}^{+\infty}\epsilon^{-\frac{x^2}{h^2}}dx=1,$$

whence $A=\frac{1}{\sqrt{\pi}h}$, and the law of error is

$$\frac{1}{\sqrt{\pi}}\epsilon^{-\frac{x^2}{h^2}}\frac{\delta x}{h}.$$ Law of error.

The law of error, as regards *distance from the mark, without reference to the direction* of error, is evidently

$$\iint\phi(x^2)\phi(y^2)dxdy,$$

taken through the space between concentric circles whose radii are r and $r+\delta r$, and is therefore

$$\frac{2}{h^2}\epsilon^{-\frac{r^2}{h^2}}r\delta r,$$

which is of the same form as the law of error to the right or left of a line, with the additional factor r for the greater space for error at greater distances from the centre. As a verification, we see at once that

$$\frac{2}{h^2}\int_0^{\infty}\epsilon^{-\frac{r^2}{h^2}}rdr=1$$

as was to be expected.

392. The *Probable Error* of an observation is a numerical quantity such that the error of the observation is as likely to exceed as to fall short of it in magnitude. Probable error.

Probable error.

If we assume the law of error just found, and call P the probable error in one trial,

$$\int_0^P \epsilon^{-\frac{x^2}{h^2}} dx = \int_P^\infty \epsilon^{-\frac{x^2}{h^2}} dx.$$

The solution of this equation by trial and error leads to the approximate result

$$P = 0{\cdot}477\, h.$$

Probable error of a sum, difference, or multiple.

393. The probable error of any given multiple of the value of an observed quantity is evidently the same multiple of the probable error of the quantity itself.

The probable error of the sum or difference of two quantities, affected by *independent* errors, is the square root of the sum of the squares of their separate probable errors.

To prove this, let us investigate the *law* of error of

$$X \pm Y = Z$$

where the laws of error of X and Y are

$$\frac{1}{\sqrt{\pi}} \epsilon^{-\frac{x^2}{a^2}} \frac{dx}{a}, \text{ and } \frac{1}{\sqrt{\pi}} \epsilon^{-\frac{y^2}{b^2}} \frac{dy}{b},$$

respectively. The chance of an error in Z, of a magnitude not exceeding the limits $[z, z + \delta z]$, is evidently

$$\frac{1}{\pi ab} \int_{-\infty}^{+\infty} \epsilon^{-\frac{x^2}{a^2}} dx \int_{z-x}^{z+\delta z-x} \epsilon^{-\frac{y^2}{b^2}} dy.$$

For, whatever value is assigned to x, the value of y is given by the limits $z - x$ and $z + \delta z - x$ [or $z + x$, $z + \delta z + x$; but the chances of $\pm x$ are the same, and both are included in the limits ($\pm \infty$) of integration with respect to x].

The value of the above integral becomes, by effecting the integration with respect to y,

$$\frac{\delta z}{\pi ab} \int_{-\infty}^{+\infty} \epsilon^{-\frac{x^2}{a^2}} \epsilon^{-\frac{(z-x)^2}{b^2}} dx,$$

and this is easily reduced to

$$\frac{1}{\sqrt{\pi}} \epsilon^{-\frac{z^2}{a^2+b^2}} \frac{\delta z}{\sqrt{a^2+b^2}}.$$

Thus the probable error is $0{\cdot}477\sqrt{a^2+b^2}$, whence the proposition. And the same theorem is evidently true for *any* number of quantities.

Practical application.

394. As above remarked, the principal use of this theory is in the deduction, from a large series of observations, of the values of the quantities sought in such a form as to be liable to the smallest probable error. As an instance—by the principles of physical astronomy, the place of a planet is calculated from assumed values of the elements of its orbit, and tabulated in the *Nautical Almanac.* The *observed* places do not exactly agree with the predicted places, for two reasons—first, the data for calculation are not exact (and in fact the main object of the observation is to correct their assumed values); second, the observation is in error to some unknown amount. Now the difference between the observed, and the calculated, places depends on the errors of assumed elements and of observation. Our methods are applied to eliminate as far as possible the second of these, and the resulting equations give the required corrections of the elements. Practical application. Method of least squares.

Thus if θ be the calculated R.A. of a planet: δa, δe, $\delta\varpi$, etc., the corrections required for the assumed elements—the true R.A. is

$$\theta + A\delta a + E\delta e + \Pi\delta\varpi + \text{etc.},$$

where A, E, Π, etc., are approximately known. Suppose the observed R.A. to be Θ, then

$$\theta + A\delta a + E\delta e + \Pi\delta\varpi + \ldots = \Theta$$

or

$$A\delta a + E\delta e + \Pi\delta\varpi + \ldots = \Theta - \theta,$$

a known quantity, subject to error of observation. Every observation made gives us an equation of the same *form* as this, and in general the number of observations greatly exceeds that of the quantities δa, δe, $\delta\varpi$, etc., to be found. But it will be sufficient to consider the simple case where only *one* quantity is to be found.

Suppose a number of observations, of the same quantity x, lead to the following equations:—

$$x = B_1, \quad x = B_2, \text{ etc.,}$$

and let the probable errors be E_1, E_2, ... Multiply the terms of each equation by numbers inversely proportional to E_1, E_2, ... This will make the probable errors of the second members of all the equations the same, e suppose. The equations have now the general form

$$ax = b,$$

and it is required to find a system of linear factors, by which these equations, being multiplied in order and added, shall lead to a final equation giving the value of x with the probable error a minimum. Let them be f_1, f_2, etc. Then the final equation is

$$(\Sigma af)x = \Sigma(bf)$$

and therefore

$$P^2(\Sigma af)^2 = e^2\Sigma(f^2)$$

by the theorems of § 393, if P denote the probable error of x.

Method of least squares.

Hence $\frac{\Sigma(f^2)}{(\Sigma af)^2}$ is a minimum, and its differential coefficients with respect to each separate factor f must vanish.

This gives a series of equations, whose general form is

$$f\Sigma(af) - a\Sigma(f^2) = 0,$$

which give evidently $f_1 = a_1$, $f_2 = a_2$, etc.

Hence the following rule, which may easily be seen to hold for any number of linear equations containing a smaller number of unknown quantities,

Make the probable error of the second member the same in each equation, by the employment of a proper factor; multiply each equation by the coefficient of x in it and add all, for one of the final equations; and so, with reference to y, z, etc., for the others. The probable errors of the values of x, y, etc., found from these final equations will be less than those of the values derived from any other *linear* method of combining the equations.

This process has been called the method of *Least Squares*, because the values of the unknown quantities found by it are such as to render the sum of the squares of the errors of the original equations a minimum.

That is, in the simple case taken above,

$$\Sigma(ax - b)^2 = \text{minimum}.$$

For it is evident that this gives, on differentiating with respect to x, $$\Sigma a(ax - b) = 0,$$ which is the law above laid down for the formation of the single equation.

Methods of representing experimental results.

395. When a series of observations of the same quantity has been made at different times, or under different circumstances, the law connecting the value of the quantity with the time, or some other variable, may be derived from the results in several ways—all more or less approximate. Two of these methods, however, are so much more extensively used than the others, that we shall devote a page or two here to a preliminary notice of them, leaving detailed instances of their application till we come to Heat, Electricity, etc. They consist in (1.) a *Curve*, giving a graphic representation of the relation between the ordinate and abscissa, and (2.) an *Empirical Formula* connecting the variables.

396. Thus if the abscissæ represent intervals of time, and the ordinates the corresponding height of the barometer, we may construct curves which show at a glance the dependence of barometric pressure upon the time of day; and so on. Such curves may be accurately drawn by photographic processes on a sheet of sensitive paper placed behind the mercurial column, and made to move past it with a uniform horizontal velocity by clockwork. A similar process is applied to the Temperature and Electricity of the atmosphere, and to the components of terrestrial magnetism. Curves.

397. When the observations are not, as in the last section, continuous, they give us only a series of points in the curve, from which, however, we may in general approximate very closely to the result of continuous observation by drawing, *liberâ manu*, a curve passing through these points. This process, however, must be employed with great caution; because, unless the observations are sufficiently close to each other, most important fluctuations in the curve may escape notice. It is applicable, with abundant accuracy, to all cases where the quantity observed changes very slowly. Thus, for instance, weekly observations of the temperature at depths of from 6 to 24 feet underground were found by Forbes sufficient for a very accurate approximation to the law of the phenomenon.

398. As an instance of the processes employed for obtaining an empirical formula, we may mention methods of *Interpolation*, to which the problem can always be reduced. Thus from sextant observations, at known intervals, of the altitude of the sun, it is a common problem of astronomy to determine at what instant the altitude is greatest, and what is that greatest altitude. The first enables us to find the true solar time at the place, and the second, by the help of the *Nautical Almanac*, gives the latitude. The differential calculus, and the calculus of finite differences, give us formulæ for any required data; and Lagrange has shown how to obtain a very useful one by elementary algebra. Interpolation and empirical formulæ.

By Taylor's Theorem, if $y=f(x)$, we have

$$y=f(x_0+\overline{x-x_0})=f(x_0)+(x-x_0)f'(x_0)+\frac{(x-x_0)^2}{1.2}f''(x_0)+\ldots$$

$$+\frac{(x-x_0)^n}{1.2\ldots n}f^{(n)}[x_0+\theta(x-x_0)] \quad (1).$$

Interpolation and empirical formulæ.

where θ is a proper fraction, and x_0 is *any* quantity whatever. This formula is useful only when the successive derived values of $f(x_0)$ diminish very rapidly.

In finite differences we have

$$f(x+h)=D^h f(x)=(1+\Delta)^h f(x)$$
$$=f(x)+h\Delta f(x)+\frac{h^2}{1.2}\Delta^2 f(x)+\dots \qquad (2).$$

a very useful formula when the higher differences are small.

(1) suggests the proper form for the required expression, but it is only in rare cases that $f'(x_0)$, $f''(x_0)$, etc., are derivable directly from observation. But (2) is useful, inasmuch as the successive differences, $\Delta f(x)$, $\Delta^2 f(x)$, etc., are easily calculated from the tabulated results of observation, provided these have been taken for equal successive increments of x.

If for values x_1, x_2, ... x_n a function takes the values y_1, y_2, y_3, ... y_n, Lagrange gives for it the obvious expression

$$\left[\frac{y_1}{x-x_1}\frac{1}{(x_1-x_2)(x_1-x_3)\dots(x_1-x_n)}+\frac{y_2}{x-x_2}\frac{1}{(x_2-x_1)(x_2-x_3)\dots(x_2-x_n)}+\dots\right](x-x_1)(x-x_2)\dots(x-$$

Here it is of course assumed that the function required is a rational and integral one in x of the $n-1^{\text{th}}$ degree; and, in general, a similar limitation is in practice applied to the other formulæ above; for in order to find the complete expression for $f(x)$ in either, it is necessary to determine the values of $f'(x_0)$, $f''(x_0)$, ... in the first, or of $\Delta f(x)$, $\Delta^2 f(x)$, ... in the second. If n of the coefficients be required, so as to give the n chief terms of the general value of $f(x)$, we must have n observed simultaneous values of x and $f(x)$, and the expressions become determinate and of the $n-1^{\text{th}}$ degree in $x-x_0$ and h respectively.

In practice it is usually sufficient to employ at most three terms of either of the first two series. Thus to express the length l of a rod of metal as depending on its temperature t, we may assume from (1)

$$l=l_0+A(t-t_0)+B(t-t_0)^2,$$

l_0 being the measured length at any temperature t_0.

CHAPTER IV.

MEASURES AND INSTRUMENTS.

399. Having seen in the preceding chapter that for the investigation of the laws of nature we must carefully watch experiments, either those gigantic ones which the universe furnishes, or others devised and executed by man for special objects—and having seen that in all such observations accurate measurements of Time, Space, Force, etc., are absolutely necessary, we may now appropriately describe a few of the more useful of the instruments employed for these purposes, and the various standards or units which are employed in them. Necessity of accurate measurements.

400. Before going into detail we may give a rapid *résumé* of the principal Standards and Instruments to be described in this chapter. As most, if not all, of them depend on physical principles to be detailed in the course of this work—we shall assume in anticipation the establishment of such principles, giving references to the future division or chapter in which the experimental demonstrations are more particularly explained. This course will entail a slight, but unavoidable, confusion—slight, because Clocks, Balances, Screws, etc., are familiar even to those who know nothing of Natural Philosophy; unavoidable, because it is in the very nature of our subject that no one part can grow alone, each requiring for its full development the utmost resources of all the others. But if one of our departments thus borrows from others, it is satisfactory to find that it more than repays by the power which its improvement affords them.

401. We may divide our more important and fundamental instruments into four classes—

Classes of instruments.

Those for measuring Time;
„ „ Space, linear or angular;
„ „ Force;
„ „ Mass.

Other instruments, adapted for special purposes such as the measurement of Temperature, Light, Electric Currents, etc., will come more naturally under the head of the particular physical energies to whose measurement they are applicable.

402. We shall now consider in order the more prominent instruments of each of these four classes, and some of their most important applications :—

Clock, Chronometer, Chronoscope, Applications to Observation and to self-registering Instruments.

Vernier and Screw-Micrometer, Cathetometer, Spherometer, Dividing Engine, Theodolite, Sextant or Circle.

Common Balance, Bifilar Balance, Torsion Balance, Pendulum, Dynamometer.

Among Standards we may mention—

1. *Time.*—Day, Hour, Minute, Second, sidereal and solar.
2. *Space.*—Yard and Métre: Degree, Minute, Second.
3. *Force.*—Weight of a Pound or Kilogramme, etc., in any particular locality (gravitation unit); kinetic unit.
4. *Mass.*—Pound, Kilogramme, etc.

403. Although without instruments it is impossible to procure or apply any standard, yet, as without the standards no instrument could give us *absolute* measure, we may consider the standards first—referring to the instruments as if we already knew their principles and applications.

Angular measure.

404. We need do no more than mention the standard of angular measure, the *Degree* or ninetieth part of a right angle, and its successive subdivisions into sixtieths called *Minutes*, *Seconds*, *Thirds*, etc. This system of division is extremely inconvenient, but it has been so long universally adopted by all Europe, that the far preferable form, the decimal division of the right angle, decreed by the French Republic when it successfully introduced other more sweeping changes, utterly failed. Seconds, however, are generally divided into decimal parts.

The decimal division is employed, of course, when *circular* measure is adopted, the unit of circular measure being the angle subtended at the centre of any circle by an arc equal in length to the radius. Thus two right angles have the circular measure π or 3·14159, so that π and 180° represent the same angle: and the unit angle, or the angle of which the arc is equal to radius, is 57°·29578..., or 57°.17'.44"·8.—(Compare § 41.) Angular measure.

Hence the number of degrees n in any angle θ given in circular measure, or the converse, will be found at once by the equation $\frac{\theta}{\pi}=\frac{n}{180}$, and therefore $n=\theta\times 57^\circ{\cdot}29578\ldots=\theta\times 57^\circ.17'.44''{\cdot}8\ldots$

405. The practical standard of time is the *Sidereal Day*, being the period, nearly constant, of the earth's rotation about its axis (§ 247). It has been calculated from ancient observations of eclipses that this has not altered by $\frac{1}{10.000.000}$ of its length from 720 B.C. From it is easily derived the *Mean Solar Day*, or the mean interval which elapses between successive passages of the sun across the meridian of any place. This is not so nearly as the former, an absolute or invariable unit; secular changes in the period of the earth's rotation about the sun affect it, though very slightly. It is divided into 24 hours, and the hour, like the degree, is subdivided into successive sixtieths, called minutes and seconds. The usual subdivision of seconds is decimal. Measure of time.

It is well to observe that seconds and minutes of time are distinguished from those of angular measure by notation. Thus we have for time $13^h\ 43^m\ 27^s{\cdot}58$, but for angular measure $13^\circ\ 43'\ 27''{\cdot}58$.

When long periods of time are to be measured, the mean solar year, consisting of 366·242203 sidereal days, or 365·242242 mean solar days, or the century consisting of 100 such years, may be conveniently employed as the unit.

406. The ultimate standard of accurate chronometry must (if the human race live on the earth for a few million years) be founded on the physical properties of some body of more constant character than the earth: for instance, a carefully arranged metallic spring, hermetically sealed in an exhausted glass vessel. The time of vibration of such a spring would be necessarily more constant from day to day than that of the balance-spring of the best possible chronometer, disturbed as this is by the train of mechanism with which it is connected: and it would almost certainly be more constant from age to age than the time of rotation of the earth (cooling and shrinking, as it certainly is, to an extent that must be very considerable in fifty million years). Necessity for a perennial standard. A spring suggested.

Measure of length, founded on artificial metallic standards.

407. The British standard of length is the *Imperial Yard*, defined as the distance between two marks on a certain metallic bar, preserved in the Tower of London, when the whole has a temperature of 60° Fahrenheit. It was not directly derived from any fixed quantity in nature, although some important relations with such have been measured with great accuracy. It has been carefully compared with the length of a second's pendulum vibrating at a certain station in the neighbourhood of London, so that if it should again be destroyed, as it was at the burning of the Houses of Parliament in 1834, and should all exact copies of it, of which several are preserved in various places, be also lost, it can be restored by pendulum observations. A less accurate, but still (except in the event of earthquake disturbance) a very good, means of reproducing it exists in the measured base-lines of the Ordnance Survey, and the thence calculated distances between definite stations in the British Islands, which have been ascertained in terms of it with a degree of accuracy sometimes within an inch per mile, that is to say, within about $\frac{1}{60000}$.

Earth's dimensions not constant,

408. In scientific investigations, we endeavour as much as possible to keep to one unit at a time, and the foot, which is defined to be one-third part of the yard, is, for British measurement, generally the most convenient. Unfortunately the inch, or one-twelfth of a foot, must sometimes be used, but it is subdivided decimally. The statute mile, or 1760 yards, is unfortunately often used when great lengths are considered. Thus it appears that the British measurement of length is more inconvenient in its several denominations than the European measurement of time, or angles.

409. A far more perfect metrical system than the British, is the French, in which the decimal division is exclusively employed. Here the standard is the *Mètre*, defined originally as the ten-millionth part of the length of the quadrant of the earth's meridian from the pole to the equator; but now defined practically by the accurate standard metres laid up in various national repositories in Europe. It is somewhat longer than the yard, as the following Table shows. Its great convenience is the decimal division. Thus in any expression the units represent mètres, the tens decamètres, etc.; the first decimal

nor easily measured with great accuracy.

place represents decimètres, the second centimètres, the third millimètres, and so on. Measure of length.

Inch = 25·39954 millimètres.	Millimètre = ·03937079 inch.
Foot = 3·047945 decimètres.	Decimètre = ·3280899 foot.
Mile = 1609·315 mètres.	Kilomètre = ·6213824 mile.

410. The unit of superficial measure is in Britain the square yard, in France the mètre carré. Of course we may use square inches, feet, or miles, as also square millimètres, kilomètres, etc, or the *Hectare* = 10,000 square mètres. Measure of surface.

Square inch	= 6·451367	square	centimètres.
„ foot	= 9·28997	„	decimètres.
„ yard	= 83·60971	„	decimètres.
Acre	= ·4046711 of a hectare.		
Square mile	= 258·9895 hectares.		
Hectare	= 2·471143 acres.		

411. Similar remarks apply to the cubic measure in the two countries, and we have the following Table:— Measure of volume.

Cubic inch	= 16·38618	cubic	centimètres.
„ foot	= 28·315312	„	decimètres, or *Litres*.
Gallon	= 4·54346 litres.		
„	= 277·274 cubic inches.		
Litre	= 0·035317 cubic feet.		

412. The British unit of mass is the Pound (defined by standards only); the French is the *Kilogramme*, defined originally as a litre of water at its temperature of maximum density; but now practically defined by existing standards. Measure of mass.

Grain = 64·79896 milligrammes.	Gramme = 15·43235 grains.
Pound = 453·5927 grammes.	Kilogram. = 2·20462125 lbs.

Professor W. H. Miller finds (*Phil. Trans.* 1857) that the "*kilogramme des Archives*" is equal in mass to 15432·34874 grains; and the "*kilogramme type laiton*," deposited in the Ministère de l'Intérieure in Paris, as standard for French commerce, is 15432·344 grains.

413. The measurement of force, whether in terms of the weight of a stated mass in a stated locality, or in terms of the *absolute* or *kinetic* unit, has been explained in Chap. II. (See §§ 220-226.) From the measures of force and length, we derive at once the measure of work or mechanical effect. That practically employed by engineers is founded on the gravita- Measure of force.

tion measure of force. Neglecting the difference of gravity at London and Paris, we see from the above tables that the following relations exist between the London and the Parisian reckoning of work :—

Foot-pound = 0·13825 kilogramme-mètre.
Kilogramme-mètre = 7·2331 foot-pounds.

Clock. **414.** A *Clock* is primarily an instrument which, by means of a train of wheels, records the number of vibrations executed by a pendulum; a *Chronometer* or *Watch* performs the same duty for the oscillations of a flat spiral spring—just as the train of wheel-work in a gas-meter counts the number of revolutions of the main shaft caused by the passage of the gas through the machine. As, however, it is impossible to avoid friction, resistance of air, etc., a pendulum or spring, left to itself, would not long continue its oscillations, and, while its motion continued, would perform each oscillation in less and less time as the arc of vibration diminished: a continuous supply of energy is furnished by the descent of a weight, or the uncoiling of a powerful spring. This is so applied, through the train of wheels, to the pendulum or balance-wheel by means of a mechanical contrivance called an *Escapement*, that the oscillations are maintained of nearly uniform extent, and therefore of nearly uniform duration. The construction of escapements, as well as of trains of clock-wheels, is a matter of *Mechanics*, with the details of which we are not concerned, although it may easily be made the subject of mathematical investigation. The means of avoiding errors introduced by changes of temperature, which have been carried out in *Compensation* pendulums and balances, will be more properly described in our chapters on Heat. It is to be observed that there is little inconvenience if a clock lose or gain *regularly*; that can be easily and accurately allowed for: irregular rate is fatal.

Electrically controlled clocks. **415.** By means of a recent application of electricity to be afterwards described, one good clock, carefully regulated from time to time to agree with astronomical observations, may be made (without injury to its own performance) to control any number of other less-perfectly constructed clocks, so as to compel their pendulums to vibrate, beat for beat, with its own.

Chronoscope. **416.** In astronomical observations, time is estimated to tenths of a second by a practised observer, who, while watching the phenomena, counts the beats of the clock. But for the *very* accurate measurement of short intervals, many instruments have been devised. Thus if a small orifice be opened in a large and deep vessel full of mercury, and if we know by trial the weight of metal that escapes say in five minutes, a simple proportion gives the interval which elapses during the escape of any given weight. It is easy to contrive an adjustment by which a vessel may be placed under, and withdrawn from, the issuing stream at the time of occurrence of any two successive phenomena. Chronoscope.

417. Other contrivances, called Stop-watches, Chronoscopes, etc., which can be read off at rest, started on the occurrence of any phenomenon, and stopped at the occurrence of a second, then again read off; or which allow of the making (by pressing a stud) a slight ink-mark, on a dial revolving at a given rate, at the instant of the occurrence of each phenomenon to be noted; and such like, are common enough. But, of late, these have almost entirely given place to the Electric Chronoscope, an instrument which will be fully described later, when we shall have occasion to refer to experiments in which it has been usefully employed.

418. We now come to the measurement of space, and of angles, and for these purposes the most important instruments are the *Vernier* and the *Screw*.

419. Elementary geometry, indeed, gives us the means of dividing any straight line into any assignable number of equal parts; but in practice this is by no means an accurate or reliable method. It was formerly used in the so-called Diagonal Scale, of which the construction is evident from the diagram. The reading is effected by a sliding piece whose edge is perpendicular to the length of the scale. Suppose that it is PQ whose position on the scale is required. This can evidently cut only *one* of the transverse lines. *Its* number gives the number of tenths of an inch [4 in the figure], and the horizontal line next above the point of intersection gives evidently the number Diagonal scale.

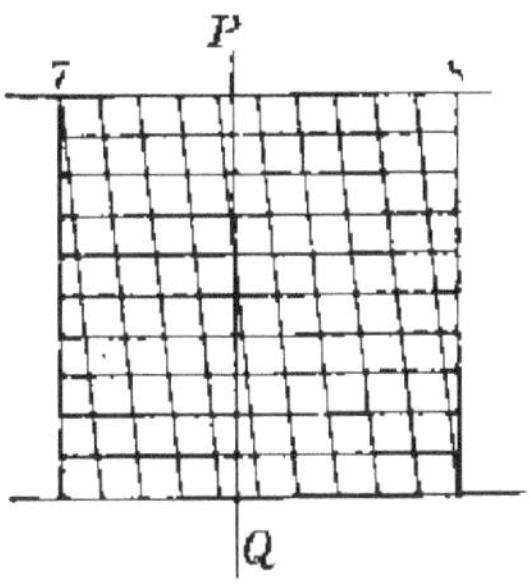

Diagonal scale.

of hundredths [in the present case 4]. Hence the reading is 7·44. As an idea of the comparative uselessness of this method, we may mention that a quadrant of 3 feet radius, which belonged to Napier of Merchiston, and is divided on the limb by this method, reads to minutes of a degree; no higher accuracy than is now attainable by the pocket sextants made by Troughton and Simms, the radius of whose arc is virtually little more than an inch. The latter instrument is read by the help of a Vernier.

Vernier.

420. The Vernier is commonly employed for such instruments as the Barometer, Sextant, and Cathetometer, while the Screw is applied to the more delicate instruments, such as Astronomical Circles, Micrometers, and the Spherometer.

421. The vernier consists of a slip of metal which slides along a divided scale, the edges of the two being coincident. Hence, when it is applied to a divided circle, its edge is circular, and it moves about an axis passing through the centre of the divided limb.

In the sketch let 0, 1, 2,... 10 be the divisions on the vernier, 0, 1, 2, etc., any set of consecutive divisions on the limb or scale

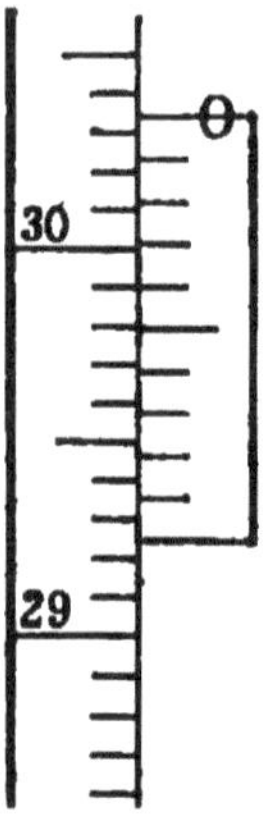

along whose edge it slides. If, when 0 and 0 coincide, 10 and 11 coincide also, then 10 divisions of the vernier are equal in length to 11 on the limb; and therefore each division on the vernier is $\frac{11}{10}$ths, or $1\frac{1}{10}$ of a division on the limb. If, then, the vernier be moved till 1 coincides with 1, 0 will be $\frac{1}{10}$th of a division of the limb beyond 0; if 2 coincide with 2, 0 will be $\frac{2}{10}$ths beyond 0; and so on. Hence to read the vernier in any position, note first the division next to 0, and behind it on the limb. This is the *integral* number of divisions to be read. For the fractional part, see which division of the vernier is in a line with one on the limb; if it be the 4th (as in the figure), that indicates an addition to the reading of $\frac{4}{10}$ths of a division of the limb; and so on. Thus, if the figure represent a barometer scale divided into inches and tenths, the reading is $30^{\text{in}}{\cdot}34$, the zero line of the vernier being adjusted to the level of the mercury.

422. If the limb of a sextant be divided, as it usually is, to third parts of a degree, and the vernier be formed by dividing 21 of these into 20 equal parts, the instrument can be read to twentieths of divisions on the limb, that is, to minutes of arc. Vernier.

If no line on the vernier coincide with one on the limb, then since the divisions of the former are the longer there will be one of the latter included between the two lines of the vernier, and it is usual in practice to take the mean of the readings which would be given by a coincidence of either pair of bounding lines.

423. In the above sketch and description, the numbers on the scale and vernier have been supposed to run *opposite* ways. This is generally the case with British instruments. In some foreign ones the divisions run in the same direction on vernier and limb, and in that case it is easy to see that to read to tenths of a scale division we must have ten divisions of the vernier equal to *nine* of the scale.

In general, to read to the nth part of a scale division, n divisions of the vernier must equal $n+1$ or $n-1$ divisions on the limb, according as these run in opposite or similar directions.

424. The principle of the *Screw* has been already noticed (§ 102). It may be used in either of two ways, *i.e.*, the nut may be fixed, and the screw advance through it, or the screw may be provented from moving longitudinally by a fixed collar, in which case the nut, if prevented by fixed guides from rotating, will move in the direction of the common axis. The advance in either case is evidently proportional to the angle through which the screw has turned about its axis, and this may be measured by means of a divided head fixed perpendicularly to the screw at one end, the divisions being read off by a pointer or vernier attached to the frame of the instrument. The nut carries with it either a tracing point (as in the dividing engine) or a wire, thread, or half the object-glass of a telescope (as in micrometers), the thread or wire, or the play of the tracing point, being at right angles to the axis of the screw. Screw.

425. Suppose it be required to divide a line into any number of equal parts. The line is placed parallel to the axis of the screw with one end exactly under the tracing point, or

Screw. under the fixed wire of a microscope carried by the nut, and the screw-head is read off. By turning the head, the tracing point or microscope wire is brought to the other extremity of the line; and the number of turns and fractions of a turn required for the whole line is thus ascertained. Dividing this by the number of equal parts required, we find at once the number of turns and fractional parts corresponding to *one* of the required divisions, and by giving that amount of rotation to the screw over and over again, drawing a line after each rotation, the required division is effected.

426. In the Micrometer, the movable wire carried by the nut is parallel to a fixed wire. By bringing them into optical contact the zero reading of the head is known; hence when another reading has been obtained, we have by subtraction the number of turns corresponding to the length of the object to be measured. The *absolute* value of a turn of the screw is determined by calculation from the number of threads in an inch, or by actually applying the micrometer to an object of known dimensions.

Spherometer. **427.** For the measurement of the thickness of a plate, or the curvature of a lens, the *Spherometer* is used. It consists of a cylindrical stem through the axis of which a good screw works. The stem is supported by three feet, equidistant from each other, and having their extremities in a plane perpendicular to the axis. The lower extremity of the screw, when worked down into this plane, is equidistant from each of the feet—and the extremities of all are delicately pointed. The number of turns, whole or fractional, of the screw, is read off by a divided head and a pointer fixed to the stem. Suppose it be required to measure the thickness of a plate of glass. The three feet of the instrument are placed upon a truly flat surface, and the screw is gradually turned until its point just touches the surface. This is determinable with the utmost accuracy, by the whole system commencing to *rock*, if slightly touched, the instant that the screw point passes below the plane of the three feet. The reason of this is, of course, that it is geometrically impossible to make a perfectly rigid body stand on four feet, without infinitely perfect fitting. At the instant at which this rocking (which is exceedingly distinct to the touch, and even to the ear) commences, the point of the screw is *just below* the plane of the feet of the instrument. The screw-head is now read off, and the screw turned backwards until room is left for the insertion, beneath its point, of the plate whose thickness is to be measured. The screw is now turned until the rocking just recommences, in which case it is evident that if the screw-point were depressed through a space equal to the thickness of the plate, it would be again *just below* the plane of the feet. From the difference of the readings of the head, we therefore easily calculate the thickness of the plate, the value of one turn of the screw having been, once for all, ascertained. Spherometer.

428. If the curvature of a lens is to be measured, the instrument is first placed, as before, on a plane surface, and the reading for the commencement of rocking is taken. The same operation is repeated on the spherical surface. The difference of the screw readings is evidently the greatest thickness of the glass which would be cut off by a plane passing through the three feet. This is sufficient, with the distance between each pair of feet, to enable us to calculate the radius of the spherical surface.

In fact if a be the distance between each pair of feet, l the length of screw corresponding to the difference of the two readings, R the radius of the spherical surface; we have at once $2R=\frac{a^2}{3l}+l$, or, as l is generally very small compared with a, the diameter is, very approximately, $\frac{a^2}{3l}$.

429. The *Cathetometer* is used for the accurate determination of differences of level—for instance, in measuring the height to which a fluid rises in a capillary tube above the exterior free surface. It consists of a divided metallic stem, which can (by means of levelling screws in its three feet) be placed very nearly vertical. Upon this slides a metallic piece, bearing a telescope whose axis is rendered horizontal by means of a level. This is, of course, perpendicular to the stem; and when the latter is made to revolve in its supports, describes a horizontal plane. The adjustments are somewhat tedious, but present no other difficulty. In using the instrument the telescope is directed first to one of the objects whose difference of level Cathetometer.

Catheto-meter. is to be found, then (with its bearing piece) it is moved by a delicate screw up or down the stem, until a horizontal wire in the focus of its eye-piece coincides with the image of the object. The vernier attached to the telescope is then read off—and, the process being repeated for the second object, a simple subtraction gives at once the required difference of level.

Balance. **430.** The principle of the *Balance* is known to everybody. We may note here a few of the precautions adopted in the best balances to guard against the various defects to which the instrument is liable; and the chief points to be attended to in its construction to secure delicacy, and rapidity of weighing.

The balance-beam should be as stiff as possible, and yet not very heavy. For this purpose it is generally formed either of tubes, or of a sort of lattice frame-work. To avoid friction, the axle consists of a knife-edge, as it is called; that is, a wedge of hard steel, which, when the balance is in use, rests on horizontal plates of polished agate. A similar contrivance is applied in very delicate balances at the points of the beam from which the scale-pans are suspended. When not in use, and just before use, the beam with its knife-edge is lifted by a lever arrangement from the agate plates. While thus secured it is loaded with weights as nearly as possible equal (this can be attained by previous trial with a coarser instrument), and the accurate determination is then readily effected. The last fraction of the required weight is determined by a rider, a very small weight, generally formed of wire, which can be worked (by a lever) from the outside of the glass case in which the balance is enclosed, and which may be placed in different positions upon one arm of the beam. This arm is graduated to tenths, etc., and thus shows at once the value of the rider in any case as depending on its moment or leverage, § 232.

431. The most important qualities of a good balance are—

1. *Sensibility.*—The beam should be sensibly deflected from a horizontal position by the smallest difference between the weights in the scale-pans. The definite measure of the sensibility is the angle through which the beam is deflected by a stated percentage of difference between the loads in the pans.

2. *Stability.*—This means rapidity of oscillation, and consequently speed in the performance of a weighing. It depends

mainly upon the depth of the centre of gravity of the whole below the knife-edge, and the length of the beam. Balance.

3. *Constancy.*—Successive weighings of the same body must give the same result—all necessary corrections (to be explained later) depending on temperature, height of barometer, etc., being allowed for.

In our Chapter on Statics we shall give the investigation of the amounts of these qualities for any given form and dimensions of the instrument.

A fine balance should turn with about a 500,000th of the greatest load which can safely be placed in either pan. In fact few measurements of any kind are correct to more than *six* significant figures.

The process of *Double Weighing*, which consists in counterpoising a mass by shot, or sand, or pieces of fine wire, and then substituting weights for it in the same pan till equilibrium is attained, is more laborious, but more accurate, than single weighing; as it eliminates all errors arising from unequal length of the arms, etc.

432. In the *Torsion-balance* invented, and used with great effect, by Coulomb, a force is measured by the torsion of a fibre of silk, a glass thread, or a metallic wire. The fibre or wire is fixed at its upper end, or at both ends, according to circumstances. In general it carries a very light horizontal rod or needle, to the extremities of which are attached the body on which is exerted the force to be measured, and a counterpoise. The upper extremity of the torsion fibre is fixed to an index passing through the centre of a divided disc, so that the angle through which that extremity moves is directly measured. If, at the same time, the angle through which the needle has turned be measured, or, more simply, if the index be always turned till the needle assumes a definite position determined by marks or sights attached to the case of the instrument—we have the amount of torsion of the fibre, and it becomes a simple statical problem to determine from the latter the force to be measured; its direction, and point of application, and the dimensions of the apparatus, being known. The force of torsion as depending on the angle of torsion was found by Coulomb to follow the law of simple proportion up to the limits of Torsion-balance.

Torsion-balance. perfect elasticity—as might have been expected from Hooke's Law (see *Properties of Matter*), and it only remains that we determine the amount for a particular angle in absolute measure. This determination is in general simple enough in theory; but in practice requires considerable care and nicety. The torsion-balance, however, being chiefly used for comparative, not absolute, measure, this determination is often unnecessary. More will be said about it when we come to its applications.

433. The ordinary spiral spring balances used for roughly comparing either small or large weights or forces, are, properly speaking, only a modified form of torsion-balance,[1] as they act almost entirely by the torsion of the wire, and not by longitudinal extension or by flexure. Spring balances we believe to be capable, if carefully constructed, of rivalling the ordinary balance in accuracy, while, for some applications, they far surpass it in sensibility and convenience. They measure directly *force*, not *mass*; and therefore if used for determining masses in different parts of the earth, a correction must be applied for the varying force of gravity. The correction for temperature must not be overlooked. These corrections may be avoided by the method of double weighing.

Pendulum. **434.** Perhaps the most delicate of all instruments for the measurement of force is the *Pendulum*. It is proved in kinetics (see Div. II.) that for any pendulum, whether oscillating about a mean vertical position under the action of gravity, or in a horizontal plane, under the action of magnetic force, or force of torsion, the square of the number of *small* oscillations in a given time is proportional to the magnitude of the force under which these oscillations take place.

For the estimation of the relative amounts of gravity at different places, this is by far the most perfect instrument. The method of coincidences by which this process has been rendered so excessively delicate will be described later.

In fact, the kinetic measure of force, as it is the true, is also far the most perfect, one—and admits of easy reduction to absolute measure.

Bifilar. **435.** Weber and Gauss, in constructing apparatus for observations of terrestrial magnetism, endeavoured so to modify

[1] J. Thomson. *Cambridge and Dublin Math. Journal* (1848).

them as to admit of their being read from some distance. For this purpose each bar, made at that time too ponderous, carried a plane mirror. By means of a scale, seen after reflection in the mirror and carefully read with a telescope, it was of course easy to compute the deviations which the mirror had experienced. But, for many reasons, it was deemed necessary that the deflections, even under considerable force, should be very small. With this view the *Bifilar* suspension was introduced. The bar-magnet is suspended horizontally by two vertical wires or fibres of equal length so adjusted as to share its weight equally between them. When the bar turns, the suspension-fibres become inclined to the vertical, and therefore the bar must rise. Hence, if we neglect the torsion of the fibres, the bifilar actually measures a force by comparing it with the weight of the suspended magnet. Bifilar.

Let a be the half length of the bar between the points of attachment of the wires, θ the angle through which the bar has been turned (in a horizontal plane) from its position of equilibrium, l the length of one of the wires, ι its inclination to the horizon.

Then $l\cos\iota$ is the difference of levels between the ends of each wire, and evidently, by the geometry of the case,

$$\tfrac{1}{2}l\sin\iota = a\sin\tfrac{1}{2}\theta.$$

Now if Q be the couple tending to turn the bar, and W its weight, the principle of mechanical effect gives

$$\begin{aligned} Qd\theta &= -Wd(l\cos\iota) \\ &= Wl\sin\iota d\iota. \end{aligned}$$

But, by the geometrical condition above,

$$l^2\sin\iota\cos\iota d\iota = a^2\sin\theta d\theta.$$

Hence

$$\frac{Q}{a^2\sin\theta} = \frac{W}{l\cos\iota},$$

or

$$Q = \frac{Wa^2}{l^2}\frac{\sin\theta}{\sqrt{1-\frac{4a^2}{l^2}\sin^2\frac{\theta}{2}}},$$

which gives the couple in terms of the deflection θ.

If the torsion of the fibres be taken into account, it will be sensibly equal to θ (since the greatest inclination to the vertical is small), and therefore the couple resulting from it will be $E\theta$, where E is some constant. This must be added to the value of Q just found in order to get the whole deflecting couple.

Dynamometer. **436.** Dynamometers are instruments for measuring energy. *White's friction brake* measures the amount of work actually performed in any time by an engine or other "prime mover," by allowing it during the time of trial to waste all its work on friction. *Morin's dynamometer* measures work without wasting any of it, in the course of its transmission from the prime mover to machines in which it is usefully employed. It consists of a simple arrangement of springs, measuring at every instant the *couple* with which the prime mover turns the shaft that transmits its work, and an integrating machine from which the work done by this couple during any time can be read off.

Let L be the couple at any instant, and ϕ the whole angle through which the shaft has turned from the moment at which the reckoning commences. The integrating machine shows at any moment the value of $\int L d\phi$, which (§ 240) is the whole work done.

Friction brakes. **437.** White's friction brake consists of a lever clamped to the shaft, but not allowed to turn with it. The moment of the force required to prevent the lever from going round with the shaft, multiplied by the whole angle through which the shaft turns, measures the whole work done against the friction of the clamp. The same result is much more easily obtained by wrapping a rope or chain several times round the shaft, or round a cylinder or drum carried round by the shaft, and applying measured forces to its two ends in proper directions to keep it nearly steady while the shaft turns round without it. The difference of the moments of these two forces round the axis, multiplied by the angle through which the shaft turns, measures the whole work spent on friction against the rope. If we remove all other resistance to the shaft, and apply the proper amount of force at each end of the dynamometric rope or chain (which is very easily done in practice), the prime mover is kept running at the proper speed for the test, and having its whole work thus wasted for the time and measured.

DIVISION II.

ABSTRACT DYNAMICS.

CHAPTER V.—INTRODUCTORY.

438. UNTIL we know thoroughly the nature of matter and the forces which produce its motions, it will be utterly impossible to submit to mathematical reasoning the *exact* conditions of any physical question. It has been long understood, however, that an approximate solution of almost any problem in the ordinary branches of Natural Philosophy may be easily obtained by a species of *abstraction*, or rather *limitation of the data*, such as enables us easily to solve the modified form of the question, while we are well assured that the circumstances (so modified) affect the result only in a superficial manner. Approximate treatment of physical questions.

439. Take, for instance, the very simple case of a crowbar employed to move a heavy mass. The accurate mathematical investigation of the action would involve the simultaneous treatment of the motions of every part of bar, fulcrum, and mass raised; and from our almost complete ignorance of the nature of matter and molecular forces, it is clear that such a treatment of the problem is impossible.

It is a result of observation that the particles of the bar, fulcrum, and mass, separately, retain throughout the process nearly the same relative positions. Hence the idea of solving, instead of the above impossible question, another, in reality quite different, but, while infinitely simpler, obviously leading to *nearly* the same results as the former.

440. The new form is given at once by the experimental result of the trial. Imagine the masses involved to be *perfectly*

Approximate treatment of physical questions.

rigid (*i.e.*, incapable of changing their form or dimensions), and the infinite series of forces, really acting, may be left out of consideration; so that the mathematical investigation deals with a finite (and generally small) number of forces instead of a practically infinite number. Our warrant for such a substitution is to be established thus.

441. The only effects of the intermolecular forces would be exhibited in alterations of the molecular form or volume of the masses involved. But as these (practically) remain almost unchanged, the forces which produce, or tend to produce, them may be left out of consideration. Thus we are enabled to investigate the action of machinery supposed to consist of separate portions whose form and dimensions are unalterable.

Further approximations.

442. If we go a little further into the question, we find that the lever *bends*, some parts of it are extended and others compressed. This would lead us into a very serious and difficult inquiry if we had to take account of the whole circumstances. But (by experience) we find that a sufficiently accurate solution of this more formidable case of the problem may be obtained by supposing (what can *never* be realized in practice) the mass to be homogeneous, and the forces consequent on a dilatation, compression, or distortion, to be proportional in magnitude, and opposed in direction, to these deformations respectively. By this further assumption, close approximations may be made to the vibrations of rods, plates, etc., as well as to the statical effect of springs, etc.

443. We may pursue the process further. Compression, in general, develops heat, and extension, cold. These *alter* sensibly the elasticity of a body. By introducing such considerations, we reach, without great difficulty, what may be called a *third* approximation to the solution of the physical problem considered.

444. We might next introduce the conduction of the heat, so produced, from point to point of the solid, with its accompanying modifications of elasticity, and so on; and we might then consider the production of thermo-electric currents, which (as we shall see) are always developed by unequal heating in a mass if it be not perfectly homogeneous. Enough, however, has been said to show, *first*, our utter ignorance as to the true and complete solution of any physical question by the only perfect method, that of the consideration of the circumstances which affect the motion of every portion, separately, of each body concerned; and, *second*, the practically sufficient manner in which practical questions may be attacked by limiting their generality, *the limitations introduced being themselves deduced from experience*, and being therefore Nature's own solution (to a less or greater degree of accuracy) of the infinite additional number of equations by which we should otherwise have been encumbered.

Further approximations.

445. To take another case: in the consideration of the propagation of waves on the surface of a fluid, it is impossible, not only on account of mathematical difficulties, but on account of our ignorance of *what* matter is, and what forces its particles exert on each other, to form the equations which would give us the separate motion of each. Our first approximation to a solution, and one sufficient for most practical purposes, is derived from the consideration of the motion of a homogeneous, incompressible, and perfectly plastic mass; a hypothetical substance which, of course, nowhere exists in nature.

446. Looking a little more closely, we find that the actual motion differs considerably from that given by the analytical solution of the restricted problem, and we introduce further considerations, such as the *compressibility* of fluids, their *internal friction*, the heat generated by the latter, and its effects in dilating the mass, etc. etc. By such successive corrections we attain, at length, to a mathematical result which (at all events in the present state of experimental science) agrees, within the limits of experimental error, with observation.

447. It would be easy to give many more instances substantiating what has just been advanced, but it seems scarcely necessary to do so. We may therefore at once say that there is no question in physical science which can be *completely and accurately* investigated by mathematical reasoning (in which, be it carefully remembered, it is *not* necessary that *symbols* should be introduced), but that there are different degrees of approximation, involving assumptions more and more nearly coincident with observation, which may be arrived at in the solution of any particular question.

Object of the present division of the work.

448. *The object of the present division of this volume is to deal with the first and second of these approximations.* In it we shall suppose all solids either RIGID, *i.e.*, unchangeable in form and volume, or ELASTIC; but in the latter case, we shall assume the law, connecting a compression or a distortion with the force which causes it, to have a particular form deduced from experiment. And we shall in the latter case neglect the thermal or electric effects which compression or distortion generally cause. We shall also suppose fluids, whether liquids or gases, to be either INCOMPRESSIBLE or compressible according to certain known laws; and we shall omit considerations of fluid friction, although we admit the consideration of friction between solids. Fluids will therefore be supposed *perfect*, *i.e.*, such that any particle may be moved amongst the others by the slightest force.

449. When we come to Properties of Matter and the Physical Forces, we shall give in detail, as far as they are yet known, the modifications which further approximations have introduced into the previous results.

Laws of friction.

450. The laws of friction between solids were very ably investigated by Coulomb; and, as we shall require them in the succeeding chapters, we give a brief summary of them here; reserving the more careful scrutiny of experimental results to our chapter on Properties of Matter.

451. To produce sliding of one solid body on another, the surfaces in contact being plane, requires a tangential force which depends—(1.) upon the nature of the bodies; (2.) upon their polish, or the species and quantity of lubricant which may have been applied; (3.) upon the normal pressure between them, to which it is in general directly proportional; (4.) upon the length of time during which they have been suffered to remain in contact.

It does not (except in extreme cases where scratching or abrasion takes place) depend sensibly upon the area of the surfaces in contact. This, which is called Statical Friction, is thus capable of opposing a tangential resistance to motion which may be of any requisite amount up to μR; where R is the whole normal pressure between the bodies; and μ (which depends mainly upon the nature of the surfaces in contact) is the *coefficient of Statical Friction*. This coefficient varies greatly with the circumstances, being in some cases as low as 0·03, in others as high as 0·80. Later, we shall give a table of its values. Where the applied forces are insufficient to produce motion, the whole amount of statical friction is not called into play; its amount then just reaches what is sufficient to equilibrate the other forces, and its direction is the opposite of that in which their resultant tends to produce motion. When the statical friction has been overcome, and sliding is produced, experiment shows that a force of friction continues to act, opposing the motion, sensibly proportional to the normal pressure, but for the same two bodies the *coefficient of Kinetic Friction* is less than that of Statical Friction, and is approximately the same whatever be the rate of motion. Laws of friction.

452. When, among the forces acting in any case of equilibrium, there are frictions of solids on solids, the circumstances would not be altered by doing away with all friction, and replacing its forces by forces of mutual action supposed to remain unchanged by any infinitely small relative motions of the parts between which they act. By this artifice all such cases may be brought under the general principle of Lagrange, § 289. Introduction of friction into the dynamical equations.

453. In the following Chapters on Abstract Dynamics we will confine ourselves strictly to such portions of this extensive subject as are likely to be useful to us in the rest of the work, or are of sufficient importance of themselves to warrant their introduction—except in special cases where results, more curious than useful, are given to show the nature of former applications of the methods, or to exhibit special methods of investigation adapted to the difficulties of peculiar problems. For a general view of the subject as a purely analytical problem the reader is referred to special mathematical treatises, such as those of Poisson, Delaunay, Duhamel, Todhunter, Tait and Steele, Griffin, etc. From these little is to be learned save dexterity in the solution of problems which are in general of no great physical interest—the objects of these treatises being professedly the mathematical analysis of the subject; while in the present work we are engaged specially with those questions which best illustrate physical principles—neither seeking, nor avoiding, difficulties of a purely mathematical kind. Rejection of merely *curious* speculations.

CHAPTER VI.

STATICS OF A PARTICLE.—ATTRACTION.

Objects of the chapter.

454. WE naturally divide Statics into two parts—the equilibrium of a particle, and that of a rigid or elastic body or system of particles whether solid or fluid. The second law of motion suffices for one part—for the other, the third, and its consequences pointed out by Newton, are necessary. In a very few sections we shall dispose of the first of these parts, and the rest of this chapter will be devoted to a digression on the important subject of Attraction.

Conditions of equilibrium of a particle.

455. By § 255, forces acting at the same point, or on the same material particle, are to be compounded by the same laws as velocities. Hence, evidently, the sum of their resolved parts in any direction must vanish if there is equilibrium. And thence the necessary and sufficient conditions.

They follow also directly from Newton's statement with regard to work, if we suppose the particle to have any velocity, constant in direction and magnitude (and § 245, this is the most general supposition we can make, since absolute rest is probably non-existent). For the work done in any time, since there is no change of kinetic energy, is the product of the displacement during that time into the algebraic sum of the effective components of the applied forces. Hence this sum must vanish for *every* direction. Practically, as any displacement may be resolved into three, in any three directions not coplanar, these three suffice for the criterion. But, in general, it is convenient to assume them in directions at right angles to each other.

Hence, for the equilibrium of a material particle, it is *necessary*, and *sufficient*, that the (algebraic) sums of the applied forces, resolved in any three rectangular directions, should vanish.

If P be one of the forces, l, m, n its direction-cosines, we have at once $\Sigma lP=0,\ \Sigma mP=0,\ \Sigma nP=0.$ Equilibrium of a particle.

If there be not equilibrium, suppose R, with direction-cosines λ, μ, ν, to be the resultant force. If reversed in direction, it will, with the other forces, produce equilibrium. Hence

$$\Sigma lP-\lambda R=0,\ \Sigma mP-\mu R=0,\ \Sigma nP-\nu R=0.$$

And $$R^2=(\Sigma lP)^2+(\Sigma mP)^2+(\Sigma nP)^2,$$

while $$\frac{\lambda}{\Sigma lP}=\frac{\mu}{\Sigma mP}=\frac{\nu}{\Sigma nP}.$$

456. We may take one or two particular cases as examples of the general results above. Thus,

(1.) If the particle rest on a smooth curve, the resolved force along the curve must vanish.

If x, y, z be the co-ordinates of the point of the curve at which the particle rests, we have evidently

$$P(l\frac{dx}{ds}+m\frac{dy}{ds}+n\frac{dz}{ds})=0.$$

When P, l, m, n are given in terms of x, y, z, this, with the *two* equations to the curve, determines the position of equilibrium.

(2.) If the curve be rough, the resultant force along it must be balanced by the friction.

If F be the friction, the condition is

$$P(l\frac{dx}{ds}+m\frac{dy}{ds}+n\frac{dz}{ds})-F=0.$$

This gives the amount of friction which will be called into play; and equilibrium will subsist until, as a limit, the friction is μ times the normal pressure on the curve. But the normal pressure is

$$P\left((m\frac{dz}{ds}-n\frac{dy}{ds})^2+(n\frac{dx}{ds}-l\frac{dz}{ds})^2+(l\frac{dy}{ds}-m\frac{dx}{ds})^2\right)^{\frac{1}{2}}$$

Hence, the limiting positions, between which equilibrium is possible, are given by the two equations to the curve, combined with

$$P(l\frac{dx}{ds}+m\frac{dy}{ds}+n\frac{dz}{ds})\pm\mu P\left((m\frac{dz}{ds}-n\frac{dy}{ds})^2+(n\frac{dx}{ds}-l\frac{dz}{ds})^2+(l\frac{dy}{ds}-m\frac{dx}{ds})^2\right)^{\frac{1}{2}}=0.$$

(3.) If the particle rest on a smooth surface, the resultant of the applied forces must evidently be perpendicular to the surface.

If $\phi(x, y, z)=0$ be the equation to the surface, we must therefore have

$$\frac{\frac{d\phi}{dx}}{lP}=\frac{\frac{d\phi}{dy}}{mP}=\frac{\frac{d\phi}{dz}}{nP},$$

and these three equations determine the position of equilibrium.

Equilibrium of a particle.

(4.) If it rest on a rough surface, friction will be called into play, resisting motion along the surface; and there will be equilibrium at any point within a certain boundary, determined by the condition that at *it* the friction is μ times the normal pressure on the surface, while within it the friction bears a less ratio to the normal pressure. When the only applied force is gravity, we have a very simple result, which is often practically useful. Let θ be the angle between the normal to the surface and the vertical at any point; the normal pressure on the surface is evidently $W\cos\theta$, where W is the weight of the particle; and the resolved part of the weight parallel to the surface, which must of course be balanced by the friction, is $W\sin\theta$. In the limiting position, when sliding is just about to commence, the greatest possible amount of statical friction is called into play, and we have

$$W\sin\theta = \mu W\cos\theta,$$

or

$$\tan\theta = \mu.$$

Angle of repose.

The value of θ thus found is called the *Angle of Repose*, and may be seen in nature in the case of sand-heaps, and slopes formed by debris from a disintegrating cliff (especially of a flat or laminated character), on which the lines of greatest slope are inclined to the horizon at an angle determined by this consideration.

Let $\phi(x, y, z) = 0$ be the surface: P, with direction-cosines l, m, n, the resultant of the applied forces. The normal pressure is

$$P.\frac{l\frac{d\phi}{dx} + m\frac{d\phi}{dy} + n\frac{d\phi}{dz}}{\sqrt{(\frac{d\phi}{dx})^2 + (\frac{d\phi}{dy})^2 + (\frac{d\phi}{dz})^2}}.$$

The resolved part of P parallel to the surface is

$$P\sqrt{\frac{(m\frac{d\phi}{dz} - n\frac{d\phi}{dy})^2 + (n\frac{d\phi}{dx} - l\frac{d\phi}{dz})^2 + (l\frac{d\phi}{dy} - m\frac{d\phi}{dx})^2}{(\frac{d\phi}{dx})^2 + (\frac{d\phi}{dy})^2 + (\frac{d\phi}{dz})^2}}.$$

Hence, for the boundary of the portion of the surface within which equilibrium is possible, we have the additional equation

$$(m\frac{d\phi}{dz} - n\frac{d\phi}{dy})^2 + (n\frac{d\phi}{dx} - l\frac{d\phi}{dz})^2 + (l\frac{d\phi}{dy} - m\frac{d\phi}{dx})^2 = \mu^2(l\frac{d\phi}{dx} + m\frac{d\phi}{dy} + n\frac{d\phi}{dz})^2.$$

457. A most important case of the composition of forces acting at one point is furnished by the consideration of the attraction of a body of any form upon a material particle anywhere situated. Experiment has shown that the attraction exerted by any portion of matter upon another is not modified by the neighbourhood, or even by the interposition, of other matter; and thus the attraction of a body on a particle is the resultant of the several attractions exerted by its parts. To treatises on applied mathematics we must refer for the examination of the consequences, often very curious, of various laws of attraction; but, dealing with Natural Philosophy, we confine ourselves to the law of gravitation, which, indeed, furnishes us with an ample supply of most interesting as well as useful results. Attraction.

458. This law, which (as a property of matter) will be carefully considered in the next Division of this Treatise, may be thus enunciated. Universal law of attraction.

Every particle of matter in the universe attracts every other particle with a force, whose direction is that of the line joining the two, and whose magnitude is directly as the product of their masses, and inversely as the square of their distance from each other.

Experiment shows (as will be seen further on) that the same law holds for electric and magnetic attractions; and it is probable that it is the fundamental law of all natural action, at least when the acting bodies are not in actual contact.

459. For the special applications of Statical principles to which we proceed, it will be convenient to use a special unit of mass, or quantity of matter, and corresponding units for the measurement of electricity and magnetism. Special unit of quantity of matter.

Thus if, in accordance with the physical law enunciated in § 458, we take as the expression for the forces exerted on each other by masses M and m, at distance D, the quantity

$$\frac{Mm}{D^2};$$

it is obvious that our *unit* force is the mutual attraction of two units of mass placed at unit of distance from each other.

460. It is convenient for many applications to speak of the

Linear, surface, and volume, densities.

density of a distribution of matter, electricity, etc., along a line, over a surface, or through a volume.

Here density of line = quantity of matter per unit of length.
" " surface = " " " area.
" " volume = " " " volume.

Electric and magnetic reckonings of *quantity*.

461. In applying the succeeding investigations to electricity or magnetism, it is only necessary to premise that M and m stand for *quantities* of free electricity or magnetism, whatever these may be, and that here the idea of *mass* as depending on *inertia* is not necessarily involved. The formula $\frac{Mm}{D^2}$ will still represent the mutual action, if we take as unit of imaginary electric or magnetic matter, such a quantity as exerts unit force on an equal quantity at unit distance. Here, however, one or both of M, m may be negative; and, as in these applications like kinds *repel* each other, the mutual action will be attraction or repulsion, according as its sign is negative or positive. With these provisos, the following theory is applicable to any of the above-mentioned classes of forces. We commence with a few simple cases which can be completely treated by means of elementary geometry.

Positive and negative masses admitted in abstract theory of attraction.

Uniform spherical shell. Attraction on *internal* point.

462. *If the different points of a spherical surface attract equally with forces varying inversely as the squares of the distances, a particle placed within the surface is not attracted in any direction.*

Let $HIKL$ be the spherical surface, and P the particle within it. Let two lines HK, IL, intercepting very small arcs HI, KL, be drawn through P; then, on account of the similar triangles HPI, KPL, those arcs will be proportional to the distances HP, LP; and any small elements of the spherical surface at HI and KL, each bounded all round by straight lines passing through P [and very nearly coinciding with HK], will be in the duplicate ratio of those lines. Hence the forces exercised by the matter of these elements on the particle P are equal; for they are as the quantities of matter directly, and the squares of the dis-

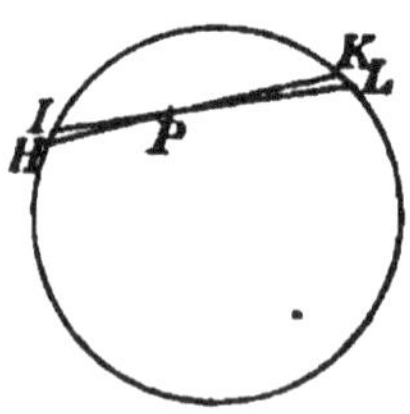

tances, inversely; and these two ratios compounded give that of equality. The attractions therefore, being equal and opposite, destroy one another: and a similar proof shows that all the attractions due to the whole spherical surface are destroyed by contrary attractions. Hence the particle P is not urged in any direction by these attractions. Uniform spherical shell. Attraction on internal point.

463. The division of a spherical surface into infinitely small elements, will frequently occur in the investigations which follow: and Newton's method, described in the preceding demonstration, in which the division is effected in such a manner that all the parts may be taken together in *pairs of opposite elements with reference to an internal point*; besides other methods deduced from it, suitable to the special problems to be examined; will be repeatedly employed. The present digression, in which some definitions and elementary geometrical propositions regarding this subject are laid down, will simplify the subsequent demonstrations, both by enabling us, through the use of convenient terms, to avoid circumlocution, and by affording us convenient means of reference for elementary principles, regarding which repeated explanations might otherwise be necessary. Digression on the division of surfaces into elements.

464. If a straight line which constantly passes through a fixed point be moved in any manner, it is said to describe, or generate, a *conical surface* of which the fixed point is the vertex. Explanations and definitions regarding cones.

If the generating line be carried from a given position continuously through any series of positions, no two of which coincide, till it is brought back to the first, the entire line on the two sides of the fixed point will generate a complete conical surface, consisting of two sheets, which are called *vertical or opposite cones*. Thus the elements HI and KL, described in Newton's demonstration given above, may be considered as being cut from the spherical surface by two *opposite cones* having P for their common vertex.

465. If any number of spheres be described from the vertex of a cone as centre, the segments cut from the concentric spherical surfaces will be similar, and their areas will be as the squares of the radii. The quotient obtained by dividing the area of one of these segments by the square of the radius of the spherical surface from which it is cut, is taken as the measure The solid angle of a cone, or of a complete conical surface.

Solid angle of cone.

of the *solid angle of the cone*. The segments of the same spherical surfaces made by the opposite cone, are respectively equal and similar to the former. Hence the solid angles of two vertical or opposite cones are equal: either may be taken as the solid angle of the complete conical surface, of which the opposite cones are the two sheets.

Sum of all the solid angles round a point $=4\pi$.

466. Since the area of a spherical surface is equal to the square of its radius multiplied by 4π, it follows that the sum of the solid angles of all the distinct cones which can be described with a given point as vertex, is equal to 4π.

Sum of the solid angles of all the complete conical surfaces $=2\pi$.

467. The solid angles of vertical or opposite cones being equal, we may infer from what precedes that the sum of the solid angles of all the complete conical surfaces which can be described without mutual intersection, with a given point as vertex, is equal to 2π.

Solid angle subtended at a point by a terminated surface.

468. The solid angle subtended at a point by a superficial area of any kind, is the solid angle of the cone generated by a straight line passing through the point, and carried entirely round the boundary of the area.

Orthogonal and oblique sections of a small cone.

469. A very small cone, that is, a cone such that any two positions of the generating line contain but a very small angle, is said to be cut at right angles, or orthogonally, by a spherical surface described from its vertex as centre, or by any surface, whether plane or curved, which touches the spherical surface at the part where the cone is cut by it.

A very small cone is said to be cut obliquely, when the section is inclined at any finite angle to an orthogonal section; and this angle of inclination is called the *obliquity of the section*.

The area of an orthogonal section of a very small cone is equal to the area of an oblique section in the same position, multiplied by the cosine of the obliquity.

Hence the area of an oblique section of a small cone is equal to the quotient obtained by dividing the product of the square of its distance from the vertex, into the solid angle, by the cosine of the obliquity.

Area of segment cut from spherical surface by small cone.

470. Let E denote the area of a very small element of a spherical surface at the point E (that is to say, an element every part of which is very near the point E), let ω denote the solid angle subtended by E at any point P, and let PE, produced if necessary, meet the surface again in E': then a denoting the radius of the spherical surface, we have Area of segment cut from spherical surface by small cone.

$$E=\frac{2a.\omega.PE^2}{EE'}.$$

For, the obliquity of the element E, considered as a section of the cone of which P is the vertex and the element E a section; being the angle between the given spherical surface and another described from P as centre, with PE as radius; is equal to the angle between the radii, EP and EC, of the two spheres. Hence, by considering the isosceles triangle ECE', we find that the cosine of the obliquity

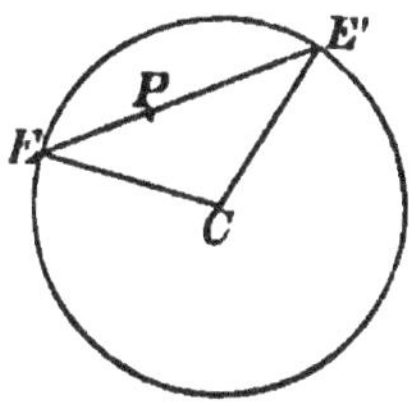

is equal to $\frac{\frac{1}{2}EE'}{EC}$ or to $\frac{EE'}{2a}$, and we arrive at the preceding expression for E.

471. *The attraction of a uniform spherical surface on an external point is the same as if the whole mass were collected at the centre.*[1] Uniform spherical shell. Attraction on external point.

Let P be the external point, C the centre of the sphere, and CAP a straight line cutting the spherical surface in A. Take I in CP, so that CP, CA, CI may be continual proportionals, and let the whole spherical surface be divided into *pairs of opposite elements with reference to the point I.*

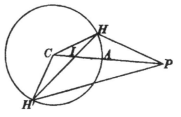

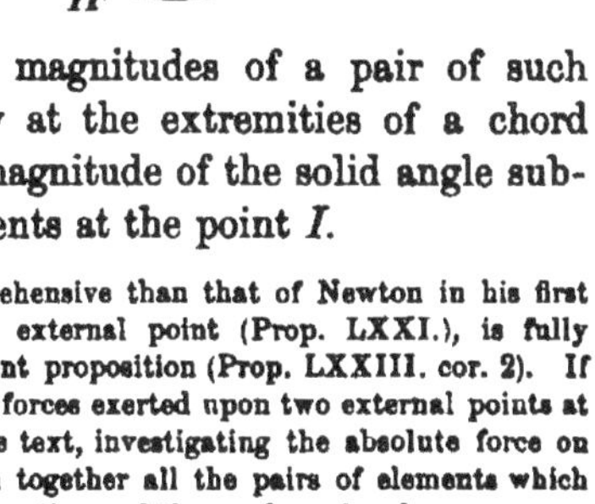

Let H and H' denote the magnitudes of a pair of such elements, situated respectively at the extremities of a chord HH'; and let ω denote the magnitude of the solid angle subtended by either of these elements at the point I.

[1] This theorem, which is more comprehensive than that of Newton in his first proposition regarding attraction on an external point (Prop. LXXI.), is fully established as a corollary to a subsequent proposition (Prop. LXXIII. cor. 2). If we had considered the proportion of the forces exerted upon two external points at different distances, instead of, as in the text, investigating the absolute force on one point, and if besides we had taken together all the pairs of elements which would constitute two narrow annular portions of the surface, in planes perpendicular to PC, the theorem and its demonstration would have coincided precisely with Prop. LXXI. of the *Principia*.

Uniform spherical shell. Attraction on *external* point.

We have (§ 469),

$$H=\frac{\omega.IH^2}{\cos CHI}, \text{ and } H'=\frac{\omega.IH'^2}{\cos CH'I}.$$

Hence, if ρ denote the density of the surface, the attractions of the two elements H and H' on P are respectively

$$\rho\frac{\omega}{\cos CHI}\cdot\frac{IH^2}{PH^2}, \text{ and } \rho\frac{\omega}{\cos CH'I}\cdot\frac{IH'^2}{PH'^2}.$$

Now the two triangles PCH, HCI have a common angle at C, and, since $PC : CH :: CH : CI$, the sides about this angle are proportional. Hence the triangles are similar; so that the angles CPH and CHI are equal, and

$$\frac{IH}{HP}=\frac{CH}{CP}=\frac{a}{CP}.$$

In the same way it may be proved, by considering the triangles PCH', $H'CI$, that the angles CPH' and $CH'I$ are equal, and that

$$\frac{IH'}{H'P}=\frac{CH'}{CP}=\frac{a}{CP}.$$

Hence the expressions for the attractions of the elements H and H' on P become

$$\rho\frac{\omega}{\cos CHI}\cdot\frac{a^2}{CP^2}, \text{ and } \rho\frac{\omega}{\cos CH'I}\cdot\frac{a^2}{CP^2},$$

which are equal, since the triangle HCH' is isosceles; and, for the same reason, the angles CPH, CPH', which have been proved to be respectively equal to the angles CHI, $CH'I$, are equal. We infer that the resultant of the forces due to the two elements is in the direction PC, and is equal to

$$2\omega.\rho.\frac{a^2}{CP^2}.$$

To find the total force on P, we must take the sum of all the forces along PC due to the pairs of opposite elements; and, since the multiplier of ω is the same for each pair, we must add all the values of ω, and we therefore obtain (§ 466), for the required resultant,

$$\frac{4\pi\rho a^2}{CP^2}.$$

The numerator of this expression; being the product of the

density, into the area of the spherical surface; is equal to the mass of the entire charge; and therefore the force on P is the same as if the whole mass were collected at C.

Cor. The force on an external point, infinitely near the surface, is equal to $4\pi\rho$, and is in the direction of a normal at the point. The force on an internal point, however near the surface, is, by a preceding proposition, equal to nothing.

472. Let σ be the area of an infinitely small element of the surface at any point P, and at any other point H of the surface let a small element subtending a solid angle ω, at P, be taken. The area of this element will be equal to

Attraction on an element of the surface.

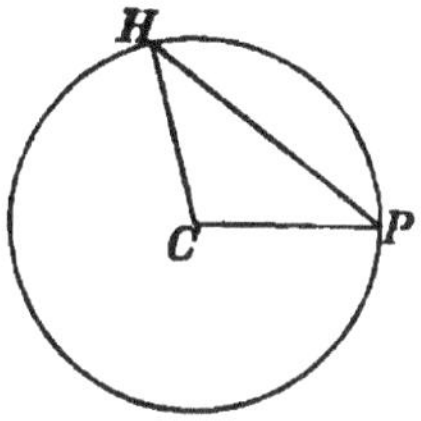

$$\frac{\omega . PH^2}{\cos CHP},$$

and therefore the attraction along HP, which it exerts on the element σ at P, will be equal to

$$\frac{\rho\omega . \rho\sigma}{\cos CHP}, \text{ or } \frac{\omega}{\cos CHP}\rho^2\sigma.$$

Now the total attraction on the element at P is in the direction CP; the component in this direction of the attraction due to the element H, is

$$\omega . \rho^2\sigma;$$

and, since all the cones corresponding to the different elements of the spherical surface lie on the same side of the tangent plane at P, we deduce, for the resultant attraction on the element σ,

$$2\pi\rho^2\sigma.$$

From the corollary to the preceding proposition, it follows that this attraction is half the force which would be exerted on an external point, possessing the same quantity of matter as the element σ, and placed infinitely near the surface.

473. In some of the most important elementary problems of the theory of electricity, spherical surfaces with densities varying inversely as the cubes of distances from eccentric points occur: and it is of fundamental importance to find the attraction of such a shell on an internal or external point. This may be done synthetically as follows; the investigation being, as we shall see below, virtually the same as that of § 462, or § 471.

Attraction of a spherical surface of which the density varies inversely as the cube of the distance from a given point.

Attraction of a spherical surface of which the density varies inversely as the cube of the distance from a given point.

474. Let us first consider the case in which the given point S and the attracted point P are separated by the spherical surface. The two figures represent the varieties of this case in which, the point S being without the sphere, P is within; and S being within, the attracted point is external. The same demonstration is applicable literally with reference to the two figures; but, for avoiding the consideration of negative quantities, some of the expressions may be conveniently modified to suit the second figure. In such instances the two expressions are given in a double line, the upper being that which is most convenient for the first figure, and the lower for the second.

Let the radius of the sphere be denoted by a, and let f be the distance of S from C, the centre of the sphere (not represented in the figures).

Join SP and take T in this line (or its continuation) so that

$$\text{(fig. 1)} \quad SP.ST = f^2 - a^2.$$
$$\text{(fig. 2)} \quad SP.TS = a^2 - f^2.$$

Through T draw any line cutting the spherical surface at K, K'. Join SK, SK', and let the lines so drawn cut the spherical surface again in E, E'.

Let the whole spherical surface be divided into pairs of opposite elements with reference to the point T. Let K and K' be a pair of such elements situated at the extremities of the chord KK', and subtending the solid angle ω at the point T; and let elements E and E' be taken subtending at S the same solid angles respectively as the elements K and K'. By this means we may divide the whole spherical surface into pairs of conjugate elements, E, E', since it is easily seen that when we have taken every pair of elements, K, K', the whole surface

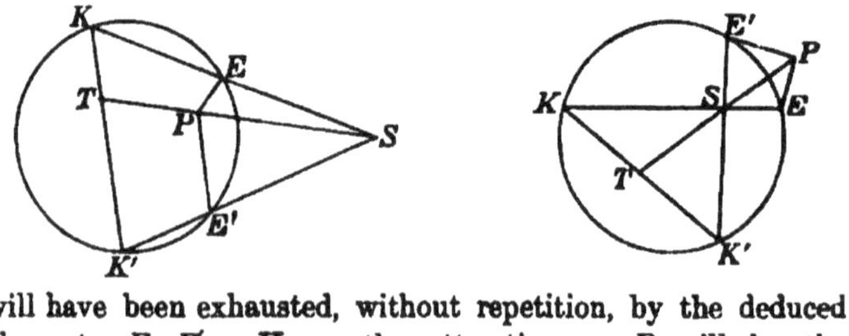

will have been exhausted, without repetition, by the deduced elements, E, E'. Hence the attraction on P will be the final resultant of the attractions of all the pairs of elements, E, E'.

Attraction of a spherical surface of which the density varies inversely as the cube of the distance from a given point.

Now if ρ be the surface density at E, and if F denote the attraction of the element E on P, we have

$$F=\frac{\rho.E}{EP^2}.$$

According to the given law of density we shall have

$$\rho=\frac{\lambda}{SE^3},$$

where λ is a constant. Again, since SEK is equally inclined to the spherical surface at the two points of intersection, we have

$$E=\frac{SE^2}{SK^2}.K=\frac{SE^2}{SK^2}\cdot\frac{2a\omega.TK^2}{KK'};$$

and hence

$$F=\frac{\dfrac{\lambda}{SE^3}\cdot\dfrac{SE^2}{SK^2}\cdot\dfrac{2a\omega.TK^2}{KK'}}{EP^2}=\lambda.\frac{2a}{KK'}\cdot\frac{TK^2}{SE.SK^2.EP^2}\cdot\omega.$$

Now, by considering the great circle in which the sphere is cut by a plane through the line SK, we find that

$$(\text{fig. } 1)\ SK.SE=f^2-a^2,$$

$$(\text{fig. } 2)\ KS.SE=a^2-f^2,$$

and hence $SK.SE=SP.ST$, from which we infer that the triangles KST, PSE are similar; so that $TK:SK::PE:SP$. Hence

$$\frac{TK^2}{SK^2.PE^2}=\frac{1}{SP^2},$$

and the expression for F becomes

$$F=\lambda.\frac{2a}{KK'}\cdot\frac{1}{SE.SP^2}\cdot\omega.$$

Modifying this by preceding expressions we have

$$(\text{fig. } 1)\ F=\lambda.\frac{2a}{KK'}\cdot\frac{\omega}{(f^2-a^2)SP^2}\cdot SK,$$

$$(\text{fig. } 2)\ F=\lambda.\frac{2a}{KK'}\cdot\frac{\omega}{(a^2-f^2)SP^2}\cdot KS.$$

Similarly, if F' denote the attraction of E' on P, we have

$$(\text{fig. } 1)\ F'=\lambda\frac{2a}{KK'}\cdot\frac{\omega}{(f^2-a^2)SP^2}\cdot SK',$$

$$(\text{fig. } 2)\ F'=\lambda\frac{2a}{KK'}\cdot\frac{\omega}{(a^2-f^2)SP^2}\cdot K'S.$$

Now in the triangles which have been shown to be similar, the angles TKS, EPS are equal; and the same may be proved of

Attraction of a spherical surface of which the density varies inversely as the cube of the distance from a given point.

the angles $K'ST$, PSE'. Hence the two sides SK, SK' of the triangle KSK' are inclined to the third at the same angles as those between the line PS and directions PE, PE' of the two forces on the point P; and the sides SK, SK' are to one another as the forces, F, F', in the directions PE, PE'. It follows, by "the triangle of forces," that the resultant of F and F' is along PS, and that it bears to the component forces the same ratios as the side KK' of the triangle bears to the other two sides. Hence the resultant force due to the two elements E and E' on the point P, is towards S, and is equal to

$$\lambda.\frac{2a}{KK'}\cdot\frac{\omega}{(f^2\sim a^2).SP^2}\cdot KK', \text{ or } \frac{\lambda.2a.\omega}{(f^2\sim a^2)SP^2}.$$

The total resultant force will consequently be towards S; and we find, by summation (§ 466) for its magnitude,

$$\frac{\lambda.4\pi a}{(f^2\sim a^2)SP^2}.$$

Hence we infer that the resultant force at any point P, separated from S by the spherical surface, is the same as if a quantity of matter equal to $\frac{\lambda.4\pi a}{f^2\sim a^2}$ were concentrated at the point S.

475. To find the attraction when S and P are either both without or both within the spherical surface.

Take in CS, or in CS produced through S, a point S_1, such that

$$CS.CS_1=a^2.$$

Then, by a well-known geometrical theorem, if E be any point on the spherical surface, we have

$$\frac{SE}{S_1E}=\frac{f}{a}.$$

Hence we have

$$\frac{\lambda}{SE^3}=\frac{\lambda a^3}{f^3.S_1E^3}.$$

Hence, ρ being the electrical density at E, we have

$$\rho=\frac{\frac{\lambda a^3}{f^3}}{S_1E^3}=\frac{\lambda_1}{S_1E^3},$$

if

$$\lambda_1=\frac{\lambda a^3}{f^3}.$$

Hence, by the investigation in the preceding section, the attraction on P is towards S_1, and is the same as if a quantity of matter equal to $\frac{\lambda_1 . 4\pi a}{f_1^2 \sim a^2}$ were concentrated at that point; Attraction of a spherical surface of which the density varies inversely as the cube of the distance from a given point.

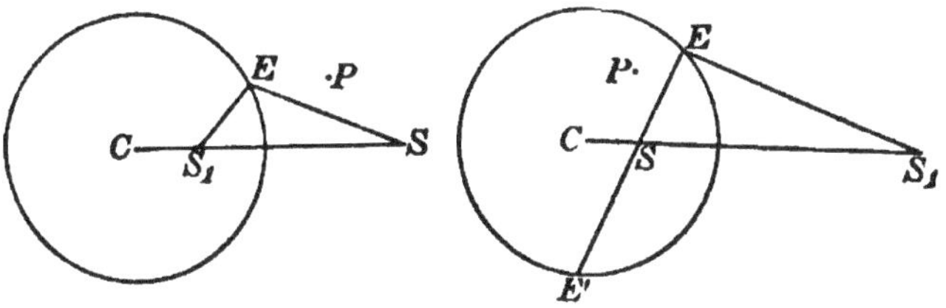

f_1 being taken to denote CS_1. If for f_1 and λ_1 we substitute their values, $\frac{a^2}{f}$ and $\frac{\lambda a^3}{f^3}$, we have the modified expression

$$\frac{\lambda \frac{a}{f} \cdot 4\pi a}{a^2 \sim f^2}$$

for the quantity of matter which we must conceive to be collected at S_1.

476. If a spherical surface be electrified in such a way that the electrical density varies inversely as the cube of the distance from an internal point S, or from the corresponding external point S_1, it will attract any external point, as if its whole electricity were concentrated at S, and any internal point, as if a quantity of electricity greater than its own in the ratio of a to f were concentrated at S_1. Uninsulated sphere under the influence of an electric point.

Let the density at E be denoted, as before, by $\frac{\lambda}{SE^3}$. Then, if we consider two opposite elements at E and E', which subtend a solid angle ω at the point S, the areas of these elements being $\frac{\omega . 2a\, SE^2}{EE'}$ and $\frac{\omega . 2a . SE'^2}{EE'}$, the quantity of electricity which they possess will be

$$\frac{\lambda . 2a . \omega}{EE'}\left(\frac{1}{SE} + \frac{1}{SE'}\right) \text{ or } \frac{\lambda . 2a . \omega}{SE . SE'}.$$

Now $SE . SE'$ is constant (Euc. III. 35) and its value is $a^2 - f^2$. Hence, by summation, we find for the total quantity of electricity on the spherical surface

$$\frac{\lambda . 4\pi a}{a^2 - f^2}.$$

Hence, if this be denoted by m, the expressions in the preceding paragraphs, for the quantities of electricity which we must

Uninsulated sphere under the influence of an electric point.

suppose to be concentrated at the point S or S_1, according as P is without or within the spherical surface, become respectively

$$m, \text{ and } \frac{a}{f}m.$$

Direct analytical calculation of attractions.

477. The *direct* analytical solution of such problems consists in the expression, by § 455, of the three components of the whole attraction as the sums of its separate parts due to the several particles of the attracting body; the transformation, by the usual methods, of these sums into definite integrals; and the evaluation of the latter. This is, in general, inferior in elegance and simplicity to the less direct mode of solution depending upon the determination of the potential energy of the attracted particle with reference to the forces exerted upon it by the attracting body, a method which we shall presently develop with peculiar care, being of incalculable value in the theories of Electricity and Magnetism as well as in that of Gravitation. But before we proceed to it, we give some instances of the direct method, beginning with the case of a spherical shell.

Uniform spherical shell.

(*a*) Let P be the attracted point, O the centre of the shell. Let any plane perpendicular to OP cut it in N, and the sphere in the small circle QR. Let $Q\hat{O}P = \theta$, $OQ = a$, $OP = D$. Then as the whole attraction is evidently along PO, we may at once resolve the parts of it in that direction. The circular band corresponding to $[\theta, \theta + d\theta]$ has for area $2\pi a^2 \sin\theta d\theta$. Hence if M be the mass of the shell, the attraction of the band on P, resolved along PO, is

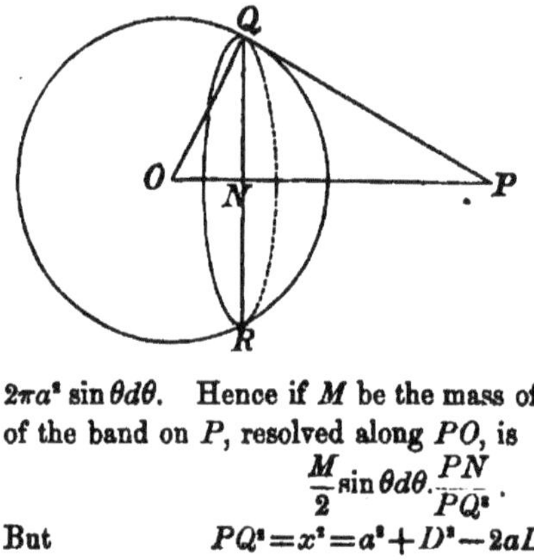

$$\frac{M}{2}\sin\theta d\theta . \frac{PN}{PQ^2}.$$

But
$$PQ^2 = x^2 = a^2 + D^2 - 2aD\cos\theta,$$
$$xdx = aD\sin\theta d\theta.$$

Also
$$PN = D - a\cos\theta = \frac{x^2 - a^2 + D^2}{2D};$$

and the attraction of the band is

$$\frac{M}{4D^2}\frac{x^2 - a^2 + D^2}{ax^2}dx.$$

This divides itself, on integration, into two cases,

Uniform spherical shell.

(1) P external, *i.e.*, $D>a$. Here the limits of x are $D-a$ and $D+a$, and the attraction is $\frac{M}{4D^2}\left[\frac{x}{a}+\frac{D^2-a^2}{ax}\right]_{D-a}^{D+a}=\frac{M}{D^2}$, as before.

(2) P internal, *i.e.*, $D<a$. Here the limits are $a-D$ and $a+D$, and the attraction is $\frac{M}{4D^2}\left[\frac{x}{a}+\frac{a^2-D^2}{ax}\right]_{a-D}^{a+D}=0$.

Uniform circular disc, on particle in its axis.

(*b*) A useful case is that of the attraction of a circular plate of uniform surface density on a point in a line through its centre, and perpendicular to its plane.

If a be the radius of the plate, h the distance of the point from it, and M its mass, the attraction (which is evidently in a direction perpendicular to the plate) is easily seen to be

$$\frac{M}{a^2}\int_0^a\frac{2hrdr}{(h^2+r^2)^{\frac{3}{2}}}=\frac{2M}{a^2}\left\{1-\frac{h}{\sqrt{h^2+a^2}}\right\}.$$

If ρ denote the surface density of the plate, this becomes

$$2\pi\rho\left(1-\frac{h}{\sqrt{h^2+a^2}}\right);$$

which, for an infinite plate, becomes

$$2\pi\rho.$$

From the preceding formula many useful results may easily be deduced: thus,

Cylinder on particle in axis.

(*c*) A uniform *cylinder* of length l, and diameter a, attracts a point in its axis at a distance x from the nearest end with a force

$$2\pi\rho\int_x^{x+l}\left(1-\frac{h}{\sqrt{h^2+a^2}}\right)dh=2\pi\rho\{l-\sqrt{(x+l)^2+a^2}+\sqrt{x^2+a^2}\}.$$

When the cylinder is of infinite length (in one direction) the attraction is therefore

$$2\pi\rho(\sqrt{x^2+a^2}-x);$$

and, when the attracted particle is in contact with the centre of the end of the infinite cylinder, this is

$$2\pi\rho a.$$

Right cone on particle at vertex.

(*d*) A right cone, of semivertical angle α, and length l, attracts a particle at its vertex. Here we have at once for the attraction, the expression

$$2\pi\rho l(1-\cos\alpha),$$

which is simply proportional to the length of the axis.

It is of course easy, when required, to find the necessarily less simple expression for the attraction on any point of the axis.

(*e*) For magnetic and electro magnetic applications a very use-

Positive and negative discs.

ful case is that of two equal discs, each perpendicular to the line joining their centres, on any point in that line—their masses (§ 461) being of opposite sign—that is, one repelling and the other attracting.

Let a be the radius, ρ the mass of a superficial unit, of either, c their distance, x the distance of the attracted point from the nearest disc. The whole action is evidently

$$2\pi\rho\left\{\frac{x+c}{\sqrt{(x+c)^2+a^2}}-\frac{x}{\sqrt{x^2+a^2}}\right\}.$$

In the particular case when c is diminished without limit, this becomes

$$2\pi\rho c\frac{a^2}{(x^2+a^2)^{\frac{3}{2}}}.$$

Variation of force in crossing an attracting surface.

478. Let P and P' be two points infinitely near one another on two sides of a surface over which matter is distributed; and let ρ be the density of this distribution on the surface in the neighbourhood of these points. Then whatever be the resultant attraction, R, at P, due to all the attracting matter, whether lodging on this surface, or elsewhere, the resultant force, R', on P' is the resultant of a force equal and parallel to R, and a force equal to $4\pi\rho$, in the direction from P' perpendicularly towards the surface. For, suppose PP' to be perpendicular to the surface, which will not limit the generality of the proposition, and consider a circular disc, of the surface, having its centre in PP', and radius infinitely small in comparison with the radii of curvature of the surface but infinitely great in comparison with PP'. This disc will [§ 477, (*b*)] attract P and P' with forces, each equal to $2\pi\rho$ and opposite to one another in the line PP'. Whence the proposition. It is one of much importance in the theory of electricity.

Uniform hemisphere on particle at edge.

(*a*) As a further example of the direct analytical process, let us find the components of the attraction exerted by a uniform *hemisphere* on a particle at its edge. Let A be the particle, AB a diameter of the base, AC the tangent to the base at A; and AD perpendicular to AC, and AB. Let RQA be a section by a plane passing through AC; AQ any radius-vector of this section;

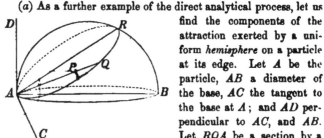

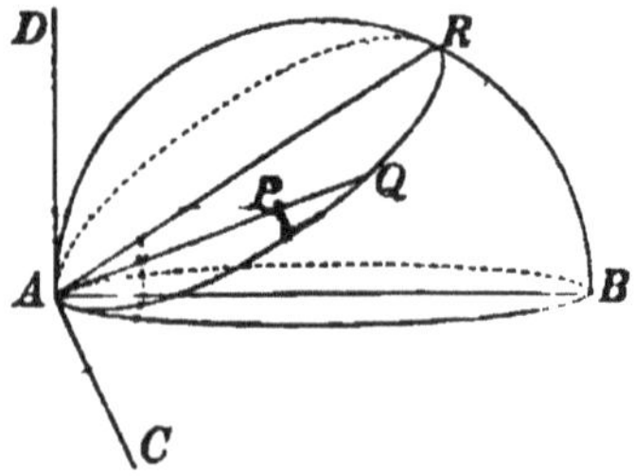

P a point in AQ. Let $AP=r$, $R\hat{A}Q=\phi$, $R\hat{A}B=\theta$. Then, evidently, the volume of an element at P is

Uniform hemisphere on particle at edge.

$$rd\phi . r\sin\theta d\theta . dr = r^2 \sin\theta d\phi d\theta dr.$$

The attraction on unit of matter at A is evidently zero along AC. Along AB it is

$$\rho \iiint \sin\theta d\phi d\theta dr \cos\phi \cos\theta,$$

between proper limits. The limits of r are 0 and $2a\cos\theta\cos\phi$, those of θ are 0 and $\frac{\pi}{2}$, and those of ϕ are $-\frac{\pi}{2}$ and $+\frac{\pi}{2}$.

Hence, Attraction along $AB = \frac{2}{3}\pi\rho a$.

Along AD the attraction is

$$\rho \int_{-\frac{\pi}{2}}^{+\frac{\pi}{2}} \int_0^{\frac{\pi}{2}} \int_0^{2a\cos\theta\cos\phi} \sin\theta d\phi d\theta dr \cos\phi \sin\theta = \frac{4}{3}\rho a.$$

(*b*) Hence at the southern base of a hemispherical hill of radius a and density ρ, the true latitude (as measured by the aid of the plumb-line, or by reflection of starlight in a trough of mercury) is diminished by the attraction of the mountain by the angle

Alteration of latitude; by hemispherical hill or cavity,

$$\frac{\frac{2}{3}\pi\rho a}{G-\frac{4}{3}\rho a}$$

where G is the attraction of the earth, estimated in the same units. Hence, if R be the radius and σ the mean density of the earth, the angle is

$$\frac{\frac{2}{3}\pi\rho a}{\frac{4}{3}\pi\sigma R - \frac{4}{3}\rho a}, \text{ or } \frac{1}{2}\frac{\rho a}{\sigma R} \text{ approximately.}$$

Hence the latitudes of stations at the base of the hill, north and south of it, differ by $\frac{a}{R}(2+\frac{\rho}{\sigma})$; instead of by $\frac{2a}{R}$, as they would do if the hill were removed.

In the same way the latitude of a place at the southern edge of a hemispherical *cavity* is increased on account of the cavity by $\frac{1}{2}\frac{\rho a}{\sigma R}$ where ρ is the density of the superficial strata.

479. As a curious additional example of the class of questions we have just considered, a deep crevasse, extending east and west, increases the latitude of places at its southern edge by (approximately) the angle $\frac{3}{4}\frac{\rho a}{\sigma R}$ where ρ is the density of

by crevasse.

Alteration of latitude; by crevasse.

the crust of the earth, and a is the width of the crevasse. Thus the north edge of the crevasse will have a *lower* latitude than the south edge if $\frac{3}{2}\frac{\rho}{\sigma} > 1$, which might be the case, as there are rocks of density $\frac{2}{3} \times 5\cdot5$ or 3·67 times that of water. At a considerable depth in the crevasse, this change of latitudes is nearly *doubled*, and then the southern side has the greater latitude if the density of the crust be not less than 1·83 times that of water.

Attraction of a sphere composed of concentric shells of uniform density.

480. It is interesting, and will be useful later, to consider as a particular case, the attraction of a sphere whose mass is composed of concentric layers, each of uniform density.

Let R be the radius, r that of any layer, $\rho = F(r)$ its density. Then, if σ be the mean density,

$$\tfrac{4}{3}\pi\sigma R^3 = 4\pi\int_0^R \rho r^2 dr,$$

from which σ may be found.

The surface attraction is $\frac{4}{3}\pi\sigma R, = G$, suppose.

At a distance r from the centre the attraction is $\frac{4\pi}{r^2}\int_0^r \rho r^2 dr$.

If it is to be the same for all points inside the sphere

$$\int_0^r \rho r^2 dr = \frac{G}{4\pi} r^2.$$

Hence $\rho = F(r) = \frac{1}{2\pi}\cdot\frac{G}{r}$ is the requisite law of density.

If the density of the upper crust be τ, the attraction at a depth h, small compared with the radius, is

$$\tfrac{4}{3}\pi\sigma_1(R-h) = G_1$$

where σ_1 is the mean density of nucleus when a shell of thickness h is removed from the sphere. Also, evidently,

$$\tfrac{4}{3}\pi\sigma_1(R-h)^3 + 4\pi\tau(R-h)^2h = \tfrac{4}{3}\pi\sigma R^3,$$

or $$G_1(R-h)^2 + 4\pi\tau(R-h)^2 h = GR^2,$$

whence $$G_1 = G\left(1+\frac{2h}{R}\right) - 4\pi\tau h.$$

The attraction is therefore unaltered at a depth h if

$$\frac{G}{R} = \tfrac{4}{3}\pi\sigma = 2\pi\tau.$$

481. Some other simple cases may be added here, as their results will be of use to us subsequently.

(*a*) The attraction of a circular arc, AB, of uniform density, on a particle at the centre, C, of the circle, lies evidently in the line CD bisecting the arc. Also the resolved part parallel to CD of the attraction of an element at P is Attraction of a uniform circular arc,

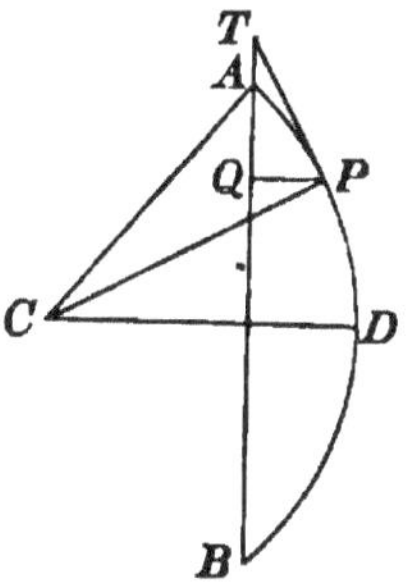

$$\frac{\text{mass of element at } P}{CD^2}\cos.P\overset{<}{C}D.$$

Now suppose the density of the chord AB to be the same as that of the arc. Then for (mass of element at $P \times \cos P\overset{<}{C}D$) we may put mass of projection of element on AB at Q; since, if PT be the tangent at P, $P\overset{<}{T}Q = P\overset{<}{C}D$.

Hence attraction along $CD = \dfrac{\text{Sum of projected elements}}{CD^2}$

$$=\frac{\rho AB}{CD^2},$$

if ρ be the density of the given arc,

$$=\frac{2\rho\sin A\overset{<}{C}D}{CD}.$$

It is therefore the same as the attraction of a mass equal to the chord, with the arc's density, concentrated at the point D.

(*b*) Again a limited straight line of uniform density attracts any external point in the same direction and with the same force as the corresponding arc of a circle of the same density, which has the point for centre, and touches the straight line. straight line.

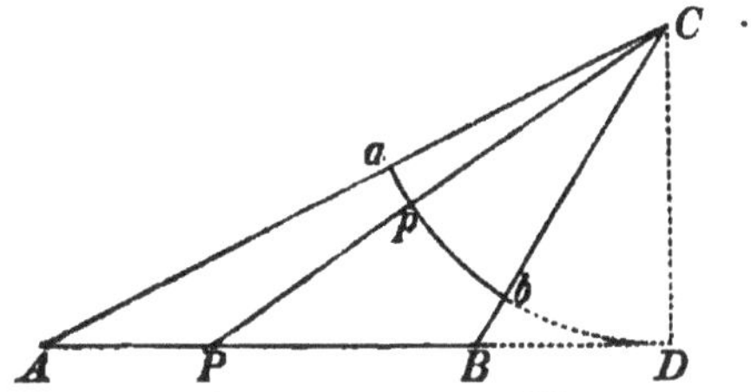

For if CpP be drawn cutting the circle in p and the line in P; Element at p : element at $P :: Cp : CP\dfrac{CP}{CD}$; that is, as $Cp^2 : CP^2$. Hence the attractions of these elements on C are equal and in the same line. Thus the arc ab attracts C as the line AB does; and, by the last proposition, the attraction of AB bisects the angle ACB, and is equal to

$$\frac{2\rho}{CD}\sin\tfrac{1}{2}A\overset{<}{C}B.$$

Attraction of a uniform straight line.

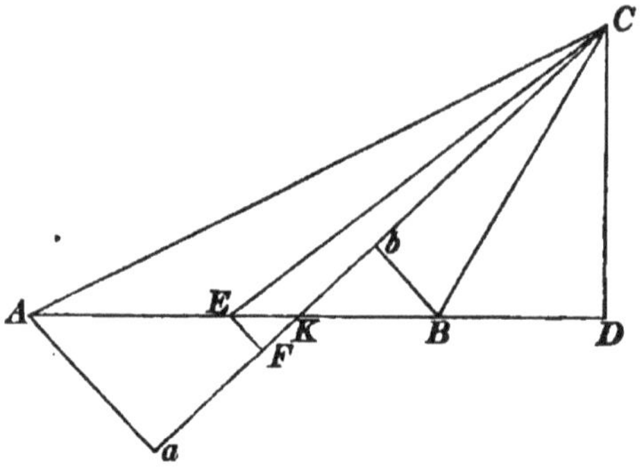

(*c*) This may be put into other useful forms—thus, let CKF bisect the angle ACB, and let Aa, Bb, EF, be drawn perpendicular to CF from the ends and middle point of AB. We have

$$\sin K\hat{C}B = \frac{KB}{CB}\sin C\hat{K}D = \frac{AB}{AC+CB}\,\frac{CD}{CK}.$$

Hence the attraction, which is along CK, is

$$\frac{2\rho AB}{(AC+CB)CK} = \frac{\rho AB}{2(AC+CB)(\overline{AC+CB}^2 - AB^2)}\cdot CF. \qquad (1)$$

For, evidently,

$$bK : Ka :: BK : KA :: BC : CA :: bC : Ca,$$

i.e., ab is divided, externally in C, and internally in K, in the same ratio. Hence, by geometry,

$$KC \cdot CF = aC \cdot Cb = \tfrac{1}{4}\{\overline{AC+CB}^2 - AB^2\},$$

which gives the transformation in (1).

(*d*) CF is obviously the tangent at C to a hyperbola, passing through that point, and having A and B as foci. Hence, if in *any* plane through AB any hyperbola be described, with foci A and B, it will be a line of force as regards the attraction of the line AB; that is, as will be more fully explained later, a curve which at every point indicates the direction of attraction.

(*e*) Similarly, if a prolate spheroid be described with foci A and B, and passing through C, CF will evidently be the normal at C: thus the force on a particle at C will be perpendicular to the spheroid; and the particle would evidently rest in equilibrium on the surface, even if it were smooth. This is an instance of (what we shall presently develop at some length) a surface of equilibrium, a level surface, or an equipotential surface.

(*f*) We may further prove, by a simple application of the preceding theorem, that the lines of force due to the attraction of two infinitely long rods in the line AB produced, one of which is attractive and the other repulsive, are the series of ellipses described from the extremities, A and B, as foci, while the surfaces of equilibrium are generated by the revolution of the confocal hyperbolas.

482. As of immense importance, in the theory not only f gravitation but of electricity, of magnetism, of fluid motion, f the conduction of heat, etc., we give here an investigation of ie most important properties of the *Potential*. Potential.

483. This function was introduced for gravitation by aplace, but the name was first given to it by Green, who may lmost be said to have created the theory, as we now have it. reen's work was neglected till 1846, and before that time most f its important theorems had been re-discovered by Gauss, 'hasles, Sturm, and Thomson.

In § 273, the *potential energy* of a conservative system in ny configuration was defined. When the forces concerned re forces acting, either really or apparently, at a distance, as ttraction of gravitation, or attractions or repulsions of electric r magnetic origin, it is in general most convenient to choose, or the zero configuration, infinite distance between the bodies oncerned. We have thus the following definition:—

484. The mutual potential energy of two bodies in any elative position is the amount of work obtainable from their nutual repulsion, by allowing them to separate to an infinite istance asunder. When the bodies attract mutually, as for nstance when no other force than gravitation is operative, heir mutual potential energy, according to the convention for ero now adopted, is negative, or (§ 547 below) their *exhaustion f potential energy* is positive.

485. The *Potential* at any point, due to any attracting or epelling body, or distribution of matter, is the mutual potential nergy between it and a unit of matter placed at that point. But in the case of gravitation, to avoid defining the potential is a negative quantity, it is convenient to change the sign. Thus the gravitation potential, at any point, due to any mass, s the quantity of work required to remove a unit of matter rom that point to an infinite distance.

486. Hence if V be the potential at any point P, and V_1 hat at a proximate point Q, it evidently follows from the above lefinition that $V-V_1$ is the work required to remove an independent unit of matter from P to Q; and it is useful to note hat this is altogether independent of the form of the path hosen between these two points, as it gives us a preliminary

Potential. idea of the power we acquire by the introduction of this mode of representation.

Suppose Q to be so near to P that the attractive forces exerted on unit of matter at these points, and therefore at any point in the line PQ, may be assumed to be equal and parallel. Then if F represent the resolved part of this force along PQ, $F.PQ$ is the work required to transfer unit of matter from P to Q. Hence

$$V - V_1 = F.PQ,$$

or

$$F = \frac{V - V_1}{PQ},$$

Force in terms of the potential. that is, the attraction on unit of matter at P in any direction PQ, is the rate at which the potential at P increases per unit of length of PQ.

Equipotential surface. **487.** A surface, at every point of which the potential has the same value, and therefore called an *Equipotential Surface*, is such that the attraction is everywhere in the direction of its normal. For in no direction along the surface does the potential change in value, and therefore there is no force in any such direction. Hence if the attracted particle be placed on such a surface (supposed smooth and rigid), it will rest in any position, and the surface is therefore sometimes called a *Surface of Equilibrium*. We shall see later, that the force on a particle of a liquid at the free surface is always in the direction of the normal, hence the term *Level Surface*, which is often used for the other terms above.

Relative intensities of force at different points of an equipotential surface. **488.** If a series of equipotential surfaces be constructed for values of the potential increasing by equal small amounts, it is evident from § 486 that the attraction at any point is inversely proportional to the normal distance between two successive surfaces close to that point; since the numerator of the expression for F is, in this case, constant.

Line of force. **489.** A line drawn from any origin, so that at every point of its length its tangent is the direction of the attraction at that point, is called a *Line of Force*; and it obviously cuts at right angles every equipotential surface which it meets.

These three last sections are true *whatever* be the law of attraction; in the next we are restricted to the law of the inverse square of the distance.

490. If, through every point of the boundary of an infinitely nall portion of an equipotential surface, the corresponding nes of force be drawn, we shall evidently have a tubular ırface of infinitely small section. The force in any direction, ; any point within such a tube, so long as it does not cut ırough attracting matter, is inversely as the section of the ıbe made by a plane passing through the point and perpendicular to the given direction. Or, more simply, the whole ›rce is at every point tangential to the direction of the tube, nd inversely as its transverse section: from which the more eneral statement above is easily seen to follow. Variation of intensity along a line of force.

This is an immediate consequence of a most important heorem, which will be proved later, § 492. *The surface integral of the attraction exerted by any distribution of matter in he direction of the normal at every point of any closed surface s $4\pi M$; where M is the amount of matter within the surface, hile the attraction is considered positive or negative according s it is inwards or outwards at any point of the surface.*

For in the present case the force perpendicular to the tubular part of the surface vanishes, and we need consider the ends only. When none of the attracting mass is within the portion of the ube considered, we have at once

$$F\varpi - F'\varpi' = 0,$$

F being the force at any point of the section whose area is ϖ. This is equivalent to the celebrated equation of Laplace—App. B (*a*); and below, § 491 (*c*).

When the attracting body is symmetrical about a point, the lines of force are obviously straight lines drawn from this point. Hence the tube is in this case a cone, and, by § 469, ϖ is proportional to the square of the distance from the vertex. Hence F is inversely as the square of the distance for points external to the attracting mass.

When the mass is symmetrically disposed about an axis in infinitely long cylindrical shells, the lines of force are evidently perpendicular to the axis. Hence the tube becomes a *wedge*, whose section is proportional to the distance from the axis, and the attraction is therefore inversely as the distance from the axis.

When the mass is arranged in infinite parallel planes, each

Variation of intensity along a line of force.

of uniform density, the lines of force are obviously perpendicular to these planes; the tube becomes a *cylinder;* and since its section is constant, the force is the same at all distances.

If an infinitely small length l of the portion of the tube considered pass through matter of density ρ, and if ω be the area of the section of the tube in this part, we have

$$F\varpi - F'\varpi' = 4\pi l\omega\rho.$$

This is equivalent to Poisson's extension of Laplace's equation [§ 491 (*c*)].

Potential due to an attracting point,

491. In estimating work done against a force which varies inversely as the square of the distance from a fixed point, the mean force is to be reckoned as the geometrical mean between the forces at the beginning and end of the path: and, whatever may be the path followed, the effective space is to be reckoned as the difference of distances from the attracting point. Thus the work done in any course is equal to the product of the difference of distances of the extremities from the attracting point, into the geometrical mean of the forces at these distances; or, if O be the attracting point, and m its force on a unit mass at unit distance, the work done in moving a particle, of unit mass, from any position P to any other position P', is

$$(OP' - OP)\sqrt{\frac{m^2}{OP^2 \cdot OP'^2}}, \text{ or } \frac{m}{OP} - \frac{m}{OP'}.$$

To prove this it is only necessary to remark, that for any infinitely small step of the motion, the effective space is clearly the difference of distances from the centre, and the working force may be taken as the force at either end, or of any intermediate value, the geometrical mean for instance: and the preceding expression applied to each infinitely small step shows that the same rule holds for the sum making up the whole work done through any finite range, and by any path.

Hence, by § 485, it is obvious that the potential at P, of a mass m situated at O, is $\frac{m}{OP}$; and thus that the potential of any

to any mass.

mass at a point P is to be found by adding the quotients of every portion of the mass, each divided by its distance from P

(*a*) For the analytical proof of these propositions, consider, first, a pair of particles, O and P, whose masses are m and unity, and co-ordinates a, b, c, x, y, z. If D be their distance

Analytical investigation of the value of the potential.

$$D^2 = \overline{x-a}^2 + \overline{y-b}^2 + \overline{z-c}^2.$$

The mutual attraction has components

$$X = -m\frac{x-a}{D^3}, \quad Y = -m\frac{y-b}{D^3}, \quad Z = -m\frac{z-c}{D^3};$$

and therefore the work required to remove P to infinity is

$$+m\int\frac{(x-a)dx+(y-b)dy+(z-c)dz}{D^3}$$

$$= +m\int\frac{dD}{D^2}$$

which, since the superior limit is $D = \infty$, is equal to

$$+\frac{m}{D}.$$

The mutual potential energy is therefore, in this case, the product of the masses divided by their mutual distance; and therefore the potential at x, y, z, due to m, is $\frac{m}{D}$.

Again, if there be more than one fixed particle m, the same investigation shows us that the potential at x, y, z is

$$\Sigma\left(\frac{m}{D}\right).$$

And if the particles form a continuous mass, whose density at a, b, c is ρ, we have of course for the potential the expression

$$\iiint\rho\frac{dadbdc}{D},$$

the limits depending on the form of the mass.

If we call V the potential at any point P (x, y, z), it is evident (from the way in which we have obtained its value) that the components of the attraction on unit of matter at P are

Force at any point.

$$X = \frac{dV}{dx}, \quad Y = \frac{dV}{dy}, \quad Z = \frac{dV}{dz}.$$

Hence the force, resolved along any curve of which s is the arc, is

$$X\frac{dx}{ds} + Y\frac{dy}{ds} + Z\frac{dz}{ds} = \frac{dV}{dx}\cdot\frac{dx}{ds} + \frac{dV}{dy}\cdot\frac{dy}{ds} + \frac{dV}{dz}\cdot\frac{dz}{ds}$$

$$= \frac{dV}{ds}.$$

All this is evidently independent of the question whether P lies within the attracting mass or not.

Force within a homogeneous sphere.

(*b*) If the attracting mass be a sphere of density ρ, and centre a, b, c, and if P be within its surface, we have, since the external shell has no effect,

$$X=\frac{dV}{dx}=-\frac{4}{3}\pi\rho D^3.\frac{x-a}{D^3}$$

$$=-\frac{4}{3}\pi\rho(x-a).$$

Rate of increase of the force in any direction.

Hence $$\frac{dX}{dx}=\frac{d^2V}{dx^2}=-\frac{4}{3}\pi\rho.$$

(*c*) Now, in general, if

$$\nabla^2=\frac{d^2}{dx^2}+\frac{d^2}{dy^2}+\frac{d^2}{dz^2},$$

we have $\nabla^2\frac{1}{D}=0$, as was proved before, App. B (*g*), (14). For

$$\frac{d}{dx}\left(\frac{1}{D}\right)=-\frac{x-a}{D^3}$$

$$\frac{d^2}{dx^2}\left(\frac{1}{D}\right)=-\frac{1}{D^3}+\frac{3(x-a)^2}{D^5},$$

and from this, and the similar expressions for the second differentials in y and z, the theorem follows at once.

Hence as $$V=\iiint\rho\frac{dadbdc}{D}$$

and ρ does not involve x, y, z, we see that *as long as D does not vanish within the limits of integration*, *i.e.*, as long as P is not a point of the attracting mass

$$\nabla^2V=0;$$

or, in terms of the components of the force,

Laplace's equation.

$$\frac{dX}{dx}+\frac{dY}{dy}+\frac{dZ}{dz}=0.$$

If P be within the attracting mass, suppose a small sphere to be described so as to contain P. Divide the potential into two parts, V_1 that of the sphere, V_2 that of the rest of the body.

The expression above shows that

$$\nabla^2V_2=0.$$

Also the expressions for $\frac{d^2V}{dx^2}$, etc., in the case of a sphere (*b*) give

$$\nabla^2V_1=-4\pi\rho,$$

where ρ is the density of the sphere.

Hence as $$V=V_1+V_2$$

Poisson's extension of Laplace's equation.

$$\nabla^2V=-4\pi\rho,$$

which is the general equation of the potential, and includes the case of P being wholly external to the attracting mass, since then $\rho=0$. In terms of the components of the force, this equation becomes Poisson's extension of Laplace's equation.

$$\frac{dX}{dx}+\frac{dY}{dy}+\frac{dZ}{dz}=-4\pi\rho.$$

(*d*) We have already, in these most important equations, the means of verifying various former results, and also of adding new ones.

Thus, to find the attraction of a hollow sphere composed of concentric shells, each of uniform density, on an external point (by which we mean a point *not* part of the mass). In this case symmetry shows that V must depend upon the distance from the centre of the sphere alone. Let the centre of the sphere be origin, and let Potential of matter arranged in concentric spherical shells of uniform density.

$$r^2=x^2+y^2+z^2.$$

Then V is a function of r alone, and consequently

$$\frac{dV}{dx}=\frac{dV}{dr}\frac{dr}{dx}=\frac{x}{r}\frac{dV}{dr},$$

$$\frac{d^2V}{dx^2}=\frac{1}{r}\frac{dV}{dr}-\frac{x^2}{r^3}\frac{dV}{dr}+\frac{x^2}{r^2}\frac{d^2V}{dr^2},$$

and

$$\nabla^2V=\frac{2}{r}\frac{dV}{dr}+\frac{d^2V}{dr^2}.$$

Hence, when P is outside the sphere, or in the hollow space within it,

$$\frac{2}{r}\frac{dV}{dr}+\frac{d^2V}{dr^2}=0.$$

A first integral of this is $r^2\frac{dV}{dr}=C.$

For a point outside the shell C has a finite value, which is easily seen to be $-M$, where M is the mass of the shell.

For a point in the internal cavity $C=0$, because evidently at the centre there is no attraction—*i.e.* there $r=0$, $\frac{dV}{dr}=0$ together. Hence there is no attraction on *any* point in the cavity.

We need not be surprised at the apparent discontinuity of this solution. It is owing to the *discontinuity of the given distribution of matter*. Thus it appears, by § 491 (*c*), that the true general equation to the potential is not what we have taken above, but

$$\frac{d^2V}{dr^2}+\frac{2}{r}\frac{dV}{dr}=-4\pi\rho,$$

where ρ, the density of the matter at distance r from the centre, is zero when $r<a$ the radius of the cavity: has a finite value σ, which for simplicity we may consider constant, when $r>a$ and $<a'$ the radius of the outer bounding surface: and is zero, again,

Potential of matter arranged in concentric spherical shells of uniform density.

for all values of r exceeding a'. Hence, integrating from $r=0$ to $r=r$, any value, we have (since $r^2\frac{dV}{dr}=0$ when $r=0$),

$$r^2\frac{dV}{dr}=-4\pi\int_0^r \rho r^2 dr=-M_1,$$

if M_1 denote the whole amount of matter within the spherical surface of radius r; which is the discontinuous function of r specified as follows:—

From $r=0$ to $r=a$, $\quad r=a$ to $r=a'$, $\quad r=a'$ to $r=\infty$,

$$M_1=0, \qquad M_1=\frac{4\pi\sigma}{3}(r^3-a^3), \quad M_1=\frac{4\pi\sigma}{3}(a'^3-a^3).$$

We have entered thus into detail in this case, because such apparent anomalies are very common in the analytical solution of physical questions. To make this still more clear, we subjoin a graphic representation of the values of V, $\frac{dV}{dr}$, and $\frac{d^2V}{dr^2}$ for this case. $ABQC$, the curve for V, is partly a straight line, and has a point of inflection at Q: but there is no discontinuity and no abrupt change of direction. $OEFD$, that for $\frac{dV}{dr}$, is continuous, but its direction twice changes abruptly. That for $\frac{d^2V}{dr^2}$ consists of three detached portions, OE, GH, KL.

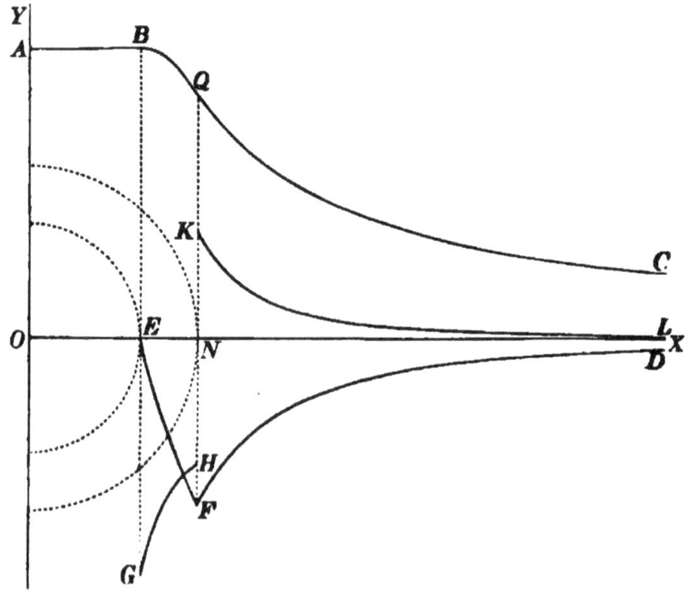

(*e*) For a mass disposed in infinitely long concentric cylindrical shells, each of uniform density, if the axis of the cylinders be z, we must evidently have V a function of x^2+y^2 only.

Coaxal right cylinders of uniform density and infinite length.

Hence $\frac{dV}{dz}=0$, or the attraction is wholly perpendicular to the axis.

Also, $\frac{d^2V}{dz^2}=0$; and therefore by (*d*)

$$\nabla^2 V=\frac{d^2V}{dr^2}+\frac{1}{r}\frac{dV}{dr}=-4\pi\rho.$$

Hence
$$r\frac{dV}{dr}=C-4\pi\int\rho r dr,$$

from which conclusions similar to the above may be drawn.

(*f*) If, finally, the mass be arranged in infinite parallel planes, each of uniform density, and perpendicular to the axis of x; the resultant force must be parallel to this direction: that is to say, $Y=0$, $Z=0$, and therefore

$$\frac{dX}{dx}=-4\pi\rho,$$

Matter arranged in infinite parallel planes of uniform density.

which, if ρ is known in terms of x, is completely integrable.

Outside the mass, $\rho=0$, and therefore

$$X=C,$$

or the attraction is the same at all distances—a result easily verified by the direct methods.

If the mass consist of an infinite plane lamina of thickness t, and constant density ρ; then, supposing the origin to be half-way between its faces, $X=C-4\pi\rho x$

so long as x is between $+\frac{t}{2}$ and $-\frac{t}{2}$. But for $x=0$ we must evidently have $X=0$, and therefore $C=0$. Hence

$$X=-4\pi\rho x.$$

Outside the lamina $X=C_1$ (since $\rho=0$). At the positive surface, and everywhere beyond it, $C_1=-2\pi\rho t$, and at and beyond the negative it is $+2\pi\rho t$. The difference of these is $-4\pi\rho t$ (§ 478).

(*g*) Since in any case $\frac{dV}{ds}$ is the attraction resolved in the direction of the tangent to the arc s, it will be wholly perpendicular to that arc if

Equipotential surface.

$$\frac{dV}{ds}=0,$$

or
$$V=C.$$

This is the equation to an *equipotential* surface.

Equi-potential surface.

If n be the normal to such a surface, measured outwards, the whole attraction at any point is evidently

$$\frac{dV}{dn},$$

and its direction is that in which V increases.

Integral of normal attraction over a closed surface.

492. Let S be any closed surface, and let O be a point, either external or internal, where a mass, m, of matter is collected. Let N be the component of the attraction of m in the direction of the normal drawn inwards from any point P, of S. Then, if $d\sigma$ denotes an element of S, and $\iint$ integration over the whole of it,

$$\iint N d\sigma = 4\pi m, \text{ or } = 0,$$

according as O is internal or external.

Equivalent to Poisson's extension of Laplace's equation.

Case 1, *O internal.* Let $OP_1P_2P_3\ldots$ be a straight line drawn in any direction from O, cutting S in P_1, P_2, P_3, etc., and therefore passing out at P_1, in at P_2, out again at P_3, in again at P_4 and so on. Let a conical surface be described by lines through O, all infinitely near $OP_1P_2\ldots$, and let ω be its solid angle (§ 465). The portions of $\iint N d\sigma$ corresponding to the elements cut from S by this case will be clearly each equal in absolute magnitude to ωm, but will be alternately positive and negative. Hence as there is an odd number of them their sum is $+\omega m$. And the sum of these, for all solid angles round O is (§ 466) equal to $4\pi m$; that is to say, $\iint N d\sigma = 4\pi m$.

Equivalent to Laplace's equation.

Case 2, *O external.* Let $OP_1P_2P_3\ldots$ be a line drawn from O passing across S, inwards at P_1, outwards at P_2, and so on. Drawing, as before, a conical surface of infinitely small solid angle, ω, we have still ωm for the absolute value of each of the portions of $\iint N d\sigma$ corresponding to the elements which it cuts from S; but their signs are alternately negative and positive: and therefore as their number is even, their sum is zero. Hence $\iint N d\sigma = 0$.

From these results it follows immediately that if there be any continuous distribution of matter, partly within and partly without a closed surface S, and N and $d\sigma$ be still used with the same signification, we have

$$\iint N d\sigma = 4\pi M$$

if M denote the whole amount of matter within S.

This is only a particular case of the analytical theorem of Chap. I. App. A (a). For if $\alpha=1$, and $U'=1$, it becomes Integral of normal attraction over a closed surface.

$$0=\iint d\sigma\partial U-\iiint\nabla^2 Udxdydz.$$

Now let U be the potential at (x, y, z), due to the distribution of matter in question. Then, according to the meaning of ∂, we have $\partial U=-N$. Also, let ρ be the density of the matter at (x, y, z). Then [§ 491 (c)] we have

$$\nabla^2 U=-4\pi\rho.$$

Hence the preceding equation gives

$$\iint Nd\sigma=4\pi\iiint\rho dxdydz=4\pi M.$$

493. From this it follows that the potential cannot have a maximum or minimum value at a point in free space. For if it were so, a closed surface could be described about the point, and indefinitely near it, so that at every point of it the value of the potential would be less than, or greater than, that at the point; so that N would be negative or positive all over the surface, and therefore $\iint Nd\sigma$ would be finite, which is impossible, as the surface contains none of the attracting mass. Potential in free space cannot have a maximum or minimum value.

494. It is also evident that N must have positive values at some parts of this surface, and negative values at others, unless it is zero all over it. Hence in free space the potential, if not constant round any point, increases in some directions from it, and diminishes in others; and therefore a material particle placed at a point of zero force under the action of any attracting bodies, and free from all constraint, is in unstable equilibrium, a result due to Earnshaw.[1] Consequences.

495. If the potential be constant over a closed surface which contains none of the attracting mass, it has the same constant value throughout the interior. For if not, it must have a maximum or minimum value somewhere within, which is impossible.

496. The mean potential over any spherical surface, due to matter entirely without it, is equal to the potential at its centre; a theorem apparently first given by Gauss. See also Cambridge *Mathematical Journal*, Feb. 1845 (vol. iv. p. 225). It is one of Mean potential over a spherical surface equal to that at its centre.

[1] Cambridge *Phil. Trans.*, March 1839.

Mean potential over a spherical surface equal to that at its centre.

the most elementary propositions of spherical harmonic analysis applied to potentials, found by applying App. B. (16) to the formulæ of § 539, below. But the proof in the paper now referred to is noticeable as independent of the harmonic expansion.

Let, in Chap. I. App. B. (a), S be a spherical surface, of radius a; and let U be the potential at (x, y, z), due to matter altogether external to it; let U' be the potential of a unit of matter uniformly distributed through a smaller concentric spherical surface; so that, outside S and to some distance within it, $U' = \frac{1}{r}$; and lastly, let $a = 1$. The middle member of App. B (a) (1) becomes

$$\frac{1}{a}\iint \partial U d\sigma - \iiint U' \nabla^2 U dx dy dz,$$

which is equal to zero, since $\nabla^2 U = 0$ for the whole internal space, and (§ 492) $\iint \partial U d\sigma = 0$. Equating therefore the third member to zero we have

$$\iint d\sigma U \partial U' = \iiint U \nabla^2 U' dx dy dz.$$

Now at the surface, S, $\partial U' = -\frac{1}{a^2}$; and for all points external to the sphere of matter to which U' is due, $\nabla^2 U' = 0$, and for all internal points $\nabla^2 U' = -4\pi\rho'$, if ρ' be the density of the matter. Hence the preceding equation becomes

$$\frac{1}{a^2}\iint U d\sigma = 4\pi \iiint \rho' U dx dy dz.$$

Let now the density ρ' increase without limit, and the spherical space within which the triple internal extends, therefore become infinitely small. If we denote by U_0 the value of U at its centre, which is also the centre of S, we shall have

$$\iiint \rho' U dx dy dz = U_0 \iiint \rho' dx dy dz = U_0.$$

Hence the equation becomes

$$\frac{\iint U d\sigma}{4\pi a^2} = U_0.$$

which was to be proved.

Theorem of Gauss.

497. If the potential of any masses has a constant value, V, through any finite portion, K, of space, unoccupied by matter, it is equal to V through every part of space which can be reached in any way without passing through any of those masses: a very remarkable proposition, due to Gauss. For, if the potential differ from V in space contiguous to K, it must (§ 495) be greater in some parts and less in others. Theorem of Gauss.

From any point C within K, as centre, in the neighbourhood of a place where the potential is greater than V, describe a spherical surface not large enough to contain any part of any of the attracting masses, nor to include any of the space external to K except such as has potential greater than V. But this is impossible, since we have just seen (§ 497) that the mean potential over the spherical surface must be V. Hence the supposition that the potential is greater than V in some places and less in others, contiguous to K and not including masses, is false.

498. Similarly we see that in any case of symmetry round an axis, if the potential is constant through a certain finite distance, however short, along the axis, it is constant throughout the whole space that can be reached from this portion of the axis, without crossing any of the masses. (See § 546, below.)

499. Let S be any finite portion of a surface, or complete closed surface, or infinite surface, and let E be any point on S. (a.) It is possible to distribute matter over S so as to produce potential equal to $F(E)$, any arbitrary function of the position of E, over the whole of S. (b.) There is only one whole quantity of matter, and one distribution of it, which can satisfy this condition. Green's problem.

In Chap. I. App. A. (b) (e), etc., let $\alpha = 1$. By (e) we see that there is one, and that there is only one, solution of the equation

$$\nabla^2 U = 0$$

for all points not belonging to S, subject to the condition that U shall have a value arbitrarily given over the whole of S. Continuing to denote by U the solution of this problem, and considering first the case of S an open shell, that is to say, a finite portion of curved surface (including a plane, of course, as a particular case), let, in Chap. I. A. (a), U' be the potential at (x, y, z) due to a distribution of matter, having $\varpi(Q)$ for density at any point, Q. Let the triple integration extend throughout infinite space, exclusive of the infinitely thin shell S. Although in the investigation referred to [A. (a)] the triple integral extended only through the finite space contained within a closed surface, the same process shows that we have now, instead of the second and third

Green's problem;

members of (1) of that investigation, the following equated expressions :—

$$\iint d\sigma U'\{[\partial U]-(\partial U)\}-\iiint dxdydz U'\nabla^2 U$$
$$=\iint d\sigma U\{[\partial U']-(\partial U')\}-\iiint dxdydz U\nabla^2 U'$$

where $[\partial U]$ denotes the rate of variation of U on either side of S, infinitely near E, reckoned per unit of length from S; and (∂U) denotes the rate of variation of U infinitely near E, on the other side of S, reckoned per unit of length *towards* S; and $[\partial U']$, $(\partial U')$ denote the same for U'. Now we shall suppose the matter of which U' is the potential not to be condensed in finite quantities in any finite areas of S, which will make

$$[\partial U']=(\partial U'):$$

and the conditions defining U and U' give, throughout the space of the triple integral,

$$\nabla^2 U=0, \text{ and } \nabla^2 U'=-4\pi\varpi;$$

ϖ denoting the value of ϖ (Q) when Q is the point (x, y, z). Hence the preceding equation becomes

$$\iint d\sigma U'\{[\partial U]-(\partial U)\}=4\pi\iiint dxdydz\varpi U.$$

Let now the matter of which U' is the potential be equal in amount to unity and be confined to an infinitely small space round a point Q. We shall have

$$U'=\frac{1}{EQ}$$

$$\iiint dxdydz\varpi U=U(Q)\iiint\varpi dxdydz=U(Q),$$

if we denote the value of U at (Q) by $U(Q)$. The equation becomes

$$\iint\frac{[\partial U]-(\partial U)}{EQ}d\sigma=4\pi U(Q) \qquad (1).$$

reduced to the proper general solution of Laplace's equation.

Hence a distribution of matter over S, having

$$\frac{1}{4\pi}\{[\partial U]-(\partial U)\} \qquad (2)$$

for density at the point E, gives U as its potential at (x, y, z). We conclude, therefore, that it is possible to find one, but only one, distribution of matter over S which shall produce an arbitrarily given potential over the whole of S; and in (2) we have the solution of this problem, when the problem of finding U to fulfil the conditions stated above, has been solved.

If S is any finite closed surface, any group of surfaces, open or closed, or an infinite surface, the same conclusions clearly hold. The triple integration used in the investigation must then be separately carried out through all the portions of space separated from one another by S, or by portions of S.

Green's problem;

If the solution, ρ, of the problem has been obtained for the case in which the arbitrary function is the potential at any point of S, due to a unit of matter at any point P not belonging to S, that is to say, for the case of $F(E)=\frac{1}{EP}$, the solution of the general problem was shown by Green to deducible from it thus:—

solved synthetically in terms of particular solution of Laplace's equation.

$$U=\iint\rho F(E)d\sigma \qquad (3).$$

The proof is obvious: For let, for a moment, ${}_{,}\rho$ denote the superficial density required to produce U, then ${}_{,}\rho'$ denoting the value of ${}_{,}\rho$ for any other element, E', of S, we have

$$F(E)=\iint\frac{{}_{,}\rho' d\sigma'}{E'E}.$$

Hence the preceding double integral becomes

$$\iint d\sigma\rho\iint d\sigma'\frac{{}_{,}\rho'}{E'E}, \text{ or } \iint d\sigma'{}_{,}\rho'\iint d\sigma\frac{\rho}{E'E}.$$

But, by the definition of ρ,

$$\iint d\sigma\frac{\rho}{E'E}=\frac{1}{EP};$$

and therefore the expression becomes

$$\iint d\sigma'\frac{{}_{,}\rho'}{E'P},$$

which is equal to U, according to the definition of ${}_{,}\rho$.

The expression (46) of App. B., from which the spherical harmonic expansion of an arbitrary function was derived, is a case of the general result (3) now proved.

500. It is important to remark that, if S consist, in part, of a closed surface, Q, the determination of U within it will be independent of those portions of S, if any, which lie without it; and, *vice versa*, the determination of U through external space will be independent of those portions of S, if any, which lie within the part Q. Or if S consist, in part, of a surface Q, extending infinitely in all directions, the determination of U through all space on either side of Q, is independent of those portions of S, if any, which lie on the other side. This follows from the preceding investigation, modified by confining the triple integration to one of the two portions of space separated completely from one another by Q.

Isolation of effect by closed portion of surface.

Green's problem; applied to a given distribution of electricity, M, influencing a group, S, of conducting surfaces.

501. Another remark of extreme importance is this:—If $F(E)$ be the potential at E of any distribution, M, of matter, and if S be such as to separate perfectly any portion or portions of space, H, from all of this matter; that is to say, such that it is impossible to pass into H from any part of M without crossing S; then, throughout H, the value of U will be the potential of M.

For if V denote this potential, we have, throughout H, $\nabla^2 V=0$; and at every point of the boundary of H, $V=F(E)$. Hence, considering the theorem of Chap. I. App. A (c), for the space H alone, and its boundary alone, instead of S, we see that, through this space, V satisfies the conditions prescribed for U, and therefore, through this space, $U=V$.

502. Thus, for instance, if S consist of three detached surfaces, S_1, S_2, S_3, as in the diagram, of which S_1, S_2 are closed, and S_3 is an open shell, and if $F(E)$ be the potential due to M, at any point, E, of any of these portions of S; then throughout H_1, and H_2, the spaces within S_1 and without S_2, the value of U is simply the potential of M. The value of U through K, the remainder of space, depends, of course, on the character of the composite surface S, and is a case of the general problem of which the solution was proved to be possible and single in Chap. I. App. A.

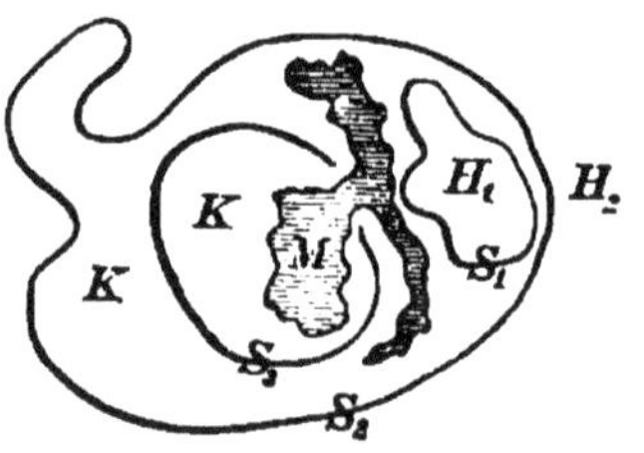

General problem of electric influence possible and determinate.

503. From § 500 follows the grand proposition:—*It is possible to find one, but no other than one, distribution of matter over a surface S which shall produce over S, and throughout all space H separated by S from every part of M, the same potential as any given mass M.*

Thus, in the preceding diagram, it is possible to find one, and but one, distribution of matter over S_1, S_2, S_3 which shall produce over S_3 and through H_1 and H_2 the same potential as M.

The statement of this proposition most commonly made is: *It is possible to distribute matter over any surface, S, completely enclosing a mass M, so as to produce the same potential as M through all space outside M*; which, though seemingly more limited, is, when interpreted with proper mathematical comprehensiveness, equivalent to the foregoing.

Green's problem.

504. If S consist of several closed or infinite surfaces, S_1, S_2, S_3, respectively separating certain isolated spaces H_1, H_2, H_3, from H, the remainder of all space, and if $F(E)$ be the potential of masses m_1, m_2, m_3, lying in the spaces H_1, H_2, H_3; the portions of U due to S_1, S_2, S_3, respectively will throughout H be equal respectively to the potentials of m_1, m_2, m_3, separately.

Simultaneous electric influences in spaces separated by infinitely thin conducting surfaces.

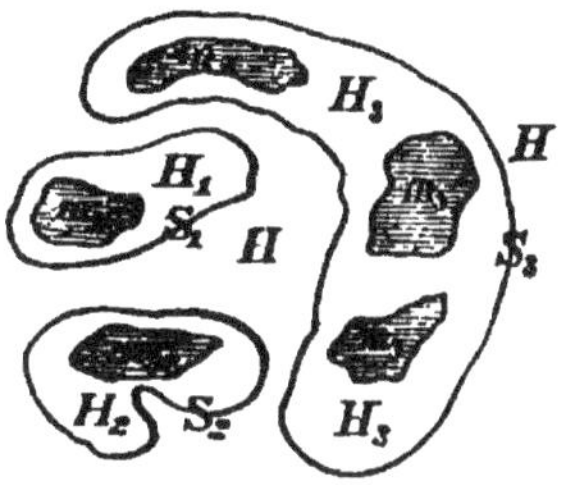

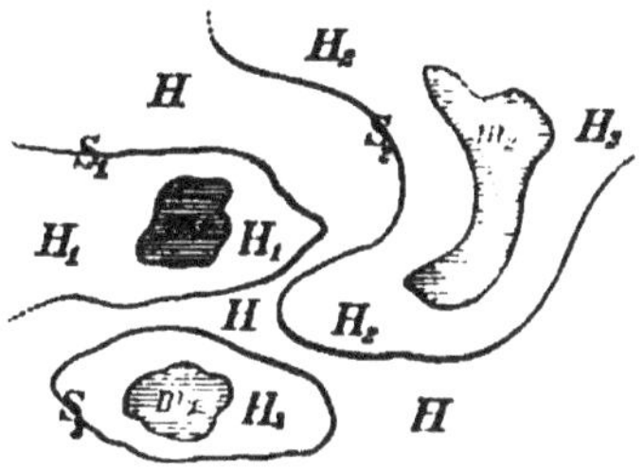

For as we have just seen, it is possible to find one, but only one, distribution of matter over S_1 which shall produce the potential of m_1, throughout all the space H, H_2, H_3, etc., and one, but only one, distribution over S_2 which shall produce the potential of m_2 throughout H, H_1, H_3, etc.; and so on. But these distributions on S_1, S_2, etc., jointly constitute a distribution producing the potential $F(E)$ over every part of S, and therefore the sum of the potentials due to them all, at any point, fulfils the conditions presented for U. This is therefore (§ 500) *the* solution of the problem.

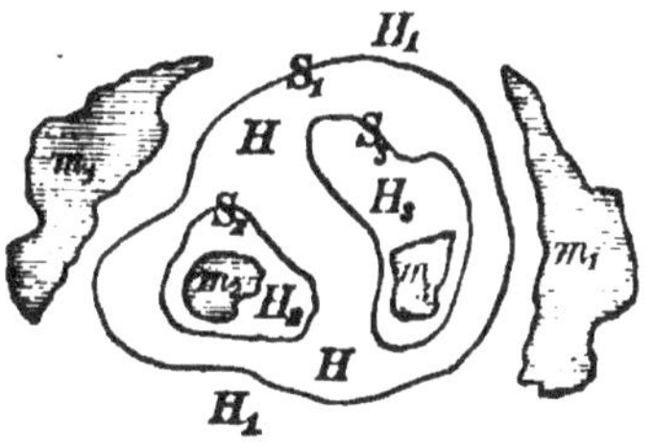

505. Considering still the case in which $F(E)$ is prescribed to be the potential of a given mass, M: let S be an equipotential surface enclosing M, or a group of isolated surfaces enclosing all the parts of M, and each equipotential for the whole of M. The potential due to the supposed distribution over S will be the same as that of M, through all external space, and will be constant (§ 496) through each enclosed portion of space. Its

Reducible case of Green's problem.

Reducible case of Green's problem;

resultant attraction will therefore be the same as that of M on all external points, and zero on all internal points. Hence we see at once that the density of the matter distributed over it, to produce $F(E)$, is equal to $\frac{R}{4\pi}$ where R denotes the resultant force of M, at the point E.

We have $[\partial U] = -R$ and $(\partial U) = 0$. Whence, by § 500 (2), the law of density

applied to the invention of solved problems of electric influence.

506. When M consists of two portions m_1 and m' separated by an equipotential S_1, and S consists of two portions, S_1, and S', of which the latter separate the former perfectly from m'; we see by § 504 that the distribution over S_1 produces through all space on the side of it on which S' lies, the same potential, V_1, as m_1, and the distribution on S' produces through space on the side of it on which S_1 lies, the same potential, V', as m'. But the supposed distribution on the whole of S is such as to produce a constant potential, C_1 over S_1, and consequently the same at every point within S_1. Hence the internal potential due to S_1 alone, is $C_1 - V'$.

Thus, passing from potentials to attractions, we see that the resultant attraction of S_1 alone, on all points on one side of it is the same as that of m_1; and on the other side is equal and opposite to that of the remainder m' of the whole mass. The most direct and simple complete statement of this result is as follows:—

If masses m, m', in portions of space, H, H', completely separated from one another by one continuous surface S, whether closed or infinite, are known to produce tangential forces equal and in the same direction at each point of S, one and the same distribution of matter over S will produce the force of m throughout H', and that of m' throughout H. The density of this distribution is equal to $\frac{R}{4\pi}$, if R denote the resultant force due to one of the masses, and the other with its *sign* changed. And it is to be remarked that the direction of this resultant force is, at every point, E, of S, perpendicular to S, since the potential due to one mass, and the other with its sign changed, is constant over the whole of S.

507. Green, in first publishing his discovery of the result stated in § 505, remarked that it shows a way to find an infinite variety of closed surfaces for any one of which we can solve the problem of determining the distribution of matter over it which shall produce a given uniform potential at each point of its surface, and consequently the same also throughout its interior. Thus, an example which Green himself gives, let M be a uniform bar of matter, AA'. The equipotential surfaces round it are, as we have seen above (§ 481), prolate ellipsoids of revolution, each having A and A' for its foci; and the resultant force at C was found to be Reducible case of Green's problem:—examples.

$$\frac{m}{l(l^2-a^2)}\cdot CF,$$

the whole mass of the bar being denoted by m, its length by $2a$, and $A'C+AC$ by $2l$. We conclude that a distribution of matter over the surface of the ellipsoid, having

$$\frac{1}{4\pi}\frac{m.CF}{l(l^2-a^2)}$$

for density at C, produces on all external space the same resultant force as the bar, and zero force or a constant potential through the internal space. This is a particular case of the general result regarding ellipsoidal shells, proved below, in §§ 520, 521.

508. As a second example, let M consist of two equal particles, at points I, I'. If we take the mass of each as unity, the potential at P is $\frac{1}{IP}+\frac{1}{I'P}$; and therefore

$$\frac{1}{IP}+\frac{1}{I'P}=C$$

is the equation of an equipotential surface; it being understood that negative values of IP and $I'P$ are inadmissible, and that any constant value, from ∞ to 0, may be given to C. The curves in the annexed diagram have been drawn, from this equation, for the cases of C equal respectively to 10, 9, 8, 7, 6, 5, 4·5, 4·3, 4·2, 4·1, 4, 3·9, 3·8, 3·7, 3·5, 3, 2·5, 2; the value of II' being unity.

Reducible case of Green's problem:—examples.

The corresponding equipotential surfaces are the surfaces traced by these curves, if the whole diagram is made to rotate

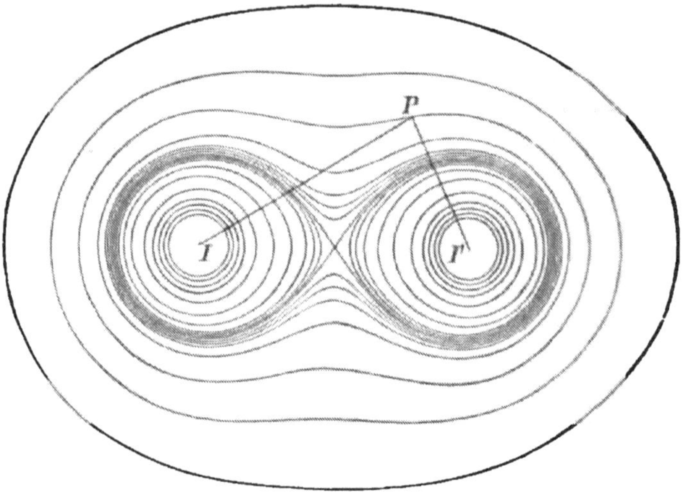

round II' as axis. Thus we see that for any values of C less than 4 the equipotential surface is one closed surface. Choosing any one of these surfaces, let R denote the resultant of forces equal to $\frac{1}{IP^2}$ and $\frac{1}{I'P^2}$ in the lines PI and PI'. Then if matter be distributed over this surface, with density at P equal to $\frac{R}{4\pi}$, its attraction on any internal point will be zero; and on any external point, will be the same as that of I and I'.

509. For each value of C greater than 4, the equipotential surface consists of two detached ovals approximating (the last three or four in the diagram, very closely) to spherical surfaces, with centres lying between the points I and I', but approximating more and more closely to these points, for larger and larger values of C.

Considering one of these ovals alone, one of the series enclosing I', for instance, and distributing matter over it according to the same law of density, $\frac{R}{4\pi}$, we have a shell of matter which exerts (§ 507) on external points the same force as I'; and on internal points a force equal and opposite to that of I.

Reducible case of Green's problem:—examples. Electric images.

510. As an example of exceedingly great importance in the theory of electricity, let M consist of a positive mass, m, concentrated at a point I, and a negative mass, $-m'$, at I'; and let S be a spherical surface cutting II', and II' produced in points A, $A_{,}$, such that $IA : AI' :: IA_{,} : I'A_{,} :: m : m'$. Then, by a well-known geometrical proposition, we shall have $IE : I'E :: m : m'$; and therefore

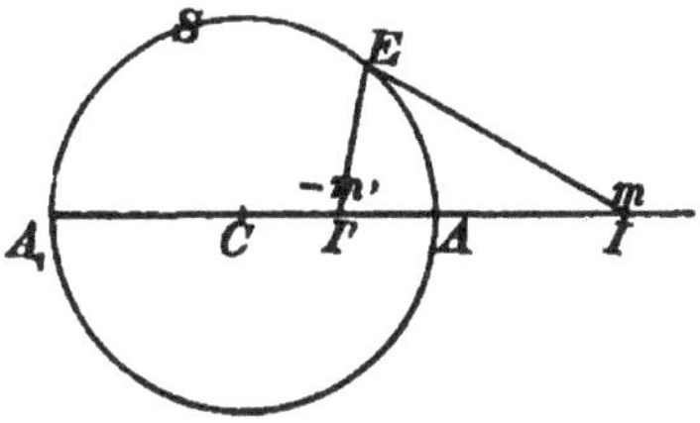

$$\frac{m}{IE} = \frac{m'}{I'E}.$$

Hence, by what we have just seen, one and the same distribution of matter over S will produce the same force as m' through all external space, and the same as m through all the space within S. And, finding the resultant of the forces $\frac{m}{IE^2}$ in EI, and $\frac{m'}{I'E^2}$ in $I'E$ produced, which, as these forces are inversely as IE to $I'E$, is (§ 256) equal to

$$\frac{m}{IE^2.I'E} II', \text{ or } \frac{m^2 II'}{m'} \frac{1}{IE^3},$$

we conclude that the density in the shell at E is

$$\frac{m^2 II'}{4\pi m'} \cdot \frac{1}{IE^3}.$$

That the shell thus constituted does attract external points as if its mass were collected at I', and internal points as a certain mass collected at I, was proved geometrically in § 474 above.

511. If the spherical surface is given, and one of the points, I, I', for instance I, the other is found by taking $CI' = \frac{CA^2}{CI}$; and for the mass to be placed at it we have

$$m' = m\frac{I'A}{AI} = m\frac{CA}{CI} = m\frac{CI'}{CA}.$$

Electric images.

Hence if we have any number of particles m_1, m_2, etc., at points I_1, I_2, etc., situated without S, we may find in the same way corresponding internal points I'_1, I'_2, etc., and masses m'_1, m'_2 etc.; and, by adding the expressions for the density at E given for each pair by the preceding formula, we get a spherical shell of matter which has the property of acting on all external space with the same force as $-m'_1$, $-m'_2$, etc., and on all internal points with a force equal and opposite to that of m_1, m_2, etc.

512. An infinite number of such particles may be given constituting a continuous mass M; when of course the corresponding internal particles will constitute a continuous mass, $-M'$, of the opposite kind of matter; and the same conclusion will hold. If S is the surface of a solid or hollow metal ball connected with the earth by a fine wire, and M an external influencing body, the shell of matter we have determined is precisely the distribution of electricity on S called out by the influence of M: and the mass $-M'$, determined as above, is called the *Electric Image* of M in the ball, since the electric action through the whole space external to the ball would be unchanged if the ball were removed and $-M'$ properly placed in the space left vacant. We intend to return to this subject under Electricity.

Transformation by reciprocal radius-vectors.

513. Irrespectively of the special electric application, this method of images gives a remarkable kind of transformation which is often useful. It suggests for mere geometry what has been called the transformation by reciprocal radius-vectors: that is to say, the substitution for any set of points, or for any diagram of lines or surfaces, another obtained by drawing radii to them from a certain fixed point or origin, and measuring off lengths inversely proportional to these radii along their directions. We see in a moment by elementary geometry that any line thus obtained cuts the radius-vector through any point of it at the same angle and in the same plane as the line from which it is derived. Hence any two lines or surfaces that cut one another give two transformed lines or surfaces cutting at the same angle: and infinitely small lengths, areas, and volumes transform into others whose magnitudes are altered respectively in the ratios of the first, second, and third powers of the distances of the latter from the origin, to the same powers of the distances of the former from the same. Hence the lengths, areas, and volumes in the transformed diagram, corresponding to a set of given equal infinitely small lengths, areas, and volumes, however situated, at different distances from the origin, are inversely as the squares, the fourth powers and the sixth powers of these distances. Further, it is easily proved that a straight line and a plane transform into a circle and a spherical surface, each passing through the origin; and that, generally, circles and spheres transform into circles and spheres. Transformation by reciprocal radius-vector.

514. In the theory of attraction, the transformation of masses, densities, and potentials has also to be considered. Thus, according to the foundation of the method (§ 512), equal masses, of infinitely small dimensions at different distances from the origin, transform into masses inversely as these distances, or directly as the transformed distances: and, therefore, equal densities of lines, of surfaces, and of solids, given at any stated distances from the origin, transform into densities directly as the first, the third, and the fifth powers of those distances; or inversely as the same powers of the distances, from the origin, of the corresponding points in the transformed system.

515. The statements of the last two sections, so far as proportions alone are concerned, are most conveniently expressed thus:— General summary of ratios.

Let P be any point whatever of a geometrical diagram, or of a distribution of matter, O one particular point ("the origin"), and a one particular length (the radius of the "reflecting sphere"). In OP take a point P', corresponding to P, and for any mass m, in any infinitely small part of the given distribution, place a mass m'; fulfilling the conditions

$$OP' = \frac{a^2}{OP}, \quad m' = \frac{a}{OP} m = \frac{OP'}{a} m.$$

Then if L, A, V, $\rho(L)$, $\rho(A)$, $\rho(V)$ denote an infinitely small length, area, volume, linear-density, surface-density, volume-density in the given distribution, infinitely near to P, or anywhere at the same distance, r, from O as P, and if the corresponding elements in the transformed diagram or distribution be denoted in the same way with the addition of accents, we have

General summary of ratios.

$$L'=\frac{a^2}{r^2}L=\frac{r'^2}{a^2}L;\ \ A'=\frac{a^4}{r^4}A=\frac{r'^4}{a^4}A;\ \ V'=\frac{a^6}{r^6}V=\frac{r'^6}{a^6}V.$$

$$\rho'(L)=\frac{a}{r}\rho(L)=\frac{r'}{a}\rho(L);\ \ \rho'(A)=\frac{a^3}{r^3}\rho(A)=\frac{r'^3}{a^3}\rho(A);$$

$$\rho'(V)=\frac{a^5}{r^5}\rho(V)=\frac{r'^5}{a^5}\rho(V).$$

The usefulness of this transformation in the theory of electricity and of attraction in general, depends entirely on the following theorem :—

Application to the potential.

516. (*Theorem.*)—Let ϕ denote the potential at P due to the given distribution, and ϕ' the potential at P' due to the transformed distribution : then shall

$$\phi'=\frac{r}{a}\phi=\frac{a}{r'}\phi.$$

Let a mass m collected at I be any part of the given distribution, and let m' at I' be the corresponding part in the transformed distribution. We have

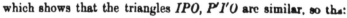

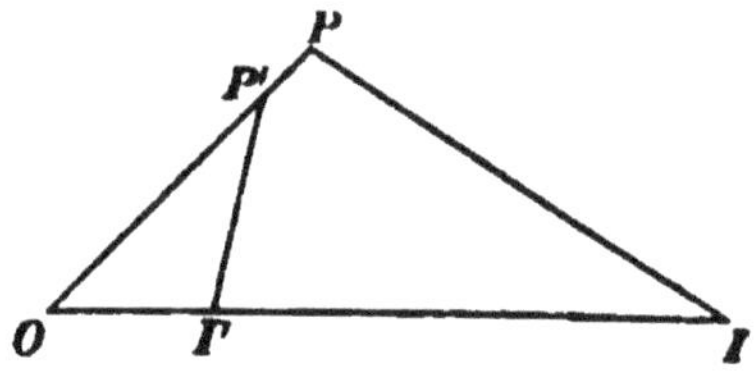

$$a^2=OI'.OI=OP'.OP,$$

and therefore

$$OI:OP::OP':OI';$$

which shows that the triangles IPO, $P'I'O$ are similar, so that

$$IP:P'I'::\sqrt{OI.OP}:\sqrt{OP'.OI'}::OI.OP:a^2.$$

We have besides

$$m:m'::OI:a,$$

and therefore

$$\frac{m}{IP}:\frac{m'}{I'P'}::OP:a.$$

Hence each term of ϕ bears to the corresponding term of ϕ' the same ratio; and therefore the sum, ϕ, must be to the sum ϕ', in that ratio, as was to be proved.

Any distribution on a spherical shell.

517. As an example, let the given distribution be confined to a spherical surface, and let O be its centre and a its own radius. The transformed distribution is the same. But the space within it becomes transformed into the space without it. Hence if ϕ be the potential due to any spherical shell at a point P, within it, the potential due to the same shell at the point P' in OP produced till $OP' = \frac{a^2}{OP}$, is equal to $\frac{a}{OP'}\phi$ (which is an elementary proposition in the spherical harmonic treatment of potentials, as we shall see presently). Thus, for instance, let the distribution be uniform. Then, as we know there is no force on an interior point, ϕ must be constant; and therefore the potential at P', any external point, is inversely proportional to its distance from the centre. Any distribution on a spherical shell.

Or let the given distribution be a uniform shell, S, and let O be any eccentric or any external point. The transformed distribution becomes (§§ 513, 514) a spherical shell, S', with density varying inversely as the cube of the distance from O. If O is within S, it is also enclosed by S', and the whole space within S transforms into the whole space without S'. Hence (§ 516) the potential of S' at any point without it is inversely as the distance from O, and is therefore that of a certain quantity of matter collected at O. Or if O is external to S, and consequently also external to S', the space within S transforms into the space within S'. Hence the potential of S' at any point within it is the same as that of a certain quantity of matter collected at O, which is now a point external to it. Thus, without taking advantage of the general theorems (§§ 499, 506), we fall back on the same results as we inferred from them in § 510, and as we proved synthetically earlier (§§ 471, 474, 475). It may be remarked that those synthetical demonstrations consist merely of transformations of Newton's demonstration, that attractions balance on a point within a uniform shell. Thus the first of them (§ 471) is the image of Newton's in a concentric spherical surface; and the second is its image in a spherical surface having its centre external to the shell, or internal but eccentric, according as the first or the second diagram is used. Uniform shell eccentrically reflected.

518. We shall give just one other application of the theorem of § 516 at present, but much use of it will be made later, in the theory of Electricity. Uniform solid sphere eccentrically reflected.

Let the given distribution of matter be a uniform solid sphere, B, and let O be external to it. The transformed system will be a solid sphere, B', with density varying inversely as the fifth power of the distance from O, a point external to it.

Uniform solid sphere eccentrically reflected.

The potential of S is the same throughout external space as that due to its mass, m, collected at its centre, C. Hence the potential of S' through space external to it is the same as that of the corresponding quantity of matter collected at C', the transformed position of C. This quantity is of course equal to the mass of B'. And it is easily proved that C' is the position of the image of O in the spherical surface of B'. We conclude that a solid sphere with density varying inversely as the fifth power of the distance from an external point, O, attracts any external point as if its mass were condensed at the image of O in its external surface. It is easy to verify this for points of the axis by direct integration, and thence the general conclusion follows according to § 490.

Attraction of an ellipsoid.

519. The determination of the attraction of an ellipsoid, or of an ellipsoidal shell, is a problem of great interest, and its results will be of great use to us afterwards, especially in Magnetism. We have left it till now, in order that we may be prepared to apply the properties of the potential, as they afford an extremely elegant method of treatment. A few definitions and lemmas are necessary.

Corresponding points on confocal ellipsoids.

Corresponding points on two confocal ellipsoids are such as coincide when either ellipsoid by a pure strain is deformed so as to coincide with the other.

And it is easily shown, as below, that if any two points, P, Q, be assumed on one shell, and their *corresponding* points, p, q, on the other, we have $Pq = Qp$.

If
$$\frac{x^2}{a^2}+\frac{y^2}{b^2}+\frac{z^2}{c^2}=1 \qquad (1)$$

and
$$\frac{x^2}{a^2+h}+\frac{y^2}{b^2+h}+\frac{z^2}{c^2+h}=1 \qquad (2)$$

be any two confocal ellipsoids; and P $[\xi, \eta, \zeta]$, a point on (1), p

$$\left[\frac{\sqrt{a^2+h}}{a}\xi,\ \frac{\sqrt{b^2+h}}{b}\eta,\ \frac{\sqrt{c^2+h}}{c}\zeta\right]$$

is evidently a point on (2), and is the *corresponding* point to P. Let Q be $[\xi', \eta', \zeta']$. Then

$$Pq^2=\left(\xi-\frac{\sqrt{a^2+h}}{a}\xi'\right)^2+\left(\eta-\frac{\sqrt{b^2+h}}{b}\eta'\right)^2+\left(\zeta-\frac{\sqrt{c^2+h}}{c}\zeta'\right)^2,$$

$$Qp^2=\left(\xi'-\frac{\sqrt{a^2+h}}{a}\xi\right)^2+\left(\eta'-\frac{\sqrt{b^2+h}}{b}\eta\right)^2+\left(\zeta'-\frac{\sqrt{c^2+h}}{c}\zeta\right)^2.$$

And $$Pq^2 - Qp^2 = \left(\frac{a^2+h}{a^2} - 1\right)(\xi^2 - \xi'^2) + \ldots$$
$$= h\left\{\frac{\xi^2}{a^2} + \frac{\eta^2}{b^2} + \frac{\zeta^2}{c^2} - \frac{\xi'^2}{a^2} - \frac{\eta'^2}{b^2} - \frac{\zeta'^2}{c^2}\right\}$$
$$= 0.$$
Corresponding points on confocal ellipsoids.

The species of shell which it is most convenient to employ in the subdivision of a homogeneous ellipsoid is bounded by similar, similarly situated, and concentric ellipsoidal surfaces; and it is evident from the properties of pure strain (§ 182) that such a shell may be produced from a spherical shell of uniform thickness by simple extensions and compressions in three rectangular directions. Unless the contrary be specified, the word "shell" will always signify an infinitely thin shell of this kind. Ellipsoidal shell,

520. Since, by § 462, a homogeneous spherical shell exerts no attraction on an internal point, a homogeneous shell (which need not be infinitely thin) bounded by similar, and similarly situated and concentric ellipsoids, exerts no attraction on an internal point. exerts no attraction upon an internal particle.

For suppose the spherical shell of § 462, by simple extensions and compressions in three rectangular directions, to be transformed into an ellipsoidal shell. In this distorted form the masses of all parts are reduced or increased in the proportion of the mass of the ellipsoid to that of the sphere. Also the ratio of the lines HP, PK is unaltered, § 158. Hence the elements IH, KL still attract P equally, and the proposition follows as in § 462.

Hence inside the shell the potential is constant.

521. Two confocal shells (§ 519) being given, the potential of the first at any point, P, of the surface of the second, is to that of the second at the corresponding point, p, on the surface of the first, as the mass of the first is to the mass of the second. This beautiful proposition is due to Chasles. Comparison of potentials of two shells

To any element of the mass of the outer shell at Q corresponds an element of mass of the inner at q, and these bear the same ratio to the whole masses of their respective shells, that the corresponding element of the spherical shell from which either may be derived bears to its whole mass. Whence, since $Pq = Qp$, the proposition is true for the corresponding elements at Q and q, and therefore for the entire shells.

Equipotential surface for ellipsoidal shell.

Also, as the potential of a shell on an internal point is constant, and as one of two confocal ellipsoids is wholly within the other: it follows that the external equipotential surfaces for any such shell are confocal ellipsoids, and therefore that the attraction of the shell on an external point is normal to a confocal ellipsoid passing through the point.

Attraction of homogeneous ellipsoid.

522. Now it has been shown (§ 478) that the attraction of a shell on an external point near its surface exceeds that on an internal point infinitely near it by $4\pi\rho$ where ρ is the surface-density of the shell at that point. Hence, as, § 520, there is no attraction on an internal point, the attraction of a shell on a point at its exterior surface is $4\pi\rho$: or $4\pi\rho t$ if ρ be now put for the volume-density, and t for the (infinitely small thickness of the shell, § 491 (f). From this we obtain immediately the determination of the whole attraction of a homogeneous ellipsoid on an external particle.

Let a_0, b_0, c_0 be the axes of the attracting ellipsoid, and let $a=a_0\theta$, $b=b_0\theta$, $c=c_0\theta$, be the axes of any similar, similarly situated, and concentric surface drawn within it; θ being thus a proper fraction. If we consider a shell bounded by surfaces corresponding to θ, and $\theta-d\theta$, respectively, its attraction on the external point P (ξ, η, ζ), is to that of a shell whose surfaces are confocal with these, and whose outer surface passes through P, as the mass of the first shell to that of the second. If A, B, C be the axes of this outer surface, we have

$$\left.\begin{aligned} A^2&=a^2+h=a_0^2\theta^2+\theta^2\phi^2\\ B^2&=b^2+h=b_0^2\theta^2+\theta^2\phi^2\\ C^2&=c^2+h=c_0^2\theta^2+\theta^2\phi^2,\end{aligned}\right\} \qquad (1),$$

where ϕ is a new variable, connected with θ by the equation

$$\frac{\xi^2}{A^2}+\frac{\eta^2}{B^2}+\frac{\zeta^2}{C^2}=1, \qquad (2).$$

or

$$\frac{\xi^2}{a_0^2+\phi^2}+\frac{\eta^2}{b_0^2+\phi^2}+\frac{\zeta^2}{c_0^2+\phi^2}=\theta^2. \qquad (3).$$

Now it is evident that, if $A-dA$, $B-dB$, $C-dC$, be the axes of the inner surface of the new shell,

$$\frac{dA}{A}=\frac{dB}{B}=\frac{dC}{C}=\frac{da}{a}=\frac{db}{b}=\frac{dc}{c}=\frac{d\theta}{\theta},$$

and that, if ϖ be the perpendicular from the centre on the tangent plane at P, and t the thickness of the shell at that point, Attraction of homogeneous ellipsoid.

$$\frac{t}{\varpi}=\frac{dA}{A}=\frac{d\theta}{\theta}.$$

Also, by geometry,

$$\frac{1}{\varpi^2}=\frac{\xi^2}{A^4}+\frac{\eta^2}{B^4}+\frac{\zeta^2}{C^4}, \qquad (4),$$

and the direction-cosines of ϖ are

$$\frac{\varpi\xi}{A^2}, \frac{\varpi\eta}{B^2}, \frac{\varpi\zeta}{C^2}.$$

Hence the attraction, parallel to axis of x, of shell $[\theta, \theta-d\theta]$ is

$$4\pi\rho\frac{a_0b_0c_0\theta^3}{ABC}\varpi\frac{d\theta}{\theta}\frac{\varpi\xi}{A^2}=4\pi\rho a_0b_0c_0\xi\frac{\varpi^2\theta^2d\theta}{A^2BC}.$$

For the whole attraction in this direction we have only to integrate this expression, as a function of θ, from $\theta=0$ to $\theta=1$.

The integration is easier if we make ϕ the variable. Thus, differentiating (3), we have

$$-\frac{1}{\varpi^2}\phi d\phi=\frac{d\theta}{\theta^3},$$

and therefore the whole attraction is

$$-4\pi\rho a_0b_0c_0\xi\int\frac{\theta^5\phi d\phi}{A^2BC}$$

$$=-2\pi\rho a_0b_0c_0\xi\int\frac{d(\phi^2)}{\sqrt{(a_0^2+\phi^2)^3(b_0^2+\phi^2)(c_0^2+\phi^2)}}.$$

The limits are given at once by (3), if we remember that θ ranges from 0 to 1; and are evidently ∞, and the positive root of

$$\frac{\xi^2}{a_0^2+\phi^2}+\frac{\eta^2}{b_0^2+\phi^2}+\frac{\zeta^2}{c_0^2+\phi^2}=1.$$

Call this root α^2, then the x component of the attraction is

$$\tfrac{3}{2}M\xi\int_{\alpha^2}^{\infty}\frac{d(\phi^2)}{\sqrt{(a_0^2+\phi^2)^3(b_0^2+\phi^2)(c_0^2+\phi^2)}}. \qquad (5)$$

where M is the mass of the ellipsoid.

It is worthy of remark that the three components depend upon the *one* elliptic integral

$$\Phi=\int_{\alpha^2}^{\infty}\frac{d(\phi^2)}{\sqrt{(a_0^2+\phi^2)(b_0^2+\phi^2)(c_0^2+\phi^2)}},$$

and are $\quad X=-\tfrac{3}{2}M\xi\dfrac{d\Phi}{d(a_0^2)},$

with similar expressions for Y and Z as partial differential coefficients with respect to b_0^2 and c_0^2 respectively, α being treated as a constant. When the attracted point lies on the surface of the ellipsoid, the only requisite change is to put $\alpha=0$.

Attraction of a spheroid of revolution.

If we put $c_0=b_0$ the ellipsoid becomes a spheroid of revolution: and for its attraction parallel to the axis we have

$$\tfrac{3}{2}M\xi\int_{\alpha^2}^{\infty}\frac{d(\phi^2)}{(a_0^2+\phi^2)^{\frac{3}{2}}(b_0^2+\phi^2)}$$

where α^2 is the positive root of the equation

$$\frac{\xi^2}{a_0^2+\alpha^2}+\frac{\eta^2+\zeta^2}{b_0^2+\alpha^2}=1.$$

This integral is, of course, easily expressed in finite terms. But, as we shall see presently, it is sufficient to find its value for a point *on* the surface; for which we have

$$X=\tfrac{3}{2}M\xi\int_0^{\infty}\frac{d(\phi^2)}{(a_0^2+\phi^2)^{\frac{3}{2}}(b_0^2+\phi^2)} \qquad (6).$$

To work this out in real finite terms for an *oblate* spheroid, let

$$b_0^2=a_0^2+b_0^2e^2,$$

and the definite integral above becomes

$$\int_0^{\infty}\frac{d(\phi^2+a_0^2)}{(a_0^2+\phi^2)^{\frac{3}{2}}(a_0^2+b_0^2e^2+\phi^2)}=\int_{a_0^2}^{\infty}\frac{d(\omega^2)}{\omega^3(b_0^2e^2+\omega^2)}$$

$$=\frac{2}{b_0^3e^3}\left(\frac{b_0e}{a_0}-\tan^{-1}\frac{b_0e}{a_0}\right).$$

Now $$M=\tfrac{4}{3}\pi\rho b_0^3\sqrt{1-e^2}.$$

Hence we easily obtain

$$X=4\pi\rho\xi\left(\frac{1}{e^2}-\frac{\sqrt{1-e^2}}{e^3}\sin^{-1}e\right) \qquad (7).$$

For one of the components perpendicular to the axis we have

$$\tfrac{3}{2}M\eta\int_{\alpha^2}^{\infty}\frac{d(\phi^2)}{(b_0^2+\phi^2)^2\sqrt{a_0^2+\phi^2}};$$

which, when the point is on the surface, becomes

$$Y=\tfrac{3}{2}M\eta\int_0^{\infty}\frac{d(\phi^2)}{(b_0^2+\phi^2)^2\sqrt{a_0^2+\phi^2}} \qquad (8)$$

The definite integral is easily reduced to

$$2\int_{a_0}^{\infty}\frac{dz}{(b_0^2e^2+z^2)^2}=\frac{1}{b_0^3e^3}\left(\tan^{-1}\frac{b_0e}{a_0}-\frac{a_0e}{b_0}\right).$$

Hence we have

$$Y=2\pi\rho\eta\left(\frac{\sqrt{1-e^2}}{e^3}\sin^{-1}e-\frac{1-e^2}{e^2}\right) \qquad (9).$$

Maclaurin's theorem

523. From what we have already given of the analysis of this question, it is easy to deduce the following splendid theorem, due to Maclaurin:—

The attractions exerted by two homogeneous and confocal ellip-

soids on the same point external to each, or external to one and on the surface of the other, are in the same direction and proportional to their masses. Maclaurin's theorem.

The x component is, as above,

$$\tfrac{3}{2}M\xi\int_{\alpha^2}^{\infty}\frac{d(\phi^2)}{\sqrt{(a_0^2+\phi^2)^3(b_0^2+\phi^2)(c_0^2+\phi^2)}}$$

with α^2 the positive root of

$$\frac{\xi^2}{a_0^2+\phi^2}+\frac{\eta^2}{b_0^2+\phi^2}+\frac{\zeta^2}{c_0^2+\phi^2}=1.$$

For a confocal ellipsoid the axes are

$$a_1^2=a_0^2+h,\ b_1^2=b_0^2+h,\ c_1^2=c_0^2+h.$$

And the x component is

$$\tfrac{3}{2}M_1\xi\int_{\alpha_1^2}^{\infty}\frac{d(\phi^2)}{\sqrt{(a_1^2+\phi^2)^3(b_1^2+\phi^2)(c_1^2+\phi^2)}}$$

where α_1^2 is the positive root of

$$\frac{\xi^2}{a_1^2+\phi^2}+\frac{\eta^2}{b_1^2+\phi^2}+\frac{\zeta^2}{c_1^2+\phi^2}=1.$$

If we put $h+\phi^2$ for ϕ^2 in the first integral and in the equation for its limit, we reproduce the second and its limiting equation. Hence the integrals are equal, and corresponding components of the attraction are as M to M_1.

524. In a similar way we may at once prove Ivory's theorem— Ivory's theorem.

Let corresponding points P, p, be taken on the surfaces of two homogeneous confocal ellipsoids, E, e. The x component of the attraction of E on p, is to that of e on P as the area of the section of E by the plane of yz is to that of the coplanar section of e.

The x component of M on ξ, η, ζ, is already given [§ 522 (5)].

That of M_1 on $\frac{a_0}{a_1}\xi,\ \frac{b_0}{b_1}\eta,\ \frac{c_0}{c_1}\zeta$, is

$$\tfrac{3}{2}M_1\frac{a_0}{a_1}\xi\int_{\alpha_1^2}^{\infty}\frac{d(\phi^2)}{\sqrt{(a_1^2+\phi^2)^3(b_1^2+\phi^2)(c_1^2+\phi^2)}}$$

with the condition that α_1^2 is the positive root of

$$\frac{\xi^2}{a_1^2+\phi^2}+\frac{\eta^2}{b_1^2+\phi^2}+\frac{\zeta^2}{c_1^2+\phi^2}=1,$$

or

$$\frac{\xi^2}{a_0^2+h+\phi^2}+\frac{\eta^2}{b_0^2+h+\phi^2}+\frac{\zeta^2}{c_0^2+h+\phi^2}=1.$$

Now the integrals are evidently equal—and the whole expressions are as M to $M_1\frac{a_0}{a_1}$; *i.e.*, as b_0c_0 to b_1c_1.

Ivory's theorem.

Poisson showed that this theorem is true for any law of force whatever. This is easily proved by employing in the general expressions for the components of the attraction of any body, after *one* integration, the properties of corresponding points upon confocal ellipsoids (§ 519).

Law of attraction when a uniform spherical shell exerts no action on an internal point.

525. An ingenious application of Ivory's theorem, by Duhamel, must not be omitted here. Concentric spheres are a particular case of confocal ellipsoids, and therefore the attraction of any sphere on a point on the surface of an internal concentric sphere, is to that of the latter upon a point in the surface of the former as the squares of the radii of the spheres. Now *if the law of attraction be such that a homogeneous spherical shell of uniform thickness exerts no attraction on an internal point*, the action of the larger sphere on the internal point is reducible to that of the smaller. Hence *the law is that of the inverse square of the distance*, as is easily seen by making the smaller sphere less and less till it becomes a mere particle. This theorem is due originally to Cavendish.

Centre of gravity.

526. (*Definition.*) If the action of terrestrial or other gravity on a rigid body is reducible to a single force in a line passing always through one point fixed relatively to the body, whatever be its position relatively to the earth or other attracting mass, that point is called its *centre of gravity*, and the body is called a *centrobaric body*.

Centrobaric bodies, proved possible by Green.

527. One of the most startling results of Green's wonderful theory of the potential is its establishment of the existence of centrobaric bodies; and the discovery of their properties is not the least curious and interesting among its very various applications.

Properties of centrobaric bodies.

528. If a body (B) is centrobaric relatively to any one attracting mass (A), it is centrobaric relatively to every other: and it attracts all matter external to itself as if its own mass were collected in its centre of gravity.[1]

Let O be any point so distant from B that a spherical surface described from it as centre, and not containing any part of B, is large enough entirely to contain A. Let A be placed within any such spherical surface and made to rotate about any axis, OK, through O. It will always attract B in a line through G, the centre of gravity of B. Hence if every particle of its mass

[1] Thomson, Proc. R.S.E., Feb. 1864.

be uniformly distributed over the circumference of the circle that it describes in this rotation, the mass, thus obtained, will also attract B in a line through G. And this will be the case however this mass is rotated round O; since before obtaining it we might have rotated A and OK in any way round O, holding them fixed relatively to one another. We have therefore found a body, A', symmetrical about an axis, OK, relatively to which B is necessarily centrobaric. Now, O being kept fixed, let OK, carrying A' with it, be put successively into an infinite number, n, of positions uniformly distributed round O; that is to say, so that there are equal numbers of positions of OK in all equal solid angles round O: and let $\frac{1}{n}$ part of the mass of A' be left in each of the positions into which it was thus necessarily carried. B will experience from A all this distribution of matter, still a resultant force through G. But this distribution, being symmetrical all round O, consists of uniform concentric shells, and (§ 471) the mass of each of these shells might be collected at O without changing its attraction on any particle of B, and therefore without changing its resultant attraction on B. Hence B is centrobaric relatively to a mass collected at O; this being any point whatever not nearer than within a certain limiting distance from B (according to the condition stated above). That is to say, any point placed beyond this distance is attracted by B in a line through G; and hence, beyond this distance, the equipotential surfaces of B are spherical with G for common centre. B therefore attracts points beyond this distance as if its mass were collected at G: and it follows (§ 497) that it does so also through the whole space external to itself. Hence it attracts any group of points, or any mass whatever, external to it, as if its own mass were collected at G.

Properties of centrobaric bodies.

529. Hence §§ 497, 492 show that—

(*a*) *The centre of gravity of a centrobaric body necessarily lies in its interior*; or in other words, *can only be reached from external space by a path cutting through some of its mass*. And

(*b*) *No centrobaric body can consist of parts isolated from one another, each in space external to all:* in other words, *the outer boundary of every centrobaric body is a single closed surface.*

Properties of centrobaric bodies.

Thus we see, by (a), that no symmetrical ring, or hollow cylinder with open ends, can have a centre of gravity; for its centre of gravity, if it had one, would be in its axis, and therefore external to its mass.

530. *If any mass whatever, M, and any single surface, S, completely enclosing it be given, a distribution of any given amount, M', of matter on this surface may be found which shall make the whole centrobaric with its centre of gravity in any given position (G) within that surface.*

The condition here to be fulfilled is to distribute M' over S, so as by it to produce the potential

$$\frac{M+M'}{EG} - V,$$

any point, E, of S; V denoting the potential of M at this point. The possibility and singleness of the solution of this problem were proved above (§ 499). It is to be remarked, however, that if M' be not given in sufficient amount, an extra quantity must be taken, but neutralized by an equal quantity of negative matter, to constitute the required distribution on S.

The case in which there is no given body M to begin with is important; and yields the following:—

Centrobaric shell.

531. *A given quantity of matter may be distributed in one way, but in only one way, over any given closed surface, so as to constitute a centrobaric body with its centre of gravity at any given point within it.*

Thus we have already seen that the condition is fulfilled by making the density inversely as the distance from the given point, if the surface be spherical. From what was proved in §§ 501, 506 above, it appears also that a centrobaric shell may be made of either half of the lemniscate in the diagram of § 508, or of any of the ovals within it, by distributing matter with density proportional to the resultant force of m at I and m' at I'; and that the one of these points which is within it is its centre of gravity. And generally, by drawing the equipotential surfaces relatively to a mass m collected at a point I, and any other distribution of matter whatever not surrounding this point; and by taking one of these surfaces which encloses I but no other part of the mass, we learn, by Green's general theorem, and the special proposition of § 506, how to distribute matter over it so as to make it a centrobaric shell with I for centre of gravity.

Properties of centrobaric bodies.

532. Under *hydrokinetics* the same problem will be solved for a cube, or a rectangular parallelepiped in general, in terms of converging series; and under *electricity* (in a subsequent volume) it will be solved in finite algebraic terms for the surface of a lense bounded by two spherical surfaces cutting one another at any sub-multiple of two right angles, and for either part obtained by dividing this surface in two by a third spherical surface cutting each of its sides at right angles.

Centrobaric solid.

533. *Matter may be distributed in an infinite number of ways throughout a given closed space, to constitute a centrobaric body with its centre of gravity at any given point within it.*

For by an infinite number of surfaces, each enclosing the given point, the whole space between this point and the given closed surface may be divided into infinitely thin shells; and matter may be distributed on each of these so as to make it centrobaric with its centre of gravity at the given point. Both the forms of these shells and the quantities of matter distributed on them, may be arbitrarily varied in an infinite variety of ways.

Thus, for example, if the given closed surface be the pointed oval constituted by either half of the lemniscate of the diagram of § 508, and if the given point be the point I within it, a centrobaric solid may be built up of the interior ovals with matter distributed over them to make them centrobaric shells as above (§ 531). From what was proved in § 518, we see that a solid sphere, with its density varying inversely as the fifth power of the distance from an external point, is centrobaric, and that its centre of gravity is the *image* (§ 512) of this point relatively to its surface.

The centre of gravity (if it exist) is the centre of inertia.

534. The centre of gravity of a centrobaric body composed of true gravitating matter is its centre of inertia. For a centrobaric body, if attracted only by another infinitely distant body, or by matter so distributed round itself as to produce (§ 499) uniform force in parallel lines throughout the space occupied by it, experiences (§ 528) a resultant force always through its centre of gravity. But in this case this force is the resultant of parallel forces on all the particles of the body, which (see

Properties of centrobaric bodies.

Properties of Matter, below) are rigorously proportional to their masses: and in § 561 it is proved that the resultant of such a system of parallel forces passes through the point defined in § 230, as the centre of inertia.

A body is kinetically symmetrical about its centre of gravity.

535. The moments of inertia of a centrobaric body are equal round all axes through its centre of inertia. In other words (§ 285), all these axes are principal axes, and the body is kinetically symmetrical round its centre of inertia.

Let it be placed with its centre of inertia at a point O (origin of co-ordinates), within a closed surface having matter so distributed over it (§ 499) as to have xyz (which satisfies $\nabla^2(xyz)=0$) for potential at any point (x, y, z) within it. The resultant action on the body is (§ 528) the same as if it were collected at O; that is to say, zero: or, in other words, the forces on its different parts must balance. Hence (§ 551, I., below) if ρ be the density of the body at (x, y, z)

$$\iiint yz.\rho dxdydz=0,\ \iiint zx.\rho dxdydz=0,\ \iiint xy.\rho dxdydz=0.$$

Hence OX, OY, OZ are principal axes; and this, however the body is turned, only provided its centre of gravity is kept at O.

To prove this otherwise, let V denote the potential of the given body at (x, y, z); u any function of x, y, z; and ϖ the triple integral

$$\iiint\left(\frac{du}{dx}\frac{dV}{dx}+\frac{du}{dy}\frac{dV}{dy}+\frac{du}{dz}\frac{dV}{dz}\right)dxdydz,$$

extended through the interior of a spherical surface, S, enclosing all of the given body, and having for centre its centre of gravity. Then, as in Chap. I. App. A, we have

$$\varpi=\iint\partial u\,Vd\sigma-\iiint V\nabla^2 udxdydz$$
$$=\iint\partial V ud\sigma-\iiint u\nabla^2 Vdxdydz.$$

But if m be the whole mass of the given body, and a the radius of S, we have, over the whole surface of S,

$$V=\frac{m}{a},\ \text{and}\ \partial V=-\frac{m}{a^2}.$$

Also [§ 491 (c)] $\quad \nabla^2 V=-4\pi\rho,$

vanishing of course for all points not belonging to the mass of the given body. Hence from the preceding we have

$$4\pi\iiint u\rho dxdydz=\frac{m}{a^2}\iint(a\partial u+u)d\sigma-\iiint V\nabla^2 udxdydz$$

Let now u be any function fulfilling $\nabla^2 u=0$ through the whole space within S; so that, by § 492, we have $\iint\partial ud\sigma=0$, and by

§ 496, $\iint u d\sigma = 4\pi a^2 u_0$, if u_0 denote the value of u at the centre of S. Hence $\iiint u\rho dx dy dz = m u_0$. Properties of centrobaric bodies.

Let, for instance, $u = yz$. We have $u_0 = 0$, and therefore

$$\iiint yz\rho dx dy dz = 0,$$

as we found above. Or let $u = (x^2 + y^2) - (x^2 + z^2)$, which gives $u_0 = 0$; and consequently proves that

$$\iiint (x^2 + z^2)\rho dx dy dz = \iiint (x^2 + y^2)\rho dx dy dz,$$

or the moment of inertia round OY is equal to that round OX, verifying the conclusion inferred from the other result.

536. The *spherical harmonic analysis*, which forms the subject of an Appendix to Chapter I., had its origin in the theory of attraction, treated with a view especially to the figure of the earth; having been first invented for the sake of expressing in converging series the attraction of a body of nearly spherical figure. It is also perfectly appropriate for expressing the potential, or the attraction, of an infinitely thin spherical shell, with matter distributed over it according to any arbitrary law. This we shall take first, being the simpler application. Origin of spherical harmonic analysis.

Let x, y, z be the co-ordinates of P, the point in question, reckoned from O the centre, as origin of co-ordinates: ρ and ρ' the values of the density of the spherical surface at points E and E', of which the former is the point in which it is cut by OP, or this line produced: $d\sigma'$ an element of the surface at E', a its radius. Then, V being the potential at P, we have

$$V = \iint \frac{\rho' d\sigma'}{E'P} \qquad (1).$$

But, by B (48)

$$\left.\begin{aligned} \frac{1}{E'P} &= \frac{1}{a}\left\{1 + \sum_1^\infty Q_i \left(\frac{r}{a}\right)^i\right\} \text{ when } P \text{ is internal,} \\ \text{and} \qquad &= \frac{1}{r}\left\{1 + \sum_1^\infty Q_i \left(\frac{a}{r}\right)^i\right\} \quad ,, \quad ,, \text{ external,} \end{aligned}\right\} \qquad (2)$$

where Q_i is the biaxal surface harmonic of (E, E'). Hence, if

$$\rho' = S_0 + S_1 + S_2 + \text{etc.} \qquad (3)$$

be the harmonic expansion for ρ, we have, according to B (52),

$$\left.\begin{aligned} V &= 4\pi a \left\{\sum_0^\infty \frac{S_i}{2i+1}\left(\frac{r}{a}\right)^i\right\} \text{ when } P \text{ is internal,} \\ \text{and} \qquad &= \frac{4\pi a^2}{r}\left\{\sum_0^\infty \frac{S_i}{2i+1}\left(\frac{a}{r}\right)^i\right\} \quad ,, \quad ,, \text{ external,} \end{aligned}\right\} \qquad (4).$$

If, for instance, $\rho = S_i$, we have

$$V = \frac{4\pi r^i}{a^{i-1}} \frac{S_i}{2i+1} \text{ inside,}$$

Application of spherical harmonic analysis.

and $$V = \frac{4\pi a^{i+2}}{r^{i+1}} \frac{S_i}{2i+1} \text{ outside.}$$

Thus we conclude that

537. A spherical harmonic distribution of density on a spherical surface produces a similar and similarly placed spherical harmonic distribution of potential over every concentric spherical surface through space, external and internal; and so also consequently of radial component force. But the amount of the latter differs, of course (§ 478), by $4\pi\rho$, for points infinitely near one another outside and inside the surface, if ρ denote the density of the distribution on the surface between them.

If R denote the radial component of the force, we have

$$\left.\begin{aligned} R = -\frac{dV}{dr} &= -\frac{4\pi r^{i-1}}{a^{i-1}} \frac{iS_i}{2i+1} \text{ inside,} \\ \text{and} \qquad &= \frac{4\pi a^{i+2}}{r^{i+2}} \frac{(i+1)S_i}{2i+1} \text{ outside.} \end{aligned}\right\} \quad (5)$$

Hence, if $r=a$, we have

$$R\,(\text{outside}) - R\,(\text{inside}) = 4\pi S_i = 4\pi\rho.$$

538. The potential is of course a solid harmonic through space, both internal and external; and is of positive degree in the internal, and of negative in the external space. The expression for the radial component of the force, in each division of space, is reduced to the same form by multiplying it by the distance from the centre.

539. The harmonic development gives an expression in converging series, for the potential of any distribution of matter through space, which is useful in some applications.

Let x, y, z be the co-ordinates of P, the attracted point, and x', y', z' those of P' any point of the given mass. Then, if ρ be the density of the matter at P', and V the potential at P, we have

$$V = \iiint \frac{\rho' dx' dy' dz'}{[(x-x')^2 + (y-y')^2 + (z-z')^2]^{\frac{1}{2}}} \quad (6)$$

The most convenient view we can take as to the space through which the integration is to be extended is to regard it as infinite in all directions, and to suppose ρ' to be a discontinuous function of x', y', z', vanishing through all space unoccupied by matter.

Now by B (u) we have

$$\frac{1}{[(x-x')^2+(y-y')^2+(z-z')^2]^{\frac{1}{2}}}=\frac{1}{r'}\{1+\sum_1^\infty Q_i\left(\frac{r}{r'}\right)^i\} \text{ when } r'>r$$

and $$=\frac{1}{r}\{1+\sum_1^\infty Q_i\left(\frac{r'}{r}\right)^i\} \quad ,, \quad r'<r \qquad (7).$$

Application of spherical harmonic analysis.

Substituting this in (6) we have

$$V=(\iiint)\frac{\rho' dx' dy' dz'}{r'}+\frac{1}{r}[\iiint]\rho' dx' dy' dz'$$

$$+\sum_1^\infty\{r^i(\iiint)Q_i\frac{\rho' dx' dy' dz'}{r'^{i+1}}+\frac{1}{r^{i+1}}[\iiint]Q_i r'^i\rho' dx' dy' dz'\} \qquad (8),$$

where $(\iiint)$ denotes integration through all the space external to the spherical surface of radius r, and $[\iiint]$ integration through the interior space.

Potential of a distant body.

This formula is useful for expressing the attraction of a mass of any figure on a distant point in a single converging series. Thus when OP is greater than the greatest distance of any part of the body from O, the first series disappears, and the expression becomes a single converging series, in ascending powers of $\frac{1}{r}$:—

$$V=\frac{1}{r}\{\iiint\rho' dx' dy' dz'+\Sigma\frac{1}{r^i}\iiint Q_i r'^i\rho' dx' dy' dz'\} \qquad (9).$$

If we use the notation of B (u) (53), this becomes

$$V=\frac{1}{r}\{\iiint\rho' dx' dy' dz'+\sum_1^\infty r^{-2i}\iiint\rho' H_i[(x,y,z),(x',y',z')]dx' dy' dz'\} \qquad (10)$$

and we have, by B (v') and (w),

$$H_i[(x,y,z),(x',y',z')]=\frac{1.3.5\ldots(2i-1)}{1.2.3\ldots i}[\cos^i\theta-\frac{i(i-1)}{2.(2i-1)}\cos^{i-2}\theta+\frac{i(i-1)(i-2)(i-3)}{2.4.(2i-1)(2i-3)}\cos^{i-4}\theta-\text{etc.}]r^i r'^i \qquad (11)$$

where $$\cos\theta=\frac{xx'+yy'+zz'}{rr'}.$$

From this we find

$$H_1=xx'+yy'+zz'; \quad H_2=\tfrac{3}{2}[(xx'+yy'+zz')^2-\tfrac{1}{3}(x^2+y^2+z^2)(x'^2+y'^2+z'^2)];$$

and so on.

Let now M denote the mass of the body; and let O be taken at its centre of gravity. We shall have

$$\iiint\rho' dx' dy' dz'=M; \text{ and } \iiint\rho' H_1 dx' dy' dz'=0.$$

Further, let OX, OY, OZ be taken as principal axes (§§ 281, 282), so that $$\iiint\rho' y' z' dx' dy' dz'=0, \text{ etc.},$$

and let A, B, C be the moments of inertia round these axes. This will give

$$H_2\rho' dx' dy' dz'=\tfrac{1}{2}\{(3x^2-r^2)\iiint\rho' x'^2 dx' dy' dz'+\text{etc.}\}=\tfrac{1}{2}\{(3x^2-r^2)[\tfrac{1}{2}(A+B+C)-A]+\text{etc.}\}$$

$$A(r^2-3x^2)+B(r^2-3y^2)C+(r^2-3z^2)\}=\tfrac{1}{2}\{(B+C-2A)x^2+(C+A-2B)y^2+(A+B-2C)z^2\}.$$

Potential of a distant body.

Hence neglecting terms of the third and higher orders of small quantities (powers of $\frac{r'}{r}$), we have the following approximate expression for the potential:—

$$V=\frac{M}{r}+\frac{1}{2r^3}\{(B+C-2A)x^2+(C+A-2B)y^2+(A+B-2C)z^2\} \quad (12)$$

As one example of the usefulness of this result, we may mention the investigation of the disturbance in the moon's motion produced by the non-sphericity of the earth, and of the reaction of the same disturbing force on the earth, causing *lunar nutation and precession*, which will be explained later.

Differentiating, and retaining only terms of the first and second degrees of approximation, we have for the components of the mutual force between the body and a unit particle at (x, y, z),

$$X=\frac{Mx}{r^3}-\frac{(B+C-2A)x}{r^5}+\frac{5}{2}\frac{x}{r^7}[(B+C-2A)x^2+(C+A-2B)y^2+(A+B-2C)z^2]$$

$$Y=\text{etc.},\quad Z=\text{etc.}$$

whence

$$Zy-Yz=3\frac{(C-B)yz}{r^5},\quad Xz-Zx=3\frac{(A-C)zx}{r^5},\quad Yx-Xy=3\frac{(B-A)xy}{r^5}, \quad (14)$$

Comparing these with Chap. IX., below, we conclude that

Attraction of a particle on a distant body.

540. The attraction of a distant particle, P, on a rigid body if transferred (according to Poinsot's method explained below, § 555) to the centre of inertia, I, of the latter, gives a couple approximately equal and opposite to that which constitutes the resultant effect of centrifugal force, if the body rotates with a certain angular velocity about IP. The square of this angular velocity is inversely as the cube of the distance of P, irrespectively of its direction; being numerically equal to three times the reciprocal of the cube of this distance, if the unit of mass is such as to exercise the proper kinetic unit (§ 225) force on another equal mass at unit distance. The general tendency of the gravitation couple is to bring the principal axis of least moment of inertia into line with the attracting point. The expressions for its components round the principal axes will be used later (Chap. IX.) for the investigation of the phenomena of precession and nutation produced, in virtue of the earth's non-sphericity, by the attractions of the sun and moon; and (Chap. IX.) to estimate the retardation produced by tidal friction against the earth's rotation, according to the principle explained above (§ 276).

541. It appears from what we have seen that the amount of the gravitation couple is inversely as the cube of the distance between the centre of inertia and the external attracting point: and therefore that the shortest distance of the line of the resultant force from the centre of inertia varies inversely as the distance of the attracting point. We thus see *how* to a first approximation every rigid body is centrobaric relatively to a distant attracting point.

Attraction of a particle on a distant body.

Principle of the approximation used in the common theory of the centre of gravity.

542. The real meaning and value of the spherical harmonic method for a solid mass will be best understood by considering the following application:—

Let $$\rho = F(r) S_i \qquad (15)$$
where $F(r)$ denotes any function of r, and S_i a surface spherical harmonic function of order i, with coefficients independent of r. Substituting accordingly for ρ' in (8), and attending to B. (52) and (16), we find

$$V = \frac{4\pi S_i}{2i+1}\{r^i \int_r^\infty r'^{-i+1} F(r')\,dr' + r^{-i-1}\int_0^r r'^{i+2} F(r')\,dr'\} \qquad (16).$$

543. As an example, let it be required to find the potential of a solid sphere of radius a, having matter distributed through it according to a solid harmonic function, V_i.

Potential of solid sphere with harmonic distribution of density.

That is to say, let

$$\rho = V_i = r^i S_i, \text{ when } r < a,$$
and $$\rho' = 0 \qquad ,, \quad r > a.$$
Hence in the preceding formula $F(r) = r^i$ from $r=0$ to $r=a$, and $F(r)=0$, when $r>a$; and it becomes

$$\left.\begin{aligned} V &= 4\pi V_i\left\{\frac{a^2}{2(2i+1)} - \frac{r^2}{2(2i+3)}\right\} \text{ when } P \text{ is internal,} \\ \text{and } &= \frac{4\pi}{(2i+1)(2i+3)}\,\frac{a^{2i+3} V_i}{r^{2i+1}} \quad ,, \quad ,, \text{ external.}\end{aligned}\right\} \qquad (17).$$

This result may also be obtained by the aid of the algebraical formula B (12) thus, on the same principle as the potential of a uniform spherical shell was found in § 491 (*d*).

We have by § 491 (*c*)

$$\left.\begin{aligned} \nabla^2 V &= -4\pi V_i, \text{ when } r < a, \\ \text{and } &= 0 \qquad ,, \quad r > a. \end{aligned}\right\} \qquad (18).$$

But by taking $m=2$ in B (12) we have

$$\nabla^2(r^2 V_i) = 2(2i+3) V_i,$$

Potential of solid sphere with harmonic distribution of density.

and therefore the solution of the equation

$$\nabla^2 V = -4\pi V_i$$

is

$$V = -4\pi \frac{r^2 V_i}{2(2i+3)} + U \qquad (19)$$

where U is any function whatever satisfying the equation

$$\nabla^2 U = 0$$

through the whole interior of the sphere. By choosing U and the external values of V so as to make the values of V equal to one another for points infinitely near one another outside and inside the bounding surface, to fulfil the same condition for $\frac{dV}{dr}$, and to make V vanish when $r=\infty$, and when $r=0$, we find

$$U = 4\pi V_i \frac{a^2}{2(2i+1)},$$

and obtain the expression of (17) for V external. For in the first place, V external and U must clearly be $A\frac{V_i}{r^{2i+1}}$, and BV_i, where A and B are constants: and the two conditions give the equations to determine them.

Potential of any mass, in harmonic series.

544. From App. B. (52) it follows immediately that any function of x, y, z whatever may be expressed, through the whole of space, in a series of surface harmonic functions, each having its coefficients functions of the distance (r) from the origin. Hence (16), with S_i placed under the sign of integration for r', gives the harmonic development of the potential of any mass whatever; being the result of the triple integrations indicated in (8) of § 539, when the mass is specified by means of a harmonic series expressing the density.

Application to figure of the earth.

545. The most important application of the harmonic development for solid spheres hitherto made is for investigating, in the Theory of the Figure of the Earth, the attraction of a finite mass consisting of approximately spherical layers of matter equally dense through each, but varying in density from layer to layer. The result of the general analytical method explained above, when worked out in detail for this case, is to exhibit the potential as the sum of two parts, of which the first and chief is the potential due to a solid sphere, A, and the second to a spherical shell, B. The sphere, A, is obtained by reducing the given spheroid to a spherical figure by cutting away all the matter lying outside the proper mean spherical surface, and filling the space vacant inside it where the original spheroid lies within it, without altering the density anywhere. The shell, B, is a spherical surface loaded with equal quantities of positive and negative matter, so as to compensate for the transference of matter by which the given spheroid was changed into A. The analytical expression of all this may be written down immediately from the preceding formulæ (§§ 536, 537); but we reserve it until, under hydrostatics and hydrokinetics, we shall be occupied with the theory of the Figure of the Earth, and of the vibrations of liquid globes.

Application to figure of the earth.

546. The analytical method of spherical harmonics is very valuable for several practical problems of electricity, magnetism, and electro-magnetism, in which distributions of force symmetrical round an axis occur: especially in this; that if the force (or potential) at every point through some finite length along the axis be given, it enables us immediately to deduce converging series for calculating the force for points through some finite space not in the axis. (See § 498.)

Case of the potential symmetrical about an axis.

O being any conveniently chosen point of reference, in the axis of symmetry, let us have, in series converging for a portion AB of the axis,

$$U=a_0+\frac{b_0}{r}+a_1r+\frac{b_1}{r^2}+a_2r^2+\frac{b_2}{r^3}+\text{etc.} \qquad (a)$$

where U is the potential at a point, Q, in the axis, specified by $OQ=r$. Then if V be the potential at any point P, specified by $OP=r$ and $QOP=\theta$, and, as in App. B. (47), Q_1, Q_2, ... denote the axial surface harmonics of θ, of the successive integral orders, we must have, for all values of r for which the series converges,

$$V=a_0+\frac{b_0}{r}+\left(a_1r+\frac{b_1}{r^2}\right)Q_1+\left(a_2r^2+\frac{b_2}{r^3}\right)Q_2+\text{etc.} \qquad (b)$$

provided P can be reached from Q and all points of AB within some finite distance from it however small, without passing through any of the matter to which the force in question is due, or any space for which the series does not converge. For throughout this space (§ 498) $V-V'$ must vanish, if V' be the [illegible]lue of the sum of the series; since $V-V'$ is [App. B. (g)] [illegible]ntial function, and it vanishes for a finite portion of the [illegible]taining Q.

[illegible] (b) is of course convergent for all values of r which [illegible]gent, since the ultimate ratio $Q_{i+1}\div Q_i$ for in-

Potential symmetrical about an axis.

finitely great values of i, is unity, as we see from any of the expressions for these functions in App. B.

In general, that is to say unless O be a singular point, the series for U consists, according to Maclaurin's theorem, of ascending integral powers of r only, provided r does not exceed a certain limit. In certain classes of cases there are singular points, such that if O be taken at one of them, U will be expressed in a series of powers of r with fractional indices, convergent and real at least for all finite positive values of r not exceeding a certain limit. The expression for the potential in the neighbourhood of O in any such case, in terms of solid spherical harmonics relatively to O as centre, will contain harmonics [App. B. (a)] of fractional degrees.

Examples.

Examples—(I.) The potential of a circular ring of radius a and linear density ρ, at a point in the axis, distant by r from the centre, is

$$\frac{2\pi a\rho}{(a^2+r^2)^{\frac{1}{2}}}.$$

Hence $$U=2\pi\rho\left(1-\tfrac{1}{2}\frac{r^2}{a^2}+\frac{1.3}{2.4}\frac{r^4}{a^4}-\text{etc.}\right) \text{ when } r<a,$$

and $$U=\frac{2\pi a\rho}{r}\left(1-\tfrac{1}{2}\frac{a^2}{r^2}+\frac{1.3}{2.4}\frac{a^4}{r^4}-\text{etc.}\right) \text{ when } r>a,$$

from which we have

$$V=2\pi\rho\left(1-\tfrac{1}{2}\frac{r^2}{a^2}Q_2+\frac{1.3}{2.4}\frac{r^4}{a^4}Q_4-\text{etc.}\right) \text{ when } r<a,$$

and $$V=2\pi\rho\left(\frac{a}{r}-\tfrac{1}{2}\frac{a^3}{r^3}Q_2+\frac{1.3}{2.4}\frac{a^5}{r^5}Q_4-\text{etc.}\right) \text{ when } r>a.$$

(II.) Multiplying the preceding unexpanded expression for U by da, and integrating with reference to a from $a=0$ as lower limit, and now calling U the potential of a circular disc of uniform surface density ρ, and radius a, at a point in its axis, we find

$$U=2\pi\rho\{(a^2+r^2)^{\frac{1}{2}}-r\},$$

r being positive.

Hence, expanding first in ascending, and secondly in descending powers of r, for the cases of $r<a$ and $r>a$, we find

$$V=2\pi\rho\left\{-rQ_1+a+\tfrac{1}{2}\frac{r^2}{a}Q_2-\frac{1.1}{2.4}\frac{r^4}{a^3}Q_4+\frac{1.1.3}{2.4.6}\frac{r^6}{a^5}Q_6-\text{etc.}\right\} \text{ when } r<a.$$

and $$V=2\pi\rho\left\{\tfrac{1}{2}\frac{a^2}{r}-\frac{1.1}{2.4}\frac{a^4}{r^3}Q_2+\frac{1.1.3}{2.4.6}\frac{a^6}{r^5}Q_4-\text{etc.}\right\} \text{ when } r>a.$$

It must be remarked that the first of these expressions is only continuous from $\theta=0$ to $\theta=\frac{1}{2}\pi$; and that from $\theta=\frac{1}{2}\pi$ to $\theta=\pi$ the first term of it must be made

$$+2\pi\rho rQ_1, \text{ instead of } -2\pi\rho rQ_1.$$

(III.) Again, taking $\frac{-d}{dr}$ of the expression for U in (II.), and now calling U the potential of a disc of infinitely small thickness c with positive and negative matter of surface density $\frac{\rho}{c}$ on its two sides, we have

Potential symmetrical about an axis.

$$U=2\pi\rho\{1-\frac{r}{(a^2+r^2)^{\frac{1}{2}}}\},$$

[obtainable also from § 479 (e), by integrating with reference to x, putting r for x, and ρ for ρc]. Hence for this case

$$V=2\pi\rho(1-\frac{r}{a}Q_1+\frac{1}{2}\frac{r^2}{a^2}Q_2-\frac{1.3}{2.4}\frac{r^4}{a^4}Q_4+\text{etc.})\text{ when } r<a,$$

Potential in the neighbourhood of a circular galvanometer coil.

and $$V=2\pi\rho(\frac{1}{2}\frac{a^2}{r^2}Q_1-\frac{1.3}{2.4}\frac{a^4}{r^4}Q_3+\text{etc.})\text{ when } r>a.$$

The first of these expressions also is discontinuous; and when θ is $>\frac{1}{2}\pi$ and $<\pi$, its first term must be taken as $-2\pi\rho$ instead of $2\pi\rho$.

547. If two systems, or distributions of matter, M and M', given in spaces each finite, but infinitely far asunder, be allowed to approach one another, a certain amount of work is obtained by mutual gravitation: and their mutual potential energy loses, or as we may say *suffers exhaustion*, to this amount: which amount will (§ 486) be the same by whatever paths the changes of position are effected, provided the relative initial positions and the relative final positions of all the particles are given. Hence if m_1, m_2, ... be particles of M; m'_1, m'_2, ... particles of M'; v'_1, v'_2, ... the potentials due to M' at the points occupied by m_1, m_2, ...; v_1, v_2, ... those due to M at the points occupied by m'_1, m'_2, ...; and E the exhaustion of mutual potential energy between the two systems in any actual configurations; we have

Exhaustion of potential energy

$$E=\Sigma mv'=\Sigma m'v.$$

This may be otherwise written, if ρ denote a discontinuous function, expressing the density at any point, (x, y, z) of the mass M, and vanishing at all points not occupied by matter of this distribution, and if ρ' be taken to specify similarly the other mass M'. Thus we have

$$E=\iiint\rho v'dxdydz=\iiint\rho' vdxdydz,$$

the integrals being extended through all space. The equality of the second and third members here is verified by remarking that

Exhaustion of potential energy.

$$v=\iiint\frac{{}_{,}\rho d_{,}xd_{,}yd_{,}z}{D},$$

if D denote the distance between (x, y, z) and $({}_{,}x, {}_{,}y, {}_{,}z)$, the latter being any point of space, and ${}_{,}\rho$ the value of ρ at it. A corresponding expression of course gives v': and thus we find a sextuple integral to express identically the second and third members, or the value of E, as follows:—

$$E=\iiiiiint\frac{{}_{,}\rho\rho' d_{,}xd_{,}yd_{,}zdxdydz}{D}.$$

Green's method.

548. It is remarkable that it was on the consideration of an analytical formula which, when properly interpreted with reference to two masses, has precisely the same signification as the preceding expressions for E, that Green founded his whole structure of general theorems regarding attraction.

In App. A. (a) let a be constant, and let U, U' be the potentials at (x, y, z) of two finite masses, M, M', finitely distant from one another: so that if ρ and ρ' denote the densities of M and M' respectively at the point (x, y, z), we have [§ 491 (c)]

$$\nabla^2 U=-4\pi\rho,\ \nabla^2 U'=-4\pi\rho'.$$

It must be remembered that ρ vanishes at every point not forming part of the mass M: and so for ρ' and M'. In the present merely abstract investigation the two masses may, in part or in whole, jointly occupy the same space: or they may be merely imagined subdivisions of the density of one real mass. Then, supposing S to be infinitely distant in all directions, and observing that $U\partial U'$ and $U'\partial U$ are small quantities of the order of the inverse cube of the distance of any point of S from M and M', whereas the whole area of S over which the surface integrals of App. A. (a) (1) are taken is infinitely great, only of the order of the square of the same distance, we have

$$\iint dSU'\partial U=0, \text{ and } \iint dSU\partial U'=0.$$

Hence (a) (1) becomes

$$\iiint\left(\frac{dU}{dx}\frac{dU'}{dx}+\frac{dU}{dy}\frac{dU'}{dy}+\frac{dU}{dz}\frac{dU'}{dz}\right)dxdydz=4\pi\iiint\rho U'dxdydz=4\pi\iiint\rho' Udxdydz,$$

showing that the first member divided by 4π is equal to the exhaustion of potential energy accompanying the approach of the two masses from an infinite mutual distance to the relative position which they actually occupy.

Without supposing S infinite, we see that the second member of (a) (1), divided by 4π, is the direct expression for the exhaustion of mutual energy between M' and a distribution consisting of the part of M within S and a distribution over S, of density $\frac{1}{4\pi}\partial U'$; and the third member the corresponding expression for M and derivations from M'.

Exhaustion of potential energy,

549. If, instead of two distributions, M and M', two particles, m_1, m_2 alone be given; the exhaustion of mutual potential energy in allowing them to come together from infinity, to any distance $D(1, 2)$ asunder, is

in allowing condensation of diffused matter.

$$\frac{m_1 m_2}{D(1,2)}.$$

If now a third particle m_3 be allowed to come into their neighbourhood, there is a further exhaustion of potential energy amounting to

$$\frac{m_1 m_3}{D(1,3)}+\frac{m_2 m_3}{D(2,3)}.$$

By considering any number of particles coming thus necessarily into position in a group, we find for the whole exhaustion of potential energy

$$E=\Sigma\Sigma\frac{mm'}{D}$$

where m, m' denote the masses of any two of the particles, D the distance between them, and $\Sigma\Sigma$ the sum of the expressions for all the pairs, each pair taken only once. If v denote the potential at the point occupied by m, of all the other masses, the expression becomes a simple sum, with as many terms as there are masses, which we may write thus—

$$E=\tfrac{1}{2}\Sigma mv;$$

the factor $\frac{1}{2}$ being necessary, because Σmv takes each such term as $\frac{m_1 m_2}{D(1,2)}$ twice over. If the particles form an ultimately continuous mass, with density ρ at any point (x, y, z), we have only to write the sum as an integral; and thus we have

$$E=\tfrac{1}{2}\iiint\rho v\,dx\,dy\,dz$$

as the exhaustion of potential energy of gravitation accompanying the condensation of a quantity of matter from a state of infinite diffusion (that is to say, a state in which the density is everywhere infinitely small) to its actual condition in any finite body.

Exhaustion of potential energy.

An important analytical transformation of this expression is suggested by the preceding interpretation of App. A. (a): by which we find[1]

$$E=\frac{1}{8\pi}\iiint\left(\frac{dv^2}{dx^2}+\frac{dv^2}{dy^2}+\frac{dv^2}{dz^2}\right)dxdydz,$$

or

$$E=\frac{1}{8\pi}\iiint R^2 dxdydz,$$

if R denote the resultant force at (x, y, z), the integration being extended through all space.

Detailed interpretations in connexion with the theory of energy, of the remainder of App. A., with a constant, and of its more general propositions and formulæ not involving this restriction, especially of the minimum problems with which it deals, are of importance with reference to the dynamics of incompressible fluids, and to the physical theory of the propagation of electric and magnetic force through space occupied by homogeneous or heterogeneous matter; and we intend to return to it when we shall be specially occupied with these subjects.

Gauss's method.

550. The manner in which Gauss independently proved Green's theorems is more immediately and easily interpretable in terms of energy, according to the commonly-accepted idea of forces acting simply between particles at a distance without any assistance or influence of interposed matter. Thus, to prove that a given quantity, Q, of matter is distributable in one and only one way over a given single finite surface S (whether a closed or an open shell), so as to produce equal potential over the whole of this surface, he shows (1.) that the integral

$$\iiiint\frac{\rho\rho' d\sigma d\sigma'}{PP'}$$

has a minimum value, subject to the condition

$$\iint \rho d\sigma = Q,$$

where ρ is a function of the position of a point, P, on S, ρ' its value at P', and $d\sigma$ and $d\sigma'$ elements of S at these points: and (2.) that this minimum is produced by only one determinate distribution of values of ρ. By what we have just seen (§ 549) the first of these integrals is double the potential energy of a distribution over S of an infinite number of infinitely small mutually repelling particles: and hence this minimum problem

[1] Nichol's *Cyclopædia*, 2d Ed. 1860. Magnetism, Dynamical Relations of.

is (§ 292) merely an analytical statement of the problem to find how these particles must be distributed to be in stable equilibrium.

Equilibrium of repelling particles enclosed in a rigid smooth surface.

Similarly, Gauss's second minimum problem, of which the preceding is a particular case, and which is, to find ρ so as to make

$$\iint(\tfrac{1}{2}v-\Omega)\rho d\sigma$$

a minimum, subject to

$$\iint\rho d\sigma = Q,$$

where Ω is any given arbitrary function of the position of P, and

$$v=\iint\frac{\rho' d\sigma'}{PP'},$$

is merely an analytical statement of the question, how must a given quantity of repelling particles confined to a surface S be distributed so as to make the whole potential energy due to their mutual forces, and to the forces exerted on them by a given fixed attracting or repelling body (of which Ω is the potential at P), be a minimum; in other words (§ 292), to find how the moveable particles will place themselves, under this influence.

CHAPTER VII.

STATICS OF SOLIDS AND FLUIDS.

Rigid body, **551.** As already explained, § 454, the present application requires the third law of motion and its consequences. These are embodied, sufficiently for our present purpose, in the general statement of Lagrange, § 293, with its extension in § 452, to forces of friction. We commence with the case of a *rigid* body or system, by which we understand a group of material particles which are maintained by their mutual forces in definite positions relatively to each other, though not with reference to external bodies. This is approximately true of all solid bodies, so long as the applied forces are not sufficiently powerful to overcome to a sensible extent the intense molecular reactions. We commence with a general investigation; simpler methods for special cases will be introduced later.

moved without rotation; I. Suppose the body to be displaced by *translation* merely in any direction whatever. Then if all the forces be resolved parallel to this line, each component will have wrought through the same space. But on account of the equilibrium, no work is done; and therefore the sum of the components is zero.

Thus, for equilibrium of a rigid body, it is necessary that the (algebraic) sum of the components of all the forces acting on the body, in *any* direction whatever, *vanish*.

It is sufficient, to insure this, that the sum vanish in each of any three assumed directions not in one plane; for any translation may be resolved parallel to three such lines. In practice the three directions usually selected are at right angles to each other.

rotated. II. Suppose the body to be displaced, by *rotation* merely, through an infinitely small angle about *any* axis. Then (§ 240 the whole work done by the forces is the product of the angle of rotation by the (algebraic) sum of the moments of the forces about the axis. This must vanish, on account of the equilibrium. Rigid body.

Hence, for equilibrium of a rigid body, it is necessary that the sum of the moments of the forces about *any* axis whatever shall vanish.

Since a rotation about *any* axis whatever is equivalent to translations parallel to, and rotations about, any three axes (not in one plane) passing through one point—it is sufficient, in addition to the above condition (I.) among the forces, that the sums of their moments about any three rectangular axes meeting in a point should vanish.

III. When the forces are all in one plane, the above six conditions are reduced to three, viz. :— Forces in one plane.

The sums of the components in any two (rectangular) directions must separately vanish.

The sum of the moments about any axis perpendicular to the plane of the forces must vanish.

(*a*) If X, Y, Z be the rectangular components of the force which is applied at the point P, (x, y, z) of the rigid body, and if the body be displaced by a *translation* merely, whose components parallel to the axes are ξ, η, ζ; we evidently have for the work done the expression Analytical statement.

$$\xi\Sigma(X)+\eta\Sigma(Y)+\zeta\Sigma(Z).$$

By Lagrange's form of Newton's principle, this must vanish if the body is in equilibrium under the action of the applied forces. Now ξ, η, ζ are independent, and may have any values *whatever*. Hence we must have

$$\Sigma(X)=0,\ \Sigma(Y)=0,\ \Sigma(Z)=0,$$

which are, in fact, the same as the conditions already obtained (§ 455) for the equilibrium of a particle.

(*b*) Next, suppose the body to be displaced by rotation about *any* axis. As the origin is any point whatever, we may assume the axis to pass through it. Let the rotation be equivalent to infinitely small simultaneous rotations ω_x, ω_y, ω_z about the axes of x, y, and z (§ 96). Then at once, by § 95, we see that the displacement of P has for components parallel to the axes the quantities

$$\omega_y z-\omega_z y,\quad \omega_z x-\omega_x z,\quad \omega_x y-\omega_y x.$$

General conditions of equilibrium of a rigid body.

Hence the whole work done by the applied forces is

$$\Sigma\{X(\omega_y z-\omega_z y)+Y(\omega_z x-\omega_x z)+Z(\omega_x y-\omega_y x)\}$$

or $$\omega_x\Sigma(Zy-Yz)+\omega_y\Sigma(Xz-Zx)+\omega_z\Sigma(Yx-Xy).$$

Since this must vanish, and since ω_x, ω_y, ω_z are necessarily independent, we have the three additional conditions

$$\Sigma(Zy-Yz)=0,\quad \Sigma(Xz-Zx)=0,\quad \Sigma(Yx-Xy)=0.$$

(*c*) If there be not equilibrium, we evidently have as the resultant of the forces—

I. Forces $\Sigma(X)$, $\Sigma(Y)$, $\Sigma(Z)$ applied at the origin (which may be any point whatever) in lines parallel to the axes. Let R be their resultant, and l, m, n its direction-cosines.

II. Moments, or Couples,

$$\Sigma(Zy-Yz),\quad \Sigma(Xz-Zx),\quad \Sigma(Yx-Xy)$$

about the axes respectively.

These may be represented by $G\lambda$, $G\mu$, $G\nu$, where $\lambda^2+\mu^2+\nu^2=1$. The physical meaning of this will appear presently.

(*d*) When the forces are in one plane, that of xy for instance, we have only three conditions, viz.,

$$\Sigma(X)=0,\quad \Sigma(Y)=0,\quad \text{and}\quad \Sigma(Yx-Xy)=0.$$

Resultant of any set of forces.

552. If any forces whatever act on a rigid body and do not produce equilibrium, it is evident that their resultant (whatever it may be) must, if reversed, and applied simultaneously with them, produce equilibrium. This reversed resultant must therefore, in general, satisfy *six* conditions. Hence it is only in special cases that a number of forces acting on a rigid body are equivalent to a *single* force, but in an infinite variety of ways they can generally be reduced to *two* forces, there being *four* conditions to spare.

Couples.

553. Before proceeding to apply the general conditions of equilibrium, just obtained, it will be useful to introduce the great simplification devised by Poinsot, which consists in the use of *Couples*. In § 234 we have already defined a couple, and shown that the sum of the moments of its forces is the same about all axes perpendicular to its plane. It may therefore be shifted to any new position in its own plane, or in any parallel plane, without alteration of its effect on the rigid body to which it is applied. Its arm may be turned through any angle in the plane of the forces, and the length of the arm and the magnitudes of the forces may be altered at pleasure, with ut changing its statical effect—provided the *moment* remain nchanged. Hence, as in § 234, a couple is completely re- resented by its axis. According to the convention of § 234 he axis of a couple which tends to produce rota- ion in the direction of the hands of a watch, nust be drawn through the *back* of the watch nd *vice versâ*. This may easily be remembered y the help of a simple diagram such as we ive, in which the arrow-heads indicate the direc- ions of rotation, and of the axis, respectively. Couples

554. Couples are to be compounded or resolved by treat- ng their axes by the law of the parallelogram, in a manner dentical with that which we have seen must be employed or linear and angular velocities, and forces. This follows mmediately from § 551, II. above; but may easily be proved y a synthetical process. For, to find the resolved part of couple in any plane, B, inclined to that, A, of the couple, ring, by § 553, the arm of the couple to coincide with the ntersection of the planes. The forces are therefore in A, per- endicular to this intersection, and their resolved parts in B re found by multiplying by the cosine of the angle between he planes. The moment of the resolved parts in B, since the rm is unaltered, is therefore diminished in this proportion. But the angle between two planes is the angle between perpen- liculars to them, *i.e.*, between the axes of the couple and its esolved part. And the length of the axis of the resolved part, eing its moment, is to that of the original couple as the cosine f the inclination of the axes to unity. Composition of couples.

Hence a couple G, whose axis has direction-cosines λ, μ, ν, is equivalent to the three couples $G\lambda$, $G\mu$, $G\nu$ about the axes of x, y, z respectively. And so on, precisely as at the end of § 95. Whence we see the meaning of the symbols in § 551 (c) II.

555. If a force, F, act at any point, A, of a body, it may e transferred to any other point, B, without alteration of its lirection and magnitude, if we introduce a couple whose moment s equal to that of F, when applied at A, about the point B. For, by the principle of superposition of forces, we may intro- luce at B, in the line through it parallel to the given force F, pair of equal and opposite forces F and $-F$. F at A, and Force resolved into force and couple.

Application to equilibrium of rigid body.

—F at B, form the couple just mentioned, and there remains F at B.

From this we have, at once, the conditions of equilibrium of a rigid body already twice investigated, § 551. For, each force may be transferred to any assumed point as origin, if we introduce the corresponding couple. And the forces, which now act at one point, must equilibrate according to the principles of Chap. VI.; while the resultant couple, and therefore its components about any three lines at right angles to each other, must vanish.

Forces represented by the sides of a polygon.

556. Hence forces represented, not merely in magnitude and direction, but in lines of action, by the sides of any closed polygon whether plane or gauche, taken in the same order, are equivalent to a single couple. For when transferred to any origin, they equilibrate, by the Polygon of Forces (§§ 27, 25). When the polygon is plane, twice its area is the moment of the couple; when gauche, the resolved part of the couple about any axis is twice the area of the projection on a plane perpendicular to that axis. The complete couple in this case has its axis perpendicular to the plane (§ 236) on which the projected area is a maximum.

Forces proportional and perpendicular to the sides of a triangle.

557. Lines, perpendicular to the sides of a triangle, and passing through their middle points, meet; and their inclinations are the same as those of the corresponding sides of the triangle, if we take these sides all round in the same order. Hence, if at the middle points of the sides of a triangle, and in its plane, forces be applied tending inwards; and if their magnitudes be proportional to the sides of the triangle, they produce equilibrium. The same is true of any plane polygon as may be at once seen by dividing it into triangles. And it will be shown later, that if forces be applied perpendicular to the faces of any closed polyhedron, at their centres of inertia, tending inwards and proportional to their areas, these also will form an equilibrating system.

Composition of force and couple.

558. A couple and a force in a given line inclined to its plane may be reduced to a smaller couple in a plane perpendicular to the force, and a force equal and parallel to the given force. For the couple may be resolved into two, one in a plane containing the direction of the force, and the other in a plane perpendicular to the force. And that the force and couple which are in the same plane are equivalent to the same force acting in a parallel direction but in a different line, is merely the converse of § 555. Composition of force and couple.

559. Hence any set of forces acting on a rigid body may be resolved into a force at any point and a couple (§ 551). By § 558, these may be reduced to the same force acting in a definite line in the body, and a couple whose plane is perpendicular to the force, and which is the least couple which can appear as part of the resultant of the given set of forces. The definite line in which the force then acts is called the *Central Axis*, and it is obviously the line about which the moment of the given forces is least. Composition of any set of forces acting on a rigid body. Central axis

With the notation of § 551 (c), let us suppose the origin to be changed to the point x', y', z'. The resultant force has still the components $\Sigma(X)$, $\Sigma(Y)$, $\Sigma(Z)$, or Rl, Rm, Rn, parallel to the axes. But the couples are

$$\Sigma[Z(y-y')-Y(z-z')],\quad \Sigma[X(z-z')-Z(x-x')],\quad \Sigma[Y(x-x')-X(y-y')];$$

or

$$G\lambda-R(ny'-mz'),\quad G\mu-R(lz'-nx'),\quad G\nu-R(mx'-ly').$$

The conditions that the resultant force shall be perpendicular to the plane of the resultant couple are

$$\frac{G\lambda-R(ny'-mz')}{l}=\frac{G\mu-R(lz'-nx')}{m}=\frac{G\nu-R(mx'-ly')}{n}.$$

These two equations among x', y', z' are those of the central axis.

We may also obtain them by seeking the conditions that the resultant couple

$$\sqrt{[G\lambda-R(ny'-mz')]^2+[G\mu-R(lz'-nx')]^2+[G\nu-R(mx'-ly')]^2}$$

may be a minimum subject to the independent variations of x', y', z'. This method gives three equations (by the consideration that the partial differential coefficients of the above expression with respect to x', y', z' must separately vanish), which are reducible to the two already obtained; and of which we give the first only. It is

$$n\{G\mu-R(lz'-nx')\}-m\{G\nu-R(mx'-ly')\}=0.$$

It is evident from the simplest properties of couples, but may be easily proved by the above equations, that the resultant couple has the same magnitude at all points of the central axis.

560. By combining the resultant force with one of the forces of the resultant couple, we have obviously an infinite number Reduction to two forces.

Reduction to two forces. of ways of reducing any set of forces acting on a rigid body to *two* forces whose directions do not meet. But there is one case in which the result is symmetrical, and which is therefore worthy of notice.

Symmetrical case. Thus, supposing the central axis of the system has been found, draw a line, AA', at right angles through any point C in it, so that CA may be equal to CA'. For R, acting along the central axis, substitute (by § 561) $\frac{1}{2}R$ at each end of AA'. Thus, choosing this line AA' as the arm of the couple, and calling it a, we have at each extremity of it, two forces, $\frac{G}{a}$ perpendicular to the central axis, and $\frac{1}{2}R$ parallel to the central axis. Compounding these we get two forces, each equal to $(\frac{1}{4}R^2+\frac{G^2}{a^2})^{\frac{1}{2}}$, though A and A' respectively, perpendicular to AA', and equally inclined at the angle $\tan^{-1}\frac{2G}{Ra}$ on the two sides of the plane through AA' and the central axis.

Composition of parallel forces. **561.** A very simple, but important, case, is that of any number of *parallel* forces acting at different points of a rigid body.

Here, for equilibrium, we must obviously have the (algebraic) sum of the forces equal to zero; and their moments about any two axes perpendicular to the common direction of the forces must also vanish.

If P be one of the forces, x, y, z its point of application, l, m, n the direction-cosines of all: R the resultant, acting at $\bar{x}, \bar{y}, \bar{z}$; we have $\Sigma(P)=R$, and

$$\Sigma(Pny-Pmz)=Rn\bar{y}-Rm\bar{z},$$

with two other similar equations.

From these $$\Sigma(P)=R,$$

$$\Sigma(Px)=R\bar{x},\ \Sigma(Py)=R\bar{y},\ \Sigma(Pz)=R\bar{z}.$$

The solution is definite, and indicates a particular point $\bar{x}, \bar{y}, \bar{z}$, whose position is independent of the quantities l, m, n. Hence

If there be not equilibrium, the resultant of such a set of forces is a single force equal to their (algebraic) sum, and acting at a definite point in the body, called the *Centre of Parallel Forces*, whose position depends on the relative magnitudes, and points of application, of the forces, but not upon their common direction.

562. It is obvious, from the formulæ of § 230, that if masses proportional to the forces be placed at the several points of application of these forces, the centre of inertia of these masses will be the same point in the body as the centre of parallel forces. Hence the reactions of the different parts of a rigid body against acceleration in parallel lines are rigorously reducible to one force, acting at the centre of inertia. The same is true approximately of the action of gravity on a rigid body of small dimensions relatively to the earth, and hence the centre of inertia is sometimes (§ 230) called the *Centre of Gravity.* But, except on a centrobaric body (§ 527), gravity is not in general reducible to a single force: and when it is so, this force does not pass through a point fixed relatively to the body in all positions. Centre of gravity.

563. In one case the concluding statement of § 561 must be modified, viz., when the algebraic sum of the given forces vanishes. In this case the resultant is a couple whose plane is parallel to the common direction of the forces. A good example of this is furnished by a magnetized mass of steel, of moderate dimensions, subject to the influence of the earth's magnetism only. As will be shown later, the amounts of the so-called north and south magnetisms in each element of the mass are equal, and are therefore subject to equal and opposite forces, all parallel to the line of dip. Thus a compass-needle experiences from the earth's magnetism merely a couple or *directive* action, and is not attracted or repelled as a whole. Parallel forces whose algebraic sum is zero

564. If three forces, acting on a rigid body, produce equilibrium, their directions must lie in one plane; and must all meet in one point, or be parallel. For the proof we may introduce a consideration which will be very useful to us in investigations connected with the statics of flexible bodies and fluids. Conditions of equilibrium of three forces.

If any forces, acting on a solid, or fluid, body, produce equilibrium, we may suppose any portions of the body to become fixed, or rigid, or rigid and fixed, without destroying the equilibrium. Physical axiom.

Applying this principle to the case above, suppose any two points of the body, respectively in the lines of action of two of the forces, to be fixed—the third force must have no moment about the line joining these points; that is, its direction must pass through the line joining them. As any two points in the lines of action may be taken, it follows that the three forces are coplanar. And three forces, in one plane, cannot equilibrate unless their directions are parallel, or pass through a point.

Equilibrium under the action of gravity.

565. It is easy, and useful, to consider various cases of equilibrium when no forces act on a rigid body but gravity and the pressures, normal or tangential, between it and fixed supports. Thus if one given point only of the body be fixed, it is evident that the centre of gravity must be in the vertical line through this point—else the weight, and the reaction of the support, would form an unbalanced couple. Also for stable equilibrium the centre of gravity must be *below* the point of suspension. Thus a body of any form may be made to stand in stable equilibrium on the point of a needle if we rigidly attach to it such a mass as to cause the joint centre of gravity to be below the point of the needle.

Rocking stones.

566. An interesting case of equilibrium is suggested by what are called Rocking Stones, where, whether by natural or by artificial processes, the lower surface of a loose mass of rock is worn into a convex form which may be approximately spherical, while the bed of rock on which it rests in equilibrium is, whether convex or concave, also approximately spherical, if not plane. A loaded sphere resting on a spherical surface is therefore a type of such cases.

Let O, O' be the centres of curvature of the fixed, and rocking, bodies respectively, when in the position of equilibrium. Take any two infinitely small, equal, arcs PQ, Pp; and at Q make the angle $O'QR$ equal to POp. When, by displacement, Q and p become the points in contact, QR will evidently be vertical; and, if the centre of gravity G, which must be in OPO' when the movable body is in its position of equilibrium, be to the left of QR, the equilibrium will obviously be stable. Hence, if it be below R, the equilibrium is stable, and not unless.

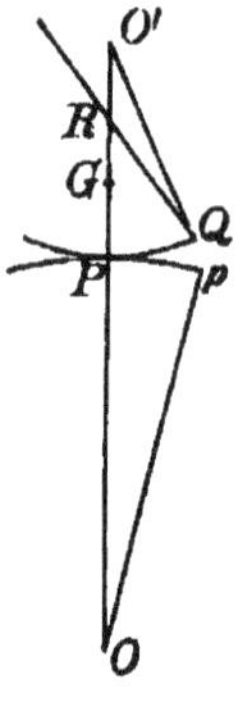

Now if ρ and σ be the radii of curvature OP, $O'P$ of the two surfaces, and θ the angle POp, the angle $QO'R$ will be equal to $\frac{\rho\theta}{\sigma}$; and we have in the triangle $QO'R$ (§ 112

$$RO' : \sigma :: \sin\theta : \sin\left(\theta + \frac{\rho\theta}{\sigma}\right)$$ Rocking stones.

$$:: \sigma : \sigma + \rho \text{ (approximately)}.$$

Hence $$PR = \sigma - \frac{\sigma^2}{\sigma+\rho} = \frac{\rho\sigma}{\rho+\sigma};$$

and therefore, for stable equilibrium,

$$PG < \frac{\rho\sigma}{\rho+\sigma}.$$

If the lower surface be plane, ρ is infinite, and the condition becomes (as in § 291)

$$PG < \sigma.$$

If the lower surface be concave the sign of ρ must be changed, and the condition becomes

$$PG < \frac{\rho\sigma}{\rho-\sigma},$$

which cannot be negative, since ρ *must* be numerically greater than σ in this case.

567. If two points be fixed, the only motion of which the system is capable is one of rotation about a fixed axis. The centre of gravity must then be in the vertical plane passing through those points, and *below* the line joining them. Equilibrium: about an axis,

568. If a rigid body rest on a fixed surface there will in general be only *three* points of contact, § 427; and the body will be in stable equilibrium if the vertical line drawn from its centre of gravity cuts the plane of these three points *within* the triangle of which they form the corners. For if one of these supports be removed, the body will obviously tend to fall towards that support. Hence each of the three prevents the body from rotating about the line joining the other two. Thus, for instance, a body stands stably on an inclined plane (if the friction be sufficient to prevent it from sliding down) when the vertical line drawn through its centre of gravity falls within the base, or area bounded by the shortest line which can be drawn round the portion in contact with the plane. Hence a body, which cannot stand on a horizontal plane, may stand on an inclined plane. on a fixed surface.

569. A curious theorem, due to Pappus, but commonly attributed to Guldinus, may be mentioned here, as it is em- Pappus' theorem.

Pappus' theorem.

ployed with advantage in some cases in finding the centre of gravity of a body—though it is really one of the geometrical properties of the Centre of Inertia. It is obvious from § 230. *If a plane closed curve revolve through any angle about an axis in its plane, the solid content of the surface generated is equal to the product of the area of either end into the length of the path described by its centre of gravity; and the area of the curved surface is equal to the product of the length of the curve into the length of the path described by its centre of gravity.*

570. The general principles upon which forces of constraint and friction are to be treated have been stated above (§§ 293, 329, 452). We add here a few examples for the sake of illustrating the application of these principles to the equilibrium of a rigid body in some of the more important practical cases of constraint.

Mechanical powers.

571. The application of statical principles to the ***Mechanical Powers***, or elementary machines, and to their combinations, however complex, requires merely a statement of their kinematical relations (as in §§ 79, 85, 102, etc.) and an immediate translation into Dynamics by Newton's principle (§ 269); or by Lagrange's Virtual Velocities (§ 289), with special attention to the introduction of forces of friction as in § 452. In no case can this process involve further difficulties than are implied in seeking the geometrical circumstances of any infinitely small disturbance, and in the subsequent solution of the equations to which the translation into dynamics leads us. We will not, therefore, stop to discuss any of these questions; but will take a few examples of no very great difficulty, before for a time quitting this part of the subject. The principles already developed will be of constant use to us in the remainder of the work, which will furnish us with ever-recurring opportunities of exemplifying their use and mode of application.

Let us begin with the case of the Balance, of which we promised (§ 431) to give an investigation.

Examples.

572. *Ex.* I. We will assume the line joining the points of attachment of the scale-pans to the arms to be at right angles to the line joining the centre of gravity of the beam with the fulcrum. It is obvious that the centre of gravity of the beam must not coincide with the knife-edge, else the beam would rest indifferently in any position. We will suppose, in the first place, that the arms are not of equal length. Examples. Balance.

Let O be the fulcrum, G the centre of gravity of the beam, M its mass; and suppose that with loads P and Q in the pans the beam rests (as drawn) in a position making an angle θ with the horizontal line.

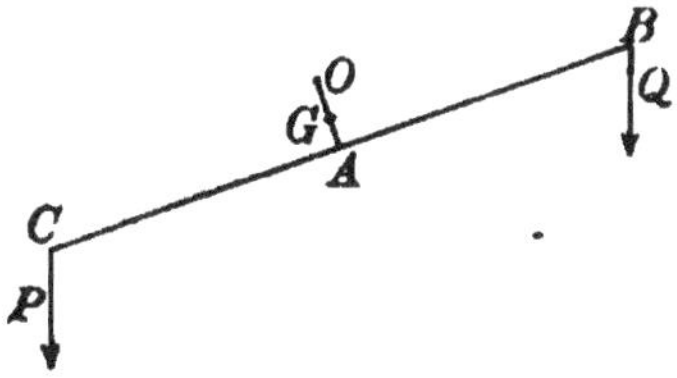

Taking moments about O, and, for convenience (see § 220), using gravitation measurement of the forces, we have Sensibility.

$$Q(AB\cos\theta + OA\sin\theta) + M.OG\sin\theta = P(AC\cos\theta - OA\sin\theta).$$

From this we find

$$\tan\theta = \frac{P.AC - Q.AB}{(P+Q)OA + M.OG}.$$

If the arms be equal we have

$$\tan\theta = \frac{(P-Q)AB}{(P+Q)OA + M.OG}.$$

Hence the Sensibility (§ 431) is greater, (1.) as the arms are longer, (2.) as the mass of the beam is less, (3.) as the fulcrum is nearer to the line joining the points of attachment of the pans, (4.) as the fulcrum is nearer to the centre of gravity of the beam. If the fulcrum be *in* the line joining the points of attachment of the pans, the sensibility is the same for the same *difference* of loads in the pans.

To determine the Stability we must investigate the time of oscillation of the balance when slightly disturbed. It will be seen, by reference to a future chapter, that the equation of motion is approximately Stability.

$$\{Mk^2 + (P+Q)AB^2\}\ddot{\theta} + Qg(AB\cos\theta + OA\sin\theta)$$
$$+ MgOG\sin\theta - Pg(AC\cos\theta - OA\sin\theta) = 0,$$

k being the radius of gyration (§ 281) of the beam. If we suppose the arms and their loads equal, we have for the time of an infinitely small oscillation

$$\pi\sqrt{\frac{Mk^2 + 2P.AB^2}{(2P.OA + M.OG)g}}.$$

Thus the stability is greater for a given load, (1.) the less the

Examples. Balance.

length of the beam, (2.) the less its mass, (3.) the less its radius of gyration, (4.) the further the fulcrum from the beam and from its centre of gravity. With the exception of the second these adjustments are the very opposite of those required for sensibility. Hence all we can do is to effect a judicious compromise; but the less the mass of the beam, the better will the balance be, in *both* respects.

The general equation, above written, shows that if the length and the radius of gyration, of one arm be diminished, the corresponding load being increased so as to maintain equilibrium —a form of balance occasionally useful—the sensibility is increased.

Rod with frictionless constraint.

Ex. II. Find the position of equilibrium of a rod AB resting on a smooth horizontal rail D, its lower end pressing against a smooth vertical wall AC parallel to the rail.

The figure represents a vertical section through the rod, which must evidently be in a plane perpendicular to the wall and rail. The equilibrium is obviously unstable.

The only forces acting are three, R the pressure of the wall on the rod, horizontal; S that of the rail on the rod, perpendicular to the rod; W the weight of the rod, acting vertically downwards at its centre of gravity. If the half-length of the rod be a, and the distance of the rail from the wall b, these are given—and all that is wanted to fix the position of equilibrium is the angle, CAB, which the rod makes with the wall. If we call it θ we have

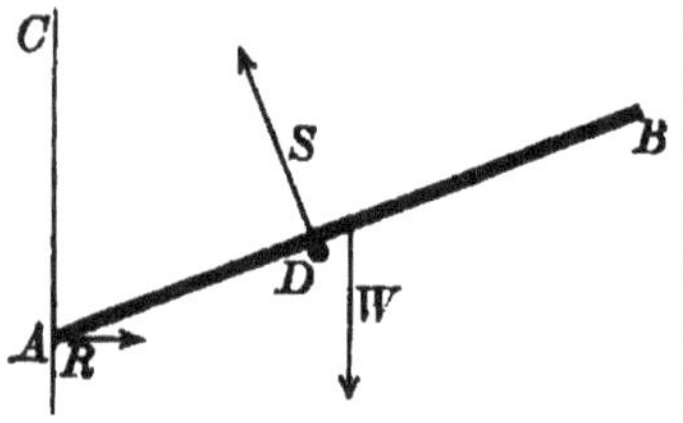

$$AD = \frac{b}{\sin\theta}.$$

Resolving horizontally, $R - S\cos\theta = 0$, (1.)

vertically, $W - S\sin\theta = 0$. (2.)

Taking moments about A

$$S.AD - W.a\sin\theta = 0$$

or $S.b - W.a\sin^2\theta = 0$ (3.)

As there are only three unknown quantities R, S, and θ, these three equations contain the complete solution of the problem. By (2) and (3)

Examples.

$$\sin^2\theta=\frac{b}{a}, \text{ which gives } \theta.$$

Hence by (2) $$S=\frac{W}{\sin\theta},$$

and by (1) $$R=S\cos\theta=W\cot\theta.$$

Rod constrained by rough surfaces.

Ex. III. As an additional example, suppose the wall and ail to be rough, and that μ is the coefficient of statical friction or both. If the rod be placed in the position of equilibrium ust investigated for the case of no friction, none will be called nto play, for there will be no tendency to motion to be over-:ome. If the end A be brought lower and lower, more and nore friction will be called into play to overcome the tend-ency of the rod to fall between the wall and the rail, until we come to a limiting position in which motion is about to com-mence. In that position the friction at A is μ times the pressure on the wall, and acts *upwards*. That at D is μ times the pressure on the rod, and acts in the direction DB. Putting $CAD=\theta_1$ in this case, our three equations become

$$R_1+\mu S_1\sin\theta_1-S_1\cos\theta_1=0 \qquad (1_1)$$

$$W-\mu R_1-S_1\sin\theta_1 \qquad =0 \qquad (2_1)$$

$$S_1 b-Wa\sin^2\theta_1 \qquad =0 \qquad (3_1).$$

The directions of both the friction-forces passing through A, neither appears in (3_1). This is why A is preferable to any other point about which to take moments.

By eliminating R_1 and S_1 from these equations we get

$$1-\frac{a}{b}\sin^2\theta_1=\mu\frac{a}{b}\sin^2\theta_1(\cos\theta_1-\mu\sin\theta_1) \qquad (4_1)$$

from which θ_1 is to be found. Then S_1 is known from (3_1), and R_1 from either of the others.

If the end A be raised above the position of equilibrium without friction, the tendency is for the rod to fall *outside* the rail; more and more friction will be called into play, till the position of the rod (θ_2) is such that the friction reaches its greatest value, μ times the pressure. We may thus find another *limiting* position for stability; and in any position between these the rod is in equilibrium.

It is useful to observe that in this second case the direction of each friction is the opposite to that in the former, and the

Examples.

same equations will serve for both if we adopt the analytic artifice of changing the *sign* of μ. Thus for θ_2, by (4_1),

$$1-\frac{a}{b}\sin^2\theta_2 = -\mu\frac{a}{b}\sin^2\theta_2\,(\cos\theta_2+\mu\sin\theta_2)$$

Block on rough plane.

Ex. IV. A rectangular block lies on a rough horizontal plane and is acted on by a horizontal force whose line of action is midway between two of the vertical sides. Find the magnitude of the force when just sufficient to produce motion, and whether the motion will be of the nature of *sliding* or *overturning.*

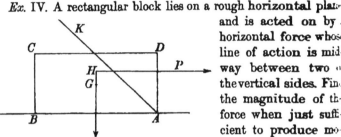

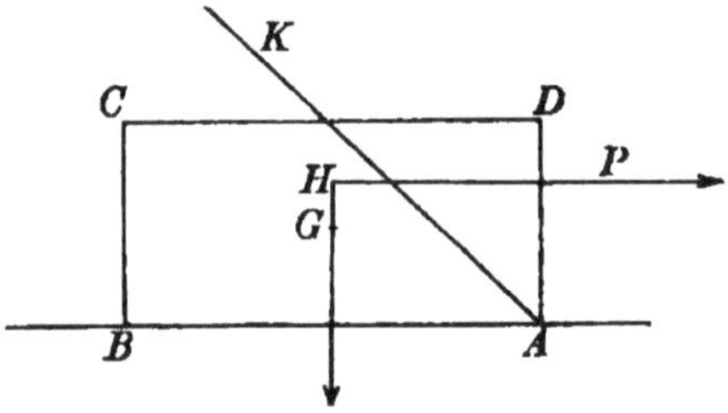

If the force P tends to overturn the body, it is evident that it will turn about the edge A, and therefore the pressure, R, of the plane and the friction, S, act at that edge. Our statical conditions are, of course,

$$R=W$$
$$S=P$$
$$Wb=Pa$$

where b is half the length of the solid, and a the distance of P from the plane. From these we have $S=\frac{b}{a}W$.

Now S cannot exceed μR, whence we must not have $\frac{b}{a}$ greater than μ, if it is to be possible to upset the body by a horizontal force in the line given for P.

A simple geometrical construction enables us to solve this and similar problems, and will be seen at once to be merely a graphic representation of the above process. Thus if we produce the directions of the applied force, and of the weight, to meet in H, and make at A the angle BAK whose co-tangent is the coefficient of friction : there will be a tendency to upset, or not, according as H is above, or below, AK.

Ex. V. A mass, such as a gate, is supported by two rings, A and B, which pass loosely round a rough vertical post. In

Examples. Mass supported by rings passing round a rough post.

uilibrium, it is obvious that at A the part of the ring earest the mass, and at B the part farthest from it, will be contact with the post. The pressures exerted on the rings, R and , will evidently have the directions C, CB, indicated in the diagram. no other force besides gravity act n the mass, the line of action of its eight, W, must pass through the oint C (§ 564). And it is obvious hat, however small be the coefficient of friction, provided there be riction at all, equilibrium is always ossible if the distance of the centre of gravity from the post e great enough compared with the distance between the rings.

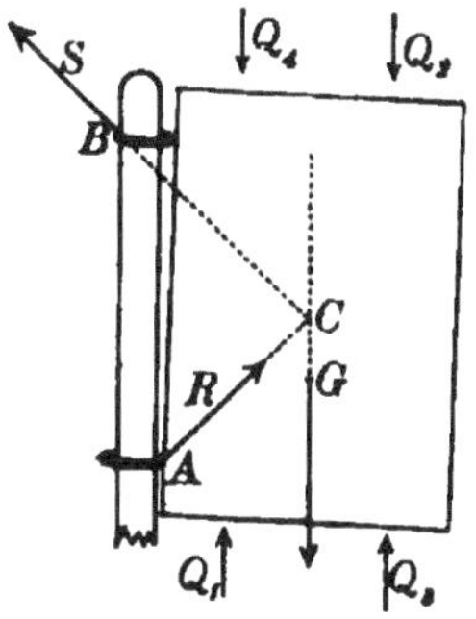

When the mass is just about to slide down, the full amount of friction is called into play, and the angles which R and S make with the horizon are each equal to the angle of repose. If we draw AC, BC according to this condition, then for equilibrium the centre of gravity G must not lie between the post and the vertical line through the point C thus determined. If, as in the figure, G lies in the vertical line through C, then a force applied upwards at Q_1, or downwards at Q_2, will remove the tendency to fall; but a force applied upwards at Q_3, or downwards at Q_4, will produce sliding at once.

A similar investigation is easily applied to the jamming of a sliding piece or drawer, and to the determination of the proper point of application of a force to move it. This we leave to the student.

573. Having thus briefly considered the equilibrium of a rigid body, we propose, before entering upon the subject of the deformation of elastic solids, to consider certain intermediate cases, in each of which we make a particular assumption the basis of the investigation, and thereby avoid a very considerable amount of analytical difficulties.

Equilibrium of a flexible and inextensible cord.

574. Very excellent examples of this kind are furnished by the statics of a flexible and inextensible cord or chain, fixed at both ends, and subject to the action of any forces. The curve in which the chain hangs in any case may be called a

Catenary. *Catenary*, although the term is usually restricted to the case of a uniform chain acted on by gravity only.

Three methods of investigation. **575.** We may consider separately the conditions of equilibrium of each element; or we may apply the general condition (§ 292) that the whole potential energy is a minimum, in the case of any conservative system of forces; or, especially when gravity is the only external force, we may consider the equilibrium of a *finite* portion of the chain treated for the time as a rigid body (§ 564).

Equations of equilibrium with reference to tangent and osculating plane. **576.** The first of these methods gives immediately the three following equations of equilibrium, for the catenary in general:—

(1.) The rate of variation of the tension per unit of length along the cord is equal to the tangential component of the applied force, per unit of length.

(2.) The plane of curvature of the cord contains the normal component of the applied force, and the centre of curvature is on the opposite side of the arc from that towards which this force acts.

(3.) The amount of the curvature is equal to the normal component of the applied force per unit of length at any point divided by the tension of the cord at the same point.

The first of these is simply the equation of equilibrium of an infinitely small element of the cord relatively to tangential motion. The second and third express that the component of the resultant of the tensions at the two ends of an infinitely small arc, along the normal through its middle point, is directly opposed and is equal to the normal applied force, and is equal to the whole amount of it on the arc. For the plane of the tangent lines in which those tensions act is (§ 8) the plane of curvature. And if θ be the angle between them (or the infinitely small angle by which the angle between their positive directions falls short of π), and T the arithmetical mean of their magnitudes, the component of their resultant along the line bisecting the angle between their positive directions is $2T\sin\frac{1}{2}\theta$, rigorously: or $T\theta$, since θ is infinitely small. Hence $T\theta = N\delta s$ if δs be the length of the arc, and $N\delta s$ the whole amount of normal force applied to it. But (§ 9) $\theta = \frac{\delta s}{\rho}$ if ρ be the radius of curvature; and therefore

$$\frac{1}{\rho}=\frac{N}{T},$$

Catenary.

hich is the equation stated in words (3) above.

577. From (1) of § 576, we see that if the applied forces a any particle of the cord constitute a conservative system, ad if any equal infinitely small lengths of the string experience e same force and in the same direction when brought into ny one position by motion of the string, the difference of the nsions of the cord at any two points of it when hanging in quilibrium, is equal to the difference of the potential (§ 485) f the forces between the positions occupied by these points. ence, whatever be the position where the potential is reckoned ero, the tension of the string at any point is equal to potential t the position occupied by it, with a constant added.

Integral for tension.

578. Instead of considering forces along and perpendicular the tangent, we may resolve all parallel to any fixed direcon: and we thus see that the component of applied force per nit of length of the chain at any point of it, must be equal to he rate of diminution per unit of length of the cord, of the omponent of its tension parallel to the fixed line of this component. By choosing any three fixed rectangular directions we hus have the three differential equations convenient for the nalytical treatment of catenaries by the method of rectangular o-ordinates.

Cartesian equations of equilibrium

These equations are

$$\left.\begin{aligned}\frac{d}{ds}\left(T\frac{dx}{ds}\right)&=-\sigma X\\ \frac{d}{ds}\left(T\frac{dy}{ds}\right)&=-\sigma Y\\ \frac{d}{ds}\left(T\frac{dz}{ds}\right)&=-\sigma Z\end{aligned}\right\}\qquad(1),$$

if s denote the length of the cord from any point of it, to a point P; x, y, z the rectangular co-ordinates of P; X, Y, Z the components of the applied forces at P, per unit mass of the cord; σ the mass of the cord per unit length at P; and T its tension at this point.

These equations afford analytical proofs of § 576, (1), (2), and (3) thus:—Multiplying the first by dx, the second by dy, and the third by dz, adding and observing that

Catenary.

$$\frac{dx}{ds}d\frac{dx}{ds}+\frac{dy}{ds}d\frac{dy}{ds}+\frac{dz}{ds}d\frac{dz}{ds}=\tfrac{1}{2}d\frac{dx^2+dy^2+dz^2}{ds^2}=0,$$

we have

$$dT=-\sigma(Xdx+Ydy+Zdz)=-\sigma(X\frac{dx}{ds}+Y\frac{dy}{ds}+Z\frac{dz}{ds})ds, \quad (2$$

which is (1) of § 576. Again, eliminating dT and T, we hav[e]

$$X(\frac{dy}{ds}d\frac{dz}{ds}-\frac{dz}{ds}d\frac{dy}{ds})+Y(\frac{dz}{ds}d\frac{dx}{ds}-\frac{dx}{ds}d\frac{dz}{ds})+Z(\frac{dx}{ds}d\frac{dy}{ds}-\frac{dy}{ds}d\frac{dx}{ds})=0, \quad (3$$

which (§§ 9, 26) shows that the resultant of X, Y, Z is in t[he] osculating plane, and therefore is the analytical expression o[f] § 576 (2). Lastly, multiplying the first by $d\frac{dx}{ds}$, the second b[y] $d\frac{dy}{ds}$, and the third by $d\frac{dz}{ds}$, and adding, we find

$$T=-\sigma\frac{(Xd\frac{dx}{ds}+Yd\frac{dy}{ds}+Zd\frac{dz}{ds})ds}{(d\frac{dx}{ds})^2+(d\frac{dy}{ds})^2+(d\frac{dz}{ds})^2}, \quad (4$$

which is the analytical expression of § 576 (3).

Method of energy.

579. The same equations of equilibrium may be deriv[ed] from the energy condition of equilibrium; analytically wit[h] ease by the methods of the calculus of variations.

Let V be the potential at (x, y, z) of the applied forces pe[r] unit mass of the cord. The potential energy of any given length of the cord, in any actual position between two given fixed points, will be

$$\int V\sigma ds.$$

This integral, extended through the given length of the cord between the given points, must be a minimum; while the indefinite integral, s, from one end up to the point (x, y, z) remains unchanged by the variations in the positions of this point. Hence, by the calculus of variations,

$$\delta\int V\sigma ds+\int\lambda\delta ds=0,$$

where λ is a function of x, y, z to be eliminated.

Now σ is a function of s, and therefore as s does not vary when x, y, z are changed into $x+\delta x$, $y+\delta y$, $z+\delta z$, the co-ordinates of the same particle of the chain in another position, we have

$$\delta(\sigma V)=\sigma\delta V=-\sigma(X\delta x+Y\delta y+Z\delta z).$$

Using this, and

$$\delta ds=\frac{dxd\delta x+dyd\delta y+dzd\delta z}{ds},$$

in the variational equation; and integrating the last term by parts according to the usual rule; we have

$$s\{[\sigma X+\frac{d}{ds}(\overline{V\sigma+\lambda}\frac{ds}{dx})]\delta x+[\sigma Y+\frac{d}{ds}(\overline{V\sigma+\lambda}\frac{ds}{dy})]\delta y+[\sigma Z+\frac{d}{ds}(\overline{V\sigma+\lambda}\frac{ds}{dz})]\delta z\}=0:$$ Catenary. Energy equation of equilibrium.

whence finally

$$\frac{d}{ds}\{(V\sigma+\lambda)\frac{dx}{ds}\}+X\sigma=0,$$

$$\frac{d}{ds}\{(V\sigma+\lambda)\frac{dy}{ds}\}+Y\sigma=0,$$

$$\frac{d}{ds}\{(V\sigma+\lambda)\frac{dz}{ds}\}+Z\sigma=0,$$

which, if T be put for $V\sigma+\lambda$, are the same as the equations (1) of § 578.

580. The form of the common catenary (§ 574) may be of ourse investigated from the differential equations (§ 578) of ie catenary in general. It is convenient and instructive, owever, to work it out *ab initio* as an illustration of the third iethod explained in § 575. Common catenary.

Third method.—The chain being in equilibrium, *any* arc of it may be supposed to become rigid without disturbing the equilibrium. The only forces acting on this rigid body are the tensions at its ends, and its weight. These forces being three in number, must be in one plane (§ 564), and hence, since one of them is vertical, the whole curve lies in a vertical plane. In this plane let x_0, z_0, s_0, x_1, z_1, s_1, belong to the two ends of the arc which is supposed rigid, and T_0, T_1, the tensions at those points. Resolving horizontally we have

$$T_0(\frac{dx}{ds})_0=T_1(\frac{dx}{ds})_1.$$

Hence $T\frac{dx}{ds}$ is constant throughout the curve. Resolving vertically we have

$$T_1(\frac{dz}{ds})_1-T_0(\frac{dz}{ds})_0=\sigma(s_1-s_0),$$

the weight of unit of mass being now taken as the unit of force.

Hence if T_0 be the tension at the lowest point, where $\frac{dz}{ds}=0$, $s=0$, and T the tension at any point (x, z) of the curve, we have

$$T=T_0\frac{ds}{dx}=\sigma s\frac{ds}{dz} \qquad (1).$$

Hence

$$T_0\frac{d}{ds}(\frac{dz}{dx})=\sigma,$$

Catenary; common.

or $$T_0\frac{d^2z}{dx^2}=\sigma\frac{ds}{dx}=\sigma\sqrt{1+(\frac{dz}{dx})^2}. \qquad (2$$

Integrating we have

$$\log\{\frac{dz}{dx}+\sqrt{1+(\frac{dz}{dx})^2}\}=\frac{\sigma}{T_0}x+C',$$

and the constant is zero if we take the origin so that $x=0$, whe[n] $\frac{dz}{dx}=0$, *i.e.*, where the chain is horizontal.

Hence

$$\frac{dz}{dx}+\sqrt{1+(\frac{dz}{dx})^2}=\epsilon^{\frac{\sigma}{T_0}x} \qquad (3$$

whence $$\frac{dz}{dx}=\tfrac{1}{2}(\epsilon^{\frac{\sigma}{T_0}x}-\epsilon^{-\frac{\sigma}{T_0}x});$$

and by integrating again

$$z+C''=\frac{T_0}{2\sigma}(\epsilon^{\frac{\sigma}{T_0}x}+\epsilon^{-\frac{\sigma}{T_0}x}).$$

This may be written

$$z=\frac{a}{2}(\epsilon^{\frac{x}{a}}+\epsilon^{-\frac{x}{a}}) \qquad (4$$

the ordinary equation of the catenary, the axis of x being take[n] at a distance a or $\frac{T_0}{\sigma}$ below the horizontal element of the chain.

The co-ordinates of that element are therefore $x=0$, $z=\frac{T_0}{\sigma}=a$. The latter shows that

$$T_0=\sigma a,$$

or the tension at the lowest point of the chain (and therefore al[so] the horizontal component of the tension throughout) is the weig[ht] of a length a of the chain.

Now, by (1), $T=T_0\frac{ds}{dx}=\sigma z$, by (4), and therefore

the tension at any point is equal to the weight of a portion o[f] the chain equal to the vertical ordinate at that point.

Relative kinetic problem.

581. From § 576 it follows immediately that if a material particle of unit mass be carried along any catenary with a velocity, $\dot{s}$, equal to T, the numerical measure of the tension at any point, the force upon it by which this is done is in the same direction as the resultant of the applied force on the catenary at this point, and is equal to the amount of this force per unit of ength, multiplied by T. For, denoting by S the tangential nd (as before) by N the normal component of the applied orce per unit of length at any point P of the catenary, we iave, by § 576 (1), S for the rate of variation of $\dot{s}$ per unit ength, and therefore $S\dot{s}$ for its variation per unit of time. That s to say, Kinetic question relative to catenary.

$$\ddot{s}=S\dot{s}=ST,$$

ɪr (§ 259) the tangential component force on the moving ɪarticle is equal to ST. Again, by § 576 (3),

$$NT=\frac{T^2}{\rho}=\frac{\dot{s}^2}{\rho},$$

ɪr the centrifugal force of the moving particle in the circle of urvature of its path, that is to say, the normal component of he force on it, is equal to NT. And lastly, by (2) this force s in the same direction as N. We see therefore that the lirection of the whole force on the moving particle is the ame as that of the resultant of S and N; and its magnitude s T times the magnitude of this resultant.

Or, by taking
$$\frac{ds}{T}=dt,$$
in the differential equation of § 578, we have
$$\frac{d^2x}{dt^2}=-T\sigma X,\quad \frac{d^2y}{dt^2}=-T\sigma Y,\quad \frac{d^2z}{dt^2}=-T\sigma Z,$$
which proves the same conclusion.

When σ is constant, and the forces belong to a conservative system, if V be the potential at any point of the cord, we have, by § 578 (2), $T=\sigma V+C$.
Hence, if $U=\frac{1}{2}(\sigma V+C)^2$, these equations become
$$\frac{d^2x}{dt^2}=-\frac{dU}{dx},\quad \frac{d^2y}{dt^2}=-\frac{dU}{dy},\quad \frac{d^2z}{dt^2}=-\frac{dU}{dz}.$$
The integrals of these equations which agree with the catenary, are those only for which the energy constant is such that $\dot{s}^2=2U$.

582. Thus we see how, from the more familiar problems of the kinetics of a particle, we may immediately derive curious cases of catenaries. For instance: a particle under the influence of a constant force in parallel lines moves (Chap. VIII.) in a parabola with its axis vertical, with velocity at each point equal to that generated by the force acting through a space Examples.

Kinetic question relative to catenary. Examples.

equal to its distance from the directrix. Hence, if z denote this distance, and f the constant force,

$$T=\sqrt{2fz}$$

in the allied parabolic catenary; and the force on the catenary is parallel to the axis, and is equal in amount per unit of length, to

$$\frac{f}{\sqrt{2fz}} \text{ or } \sqrt{\frac{f}{2z}}.$$

Hence if the force on the catenary be that of gravity, it must have its axis vertical (its vertex downwards of course for stable equilibrium) and its mass per unit length at any point must be inversely as the square root of the distance of this point above the directrix. From this it follows that the whole weight of any arc of it is proportional to its horizontal projection. Or, again, as will be proved later with reference to the motions of comets, a particle moves in a parabola under the influence of a force towards a fixed point varying inversely as the square of the distance from this point, if its velocity be that due to falling from rest at an infinite distance. This velocity being $\sqrt{\frac{2\mu}{r}}$, at distance r, it follows, according to § 581, that a cord will hang in the same parabola, under the influence of a force towards the same centre, and equal to

$$\frac{\mu}{r^2} \div \sqrt{\frac{2\mu}{r}}, \text{ or } \sqrt{\frac{\mu}{2r^3}}.$$

If, however, the length of the cord be varied between two fixed points, the central force still following the same law, the altered catenary will no longer be parabolic: but it will be the path of a particle under the influence of a central force equal to

$$\left(C+\sqrt{\frac{2\mu}{r}}\right)\sqrt{\frac{\mu}{2r^3}},$$

since the tension would (§ 581) be $C+\sqrt{\frac{2\mu}{r}}$, instead of $\sqrt{\frac{2\mu}{r}}$, at distance r from the origin.

583. Or if the question be, to find what force towards a given fixed point, will cause a cord to hang in any given plane curve with this point in its plane; it may be answered imnediately from the solution of the corresponding problem in central forces." But the general equations, § 578, are always asily applicable; as, for instance, to the following useful in-'erse case of the gravitation catenary:— Catenary. Inverse problem.

Find the section, at each point, of a chain of uniform naterial, so that when its ends are fixed the tension at each point lue to its weight may be proportional to its strength (or section) at hat point. Find also the form of the Curve, called the Catenary f Uniform Strength, in which it will hang. Catenary of uniform strength.

Here, as the only external force is gravity, the chain is in a vertical plane—in which we may assume the horizontal axis of x to lie. If μ be the weight of the chain at the point (x, z) reckoned per unit of length; our equations [§ 578 (1)] become .

$$\frac{d}{ds}\left(T\frac{dx}{ds}\right)=0,\quad \frac{d}{ds}\left(T\frac{dz}{ds}\right)=\mu.$$

But, by hypothesis $T\propto\mu$. Let it be $b\mu$. Hence, by the first equation, if μ_0 be the value of μ at the lowest point

$$\mu=\mu_0\frac{ds}{dx};$$

whence, by the second equation,

$$\frac{d}{ds}\left(\frac{dz}{dx}\right)=\frac{1}{b}\frac{ds}{dx},$$

or

$$\frac{d^2z}{dx^2}=\frac{1}{b}\left[1+\left(\frac{dz}{dx}\right)^2\right].$$

Integrating we find

$$\tan^{-1}\frac{dz}{dx}=\frac{x}{b},$$

no constant being required if we take the axis of x so as to touch the curve at its lowest point. Integrating again we have

$$\frac{z}{b}=-\log\cos\frac{x}{b},$$

no constant being added, if the origin be taken at the lowest point. We may write the equation in the form

$$\sec\frac{x}{b}=\epsilon^{\frac{z}{b}}.$$

From this form of the equation we see that the curve has vertical asymptotes at a horizontal distance πb from each other. And thus it is easy to calculate for any given data, as to the tensile strength and specific gravity of the material employed, the greatest span which can be attained in any case.

Flexible string on smooth surface.

584. When a perfectly flexible string is stretched over a smooth surface, and acted on by no other force throughout its length than the resistance of this surface, it will, when in stable equilibrium, lie along a line of minimum length on the surface, between any two of its points. For (§ 564) its equilibrium can be neither disturbed nor rendered unstable by placing staples over it, through which it is free to slip, at any two points where it rests on the surface: and for the intermediate part the energy criterion of stable equilibrium is that just stated.

There being no tangential force on the string in this case, and the normal force upon it being along the normal to the surface, its osculating plane (§ 576) must cut the surface everywhere at right angles. These considerations, easily translated into pure geometry, establish the fundamental property of the geodetic lines on any surface. The analytical investigations of §§ 578, 579, when adapted to the case of a chain of *not* given length, stretched between two given points on a given smooth surface, constitute the direct analytical demonstration of this property.

In this case it is obvious that the tension of the string is the same at every point, and the pressure of the surface upon it is [§ 576 (3)] at each point proportional to the curvature of the string.

On rough surface.

585. No real surface being perfectly smooth, a cord or chain may rest upon it when stretched over so great a length of a geodetic on a convex rigid body as to be not of minimum length between its extreme points: but practically, as in tying a cord round a ball, for permanent security it is necessary, by staples or otherwise, to constrain it from lateral slipping at successive points near enough to one another to make each free portion a true minimum on the surface.

Rope coiled about rough cylinder.

586. A very important practical case is supplied by the consideration of a rope wound round a rough cylinder. We may suppose it to lie in a plane perpendicular to the axis, as we thus simplify the question very considerably without sensibly injuring the utility of the solution. To simplify still further, we shall suppose that no forces act on the rope, but tensions and the reaction of the cylinder. In practice this is equivalent to the supposition that the tensions and reactions are very large compared with the weight of the rope or chain; which, however, is inadmissible in some important cases; especially such as occur in the application of the principle to brakes for laying submarine cables, to dynamometers, and to windlasses (or capstans with horizontal axes). Rope coiled about rough cylinder.

If R be the normal reaction of the cylinder per unit of length of the cord, at any point; T and $T+\delta T$ the tensions at the extremities of an arc δs; $\delta\theta$ the inclination of these lines; we have, as in § 576,

$$T\delta\theta = R\delta s.$$

And the friction called into play is evidently equal to δT. When the rope is about to slip, the friction has its greatest value, and then

$$\delta T = \mu R\delta s = \mu T\delta\theta.$$

This gives, by integration,

$$T = T_0 \epsilon^{\mu\theta},$$

showing that, for equal successive amounts of integral curvature (§ 10), the tension of the rope augments in *geometrical* progression. To give an idea of the magnitudes involved, suppose $\mu = \cdot 5$, $\theta = \pi$, then

$$T = T_0 \epsilon^{\cdot 5\pi} = 4\cdot 81 T_0 \text{ roughly}.$$

Hence if the rope be wound three times round the post or cylinder the ratio of the tensions of its ends, when motion is about to commence, is

$$(4\cdot 81)^6 : 1 \text{ or about } 12{,}390 : 1.$$

Thus we see how, by the aid of friction, one man may easily check the motion of the largest vessel, by the simple expedient of coiling a rope a few times round a post. This application of friction is of great importance in many other applications, especially to Dynamometers.

587. With the aid of the preceding investigations, the student may easily work out for himself the formulæ expressing the solution of the general problem of a cord under the action of any forces, and constrained by a rough surface; they are not of sufficient importance or interest to find a place here.

588. An elongated body of elastic material, which for brevity we shall generally call a *Wire*, bent or twisted to any Elastic wire.

Elastic wire, fibre, bar, rod, lamina, or beam.

degree, subject only to the condition that the radius of curvature and the reciprocal of the twist (§ 119) are everywhere very great in comparison with the greatest transverse dimension, presents a case in which, as we shall see, the solution of the general equations for the equilibrium of an elastic solid is either obtainable in finite terms, or is reducible to comparatively easy questions agreeing in mathematical conditions with some of the most elementary problems of hydrokinetics, electricity, and thermal conduction. And it is only for the determination of certain constants depending on the section of the wire and the elastic quality of its substance, which measure its flexural and torsional rigidity, that the solutions of these problems are required. When the constants of flexure and torsion are known, as we shall now suppose them to be, whether from theoretical calculation or experiment, the investigation of the form and twist of any length of the wire, under the influence of any forces which do not produce a violation of the condition stated above, becomes a subject of mathematical analysis involving only such principles and formulæ as those that constitute the theory of curvature (§§ 5-13) and twist (§§ 119-123) in geometry or kinematics.

589. Before entering on the general theory of elastic solids, we shall therefore, according to the plan proposed in § 573, examine the dynamic properties and investigate the conditions of equilibrium of a perfectly elastic wire, without admitting any other condition or limitation of the circumstances than what is stated in § 588, and without assuming any special quality of isotropy, or of crystalline, fibrous or laminated structure in the substance. The following short geometrical digression is a convenient preliminary:—

Composition and resolution of curvatures in a curve line.

590. The geometrical composition of curvatures with one another, or with rates of twist, is obvious from the definition and principles regarding curvature given above in §§ 5-13 and twist in §§ 119-123, and from the composition of angular velocities explained in § 96. Thus if one line, OT, of a rigid body be always held parallel to the tangent, PT, at a point P moving with unit velocity along a curve, whether plane or tortuous, it will have, round an axis perpendicular to OT and to the radius of curvature (that is to say, perpendicular to the osculating plane), an angular velocity numerically equal to the curvature. The body may besides be made to rotate with any angular velocity round OT. Thus, for instance, if a line of it, OA, be kept always parallel to a transverse (§ 120) PA, the component angular velocity of the rigid body round OT will at every instant be equal to the "rate of twist" (§ 120) of the transverse round the tangent to the curve. Again, the angular velocity round OA may be resolved into components round two lines OK, OL, perpendicular to one another and to OT; and the whole curvature of the curve may be resolved accordingly into two component curvatures in planes perpendicular to those two lines respectively. The amounts of these component curvatures are of course equal to the whole curvature multiplied by the cosines of the inclinations of the osculating plane to their respective planes. And it is clear that each component curvature is simply the curvature of the projection of the actual curve on its plane.[1]

Composition and resolution of curvature in a curve line.

591. Besides showing how the constants of flexural and torsional rigidity are to be determined theoretically from the form of the transverse section of the wire, and the proper data as to the elastic qualities of its substance, the complete theory simply indicates that, provided the conditional limit (§ 588) of deformation is not exceeded, the following laws will be obeyed by the wire under stress:—

Laws of flexure and torsion.

Let the whole mutual action between the parts of the wire on the two sides of the cross section at any point (being of course the action of the matter infinitely near this plane on one side, upon the matter infinitely near it on the other side), be reduced to a single force through any point of the section and a single couple. Then—

I. The twist and curvature of the wire in the neighbourhood of this section are independent of the force, and depend solely on the couple.

[1] The curvature of the projection of a curve on a plane inclined at an angle α to the osculating plane, is $\frac{1}{\rho}\cos\alpha$ if the plane be parallel to the tangent; and $\frac{1}{\rho\cos^2\alpha}$ if it be parallel to the principal normal (or radius of absolute curvature). There is no difficulty in proving either of these expressions.

Laws of flexure and torsion.

II. The curvatures and rates of twist producible by all several couples separately, constitute, if geometrically compounded, the curvature and rate of twist which are actually produced by a mutual action equal to the resultant of these couples.

592. It may be added, although not necessary for our present purpose, that there is one determinate point in the cross section such that if it be chosen as the point to which the forces are transferred, a higher order of approximation is obtained for the fulfilment of these laws than if any other point of the section be taken. That point, which in the case of a wire of substance uniform through its cross section is the centre of inertia of the area of the section, we shall generally call the elastic centre, or the centre of elasticity, of the section. It has also the following important property:—The line of elastic centres, or, as we shall call it, the elastic central line remains sensibly unchanged in length to whatever stress within our conditional limits (§ 588) the wire be subjected. The elongation or contraction produced by the neglected resultant force if this is in such a direction as to produce any, will cause the line of *rigorously no elongation* to deviate only infinitesimally from the elastic central line, in any part of the wire finitely curved. It will, however, clearly cause there to be no line of *rigorously unchanged length,* in any straight part of the wire; but as the whole elongation would be infinitesimal in comparison with the effective actions with which we are concerned, this case constitutes no exception to the preceding statement.

Warping of normal section by torsion and flexure, infinitesimal.

593. Considering now a wire of uniform constitution and figure throughout, and naturally straight; let any two planes of reference perpendicular to one another through its elastic central line when straight, cut the normal section through P in the lines PK and PL. These two lines (supposed to belong to the substance, and move with it) will remain infinitely nearly at right angles to one another, and to the tangent, PT, to the central line, however the wire may be bent or twisted within the conditional limits. Let κ and λ be the component curvatures (§ 590) in the two planes perpendicular to PK and PL through PT, and let τ be the twist (§ 120) of the wire at P. We have just seen (§ 590) that if P be moved

Rotations corresponding to flexure and torsion.

at a unit rate along the curve, a rigid body with three rectangular axes of reference $O_{,}K$, $O_{,}L$, $O_{,}T$ kept always parallel to PK, PL, PT, will have angular velocities κ, λ, τ round those axes respectively. Hence if the point P and the lines PT, PK, PL be at rest while the wire is bent and twisted from its unstrained to its actual condition, the lines of reference $P'K'$, $P'L'$, $P'T'$ through any point P' infinitely near P, will experience a rotation compounded of $\kappa.PP'$ round $P'K'$, $\lambda.PP'$ round $P'L'$, and $\tau.PP'$ round $P'T'$.

Rotations corresponding to flexure and torsion.

594. Considering now the elastic forces called into action, we see that if these constitute a conservative system, the work required to bend and twist any part of the wire from its unstrained to its actual condition, depends solely on its figure in these two conditions. Hence if $w.PP'$ denote the amount of this work, for the infinitely small length PP' of the rod, w must be a function of κ, λ, τ; and therefore if K, L, T denote the components of the couple-resultant of all the forces which must act on the section through P' to hold the part PP' in its strained state, it follows, from §§ 240, 272, 274, that

Potential energy of elastic force in bent and twisted wire.

$$K\delta\kappa = \delta_\kappa w,\ \ L\delta\lambda = \delta_\lambda w,\ \ T\delta\tau = \delta_\tau w \qquad (1)$$

where $\delta_\kappa w$, $\delta_\lambda w$, $\delta_\tau w$ denote the augmentations of w due respectively to infinitely small augmentations $\delta\kappa$, $\delta\lambda$, $\delta\tau$, of κ, λ, τ.

595. Now however much the shape of any finite length of the wire may be changed, the condition of § 588 requires clearly that the changes of shape in each infinitely small part, that is to say, the strain (§ 154) of the substance, shall be everywhere very small (infinitely small in order that the theory may be rigorously applicable). Hence the principle of superposition [§ 591, II.] shows that if κ, λ, τ be each increased or diminished in one ratio, K, L, T will be each increased or diminished in the same ratio: and consequently w in the duplicate ratio, since the angle through which each couple acts is altered in the same ratio as the amount of the couple; or, in algebraic language, w is a homogeneous quadratic function of κ, λ, τ.

Thus if A, B, C, a, b, c denote six constants, we have

$$w = \tfrac{1}{2}(A\kappa^2 + B\lambda^2 + C\tau^2 + 2a\lambda\tau + 2b\tau\kappa + 2c\kappa\lambda) \qquad (2).$$

Hence, by § 594 (1),

Components of restituent couple.

$$\left.\begin{aligned} K &= A\kappa + c\lambda + b\tau \\ L &= c\kappa + B\lambda + a\tau \\ T &= b\kappa + a\lambda + C\tau \end{aligned}\right\} \quad (3)$$

By the known reduction of the homogeneous quadratic function these expressions may of course be reduced to the following simple forms:—

$$\left.\begin{aligned} w &= \tfrac{1}{2}(A_1\vartheta_1^2 + A_2\vartheta_2^2 + A_3\vartheta_3^2) \\ L_1 &= A_1\vartheta_1, \; L_2 = A_2\vartheta_2, \; L_3 = A_3\vartheta_3 \end{aligned}\right\} \quad (4)$$

where ϑ_1, ϑ_2, ϑ_3 are linear functions of κ, λ, τ. And if these functions are restricted to being the expressions for the components round three rectangular axes, of the rotations κ, λ, τ viewed as angular velocities round the axes PK, PL, PT, the positions of the new axes, PQ_1, PQ_2, PQ_3, and the values of A_1, A_2, A_3 are determinate; the latter being the roots of the determinant cubic [§ 181 (10)] founded on (A, B, C, a, b, c). Hence we conclude that

Three principal or normal axes of torsion and flexure.

596. There are in general three determinate rectangular directions, PQ_1, PQ_2, PQ_3, through any point P of the middle line of a wire, such that if opposite couples be applied to any two parts of the wire in planes perpendicular to any one of them, every intermediate part will experience rotation in a plane parallel to those of the balanced couples. The moments of the couples required to produce unit rate of rotation round these three axes are called the *principal torsion-flexure* rigidities of the wire. They are the elements denoted by A_1, A_2, A_3 in the preceding analysis.

Three principal torsion-flexure rigidities.

597. If the rigid body imagined in § 593 have moments of inertia equal to A_1, A_2, A_3 round three principal axes through O kept always parallel to the principal torsion-flexure axes through P while P moves at unit rate along the wire, its moment of momentum round any axis will be equal to the moment of the component torsion-flexure couple round the parallel axes through P. This is shown by the agreement of the preceding formulæ with those for the moment of momentum of a rotating rigid body given below (Chap. IX.)

Three principal or normal spirals.

598. The form assumed by the wire when balanced under the influence of couples round one of the three principal axes is of course a uniform helix having a line parallel to it for axis, and lying on a cylinder whose radius is determined by the condition that the whole rotation of one end of the wire from its unstrained position, the other end being held fixed, is equal to the amount due to the couple applied.

Three principal or normal spirals.

Let l be the length of the wire from one end, E, held fixed, to the other end, E', where a couple, L, is applied in a plane perpendicular to the principal axes PQ_1 through any point of the wire. The rotation being [§ 595 (4)] at the rate $\frac{L}{A_1}$, per unit of length, amounts on the whole to $l\frac{L}{A_1}$. This therefore is the angular space occupied by the helix on the cylinder on which it lies. Hence if r denote the radius of this cylinder, and i_1 the inclination of the helix to its axis (being the inclination of PQ_1 to the length of the wire), we have

$$r\frac{Ll}{A_1}=l\sin i_1;$$

whence

$$r=\frac{A_1\sin i_1}{L} \qquad (5).$$

599. In the most important practical cases, as we shall see later, those namely in which the substance is either "isotropic," as is the case sensibly with common metallic wires, or, as in rods or beams of fibrous or crystalline structure with an axis of elastic symmetry along the length of the piece, one of the three normal axes of torsion and flexure coincides with the length of the wire, and the two others are perpendicular to it; the first being an axis of pure torsion, and the two others axes of pure flexure. Thus opposing couples round the axis of the wire twist it simply without bending it; and opposing couples in either of the two principal planes of flexure, bend it into a circle. The unbent straight line of the wire, and the circular arcs into which it is bent by couples in the two principal planes of flexure, are what the three principal spirals of the general problem become in this case.

Case in which elastic central line is a normal axis of torsion.

600. In the more particular case in which two principal rigidities against flexure are equal, every plane through the length of the wire is a principal plane of flexure, and the rigidity against flexure is equal in all. This is clearly the case with a common round wire, or rod: or with one of square section. It will be shown later to be the case for a rod of iso-

Equal flexibility in all directions.

Case of equal flexibility in all directions.

tropic material and of any form of normal section whic. "kinetically symmetrical," § 285, round all axes in its pl through its centre of inertia.

601. In this case, if one end of the rod or wire be he fixed, and a couple be applied in any plane to the ot end, a uniform spiral form will be produced round an a perpendicular to the plane of the couple. The lines of t substance parallel to the axis of the spiral are not, howev parallel to their original positions, as (§ 598) in each of t three principal spirals of the general problem : and li traced along the surface of the wire parallel to its len when straight, become as it were secondary spirals, circ round the main spiral formed by the central line of t deformed wire; instead of being all spirals of equal step, as each one of the principal spirals of the general problem. Last in the present case, if we suppose the normal section of t wire to be circular, and trace uniform spirals along its surfa when deformed in the manner supposed (two of which, f instance, are the lines along which it is touched by the in scribed and the circumscribed cylinder), these lines do n become straight, but become spirals laid on as it were roun the wire, when it is allowed to take its natural straight a untwisted condition.

Let, in § 595, PQ_1 coincide with the central line of the wir and let $A_1=A$, and $A_2=A_3=B$; so that A measures the rigidi of torsion and B that of flexure. One end of the wire bein held fixed, let a couple G be applied to the other end, round a axis inclined at an angle θ to the length. The rates of twist an of flexure each per unit of length, according to (4.) of § 59 will be

$$\frac{G\cos\theta}{A}, \text{ and } \frac{G\sin\theta}{B},$$

respectively. The latter being (§ 9) the same thing as th curvature, and the inclination of the spiral to its axis being θ, it follows (§ 126, or § 590, foot-note) that $\frac{B\sin\theta}{G}$ is the radius of curvature of its projection on a plane perpendicular to this lin that is to say, the radius of the cylinder on which the spiral lie

602. A wire of equal flexibility in all directions may clearl be held in any specified spiral form, and twisted to any stated degree, by a determinate force and couple applied at one end, the other end being held fixed. The direction of the force must be parallel to the axis of the spiral, and, with the couple, must constitute a system of which this line is (§ 559) the *central axis:* since otherwise there could not be the same system of balancing forces in every normal section of the spiral. All this may be seen clearly by supposing the wire to be first brought by any means to the specified condition of strain; then to have rigid planes rigidly attached to its two ends perpendicular to its axis, and these planes to be rigidly connected by a bar lying in this line. The spiral wire now left to itself cannot but be in equilibrium: although if it be too long (according to its form and degree of twist) the equilibrium may be unstable. The force along the central axis, and the couple, are to be determined by the condition that, when the force is transferred after Poinsot's manner to the elastic centre of any normal section, they give two couples together equivalent to the elastic couples of flexure and torsion.

Wire strained to any given spiral and twist.

Let α be the inclination of the spiral to the plane perpendicular to its axis; r the radius of the cylinder on which it lies; τ the rate of twist given to the wire in its spiral form. The curvature is (§ 126) equal to $\frac{\cos^2\alpha}{r}$; and its plane, at any point of the spiral, being the plane of the tangent to the spiral and the diameter of the cylinder through that point, is inclined at the angle α to the plane perpendicular to the axis. Hence the components in this plane, and in the plane through the axis of the cylinder of the flexural couple are respectively

$$\frac{B\cos^2\alpha}{r}\cos\alpha, \text{ and } \frac{B\cos^2\alpha}{r}\sin\alpha.$$

Also, the components of the torsional couple, in the same planes, are $A\tau\sin\alpha$, and $-A\tau\cos\alpha$.

Hence, for equilibrium

$$\left.\begin{aligned} G &= \frac{B\cos^2\alpha}{r}\cos\alpha + A\tau\sin\alpha \\ -Rr &= \frac{B\cos^2\alpha}{r}\sin\alpha - A\tau\cos\alpha \end{aligned}\right\} \qquad (6),$$

which give explicitly the values, G and R, of the couple and force required, the latter being reckoned as positive when its direction

Wire strained to any given spiral and twist.

is such as to pull *out* the spiral, or when the ends of the rigid b supposed above are pressed *inwards* by the plates attached to t ends of the spiral.

If we make $R=0$, we fall back on the case considered pre ously (§ 601). If, on the other hand, we make $G=0$, we hav

$$\tau = -\frac{1}{r}\frac{B}{A}\frac{\cos^2\alpha}{\sin\alpha},$$

and
$$R = -\frac{B}{r^2}\frac{\cos^2\alpha}{\sin\alpha} = \frac{A\tau}{r\cos\alpha},$$

from which we conclude that

Twist determined to reduce the action to a single force.

603. A wire of equal flexibility in all directions may b held in any stated spiral form by a simple force along its ax between rigid pieces rigidly attached to its two ends, provid that, along with its spiral form, a certain degree of twist i given to it. The force is determined by the condition that i moment round the perpendicular through any point of th spiral to its osculating plane at that point, must be equa and opposite to the elastic unbending couple. The degree o twist is that due (by the simple equation of torsion) to th moment of the force thus determined, round the tangent a any point of the spiral. The direction of the force being according to the preceding condition, such as to press togethe the ends of the spiral, the direction of the twist in the wire i opposite to that of the tortuosity (§ 9) of its central curve.

Spiral springs.

604. The principles and formulæ (§§ 598, 603) with which we have just been occupied are immediately applicable to th theory of spiral springs; and we shall therefore make a shor digression on this curious and important practical subject before completing our investigation of elastic curves.

A common spiral spring consists of a uniform wire shaped permanently to have, when unstrained, the form of a regular helix, with the principal axes of flexure and torsion everywhere similarly situated relatively to the curve. When used in the proper manner, it is acted on, through arms or plates rigidly attached to its ends, by forces such that its form as altered by them is still a regular helix. This condition is obviously fulfilled if (one terminal being held fixed) an infinitely small force and infinitely small couple be applied to the other terminal along the axis and in a plane perpendicular to it, and if the force and uple be increased to any degree, and always kept along and the plane perpendicular to the axis of the altered spiral. It ould, however, introduce useless complication to work out the tails of the problem except for the case (§ 599) in which one the principal axes coincides with the tangent to the central ne, and is therefore an axis of pure torsion; as spiral springs practice always belong to this case. On the other hand, a very teresting complication occurs if we suppose (what is easily alized in practice, though to be avoided if merely a good ring is desired) the normal section of the wire to be of such a gure, and so situated relatively to the spiral, that the planes greatest and least flexural rigidity are oblique to the tangent lane of the cylinder. Such a spring when acted on in the gular manner at its ends must experience a certain degree of irning through its whole length round its elastic central curve order that the flexural couple developed may be, as we shall nmediately see it must be, precisely in the osculating plane of he altered spiral. But all that is interesting in this very urious effect will be illustrated later (§ 624) in full detail in the ase of an open circular arc altered by a couple in its own plane, nto a circular arc of greater or less radius; and for brevity nd simplicity we shall confine the detailed investigation of piral springs on which we now enter, to the cases in which ither the wire is of equal flexural rigidity in all directions, or he two principal planes of (greatest and least or least and reatest) flexural rigidity coincide respectively with the tangent lane to the cylinder, and the normal plane touching the central urve of the wire, at any point.

Spiral springs.

605. The axial force, on the movable terminal of the pring, transferred according to Poinsot (§ 555) to any point n the elastic central curve, gives a couple in the plane through hat point and the axis of the spiral. The resultant of this and he couple which we suppose applied to the terminal in the lane perpendicular to the axis of the spiral is the effective ending and twisting couple: and as it is in a plane perpendicular to the tangent plane to the cylinder, the component of it to which bending is due must be also perpendicular to this plane, and therefore is in the osculating plane of the spiral. This component couple therefore simply maintains a curvature

Spiral springs.

different from the natural curvature of the wire, and the ot that is, the couple in the plane normal to the central cu pure torsion. The equations of equilibrium merely exp this in mathematical language.

Resolving as before (§ 602) the flexural and the tor couples each into components in the planes through the axi the spiral, and perpendicular to it, we have

$$\left.\begin{aligned} G&=B\left(\frac{\cos^2\alpha}{r}-\frac{\cos^2\alpha_0}{r_0}\right)\cos\alpha'+A\tau\sin\alpha', \\ -Rr&=B\left(\frac{\cos^2\alpha}{r}-\frac{\cos^2\alpha_0}{r_0}\right)\sin\alpha'-A\tau\cos\alpha', \\ \text{and, by § 126,}\quad \tau&=\frac{\cos\alpha\sin\alpha}{r}-\frac{\cos\alpha_0\sin\alpha_0}{r_0}, \end{aligned}\right\}$$

where A denotes the torsional rigidity of the wire, and B flexural rigidity in the osculating plane of the spiral; α_0 the i clination, and r_0 the radius of the cylinder, of the spiral wh unstrained; α and r the same parameters of the spiral wh under the influence of the axial force R and couple G; and τ degree of twist in the change from the unstrained to the strain condition.

These equations give explicitly the force and couple requir to produce any stated change in the spiral; or if the force a couple are given they determine α', r' the parameters of altered curve.

As it is chiefly the external action of the spring that we concerned with in practical applications, let the parameters of the spiral be eliminated by the following assumptions:—

$$\left.\begin{aligned} x&=l\sin\alpha, & \phi&=\frac{l\cos\alpha}{r} \\ x_0&=l\sin\alpha_0, & \phi_0&=\frac{l\cos\alpha_0}{r_0} \end{aligned}\right\}$$

where l denotes the length of the wire, ϕ the angle betw planes through the two ends of the spiral, and its axis, and x th distance between planes through the ends and perpendicular the axis in the strained condition; and, similarly, ϕ_0, x_0 for th unstrained condition; so that we may regard (ϕ, x) and $(\phi_0, x$ as the co-ordinates of the movable terminal relatively to th fixed in the two conditions of the spring. Thus the precedin equations become

Spiral springs.

$$\left.\begin{aligned} &= \frac{B}{l^3}\{\sqrt{(l^2-x^2)}\phi - \sqrt{(l^2-x_0^2)}\phi_0\}\sqrt{(l^2-x^2)} + \frac{A}{l^3}(x\phi - x_0\phi_0)x \\ &= -\frac{B}{l^3}\{\sqrt{(l^2-x^2)}\phi - \sqrt{(l^2-x_0^2)}\phi_0\}\frac{x\phi}{\sqrt{(l^2-x^2)}} + \frac{A}{l^3}(x\phi - x_0\phi_0)\phi \end{aligned}\right\} \quad (9).$$

Here we see that $Ld\phi + Rdx$ is the differential of a function of the two independent variables, x, ϕ. Thus if we denote this function by E, we have

$$\left.\begin{aligned} E &= \tfrac{1}{2}\frac{B}{l^3}\{\sqrt{(l^2-x^2)}\phi - \sqrt{(l^2-x_0^2)}\phi_0\}^2 + \tfrac{1}{2}\frac{A}{l^3}(x\phi - x_0\phi_0)^2 \\ L &= \frac{dE}{d\phi}, \quad R = \frac{dE}{dx} \end{aligned}\right\} \quad (10),$$

a conclusion which might have been inferred at once from the general principle of energy, thus:—

606. The potential energy of the strained spring is easily seen from § 595 (4), above, to be

$$\tfrac{1}{2}[B(\varpi - \varpi_0)^2 + A\tau^2]l,$$

if A denote the torsional rigidity, B the flexural rigidity in the plane of curvature, ϖ and ϖ_0 the strained and unstrained curvatures, and τ the torsion of the wire in the strained condition, the torsion being reckoned as zero in the unstrained condition. The axial force, and the couple, required to hold the spring to any given length reckoned along the axis of the spiral, and to any given angle between planes through its ends and the axes, are of course (§ 272) equal to the rates of variation of the potential energy, per unit of variation of these co-ordinates respectively. It must be carefully remarked, however, that, if the terminal rigidly attached to one end of the spring be held fast so as to fix the tangent at this end, and the motion of the other terminal be so regulated as to keep the figure of the intermediate spring always truly spiral, this motion will be somewhat complicated; as the radius of the cylinder, the inclination of the axis of the spiral to the fixed direction of the tangent at the fixed end, and the position of the point in the axis in which it is cut by the plane perpendicular to it through the fixed end of the spring, all vary as the spring changes in figure. The *effective components* of any infinitely small motion of the movable terminal are its component translation along, and rotation round, the instantaneous position of the axis of the spiral [two degrees of freedom], along with which it will

Spiral springs.

generally have an infinitely small translation in some dir... and rotation round some line, each perpendicular to this ... and determined from the two degrees of arbitrary moti... the condition that the curve remains a true spiral.

607. In the practical use of spiral springs, this condi... not rigorously fulfilled: but, instead, either of two pla... generally followed:—(1.) Force, without any couple, is ap... pulling out or pressing together two definite points of the ... terminals, each as nearly as may be in the axis of the unstra... spiral; or (2.) One terminal being held fixed, the oth... allowed to slide, without any turning, in a fixed direction, b... as nearly as may be the direction of the axis of the spiral w... unstrained. The preceding investigation is applicable to ... infinitely small displacement in either case: the couple b... put equal to zero for case (1.), and the instantaneous rota... motion round the axis of the spiral equal to zero for case (2.)

For infinitely small displacements let $\phi=\phi_0+\delta\phi$, a... $x=x_0+\delta x$, in (10), so that now

$$L=\frac{dE}{d\delta\phi},\quad R=\frac{dE}{d\delta x}.$$

Then, retaining only terms of the lowest degree relative to ... and $\delta\phi$ in each formula, and writing x and ϕ instead of x_0 a... ϕ_0, we have

$$E=\frac{1}{2l^3}\{(B\frac{x^2}{l^2-x^2}+A)\phi^2\delta x^2+2(A-B)x\phi\delta x\delta\phi+[B(l^2-x^2)+Ax^2]\delta\phi^2$$

$$R=\frac{1}{l^3}\{(B\frac{x^2}{l^2-x^2}+A)\phi^2\delta x+(A-B)x\phi\delta\phi\}$$

$$L=\frac{1}{l^3}\{(A-B)x\phi\delta x+[B(l^2-x^2)+Ax^2]\delta\phi\}$$

Example 1.—For a spiral of 45° inclination we have

$$x^2=\tfrac{1}{2}l^2 \text{ and } \phi^2=\tfrac{1}{2}\frac{l^2}{r^2}:$$

and the formulæ become

$$\left.\begin{aligned}R&=\tfrac{1}{2}\frac{1}{lr^2}[(A+B)\delta x+(A-B)r\delta\phi]\\ L&=\tfrac{1}{2}\frac{1}{lr}[(A-B)\delta x+(A+B)r\delta\phi]\end{aligned}\right\}\quad (12)$$

A careful study of this case, illustrated if necessary by a m...l easily made out of ordinary iron or steel wire, will be found ... instructive.

Example 2.—Let $\frac{x}{l}$ be very small. Neglecting, therefore, its square, we have $\phi=\frac{l}{r}$, and $L=\frac{B}{l}\delta\phi=B\delta\frac{1}{r}$; and $R=\frac{A}{lr^2}\delta x$. Spiral spring of infinitely small inclination:

The first of these is simply the equation of direct flexure (§ 595). The interpretation of the second is as follows:—

608. In a spiral spring of infinitely small inclination to the plane perpendicular to its axis, the displacement produced in the movable terminal by a force applied to it in the axis of the spiral is a simple rectilineal translation in the direction of the axis, and is equal to the length of the circular arc through which an equal force carries one end of a rigid arm or crank equal in length to the radius of the cylinder, attached perpendicularly to one end of the wire of the spring supposed straightened and held with the other end absolutely fixed, and the end which bears the crank, free to turn in a collar. This statement is due to J. Thomson,[1] who showed that in pulling out a spiral spring of infinitely small inclination the action exercised and the elastic quality used are the same as in a torsion-balance with the same wire straightened (§ 433). This theory is, as he proved experimentally, sufficiently approximate for most practical applications; spiral springs, as commonly made and used, being of very small inclination. There is no difficulty in finding the requisite correction, for the actual inclination in any case, from the preceding formulæ. The fundamental principle that spiral springs act chiefly by torsion seems to have been first discovered by Binet in 1814.[2] virtually a torsion-balance.

609. Returning to the case of a uniform wire straight and untwisted (that is, cylindrical or prismatic) when free from stress; let us suppose one end to be held fixed in a given direction, and no other force from without to influence it except that of a rigid frame attached to its other end acted on by a force, R, in a given line, AB, and a couple, G, in a plane perpendicular to this line. The form and twist it will have when in equilibrium are determined by the condition that the torsion and flexure at any point, P, of its length are those due to the couple G compounded with the couple obtained by bringing R Elastic curve transmitting force and couple.

[1] *Camb. & Dub. Math. Jour.* 1848. [2] St. Venant, *Comptes Rendus.* Sept. 1864.

Kirchhoff's kinetic comparison.

to P. It follows that if the rigid body of § 597 be left itself at any instant, moving in the manner prescribed round the fixed point O, and subjected only to a constant force equal to R acting on it at the point, $_{,}T$, in a line, $_{,}TD$, parallel to AB, it will continue moving in the prescribed manner.

To prove this let the body be compelled to move in the prescribed manner, and at the same time let the force R act on it in the line, $_{,}TD$. Then, taking the co-ordinate axis, OX, parallel to this line, and calling x, y, z the co-ordinates of P at any time t, we have $\frac{dx}{ds}$, $\frac{dy}{ds}$, $\frac{dz}{ds}$ for the direction-cosines of $O_{,}T$: and as the length of $O_{,}T$ is unity, the moments of R in $_{,}TD$, round OX, OY, OZ are respectively

$$0,\ R\frac{dz}{ds},\ -R\frac{dy}{ds}.$$

Hence the moments of momentum generated by these in any time (being simply their time integrals) are (since $s=t$)

$$0,\ R(z-z_0),\ -R(y-y_0),$$

if (y_0, z_0), (y, z) be the co-ordinates of P at the beginning and end of the time. But these are precisely the additions to the component flexure-torsion couples in the wire from the first to the second position of P; and therefore if the component moments of momentum of the body are equal at the beginning of the time to the component flexure-torsion couples of the wire at the point (x_0, y_0, z_0) the simple action of the force R upon it with the point O held fixed, will keep its moment of momentum constantly in agreement with the flexure-torsion couple of the wire, and consequently its lines $O_{,}T$, $O_{,}K$, $O_{,}L$ constantly parallel to the corresponding lines through P in the wire, as P moves along it at unit velocity.

This very remarkable theorem was discovered by Kirchhoff, to whom also the first thoroughly general investigation of the equations of equilibrium and motion of an elastic wire is due.[1]

610. The comparison thus established between the static problem of the bending and twisting of a wire, and the kinetic problem of the rotation of a rigid body, affords highly interesting illustrations, and, as it were, graphic representations, of the

[1] *Crelle's Journal*. 1859. Ueber das Gleichgewicht und die Bewegung e… unendlich dünnen elastischen Stabes.

circumstances of either by aid of the other; the usefulness of which in promoting a thorough mental appropriation of both must be felt by every student who values rather the physical subject than the mechanical process of working through mathematical expressions, to which so many minds able for better things in science have unhappily been devoted of late years. Kirchhoff's kinetic comparison.

When particularly occupied with the kinetic problem in chap. IX., we shall have occasion to examine the rotations corresponding to the spirals of § 601-603, and to point out also the general character of the elastic curves corresponding to some of the less simple cases of rotatory motion.

611. For the present we confine ourselves to one example, which, so far as the comparison between the static and kinetic problems is concerned, is the simplest of all—the *Elastic Curve* of James Bernoulli, and the common pendulum. A uniform straight wire, either equally flexible in all planes through its length, or having its directions of maximum and minimum flexural rigidity in two planes through its whole length, is acted on by a force and couple in one of these planes, applied either directly to one end, or by means of an arm rigidly attached to it, the other end being held fast. The force and couple may, of course (§ 558), be reduced to a single force, the extreme case of a couple being mathematically included as an infinitely small force at an infinitely great distance. To avoid any restriction of the problem, we must suppose this force applied to an arm rigidly attached to the wire, although in any case in which the line of the force cuts the wire, the force may be applied directly at the point of intersection, without altering the circumstances of the wire between this point and the fixed end. The wire will, in these circumstances, be bent into a curve lying throughout in the plane through its fixed end and the line of the force, and (§ 599) its curvatures at different points will, as was first shown by James Bernoulli, be simply as their distances from this line. The curve fulfilling this condition has clearly just two independent parameters, of which one is conveniently regarded as the mean proportional, a, between the radius of curvature at any point and its distance from the line of force, and the other, the maximum distance, b, of the wire from the line of force. By choosing any value for each of these para- Common pendulum and plane elastic curve.

Graphic construction of elastic curve transmitting force in one plane.

meters it is easy to trace the corresponding curve with a very high approximation to accuracy, by commencing with a small circular arc touching at one extremity a straight line at the given maximum distance from the line of force, and continuing by small circular arcs, with the proper increasing radii, according to the diminishing distances of their middle points from the line of force. The annexed diagrams are, however, not so drawn; but are simply traced from the forms actually assumed by a flat steel spring, of small enough breadth not to be much disturbed by tortuosity in the cases in which different parts of it cross one another. The mode of application of the force is sufficiently explained by the indications in the diagram.

Equation of the plane elastic curve.

Let the line of force be axis of x, and let ρ be the radius of curvature at any point (x, y) of the curve. The dynamical condition stated above becomes

$$\rho y = \frac{B}{T} = a^2 \qquad (1),$$

where B denotes the flexural rigidity, T the tension of the cord, and a a linear parameter of the curve depending on these elements. Hence, by the ordinary formula for ρ^{-1},

$$y = \frac{a^2\frac{d^2y}{dx^2}}{\left(1+\frac{dy^2}{dx^2}\right)^{\frac{3}{2}}} \qquad (2).$$

Multiplying by $2dy$ and integrating, we have

$$y^2 = C - \frac{2a^2}{\left(1+\frac{dy^2}{dx^2}\right)^{\frac{1}{2}}} \qquad (3);$$

and finally,

$$x = \int\frac{(y^2-C)dy}{(4a^4-C^2+2Cy^2-y^4)^{\frac{1}{2}}} \qquad (4),$$

which is the equation of the curve expressed in terms of an elliptic integral.

If, in the first integral, (3), we put $\frac{dy}{dx}=0$, we find

$$y = \pm(C\pm 2a^2)^{\frac{1}{2}} \qquad (5),$$

the upper sign within the bracket giving points of maximum, and the lower, points, if any real, of minimum distance from the axis. Hence there are points of equal maximum distance from the line of

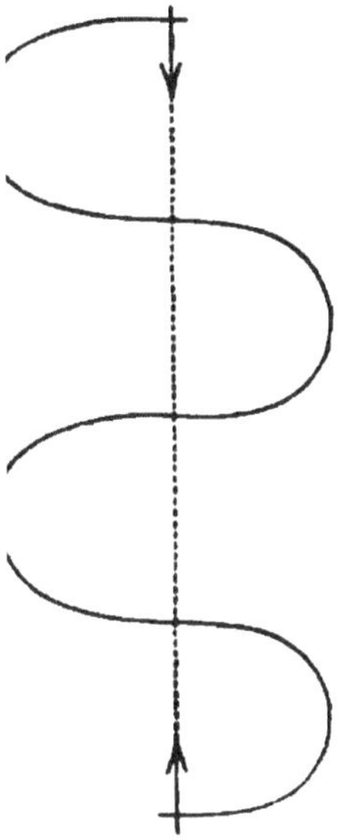

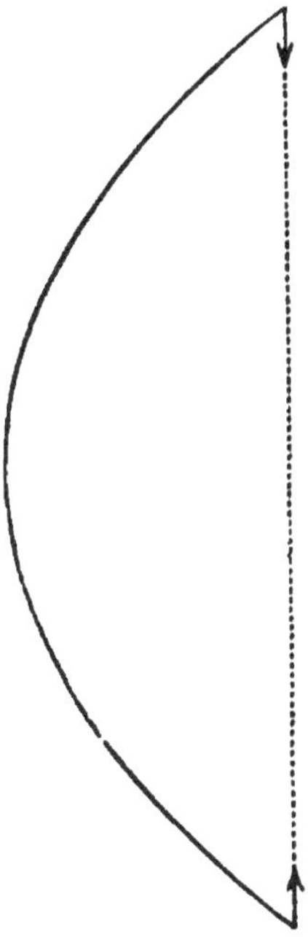

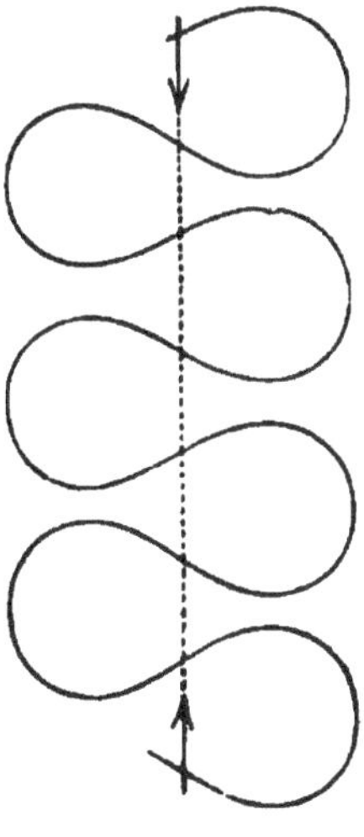

Fig. 3.

Fig. 4.

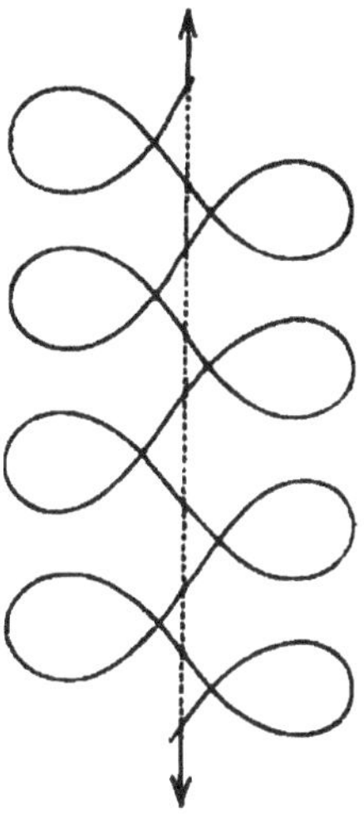

Fig. 5.

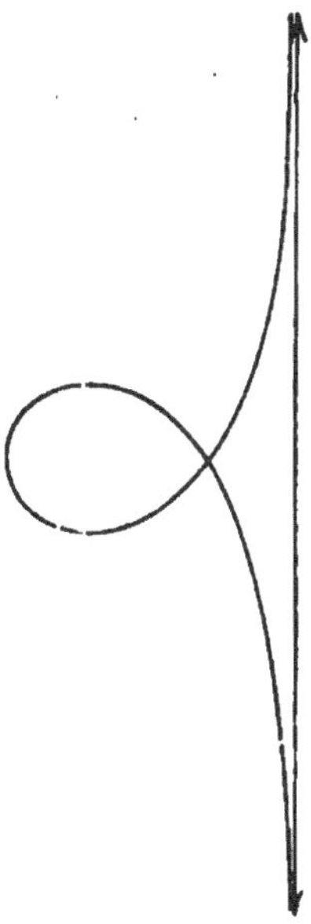

Fig. 6.

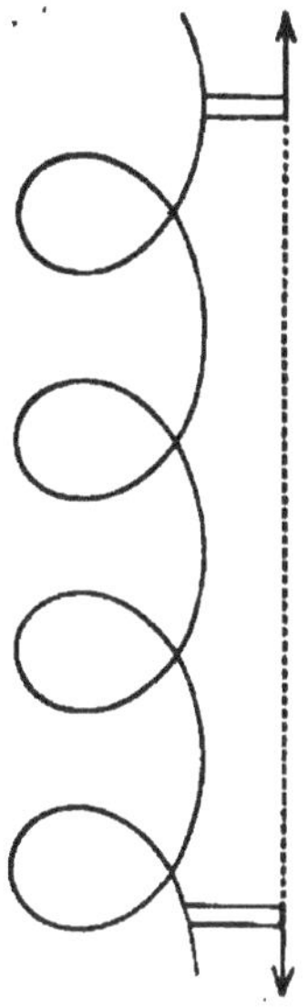

Fig. 7.

Equation of the plane elastic curve.

force on its two sides, but no real minima when $C < 2a^2$; wh therefore comprehends the cases of diagrams 1...5. But there real minima as well as maxima when $C > 2a^2$, which is theref the case of diagram 7. In this case it may be remarked the analytical equations comprehend two equal and similar tached curves symmetrically situated on the two sides of the of force; of which one only is shown in the diagram.

The intermediate case, $C = 2a^2$, is that of diagram 6. For it the final integral degrades into a logarithmic form, as follows:—

$$x = \int \frac{y\,dy}{(4a^2 - y^2)^{\frac{1}{2}}} - \int \frac{2a^2\,dy}{y(4a^2 - y^2)^{\frac{1}{2}}};$$

or, with the integrations effected, and the constant assigned to make the axis of y be that of symmetry,

$$x = -(4a^2 - y^2)^{\frac{1}{2}} + a \log \frac{2a + (4a^2 - y^2)^{\frac{1}{2}}}{y} \qquad (6).$$

This equation, when the radical is taken with the signs indicated represents the branch proceeding from the vertex, first to the negative side of the axis of y, crossing it at the double point, and going to infinity towards the positive axis of x as an asymptote The other branch is represented by the same equation with the sign of the radical reversed in each place.

It need scarcely be remarked that in (3) the sign of $(1 + \frac{dy^2}{dx^2})$ can only change, for a point moving continuously along the curve, when $\frac{dy}{dx}$ becomes infinite. The interpretation is facilitated by putting

$$\frac{dy}{dx} = \tan\theta, \text{ or } (1 + \frac{dy^2}{dx^2})^{\frac{1}{2}} = -\cos\theta,$$

which reduces (3) to

$$y^2 = 2a^2\cos\theta + C \qquad (7)$$

Here, when $C > 2a^2$ (the case in which, as we have seen above there are minimum as well as maximum values of y on one side of the line of force), there is no limit to the value of θ. It increases, of course, continuously for a point moving continuously along the curve; the augmentation being 2π for one complete period (diagram 7).

When $C < 2a^2$, θ has equal positive and negative values at the points in which the curve cuts the line of force. These values being given by the equation

$$\cos\theta = -\frac{C}{2a^2} \qquad (8).$$

are obtuse when C is positive (diagram 3), and acute when C is negative (diagram 1). The extreme negative value of C is of course $-2a^2$. Equation of the plane elastic curve.

If we take $C=-2a^2+b^2$,

$\pm b$ will be the maximum positive or negative value of y, as we see by (7); and if we suppose b to be small in comparison with a, we have the case of a uniform spring bent, as a bow, but slightly, by a string stretched between its ends.

612. An important particular case is that of figure 1, which corresponds to a bent bow having the same flexural rigidity throughout. If the amount of bending be small, the equation is easily integrated to any requisite degree of approximation. We will merely sketch the process of investigation. Bow slightly bent.

Let e be the maximum distance from the axis, corresponding to $x=0$. Then $y=e$ gives $\frac{dy}{dx}=0$, and (3) becomes

$$e^2-y^2=2a^2\left(1-\frac{1}{\sqrt{1+\frac{dy^2}{dx^2}}}\right);$$

whence

$$\frac{dy}{dx}=\frac{\sqrt{e^2-y^2}\sqrt{4a^2-e^2+y^2}}{2a^2-e^2+y^2} \qquad (9).$$

For a first approximation, omit e^2-y^2 in comparison with a^2 where they occur in the same factors, and we have

$$\frac{dy}{dx}=\frac{\sqrt{e^2-y^2}}{a},$$

or, since $y=e$ when $x=0$,

$$y=e\cos\frac{x}{a} \qquad (10)$$

the harmonic curve, or curve of sines, which is the simplest form assumed by a vibrating cord or pianoforte-wire.

For a closer approximation we may substitute for y, in those factors where it was omitted, the value given by (10); and so on. Thus we have

$$\frac{dy}{dx}=\frac{\sqrt{e^2-y^2}}{a}\left(1+\frac{3e^2}{8a^2}\sin^2\frac{x}{a}\right), \text{ nearly,}$$

or

$$\frac{dy}{\sqrt{e^2-y^2}}=\frac{dx}{a}\left(1+\frac{3e^2}{16a^2}-\frac{3e^2}{16a^2}\cos\frac{2x}{a}\right),$$

from which, by integration,

$$\cos^{-1}\frac{y}{e}=\frac{x}{a}\left(1+\frac{3e^2}{16a^2}\right)-\frac{3e^2}{32a^2}\sin\frac{2x}{a}$$

and

$$y=e\cos\left\{\frac{x}{a}\left(1+\frac{3e^2}{16a^2}\right)\right\}+\frac{3e^3}{32a^2}\sin\frac{x}{a}\sin\frac{2x}{a}.$$

Plane elastic curve and common pendulum.

613. As we choose particularly the common pendulum for the corresponding kinetic problem, the force acting on the rigid body in the comparison must be that of gravity in the vertical through its centre of gravity. It is convenient, accordingly, not to take *unity* as the velocity for the point of comparison along the bent wire, but the velocity gravity would generate in a body falling through a height equal to half the constant, a, of § 611 : and this constant, a, will then be the length of the isochronous simple pendulum. Thus if an elastic curve be held with its line of force vertical, and if a point, P be moved along it with a constant velocity equal to $\sqrt{gr}$, (r denoting the mean proportional between the radius of curvature at any point and its distance from the line of force,) the tangent at P will keep always parallel to a simple pendulum, of length a, placed at any instant parallel to it, and projected with the same angular velocity. Diagrams 1...5, correspond to *vibrations* of the pendulum. Diagram 6 corresponds to the case in which the pendulum would just reach its position of unstable equilibrium in an infinite time. Diagram 7 corresponds to cases in which the pendulum flies round continuously in one direction, with periodically increasing and diminishing velocity. The extreme case, of the circular elastic curve, corresponds to an infinitely long pendulum flying round with finite angular velocity, which of course experiences only infinitely small variation in the course of the revolution. A conclusion worthy of remark is, that the rectification of the elastic curve is the same analytical problem as finding the time occupied by a pendulum in describing any given angle.

Wire of any shape disturbed by forces and couples applied through its length.

614. Hitherto we have confined our investigation of the form and twist of a wire under stress to a portion of the whole wire not itself acted on by force from without, but merely engaged in transmitting force between two equilibrating systems applied to the wire beyond this portion; and we have, thus, not included the very important practical cases of a curve deformed by its own weight or centrifugal force, or fulfilling such conditions of equilibrium as we shall have to use afterwards in finding its equations of motion according to D'Alembert's principle. We therefore proceed now to a perfectly general investigation of the equilibrium of a curve, uniform or not uniform throughout its length; either straight, or bent and twisted in any way, when free from stress; and not restricted by any condition as to the positions of the three principal flexure-torsion axes (§ 596); under the influence of any distribution whatever of force and couple through its whole length. Wire of any shape disturbed by forces and couples applied through its length.

Let α, β, γ be the components of the mutual force, and ξ, η, ζ those of the mutual couple, acting between the matter on the two sides of the normal section through (x, y, z). Those for the normal section through $(x+\delta x, y+\delta y, z+\delta z)$ will be

$$\alpha+\frac{d\alpha}{ds}\delta s,\ \beta+\frac{d\beta}{ds}\delta s,\ \gamma+\frac{d\gamma}{ds}\delta s$$

$$\xi+\frac{d\xi}{ds}\delta s,\ \eta+\frac{d\eta}{ds}\delta s,\ \zeta+\frac{d\zeta}{ds}\delta s.$$

Hence, if $X\delta s$, $Y\delta s$, $Z\delta s$, and $L\delta s$, $M\delta s$, $N\delta s$ be the components of the applied force, and applied couple, on the portion δs of the wire between those two normal sections, we have (§ 551) for the equilibrium of this part of the wire

$$-X=\frac{d\alpha}{ds},\ -Y=\frac{d\beta}{ds},\ -Z=\frac{d\gamma}{ds} \qquad (1),$$

and (neglecting, of course, infinitely small terms of the second order, as $\delta y\delta s$)

$$-L\delta s=\frac{d\xi}{ds}\delta s+\gamma\delta y-\beta\delta z,\ \text{etc.};$$

or

$$L=\frac{d\xi}{ds}+\gamma\frac{dy}{ds}-\beta\frac{dz}{ds},\ -M=\frac{d\eta}{ds}+\alpha\frac{dz}{ds}-\gamma\frac{dx}{ds},\ -N=\frac{d\zeta}{ds}+\beta\frac{dx}{ds}-\alpha\frac{dy}{ds} \qquad (2).$$

We may eliminate α, β, γ from these six equations by means of the following convenient assumption—

$$\alpha\frac{dx}{ds}+\beta\frac{dy}{ds}+\gamma\frac{dz}{ds}=T \qquad (3),$$ Longitudinal tension.

T meaning the component of the force acting across the normal section, along the tangent to the middle line. From this, and the second and third of (2), we have

$$\alpha=T\frac{dx}{ds}-(M+\frac{d\eta}{ds})\frac{dz}{ds}+(N+\frac{d\zeta}{ds})\frac{dy}{ds}.$$

This, and the symmetrical expressions for β and γ, used in (1), give

Wire of any shape disturbed by forces and couples applied through its length.

$$\left.\begin{aligned} X&=-\frac{d}{ds}\left\{T\frac{dx}{ds}-(M+\frac{d\eta}{ds})\frac{dz}{ds}+(N+\frac{d\zeta}{ds})\frac{dy}{ds}\right\} \\ Y&=-\frac{d}{ds}\left\{T\frac{dy}{ds}-(N+\frac{d\zeta}{ds})\frac{dx}{ds}+(L+\frac{d\xi}{ds})\frac{dz}{ds}\right\} \\ Z&=-\frac{d}{ds}\left\{T\frac{dz}{ds}-(L+\frac{d\xi}{ds})\frac{dy}{ds}+(M+\frac{d\eta}{ds})\frac{dx}{ds}\right\} \end{aligned}\right\} \quad (4)$$

We have besides, from (2)

$$0=\frac{dx}{ds}(L+\frac{d\xi}{ds})+\frac{dy}{ds}(M+\frac{d\eta}{ds})+\frac{dz}{ds}(N+\frac{d\zeta}{ds}) \quad (5)$$

To complete the mathematical expression of the circumstances it only remains to introduce the equations of torsion-flexure. For this purpose, let any two lines of reference for the substance of the wire, PK, PL, be chosen at right angles to one another in the normal section through P. Let κ_0, λ_0 be the components of the curvature (§ 589) in the planes perpendicular to these lines and through the tangent, PT, when the wire is unstrained; and κ, λ what they become under the actual stress. Let τ_0 denote the rate of twist (§ 119) of either line of reference round the tangent from point to point along the wire in the unstrained condition, and τ in the strained, so that $\tau-\tau_0$ is the rate of twist produced at P by the actual stress. Thus [§ 595 (3)], we have

Equations of torsion-flexure.

$$\left.\begin{aligned} &\xi l+\eta m+\zeta n=A(\kappa-\kappa_0)+c(\lambda-\lambda_0)+b(\tau-\tau_0) \\ &\xi l'+\eta m'+\zeta n'=c(\kappa-\kappa_0)+B(\lambda-\lambda_0)+a(\tau-\tau_0) \\ &\xi\frac{dx}{ds}+\eta\frac{dy}{ds}+\zeta\frac{dz}{ds}=b(\kappa-\kappa_0)+a(\lambda-\lambda_0)+C(\tau-\tau_0) \end{aligned}\right\} \quad (6)$$

where (l, m, n), (l', m', n'), $(\frac{dx}{ds}, \frac{dy}{ds}, \frac{dz}{ds})$ denote the directions of PK, PL, PT; so that

$$\left.\begin{aligned} l\frac{dx}{ds}+m\frac{dy}{ds}+n\frac{dz}{ds}=0,\quad l'\frac{dx}{ds}+m'\frac{dy}{ds}+n'\frac{dz}{ds}=0 \\ ll'+mm'+nn'=0, \\ l^2+m^2+n^2=1,\qquad l'^2+m'^2+n'^2=1 \end{aligned}\right\} \quad (7)$$

Now if lines $O_{\prime}K$, $O_{\prime}L$, $O_{\prime}T$, each of unit length, be drawn, as in § 593, always parallel to PK, PL, PT, and if P be carried at unit velocity along the curve, the component velocity of ${}_{\prime}L$ parallel to $O_{\prime}T$, or that of ${}_{\prime}T$ parallel to $O_{\prime}K$ with its sign changed, is (§ 593) equal to κ; and similar statements apply to λ and τ. Hence,

$$\left.\begin{aligned}\kappa&=-\left(l'\frac{d\frac{dx}{ds}}{ds}+m'\frac{d\frac{dy}{ds}}{ds}+n'\frac{d\frac{dz}{ds}}{ds}\right)\\ \lambda&=+\left(l\frac{d\frac{dx}{ds}}{ds}+m\frac{d\frac{dy}{ds}}{ds}+n\frac{d\frac{dz}{ds}}{ds}\right)\\ \tau&=+\left(l'\frac{dl}{ds}+m'\frac{dm}{ds}+n'\frac{dn}{ds}\right)\end{aligned}\right\}\quad(8).$$

Torsion, and two components of curvature, of wire (or component angular velocities of rotating solid).

Equations (7) reduce (l, m, n), (l', m', n') to one variable element, being the co-ordinate by which the position of the substance of the wire, round the tangent at any point of the central curve, is specified: and (8) express κ, λ, τ in terms of this co-ordinate, and the three Cartesian co-ordinates x, y, z of P. The specification of the constrained condition of the wire gives κ_0, λ_0, τ_0 as functions of s. Thus (6) gives ξ, η, ζ each in terms of s, and the four co-ordinates, and their differential coefficients relatively to s. Substituting these in (4) and (5) we have four differential equations which, with

$$\frac{dx^2}{ds^2}+\frac{dy^2}{ds^2}+\frac{dz^2}{ds^2}=1 \quad (9),$$

constitute the five equations by which the five unknown functions (the four co-ordinates, and the tension, T,) are to be determined in terms of s, or by means of which, with s and T eliminated, the two equations of the curve may be found, and the co-ordinate for the position of the normal section round the tangent determined in terms of x, y, z.

Terminal conditions.

The terminal conditions for any specified circumstances are easily expressed in the proper mathematical terms, by aid of equations (2). Thus, for instance, if a given force and a given couple be directly applied to a free end, or if the problem be limited to a portion of the wire terminated in one direction at a point Q, and if, in virtue of actions on the wire beyond, we have a given force $(\alpha_0, \beta_0, \gamma_0)$, and a given couple (ξ_0, η_0, ζ_0) acting on the normal section through Q of the portion under consideration, and if s_0 is the length of the wire from the zero of reckoning for s up to the point Q, and L_0, M_0, N_0 the values of L, M, N at this point, the equations expressing the terminal conditions will be

$$\left.\begin{aligned}\xi&=\xi_0, & -\frac{d\xi}{ds}&=L_0+\left(\gamma_0\frac{dy}{ds}-\beta_0\frac{dz}{ds}\right)\\ \eta&=\eta_0, & -\frac{d\eta}{ds}&=M_0+\left(\alpha_0\frac{dz}{ds}-\gamma_0\frac{dx}{ds}\right)\\ \zeta&=\zeta_0, & -\frac{d\zeta}{ds}&=N_0+\left(\beta_0\frac{dx}{ds}-\alpha_0\frac{dy}{ds}\right)\end{aligned}\right|\ \text{when } s=s_0\ \Bigg\}\quad(10).$$

Terminal conditions.

From these we see, by taking $L_0=0$, $M_0=0$, $N_0=0$, $\alpha_0=\beta_0=0$, $\gamma_0=0$, $\xi_0=0$, $\eta_0=0$, $\zeta_0=0$, that

615. For the simple and important case of a natur… straight wire, acted on by a distribution of force, but n… couple, through its length, the condition fulfilled at a perf… free end, acted on by neither force nor couple, is that the cur… ture is zero at the end, and its rate of variation from zero… unit of length from the end, is, at the end, zero. In other w… the curvatures at points infinitely near the end are as … squares of their distances from the end in general (or, as … higher power of these distances, in singular cases). The … statements hold for the *change* of curvature produced by … stress, if the unstrained wire is not straight, but the ot… circumstances the same as those just specified.

Straight beam infinitely little bent.

616. As a very simple example of the equilibrium of … wire subject to forces through its length, let us suppose … natural form to be straight, and the applied forces to be … lines, and the couples to have their axes, all perpendicul… its length, and to be not great enough to produce more th… an infinitely small deviation from the straight line. Furt… in order that these forces and couples may produce no tors… let the three flexure-torsion axes be perpendicular to … along the wire. But we shall not limit the problem furt… by supposing the section of the wire to be uniform, as w… should thus exclude some of the most important pract… applications, as to beams of balances, levers in machine… beams in architecture and engineering. It is more instruct… to investigate the equations of equilibrium directly for … case than to deduce them from the equations worked out ab… for the much more comprehensive general problem. The par… ticular principle for the present case is simply that the rate of variation of the rate of variation, per unit of length along th… wire, of the bending couple in any plane through the length, is equal, at any point, to the applied force per unit of length, wit… the simple rate of variation of the applied couple subtract… This, together with the direct equations (§ 599) between th… component bending couples, gives the required equations of equilibrium.

The diagram representing a section of the wire in the plane xy, let $OP = x$, $PP' = \delta x$. Let Y and N be the components Straight beam infinitely little bent.

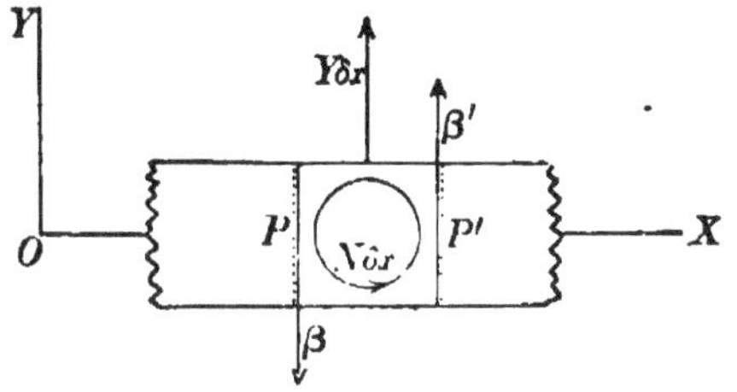

in the plane of the diagram, of the applied force and couple, each reckoned per unit of length of the wire; so that $Y\delta x$ and $N\delta x$ will be the amounts of force and couple in this plane, actually applied to the portions of the wire between P and P'.

Let, as before (§ 614), β and γ denote the components parallel to OY and OZ of the mutual force,[1] and ζ and η the components in the plane XOY, XOZ, of the mutual couple, between the portions of matter on the two sides of the normal section through P; and β', γ', and ζ', η' the same for P'. The matter between these two sections is balanced under these actions from the matter contiguous to it beyond them, and the force and couple applied to it from without. These last have, in the plane XOY, components respectively equal to $Y\delta x$ and $N\delta x$: and hence for the equilibrium of the portion PP',

$-\beta + Y\delta x + \beta' = 0$, by forces parallel to OY,

and $-\zeta + N\delta x + \zeta' + \beta\delta x = 0$, by couples in plane XOY,

the term $\beta\delta x$ in this second equation being the moment of the couple formed by the infinitely nearly equal forces β, β' in the dissimilar parallel directions through P and P'. Now

$$\beta' - \beta = \frac{d\beta}{dx}\delta x, \text{ and } \zeta' - \zeta = \frac{d\zeta}{dx}\delta x.$$

Hence the preceding equations give

$$\left.\begin{aligned} \frac{d\beta}{dx} &= -Y \\ \frac{d\zeta}{dx} &= -N - \beta \end{aligned}\right\}. \qquad (1);$$

[1] These forces, being each *in* the plane of section of the solid separating the portions of matter between which they act, are of the kind called *shearing forces*. See below, § 662.

Straight beam infinitely little bent.

and these, by the elimination of β,

$$\frac{d^2\zeta}{dx^2}=-\frac{dN}{dx}+Y$$

Similarly, by forces and couples in the plane XOZ,

$$\frac{d^2\eta}{dx^2}=-\frac{dM}{dx}+Z$$

couples in this plane being reckoned positive when they t... turn from the direction of OX to that of OZ [which is ... to the convention (551) generally adopted as being proper ... the three axes are dealt with symmetrically].

Since the wire deviates infinitely little from the straig... OX, the component curvatures are

$$\frac{d^2y}{dx^2} \text{ in the plane } XOY,$$

and $$\frac{d^2z}{dx^2} \quad " \quad " \quad XOZ.$$

Hence the equations of flexure are

$$\left.\begin{aligned} \zeta &= B\frac{d^2y}{dx^2}+a\frac{d^2z}{dx^2} \\ \eta &= a\frac{d^2y}{dx^2}+C\frac{d^2z}{dx^2} \end{aligned}\right\}$$

where B and C are the flexural rigidities (§ 596) in the ... xy and xz, and a the coefficient expressing the couple in ... produced by unit curvature in the other; three quantiti... are to be regarded, in general, as given functions of x. S... tuting these expressions for ζ and η, in (2) and (3), w... the required equations of equilibrium.

Case of independent flexure in two planes.

617. If the directions of maximum and minimum fl... rigidity lie throughout the wire in two planes, the equ... of equilibrium become simplified by these planes being ... as planes of reference, XOY, XOZ. The flexure in either ... then depends simply on the forces in it, and thus the pr... divides itself into the two quite independent problems ... tegrating the equations of flexure in the two principal p... and so finding the projections of the curve on two fixed p... agreeing with their position when the rod is straight.

In this case, and with XOY, XOZ so chosen, we have $a=$ Hence the equations of flexure (4) become simply

$$\zeta=B\frac{d^2y}{dx^2},\quad \eta=C\frac{d^2z}{dx^2};$$

and the differential equations of the curve, found by using these in (2) and (3),

Straight beam infinitely little bent. Case of independent flexure in two planes.

$$\frac{d^2\left(B\frac{d^2y}{dx^2}\right)}{dx^2}=\mathfrak{Y},\quad \frac{d^2\left(C\frac{d^2z}{dx^2}\right)}{dx^2}=\mathfrak{Z} \qquad (5),$$

where $$\mathfrak{Y}=-\frac{dN}{dx}+Y,\quad \mathfrak{Z}=-\frac{dM}{dx}+Z \qquad (6).$$

Here $\mathfrak{Y}$ and $\mathfrak{Z}$ are to be generally regarded as known functions of x, given explicitly by (6), being the amounts of component simple forces perpendicular to the wire, reckoned per unit of its length, that would produce the same figure as the distribution of force and couple we have supposed actually applied throughout the length. Later, when occupied with the theory of magnetism, we shall meet with a curious instance of the relation expressed by (6) In the meantime it may be remarked that although the figure of the wire does not sensibly differ when the simple distribution of force is substituted for any given distribution of force and couple, the shearing forces in normal sections become thoroughly altered by this change of circumstances, as is shown by (1). When the wire is uniform, B and C are constant, and the equations of equilibrium become

$$\frac{d^4y}{dx^4}=\frac{\mathfrak{Y}}{B},\quad \frac{d^4z}{dx^4}=\frac{\mathfrak{Z}}{C} \qquad (7).$$

Plank bent by its own weight.

The simplest example is had by taking $\mathfrak{Y}$ and $\mathfrak{Z}$ each constant, a very interesting and useful case, being that of a uniform beam influenced only by its own weight, except where held or pressed by its supports. Confining our attention to flexure in the one principal plane, XOY, and supposing this to be vertical, so that $\mathfrak{Y}=gw$, if w be the mass per unit of length; we have, for the complete integral, of course

$$y=\frac{gw}{B}\left(\tfrac{1}{24}x^4+Kx^3+K'x^2+K''x+K'''\right) \qquad (8),$$

where K, K', etc., denote constants of integration. These, four in number, are determined by the terminal conditions; which, for instance, may be that the value of y and of $\frac{dy}{dx}$ is given for each end. Or, as for instance in the case of a plank simply resting with its ends on two edges or tressles, and free to turn round either, the condition may be that the curvature vanishes at each end: so that if OX be taken as the line through the points of support, we have

Plank supported by its ends;

$$\left.\begin{array}{l} y=0 \\ \dfrac{d^2y}{dx^2}=0 \end{array}\right\} \text{ when } x=0 \text{ and when } x=l,$$

l being the length of the plank. The solution then is

$$y=\frac{gw}{B}\cdot\tfrac{1}{24}(x^4-2lx^3+l^3x) \qquad (9)$$

Hence, by putting $x=\frac{1}{2}l$, we find $y=\dfrac{gw}{B}\cdot\dfrac{5l^4}{16\times 24}$ for the distance by which the middle point is deflected from the straight line joining the points of support.

Or, as in the case of a plank balanced on a tressle at its middle (taken as zero of x), or hung by a rope tied round it there, we may have

by its middle.

$$\left.\begin{array}{l} y=0 \\ \dfrac{dy}{dx}=0 \end{array}\right\} \text{ when } x=0,$$

and

$$\left.\begin{array}{l} \dfrac{d^2y}{dx^2}=0 \\ \dfrac{d^3y}{dx^3}=0 \end{array}\right\} \text{ when } x=\tfrac{1}{2}l \text{ [see above, § 614 (10)].}$$

The solution in this case is, for the positive half of the plank,

$$y=\frac{gw}{B}\cdot\tfrac{1}{24}(x^4-2lx^3+\tfrac{3}{2}l^2x^2) \qquad (10)$$

By putting $x=\frac{1}{2}l$, we find $y=\dfrac{gw}{B}\cdot\dfrac{3l^4}{16.24}$. Hence

Droops compared.

618. When a uniform bar, beam, or plank is balanced on a single tressle at its middle, the droop of its ends is only $\frac{3}{5}$ of the droop which its middle has when the bar is supported on tressles at its ends. From this it follows that the former is $\frac{3}{8}$ and the latter $\frac{5}{8}$ of the droop or elevation produced by a force equal to half the weight of the bar, applied vertically downwards or upwards to one end of it, if the middle is held fast in a horizontal position. For let us first suppose the whole to rest on a tressle under its middle, and let two tressles be placed under its ends and gradually raised till the pressure is entirely taken off from the middle. During this operation the middle remains fixed and horizontal, while a force increasing to half the weight, applied vertically upwards on each end, raises it through a height equal to the sum of the droops in the two cases above referred to. This result is of course proved directly by comparing the absolute values of the droop in those two cases as found above, with the deflection from the tangent at the end of the cord in the elastic curve, figure 2, of § 611, which is cut by the cord at right angles. It may be stated otherwise thus: the droop of the middle of a uniform beam resting on tressles at its ends is increased in the ratio of 5 to 13 by laying a mass equal in weight to itself on its middle: and, if the beam is hung by its middle, the droop of the ends is increased in the ratio of 3 to 11 by hanging on each of them a mass equal to half the weight of the beam. Plank supported by its ends or middle;

619. The important practical problem of finding the distribution of the weight of a solid on points supporting it, when more than two of these are in one vertical plane, or when there are more than three altogether, which (§ 568) is indeterminate[1] if the solid is perfectly rigid, may be completely solved for a uniform elastic beam, naturally straight, resting on three or more points in rigorously fixed positions all nearly in one horizontal line, by means of the preceding results. by three or more points.

If there are i points of support, the $i-1$ parts of the rod between them in order and the two end parts will form $i+1$ curves expressed by distinct algebraic equations [§ 617 (8)], each involving four arbitrary constants. For determining these constants we have $4i+4$ equations in all, expressing the following conditions:—

I. The ordinates of the inner ends of the projecting parts of the rod, and of the two ends of each intermediate part, are respectively equal to the given ordinates of the corresponding points of support [$2i$ equations].

II. The curves on the two sides of each support have coincident tangents and equal curvatures at the point of transition from one to the other [$2i$ equations].

III. The curvature and its rate of variation per unit of length along the rod, vanish at each end [4 equations].

Thus the equation of each part of the curve is completely determined: and then, by § 616, we find the shearing force in any normal section. The difference between these in the

[1] It need scarcely be remarked that indeterminateness does not exist in nature. How it may occur in the problems of abstract dynamics, and is obviated by taking something more of the properties of matter into account, is instructively illustrated by the circumstances referred to in the text.

neighbouring portions of the rod on the two sides of a point of support, is of course equal to the pressure on this point.

Plank supported by its ends and middle.

620. The solution for the case of this problem in which two of the points of support are at the ends, and the third midway between them either exactly in the line joining them, or at any given very small distance above or below it, is found at once, without analytical work, from the particular result stated in § 618. Thus if we suppose the beam, after being first supported wholly by tressles at its ends, to be gradually pressed up by a tressle under its middle, it will bear a force simply proportional to the space through which it is raised from the zero point, until all the weight is taken off the ends and borne by the middle. The whole distance through which the middle rises during this process is, as we found, $\frac{gw}{B}\cdot\frac{8a^4}{16\cdot 24}$, and this whole elevation is $\frac{8}{5}$ of the droop of the middle in the first position. If therefore, for instance, the middle tressle be fixed exactly in the line joining those under the ends, it will bear $\frac{5}{8}$ of the whole weight, and leave $\frac{3}{16}$ to be borne by each end. And if the middle tressle be lowered from the line joining the end ones by $\frac{7}{15}$ of the space through which it would have to be lowered to relieve itself of all pressure, it will bear just $\frac{1}{3}$ of the whole weight, and leave the other two thirds to be equally borne by the two ends.

Rotation of a wire round its elastic central line.

Elastic universal flexure joint; § 109.

621. A wire of equal flexibility in all directions, and straight when freed from stress, offers, when bent and twisted in any manner whatever, not the slightest resistance to being turned round its elastic central curve, as its conditions of equilibrium are in no way affected by turning the whole wire thus equally throughout its length. The useful application of this principle, to the maintenance of equal angular motion in two bodies rotating round different axes, is rendered somewhat difficult in practice by the necessity of a perfect attachment and adjustment of each end of the wire, so as to have the tangent to its elastic central curve exactly in line with the axis of rotation. But if this condition is rigorously fulfilled, and the wire is of exactly equal flexibility in every direction, and exactly straight when free from stress, it will give, against any constant resistance, an accurately uniform motion from one to another of two bodies rotating round axes which may be inclined to one another at any angle, and need not be in one plane. If they are in one plane, if there is no resistance to the rotatory motion, and if the action of gravity on the wire is insensible, it will take some of the varieties of form (§ 612) of the plane elastic curve of James Bernoulli. But however much it is altered from this; whether by the axes not being in one plane; or by the torsion accompanying the transmission of a couple from one shaft to the other, and necessarily, when the axes are in one plane, twisting the wire out of it; or by gravity; the elastic central curve will remain at rest, the wire in every normal section rotating round it with uniform angular velocity, equal to that of each of the two bodies which it connects. Under Properties of Matter, we shall see, as indeed may be judged at once from the performances of the vibrating spring of a chronometer for twenty years, that imperfection in the elasticity of a metal wire does not exist to any such degree as to prevent the practical application of this principle, even in mechanism required to be durable. **Equable elastic rotating joint.**

It is right to remark, however, that if the rotation be too rapid, the equilibrium of the wire rotating round its unchanged elastic central curve may become unstable, as is immediately discovered by experiments (leading to very curious phenomena), when, as is often done in illustrating the kinetics of ordinary rotation, a rigid body is hung by a steel wire, the upper end of which is kept turning rapidly.

622. If the wire is not of rigorously equal flexibility in all directions, there will be a periodic inequality in the communicated angular motion, having for period a half turn of either body: or if the wire, when unstressed, is not exactly straight, there will be a periodic inequality, having the whole turn for its period. In other words, if ϕ and ϕ' be angles simultaneously turned through by the two bodies, with a constant working couple transmitted from one to the other through the wire, $\phi - \phi'$ will not be zero, as in the proper elastic universal flexure joint, but will be a function of $\sin 2\phi$ and $\cos 2\phi$ if the first defect alone exists; or it will be a function of $\sin\phi$ and $\cos\phi$ if there is the second defect whether alone or along with the first. It is probable that, if the bend in the wire when **Practical inequalities.**

Elastic rotating joint.

unstressed is not greater than can be easily provided against in actual construction, the inequality of action caused by it may be sufficiently remedied without much difficulty in practice, by setting it at one or at each end, somewhat inclined to the axis of the rotating body to which it is attached. But these considerations lead us to a subject of much greater interest in itself than any it can have from the possibility of usefulness in practical applications. The simple cases we shall choose illustrate three kinds of action which may exist, each either alone or with one or both the others, in the equilibrium of a wire not equally flexible in all directions, and straight when unstressed.

Rotation round its elastic central circle, of a straight wire made into a hoop.

623. A uniform wire, straight when unstressed, is bent till its two ends meet, which are then attached to one another, with the elastic central curve through each touching one straight line: so that whatever be the form of the normal section, and the quality, crystalline or non-crystalline, of the substance, the whole wire must become, when in equilibrium, an exact circle (gravity being not allowed to produce any disturbance). It is required to find what must be done to turn the whole wire uniformly through any angle round its elastic central circle.

If the wire is of exactly equal flexibility in all directions,[1] it will, as we have seen (§ 621), offer no resistance at all to this action, except of course by its own inertia; and if it is once set to rotate thus uniformly with any angular velocity, great or small, it would continue so for ever were the elasticity perfect and were there no resistance from the air or other matter touching the axis.

To avoid restricting the problem by any limitation, we must suppose the wire to be such that, if twisted and bent in any way, the potential energy of the elastic action developed, per unit of length, is a quadratic function of the twist, and two components of the curvature (§§ 590, 595), with six arbitrarily given coefficients. But as the wire has no twist,[2] three terms of this function disappear in the case before us, and there remain only

[1] In this case, clearly it might have been twisted before its ends were put together, without altering the circular form taken when left with its ends joined.

[2] Which we have supposed, in order that it may take a circular form: although in the important case of equal flexibility in all directions this condition would obviously be fulfilled, even with twist.

three terms,—those involving the squares and the product of the components of curvature in planes perpendicular to two rectangular lines of reference in the normal section through any point. The position of these lines of reference may be conveniently chosen so as to make the product of the components of curvature disappear: and the planes perpendicular to them will then be the planes of maximum and minimum flexural rigidity when the wire is kept free from twist.[1] There is no difficulty in applying the general equations of § 614 to express these circumstances and answer the proposed question. Leaving this as an analytical exercise to the student, we take a shorter way to the conclusion by a direct application of the principle of energy.

Rotation round its elastic central circle, of a straight wire made into a hoop.

Let the potential energy per unit of length be $\frac{1}{2}(B\kappa^2+C\lambda^2)$, when κ and λ are the component curvatures in the planes of maximum and minimum flexural rigidity: so that, as in § 617, B and C are the measures of the flexural rigidities in these planes. Now if the wire be held in any way at rest with these planes through each point of it inclined at the angles ϕ and $\frac{\pi}{2}-\phi$ to the plane of its elastic central circle, the radius of this circle being r, we should have $\kappa=\frac{1}{r}\cos\phi$, $\lambda=\frac{1}{r}\sin\phi$. Hence, since $2\pi r$ is the whole length,

$$E=\pi\left(\frac{B}{r}\cos^2\phi+\frac{C}{r}\sin^2\phi\right) \qquad (1).$$

Let us now suppose every infinitely small part of the wire to be acted on by a couple in the normal plane, and let L be the amount of this couple per unit of length, which must be uniform all round the ring in order that the circular form may be retained, and let this couple be varied so that, rotation being once commenced, ϕ may increase at any uniform angular velocity. The equation of work done per unit of time (§§ 240, 287) is

$$2\pi r L\dot{\phi}=\frac{dE}{dt}=\frac{dE}{d\phi}\dot{\phi}.$$

[1] When, as in ordinary cases, the wire is either of isotropic material (see § 677 below), or has a normal axis (§ 596) in the direction of its elastic central line, flexure will produce no tendency to twist: in other words, the products of twist into the components of curvature will disappear from the quadratic expressing the potential energy: or the elastic central line is an axis of pure torsion. But, as shown in the text, the case under consideration gains no simplicity from this restriction.

Rotation round its elastic central circle, of a straight wire made into a hoop.

And therefore, by (1),

$$-L=\frac{B-C}{r^2}\sin\phi\cos\phi=\frac{B-C}{2r^2}\sin 2\phi,$$

which shows that the couple required in the normal plane through every point of the ring, to hold it with the planes of greatest flexural rigidity touching a cone inclined at any angle ϕ, to the plane of the circle, is proportional to $\sin 2\phi$; is in the direction to prevent ϕ from increasing; and when $\phi=\frac{1}{4}\pi$ amounts to $\frac{B-C}{2r^2}$ per unit length of the circumference. From this we see that there are two positions of stable equilibrium—being those in which the plane of least flexural rigidity lies in the plane of the ring; and two positions of unstable equilibrium,—being those in which the plane of greatest flexural rigidity is in the plane of the ring.

Rotation round its elastic central circle, of a hoop of wire equally flexible in all directions, but circular when unstrained.

624. A wire of uniform flexibility in all directions, so shaped as to be a circular arc of radius a when free from stress, is bent till its ends meet, and these are joined as in § 623, so that the whole becomes a circular ring of radius r. It is required to find the couple which will hold this ring turned round the central curve through any angle ϕ in every normal section from the position of stable equilibrium (which is of course that in which the naturally concave side of the wire is on the concave side of the ring, the natural curvature being either increased or diminished, but not reversed, when the wire is bent into the ring). Applying the principle of energy exactly as in the preceding section, we find that in this case the couple is proportional to $\sin\phi$, and that when $\phi=\frac{1}{2}\pi$, its amount per unit of length of the circumference is $\frac{B}{ar}$, if B denote the flexural rigidity.

For in this case we have the potential energy

$$E=\pi rB\left\{\left(\frac{1}{a}-\frac{1}{r}\cos\phi\right)^2+\left(\frac{1}{r}\sin\phi\right)^2\right\}=\pi rB\left(\frac{1}{a^2}-\frac{2}{ar}\cos\phi+\frac{1}{r^2}\right) \quad (2),$$

and

$$L=\frac{1}{2\pi r}\frac{dE}{d\phi}=\frac{B}{ar}\sin\phi \quad (3).$$

If every part of the ring is turned half round, so as to bring the naturally concave side of the wire to the convex side of the ring, we have of course a position of unstable equilibrium.

625. A wire of unequal flexibility in different directions is ormed so that, when free from stress, it constitutes a circular rc of radius a, with the plane of greatest flexural rigidity at ach point touching a cone inclined to its plane at an angle α. ts ends are then brought together and joined, as in §§ 623, 624, o that the whole becomes a closed circular ring, of any given adius r. It is required to find the changed inclination, ϕ, to he plane of the ring, which the plane of greatest flexural igidity assumes, and the couple, G, in the plane of the ring, vhich acts between the portions of matter on each side of any ormal section.

Wire unequally flexible in different directions, and circular when unstrained, bent to another circle by balancing couples applied to its ends.

The two equations between the components of the couple nd the components of the curvature in the planes of greatest nd least flexural rigidity determine the two unknown quantities f the problem.

These equations are

$$\left.\begin{aligned} B\left(\frac{1}{r}\cos\phi-\frac{1}{a}\cos\alpha\right)&=G\cos\phi \\ C\left(\frac{1}{r}\sin\phi-\frac{1}{a}\sin\alpha\right)&=G\sin\phi \end{aligned}\right\} \qquad (4),$$

since $\frac{1}{a}\cos\alpha$ and $\frac{1}{a}\sin\alpha$ are the components of natural curvature in the principal planes, and therefore $\frac{1}{r}\cos\phi-\frac{1}{a}\cos\alpha$, and $\frac{1}{r}\sin\phi-\frac{1}{a}\sin\alpha$, are the changes from the natural to the actual curvatures in these planes maintained by the corresponding components $G\cos\phi$ and $G\sin\phi$ of the couple G.

The problem, so far as the position into which the wire turns round its elastic central curve, may be solved by an application of the principle of energy, comprehending those of §§ 623, 624 as particular cases.

Let L be the amount, per unit of length of the ring, of the couple which must be applied from without, in each normal section, to hold it with the plane of maximum flexural rigidity at each point inclined at any given angle, ϕ, to the plane of the ring. We have, as before (§§ 623, 624), for the potential energy of the elastic action in the ring when held so,

$$E=\pi r\left\{B\left(\frac{\cos\phi}{r}-\frac{\cos\alpha}{a}\right)^2+C\left(\frac{\sin\phi}{r}-\frac{\sin\alpha}{a}\right)^2\right\} \qquad (5).$$

Wire unequally flexible in different directions, and circular when unstrained, bent to another circle by balancing couples applied to its ends.

Hence

$$L=\frac{1}{2\pi r}\frac{dE}{d\phi}=\left\{-B\left(\frac{\cos\phi}{r}-\frac{\cos\alpha}{a}\right)\frac{\sin\phi}{r}+C\left(\frac{\sin\phi}{r}-\frac{\sin\alpha}{a}\right)\frac{\cos\phi}{r}\right\}$$

This equated to zero, is the same as (4) with G eliminated, it determines the relation between ϕ and r, in order that the ring when altered to radius r instead of a may be in equilibrium by itself (that is, without any application of couple in the normal section). The present method has the advantage of facilitating the distinction between the solutions, as regards stability or instability of the equilibrium, since (§ 291) for stable equilibrium E is a minimum, and for unstable equilibrium a maximum.

As a particular case, let $C=\infty$, which simplifies the problem very much. The terms involving C as a factor in (5) and (6) become nugatory in this case, and require of course that

$$\frac{\sin\phi}{r}-\frac{\sin\alpha}{a}=0.$$

But the former method is clearer and better for the present case as this result is at once given by the second of equations (4); and then the value of G, if required, is found from the first. We conclude what is stated in the following section:—

Conical bendings of developable surface.

626. Let a uniform hoop, possessing flexibility only in one tangent plane to its elastic central line at each point, be given, so shaped that when under no stress (for instance, when cut through in any normal section and uninfluenced by force from other bodies) it rests in the form of a circle of radius a, with its planes of no flexibility all round touching a cone inclined to the plane of this circle. This is very nearly the case with a common hoop of thin sheet-iron fitted upon a conical vat or on either end of a barrel of ordinary shape. Let such a hoop be shortened (or lengthened), made into a circle of radius a by riveting its ends together (§ 623) in the usual way, and left with no force acting on it from without. It will rest with its plane of no flexibility inclined at the angle $\phi=\sin^{-1}(\frac{r}{a}\sin\alpha)$ to the plane of its circular form, and the elastic couple acting in this plane between the portions of matter on the two sides of any normal section will be

$$G=\frac{B}{\cos\phi}\left(\frac{\cos\phi}{r}-\frac{\cos\alpha}{a}\right).$$

These results we see at once, by remarking that the component rvature in the plane of inflexibility at each point must be variably of the same value, $\frac{\sin a}{a}$, as in the given unstressed ndition of the hoop: and that the component couple, $G\cos\phi$, the plane perpendicular to that of no flexibility at each oint, must be such as to change the component curvature in is plane from $\frac{\cos a}{a}$ to $\frac{\cos\phi}{r}$.

Conical bendings of developable surface.

The greatest circle to which such a hoop can be changed is f course that whose radius is $\frac{a}{\sin a}$: and for this $\phi = \frac{1}{2}\pi$, or the urface of no flexibility at each point (the surface of the sheet-metal in the practical case) becomes the plane of the circle: nd therefore $G = \infty$, showing that if a hoop approaching nfinitely nearly to this condition be made, in the manner ex-lained, the internal couple acting across each normal section vill be infinitely great, which is obviously true.

627. Another very important and interesting case readily lealt with by a method similar to that which we have applied o the elastic wire, is the equilibrium of a plane elastic plate ent by any forces subject to certain conditions stated below § 632). Some definitions and preliminary considerations may e conveniently taken first.

Flexure of a plane elastic plate.

(1.) A *surface of a solid* is a surface passing through always he same particles of the solid, however it is strained.

Definitions.

(2.) The middle surface of a plate is the surface passing hrough all those of its particles which, when it is free from stress, lie in a plane midway between its two plane sides.

(3.) A normal section of a plate, or a surface normal to a plate, is a surface which, when the plate is free from stress, cuts its sides and all planes parallel to them at right angles, being therefore, when unstrained, necessarily either a single plane or a cylindrical (or prismatic) surface.

(4.) The *deflection* of any point or small part of the plate, is the distance of its middle surface there from the tangent plane to the middle surface at any conveniently chosen point of reference in it.

(5.) The *inclination* of the plate, at any point, is the inclination of the tangent plane of the middle surface there to the tangent plane at the point of reference.

Flexure of a plane elastic plate.

(6.) The *curvature of a plate* at any point, or in any pa... the curvature of its middle surface there.

(7.) In a surface infinitely nearly plane the curvature is ... to be *uniform*, if the curvatures in every two parallel n... sections are equal.

(8.) Any diameter of a plate, or distance in a plate infi... nearly plane, is called finite, unless it is an infinitely great m... tiple of the least radius of curvature multiplied by the gre... inclination.

Geometrical preliminaries.

Choosing XOY as the tangent plane at the point of referen... let (x, y, z) be any point of its middle surface, i its inclina... there, and $\frac{1}{r}$ its curvature in a normal section through t... point, inclined at an angle ϕ to ZOX. We have

$$\tan i = \sqrt{\left(\frac{dz^2}{dx^2} + \frac{dz^2}{dy^2}\right)}$$

and, if i be infinitely small,

$$\frac{1}{r} = \frac{d^2z}{dx^2}\cos^2\phi + 2\frac{d^2z}{dxdy}\sin\phi\cos\phi + \frac{d^2z}{dy^2}\sin^2\phi$$

To prove these, let ξ, η, ζ be the co-ordinates of any point of t... surface infinitely near (x, y, z). Then, by the elements of th... differential calculus,

$$\zeta = \frac{dz}{dx}\xi + \frac{dz}{dy}\eta + \tfrac{1}{2}\left(\frac{d^2z}{dx^2}\xi^2 + 2\frac{d^2z}{dxdy}\xi\eta + \frac{d^2z}{dy^2}\eta^2\right).$$

Let $\xi = \rho\cos\phi,\ \eta = \rho\sin\phi,$

so that we have

$$\left.\begin{aligned} \zeta &= A\rho + \tfrac{1}{2}B\rho^2, \text{ where } A = \frac{dz}{dx}\cos\phi + \frac{dz}{dy}\sin\phi \\ \text{and}\quad B &= \frac{d^2z}{dx^2}\cos^2\phi + 2\frac{d^2z}{dxdy}\sin\phi\cos\phi + \frac{d^2z}{dy^2}\sin^2\phi \end{aligned}\right\} \quad (3$$

Then by the formula for the curvature of a plane curve (§ 9),

$$\frac{1}{r} = \frac{B}{(1+A^2)^{\frac{3}{2}}}, \text{ or, as } A \text{ is infinitely small, } \frac{1}{r} = B,$$

and thus (2) is proved.

It follows that the surface represented by

$$z = \tfrac{1}{2}(Ax^2 + 2cxy + By^2) \quad (4).$$

is a surface of uniform curvature if A, B, c be constant throughout the admitted range of values of (x, y); these being limited by the condition that $Ax + cy$, and $cx + By$ must be everywhere infinitely small.

628. When a plane surface is bent to any other shape than a developable surface (§ 139), it must experience some degree stretching or contraction. But an essential condition for the eory of elastic plates on which we are about to enter, is that e amount of the stretching or contraction thus *necessary* in e middle surface is at most incomparably smaller than the etching and contraction of the two sides (§ 141) due to curture. This condition, if we exclude the case of bending into surface differing infinitely little from a developable surface, is uivalent to the following:—

Limitation of flexure to not imply stretching of middle surface comparable with that of either side.

The deflection [§ 627 (4)] *is, at all places finitely* [§ 627 (8)] *stant from the point of reference, incomparably smaller than e thickness.*

And if we extend the signification of "deflection" from that fined in (4) of § 627, to distance from some true developable surce, the excluded case is of course brought under the statement.

Although the truth of this is obvious, it is satisfactory to rove it by investigating the actual degrees of stretching and ntraction referred to.

629. Let us suppose a given plane surface to be bent to me curved form without any stretching or contracting of nes radiating from some particular point of it, O; and let it e required to find the stretching or contraction in the circumference of a circle described from O as centre, with any dius a, on the unstrained plane. If the stretching in each art of the circumference, and not merely on the whole, is to be und, something more as to the mode of the bending must be pecified; which, for simplicity, in the first place, we shall uppose to be, that any point P of the given surface moves in plane perpendicular to the tangent plane through O, during he straining.

Stretching of a plane by synclastic or anticlastic flexure.

Let a, θ be polar co-ordinates of P in its primitive position, and r, θ those of the projection on the tangent plane through O, of its position in the bent surface, and let z be the distance of this position from the tangent plane through O. An element, $ad\theta$, of the unstrained circle, becomes

$$(r^2d\theta^2+dr^2+dz^2)^{\frac{1}{2}}$$

on the bent surface; and, therefore, for the stretching[1] of this element we have

$$\epsilon=\left(\frac{r^2}{a^2}+\frac{dr^2}{a^2d\theta^2}+\frac{dz^2}{a^2d\theta^2}\right)^{\frac{1}{2}}-1 \qquad (1).$$

[1] Ratio of the lengthening to the unstretched length.

Stretching of a plane by synclastic or anticlastic flexure.

Hence if e denote the ratio of the elongation of the wh… circumference to its unstretched length, or the mean stretch… the circumference,

$$e=\frac{1}{2\pi}\int_0^{2\pi}d\theta\{(\frac{r^2}{a^2}+\frac{dr^2}{a^2d\theta^2}+\frac{dz^2}{a^2d\theta^2})^{\frac{1}{2}}-1\}$$

where we must suppose z and r known functions of θ. Con… ourselves now to distances from O within which the curvatu… the surface is sensibly uniform, we have

$$z=\frac{a^2}{2\rho},\text{ and } r=\rho\sin\frac{a}{\rho}=a(1-\tfrac{1}{6}\frac{a^2}{\rho^2}+\text{etc.})$$

if ρ be the radius of curvature of the normal section throug… and P: and, if we take as the zero line for θ that in which … tangent plane is cut by one of the principal normal planes (§ 1…

$$\frac{1}{\rho}=\frac{1}{\rho_1}\cos^2\theta+\frac{1}{\rho_2}\sin^2\theta=\tfrac{1}{2}(\frac{1}{\rho_1}+\frac{1}{\rho_2})+\tfrac{1}{2}(\frac{1}{\rho_1}-\frac{1}{\rho_2})\cos 2\theta$$

where ρ_1, ρ_2 are the principal radii of curvature. Hence … term $\frac{dr^2}{a^2d\theta^2}$ under the radical sign disappears if we include … terms involving higher powers than the first, of the small fract… $\frac{a^2}{\rho^2}$; and, to this degree of approximation

$$\epsilon=\{1-\tfrac{1}{3}\frac{a^2}{\rho^2}+a^2(\frac{1}{\rho_2}-\frac{1}{\rho_1})^2\sin^2\theta\cos^2\theta\}^{\frac{1}{2}}-1=-\tfrac{1}{6}\frac{a^2}{\rho^2}+\frac{a^2}{2}(\frac{1}{\rho_2}-\frac{1}{\rho_1})^2\sin^2\theta\ldots$$

or, by (4), and reductions, finally

$$\epsilon=-\tfrac{1}{6}a^2\{\frac{1}{\rho_1\rho_2}+\tfrac{1}{2}(\frac{1}{\rho_1^2}-\frac{1}{\rho_2^2})\cos 2\theta+\tfrac{1}{2}(\frac{1}{\rho_1}-\frac{1}{\rho_2})^2\cos 4\theta\}\quad ;$$

Using this in (2) we find

$$e=-\tfrac{1}{6}\frac{a^2}{\rho_1\rho_2}\quad (6).$$

The whole amount of stretching thus expressed will, it follo… from (5), be distributed uniformly through the circumference. … instead of compelling each point P to remain in the plane throug… O, perpendicular to XOY, we allow it to yield in the direction of the circumference through a space equal to

$$\frac{a^3}{24}\{(\frac{1}{\rho_1^2}-\frac{1}{\rho_2^2})\sin 2\theta+\tfrac{1}{2}(\frac{1}{\rho_1}-\frac{1}{\rho_2})^2\sin 4\theta\}\quad (\ldots$$

From (6) we conclude that

630. If a plane area be bent to a uniform degree of curva- ture throughout, without any stretching in any radius throug… a certain point of it, and with uniform stretching or contraction over the circumference of every circle described from the same int as centre, the amount of this contraction (reckoned gative where the actual effect is stretching) is equal to the tio of one-sixth of the square of the radius of the circle, to e rectangle under the maximum and minimum radii of curture of normal sections of the surface; or which is the same ing, the ratio of two-thirds of the rectangle under the maxium and minimum deflections of the circumference from the ngent plane of the surface at the centre, to the square of the dius; or, which is the same, the ratio one-third of the maxium deflection to the maximum radius of curvature.

Stretching of a plane by synclastic or anticlastic flexure.

If the surface thus bent be the middle surface of a plate of niform thickness, and if each line of particles perpendicular this surface in the unstrained plate remain perpendicular to when bent, the stretching on the convex side, and the conaction on the concave side, in any normal section, is obviously qual to the ratio of half the thickness, to the radius of curvaire. The comparison of this, with the last form of the preeding statement, proves that the second of the two conditions ated in § 628 secures the fulfilment of the first.

631. If a surface already bent as specified, be again bent to different shape still fulfilling the prescribed conditions, or if surface given curved be altered to any other shape by bendig according to the same conditions, the contraction proluced in the circumferences of the concentric circles by this ending, will of course be equal to the increment in the value f the ratio stated in the preceding section. Hence if a curved urface be bent to any other figure, without stretching in any art of it, the rectangle under the two principal radii of curvaure at every point remains unchanged. This is Gauss's celerated theorem regarding the bending of curved surfaces, of which we gave a more analytical demonstration in our introluctory Chapter (see § 150).

Stretching of a curved surface by flexure not fulfilling Gauss's condition.

Gauss's theorem regarding flexure.

632. Without further preface we now commence the theory of the flexure of a plane elastic plate with the promised (§ 627) tatement of restricting conditions.

Limitations as to the forces and flexures to be admitted in elementary theory of elastic plate.

(1.) Of the forces applied from without to any part of the plate, bounded by a normal surface [§ 627 (3)], the components parallel to any line in the plane of the plate are either evanescent or are reducible to *couples*. In other words the algebraic

Limitations as to the forces and flexures to be admitted in elementary theory of elastic plate.

sum of such components, for any part of the plate boun… a normal surface is zero.

(2.) The principal radii of curvature of the middle surf… everywhere infinitely great multiples of the thickness of th…

(3.) The deflection is nowhere, within finite distance fro… point of reference, more than an infinitely small fraction o… thickness.

(4.) Neither the thickness of the plate nor the coefficien… elasticity of its substance need be uniform throughout, b… they vary at all they must vary continuously from pl… place; and must not any of them be incomparably greate… one place than in another within any finite area of the plat…

Results of general theory stated in advance.

633. The general theory of elastic solids investigated … shows that when these conditions are fulfilled the distrib… of strain through the plate possesses the following prop… the statement of which at present, although not necessary … the particular problem on which we are entering, will pro… a thorough understanding and appreciation of the princ… involved.

(1.) The stretching of any part of the middle surface is … finitely small in comparison with that of either side, in eve… part of the plate where the curvature is finite.

(2.) The particles in any straight line perpendicular to t… plate when plane, remain in a straight line perpendicular … the curved surfaces into which its sides, and parallel plane… the substance between them, become when it is bent. A… hence the curves in which these surfaces are cut by any pla… through that line, have one point in it for centre of curvatu… of them all.

(3.) The whole thickness of the plate remains unchanged, a… every point; but the half thickness on one side (which wh… the curvature is synclastic is the convex side) of the midd… surface becomes diminished and on the other side increased, b… equal amounts comparable with the elongations and shorten… ings of lengths equal to the half thickness, measured on th… two side surfaces of the plate.

634. The conclusions from the general theory on which w… shall found the equations of equilibrium and motion of a… elastic plate are as follows :—

Laws for flexure of elastic plate assumed in advance.

Let a naturally plane plate be bent to any surface of uniform curvature [§ 627 (7)] throughout, the applied forces and the extents of displacement fulfilling the conditions and restrictions of § 632 : Then—

(1.) The force across any section of the plate is, at each point of it, in a line parallel to the tangent plane to the middle surface in the neighbourhood.

(2.) The forces across any set of parallel normal sections are equally inclined to the directions of the normal sections at all points (that is to say, are in directions which would be parallel if the plate were bent, and which deviate actually from parallelism only by the infinitely small deviations produced in the normal sections by the flexure).

(3.) The amounts of force across one normal section, or any set of parallel normal sections, on equal infinitely small areas, are simply proportional to the distances of these areas from the middle surface of the plate.

(4.) The component forces in the tangent planes of the normal sections are equal and in dissimilar directions in sections which are perpendicular to one another. For proof, see § 661. [The meaning of "dissimilar directions" in this expression is explained by the diagram; where the arrow-heads indicate the directions in which the portions of matter on the two sides of each normal section would yield if the substance were actually divided, half way through the plate from one side, by each of the normal sections indicated by dotted lines.]

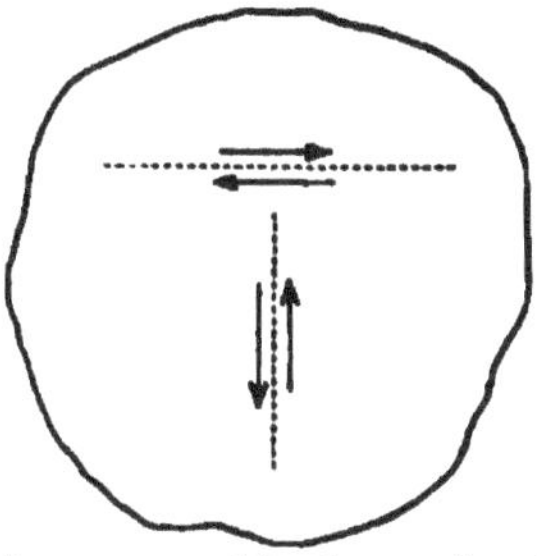

(5.) By the law of superposition, we see that if the applied forces be all doubled, or altered in any other ratio, the curvature in every normal section, and all the internal forces specified in (1), (2), (3), (4), are changed in the same ratio; and the potential energy of the internal forces becomes changed according to the square of the same ratio.

635. From § 634 (3) it follows immediately that the forces experienced by any portion of the plate bounded by a normal section through the circumference of a closed polygon or curve of the middle surface, from the action of the contiguous matter

Stress-couple acting across a normal section.

of the plate all round it, may be reduced to a set of cou[...] by taking them in groups over infinitely small recta[...] into which the bounding normal section may be imagine[...] divided by normal lines. From § 634 (2) it follows that [...] distribution of couple thus obtained is uniform along e[...] straight portion, if any there is, of the boundary, and e[...] per equal lengths in all parallel parts of the boundary.

Twisting components proved equal round any two perpendicular axes.

636. From § 634 (4) it follows that the component cou[...] round axes perpendicular to the boundary are equal in p[...] of the boundary at right angles to one another, and are [...] directions related to one anot[...] in the manner indicated by t[...] circular arrows in the diagra[...] that is to say, in such directi[...] that if the axis is, according [...] the rule of § 234, drawn *outw[...]* from the portion of the pla[...] under consideration, for one po[...] of the boundary, it must be draw[...] *inwards* for every point where the boundary is perpendicular t[...] its direction at that point.

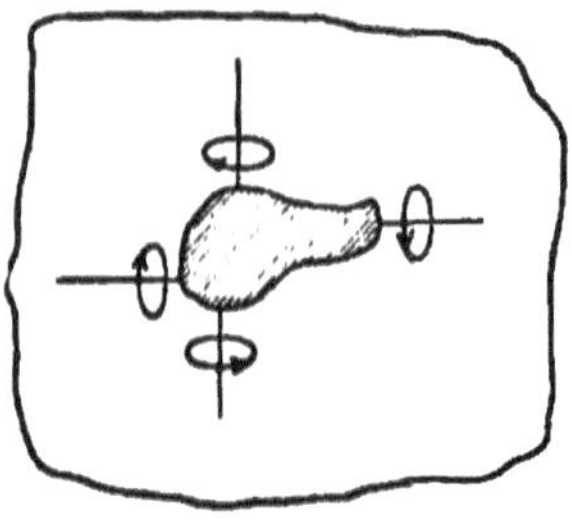

Principal axes of bending stress.

637. We may now prove that there are two normal sectio[...] at right angles to one another, in which the component coupl[...] round axes perpendicular to them vanish, and that in th[...] sections the component couples round axes coincident with t[...] sections are of maximum and minimum values.

Let OAB be a right-angled triangle of the plate. Let Λ and [...] be the two component coupl[...] acting on t[...] side OA; K a[...] Π those on t[...] side OB; and G and H those o[...] the side AB; the amount o[...] each couple being reckon[...] per unit o[...] length of the side on which it acts, and the axes and direction[...] of the several couples being as indicated by the circular arrows wh[...]

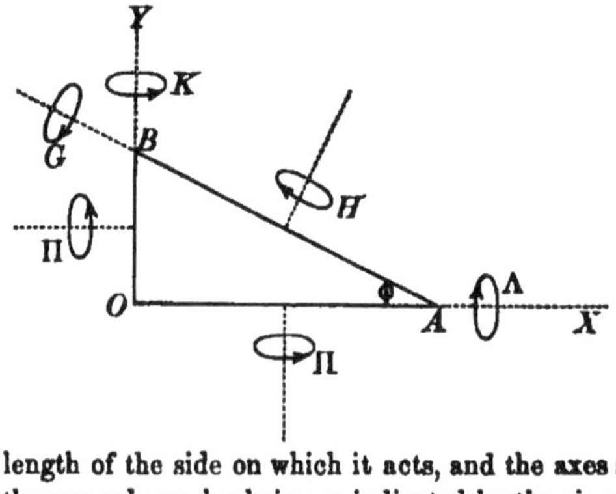

each is reckoned as positive. Then, if $AB=a$, and $BAO=\phi$, the whole amounts of the couples on the three sides are respectively

Principal axes of bending stress investigated.

$$\Lambda a\cos\phi,\ \Pi a\cos\phi,$$
$$\mathrm{K}a\sin\phi,\ \Pi a\sin\phi,$$
$$Ga,\qquad Ha.$$

Resolving the two latter round OX and OY, we have

$$Ga\cos\phi - Ha\sin\phi \text{ round } OX,$$

and $$Ga\sin\phi + Ha\cos\phi \quad ,, \quad OY.$$

But if the portion in question, of the plate, were to become rigid, its equilibrium would not be disturbed (§ 564); and therefore we must have

$$\left.\begin{array}{l} Ga\cos\phi - Ha\sin\phi = \Lambda a\cos\phi + \Pi a\sin\phi \text{ by couples round } OX \\ \text{and} \\ Ga\sin\phi + Ha\cos\phi = \mathrm{K}a\sin\phi + \Pi a\cos\phi \quad ,, \quad ,, \quad OY \end{array}\right\} \quad (1).$$

From these we find immediately

$$\left.\begin{array}{l} G=\Lambda\cos^2\phi + 2\Pi\sin\phi\cos\phi + \mathrm{K}\sin^2\phi, \\ H=(\mathrm{K}-\Lambda)\sin\phi\cos\phi + \Pi(\cos^2\phi - \sin^2\phi) \end{array}\right\} \quad (2).$$

Hence the values of ϕ, which make H vanish, give to G its maximum and minimum values, and, being determined by the equation

$$\tan 2\phi = -\frac{\Pi}{\frac{1}{2}(\mathrm{K}-\Lambda)} \quad (3),$$

differ from one another by $\frac{1}{2}\pi$.

A modification of these formulæ, which we shall find valuable, is had by putting

$$\Sigma=\tfrac{1}{2}(\mathrm{K}+\Lambda),\ \Theta=\tfrac{1}{2}(\mathrm{K}-\Lambda) \quad (4).$$

This reduces (2) to

$$\left.\begin{array}{l} G=\Sigma+\Pi\sin 2\phi - \Theta\cos 2\phi \\ H=\qquad \Pi\cos 2\phi + \Theta\sin 2\phi \end{array}\right\} \quad (5),$$

which again become

$$\left.\begin{array}{l} G=\Sigma+\Omega\cos 2(\phi-a) \\ H=-\Omega\sin 2(\phi-a) \end{array}\right\} \quad (6),$$

where a [being a value of ϕ given by (3)], and Ω are taken so that $$\left.\begin{array}{l} \Pi=\Omega\sin 2a,\ \Theta=-\Omega\cos 2a, \\ \text{so that, of course,}\quad \Omega=(\Pi^2+\Theta^2)^{\frac{1}{2}} \end{array}\right\} \quad (7).$$

This analysis demonstrates the following convenient synthesis of the whole system of internal force in question:—

638. The action experienced by each part of the plate, in virtue of the internal forces between it and the surrounding contiguous matter of the plate, being called a *stress* [in accordance with the general use of this term defined below (§ 658)], may be regarded as made up of two distinct elements—(1.) a synclastic stress, and (2.) an anticlastic stress; as we shall call them.

Synclastic and anticlastic stresses defined.

Synclastic and anticlastic stresses defined.

(1.) Synclastic stress consists of equal direct bending action round every straight line in the plane of the plate. Its amount may be conveniently regarded as measured by the amount, Σ, of the mutual couple between the portions of matter on the two sides of any straight normal section of unit length. Its effect would be to produce equal curvature in all normal sections (that is to say, a spherical figure) if the plate were equally flexible in all directions.

Anticlastic stress referred to its principal axes;

(2.) Anticlastic stress consists of two simple bending stresses of equal amounts in opposite directions round two sets of parallel straight lines perpendicular to one another in the plane of the plate. Its effect would be uniform anticlastic curvature, with equal convexities and concavities, if the plate were equally flexible in all directions. Its amount is reckoned as the amount, Ω, of the mutual couple between the portions of matter on the two sides of a straight normal section of unit length, parallel to either of these two sets of lines. It gives rise to couples of the same amount, Ω, between the portions of matter on each side of a normal section of unit length parallel

referred to axes inclined to them at 45°.

to either of the sets of lines bisecting the right angles between those; but the couples now referred to are *in* the plane of the normal section instead of perpendicular to it. This is proved and illustrated by the annexed diagram, representing [a particular case of the diagram and equations (1) of § 637] the equilibrium of an isosceles right-angled triangle under the influence of couples,

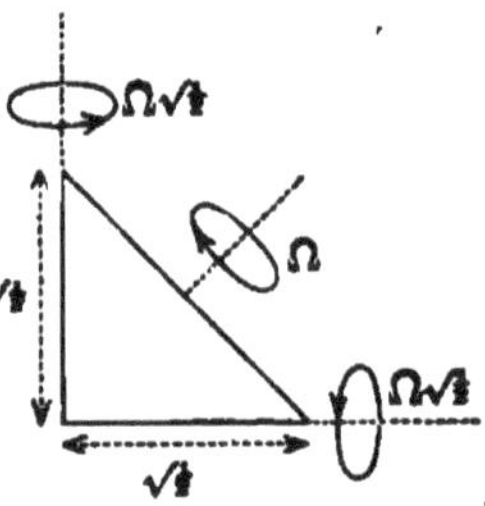

each equal to $\Omega\sqrt{\frac{1}{2}}$, applied to it round axes coinciding with its legs, and a third couple, Ω, round an axis perpendicular to its hypotenuse.

Octantal resolution and composition of anticlastic stress.

If two pairs of rectangular axes, each bisecting the right angles formed by the other, be chosen as axes of reference, an anticlastic stress having any third pair of rectangular lines for its axes may, as the preceding formulæ [§ 637 (5)] show, be resolved into two having their axes coincident with the two pairs of axes of reference respectively, by the ordinary cosine formula with each angle doubled. Hence it follows that any two anticlastic stresses may be compounded into one by the same geometrical construction as the parallelogram of forces, made upon lines inclined to one another at an angle equal to twice that between the corresponding axes of the two given stresses; and the position of the axes of the resultant stress will be indicated by the angles of this diagram each halved. Construction by parallelogram.

639. Precisely the same set of statements are of course applicable to the curvature of a surface. Thus the proposition proved in § 637 (3) for bending stresses has, for its analogue in curvature, Euler's theorem proved formerly in § 130; and analogues to the series of definitions and propositions founded on it and derived from it may be at once understood without more words or proof. Geometrical analogues.

Let $$z=\tfrac{1}{2}(\kappa x^2+2\varpi xy+\lambda y^2) \qquad (1)$$ be the equation of a curved surface infinitely near a point O at which it is touched by the plane YOX. Its curvature may be regarded as compounded of a cylindrical curvature, λ, with axis parallel to OX, a cylindrical curvature, κ, with axis parallel to OY, and an anticlastic curvature, ϖ, with axis bisecting the angles XOY, YOX'. Thus, if ϖ and λ each vanished, the surface would be cylindrical, with $\frac{1}{\kappa}$ for radius of curvature and generating lines parallel to OY. Or, if κ and λ each vanished, there would be anticlastic curvature, with sections of equal maximum curvature in the two directions, bisecting the angles XOY and YOX', and radius of curvature in those sections equal to $\frac{1}{\varpi}$. Two cylindrical curvatures round perpendicular axes, and an anticlastic curvature round axis bisecting their right angles;

If now we put $$\sigma=\tfrac{1}{2}(\kappa+\lambda),\quad \vartheta=\tfrac{1}{2}(\kappa-\lambda) \qquad (2),$$ or a spherical curvature and two anticlastic curvatures;
the equation of the surface becomes $$z=\tfrac{1}{2}\{\sigma(x^2+y^2)+\vartheta(x^2-y^2)+2\varpi xy\} \qquad (3);$$
or, if $$\left.\begin{array}{l} x=r\cos\phi,\ y=r\sin\phi, \\ z=\tfrac{1}{2}\{\sigma+\vartheta\cos 2\phi+\varpi\sin 2\phi\}r^2 \end{array}\right\} \qquad (4);$$
or, lastly, $$\left.\begin{array}{l} z=\tfrac{1}{2}\{\sigma+\omega\cos 2(\phi-a)\}r^2, \\ \vartheta=\omega\cos 2a,\ \varpi=\omega\sin 2a \end{array}\right\} \qquad (5).$$ or a spherical and one anticlastic curvature.

In these formulæ σ measures the spherical curvature; and ϑ and ϖ two components of anticlastic curvature, referred to the pair of axes $X'X$, $Y'Y$, and the other pair bisecting their angles. The resultant of ϑ and ϖ is an anticlastic curvature ω, with axes inclined, in the angle XOY at angle a to OX, and in YOX', at angle a to OY.

Work done in bending.

640. The notation of §§ 637, 639 being retained, the work done on any area A of the plate experiencing a change of curvature $(\delta\kappa, \delta\lambda, \delta\varpi)$, under the action of a stress (K, A, Π), is

$$(K\delta\kappa+\Lambda\delta\lambda+2\Pi\delta\varpi)A \qquad (1)$$

or

$$(2\Sigma\delta\sigma+2\Theta\delta\vartheta+2\Pi\delta\varpi)A \qquad (2)$$

if, as before,

$$\Sigma=\tfrac{1}{2}(K+\Lambda),\ \Theta=\tfrac{1}{2}(K-\Lambda),\ \sigma=\tfrac{1}{2}(\kappa+\lambda),\ \vartheta=\tfrac{1}{2}(\kappa-\lambda) \qquad (3)$$

Let $PQP'Q'$ be a rectangular portion of the plate with its centre at O, and its sides $Q'P$, $P'Q$ parallel to OX, and $Q'P'$, PQ parallel to OY. If

$$z=\tfrac{1}{2}(\kappa x^2+2\varpi xy+\lambda y^2)$$

be the equation of the curved surface, we have

$$\frac{dz}{dx}=\kappa x+\varpi y,\quad \frac{dz}{dy}=\varpi x+\lambda y;$$

and therefore the tangent plane at (x, y) deviates in direction from XOY by an infinitely small rotation

$$\left.\begin{array}{ll} & \kappa x+\varpi y \text{ round } OY \\ \text{and} & \varpi x+\lambda y \quad ,, \quad OX \end{array}\right\} \qquad (4)$$

Hence the rotation from XOY to the mean tangent plane for all points of the side PQ or $Q'P'$ is

$$\mp\tfrac{1}{2}Q'P.\kappa \text{ round } OY,$$

and

$$\mp\tfrac{1}{2}Q'P.\varpi \quad ,, \quad OX.$$

Hence if the tangent plane, XOY, at O remains fixed, while the curvature changes from $(\kappa, \varpi, \lambda)$ to $(\kappa+\delta\kappa,\ \varpi+\delta\varpi,\ \lambda+\delta\lambda)$, the work done by the couples $PQ.K$ round OY, and $PQ.\Pi$ round OX, distributed over the side PQ, will be

$$\tfrac{1}{2}Q'P.PQ.(K\delta\kappa+\Pi\delta\varpi),$$

and an equal amount will be done by the equal and opposite couples distributed over the side $Q'P'$ undergoing an equal and opposite rotation. Similarly, we find for the whole work done on the sides $P'Q$ and $Q'P$,

$$PQ.Q'P.(\Pi\delta\varpi+K\delta\kappa).$$

Hence the whole work done on all the four sides of the rectangle is

$$PQ.Q'P.(K\delta\kappa+2\Pi\delta\varpi+\Lambda\delta\lambda):$$

whence the proposition to be proved, since any given area of the plate may be conceived divided into infinitely small rectangles.

It is an instructive exercise to verify the result by beginning with the consideration of a portion of plate bounded by any given curve, and using the expressions (1) of § 637, by which we find, for the couples on any infinitely short portion, ds, of its boundary, specified in position by (x, y),

Work done in bending.

$$\left.\begin{array}{ll} & (-\Lambda\frac{dx}{ds}+\Pi\frac{dy}{ds})ds \text{ round } OX \\ \text{and} & (K\frac{dy}{ds}-\Pi\frac{dx}{ds})ds \quad \text{,,} \quad OY \end{array}\right\} \quad (5).$$

But, as we have just seen (4), the rotation experienced by the tangent plane to the plate at (x, y), when the curvature changes from $(\kappa, \varpi, \lambda)$ to $(\kappa+\delta\kappa, \varpi+\delta\varpi, \lambda+\delta\lambda)$, is

$$\left.\begin{array}{ll} & x\delta\kappa+y\delta\varpi \text{ round } OY \\ \text{and} & x\delta\varpi+y\delta\lambda \quad \text{,,} \quad OX \end{array}\right\} \quad (6),$$

the tangent plane to the plate at O being supposed to remain unchanged in position; and therefore the work done on the portion ds of the edge is

$$\{(K\frac{dy}{ds}-\Pi\frac{dx}{ds})(x\delta\kappa+y\delta\varpi)+(\Pi\frac{dy}{ds}-\Lambda\frac{dx}{ds})(x\delta\varpi+y\delta\lambda)\}ds.$$

The required work, being the integral of this over the whole of the bounding curve, is therefore

$$(K\delta\kappa+2\Pi\delta\varpi+\Lambda\delta\lambda)A;$$

since

$$\int x\frac{dy}{ds}ds=-\int y\frac{dx}{ds}ds=A,$$

and

$$\int x\frac{dx}{ds}ds=0, \int y\frac{dy}{ds}ds=0,$$

each integral being round the whole closed curve.

641. Considering now the elastic forces called into action by the flexure $(\kappa, \varpi, \lambda)$ reckoned from the unstressed condition of the plate (plane, or infinitely nearly plane), and denoting by w the whole amount of their potential energy, per unit area of the plate, we have, as in the case of the wire treated in § 594,

Partial differential equations work done in bending an elastic plate.

$$K\delta\kappa=\delta_\kappa w, \ \Lambda\delta\lambda=\delta_\lambda w, \ 2\Pi\delta\varpi=\delta_\varpi w \quad (7);$$

or, according to the other notation,

$$2\Sigma\delta\sigma=\delta_\sigma w, \ 2\Theta\delta\vartheta=\delta_\vartheta w, \ 2\Pi\delta\varpi=\delta_\varpi w \quad (8);$$

where, as above explained, K and Λ denote the simple bending stresses (measured by the amount of bending couple, per unit of length) round lines parallel to OY and OX respectively: Π the anticlastic stress with axes at 45° to OX and OY: and Σ and Θ the synclastic stress and the anticlastic stress with OX and OY for axes, together equivalent to K and Λ. Also, as in § 595, we see that whatever be the character, aeolotropic or isotropic, § 677, of the substance of the plate, it must be a homo-

Potential energy of an elastic plate held bent. geneous quadratic function of the three components of curvature, whether (κ, λ, ϖ) or ($\sigma, \vartheta, \varpi$). From this and (7), (8) it follows that the coefficients in the linear functions of the three components of curvature which express the components of the stress required to maintain it, must fulfil the ordinary conservative relations of equality in three pairs, reducing the whole number from nine to six.

Thus A, B, C, a, b, c denoting six constants depending on the quality of the solid substance and the thickness of the plate, we have $w=\frac{1}{2}(A\kappa^2+B\lambda^2+C\varpi^2+2a\lambda\varpi+2b\varpi\kappa+2c\kappa\lambda)$, (9)
and hence, by (7),

$$\left.\begin{aligned} \mathrm{K}&=A\kappa+c\lambda+b\varpi\\ \Lambda&=c\kappa+B\lambda+a\varpi\\ 2\Pi&=b\kappa+a\lambda+C\varpi \end{aligned}\right\} \quad (10)$$

Transforming these by § 640 (8) we have, in terms of $\sigma, \vartheta, \varpi$

$$w=\tfrac{1}{2}\{(A+B+2c)\sigma^2+(A+B-2c)\vartheta^2+C\varpi^2+2(b-a)\vartheta\varpi+2(b+a)\sigma\varpi+2(A-B)\sigma\vartheta\} \quad (11)$$

and

$$\left.\begin{aligned} 2\Sigma&=(A+B+2c)\sigma+(A-B)\vartheta+(b+a)\varpi\\ 2\Theta&=(A-B)\sigma+(A+B-2c)\vartheta+(b-a)\varpi\\ 2\Pi&=(b+a)\sigma+(b-a)\vartheta+C\varpi \end{aligned}\right\} \quad (12)$$

These second forms are chiefly useful as showing immediately the relations which must be fulfilled among the coefficients for the important case considered in the following section.

Case of equal flexibility in all directions **642.** If the plate be equally flexible in all directions, a synclastic stress must produce spherical curvature: an anticlastic stress having any pair of rectangular lines in the plate for its axes must produce anticlastic curvature having these lines for sections of equal greatest curvature on the opposite sides of the tangent plane: and in either action the amount of the curvature is simply proportional to the amount of the stress. **Synclastic and anticlastic rigidities of a plate.** Hence if $\mathfrak{h}$ and $\mathfrak{k}$ denote two coefficients depending on the compressibility and rigidity of the substance if isotropic (§§ 677, 680, below), and on the thickness of the plate, we have

$$\Sigma=\mathfrak{h}\sigma,\quad \Theta=\mathfrak{k}\vartheta,\quad \Pi=\mathfrak{k}\varpi \quad (13)$$

And therefore [§ 640 (2)]

$$w=\mathfrak{h}\sigma^2+\mathfrak{k}(\vartheta^2+\varpi^2) \quad (14)$$

Hence the coefficients in the general expressions of § 641 fulfil in the case of equal flexibility in all directions, the following conditions:—

$$a=0,\ b=0,\ A=B,\ 2(A-c)=C \qquad (15);$$

and the newly-introduced coefficients $\mathfrak{h}$ and $\mathfrak{k}$ are related to them thus:—

$$A+c=\mathfrak{h},\ \tfrac{1}{2}C=A-c=\mathfrak{k} \qquad (16).$$

643. Let us now consider the equilibrium of an infinite ite, disturbed from its natural plane by forces applied to it any way, subject only to the conditions of § 632. The sub- ınce may be of any possible quality as regards elasticity in fferent directions: and the plate itself need not be homo- neous either as to this quality, or as to its thickness, in fferent parts; provided only that round every point it is in th respects sensibly homogeneous [§ 632 (4)] to distances great comparison with the thickness at that point. Plate bent by any forces.

644. Let OX, OY be rectangular axes of reference in the ane of the undisturbed plate; and let z be the infinitely small splacement from this plane, of the point (x, y) of the plate, hen disturbed by any forces, specified in their effective com- onents as follows:—A portion, E, of the plate bounded by a ormal surface cutting the middle surface in a line enclosing n infinitely small area σ in the neighbourhood of the point $c, y)$, being considered, let $Z\sigma$ denote the sum of the compon- nt forces perpendicular to XOY on all the matter of E in the eighbourhood of the point (x, y): and $L\sigma$, $M\sigma$ the component ouples round OX and OY obtained by transferring, according o Poinsot, the forces from all points of the portion E, supposed or the moment rigid, to one point of it which it is convenient o take at the centre of inertia of the area, σ, of the part of the niddle surface belonging to it. This force and these couples, Conditions of equilibrium. long with the internal forces of elasticity exerted on the matter, of E, across its boundary, by the matter surrounding it, must (§ 564) fulfil the conditions of equilibrium for E treated as a rigid body. And E, being not really rigid, must have the curvature due, according to § 641, to the bending stress con- stituted by the last-mentioned forces. These conditions ex- pressed mathematically supply five equations from which, four elements specifying the internal forces being eliminated, we have a single partial differential equation for z in terms of x and y, which is the required equation of equilibrium.

Equations of equilibrium of plate bent by any forces, investigated.

Let σ be a rectangle $PQP'Q'$, with sides δx parallel ... and δy parallel t... Let $a\delta y$, $a'\delta y$ be the ... finitely nearly equa... ing forces perpendic... the plate in the norm... faces through P... QP' respectively: a... β, β' be the corresp... notation for PQ...

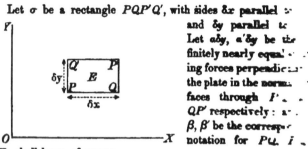

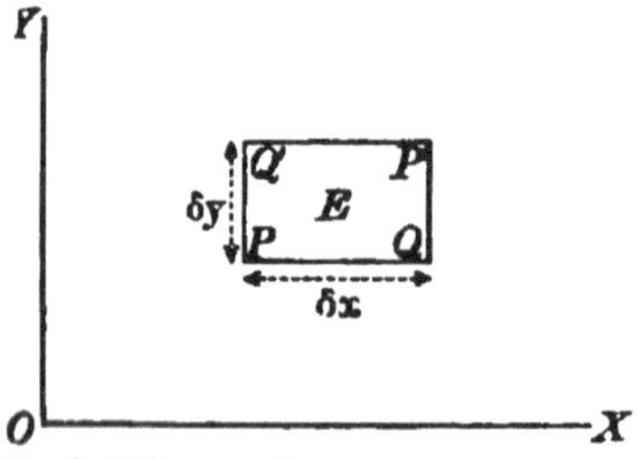

We shall have, of course,

$$a'-a=\frac{da}{dx}\delta x, \text{ and } \beta'-\beta=\frac{d\beta}{dy}\delta y.$$

The effect of these actions on the portion, E, of the plat... sidered as rigid, is forces $a'\delta y$, $\beta'\delta x$ through the middle p... QP', $Q'P'$, in the direction of z positive, and forces ... through the middle points of PQ', PQ, in the directi... negative. Hence, towards the equilibrium of E as a rigid ... they contribute

$$(a'-a)\delta y+(\beta'-\beta)\delta x, \text{ or } \left(\frac{da}{dx}+\frac{d\beta}{dy}\right)\delta x\delta y, \text{ component force paral...}$$

$$a\delta y.\delta x \text{ couple round } OY,$$

and $\quad \beta\delta x.\delta y \quad$,, $\quad$,, $\quad OX$;

(in these two last expressions the difference between a and ... between β and β' being of course neglected). Again, ... Λ, Π specify, according to the system of § 637, the b... stress at (x, y), we shall have couples infinitely nearly ... and opposite, on the pairs of opposite sides, of which, estim... in components round OX and OY, the differences, represe... the residual turning tendencies on E as a rigid body, ar... follows:—

round OX, $\begin{cases} \text{from sides } PQ, Q'P', \dfrac{d\Lambda}{dy}\delta y.\delta x, \\ \text{,, ,, } PQ', QP', \dfrac{d\Pi}{dx}\delta x.\delta y. \end{cases}$

round OY, $\begin{cases} \text{from sides } PQ, Q'P', \dfrac{d\Pi}{dy}\delta y.\delta x, \\ \text{,, ,, } PQ', QP', \dfrac{dK}{dx}\delta x.\delta y; \end{cases}$

or in all, $\quad$ round OX, $\left(\dfrac{d\Lambda}{dy}+\dfrac{d\Pi}{dx}\right)\delta x\delta y$,

and $\quad$,, $\quad OY$, $\left(\dfrac{d\Pi}{dy}+\dfrac{dK}{dx}\right)\delta x\delta y$.

The equations of equilibrium, therefore, between these and the applied forces on E as a rigid body give, if we remove the common factor, $\delta x \delta y$,

Equations of stress in plate bent by any forces.

$$\left.\begin{aligned} Z+\frac{d\alpha}{dx}+\frac{d\beta}{dy}=0 \\ L+\beta+\frac{d\Lambda}{dy}+\frac{d\Pi}{dx}=0 \\ M+\alpha+\frac{d\Pi}{dy}+\frac{dK}{dx}=0 \end{aligned}\right\} \quad (1).$$

The first of these, with α and β replaced in it by their values from the second and third, becomes

$$\frac{d^2K}{dx^2}+2\frac{d^2\Pi}{dxdy}+\frac{d^2\Lambda}{dy^2}=Z-\frac{dM}{dx}-\frac{dL}{dy} \quad (2).$$

Now κ, λ, ϖ denoting component curvatures of the plate, according to the system of § 639, we have of course

$$\kappa=\frac{d^2z}{dx^2},\quad \lambda=\frac{d^2z}{dy^2},\quad \varpi=\frac{d^2z}{dxdy} \quad (3),$$

and hence (10) of § 641 give

Equations connecting stress and curvature.

$$\left.\begin{aligned} K=A\frac{d^2z}{dx^2}+c\frac{d^2z}{dy^2}+b\frac{d^2z}{dxdy} \\ \Lambda=c\frac{d^2z}{dx^2}+B\frac{d^2z}{dy^2}+a\frac{d^2z}{dxdy} \\ 2\Pi=b\frac{d^2z}{dx^2}+a\frac{d^2z}{dy^2}+C\frac{d^2z}{dxdy} \end{aligned}\right\} \quad (4).$$

Using these in (2) we find the required differential equation of the disturbed surface. On the general supposition (§ 643) we must regard A, B, C, a, b, c as given functions of x and y. In the important practical case of a homogeneous plate they are constants; and the required equation becomes the linear partial differential equation of the fourth degree with constant coefficients, as follows:—

$$A\frac{d^4z}{dx^4}+2b\frac{d^4z}{dx^3dy}+(C+2c)\frac{d^4z}{dx^2dy^2}+2a\frac{d^4z}{dxdy^3}+B\frac{d^4z}{dy^4}=Z-\frac{dM}{dx}-\frac{dL}{dy} \quad (5).$$

For the case of equal flexibility in all directions, according to § 642 (13), this becomes

Partial differential equation of the bent surface.

$$\left.\begin{aligned} A\left(\frac{d^4z}{dx^4}+2\frac{d^4z}{dx^2dy^2}+\frac{d^4z}{dy^4}\right)=Z-\frac{dM}{dx}-\frac{dL}{dy} \\ \text{or}\quad A\left(\frac{d^2}{dx^2}+\frac{d^2}{dy^2}\right)^2z=Z-\frac{dM}{dx}-\frac{dL}{dy} \end{aligned}\right\} \quad (6).$$

645. To investigate the boundary conditions for a plate of limited dimensions, we may first consider it as forming part of

Boundary conditions; an infinite plate bounded by a normal surface drawn th... closed curve traced on its middle surface. The prece... vestigation leads immediately to expressions for the f... couple on any portion of the normal bounding surfac... the portion in question be actually cut out from the su... ing sheet, and if a distribution of force and couple ... with that so found be applied to its edge, its elastic ... will remain absolutely unchanged throughout up to t... normal edge. To fulfil this condition requires three ... expressing (1.) that the shearing force applied to the e... is, the applied tangential force in the normal surface ... **Poisson's three:** ing the edge), which is necessarily in the direction ... normal line to the plate, must be equal to the required ... and (2. and 3.) that the couple applied to any small part ... edge must have components of the proper amounts ro... two lines in the plane of the plate. These three ... were given by Poisson as necessary for the full expres... **two sufficient, proved by Kirchhoff.** the boundary condition; but Kirchhoff has demonstra... they express too much, and has shown that two ... suffice. This we shall prove by showing that when ... plate is given in any condition of stress, or free from st... may apply, round axes everywhere perpendicular to its ... surface-edge, any arbitrary distribution of couple witho... ducing any change except at infinitely small distan... the edge, provided a certain distribution of force, al... lated from the distribution of couple, be applied to th... perpendicularly to the plate.

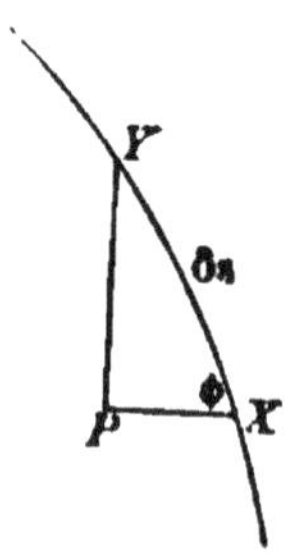

Let $XY, =\delta s$, be an infinitely small element at a point ... of a curve traced on the middle surface ... infinite plate; and, PX and PY being pa... to the axes of x and y, let $YXP=\phi$ ... if $\zeta\delta s$ denote the shearing force in the ... surface to the plate through δs, and, a... (§ 644), $\alpha.PY$ and $\beta.PX$ be those in ... surfaces through PY and PX, we must ... for the equilibrium of the triangle YPX ... posed rigid (§ 564),

$\zeta\delta s=\alpha.PY+\beta.PX$, whence $\zeta=\alpha\sin\phi+$ ...

Using here for α and β their values by (1) of § 644, we ha...

$$\zeta = -\left(M+\frac{d\Pi}{dy}+\frac{dK}{dx}\right)\sin\phi-\left(L+\frac{d\Lambda}{dy}+\frac{d\Pi}{dx}\right)\cos\phi \qquad (1).$$

Kirchhoff's boundary equations investigated.

Next, if $G\delta s$ and $H\delta s$ denote the components round XY, and round an axis perpendicular to it in the plane of the plate, of the couple acting across the normal surface through δs, we have [(2) of § 637],

$$G=\Lambda\cos^2\phi+2\Pi\sin\phi\cos\phi+K\sin^2\phi \qquad (2),$$

$$H=(K-\Lambda)\sin\phi\cos\phi+\Pi(\cos^2\phi-\sin^2\phi) \qquad (3).$$

If (ζ, G, H) denoted the action experienced by the edge in virtue of applied forces, all the plate outside a closed curve, of which δs is an element, being removed, these three equations would express the same as the three boundary equations given by Poisson. Lastly, let $\mathfrak{K}\delta s$, $G\delta s$, $\mathfrak{H}\delta s$ denote the force perpendicular to the plate, and the components of couple, actually applied at any point (x, y) of a free edge on the length δs of the middle curve. As we shall immediately see (§ 648), if

$$\mathfrak{K}-\zeta+\frac{d(\mathfrak{H}-H)}{ds}=0 \qquad (4),$$

the plate will be in the same condition of stress throughout, except infinitely near the edge, as with (ζ, G, H) for the action on the edge. Hence, eliminating ζ and H between these four equations, there remain to us (2) unchanged and another, or in all these two—

$$\left.\begin{array}{l} G=\Lambda\cos^2\phi+2\Pi\sin\phi\cos\phi+K\sin^2\phi, \text{ and} \\ =-\left(M+\frac{d\Pi}{dy}+\frac{dK}{dx}\right)\sin\phi-\left(L+\frac{d\Lambda}{dy}+\frac{d\Pi}{dx}\right)\cos\phi+\frac{d}{ds}[(K-\Lambda)\sin\phi\cos\phi+\Pi(\cos^2\phi-\sin^2\phi)] \end{array}\right\} (5),$$

which are Kirchhoff's boundary equations.

646. The proposition stated at the end of last section is uivalent to this:—That a certain distribution of normal earing force on the bounding edge of a finite plate may be etermined which shall produce the same effect as any given istribution of couple, round axes everywhere perpendicular to e normal surface supposed to constitute the edge. To prove his let equal forces act in opposite directions in lines EF, $E'F'$ n each side of the middle line and parallel to it, constituting he supposed distribution of couple. It must be understood hat the forces are actually distributed along their lines of ction, and not, as in the abstract dynamics of ideal rigid bodies, pplied indifferently at any points of these lines; but the mount of the force per unit of length, though equal in the

Distribution of shearing force determined, to produce same flexure as a given distribution of couple round axes perpendicular to boundary.

Distribution of shearing force determined, to produce same flexure as a given distribution of couple round axes perpendicular to boundary.

neighbouring parts of the two lines, must differ from p
point along the edge, to constitute any other than a un
distribution of couple. L
we may suppose the for
the opposite directions to b
confined to two lines, as
in the diagram, but to be di
over the two halves of th
on the two sides of its m
line; and further, the amo
them in equal infinitely s
breadths at different dis
from the middle line mu
proportional to these dis
as stated in § 634 (3), if the
distribution of couple is to be thoroughly such as H of §

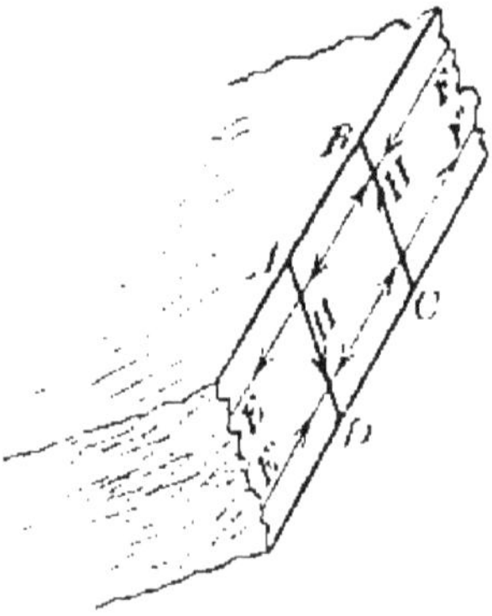

Let now the whole edge be divided into infinitely s
rectangles, such as $ABCD$ in the diagram, by lines drawn
pendicularly across it. In one of these rectangles app
balancing system of couples consisting of a diffused co
equal and opposite to the part of the given distributio
couple belonging to the area of the rectangle, and a co
of single forces in the lines AD, CB, of equal and opp
moment. This balancing system obviously cannot cause
sensible disturbance (stress or strain) in the plate, ex
within a distance comparable with the sides of the recta
and, therefore, when the same thing is done in all the rectan
into which the edge is divided, the plate is only disturb
an infinitely small distance from the edge inwards all r
But the given distribution of couple is thus removed (b
directly balanced by a system of diffused force equal
opposite everywhere to that constituting it), and there rem
only the set of forces applied in the cross lines. Of these t
are two in each cross line, derived from the operations p
formed in the two rectangles of which it is a common side,
their difference alone remains effective. Thus we see tha
the given distribution of couple be uniform along the edge
may be removed without disturbing the condition of the p
except infinitely near the edge: in other words,

647. *A uniform distribution of couple along the whole edge* *n finite plate, everywhere round axes in the plane of the plate,* *l perpendicular to the edge, produces distortion, spreading to* *'y infinitely small distances inwards from the edge all round,* *d no stress or distortion of the plate as a whole.* The truth of s remarkable proposition is also obvious when we consider at the tendency of such a distribution of couple can only be drag the two sides of the edge infinitesimally in opposite rections round the area of the plate. Later we shall investite strictly the strain, in the neighbourhood of the edge, proiced by it, and we shall find that it diminishes with extreme pidity inwards from the edge, becoming practically insensible distances exceeding twice the thickness of the plate.

Uniform distribution of twisting couple produces no flexure.

648. *A distribution of couple on the edge of a plate, round* *'es everywhere in the plane of the plate, and perpendicular to* *e edge, of any given amount per unit of length of the edge, may* *removed, and, instead, a distribution of force perpendicular to* *e plate, equal in amount per unit length of the edge, to the rate* *' variation per unit length of the amount of the couple, without* *ltering the flexure of the plate as a whole, or producing any dis-* *ırbance in its stress or strain except infinitely near the edge.*

The distribution of shearing force that produces same flexure as from distribution of twisting couple.

In the diagram of § 646 let $AB = \delta s$. Then if H be the amount of the given couple per unit length along the edge, between AD, BC, the amount of it on the rectangle $ABCD$ is $H\delta s$, and therefore H must be the amount of the forces introduced along AD, CB, in order that they may constitute a couple of the requisite moment. Similarly, if $H'\delta s$ denote the amount of the couple in the contiguous rectangle on the other side of BC, the force in BC derived from it will be H' in the direction opposite to H. There remains effective in BC a single force equal to the difference, $H' - H$.

If from A to B be the direction in which we suppose s, a length measured along the edge from any zero point, to increase, we have

$$H' - H = \frac{dH}{ds}\delta s.$$

Thus we are left with single forces, equal to $\frac{dH}{ds}\delta s$, applied in lines perpendicularly across the edge, at consecutive distances δs from one another; and for this we may substitute, without causing disturbance except infinitely near the edge, a continuous distribution of transverse force, amounting to $\frac{dH}{ds}$ per length; which is the proposition to be proved. The direc- this force, when $\frac{dH}{ds}$ is positive, is that of z negative w- immediately the form of it expressed in (4), of § 645.

Case of circular strain. **649.** As a first example of the application of the- .. tions, we shall consider the very simple case of a u plate of finite or infinite extent, symmetrically influ- :. concentric circles by a load distributed symmetrically, a: proper boundary appliances if required.

Let the origin of co-ordinates be chosen at the centre of metry, and let r, θ be polar co-ordinates of any point P, s- ·

$$x=r\cos\theta,\ y=r\sin\theta.$$

The second number of (6), § 644, will be a function of r, w for brevity we may now denote simply by Z (being the a: of load per unit area when the applied forces on each sma.. · are reducible to a single normal force through some point : Since z is now a function of r, and, as we have seen b- [§ 491 (e)],

$$\nabla^2 u=\frac{1}{r}\frac{d}{dr}\left(r\frac{du}{dr}\right)$$

when u is any function of r, equation (6) of § 644 becomes

$$\frac{A}{r}\frac{d}{dr}\left\{r\frac{d}{dr}\left[\frac{1}{r}\frac{d}{dr}\left(r\frac{dz}{dr}\right)\right]\right\}=Z$$

Hence

$$z=\frac{1}{A}\int\frac{dr}{r}\int rdr\int\frac{dr}{r}\int rZdr+\tfrac{1}{4}C(\log r-1)r^2+\tfrac{1}{4}C'r^2+C''\log r+C'''$$

which is the complete integral, with the four arbitrary con- explicitly shown. The following expressions, founded on . mediate integrals, deserve attention now, as promoting a th- comprehension of the solution; and some of them will be r- later for expressing the boundary conditions. The nota: (7) will be explained in § 650:—

(inclination, divided by radius; or curvature in normal section perpendicular to radius)

$$\frac{1}{r}\frac{dz}{dr}=\frac{1}{Ar^2}\int rdr\int\frac{dr}{r}\int rZdr+\tfrac{1}{2}C(\log r-\tfrac{1}{2})+\tfrac{1}{2}C'+\frac{C''}{r^2}$$

(curvature in radial section)

$$\frac{d^2z}{dr^2}=-\frac{1}{Ar^2}\int rdr\int\frac{dr}{r}\int rZdr+\frac{1}{A}\int\frac{dr}{r}\int rZdr+\tfrac{1}{2}C(\log r+\tfrac{1}{2})+\tfrac{1}{2}C'-\ldots$$

Plate circularly strained.

$$\left.\begin{array}{c}\text{(sum of curvatures in rectangular sections)}\\ \nabla^2 z=\frac{1}{A}\int\frac{dr}{r}\int rZdr+C\log r+C'\end{array}\right\}\quad(5)$$

$$\left.\begin{array}{l} A\frac{d^2z}{dr^2}+c\frac{dz}{rdr}=G \\ =-\frac{A-c}{Ar^2}\int rdr\int\frac{dr}{r}\int rZdr+\int\frac{dz}{r}\int rZdr+\frac{1}{2}C\{(A+c)\log r+\frac{1}{2}(A-c)\} \\ \qquad\qquad\qquad\qquad +\frac{1}{2}C'(A+c)-C''(A-c)\frac{1}{r^2} \\ H=0 \end{array}\right\}\quad(6)$$

$$L=c\frac{d^2z}{dr^2}+A\frac{dz}{rdr}\quad(7)$$

$$\left.\begin{array}{l}(A-c)\frac{d}{dr}(\frac{1}{r}\frac{dz}{dr})+\frac{dG}{dr}=A\frac{d}{dr}\nabla^2 z=-\zeta \\ =\frac{1}{r}\int rZdr+C\frac{A}{r}\end{array}\right\}\quad(8).$$

Of these (6) and (8) express, according to the notation of § 645, the couple and the shearing force acting on the normal surface cutting the middle surface of the plate in the circle of radius r. They are derivable analytically from our solution (2) by means of (2), (3), and (1) of § 645, with (4) of § 644, and (15) of § 642. The work is of course much shortened by taking $y=0$, and $x=r$, and using (3) and (4) of the present section. The student may go through this process, with or without the abbreviation, as an analytical exercise; but it is more instructive, as well as more direct, to investigate *ab initio* the equilibrium of a plate symmetrically strained in concentric circles, and so, in the course of an independent demonstration of (6) § 644, for this case, or (1) § 649, to find expressions for the flexural and shearing stresses.

Independent investigation for circular strain.

650. It is clear that, in every part of the plate, the normal sections (§ 637) of maximum and minimum, or minimum and maximum bending couples are those through and perpendicular to the radius drawn from O the centre of symmetry. At distance r from O, let L and G be the bending couples in the section through the radius, and in the section perpendicular to it; so that, if λ and κ be the curvatures in these sections, we have, by (10) of § 641 and (15) of § 642,

$$\left.\begin{array}{l}L=A\lambda+c\kappa\\ G=c\lambda+A\kappa\end{array}\right\}\quad(9).$$

Let also ζ be the shearing force (§ 616, foot-note) in the

Independent investigation for circular strain.

circular normal section of radius r. The symmetry requires that there be no shearing force in radial normal sections.

Considering now an element, E, bounded by two radii making an infinitely small angle $\delta\theta$ with one another, and two concentric circles of radii $r - \frac{1}{2}\delta r$ and $r + \frac{1}{2}\delta r$; we see that the equal couples, $L\delta r$ on its radial normal sections, round axes falling short of direct opposition by the infinitely small angle $\delta\theta$, have a resultant equal to $L\delta r\delta\theta$ round an axis perpendicular to the middle radius, in the negative direction when L is positive; and the infinitely nearly equal couples on its outer and inner circular edges have a resultant round the same axis, equal to $\frac{d(Gr\delta\theta)}{dr}\delta r$, being the difference of the values taken by $Gr\delta\theta$ when $r - \frac{1}{2}\delta r$ and $r + \frac{1}{2}\delta r$ are put for r. There is also the couple of the shearing forces on the outer and inner edges, each infinitely nearly equal to $\zeta r\delta\theta$; of which the moment is $\zeta r\delta\theta\delta r$. Hence, for the equilibrium of E under the action of these couples,

$$-L\delta r\delta\theta + \frac{d(Gr)}{dr}\delta r\delta\theta + \zeta r\delta\theta\delta r = 0,$$

or

$$-L + \frac{d(Gr)}{dr} + \zeta r = 0 \qquad (10)$$

if, as we may now conveniently do, we suppose no couples to be applied from without to any part of the plate except its bounding edges. Again, considering normal forces on E, we have $\frac{d(\zeta r\delta\theta)}{dr}\delta r$ for the sum of those acting on it from the contiguous matter of the plate, and $Zr\delta\theta\delta r$ from external matter if, as above, Z denote the amount of applied normal force per unit area of the plate. Hence, for the equilibrium of these forces,

$$\frac{d(\zeta r)}{dr} + Zr = 0 \qquad (11)$$

Substituting for ζ in (11) by (10); for L and G in the result by (9); and, in the result of this, for λ and κ their expressions by the differential calculus, which are $\frac{dz}{rdr}$ and $\frac{d^2z}{dr^2}$, since the plate is a surface of revolution differing infinitely little from plane perpendicular to the axis, we arrive finally at (1) the ifferential equation of the problem. Of the other formulæ of f § 649, (6), (7), (8) follow immediately from (9) and (10) now roved: except $H=0$, which follows from the fact that the ıdial and circular normal sections are the sections of maximum nd minimum, or minimum and maximum, curvature.

Independent investigation for circular strain.

651. We are now able to perceive the meaning of each of he four arbitrary constants.

Interpretation of terms in integral.

(1.) C''' is of course merely a displacement of the plate withut strain.

(2.) $C'' \log r$ is a displacement which produces anticlastic urvature throughout, with $\pm\frac{C''}{r^2}$ for the curvatures in the two rincipal sections: corresponding to which the bending couples, ., G, are equal to $\pm(A-c)\frac{C''}{r^2}$. An infinite plane plate, with circular aperture, and a uniform distribution of bending couple pplied to the edge all round, in each part round the tangent as xis, would experience this effect; as we see from the fact that he stress in the plate, due to C'', diminishes according to the nverse square of the distance from the centre of symmetry. t is remarkable that although the absolute value of the deflecion, $C'' \log r$, is infinite for infinite values of r, the restrictive ondition (3) of § 632 is not violated provided C'' is infinitely mall in comparison with the thickness: and it may be readily roved that the law (1) of § 633 is, in point of fact, fulfilled by his deflection, even if the whole displacement has rigorously his value, $C'' \log r$, and is precisely in the direction perpenicular to the undisturbed plane. For this case $\zeta=0$, or there s no shear.

(3.) $\frac{1}{4}C'r^2$ is a displacement corresponding to spherical urvature: and therefore involving simply a uniform synclastic tress [§ 638 (2)], of which the amount is of course [§ 641 10) or (11)] equal to $A+c$ divided by the radius of curvaure, or $(A+c)\times\frac{1}{2}C'$, agreeing with the equal values given or L and G by (6) and (7) of § 649. In this case also $\zeta=0$, or here is no shearing force. A finite plate of any shape, acted n by a uniform bending couple all round its edge, becomes ent thus spherically.

Interpretation of terms in integral.

(4.) $\frac{1}{4}C(\log r-1)r^2$ is a deflection involving a shearing force equal to $-A\frac{C}{r}$, and a bending couple,

$$\tfrac{1}{2}C\{(A+c)\log r+\tfrac{1}{2}(A-c)\},$$

in the circle of distance r from the centre of symmetry.

Symmetrical flexure of flat ring.

652. It is now a problem of the merest algebra to find the flexure of a flat ring, or portion of plane plate bounded by two concentric circles, when acted on by any given bending couples and transverse forces applied uniformly round the outer and inner edges. For equilibrium, the forces on the outer and inner edges must be in contrary directions, and equal amounts. Thus we have three arbitrary data: the amounts of the couple applied to the two edges, each reckoned per unit of length, and the whole amount, F, of the force on either edge. By (4), § 651, or (8) of § 649, we see that

$$-C=\frac{F}{2\pi A} \quad (12)$$

and there remain unknown the two constants, C' and C'', to be determined from the two equations given by putting the expression for G [(6) of § 649] equal to the given values for the values of r at the outer and inner edges respectively.

Example.—A circular table (of isotropic material), with a concentric circular aperture, is supported by its outer edge which rests simply on a horizontal circle; and is deflected by a load uniformly distributed over its inner edge (or *vice versa*, inner for outer). To find the deflection due to this load (which of course is simply added to the deflection due to the weight, determined below). Here G must vanish at each edge.

The radii of the outer and inner edges being a and a', the equations are

$$\tfrac{1}{2}C\{(A+c)\log a+\tfrac{1}{2}(A-c)\}+\tfrac{1}{2}C'(A+c)-C''(A-c)\frac{1}{a^2}=0,$$

and the same with a' for a. Hence

$$C''(A-e)\left(\frac{1}{a'^2}-\frac{1}{a^2}\right)=-\tfrac{1}{2}C(A+c)\log\frac{a}{a'},$$

and

$$\tfrac{1}{2}C'(A+c)(a^2-a'^2)=-\tfrac{1}{2}C[(A+c)(a^2\log a-a'^2\log a')+\tfrac{1}{2}(A-c)(a^2-a'^2$$

and thus, using for C its value (12), we find [(2) § 649]

$$=\frac{F}{2\pi A}\left[\tfrac{1}{4}\left(-\log r+1+\frac{a^2\log a-a'^2\log a'}{a^2-a'^2}+\tfrac{1}{2}\frac{A-c}{A+c}\right)r^2+\tfrac{1}{2}\frac{A+c}{A-c}\frac{a^2a'^2\log\frac{a}{a'}}{a^2-a'^2}\log r+C''\right].$$

Flexure of flat ring equilibrated by forces symmetrically distributed over its edges;

Putting the factor of r^2 into a more convenient form, and assigning C''' so that the deflection may be reckoned from the level of the inner edge, we have finally

$$z=\frac{F}{2\pi A}\left\{\tfrac{1}{4}\left(-\log\frac{r}{a'}+\frac{a^2}{a^2-a'^2}\log\frac{a}{a'}+\tfrac{1}{2}\frac{3A+c}{A+c}\right)r^2\right.$$

$$\left.+\tfrac{1}{2}\frac{A+c}{A-c}\frac{a^2a'^2\log\frac{a}{a'}}{a^2-a'^2}\log\frac{r}{a'}-\tfrac{1}{4}\frac{a^2a'^2}{a^2-a'^2}\log\frac{a}{a'}-\tfrac{1}{8}\frac{3A+c}{A+c}a'^2\right\}\quad(13).$$

Towards showing the distribution of stress through the breadth of the ring, we have from this, by § 649 (6),

$$G=\frac{F}{2\pi A}\cdot\tfrac{1}{2}(A+c)\left(\frac{a^2}{a^2-a'^2}\log\frac{a}{a'}-\log\frac{r}{a'}-\frac{a^2a'^2}{a^2-a'^2}\log\frac{a}{a'}\frac{1}{r^2}\right)\quad(14),$$

which, as it ought to do, vanishes when $r=a'$, and when $r=a$.

Further, by § 649 (8), $$\zeta=\frac{F}{2\pi r}\quad(15)$$

(which shows that, as is obviously true, the whole amount of the transverse force in any concentric circle of the ring is equal to F).

653. The problem of § 652, extended to admit load distributed in any symmetrical manner over the surface of the ing instead of merely confined to one edge, is solved lgebraically in precisely the same manner, when the terms lependent on Z, and exhibited in the several expressions of § 649, are found by integration. One important remark we iave to make however: that much needless labour is avoided y treating Z as a discontinuous function in these integrations n cases in which one continuous algebraic or transcendental unction does not express the distribution of load over the whole portion of plate considered. Unless this plan were ollowed, the expressions for z, $\frac{dz}{dr}$, G, and ζ, would have to be worked out separately for each annular portion of plate through which Z is continuous, and their values equated on each side of each separating circle. Hence if there were i annular portions to be thus treated separately there would be $4i$ arbitrary constants, to be determined by the $4(i-1)$ equations so obtained, and the 4 equations expressing that at the outer and inner bounding circular edges G has the prescribed values

and with load symmetrically spread over its area.

Flexure of flat ring equilibrated by forces symmetrically distributed over its edges; and with load symmetrically spread over its area.

(whether zero or not) of the applied bending couples, an[illegible] z and ζ have each a prescribed value at one or other of [illegible] circles. But by the more artful method, the multiplicati[illegible] detail required in virtue of the discontinuity of Z is con[illegible] to the successive integrations; and the arbitrary constan[illegible] which there are now but four, are determined by the condit[illegible] for the two extreme bounding edges.

Example.—A circular table (of isotropic material) wi[illegible] concentric circular aperture, is borne by its outer or inner [illegible] which rests simply on a horizontal circular support, an[illegible] loaded by matter uniformly distributed over an annular ar[illegible] its surface, extending from its inner edge outwards to a [illegible] centric circle of given radius, c. It is required to find [illegible] flexure.

First, supposing the aperture filled up, and the plate uni[illegible] from outer edge to centre, let the whole circle of radius [illegible] uniformly loaded at the rate w, a constant, per unit of its ar[illegible] We have

	$Z=$	$\int rZdr=$	$\int\frac{dr}{r}\int rZdr=$	$\int rdr\int\frac{dr}{r}\int rZdr=$	$\int\frac{dr}{r}\int rdr\int\frac{dr}{r}\int r$[illegible]
When $r=0$	w	0	0	0	0
,, $<c$	w	$\frac{1}{2}wr^2$	$\frac{1}{4}wr^2$	$\frac{1}{16}wr^4$	$\frac{1}{64}wr^4$
,, $>c$	0	$\frac{1}{2}wc^2$	$\frac{wc^2}{4}(2\log\frac{r}{c}+1)$	$\frac{wc^2}{16}(4r^2\log\frac{r}{c}+c^2)$	$\frac{wc^2}{16}(2r^2\log\frac{r}{c}-r^2+c^2$[illegible]
	I.	II.	III.	IV.	V.

Of these results, V. used in (2) gives the general solution; [illegible] IV., III., and II. in (6) and (8) give the corresponding express[illegible] for G and ζ. If, first, we suppose the value of G thus found [illegible] have any given value for each of two values, r', r'', of r, and [illegible] have a given value for one of these values of r, we have th[illegible] simple algebraic equations to find C, C', C''; and we solve a m[illegible] general problem than that proposed; to which we descend b[illegible] making the prescribed values of G and ζ zero. The power [illegible] mathematical expression and analysis in dealing with disc[illegible] tinuous functions, is strikingly exemplified in the applicab[illegible] of the result not only to the contemplated case, in which c is [illegible] termediate between r' and r''; but also to cases in which c is l[illegible] than either (when we fall back on the previous case, of § [illegible] or c greater than either (when we have a solution more dire[illegible] obtainable by taking $Z=w$ for all values of r).

priate for our present problem as for all involving the expression of arbitrary functions of θ for particular values of ϑ, is

$$v=\sum_{0}^{\infty}\{(A_i\cos i\theta+B_i\sin i\theta)\epsilon^{i\vartheta}+(\mathfrak{A}_i\cos i\theta+\mathfrak{B}_i\sin i\theta)\epsilon^{-i\vartheta}\} \qquad (12),$$

where A_i, B_i, $\mathfrak{A}_i$, $\mathfrak{B}_i$ are constants. That this is a solution, is of course verified in a moment by differentiation. From it we readily find (and the result of course is verified also by differentiation),

$$v=\sum_{i=0}^{i=\infty}\{\frac{1}{(i+2)^2-i^2}(A_i\cos i\theta+B_i\sin i\theta)\,\epsilon^{(i+2)\vartheta}\}+\sum_{i=2}^{i=\infty}\{\frac{-1}{i^2-(i-2)^2}(\mathfrak{A}_i\cos i\theta$$
$$+\mathfrak{B}_i\sin i\theta)\,\epsilon^{-(i-2)\vartheta}\}-\tfrac{1}{2}(\mathfrak{A}_1\cos\theta+\mathfrak{B}_1\sin\theta)\vartheta\,\epsilon^{\vartheta}+v' \qquad (13),$$

v' being any solution of (11), which may be conveniently taken as given by (12) with accented letters A_i', etc., to denote four new constants. If now the arbitrary periodic functions of θ, with 2π for period, given as the values whether of displacement, or shearing force, or couple, for the outer and inner circular edges, be expressed by Fourier's theorem [§ 77 (14)] in simple harmonic series; the two equations [§ 645 (5)] for each edge, applied separately to the coefficients of $\cos i\theta$ and $\sin i\theta$ in the expressions thus obtained, give eight equations for determining the eight constants A_i, $\mathfrak{A}_i$, B_i, $\mathfrak{B}_i$, A_i', $\mathfrak{A}_i'$, B_i', $\mathfrak{B}_i'$.

656. Although the problem of fulfilling arbitrary boundary onditions has not yet been solved for rectangular plates, there one remarkable case of it which deserves particular notice; ot only as interesting in itself, and important in practical pplication, but as curiously illustrating one of the most ifficult points §§ 646, 648] of the eneral theory. A ectangular plate cted on perpendicularly by a alancing system of four equal parallel forces applied at its four corners, becomes strained to a condition of uniform anticlastic curvature throughout, with the sections of no-flexure parallel to its sides, and therefore with sections of equal opposite maximum curvature in the normal planes inclined to the

Rectangular plate, held and loaded by diagonal pairs of corners.

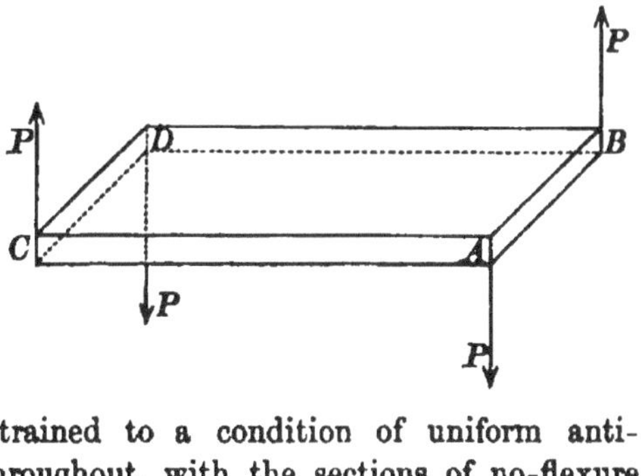

Rectangular plate, held and loaded by diagonal pairs of corners.

sides at 45°. This follows immediately from § 648, if we suppose the corners rounded off ever so little, and the f... diffused over them.

Or, in each of an infinite number of normal lines in the ed... AB, let a pair of opposite forces each equal to $\frac{1}{2}P$ be appl... which cannot disturb the plate. These, with halves of the ... forces P in the dissimilar directions at the corners A and B, con... stitute a diffused couple over the whole edge AB, amounting ... moment per unit of length to $\frac{1}{2}P$, round axes perpendic... to the plane of the edge. Similarly, the other halves of ... forces P at the corners AB, with halves of those at C a... D and introduced balancing forces, constitute diffused cou... over the edges CA and DB; and the remaining halves of ... corner forces at C and D, with introduced balancing forces, co... stitute a diffused couple over CD; each having $\frac{1}{2}P$ for ... amount of moment per unit length of the edge over which it ... diffused. Their directions are mutually related in the ma... specified in § 638 (2), and thus taken all together, they consti... an anticlastic stress of value $\Omega = \frac{1}{2}P$. Hence (§ 642) the re... is uniform anticlastic strain amounting to $\frac{1}{2}\frac{P}{k}$, and having ... axes inclined 45° to the edges; that is to say (§ 639), a flexu... with maximum curvatures on the two sides of the tangent pla... each equal to $\frac{1}{2}\frac{P}{k}$, and in normal sections in the positions stated.

Transition to finite flexures indicated.

657. Few problems of physical mathematics are m... curious than that presented by the transition from this s... tion, founded on the supposition that the greatest deflect... is but a small fraction of the thickness of the plate, to th... solution for larger flexures, in which corner portions will bend approximately as developable surfaces (cylindrical, in fact), an... a central quadrilateral part will remain infinitely nearly plane and thence to the extreme case of an infinitely thin perfectly flexible rectangle of inextensible fabric. This extreme case may be easily observed and experimented on by taking a carefully cut (§ 145) rectangle of paper, supporting it by fine threads attached to two opposite corners, and kept parallel, while two equal weights are hung by threads from the other corners.

Transmission of force through elastic solid.

658. The definitions and investigations regarding strain of §§ 154-190 constitute a kinematical introduction to the theory f elastic solids. We must now, in commencing the elementary lynamics of the subject, consider the forces called into play hrough the interior of a solid when brought into a condition of train. We adopt, from Rankine,[1] the term *stress* to designate such orces, as distinguished from strain defined (§ 154) to express he merely geometrical idea of a change of volume or figure.

Transmission of force through elastic solid.

659. When through any space in a body under the action of force, the mutual force between the portions of matter on the two sides of any plane area is equal and parallel to the mutual force across any equal, similar, and parallel plane area, the stress is said to be homogeneous through that space. In other words, the stress experienced by the matter is homogeneous through any space if all equal, similar, and similarly turned portions of matter within this space are similarly and equally influenced by force.

Homogeneous stress.

660. To be able to find the distribution of force over the surface of any portion of matter homogeneously stressed, we must know the direction, and the amount per unit area, of the force across a plane area cutting through it in any direction. Now if we know this for any three planes, in three different directions, we can find it for a plane in any direction as we see in a moment by considering what is necessary for the equilibrium of a tetrahedron of the substance. The resultant force on one of its sides must be equal and opposite to the resultant of the forces on the three others, which is known if these sides are parallel to the three planes for each of which the force is given.

Force transmitted across any surface in elastic solid.

661. Hence the stress, in a body homogeneously stressed, is completely specified when the direction, and the amount per unit area, of the force on each of three distinct planes is given. It is, in the analytical treatment of the subject, generally convenient to take these planes of reference at right angles to one another. But we should immediately fall into error did we not remark that the specification here indicated consist not of nine but in reality only of six, independent elements. For if the equilibrating forces on the six faces of a cube be each resolved into three components parallel to its three edges OX, OY, OZ, we have in all 18 forces; of which each pair acting perpendicularly on a pair of opposite faces, being equal and directly opposed,

Specification of a stress; by six independent elements.

[1] *Cambridge and Dublin Mathematical Journal*, 1850.

Relations
between
pairs of
tangential
tractions
necessary
for equili-
brium.

balance one another. The twelve tangential compon⸗⸗ ⸗ ⸗
remain constitute three pair of couples having their ax⸗⸗ ⸗⸗

direction of the three e⸗⸗ ⸗
of which must separa:⸗⸗ ⸗
equilibrium. The diagram ⸗⸗
the pair of equilibrating ⸗
having OY for axis; fr⸗⸗ ⸗
consideration of which w⸗ ⸗
that the forces on the fa⸗⸗ ⸗
parallel to OZ, are equal ⸗ ⸗
forces on the faces yx, ⸗⸗⸗
to OX. Similarly, we s⸗ ⸗
the forces on the faces
parallel to OY, are equal to those on the faces (xz), para⸗⸗⸗
OZ; and that the forces on (xx), parallel to OX, are ⸗⸗⸗ ⸗
those on (zy), parallel to OY.

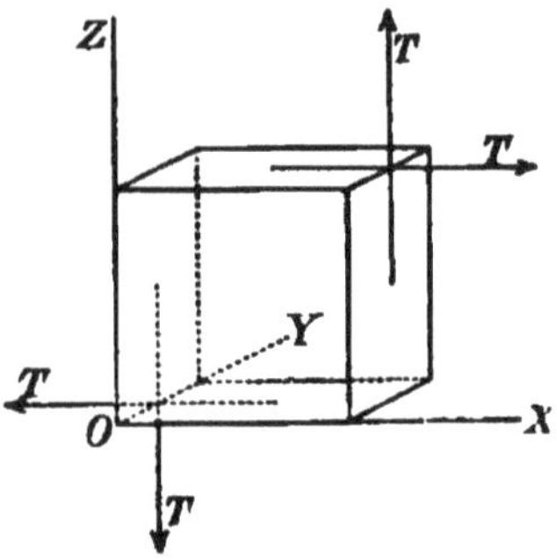

Specification
of a stress;
by six in-
dependent
elements:
three simple
longitudinal
stresses, and
three simple
distorting
stresses.

662. Thus, any three rectangular planes of reference ⸗⸗
chosen, we may take six elements thus, to specify a st⸗⸗⸗
P, Q, R the normal components of the forces on these p⸗⸗ ⸗
and S, T, U the tangential components, respectively p⸗⸗⸗⸗
dicular to OX, of the forces on the two planes meeting in ⸗⸗
perpendicular to OY, of the forces on the planes me⸗⸗⸗ ⸗
OY, and perpendicular to OZ, of the forces on the planes ⸗ ⸗
ing in OZ; each of the six forces being reckoned, per u⸗ ⸗
area. A normal component will be reckoned as positive w⸗
it is a traction tending to separate the portions of matter ⸗⸗ ⸗

Simple longi-
tudinal, and
shearing,
stresses.

two sides of its plane. P, Q, R are sometimes called ⸗⸗
longitudinal stresses, and S, T, U simple shearing str⸗⸗⸗⸗

Force across
any surface
in terms of
rectangular
specification
of stress.

From these data, to find in the manner explained in § ⸗⸗⸗ ⸗
force on any plane, specified by l, m, n, the direction-cosine⸗
its normal; let such a plane cut OX, OY, OZ in the three p⸗ ⸗
X, Y, Z. Then, if the area XYZ be denoted for a mom⸗⸗ ⸗
A, the areas YOZ, ZOX, XOY, being its projections on the u⸗⸗
rectangular planes, will be respectively equal to $Al, Am, A⸗$
Hence, for the equilibrium of the tetrahedron of matter bou⸗⸗
by those four triangles, we have, if F, G, H denote the c⸗
ponents of the force experienced by the first of them, XYZ p⸗
unit of its area, $F.A = P.lA + U.mA + T.nA,$
and the two symmetrical equations for the components parall⸗⸗ ⸗
OY and OZ. Hence, dividing by A, we conclude

$$\left.\begin{aligned} F &= Pl + Um + Tn \\ G &= Ul + Qm + Sn \\ H &= Tl + Sm + Rn \end{aligned}\right\} \quad (1).$$

Force across any surface in terms of rectangular specification of stress.

These expressions stand in the well-known relation to the ellipsoid $Px^2 + Qy^2 + Rz^2 + 2(Syz + Tzx + Uxy) = 1$ (2), according to which, if we take

$$x = lr,\ y = mr,\ z = nr,$$

and if λ, μ, ν denote the direction-cosines and p the length of the perpendicular from the centre to the tangent plane at (x, y, z) of the ellipsoidal surface, we have

$$F = \frac{\lambda}{pr},\quad G = \frac{\mu}{pr},\quad H = \frac{\nu}{pr}.$$

We conclude that

663. For any fully specified state of stress in a solid, a quadratic surface may always be determined, which shall represent the stress graphically in the following manner: Stress quadric.

To find the direction, and the amount per unit area, of the force acting across any plane in the solid, draw a line perpendicular to this plane from the centre of the quadratic to its surface. The required force will be equal to the reciprocal of the product of the length of this line into the perpendicular from the centre to the tangent plane at the point of intersection, and will be perpendicular to the latter plane.

664. From this it follows that for any stress whatever there are three determinate planes at right angles to one another such that the force acting in the solid across each of them is precisely perpendicular to it. These planes are called the principal or normal planes of the stress; the forces upon them, per unit area,—its principal or normal tractions; and the lines perpendicular to them,—its principal or normal axes, or simply its axes. The three principal semi-diameters of the quadratic surface are equal to the reciprocals of the square roots of the normal tractions. If, however, in any case each of the three normal tractions is negative, it will be convenient to reckon them rather as *positive pressures*; the reciprocals of the square roots of which will be the semi-axes of a real stress-ellipsoid representing the distribution of force in the manner explained above, with pressure substituted throughout for traction. Normal planes and axes of a stress.

of revolution and a sphere being included, as those in

two, or all three, are equal. When one of the three is neg-
and the two others positive, the surface is a hyperboloid of
sheet. When one of the normal tractions is positive and
two others negative, the surface is a hyperboloid of two sh-

666. When one of the three principal tractions vani-
while the other two are finite, the stress-quadratic becom-
cylinder, circular, elliptic, or hyperbolic, according as the
two are equal, unequal of one sign, or of contrary signs. W-
two of the three vanish, the quadratic becomes two planes:
the stress in this case is (§ 662) called a simple longitudinal st-
The theory of principal planes, and normal tractions just st-
(§ 664), is then equivalent to saying that any stress what-
may be regarded as made up of three simple longitud-
stresses in three rectangular directions. The geometrical in-
pretations are obvious in all these cases.

Composition of stresses.

667. The composition of stresses is of course to be effect-
adding the component tractions thus:—If (P_1, Q_1, R_1, S_1, T_1, U
(P_2, Q_2, R_2, S_2, T_2, U_2), etc., denote, according to § 662, a-
given set of stresses acting simultaneously in a substance, th-
joint effect is the same as that of a single resultant stress
which the specification in corresponding terms is (ΣP, ΣQ, Σ
ΣS, ΣT, ΣU).

Laws of strain and stress compared.

668. Each of the statements that have now been ma-
(§§ 659, 667) regarding stresses, is applicable to *infinitely* sm-
strains, if for traction perpendicular to any plane, reckoned p-
unit of its area, we substitute *elongation*, in the lines of the tra-
tion, reckoned per unit of length; and for *half the tangential* -
tion parallel to any direction, *shear* in the same direction reckon-
in the manner explained in § 175. The student will find it a
useful exercise to study in detail this transference of each o-
of those statements, and to justify it by modifying in the prop-
manner the results of §§ 171, 172, 173, 174, 175, 185, to adap-
them to infinitely small strains. It must be remarked that th-
strain-quadratic thus formed according to the rule of § 66-
which may have any of the varieties of character mentioned in
§§ 665, 666, is not the same as the strain-ellipsoid of § 16-

Laws of strain and stress compared.

/hich is always essentially an ellipsoid, and which, for an in- nitely small strain, differs infinitely little from a sphere.

The comparison of § 172, with the result of § 661 regarding angential tractions, is particularly interesting and important.

669. The following tabular synopsis of the meaning of the lements constituting the corresponding rectangular specifica- ions of a strain and stress explained in preceding sections, vill be found convenient:—

Rectangular elements of strains and stresses.

Components of the strain.	Components of the stress.	Planes; of which relative motion, or across which force, is reckoned.	Direction of relative motion or of force.
e	P	yz	x
f	Q	zx	y
g	R	xy	z
a	S	$\begin{cases} yx \\ zx \end{cases}$	y z
b	T	$\begin{cases} zy \\ xy \end{cases}$	z x
c	U	$\begin{cases} xz \\ yz \end{cases}$	x y

Work done by a stress within a varying solid.

670. If a unit cube of matter under any stress (P, Q, R, S, ', U) experience the infinitely small simple longitudinal strain alone, the work done on it will be Pe; since, of the com- onent forces P, U, T parallel to OX, U and T do no work in irtue of this strain. Similarly Qf, Rg are the works done if, he same stress acting, the simple longitudinal strains f or g re experienced, either alone. Again, if the cube experiences simple shear, a, whether we regard it (§ 172) as a differential liding of the planes yx, parallel to y, or of the planes zx, arallel to z, we see that the work done is Sa: and similarly, Tb if the strain is simply a shear b, parallel to OZ, of planes y, or parallel to OX, of planes xy: and Uc if the strain is a hear c, parallel to OX, of planes xz, or parallel to OY, of planes z. Hence the whole work done by the stress (P, Q, R, S, T, U) n a unit cube taking the strain (e, f, g, a, b, c), is

$$Pe + Qf + Rg + Sa + Tb + Uc \qquad (3).$$

t is to be remarked that, inasmuch as the action called a stress s a system of forces which balance one another if the portion f matter experiencing it is rigid, it cannot (§ 551) do any work

work done different from that just found.

If the side of the cube be of any length p, instead of … each force will be p^2 times, and each relative displacem… times; and therefore the work done p^3 times the resp… amounts reckoned above. Hence a body of any shap… of cubic content C, subjected throughout to a uniform st… (P, Q, R, S, T, U) while taking uniformly throughout a st… (e, f, g, a, b, c), experiences an amount of work equal to

$$(Pe + Qf + Rg + Sa + Tb + Uc)C$$

Work done on the surface of a varying solid.

It is to be remarked that this is necessarily equal t… work done on the bounding surface of the body by f… applied to it from without. For the work done on any p… of matter within the body is simply that done on its surfa… the matter touching it all round, as no force acts at a dis… from without on the interior substance. Hence if we ima… the whole body divided into any number of parts, each of … shape, the sum of the works done on all these parts is, by … disappearance of equal positive and negative terms expre… the portions of the work done on each part by the conti… parts on all its sides, and spent by these other parts in t… action, reduced to the integral amount of work done by f… from without applied all round the outer surface.

The analytical verification of this is instructive with regar… the syntax of the mathematical language in which the theory … the transmission of force is expressed. Let x, y, z be the co… ordinates of any point within the body; W the whole amou… of work done in the circumstances specified above; and $\iiint$ in… tegration extended throughout the space occupied by the body so that

$$W = \iiint (Pe + Qf + Rg + Sa + Tb + Uc)\,dx\,dy\,dz \qquad 5$$

If now we denote by α, β, γ the component displacements of any point of the matter infinitely near the point (x, y, z), experienced when the strain (e, f, g, a, b, c) takes place, whether non-rotationally (§ 182) and with some point of the body fixed, or with any motion of translation whatever and any infinitely small rotation, by adapting § 181 (5) to infinitely small strains according to our present notation (§ 669), and using in it § 190 (e), we have

Strain-components in terms of displacement.

$$\left.\begin{aligned} e&=\frac{d\alpha}{dx}, & f&=\frac{d\beta}{dy}, & g&=\frac{d\gamma}{dz}, \\ a&=\frac{d\beta}{dz}+\frac{d\gamma}{dy}, & b&=\frac{d\gamma}{dx}+\frac{d\alpha}{dz}, & c&=\frac{d\alpha}{dy}+\frac{d\beta}{dx} \end{aligned}\right\} \quad (6).$$

With these, (5) becomes

Work done through interior;

$$W=\iiint\left(P\frac{d\alpha}{dx}+U\frac{d\beta}{dx}+T\frac{d\gamma}{dx}+U\frac{d\alpha}{dy}+Q\frac{d\beta}{dy}+S\frac{d\gamma}{dy}+T\frac{d\alpha}{dz}+S\frac{d\beta}{dz}+R\frac{d\gamma}{dz}\right)dxdydz \quad (7).$$

Hence by integration

$$W=\iint[(P\alpha+U\beta+T\gamma)dydz+(U\alpha+Q\beta+S\gamma)dzdx+(T\alpha+S\beta+R\gamma)dxdy] \quad (8),$$

the limits of the integrations being so taken that, if $d\sigma$ denote an element of the bounding surface, $\iint$ integration all over it, and l, m, n the direction-cosines of the normal at any point of it, the expression means the same as

$$W=\iint\{(P\alpha+U\beta+T\gamma)l+(U\alpha+Q\beta+S\gamma)m+(T\alpha+S\beta+R\gamma)n\}d\sigma \quad (9);$$

which, with the terms grouped otherwise, becomes

$$W=\iint\{(Pl+Um+Tn)\alpha+(Ul+Qm+Sn)\beta+(Tl+Sm+Rn)\gamma\}d\sigma \quad (10).$$

The second member of this, in virtue of (1), expresses directly the work done by the forces applied from without to the bounding surface.

agrees with work done on surface.

671. If, now, we suppose the body to yield to a stress (P, Q, R, S, T, U), and to oppose this stress only with its innate resistance to change of shape, the differential equation of work done will [by (4) with de, df, etc., substituted for e, f, etc.] be

Differential equation of work done by a stress

$$dw=Pde+Qdf+Rdg+Sda+Tdb+Udc \quad (11),$$

if w denote the whole amount of work done per unit of volume in any part of the body while the substance in this part experiences a strain (e, f, g, a, b, c) from some initial state regarded as a state of no strain. This equation, as we shall see later, under Properties of Matter, expresses the work done in a natural fluid, by distorting stress (or difference of pressure in different directions) working against its innate viscosity; and w is then, according to Joule's discovery, the dynamic value of the heat generated in the process. The equation may also be pplied to express the work done in straining an imperfectly lastic solid, or an elastic solid of which the temperature varies uring the process. In all such applications the stress will epend partly on the speed of the straining motion, or on the arying temperature, and not at all, or not solely, on the state

Physical application.

Perfectly elastic body defined, in abstract dynamics.

672. *Definition.*—A perfectly elastic body is a body w... when brought to any one state of strain, requires at al... the same stress to hold it in this state; however long ... kept strained, or however rapidly its state be altered fro... other strain, or from no strain, to the strain in question. ... according to our plan (§§ 443, 448) for Abstract Dyn... we ignore variation of temperature in the body. If, how... we add a condition of absolutely no variation of temper... or of recurrence to one specified temperature after chan... strain, we have a definition of that property of perfect ela...

Its conditional fulfilment in nature.

towards which highly elastic bodies in nature approximate; which is rigorously fulfilled by all fluids, and may be ... some real solids, as homogeneous crystals. But inasmu... the elastic reaction of every kind of body against strain ... with varying temperature, and (a thermodynamic conseq... of this, as we shall see later) any increase or diminut... strain in an elastic body is necessarily[1] accompanied ... change of temperature; even a perfectly elastic body could ... in passing through different strains, act as a rigorously c... servative system, but, on the contrary, must give rise to ... sipation of energy in consequence of the conduction or radi... of heat induced by these changes of temperature.

But by making the changes of strain quickly enough to p... vent any sensible equalization of temperature by conduction ... radiation (as, for instance, Stokes has shown, is done in so... of musical notes travelling through air); or by making th... slowly enough to allow the temperature to be mainta... sensibly constant[2] by proper appliances; any highly elastic ... perfectly elastic body in nature may be got to act very ne... as a conservative system.

Potential energy of an elastic solid held strained.

673. In nature, therefore, the integral amount, ..., of wo... defined as above, is for a perfectly elastic body, independ... (§ 274) of the series of configurations, or states of stra... through which it may have been brought from the first ... the second of the specified conditions, provided it has ...

[1] "On the Thermoelastic and Thermomagnetic Properties of Matter" (W. T... son). *Quarterly Journal of Mathematics.* April 1857. [2] *Ibid.*

been allowed to change sensibly in temperature during the process.

Potential energy of an elastic solid held strained

The analytical statement is that the expression (11) for dw must be the differential of a function of e, f, g, a, b, c, regarded as independent variables; or, which means the same, w is a function of these elements, and

$$\left.\begin{array}{l} P=\frac{dw}{de},\quad Q=\frac{dw}{df},\quad R=\frac{dw}{dg}, \\ S=\frac{dw}{da},\quad T=\frac{dw}{db},\quad U=\frac{dw}{dc}. \end{array}\right\} \qquad (12).$$

In Appendix C. we shall return to the comprehensive analytical treatment of this theory, not confining it to infinitely small strains for which alone the notation $(e, f, \ldots)$, as defined in § 669, is convenient. In the meantime, we shall only say that when the whole amount of strain is infinitely small, and the stress-components are therefore all altered in the same ratio as the strain-components if these are altered all in any one ratio; w must be a homogeneous quadratic function of the six variables e, f, g, a, b, c, which, if we denote by $(e, e), (f, f)\ldots(e, f)\ldots$ constants depending on the quality of the substance and on the directions chosen for the axes of co-ordinates, we may write as follows:—

$$\left.\begin{array}{r} w=\tfrac{1}{2}\{(e,e)e^2+(f,f)f^2+(g,g)g^2+(a,a)a^2+(b,b)b^2+(c,c)c^2 \\ +2(e,f)ef+2(e,g)eg+2(e,a)ea+2(e,b)eb+2(e,c)ec \\ +2(f,g)fg+2(f,a)fa+2(f,b)fb+2(f,c)fc \\ +2(g,a)ga+2(g,b)gb+2(g,c)gc \\ +2(a,b)ab+2(a,c)ac \\ +2(b,c)bc\} \end{array}\right\} \quad (13).$$

The 21 coefficients $(e, e), (f, f)\ldots(b, c)$, in this expression constitute the 21 "coefficients of elasticity," which Green first showed to be proper and essential for a complete theory of the dynamics of an elastic solid subjected to infinitely small strains. The only condition that can be theoretically imposed upon these coefficients is that they must not permit w to become negative for any values, positive or negative, of the strain-components $e, f, \ldots$ Under Properties of Matter, we shall see that a false theory (Boscovich's), falsely worked out by mathematicians, has led to relations among the coefficients of elasticity which experiment has proved to be false.

Eliminating w from (12) by (13) we have

Stress-components expressed in terms of strain.

$$\left.\begin{aligned} P&=(e,e)e+(e,f)f+(e,g)g+(e,a)a+(e,b)b+(e,c)c \\ Q&=(e,f)e+(f,f)f+(f,g)g+(f,a)a+(f,b)b+(f,c)c \\ &\quad\text{etc.} \qquad\qquad \text{etc.} \\ &\quad\text{etc.} \qquad\qquad \text{etc.} \end{aligned}\right\} \quad (14)$$

These equations express the six components of stress (P, Q, R, S, T, U) as linear functions of the six components of strain (e, f, g, a, b, c) with 15 equalities[1] among their 36 coefficients, which leave only 21 of them independent. The mere principle of superposition (which we have used above in establishing the *quadratic* form for w) might have been directly applied to demonstrate linear formulæ for the stress-components. Thus it is that some authors have been led to lay down, as the foundation of the most general possible theory of elasticity, six equations involving 36 coefficients supposed to be independent. But it is only by the principle of energy that, as first discovered by Green, the fifteen pairs of these coefficients are proved to be equal.

Strain-components expressed in terms of stress.

The algebraic transformation of equations (14) to express the strain-components singly, by linear functions of the stress-components, may be directly effected of course by forming the proper determinants from the 36 coefficients, and taking the 36 proper quotients. From a known determinantal theorem, used as above [§ 313 (*d*)], it follows that there are 15 equalities between pairs of these 36 quotients, because of the 15 equalities in pairs of the coefficients of e, f, etc., in (14). Thus, if we denote by

$$[P,P],\ [Q,Q],\ \ldots\ [P,Q],\ \ldots\ [Q,P]\ \ldots$$

the set of 36 determinantal quotients found by that process (and, therefore, known algebraic functions of the original coefficients (e,e), (f,f), ... etc.], we have

$$\left.\begin{aligned} e&=[P,P]P+[P,Q]Q+[P,R]R+[P,S]S+[P,T]T+[P,U]U \\ f&=[Q,P]P+[Q,Q]Q+[Q,R]R+[Q,S]S+[Q,T]T+[Q,U]U \\ &\quad\text{etc.} \qquad\qquad \text{etc.} \end{aligned}\right\} \quad (15)$$

and these new coefficients satisfy 15 equations

$$[P,Q]=[Q,P],\ [P,R]=[R,P]\ldots \quad (16)$$

By what we proved in § 313 (*d*) when engaged with precisely the same algebraic transformation, we see that $[P,P]$, $[Q,Q]$, ... $[P,Q]$, ... are simply the coefficients of P^2, Q^2, ..., $2PQ$, ... in the expression for $2w$ obtained by eliminating e, f, ... from (13), so that

$$w=\tfrac{1}{2}\{[P,P]P^2+[Q,Q]Q^2+\ldots+2[P,Q]PQ+2[P,R]PR+\ldots\} \quad (17)$$

[1] Viz., $(e,f)=(f,e)$, ... $(e,a)=(a,e)$, ... $(b,c)=(c,b)$.

and

$$e=\left[\frac{dw}{dP}\right],\ f=\left[\frac{dw}{dQ}\right],\ g=\left[\frac{dw}{dR}\right],\quad a=\left[\frac{dw}{dS}\right],\ b=\left[\frac{dw}{dT}\right],\ c=\left[\frac{dw}{dU}\right] \qquad (19);$$

Strain-components expressed in terms of stress.

where the brackets [ ] denote the partial differential coefficients taken on the supposition that w is expressed as a function of P, Q, etc., as in (19); to distinguish them from those of equations (12) which were taken on the supposition that w is expressed as a function of e, f, etc., as in (13). We have also, as in § 313 (d),

$$w=\tfrac{1}{2}(Pe+Qf+Rg+Sa+Tb+Uc) \qquad (20);$$

which might have been put down in the beginning, as it simply expresses that

674. The average stress, due to elasticity of the solid, when strained from its natural condition to that of strain (e, f, g, a, b, c) is (as from the assumed applicability of the principle of superposition we see it must be) just half the stress required to keep it in this state of strain.

Average stress through any changing strain.

675. A body is called homogeneous when any two equal, similar parts of it, with corresponding lines parallel and turned towards the same parts, are undistinguishable from one another by any difference in quality. The perfect fulfilment of this condition without any limit as to the smallness of the parts, though conceivable, is not generally regarded as probable for any of the real solids or fluids known to us, however seemingly homogeneous. It is, we believe, held by all naturalists that there is a *molecular structure*, according to which, in *compound* bodies such as water, ice, rock-crystal, etc., the constituent substances lie side by side, or arranged in groups of finite dimensions, and even in bodies called *simple* (*i.e.*, not known to be chemically resolvable into other substances) there is no ultimate homogeneousness. In other words, the prevailing belief is that every kind of matter with which we are acquainted has a more or less *coarse-grained* texture, whether having visible molecules, as great masses of solid stone or brick-building, or natural granite or sandstone rocks; or, molecules too small to be visible or directly measurable by us (but *not infinitely small*)[1] in seemingly homogeneous metals, or continuous crystals, or liquids, or gases. We must of course return to this subject

Homogeneousness defined.

Molecular hypothesis

assumes a very fine-grained texture in crystals, but no ultimate homogeneousness.

[1] Probably not *undiscoverably* small, although of dimensions not yet known to us.

say that the definition of homogeneousness may be applied

practically on a very large scale to masses of building or coar-
grained conglomerate rock, or on a more moderate scale
blocks of common sandstone, or on a very small scale to see-
ingly homogeneous metals;[1] or on a scale of extreme, uncovered fineness, to vitreous bodies, continuous crystals, solidit-
gums, as India rubber, gum-arabic, etc., and fluids.

676. The substance of a homogeneous solid is called *is-tropic* when a spherical portion of it, tested by any phys-
agency, exhibits no difference in quality however it is tur-
Or, which amounts to the same, a cubical portion cut from
position in an isotropic body exhibits the same qualities r-
tively to each pair of parallel faces. Or two equal and sim-
portions cut from *any* positions in the body, not subject to
condition of parallelism (§ 675), are undistinguishable from
another. A substance which is not isotropic, but exhibits d-
ferences of quality in different directions, is called *aeolotrop*

677. An individual body, or the substance of a homogeneous solid, may be isotropic in one quality or class
qualities, but aeolotropic in others.

Thus in abstract dynamics a rigid body, or a group of
rigidly connected, contained within and rigidly attached to
rigid spherical surface, is kinetically symmetrical (§ 285) if
centre of inertia is at the centre of the sphere, and if its mome-
of inertia are equal round all diameters. It is also isot-
relatively to gravitation if it is centrobaric (§ 526), so that
centre of figure is not merely a centre of inertia, but a
centre of gravity. Or a transparent substance may trans-
light at different velocities in different directions through
(that is, be *doubly refracting*), and yet a cube of it may a-
generally does in natural crystals) absorb the same part of
beam of white light transmitted across it perpendicularly
any of its three pairs of faces. Or (as a crystal which exh-
dichroism) it may be aeolotropic relatively to the latter, or t-
either, optic quality, and yet it may conduct heat equally
all directions.

---

[1] Which, however, we know, as recently proved by Deville and Van Tro-
porous enough at high temperatures to allow very free percolation of gases.

Isotropic
and aeolo-
tropic
substances
defined.

Isotropy
and aeolo-
tropy of
different
sets of
properties.

Practical limitation of isotropy, and homogeneousness of aeolotropy, to the average in the aggregate of molecules.

678. The remarks of § 675 relative to homogeneousness in the aggregate, and the supposed ultimately heterogeneous texture of all substances however seemingly homogeneous, indicate corresponding limitations and non-rigorous practical interpretations of isotropy.

Conditions fulfilled in elastic isotropy.

679. To be elastically isotropic, we see first that a spherical or cubical portion of any solid, if subjected to uniform normal pressure (positive or negative) all round, must, in yielding, experience no deformation: and therefore must be equally compressed (or dilated) in all directions. But, further, a cube cut from any position in it, and acted on by *tangential* or distorting stress (§ 662) in planes parallel to two pairs of its sides, must experience simple deformation, or shear (§ 171), in the same direction, unaccompanied by condensation or dilatation,[1] and the same in amount for all the three ways in which a stress may be thus applied to any one cube, and for different cubes taken from any different positions in the solid.

Measures of resistance to compression and resistance to distortion.

680. Hence the elastic quality of a perfectly elastic, homogeneous, isotropic solid is fully defined by two elements;—its resistance to compression, and its resistance to distortion. The amount of uniform pressure in all directions, per unit area of its surface, required to produce a stated very small compression, measures the first of these, and the amount of the distorting stress required to produce a stated amount of distortion measures the second. The numerical measure of the first is the compressing pressure divided by the diminution of the bulk of a portion of the substance which, when uncompressed, occupies the unit volume. It is sometimes called the *elasticity of volume*, or the *resistance to compression*. Its reciprocal, or the amount of compression on unit of volume divided by the compressing pressure, or, as we may conveniently say, the compression per unit of volume, per unit of compressing pressure, is commonly called the *compressibility*. The second, or resist-

Elasticity of volume.

Compressibility.

1 It must be remembered that the changes of figure and volume we are concerned with are so small that the principle of superposition is applicable; so that if any distorting stress produced a condensation, an opposite distorting stress would produce a dilatation, which is a violation of the isotropic condition. But it is possible that a distorting stress may produce, in a truly isotropic solid, condensation or dilatation in proportion to the square of its value: and it is probable that such effects may be sensible in India rubber, or cork, or other bodies susceptible of great deformations or compressions, with persistent elasticity.

defined.

or shear (§ 175) which it produces, and is called the rig[illegible] the substance, or its *elasticity of figure.*

Discrepant reckonings of shear and distorting stress, from the simple longitudinal strains or stresses respectively involved.

681. From § 169 it follows that a strain compound[illegible] simple extension in one set of parallels, and a simple co[illegible] tion of equal amount in any other set perpendicular t[illegible] is the same as a simple shear in either of the two [illegible] planes cutting the two sets of parallels at 45°. And [illegible] numerical measure (§ 175) of this shear, or simple dist[illegible] is equal to *double* the amount of the elongation or contr[illegible] (each measured, of course, per unit of length). Similarly see (§ 668) that a longitudinal traction (or negative pre[illegible] parallel to one line, and an equal longitudinal positive pr[illegible] parallel to any line at right angles to it, is equivalent to [illegible] torting stress of tangential tractions (§ 661) parallel t[illegible] planes which cut those lines at 45°. And the num[illegible] measure of this distorting stress, being (§ 662) the amou[illegible] the tangential traction in either set of planes, is equal t[illegible] amount of the positive or negative normal pressure, *not d*[illegible]

Strain produced by a single longitudinal stress.

682. Since then any stress whatever may be made u[illegible] simple longitudinal stresses, it follows that, to find the rel[illegible] between any stress and the strain produced by it, we have [illegible] to find the strain produced by a single longitudinal str[illegible] which we may do at once thus:—A simple longitudinal str[illegible]

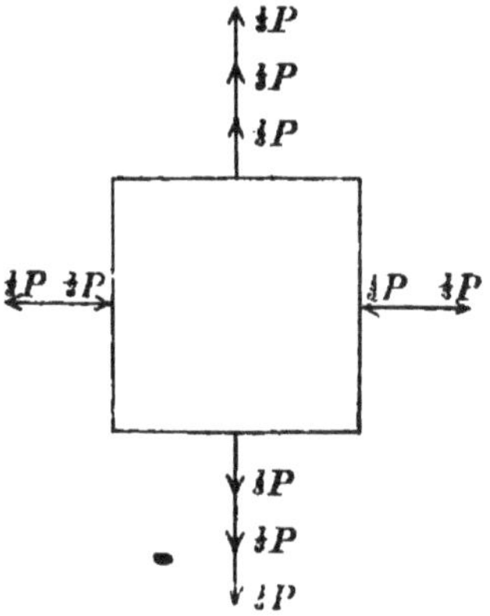

P, is equivalent to a uni[illegible] dilating tension ⅓P in all di[illegible] tions, compounded with [illegible] distorting stresses, each eq[illegible] to ⅓P, and having a comm[illegible] axis in the line of the gi[illegible] longitudinal stress, and th[illegible] other two axes any two lines [illegible] right angles to one another a[illegible] to it. The diagram, drawn i[illegible] plane through one of these la[illegible] lines, and the former, suffici[illegible] ly indicates the synthesis; [illegible] only forces not shown being those perpendicular to its plan[illegible]

Ratio of lateral contraction to longitudinal extension,

rigidity to be $\frac{3}{5}$ of the resistance to compression, for all s... and which was first shown to be false by Stokes[1] from ... obvious observations, proving enormous discrepancies fr... in many well-known bodies, and rendering it most impr... that there is any approach to a constancy of ratio betw... rigidity and resistance to compression in any class of s... Thus clear elastic jellies, and India rubber, present fa... specimens of isotropic homogeneous solids, which, while d... ing very much from one another in rigidity ("stiffness" ... probably all of very nearly the same compressibility as w... This being $\frac{1}{308000}$ per pound per square inch; the resistan... compression, measured by its reciprocal, or, as we may rea... "308000 lbs. per square inch," is obviously many hun... times the absolute amount of the rigidity of the stiffest of ... substances. A column of any of them, therefore, when p... together or pulled out, within its limits of elasticity, by bal... ing forces applied to its ends (or an India-rubber band ... pulled out), experiences no sensible change of volume, th... very sensible change of length. Hence the proportionate ... tension or contraction of any transverse diameter must ... sensibly equal to ½ the longitudinal contraction or extens... and for all ordinary stresses, such substances may be practi... regarded as incompressible elastic solids. Stokes gave re... for believing that metals also have in general greater res... ance to compression, in proportion to their rigidities, ... according to the fallacious theory, although for them the dis... crepancy is very much less than for the gelatinous bodies. T... probable conclusion was soon experimentally demonstrated ... Wertheim, who found the ratio of lateral to longitudinal cha... of lineal dimensions, in columns acted on solely by longitu... force, to be about ⅓ for glass and brass; and by Kirchhoff, w... by a very well-devised experimental method, found ·387 as ... value of that ratio for brass, and ·294 for iron. For copper ... find that it probably lies between ·226 and ·441, by ... experiments[2] of our own, measuring the torsional and lon... tudinal rigidities (§§ 596, 599, 686) of a copper wire.

different for different substances from ½ for jelly to 0 for cork.

[1] On the Friction of Fluids in Motion, and the Equilibrium and Motion of E... Solids.—*Trans. Camb. Phil. Jour.*, April 1845. See also *Camb. and Dub. M... Jour.*, March 1848.

[2] On the Elasticity and Viscosity of Metals (W. Thomson). Proc. R. S., May ...

Velocity of transmission of a simple longitudinal stress through a rod.

by the weight of the unit length. It is useful in many ap[illegible] tions of the theory of elasticity; as, for instance, in this [illegible] which will be proved later:—the velocity of transmis[illegible] longitudinal vibrations (as of sound) along a bar or [illegible] equal to the velocity acquired by a body in falling fr[illegible] height equal to half the length of the modulus.[1]

Specific modulus of an isotropic body.

688. The *specific modulus of elasticity of an isotrop[illegible] stance*, or, as it is most often called, simply *the mod[illegible] elasticity of the substance*, is the modulus of elasticity of [illegible] of it having some definitely specified sectional area. If [illegible] be such that the weight of unit length is unity, the *modul[illegible] the substance* will be the same as the length of the modul[illegible] any bar of it: a system of reckoning which, as we have [illegible] has some advantages in application. It is, however, more [illegible] to choose a common unit of area as the sectional area [illegible] bar referred to in the definition. There must also be a d[illegible] understanding as to the unit in terms of which the fo[illegible] measured, which may be either the *absolute unit* (§ 223 the gravitation unit for a specified locality; that is (§ 22[illegible] weight in that locality of the unit of mass. Experimen[illegible] hitherto have stated their results in terms of the gravita[illegible] unit, each for his own locality; the accuracy hitherto atta[illegible] being scarcely in any cases sufficient to require corre[illegible] for the different forces of gravity in the different plac[illegible] observation.

In terms of the absolute unit; or of the force of gravity on the unit of mass in any particular locality.

689. The most useful and generally convenient spec[illegible] tion of the modulus of elasticity of a substance is in gram[illegible] weight per square centimetre. This has only to be divid[illegible] the specific gravity of the substance to give the *length [illegible] modulus*. British measures, however, being still unhap[illegible] sometimes used in practical and even in high scientific [illegible]

[1] It is to be understood that the vibrations in question are so much spre[illegible] through the *length* of the body, that inertia does not sensibly influence the [illegible] verse contractions and dilatations which (unless the substance have in thi[illegible] the peculiar character presented by cork, § 684) take place along with them. [illegible] under thermodynamics, we shall see that changes of temperature produced [illegible] varying stresses cause changes of temperature which, in ordinary solids, ren[illegible] velocity of transmission of longitudinal vibrations sensibly greater than that [illegible] lated by the rule stated in the text, if we use the *static modulus* as underst[illegible] the definition there given; and we shall learn to take into account the therm[illegible] by using a definite *static modulus*, or *kinetic modulus*, according to the circumst[illegible] of any case that may occur.

Stress required to maintain a simple longitudinal strain.

692. In §§ 681, 682 we examined the effect of a s… longitudinal stress, in producing elongation in its own l… tion, and contraction in lines perpendicular to it. With st… substituted for strains, and strains for stresses, we may a… the same process to investigate the longitudinal and l… tractions required to produce a simple longitudinal strain (… is, an elongation in one direction, with no change of dimen… perpendicular to it) in a rod or solid of any shape.

Thus a simple longitudinal strain e is equivalent to a c… dilatation e without change of figure (or linear dilatati… equal in all directions), and two distortions consisting e… dilatation $\frac{1}{3}e$ in the given direction, and contraction $\frac{1}{3}e$ in e… of two directions perpendicular to it and to one another. … produce the cubic dilatation, e, alone requires (§ 680) a no… traction ke equal in all directions. And, to produce eith… the distortions simply, since the measure (§ 175) of each i… requires a distorting stress equal to $n\times\frac{2}{3}e$, which consist… tangential tractions each equal to this amount, positive … drawing outwards) in the line of the given elongation, … negative (or pressing inwards) in the perpendicular direct… Thus we have in all

$$\left.\begin{array}{l}\text{normal traction} = (k+\frac{4}{3}n)e, \text{ in the direction of the given strain, and}\\ \text{normal traction} = (k-\frac{2}{3}n)e, \text{ in every direction perpendicular to the given strain.}\end{array}\right\} \quad (4)$$

Stress-components in terms of strain for isotropic body.

693. If now we suppose any possible infinitely small str… (e, f, g, a, b, c), according to the specification of § 669, t… given to a body, the stress (P, Q, R, S, T, U) required t… maintain it will be expressed by the following formulæ, o… tained by successive applications of § 692 (4) to the com… ponents e, f, g separately, and of § 680 to a, b, c:—

$$\left.\begin{array}{l} S=na,\ T=nb,\ U=nc,\\ P=\mathfrak{A}e+\mathfrak{B}(f+g),\\ Q=\mathfrak{A}f+\mathfrak{B}(g+e),\\ R=\mathfrak{A}g+\mathfrak{B}(e+f),\\ \text{where}\quad \left.\begin{array}{l}\mathfrak{A}=k+\frac{4n}{3},\\ \mathfrak{B}=k-\frac{2n}{3},\end{array}\right\}\ n=\frac{1}{2}(\mathfrak{A}-\mathfrak{B})\end{array}\right\} \quad (5)$$

694. Similarly, by § 680 and § 682 (3), we have

Strain-components in terms of stress for isotropic body.

$$\left.\begin{aligned}
&a=\frac{1}{n}S,\ b=\frac{1}{n}T,\ c=\frac{1}{n}U,\\
&Me=\{P-\sigma(Q+R)\},\\
&Mf=\{Q-\sigma(R+P)\},\\
&Mg=\{R-\sigma(P+Q)\},\\
\text{iere}\quad &M=\frac{9nk}{3k+n},\\
\text{d}\quad &\frac{m-n}{2m}=\sigma=\frac{3k-2n}{2(3k+n)}=\tfrac{1}{2}\frac{M}{n}-1,
\end{aligned}\right\}\quad(6),$$

the formulæ expressing the strain (e, f, g, a, b, c) in terms of e stress (P, Q, R, S, T, U). They are of course merely the gebraic inversions of (5); and (§ 673) they might have been und by solving these for e, f, g, a, b, c, regarded as the unnown quantities. M is here introduced to denote Young's odulus (§ 683). and m from § 698 (5).

695. To express the equation of energy for an isotropic ıbstance, we may take the general formula,

Equation of energy for the same.

$$w=\tfrac{1}{2}(Pe+Qf+Rg+Sa+Tb+Uc)\ldots[\S\ 673\ (20)],$$

ıd eliminate from it P, Q, etc., by (5) of § 693, or, again, e, f, tc., by (6) of § 694, we thus find

$$\left.\begin{aligned}
&=(k+\frac{4n}{3})(e^2+f^2+g^2)+2(k-\frac{2n}{3})(fg+ge+ef)+n(a^2+b^2+c^2)\\
&=\tfrac{1}{2}\{(\frac{1}{n}+\frac{1}{3k})(P^2+Q^2+R^2)-2(\frac{1}{2n}-\frac{1}{3k})(QR+RP+PQ)\}+\frac{1}{n}(S^2+T^2+U^2)
\end{aligned}\right\}\ (7).$$

696. The mathematical theory of the equilibrium of an lastic solid presents the following general problems:—

Fundamental problems of mathematical theory.

A solid of any given shape, when undisturbed, is acted on in ts substance by force distributed through it in any given manner, nd displacements are arbitrarily produced, or forces arbitrarily pplied, over its bounding surface. It is required to find the lisplacement of every point of its substance.

This problem has been thoroughly solved for a shell of ıomogeneous isotropic substance bounded by surfaces which, when undisturbed, are spherical and concentric (§ 735); but ıot hitherto for a body of any other shape. The limitations ınder which solutions have been obtained for other cases (thin plates, and rods), leading, as we have seen, to important practical results, have been stated above (§§ 588, 632). To demonstrate the laws (§§ 591, 633) which were taken in an- ticipation will also be one of our applications of the gen- equations for interior equilibrium of an elastic solid, which now proceed to investigate.

Conditions of internal equilibrium,

697. Any portion in the interior of an elastic solid may regarded as becoming perfectly rigid (§ 564) without distur- ing the equilibrium either of itself or of the matter round Hence the traction exerted by the matter all round it, rega- as a distribution of force applied to its surface, must, with applied forces acting on the substance of the portion conside- fulfil the conditions of equilibrium of forces acting on a ri- body. This statement, applied to an infinitely small rectang- parallelepiped of the body, gives the general differential e- tions of internal equilibrium of an elastic solid. It is to remarked that *three* equations suffice; the conditions of equi- brium for the *couples* being secured by the relation establish- above (§ 661) among the six pairs of tangential compon- tractions on the six faces of the figure.

expressed by three equations.

Let (x, y, z) be any point within the solid, and δx, δy, δz edg- respectively parallel to the rectangular axes of reference, of an infinitely small parallelepiped of the solid having that point f- its centre.

If P, Q, R, S, T, U denote (§ 662) the stress at (x, y, z), the average amounts of the component tractions (see table, § 66- the faces of the parallelepiped will be

$$\text{on the two faces } \delta y\delta z \left\{ \begin{array}{ll} \pm\left(P\pm\frac{dP}{dx}\cdot\frac{1}{2}\delta x\right)\delta y\delta z, & \text{parallel to } OX. \\ \pm\left(U\pm\frac{dU}{dx}\cdot\frac{1}{2}\delta x\right)\delta y\delta z, & \quad ,, \quad ,, \quad OY. \\ \pm\left(T\pm\frac{dT}{dx}\cdot\frac{1}{2}\delta x\right)\delta y\delta z, & \quad ,, \quad ,, \quad OZ. \end{array} \right.$$

Taking the symmetrical expressions for the tractions on the tw- other pairs of faces, and summing for all the faces all the c- ponents parallel to the three axes separately, we have

$$\left(\frac{dP}{dx}+\frac{dU}{dy}+\frac{dT}{dz}\right)\delta x\delta y\delta z, \text{ parallel to } OX,$$

$$\left(\frac{dU}{dx}+\frac{dQ}{dy}+\frac{dS}{dz}\right)\delta x\delta y\delta z, \quad ,, \quad ,, \quad OY,$$

$$\left(\frac{dT}{dx}+\frac{dS}{dy}+\frac{dR}{dz}\right)\delta x\delta y\delta z, \quad ,, \quad ,, \quad OZ.$$

Let now X, Y, Z denote the components of the applied force on the substance at (x, y, z), reckoned per unit of volume; so that $X\delta x\delta y\delta z$, $Y\delta x\delta y\delta z$, $Z\delta x\delta y\delta z$ will be their amounts on the small portion in question. Adding these to the corresponding components just found for the tractions, equating to zero, and omitting the factor $\delta x\delta y\delta z$, we have

General equations of interior equilibrium.

$$\left.\begin{array}{l}\frac{dP}{dx}+\frac{dU}{dy}+\frac{dT}{dz}+X=0\\ \frac{dU}{dx}+\frac{dQ}{dy}+\frac{dS}{dz}+Y=0\\ \frac{dT}{dx}+\frac{dS}{dy}+\frac{dR}{dz}+Z=0\end{array}\right\}\quad(2);$$

which are the general equations of internal stress required for equilibrium.

If for P, Q, R, S, T, U we substitute the linear functions of e, f, g, a, b, c in terms of which they are expressed by (14) of § 673, we have the equations of internal strain. And if we eliminate e, f, g, a, b, c by (6) of § 670 we have, for (α, β, γ) the components of the displacement of any interior point in terms of (x, y, z) its undisplaced position in the solid, three linear partial differential equations of the second degree, which are the equations of internal equilibrium in their ultimate form. It is to be remarked that, by supposing the coefficients (e, e), (e, f), etc., to be not constant, but given functions of (x, y, z), we avoid limiting the investigation to a homogeneous body.

698. These equations being sufficient as well as necessary or the equilibrium of the body, they must secure that the con- lition of § 697 is fulfilled for any and every finite portion of t: as is easily verified.

Being sufficient, they imply that the forces on any part supposed rigid fulfil the six equations of equilibrium in a rigid body.

Let $\iiint$ denote integration throughout any particular part of the solid, $d\sigma$ an element of the surface bounding this part, and $[\iint]$ integration over the whole of this surface. We have

$$\iiint X dxdydz = -\iiint\left(\frac{dP}{dx}+\frac{dU}{dy}+\frac{dT}{dz}\right)dxdydz.$$

Hence, integrating each term once, attending to the limits as in Appendix A., and denoting by l, m, n the direction-cosines of the normal through $d\sigma$,

$$\iiint X dxdydz = -\left[\iint(Pdydz+Udzdx+Tdxdy)\right] = -\left[\iint(Pl+Um+Tn)d\sigma\right],$$

and therefore [§ 662 (1)] $\iiint X dxdydz+\left[\iint F d\sigma\right]=0$ (3).

Again we have

$$\iiint(yZ-zY)dxdydz = -\iiint\left\{y\left(\frac{dT}{dx}+\frac{dS}{dy}+\frac{dR}{dz}\right)-z\left(\frac{dU}{dx}+\frac{dQ}{dy}+\frac{dS}{dz}\right)\right\}dxdydz.$$

Verification of equations of equilibrium for any part supposed rigid.

Now, integrating by parts, etc., as in Appendix A., we have

$$\iiint y\frac{dS}{dy}dxdydz=\left[\iint ySmd\sigma\right]-\iiint Sdxdydz,$$

and $$\iiint z\frac{dS}{dz}dxdydz=\left[\iint zSnd\sigma\right]-\iiint Sdxdydz.$$

Hence $$\iiint\left(y\frac{dS}{dy}-z\frac{dS}{dz}\right)dxdydz=\left[\iint(ySm-zSn)d\sigma\right].$$

Using this in the preceding expression, integrating the other terms each once simply as before, and using § 662 (1), we find

$$\iiint(yZ-zY)dxdydz+\left[\iint(yH-zG)d\sigma\right]=0$$

The six equations of equilibrium being (3), (4), and the symmetrical equations relative to y and z; are thus proved.

Simplified equations for isotropic solid.

For an isotropic solid, the equations (2) become of course much simpler. Thus, using (5) of § 693, eliminating e, f, g, a, b, c by (6) of § 670, grouping conveniently the terms which result, and putting $$m=(k+\tfrac{1}{3}n)$$ we find

$$\left.\begin{aligned}
m\frac{d}{dx}\left(\frac{d\alpha}{dx}+\frac{d\beta}{dy}+\frac{d\gamma}{dz}\right)+n\left(\frac{d^2\alpha}{dx^2}+\frac{d^2\alpha}{dy^2}+\frac{d^2\alpha}{dz^2}\right)+X=0\\
m\frac{d}{dy}\left(\frac{d\alpha}{dx}+\frac{d\beta}{dy}+\frac{d\gamma}{dz}\right)+n\left(\frac{d^2\beta}{dx^2}+\frac{d^2\beta}{dy^2}+\frac{d^2\beta}{dz^2}\right)+Y=0\\
m\frac{d}{dz}\left(\frac{d\alpha}{dx}+\frac{d\beta}{dy}+\frac{d\gamma}{dz}\right)+n\left(\frac{d^2\gamma}{dx^2}+\frac{d^2\gamma}{dy^2}+\frac{d^2\gamma}{dz^2}\right)+Z=0
\end{aligned}\right\} \quad (6)$$

or, as we may write them short,

$$m\frac{d\delta}{dx}+n\nabla^2\alpha+X=0,\ m\frac{d\delta}{dy}+n\nabla^2\beta+Y=0,\ m\frac{d\delta}{dz}+n\nabla^2\gamma+Z=0 \quad (7)$$

if we put $$\frac{d\alpha}{dx}+\frac{d\beta}{dy}+\frac{d\gamma}{dz}=\delta \quad (8)$$

and $$\frac{d^2}{dx^2}+\frac{d^2}{dy^2}+\frac{d^2}{dz^2}=\nabla^2 \quad (9)$$

so that δ shall denote the amount of dilatation in volume experienced by the substance; and ∇^2 the same symbol of operation as formerly [Appendix A. and B., and §§ 491, 492, 499, etc.]

St. Venant's application to torsion problems.

699. One of the most beautiful applications of the general equations of internal equilibrium of an elastic solid hitherto made is that of M. de St. Venant to "the torsion of prisms."[1] To one end of a long straight prismatic rod, wire, or solid or hollow cylinder of any form, a given couple is applied in a plane per

[1] *Mémoires des Savants Etrangers.* 1855. "De la Torsion des Prismes, avec des Considérations sur leur Flexion," etc.

pendicular to the length, while the other end is held fast: it is required to find the degree of twist (§ 120) produced, and the distribution of strain and stress throughout the prism. The conditions to be satisfied here are that the resultant action between the substance on the two sides of any normal section is a couple in the normal plane, equal to the given couple. Our work for solving the problem will be much simplified by first establishing the following preliminary propositions:— Torsion problem stated.

700. Let a solid (whether aeolotropic or isotropic) be so acted on by force applied from without to its boundary, that throughout its interior there is no normal traction on any plane parallel or perpendicular to a given plane, XOY, which implies, of course, that there is no distorting stress with axes in or parallel to this plane, and that the whole stress at any point of the solid is a simple distorting stress of tangential forces in some direction in the plane parallel to XOY, and in the plane perpendicular to this direction. Then— Lemma.

(1.) The interior distorting stress must be equal, and similarly directed, in all parts of the solid lying in any line perpendicular to the plane XOY.

(2.) It being premised that the traction at every point of any surface perpendicular to the plane XOY is, by hypothesis, a distribution of force in lines perpendicular to this plane; the integral amount of it on any closed prismatic or cylindrical surface perpendicular to XOY, and bounded by planes parallel to it, is zero.

(3.) The matter within the prismatic surface and terminal planes of (2.) being supposed for a moment (§ 564) to be rigid, the distribution of tractions referred to in (2.) constitutes a couple whose moment, divided by the distance between those terminal planes, is equal to the resultant force of the tractions on the area of either, and whose plane is parallel to the lines of these resultant forces. In other words, the

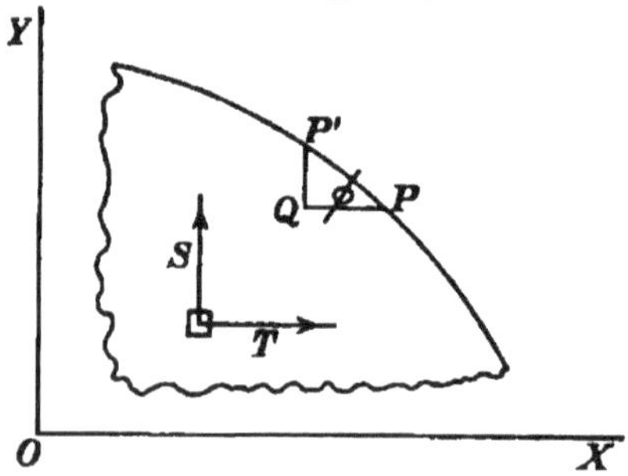

Lemma. moment of the distribution of forces over the prismatic surface referred to in (2.) round any line (OY or OX) in the plane XOY is equal to the sum of the components (T or S), perpendicular to the same line, of the traction in either of the terminal planes multiplied by the distance between these planes.

To prove (1.) consider for a moment as rigid (§ 564) an infinitesimal prism, AB (of sectional area ω), perpendicular to XOY, and having plane ends, A, B, parallel to it. There being no forces on its sides (the cylindrical boundary) perpendicular to its length, its equilibrium so far as motion in the direction of any line (OX), perpendicular to its length, requires (§ 551, I.) that the components of the tractions on its ends be equal and in opposite directions. Hence, in the notation of § 662, the distorting stress components, T, must be equal at A and B; and so must the stress components S, for the same reason.

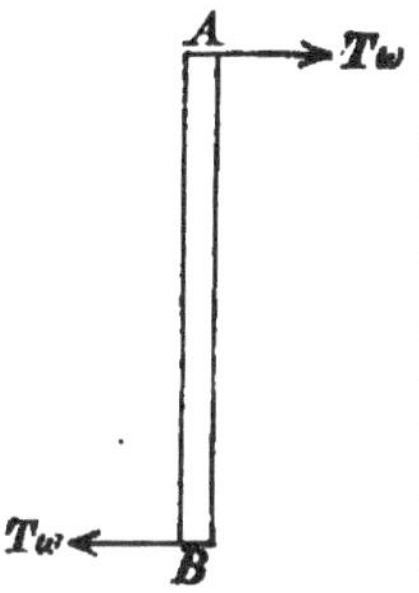

To prove (2.) and (3.) we have only to remark that they are required, according to § 551, I. and II., for the equilibrium of the rigid prism referred to in (3.)

Or, analytically, by the general equations (2) of § 697, since $X=0$, $Y=0$, $Z=0$, $P=0$, $Q=0$, $R=0$, $U=0$, by hypothesis; we have

$$\frac{dT}{dz}=0,\ \frac{dS}{dz}=0 \qquad (1.)$$

and

$$\frac{dT}{dx}+\frac{dS}{dy}=0 \qquad (2.)$$

Of these (1) prove that S and T are functions of x and y without z, or, in words, (1.) And if $\iint$ denote integration over the whole of any closed area of XOY, we have

$$\iint\left(\frac{dT}{dx}+\frac{dS}{dy}\right)dxdy=\left[\int(Tdy+Sdx)\right],^*$$

of which the second member, when the limits of the effected and indicated integrations are properly assigned, is found to be the same as $\int(T\sin\phi+S\cos\phi)ds$,

where $\int$ denotes integration over the whole bounding curve, ds

* The brackets [ ], as here used, denote integrals assigned properly for the bounding curve.

an element of its length, and ϕ the inclination of ds to XO. Lemma.
But, by (1) § 662, with $l=\sin\phi$, $m=\cos\phi$, $n=0$, we have

$$H=T\sin\phi+S\cos\phi \qquad (3),$$

if H denote the traction (parallel to OZ), reckoned as usual per unit of area, experienced by the bounding prismatic surface. Hence

$$\iint\left(\frac{dT}{dx}+\frac{dS}{dy}\right)dxdy=\int Hds \qquad (4);$$

and therefore, because of (2),

$$\int Hds=0 \qquad (5),$$

which is (2.) in symbols. Again we have, by integration by parts, and substitution, (2), of $\frac{dS}{dy}$ for $-\frac{dT}{dx}$,

$$\iint Tdxdy=[\int Txdy]^{*}-\iint x\frac{dT}{dx}dxdy$$

$$=[\int Txdy]^{*}+\iint x\frac{dS}{dy}dxdy=[\int Txdy]^{*}+[\int Sxdx]^{*}$$

$$=\int x(T\sin\phi+S\cos\phi)ds=\int xHds \qquad (6),$$

which proves (3.)

701. For a solid or hollow circular cylinder, the solution of § 699 (given first, we believe, by Coulomb) obviously is that each circular normal section remains unchanged in its own dimensions, figure, and internal arrangement (so that every straight line of its particles remains a straight line of unchanged length), but is turned round the axis of the cylinder through such an angle as to give a uniform *rate of twist* (§ 120) equal to the applied couple divided by the product of the moment of inertia of the circular area (whether annular or complete to the centre) into the rigidity of the substance. Torsion of circular cylinder.

For, if we suppose the distribution of strain thus specified to be actually produced, by whatever application of stress is necessary, we have, in every part of the substance, a simple shear parallel to the normal section, and perpendicular to the radius through it. The elastic reaction against this requires, to balance it (§§ 679, 682), a simple distorting stress consisting of forces in the normal section, directed as the shear, and others in planes through the axis, and directed parallel to the axis. The amount of the shear is, for parts of the substance at distance r from the axis, equal obviously to τr, if τ be the rate of twist. Hence the

* The brackets [ ], as here used, denote integrals assigned properly for the bounding curve.

Torsional rigidity of circular cylinder.

amount of the tangential force in either set of planes is $n\tau r$ per unit of area, if n be the rigidity of the substance. Hence there is no force between parts of the substance lying on the two sides of any element of any circular cylinder coaxal with the bounding cylinder or cylinders; and consequently no force is required on the cylindrical boundary to maintain the supposed state of strain. And the mutual action between the parts of the substance on the two sides of any normal plane section consists of force in the plane, directed perpendicular to the radius through each point, and amounting to $n\tau r$ per unit of area. The moment of this distribution of force round the axis of the cylinder is (if $d\sigma$ denote an element of the area) $n\tau\iint d\sigma r^2$, or the product of $n\tau$ into the moment of inertia of the area round the perpendicular to its plane through its centre, which is therefore equal to the moment of the couple applied at either end.

Prism of any shape constrained to a simple twist,

702. Similarly, we see that if a cylinder or prism of any shape be compelled to take exactly the state of strain above specified (§ 701) with the line through the centres of inertia of the normal sections, taken instead of the axis of the cylinder, the mutual action between the parts of it on the two sides of any normal section will be a couple of which the moment will be expressed by the same formula, that is, the product of the rigidity, into the rate of twist, into the moment of inertia of the section round its centre of inertia.

The only additional remark required to prove this is, that if the forces in the normal section be resolved in any two rectangular directions, OX, OY, the sums of the components, being respectively $n\tau\iint x d\sigma$ and $n\tau\iint y d\sigma$, each vanish by the property (§ 230) of the centre of inertia.

requires tractions on its sides.

703. But for any other shape of prism than a solid or symmetrical hollow circular cylinder, the supposed state of strain will require, besides the terminal opposed couples, force parallel to the length of the prism, distributed over the prismatic boundary, in proportion to the distance along the tangent, from each point of the surface, to the point in which this line is cut by a perpendicular to it from the centre of inertia of the normal section. To prove this let a normal section of the prism be represented in the annexed diagram. Let PK, representing the shear at any point, P, close to the prismatic boundary, be resolved into PN and PT respectively along the normal and tangent. The whole shear, PK, being equal to τr, its component, PN, is equal to $\tau r \sin \omega$ or $\tau . PE$. The corresponding component of the required stress is $n\tau . PE$, and involves (§ 661) equal forces in the plane of the diagram, and in the plane through TP perpendicular to it, each amounting to $n\tau . PE$ per unit of area.

Traction on sides of prism constrained to a simple twist.

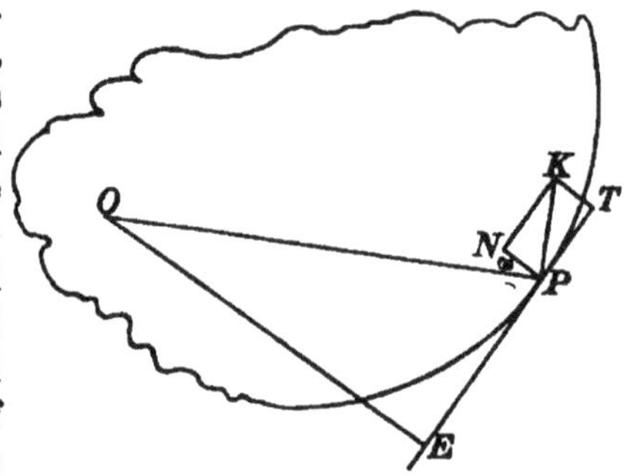

An application of force equal and opposite to the distribution thus found over the prismatic boundary, would of course alone produce in the prism, otherwise free, a state of strain which, compounded with that supposed above, would give the state of strain actually produced by the sole application of balancing couples to the two ends. The result, it is easily seen (and it will be proved below), consists of an increased twist, together with a warping of naturally plane normal sections, by infinitesimal displacements perpendicular to themselves, into certain surfaces of anticlastic curvature, with equal opposite curvatures in the principal sections (§ 130) through every point. This theory is due to St. Venant, who not only pointed out the falsity of the supposition admitted by several previous writers, that Coulomb's law holds for other forms of prism than the solid or hollow circular cylinder, but discovered fully the nature of the requisite correction, reduced the determination of it to a problem of pure mathematics, worked out the solution for a great variety of important and curious cases, compared the results with observation in a manner satisfactory and interesting to the naturalist, and gave conclusions of great value to the practical engineer.

St. Venant's correction to give the strain produced by mere twisting couples applied to the ends.

704. We take advantage of the identity of mathematical conditions in St. Venant's torsion problem, and a hydrokinetic problem first solved a few years earlier by Stokes,[1] to give

[1] "On some cases of Fluid Motion."—*Camb. Phil. Trans.* 1843.

Hydro-kinetic analogue to torsion problem.

the following statement, which will be found very useful in estimating deficiencies in torsional rigidity below the amount calculated from the fallacious extension of Coulomb's law:—

705. Conceive a liquid of density n completely filling a closed infinitely light prismatic box of the same shape within as the given elastic prism and of length unity, and let a couple be applied to the box in a plane perpendicular to its length. The *effective* moment of inertia of the liquid[1] will be equal to the correction by which the torsional rigidity of the elastic prism calculated by the false extension of Coulomb's law, must be diminished to give the true torsional rigidity.

Further, the actual *shear* of the solid, in any infinitely thin plate of it between two normal sections, will at each point be, when reckoned as a differential sliding (§ 172) parallel to these planes, equal to and in the same direction as the velocity of the liquid relatively to the containing box.

Solution of torsion problem.

706. To prove these propositions and investigate the mathematical equations of the problem, we first show that the conditions of the case (§ 699) are verified by a state of strain compounded of (1.) a simple twist round the line through the centres of inertia, and (2.) a distorting of each normal section by infinitesimal displacements perpendicular to its plane: then find the interior and surface equations to determine this warping: and lastly, calculate the actual moment of the couple to which the mutual action between the matter on the two sides of any normal section is equivalent.

Taking OX, OY in any normal section through O any convenient point (not necessarily its centre of inertia), and OZ perpendicular to them, let $x+\alpha$, $y+\beta$, $z+\gamma$ be the co-ordinates of the position to which a point (x, y, z) of the unstrained solid is displaced, in virtue of the compound strain just described. Thus γ will be a function of x and y, without z; and, if the twist (1.) be denoted by τ according to the simple twist reckoning of § 120, we shall have

$$x+\alpha=x\cos(\tau z)-y\sin(\tau z),\quad y+\beta=x\sin(\tau z)+y\cos(\tau z) \qquad (7)$$

Hence, for infinitely small values of z,

$$\alpha=-\tau yz,\quad \beta=\tau xz \qquad (8)$$

[1] That is, the moment of inertia of a rigid solid which, as will be proved in Vol. II., may be fixed within the box, if the liquid be removed, to make its motions the same as they are with the liquid in it.

Adhering to the notation of §§ 670, 693, only changing to Saxon letters, we have

Equations of strain, stress, and internal equilibrium.

$$\mathfrak{e}=0,\ \mathfrak{f}=0,\ \mathfrak{g}=0,\ \mathfrak{a}=\tau x+\frac{d\gamma}{dy},\ \mathfrak{b}=-\tau y+\frac{d\gamma}{dx},\ \mathfrak{c}=0 \qquad (9).$$

Hence [§ 693 (5)]

$$P=0,\ Q=0,\ R=0,\ S=n(\tau x+\frac{d\gamma}{dy}),\ T=n(-\tau y+\frac{d\gamma}{dx}),\ U=0 \qquad (10).$$

And with the notation of § 698, (8) and (9),

$$\delta=0,\ \nabla^2\alpha=0,\ \nabla^2\beta=0 \qquad (11).$$

Hence if also

$$\frac{d^2\gamma}{dx^2}+\frac{d^2\gamma}{dy^2}=0 \qquad (12),$$

the equations of internal equilibrium [§ 698 (6)] are all satisfied.

For the surface traction, with the notation of §§ 662, 700, we have, by § 662 (1),

Surface traction to be made zero.

$$F=0,\ G=0,\ H=T\sin\phi+S\cos\phi \qquad (13);$$

or eliminating T and S by (10), and introducing $\frac{d\gamma}{dp}$ to denote the rate of variation of γ in the direction perpendicular to the prismatic surface, and q (PE of § 703) the distance from the point of the surface for which H is expressed, to the intersection of the tangent plane with a perpendicular from O,

$$\left.\begin{array}{l} H=n\{(\frac{d\gamma}{dy}\cos\phi+\frac{d\gamma}{dx}\sin\phi)-\tau(y\sin\phi-x\cos\phi)\} \\ \text{or } H=n(\frac{d\gamma}{dp}-\tau q). \end{array}\right\} \qquad (14).$$

To find the mutual action between the matter on the two sides of a normal section, we first remark that, inasmuch as each of the two parts of the compound strain considered (the twist and the warping) separately fulfils the conditions of § 700, we must have

Couple resultant of traction in normal section.

$$\iint Tdxdy=\int xHds, \text{ and } \iint Sdxdy=\int yHds \qquad (15).$$

Hence when the prescribed surface condition $H=0$ is fulfilled, we have

$$\iint Tdxdy=0,\ \iint Sdxdy=0 \qquad (16),$$

and there remains only a couple

$$N=\iint(Sx-Ty)dxdy=n\tau\iint(x^2+y^2)\,dxdy-n\iint(y\frac{d\gamma}{dx}-x\frac{d\gamma}{dy})dxdy \ (17),$$

in the plane of the normal section. That condition, by (14), gives

$$\frac{d\gamma}{dp}=\tau q, \text{ or } \frac{d\gamma}{dy}\cos\phi+\frac{d\gamma}{dx}\sin\phi=\tau(y\sin\phi-x\cos\phi) \qquad (18),$$

for every point of the prismatic surface.

We shall see in Vol. II. that (12) and (18) are differential equations which determine a function, γ, of x, y, such that $\frac{d\gamma}{dx}$ and $\frac{d\gamma}{dy}$ are the components of the velocity of a perfect liquid initially at

Hydrokinetic application of torsional equations.

Hydro-kinetic application of torsional equations.

rest in a prismatic box as described in § 705, and set in motio[n] communicating to the box an angular velocity, τ, in the direct[ion] reckoned negative round OZ: and that the time-integral (§ … of the continuous couple by which this is done, however sudd[enly] or gradually, is $n\iint(x\frac{d\gamma}{dy}-y\frac{d\gamma}{dx})dxdy$, which is the exces[s of] $n\tau\iint(x^2+y^2)dxdy$ over N. Also, $\mathfrak{a}$ and $\mathfrak{b}$ in (9) are the com-ponents, parallel to OX and OY, of the velocity of the liq[uid] relatively to the box, since $-\tau y$ and τx are the components [of] the velocity of a point (x, y) rotating in the positive direct[ion] round OZ with the angular velocity τ. Hence the proposit[ion] (§ 705) to be proved.

707. M. de St. Venant finds solutions of these equations [in] two ways:—(A.) Taking any solution whatever of (12), he fi[nds] a series of curves for each of which (18) is satisfied, and a[ny] one of which, therefore, may be taken as the boundary of [a] prism to which that solution shall be applicable: and (B.) By the purely analytical method of Fourier, he solves (12), subj[ect] to the surface equation (18), for the particular case of a rec[t]-angular prism.

St. Venant's invention of solvable cases.

(A.) For this M. de St. Venant finds a general integral of th[e] boundary condition, viewed as a differential equation in terms [of] the two variables x, y, thus:—Multiplying (18) by ds, and re-placing $\sin\phi ds$ and $\cos\phi ds$ by their values dy and $-dx$, we hav[e]

$$\frac{d\gamma}{dx}dy-\frac{d\gamma}{dy}dx-\tfrac{1}{2}\tau d(x^2+y^2)=0 \qquad (19)$$

In this the first two terms constitute a complete differential of a function of x and y, independent variables; because γ satisfies (12). Thus, denoting this function by u, we have

$$\frac{d\gamma}{dx}=\frac{du}{dy}, \text{ and } \frac{d\gamma}{dy}=-\frac{du}{dx} \qquad (20)$$

and (19) becomes $du-\frac{1}{2}\tau d(x^2+y^2)=0$,

which requires that $u-\frac{1}{2}\tau(x^2+y^2)=C$ (21)

for every point in the boundary. It is to be remarked that, because

$$\frac{d}{dx}\frac{d\gamma}{dy}=\frac{d}{dy}\frac{d\gamma}{dx},$$

we have, from (20), $\frac{d^2u}{dx^2}+\frac{d^2u}{dy^2}=0$ (22)

or u also, as γ, fulfills the equation $\nabla^2u=0$. A function algebraically homogeneous as to x, y, which satisfies this equati[on] is [Appendix B. (a)] a spherical harmonic independent of

Hence a homogeneous solution of integral degree i can only be the part of Appendix B. (39) not containing z. This is

St. Venant's invention of solvable cases.

$$C\xi^i + C'\eta^i,$$

where [Appendix B. (26)] $\xi = x + y\sqrt{-1}$, and $\eta = x - y\sqrt{-1}$; or, if we change the constants so as to extirpate the imaginary symbol,

$$(x+y\sqrt{-1})^i + (x-y\sqrt{-1})^i\} - \sqrt{-1}B\{(x+y\sqrt{-1})^i - (x-y\sqrt{-1})^i\} \quad (23),$$

or, in terms of polar co-ordinates,

$$2r^i(A\cos i\theta + B\sin i\theta) \quad (24).$$

Using this solution for the case $i=2$ and (without loss of generality) putting $B=0$, we have

$$u = 2A(x^2 - y^2) \quad (25);$$

whence by (20)

$$\gamma = -4Axy \quad (26);$$

and the equation (21) of the series of bounding curves to which this solution is applicable is

$$\frac{x^2}{a^2} + \frac{y^2}{b^2} = 1 \quad (27),$$

if we put, for brevity,

$$\frac{-C}{\frac{1}{2}\tau - 2A} = a^2, \quad \frac{-C}{\frac{1}{2}\tau + 2A} = b^2,$$

which give $4A = \tau\dfrac{a^2-b^2}{a^2+b^2}$, so that (26) becomes

$$\gamma = -\tau\frac{a^2-b^2}{a^2+b^2}xy \quad (28).$$

Using this in (17) we have

$$N = n\tau\{\iint(x^2+y^2)dxdy - \frac{a^2-b^2}{a^2+b^2}\iint(x^2-y^2)dxdy\},$$

Solution for elliptic cylinder.

or, if I, J denote the moments of inertia of the area of the normal section, round the axes of x and y respectively,

$$N = n\tau\{J + I - \frac{a^2-b^2}{a^2+b^2}(J-I)\} \quad (29);$$

or lastly, as we have for the elliptic area (27),

$$\left.\begin{aligned} I &= \frac{\pi ab}{4}b^2, \quad J = \frac{\pi ab}{4}a^2, \\ N &= n\tau(J+I)\{1 - (\frac{a^2-b^2}{a^2+b^2})^2\} = n\tau\frac{\pi a^3 b^3}{a^2+b^2} \end{aligned}\right\} \quad (30).$$

Another very simple but most interesting case investigated by M. de St. Venant, is that arrived at by taking a harmonic of the third degree for u. Thus, introducing a factor $\frac{1}{2}\dfrac{\tau}{a}$ for the sake of homogeneousness and subsequent convenience, we have

St. Venant's invention of solvable cases.

$$\left.\begin{aligned} \tfrac{1}{2}\frac{\tau}{a}(x^3-3y^2x)-\tfrac{1}{2}\tau(x^2+y^2)=C, \\ \text{or in polar co-ordinates, } \tfrac{1}{2}\frac{\tau}{a}r^3\cos 3\theta-\tfrac{1}{2}\tau r^2=C, \end{aligned}\right\}$$

as an equation giving, by different values of C, a seri[es of] bounding lines, for which

$$\gamma=\tfrac{1}{2}\frac{\tau}{a}(y^3-3x^2y)=-\tfrac{1}{2}\frac{\tau}{a}r^3\sin 3\theta$$

Solution for equilateral triangle,

is the solution of (12), subject to (18). For the particular va[lue]

$$C=-\tfrac{2}{27}a^2\tau$$

(31) gives three straight lines, the sides of an equilateral tr[iangle] having a for perpendicular from angle to opposite side, placed relatively to x and y, as shown in the diagram (§ ... below). Thus we have the complete solution of the torsion [pro]blem for a prism of normal section an equilateral tria[ngle]. Equation (17) worked out for this area, with (32) for γ, giv[es]

$$N=n(K-\tfrac{2}{5}K)\tau.$$

But (K being the proper moment of inertia of the triangle, a[nd] A its area)

$$K=\frac{a^4}{9\sqrt{3}}=\frac{a^2}{9}A=\frac{A^2}{3\sqrt{3}};$$

and thus, for the torsional rigidity, we have the several expre[ssions]

$$N\div\tau=\tfrac{3}{5}nK=n\frac{a^4}{15\sqrt{3}}=n\frac{a^2}{15}A=n\frac{A^2}{5\sqrt{3}}=n\frac{A^4}{45K}$$

For curvilinear squares,

Similarly, taking for u a harmonic of the fourth degree a[nd] adjusting the constants to his wants, St. Venant finds equation,

$$\left.\begin{aligned} x^2+y^2-a(x^4-6x^2y^2+y^4)&=1-a \\ \text{or}\qquad r^2-ar^4\cos 4\theta&=1-a \end{aligned}\right\}$$

to give, for different values of a, a series of curvilinear squ[ares] (see diagram of § 708 (3.), below), all having rounded corn[ers] except two similar though differently turned curvilinear squa[res] with concave sides and acute angles corresponding to $a=$ and $a=-\tfrac{1}{2}(\sqrt{2}-1)$; for each of which the torsion problem [is] algebraically solved.

For star with four rounded points.

And by taking u the sum of two harmonics, of the fourth a[nd] eighth degrees respectively, and properly adjusting the constant[s] he finds

$$\frac{x^2+y^2}{r_0^2}-\tfrac{16}{49}\cdot\tfrac{16}{17}\cdot\frac{x^4-6x^2y^2+y^4}{r_0^4}+\tfrac{12}{49}\cdot\tfrac{16}{17}\cdot\frac{x^8-28x^6y^2+70x^4y^4-28x^2y^6+\ldots}{r_0^8}=1-\ldots$$

$$\text{or}\qquad \frac{r^2}{r_0^2}-\tfrac{16}{49}\cdot\tfrac{16}{17}\cdot\frac{r^4\cos 4\theta}{r_0^4}+\tfrac{12}{49}\cdot\tfrac{16}{17}\cdot\frac{r^8\cos 8\theta}{r_0^8}=1-\tfrac{32}{49}\cdot\tfrac{16}{17}$$

as the equation of the curve shown in § 709, diagram (4.), fo[r] which therefore the torsion problem is solved.

(B.) The integration (21) of the boundary equation, introduced by St. Venant for use in his synthesis, (A.) is also very useful in the analytical investigation, although he has not so applied it. First, we may remark, that the determination of u for a given form of prism is a particular case of "Green's problem" proved possible and determinate in Appendix A. (e); being to find u, a function of x, y which shall satisfy the equation St. Venant's reduction to Green's problem.

$$\frac{d^2u}{dx^2}+\frac{d^2u}{dy^2}=0,$$

for every point of the area bounded a certain given closed circuit, subject to the condition, $u=\frac{1}{2}\tau(x^2+y^2)$ (36)
for every point of the boundary.

When u is found, equations (20) and (17) with (10) complete the solution of the torsion problem.

For the case of a rectangular prism, the solution is much facilitated by taking Solution for rectangular prism,

$$\left.\begin{array}{l} u=v+A(x^2-y^2)+B, \\ \text{which gives} \quad \dfrac{d^2v}{dx^2}+\dfrac{d^2v}{dy^2}=0\,; \\ \text{and for boundary condition,} \\ v=(\frac{1}{2}\tau-A)x^2+(\frac{1}{2}\tau+A)y^2-B. \end{array}\right| \quad (37).$$

If the rectangle be not square, let its longer sides be parallel to OX; and let a, b be the lengths of each of the longer and each of the shorter sides respectively. Take, now,

$$A=\tfrac{1}{2}\tau, \text{ and } B=\tfrac{1}{4}\tau b^2 \quad (38).$$

The boundary condition becomes

$$\left.\begin{array}{ll} & v=0 \text{ when } y=\pm\frac{1}{2}b, \\ \text{and} & v=-\tau(\frac{1}{4}b^2-y^2) \text{ when } x=\pm\frac{1}{2}a \end{array}\right\} \quad (39).$$

To solve the problem by Fourier's method (compare with the more difficult problem of § 655), the requisite expansion of $\frac{1}{4}b^2-y^2$ is clearly[1] found by Fourier's analysis.

$$\tfrac{1}{4}b^2-y^2=\left(\frac{2}{\pi}\right)^3 b^2\left\{\cos\eta-\frac{1}{3^3}\cos 3\eta+\frac{1}{5^3}\cos 5\eta-\text{etc.}\right\} \quad (40);$$

$$\left.\begin{array}{ll} \text{where, for brevity,} & \eta=\dfrac{\pi y}{b}\,. \\ \text{And, for the same cause, putting} & \xi=\dfrac{\pi x}{b} \end{array}\right\} \quad (41)$$

we have, for the form of solution,

[1] Obtainable, as a matter of course, from Fourier's general theorem, but most easily by two successive integrations of the common formula

$$\tfrac{1}{4}=\frac{2}{\pi}(\cos\theta-\tfrac{1}{3}\cos 3\theta+\tfrac{1}{5}\cos 5\theta-\text{etc.})$$

Solution for rectangular prism, found by Fourier's analysis.

$$v = \Sigma (A_{2i+1} \epsilon^{-(2i+1)\xi} + B_{2i+1} \epsilon^{-(2i+1)\ldots}) \cos \ldots$$

which satisfies (37), and gives $v=0$ for $y = \pm \ldots$ [illegible] boundary condition gives, for determining A_{2i+1} [illegible]

$$[A_{2i+1} \epsilon^{-(2i+1)\frac{\pi c}{2}} + B_{2i+1} \epsilon^{\ldots}]$$

$$= [A_{2i+1} \epsilon^{+(2i+1)\frac{\pi c}{2b}} + B_{2i+1} \epsilon^{-(2i+1)\ldots}] = -\tau \ldots$$

These two equations give a common value for [illegible] quantities A_{2i+1}, B_{2i+1}; with which (42) becomes

$$v = -\tau \left(\frac{2}{\pi}\right)^3 b^2 \Sigma \frac{(-1)^i}{(2i+1)^3} \frac{\epsilon^{-(2i+1)\ldots} + \epsilon^{\ldots}}{\epsilon^{-(2i+1)\frac{\pi}{2}\ldots} + \epsilon^{\ldots}} \cos \ldots$$

From this we find, by (37), (38), and (2[illegible]),

$$\gamma = -\tau xy + \tau \left(\frac{2}{\pi}\right)^3 b^2 \Sigma \frac{(-1)^i}{(2i+1)^3} \frac{\epsilon^{\ldots} - \epsilon^{\ldots}}{\epsilon^{+(2i+1)\frac{\pi}{2}\ldots} + \epsilon^{\ldots}} \ldots$$

and (17) gives, for the torsional rigidity,

$$N \div \tau = nab^3 [\tfrac{1}{3} - \left(\frac{2}{\pi}\right)^5 \frac{b}{a} \Sigma \frac{1}{(2i+1)^5} \frac{1 - \epsilon^{\ldots}}{1 + \epsilon^{\ldots}} \ldots$$

If we had proceeded in all respects as above, only [illegible] instead of $A = \frac{1}{2}\tau$, in (37), we should have obta[illegible] sions for γ and $N \div \tau$, seemingly very different, [illegible] giving the same values. These other expressions ma[illegible] down immediately by making the interchange x, y [illegible] b, a in (45) and (46), and changing the sign of each [illegible] They obviously converge less rapidly than (45) and [illegible] we have supposed, $a > b$, and it is on this account [illegible] ceeded as above rather than in the other way. The [illegible] of the results gives astonishing theorems of pure m[illegible] such as rarely fall to the lot of those mathematicians w[illegible] themselves to pure analysis or geometry, instead of a[illegible] selves to be led into the rich and beautiful fields of ma[illegible] truth which lie in the way of physical research.

Extension to a class of curvilinear rectangles.

A relation discovered by Stokes[1] and Lamé[2] in [illegible] [which we have already used in equations (2[illegible]) [illegible] connexion with Lamé's method of curvilinear co-ordinates [illegible] us to extend the Fourier analytical method to a large [illegible] curvilineal rectangles, including the rectilineal rectan[illegible] particular case, thus:—

[1] On the Steady Motion of Incompressible Fluids. *Camb. Phil. Trans.* [illegible]

[2] Mémoire sur les lois de l'Equilibre du Fluide Ethéré. *Journal de l'É[illegible] technique*, 1834.

[3] See Thomson on the Equations of the Motion of Heat referred to [illegible] co-ordinates. *Camb. Math. Journal*, 1845.

Lamé's transformation to plane isothermal co-ordinates.

›t ξ be a function of x, y satisfying the equation

$$\frac{d^2\xi}{dx^2}+\frac{d^2\xi}{dy^2}=0 \qquad (47),$$

, as this shows that $\frac{d\xi}{dx}dy-\frac{d\xi}{dy}dx$ is a complete differential,

$$\eta=\int(\frac{d\xi}{dx}dy-\frac{d\xi}{dy}dx) \qquad (48);$$

which means the same,

$$\frac{d\eta}{dy}=\frac{d\xi}{dx}, \text{ and } \frac{d\eta}{dx}=-\frac{d\xi}{dy} \qquad (49).$$

ıis other function η also, as we see from (49), satisfies the uation

$$\frac{d^2\eta}{dx^2}+\frac{d^2\eta}{dy^2}=0 \qquad (50).$$

Theorem of Stokes and Lamé.

nd, also because of (49), two intersecting curves, whose equaons are

$$\xi=A,\ \eta=B \qquad (51),$$

ut one another at right angles. Let now, A and B being upposed given, x and y be determined by these two equations. (The point whose co-ordinates are x, y may also be regarded as ›pecified by (A, B), or by the values of ξ, η, which give curves .ntersecting in (x, y). Thus (ξ, η) with any particular values assigned to ξ and η, specifies a point in a plane. Common rectilineal co-ordinates are clearly a particular case (rectilineal orthogonal co-ordinates) of the system of curvilineal orthogonal co-ordinates thus defined. Let now u, any function of x, y, be transformed into terms of ξ, η. We have, by differentiation,

$$+\frac{d^2u}{dy^2}=\frac{d^2u}{d\xi^2}(\frac{d\xi^2}{dx^2}+\frac{d\xi^2}{dy^2})+2\frac{d^2u}{d\xi d\eta}(\frac{d\xi}{dx}\frac{d\eta}{dx}+\frac{d\xi}{dy}\frac{d\eta}{dy})+\frac{d^2u}{d\eta^2}(\frac{d\eta^2}{dx^2}+\frac{d\eta^2}{dy^2})$$

$$+\frac{du}{d\xi}(\frac{d^2\xi}{dx^2}+\frac{d^2\xi}{dy^2})+\frac{du}{d\eta}(\frac{d^2\eta}{dx^2}+\frac{d^2\eta}{dy^2}) \qquad (52),$$

which is reduced by (49) and (50) to

$$\frac{d^2u}{dx^2}+\frac{d^2u}{dy^2}=(\frac{d^2u}{d\xi^2}+\frac{d^2u}{d\eta^2})(\frac{d\xi^2}{dx^2}+\frac{d\xi^2}{dy^2}) \qquad (53).$$

Hence the equation $\quad \frac{d^2u}{dx^2}+\frac{d^2u}{dy^2}=0$

transforms into $\quad \frac{d^2u}{d\xi^2}+\frac{d^2u}{d\eta^2}=0. \qquad (54).$

Also the relations

$$\frac{du}{dy}=\frac{d\gamma}{dx},\quad \frac{du}{dx}=-\frac{d\gamma}{dy}$$

transform, in virtue of (49), into

$$\frac{du}{d\eta}=\frac{d\gamma}{d\xi},\quad \frac{du}{d\xi}=-\frac{d\gamma}{d\eta} \qquad (55).$$

Solution for rectangular prism, found by Fourier's analysis.

$$v=\Sigma\{A_{2i+1}\epsilon^{-(2i+1)\xi}+B_{2i+1}\epsilon^{+(2i+1)\xi}\}\cos(2i+1)\eta \quad (42)$$

which satisfies (37), and gives $v=0$ for $y=\pm\frac{1}{2}b$. The remaining boundary condition gives, for determining A_{2i+1} and B_{2i+1},

$$\left.\begin{aligned}&[A_{2i+1}\epsilon^{-(2i+1)\frac{\pi a}{2b}}+B_{2i+1}\epsilon^{+(2i+1)\frac{\pi a}{2b}}]\\ =&[A_{2i+1}\epsilon^{+(2i+1)\frac{\pi a}{2b}}+B_{2i+1}\epsilon^{-(2i+1)\frac{\pi a}{2b}}]=-\tau b^2\left(\frac{2}{\pi}\right)^3\frac{(-1)^i}{(2i+1)^3}\end{aligned}\right\} \quad (43)$$

These two equations give a common value for the two unknown quantities A_{2i+1}, B_{2i+1}; with which (42) becomes

$$v=-\tau\left(\frac{2}{\pi}\right)^3 b^2\Sigma\frac{(-1)^i}{(2i+1)^3}\,\frac{\epsilon^{-(2i+1)\xi}+\epsilon^{+(2i+1)\xi}}{\epsilon^{-(2i+1)\frac{\pi a}{2b}}+\epsilon^{+(2i+1)\frac{\pi a}{2b}}}\cos(2i+1)\eta \quad (44)$$

From this we find, by (37), (38), and (20),

$$\gamma=-\tau xy+\tau\left(\frac{2}{\pi}\right)^3 b^2\Sigma\frac{(-1)^i}{(2i+1)^3}\,\frac{\epsilon^{+(2i+1)\xi}-\epsilon^{-(2i+1)\xi}}{\epsilon^{+(2i+1)\frac{\pi a}{2b}}+\epsilon^{-(2i+1)\frac{\pi a}{2b}}}\sin(2i+1)\eta \quad (45)$$

and (17) gives, for the torsional rigidity,

$$N\div\tau=nab^3\left[\frac{1}{3}-\left(\frac{2}{\pi}\right)^5\frac{b}{a}\Sigma\frac{1}{(2i+1)^5}\,\frac{1-\epsilon^{-(2i+1)\frac{\pi a}{b}}}{1+\epsilon^{-(2i+1)\frac{\pi a}{b}}}\right] \quad (46)$$

If we had proceeded in all respects as above, only taking $A=-\frac{1}{2}\tau$ instead of $A=\frac{1}{2}\tau$, in (37), we should have obtained expressions for γ and $N\div\tau$, seemingly very different, but necessarily giving the same values. These other expressions may be written down immediately by making the interchange x, y, a, b for y, x, b, a in (45) and (46), and changing the sign of each term of (45). They obviously converge less rapidly than (45) and (46) if, as we have supposed, $a>b$, and it is on this account that we proceeded as above rather than in the other way. The comparison of the results gives astonishing theorems of pure mathematics, such as rarely fall to the lot of those mathematicians who confine themselves to pure analysis or geometry, instead of allowing themselves to be led into the rich and beautiful fields of mathematical truth which lie in the way of physical research.

Extension to a class of curvilinear rectangles.

A relation discovered by Stokes[1] and Lamé[2] independently [which we have already used in equations (20), (22)] taken in connexion with Lamé's method of curvilinear co-ordinates,[3] allows us to extend the Fourier analytical method to a large class of curvilineal rectangles, including the rectilineal rectangle as a particular case, thus:—

[1] On the Steady Motion of Incompressible Fluids. *Camb. Phil. Trans.*, 1842.

[2] Mémoire sur les lois de l'Equilibre du Fluide Ethéré. *Journal de l'Ecole Polytechnique*, 1834.

[3] See Thomson on the Equations of the Motion of Heat referred to Curvilinear co-ordinates. *Camb. Math. Journal*, 1845.

Lamé's transformation to plane isothermal co-ordinates.

Let ξ be a function of x, y satisfying the equation

$$\frac{d^2\xi}{dx^2}+\frac{d^2\xi}{dy^2}=0 \qquad (47),$$

and, as this shows that $\frac{d\xi}{dx}dy-\frac{d\xi}{dy}dx$ is a complete differential,

let
$$\eta=\int\left(\frac{d\xi}{dx}dy-\frac{d\xi}{dy}dx\right) \qquad (48);$$

or, which means the same,

$$\frac{d\eta}{dy}=\frac{d\xi}{dx}, \text{ and } \frac{d\eta}{dx}=-\frac{d\xi}{dy} \qquad (49).$$

This other function η also, as we see from (49), satisfies the equation

$$\frac{d^2\eta}{dx^2}+\frac{d^2\eta}{dy^2}=0 \qquad (50).$$

Theorem of Stokes and Lamé.

And, also because of (49), two intersecting curves, whose equations are

$$\xi=A,\ \eta=B \qquad (51),$$

cut one another at right angles. Let now, A and B being supposed given, x and y be determined by these two equations. The point whose co-ordinates are x, y may also be regarded as specified by (A, B), or by the values of ξ, η, which give curves intersecting in (x, y). Thus (ξ, η) with any particular values assigned to ξ and η, specifies a point in a plane. Common rectilineal co-ordinates are clearly a particular case (rectilineal orthogonal co-ordinates) of the system of curvilineal orthogonal co-ordinates thus defined. Let now u, any function of x, y, be transformed into terms of ξ, η. We have, by differentiation,

$$\frac{d^2u}{dx^2}+\frac{d^2u}{dy^2}=\frac{d^2u}{d\xi^2}\left(\frac{d\xi^2}{dx^2}+\frac{d\xi^2}{dy^2}\right)+2\frac{d^2u}{d\xi d\eta}\left(\frac{d\xi}{dx}\frac{d\eta}{dx}+\frac{d\xi}{dy}\frac{d\eta}{dy}\right)+\frac{d^2u}{d\eta^2}\left(\frac{d\eta^2}{dx^2}+\frac{d\eta^2}{dy^2}\right)$$
$$+\frac{du}{d\xi}\left(\frac{d^2\xi}{dx^2}+\frac{d^2\xi}{dy^2}\right)+\frac{du}{d\eta}\left(\frac{d^2\eta}{dx^2}+\frac{d^2\eta}{dy^2}\right) \qquad (52),$$

which is reduced by (49) and (50) to

$$\frac{d^2u}{dx^2}+\frac{d^2u}{dy^2}=\left(\frac{d^2u}{d\xi^2}+\frac{d^2u}{d\eta^2}\right)\left(\frac{d\xi^2}{dx^2}+\frac{d\xi^2}{dy^2}\right) \qquad (53).$$

Hence the equation
$$\frac{d^2u}{dx^2}+\frac{d^2u}{dy^2}=0$$

transforms into
$$\frac{d^2u}{d\xi^2}+\frac{d^2u}{d\eta^2}=0. \qquad (54).$$

Also the relations

$$\frac{du}{dy}=\frac{d\gamma}{dx},\quad \frac{du}{dx}=-\frac{d\gamma}{dy}$$

transform, in virtue of (49), into

$$\frac{du}{d\eta}=\frac{d\gamma}{d\xi},\quad \frac{du}{d\xi}=-\frac{d\gamma}{d\eta}. \qquad (55).$$

Solution for rectangle of plane isothermals.

Hence the general problem of finding u and γ has precisely the same statement in terms of ξ, η, as that given above, (22) and (20), in terms of x, y, with this exception, that we have $u=\frac{1}{2}\tau(\xi^2+\eta^2)$, but if $f(\xi,\eta)$ denote the function of ξ, η which x^2+y^2 transforms,

$u=\frac{1}{2}\tau f(\xi,\eta)$ for every point of the boundary.

The solution for the curvilineal rectangle

$$\left.\begin{array}{c|c} \xi=a & \eta=\beta \\ \xi=0 & \eta=0 \end{array}\right\}$$

is, on Fourier's plan,

$$u=\Sigma\sin\frac{i\pi\xi}{a}\left(A_i\varepsilon^{\frac{i\pi\eta}{a}}+A_i'\varepsilon^{-\frac{i\pi\eta}{a}}\right)+\Sigma\sin\frac{i\pi\eta}{\beta}\left(B_i\varepsilon^{\frac{i\pi\xi}{\beta}}+B_i'\varepsilon^{-\frac{i\pi\xi}{\beta}}\right) \quad (58)$$

where A_i, A_i' are to be determined by two equations, obtained thus:—Equate the coefficient of $\sin\frac{i\pi\xi}{a}$ when $\eta=0$ and when $\eta=\beta$ respectively to the coefficients of $\sin\frac{i\pi\xi}{a}$ in the expansions of $f(\xi,0)$ and $f(\xi,\beta)$ in series of the form

$$P_1\sin\frac{\pi\xi}{a}+P_2\sin\frac{2\pi\xi}{a}+P_3\sin\frac{3\pi\xi}{a}+\text{etc.} \quad (59)$$

by Fourier's theorem, § 77. Similarly, B_i, B_i', are determined from the expansions of $f(0,\eta)$ and $f(a,\eta)$, in series of the form

$$Q_1\sin\frac{\pi\eta}{\beta}+Q_2\sin\frac{2\pi\eta}{\beta}+Q_3\sin\frac{3\pi\eta}{\beta}+\text{etc.} \quad (60)$$

Example. Rectangle bounded by two concentric arcs and two radii.

Of one extremely simple example, very interesting in theory and valuable for practical mechanics, we shall indicate the details.

Let
$$\xi=\log\sqrt{\frac{x^2+y^2}{a^2}} \quad (61)$$

This clearly satisfies (47); and it gives, by (48),

$$\eta=\tan^{-1}\frac{y}{x} \quad (62)$$

The solution may be expressed in a series of sines of multiples of $\frac{\pi\eta}{a}$ [on the plan of (37)...(45)] by taking *

$$u=v+\frac{1}{2}\tau a^2\frac{\varepsilon^{2\xi}\cos(\beta-2\eta)}{\cos\beta} \quad (63)$$

which, with (54), gives

$$\frac{d^2v}{d\xi^2}+\frac{d^2v}{d\eta^2}=0 \quad (64)$$

* It should be noticed that this solution fails for the case of $\beta=(2i+1)\frac{\pi}{2}$.

and leaves, as boundary conditions in the solution for v,

Example. Rectangle bounded by two concentric arcs and two radii.

$$\left.\begin{array}{l} v=\tfrac{1}{2}\tau a^2\{1-\dfrac{\cos(\beta-2\eta)}{\cos\beta}\} \text{ when } \xi=0, \\ v=\tfrac{1}{2}\tau a^2\epsilon^{2a}\{1-\dfrac{\cos(\beta-2\eta)}{\cos\beta}\} \text{ when } \xi=a, \\ \text{and} \quad v=0 \text{ when } \eta=0, \text{ and when } \eta=\beta. \end{array}\right\} \quad (65).$$

The last condition shows that the B_i and B_i' part of (58) is proper for expressing v, and the first two determine B_i and B_i' as usual.

Or when it is best to have the result in series of sines of multiples of $\frac{\pi\xi}{a}$, we may take

$$u=w+\tfrac{1}{2}\tau a^2(1+\frac{\epsilon^{2a}-1}{a}\xi) \quad (66),$$

which, with (54), gives

$$\frac{d^2w}{dx^2}+\frac{d^2w}{dy^2}=0 \quad (67),$$

and leaves, as boundary conditions in the solution for w,

$$\left.\begin{array}{l} w=\tfrac{1}{2}\tau a^2\{\epsilon^{2\xi}-1-\dfrac{\epsilon^{2a}-1}{a}\xi\} \text{ when } \eta=0, \text{ and when } \eta=\beta, \\ \text{and} \quad w=0 \text{ when } \xi=0, \text{ and when } \xi=a. \end{array}\right\} \quad (68).$$

The last shows that the A_i and A_i' part of (58) is proper for w, and the two first determine A_i, A_i'.

708. St. Venant's treatise abounds in beautiful and instructive graphical illustrations of his results, from which we select the following:—

(1.) *Elliptic cylinder.*—The plain and dotted curvilineal arcs are "contour lines" (*coupes topographiques*) of the section as

Contour lines of normal section of elliptic cylinder, as warped by torsion: equilateral hyperbolas.

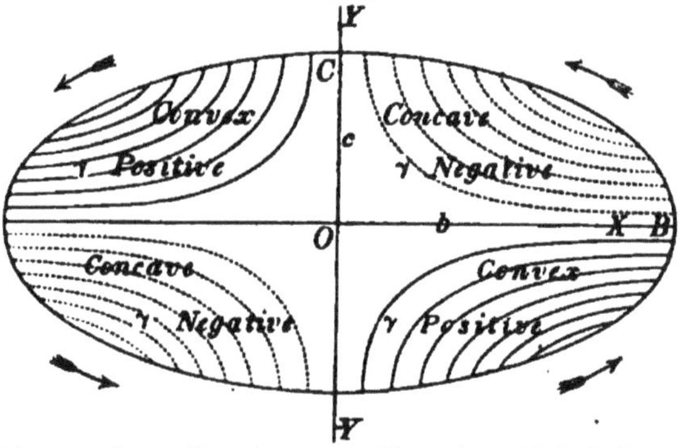

warped by torsion; that is to say, lines in which it is cut by a series of parallel planes, each perpendicular to the axis, or lines for which γ (§ 706) has different constant values. T lines are [§ 707 (28)] equilateral hyperbolas in this case. T arrows indicate the direction of rotation in the part of prism *above* the plane of the diagram.

Contour lines of normal section of triangular prism, as warped by torsion.

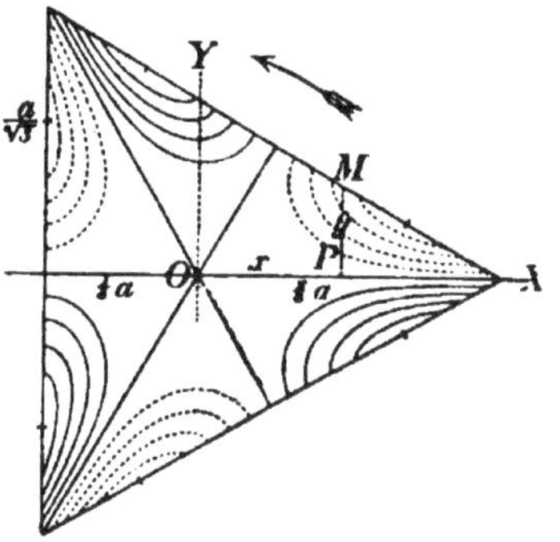

(2.) *Equilateral triang prism.*—The contour li are shown as in case the dotted curves being th where the warped secti falls *below* the plane of t diagram, the direction rotation of the part of prism above the pl being indicated by the arrow.

Diagram of St. Venant's curvilineal squares for which torsion problem is solvable.

(3.) This diagram shows the series of lines represented (34) of § 707, with the indicated values for a. It is remarka

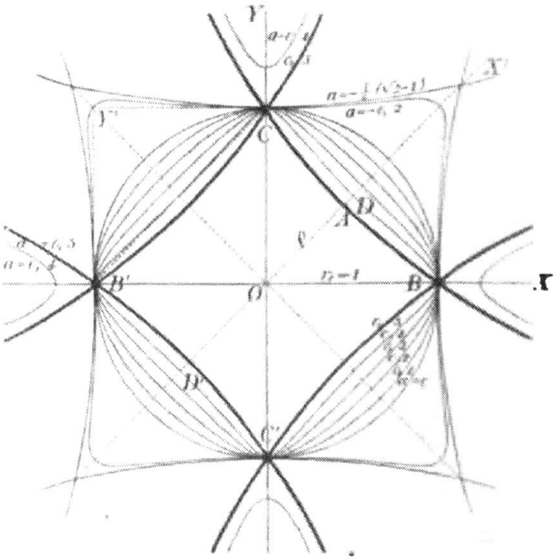

that the values $a = 0{\cdot}5$ and $a = -\frac{1}{2}(\sqrt{2} - 1)$ give similar b not equal curvilineal squares (hollow sides and acute angl one of them turned through half a right angle relatively to th other. Everything in the diagram outside the larger of these squares is to be cut away as irrelevant to the physical problem; the series of closed curves remaining exhibits figures of prisms, for any one of which the torsion problem is solved algebraically. These figures vary continuously from a circle, inwards to one of the acute-angled squares, and outwards to the other: each, except these extremes, being a continuous closed curve with no angles. The curves for $a = 0·4$ and $a = -0·2$ approach remarkably near to the rectilineal squares, partially indicated in the diagram by dotted lines.

(4.) This diagram shows the contour lines, in all respects as in the cases (1.) and (2.), for the case of a prism having for section the figure indicated. The portions of curve outside the continuous closed curve are merely indications of mathematical extensions irrelevant to the physical problem.

Contour lines for St. Venant's "étoile à quatre points arrondis."

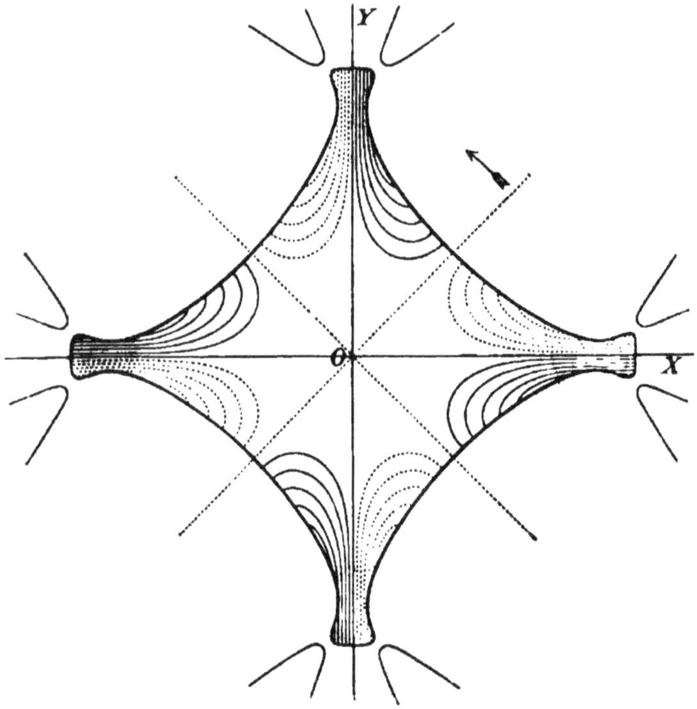

Contour lines of normal section of square prism, as warped by torsion.

(5.) This shows, as in the other cases, the contour lines for the warped section of a square prism under torsion.

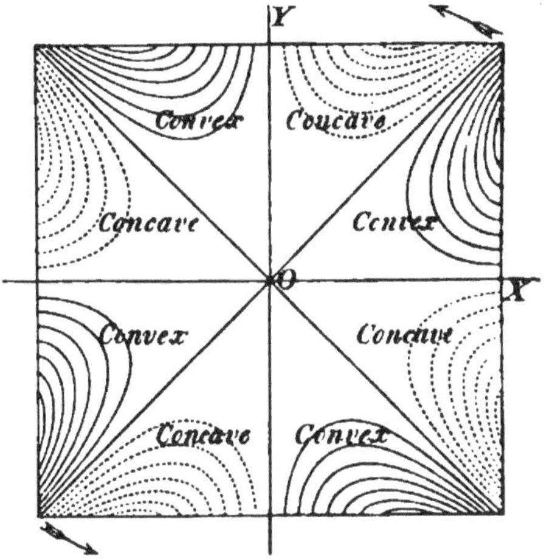

Elliptic, square, and flat rectangular bars twisted.

(6.), (7.), (8.) These are shaded drawings, showing the appearances presented by elliptic, square, and flat rectangular

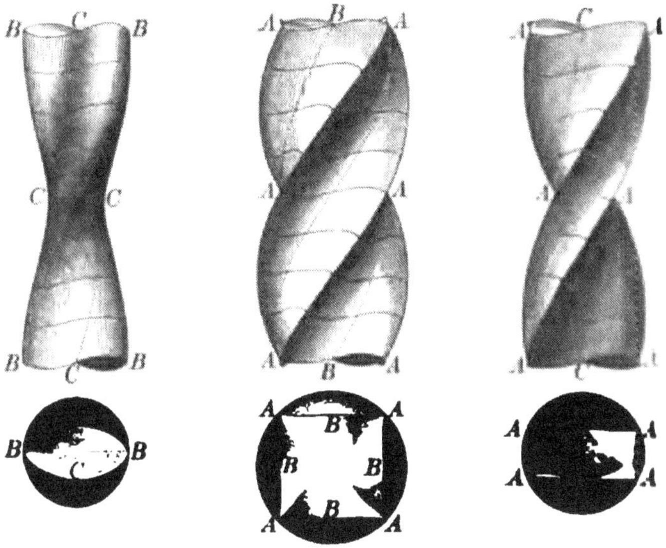

bars under exaggerated torsion, as may be realized with such a substance as India rubber.

709. Inasmuch as the moment of inertia of a plane area about an axis through its centre of inertia perpendicular to its plane is obviously equal to the sum of its moments of inertia round any two axes through the same point, at right angles to one another in its plane, the fallacious extension of Coulomb's law, referred to in § 703, would make the torsional rigidity of a bar of any section equal to $\frac{n}{M}$ (§ 694) multiplied into the sum of its flexural rigidities (see below, § 715) in any two planes at right angles to one another through its length. The true theory, as we have seen (§§ 705, 706), always gives a torsional rigidity less than this. How great the deficiency may be expected to be in cases in which the figure of the section presents projecting angles, or considerable prominences (which may be imagined from the hydrokinetic analogy we have given in § 705), has been pointed out by M. de St. Venant, with the important practical application, that strengthening ribs, or projections (see, for instance, the fourth annexed diagram), such as are introduced in engineering to give stiffness to beams, have the reverse of a good effect when *torsional* rigidity or strength is an object, although they are truly of great value in increasing the flexural rigidity, and giving strength to bear ordinary strains, which are always more or less flexural. With remarkable ingenuity and mathematical skill he has drawn beautiful illustrations of this important practical principle from his algebraic and transcendental solutions [§ 707 (32), (34), (35), (45)]. Thus

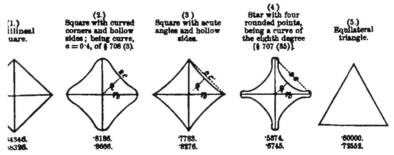

Torsional rigidity less in proportion to sum of principal flexural rigidities than according to false extension (§ 703) of Coulomb's law.

Ratios of torsional rigidities to those of solid circular rods.

(1.) ...tilineal ...uare.	(2.) Square with curved corners and hollow sides; being curve, $a = 0{\cdot}4$, of § 706 (3).	(3.) Square with acute angles and hollow sides.	(4.) Star with four rounded points, being a curve of the eighth degree [§ 707 (35)].	(5.) Equilateral triangle.

...4346.	·8186.	·7783.	·5874.	·60000.
...8326.	·8666.	·8276.	·6745.	·72552.

for an equilateral triangle, and for the rectilineal and three curvilineal squares shown in the annexed diagram, he finds for the torsional rigidities the values stated. The number im-

Ratios of torsional rigidities to those of solid circular rods (a) of same moment of inertia,

mediately below the diagram indicates in each case the fraction which the true torsional rigidity is of the old fallacious estimate (§ 703); the latter being the product of the rigidity of the substance into the moment of inertia of the cross section round an axis perpendicular to its plane through its centre of inertia. The second number indicates in each case the fraction which the torsional rigidity is of that of a solid circular cylinder of the same sectional area.

(b) of same quantity of material.

Places of greatest distortion in twisted prisms.

710. M. de St. Venant also calls attention to a conclusion from his solutions which to many may be startling, that in his simpler cases the places of greatest distortion are those points of the boundary which are nearest to the axis of the twisted prism in each case, and the places of least distortion those farthest from it. Thus in the elliptic cylinder the substance is most strained at the ends of the smaller principal diameter, and least at the ends of the greater. In the equilateral triangular and square prisms there are longitudinal lines of maximum strain through the middles of the sides. In the oblong rectangular prism there are two lines of greater maximum strain through the middles of the broader pair of sides, and two lines of less maximum strain through the middles of the narrow sides. The strain is, as we may judge from (§ 705) the hydrokinetic analogy, excessively small, but not evanescent, in the projecting ribs of a prism of the figure shown in (4) § 709. It is quite evanescent infinitely near the angle, in the triangular and rectangular prisms, and in each other case as (3) of § 709, in which there is a finite angle, whether acute or obtuse, projecting outwards.

Solid of any shape having edges, or pyramidal or conical angles, under stress.

This reminds us of a general remark we have to make, although consideration of space may oblige us to leave it without formal proof. A solid of any elastic substance, isotropic or aeolotropic, bounded by any surfaces presenting projecting edges or angles, or re-entrant angles or edges, however obtuse, cannot experience any finite stress or strain in the neighbourhood of a *projecting* angle (trihedral, polyhedral, or conical); in the neighbourhood of an edge, can only experience simple longitudinal stress parallel to the neighbouring part of the edge; and generally experiences infinite stress and strain in the neighbourhood of a *re-entrant* edge or angle; when influenced by any distribution of force, exclusive of surface tractions infinitely near the

Strain at projecting angles, evanescent.

At re-entrant angles infinite.

angles or edges in question. An important application of the last part of this statement is the practical rule, well known in mechanics, that every re-entering edge or angle ought to be rounded to prevent risk of rupture, in solid pieces designed to bear stress. An illustration of these principles is afforded by the concluding example of § 707; in which we have the complete mathematical solution of the torsion problem for prisms of fan-shaped sections, such as the annexed figures. In the cases corresponding to $a = 0$, we see, without working out the solution, that the distortion $\frac{d\gamma}{rd\eta}$ vanishes when $r = 0$, if β is $< \pi$; becomes infinite when $r = 0$, if β is $> \pi$; but is finite and determinate if $\beta = \pi$.

Liability to cracks proceeding from re-entrant angles, or any places of too sharp concave curvature.

Cases of curvilinear rectangles for which torsion problem has been solved

(1.) (2.) (3.) (4.) (5.) (6.)

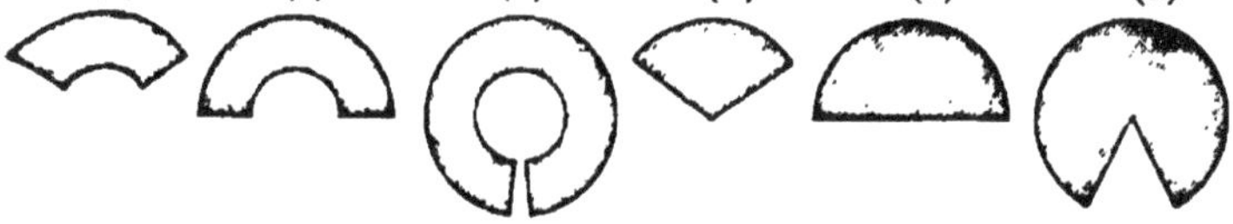

The solution indicated above determining v to satisfy (64) and (65) of § 707, if we translate it into polar co-ordinates r, η, such that $x = r\cos\eta$, and $y = r\sin\eta$, and if we put $\frac{\pi}{\beta} = \nu$, becomes merely this—

$$v = \Sigma(B_i r^{i\nu} + B_i' r^{-i\nu}) \sin i\nu\eta \, * \qquad (69),$$

where B_i, B_i' are to be determined by the equations (65) of § 707, with $r = a$ and $r = a'$ instead of $\xi = 0$ and $\xi = a$, and a'^2 instead of $a^2\epsilon^{2a}$ (a and a' denoting the radii of the concave and convex cylindrical surfaces respectively). When $a = 0$, these give $B_i' = 0$; and therefore

$$\left(\frac{dv}{rd\eta}\right)_{r=0} = 0, \quad = B_1 \cos\eta, \quad = \infty,$$

according as $\nu > 1$, $= 1$, or < 1; whence also similar results for $\left(\frac{d\gamma}{rd\eta}\right)_{r=0}$.

Distortion zero at central angle of sector (4). Infinite at central angle of sector (6); zero at all the other angles.

711. To prove the law of flexure (§§ 591, 592), and to investigate the flexural rigidity (§ 596) of a bar or wire of isotropic substance, we shall first conceive the bar to be bent into

Problem of flexure.

* Compare § 707 (23) (24); by which we see that this solution is merely the general expression in polar co-ordinates for series of spherical harmonics of x, y, with $z=0$, of degrees ι, 2ι, 3ι, etc., and $-\iota$, -2ι, -3ι, etc. These are "complete harmonics" when ι is unity or any integer.

Problem of flexure.

a circular arc, and investigate the application of force necessary to do so, subject to the following conditions:—

(1.) All lines of it parallel to its length become circular arcs in or parallel to the plane ZOX, with their centres in one line perpendicular to this plane; OZ and all parallel to it through OY being bent without change of length.

(2.) All normal sections remain plane, and perpendicular to those longitudinal lines (so that their planes come to pass through that line of centres).

(3.) No part of any normal section experiences deformation.

Forced condition of no distortion in normal sections.

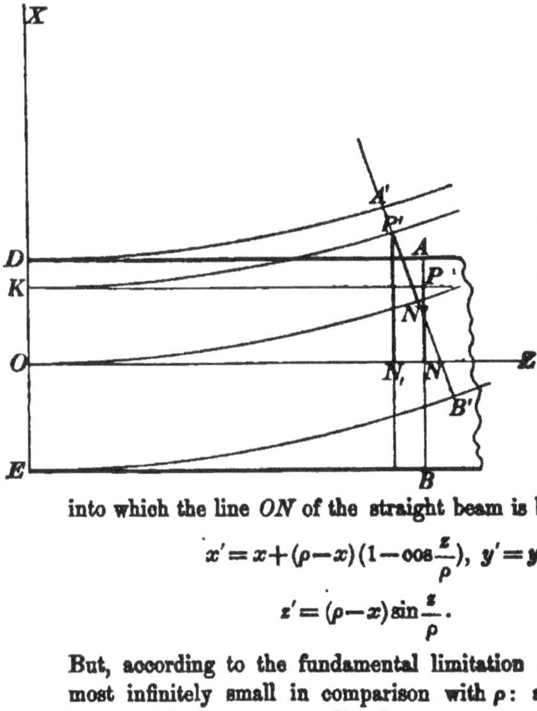

A section DOE of the beam being chosen for plane of reference, XOY, let P, (x, y, z) be any point of the unbent, and P', (x', y', z') the same point of the bent beam; each seen in projection, on the plane ZOX, in the diagram: and let ρ be the radius of the arc ON' into which the line ON of the straight beam is bent. We have

$$x' = x + (\rho - x)\left(1 - \cos\frac{z}{\rho}\right),\quad y' = y,$$

$$z' = (\rho - x)\sin\frac{z}{\rho}.$$

But, according to the fundamental limitation (§ 588), x is at most infinitely small in comparison with ρ: and through any length of the bar not exceeding its greatest transverse dimension, z is so also. Hence we neglect higher powers of $\frac{x}{\rho}$ and $\frac{z}{\rho}$ than the second in the preceding expressions; and putting $x' - x = \alpha$, $y' - y = \beta$, $z' - z = \gamma$, we have

$$\alpha = \tfrac{1}{2}\frac{z^2}{\rho},\quad \beta = 0,\quad \gamma = -\frac{xz}{\rho} \qquad (1)$$

These, substituted in § 698 (5) and § 697 (2), give

Surface traction, (P, Q), required to prevent distortion in normal section.

$$\left.\begin{array}{c} P=-(m-n)\frac{x}{\rho}, \quad Q=-(m-n)\frac{x}{\rho}, \quad R=-(m+n)\frac{x}{\rho} \\ S=0, \quad T=0, \quad U=0, \end{array}\right\} \quad (2).$$

$$X=\frac{m-n}{\rho}, \quad Y=0, \quad Z=0 \quad (3).$$

The interpretation of this result is interesting in itself, but, not requiring it for our present purpose, we leave it as an exercise to the student.

712. The problem of simple flexure supposes that no force is applied from without either as traction on the sides of the bar, or as force acting at a distance on its interior substance, but that, by opposing couples properly applied to its ends, it is kept in a circular form, with strain and stress uniform throughout its length.

Correction to do away with lateral traction, and bodily force.

To the α, β, γ of last section let corrections $\alpha'=\frac{1}{2}K(x^2-y^2)$, $\beta'=Kxy$, $\gamma'=0$, be added. This will give [by § 698 (5)]

$$P'=Q'=2mKx, \quad R'=2(m-n)Kx, \quad S'=0, \quad T'=0, \quad U'=0,$$

[and by § 698 (2)] $X'=-2mK$, $Y'=0$, $Z'=0$,

to be added to the P, $Q \ldots X$, Y, Z. Hence if we take

$$K=\frac{m-n}{2m\rho},$$

the surface tractions on the sides of the bar and the bodily forces are reduced to nothing; so that if now

St. Venant's solution of flexure problem.

$$\alpha=\frac{1}{2\rho}\{z^2+\frac{m-n}{2m}(x^2-y^2)\}, \quad \beta=\frac{1}{\rho}\frac{m-n}{2m}xy, \quad \gamma=-\frac{1}{\rho}xz \quad (1),$$

we have [§ 670 (6) and § 698 (6)]

$$\left.\begin{array}{c} e=\frac{m-n}{2\rho m}x=\frac{\sigma}{\rho}x, \quad f=\frac{m-n}{2\rho m}x=\frac{\sigma}{\rho}x, \quad g=-\frac{1}{\rho}x, \\ a=b=c=0 \end{array}\right\} \quad (2),$$

and [§ 698 (5), § 694 (6)]

$$\left.\begin{array}{c} P=0, \quad Q=0, \quad R=-\frac{(3m-n)n}{m}\frac{x}{\rho}=-M\frac{x}{\rho}, \\ X=0, \quad Y=0, \quad Z=0 \end{array}\right\} \quad (3).$$

To complete the fulfilment of the conditions, it is only necessary that the traction across each normal section be reducible to a couple. Hence $\iint Rdxdy=0$,

or, by (3), $\iint xdxdy=0$;

that is to say,

Flexure of a bar.

713. In order that no force, but only a bending cou- may be transmitted along the rod, the centre of inertia of the normal section must be in OY, that line of it in which it is cut by the surface separating longitudinally stretched fr longitudinally shortened parts of the substance.

Line through centres of inertia of normal sections remains unchanged in length.

714. In our analytical expressions only an infinitely sl part of the beam has been considered; and it has not been necessary to inquire whether the axis of the couple called in play is or is not perpendicular to the plane of flexure. B when so great a length of the beam is concerned, that the change of direction (§ 5) from one end to the other is fi the couples on the ends could not be directly opposed un their axes were both perpendicular to the plane of flexure, in- asmuch as each axis is in the proper normal section of rod. For finite flexure in a circular arc, without lateral co- straint, we must therefore have

Flexure through finite angle in one plane: must be in either of two principal planes, if produced simply by balancing couples on the two ends.

$$\iint Rydxdy = 0;\ \text{whence, by (3)},\ \iint xydxdy = 0:$$

that is to say, the plane of flexure must be perpendicular to one of the two principal axes of inertia of the normal section in its own plane. This being the case, the moment of the whole couple acting across each normal section is equal to the pro- duct of the curvature, into Young's modulus, into the moment of inertia of the area of the normal section round its principa axis perpendicular to the plane of flexure.

For we have [§ 712 (3)]

$$\iint Rxdxdy = -\frac{M}{\rho}\iint x^2 dxdy \qquad (4)$$

Principal flexural rigidities and axes.

715. Hence in a rod of isotropic substance the princip axes of flexure (§ 599) coincide with the principal axes of inertia of the area of the normal section; and the correspon ing flexural rigidities [§ 596] are the moments of inertia of this area round these axes multiplied by Young's modulus.

716. The interpretation of the results [§ 712 (2), (3)] to which the analytical investigation has led us is simply that if we imagine the whole rod divided, parallel to its length, int infinitesimal filaments (prisms when the rod is straight), each of these contracts or swells laterally with sensibly the sam freedom as if it were separated from the rest of the substan and becomes elongated or shortened in a straight line to th xtent as it is really elongated or shortened in the circular ich it becomes in the bent rod. The distortion of the cross by which these changes of lateral dimensions are neces- accompanied is illustrated in the annexed diagram, in

Geometrical interpretation of distortion in normal plane.

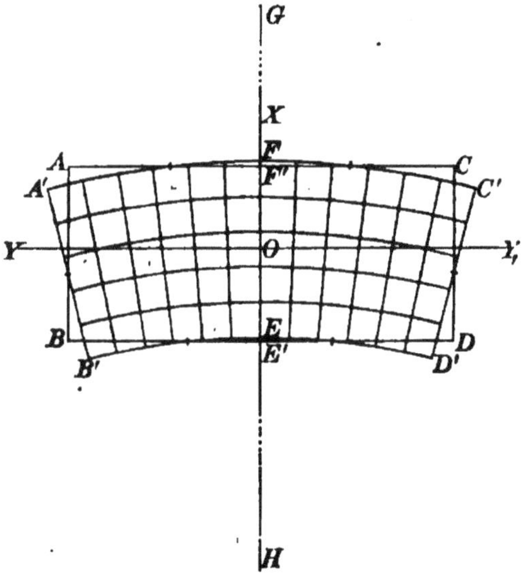

either the whole normal section of a rectangular beam, ectangular area in the normal section of a beam of any is represented in its strained and unstrained figures, the central point O common to the two. The flexure planes perpendicular to YOY, and concave upwards (or is X); G the centre of curvature, being in the direc- ndicated, but too far to be included in the diagram. straight sides AC, BD, and all straight lines parallel em, of the unstrained rectangular area become con- c arcs of circles concave in the opposite direction, their of curvature, H, being for rods of gelatinous substance, glass or metal, from 2 to 4 times as far from O on one as G is on the other. Thus the originally plane sides BD of a rectangular bar become anticlastic surfaces, of tures $\frac{1}{\rho}$ and $\frac{-\sigma}{\rho}$, in the two principal sections. A flat ıgular, or a square, rod of India rubber [for which σ

Anticlastic and conical curvatures produced in the four sides of a rectangular prism by flexure in a principal plane.

Experimental illustration.

amounts (§ 684) to very nearly ½, and which is susceptible of very great amounts of strain without utter loss of corresponding elastic action], exhibits this phenomenon remarkably well.

717. The conditional limitation (§ 588) of the curvature being very small in comparison with that of a circle of radius equal to the greatest diameter of the normal section (not previously necessary, and indeed not generally known to be necessary, we believe, when the greatest diameter is perpendicular to the plane of curvature) now receives its full explanation. For unless the *breadth*, AC, of the bar (or diameter perpendicular to the plane of flexure) be very small in comparison with the mean proportional between the radius, OH, and the thickness, AB, the distances from OY to the corners A', C' would fall short of the half thickness, OE, and the distances to B', D' would exceed it by differences comparable with its own amount. This would give rise to sensibly less and greater shortenings and stretchings in the filaments towards the corners than those expressed in our formula [§ 712 (2)], and so vitiate the solution. Unhappily mathematicians have not hitherto succeeded in solving, possibly not even tried to solve, the beautiful problem thus presented by the flexure of a broad very thin band (such as a watch spring) into a circle of radius comparable with a third proportional to its thickness and its breadth. See § 657.

Uncalculated effects of ordinary bendings of a thin flat spring.

Hence necessity for stricter limitation, § 628, of curvature than § 588 when a thin flat spring is bent in a plane perpendicular to its breadth.

718. But, provided the radius of curvature of the flexure be not only a large multiple of the *greatest* diameter, but also of a third proportional to the diameters in and perpendicular to the plane of flexure; then however great may be the ratio of the greatest diameter to the least, the preceding solution is applicable; and it is remarkable that the necessary distortion of the normal section (illustrated in the diagram of § 716) does not sensibly impede the free lateral contractions and expansions in the filaments, even in the case of a broad thin lamina (whether of precisely rectangular section, or of unequal thicknesses in different parts).

Transition to flexure of a plate.

719. Considering now a uniform thin broad lamina bent in the manner supposed in the preceding solution, we have precisely the case of a plate under the influence of a simple bending stress (§ 638). If the breadth be a, and the thickness

.he moment of inertia of the cross section is $\frac{1}{12}b^2.ab$, and re fore the flexural rigidity is $\frac{1}{12}Mab^3$, or $\frac{1}{12}Mb^3$ if the breadth unity. Hence a couple K (§ 637) would bend it to the curva- e $K \div \frac{1}{12}Mb^3$ lengthwise (or across its length), and (§ 716) uld produce the curvature $\sigma K \div \frac{1}{12}Mb^3$ breadthwise (or oss the breadth), but with concavity turned in the contrary ection. Precisely the same solution applies to the effect of oending stress, consisting of balancing couples applied to two edges, to bend it across the dimension which hitherto have been calling its breadth. And by the principle of erposition we may simultaneously apply a pair of balancing iples to each pair of parallel sides of a rectangular plate, thout altering by either balancing system the effect of the ler; so that the whole effect will be the geometrical result- t of the two effects calculated separately. Thus, a square ite of thickness b, and with each side of length unity, being ven, let pairs of balancing couples K on one pair of opposite les, and Λ on the other pair, be applied, each tending to pro- ce concavity in the same direction when positive. If κ and denote the whole curvatures produced in the planes of these uples, we shall have

Flexure of a plate: by a single bending stress; by simultaneous bending stresses in two planes at right angles to one another.

$$\kappa = \frac{1}{\frac{1}{12}Mb^3}(K - \sigma\Lambda) \qquad (1),$$

d

$$\lambda = \frac{1}{\frac{1}{12}Mb^3}(\Lambda - \sigma K) \qquad (2).$$

720. To find what the couples must be to produce simply lindrical curvature, κ, let $\lambda = 0$. We have

Stress in cylindrical curvature:

$$\Lambda = \sigma K$$

d

$$K = \frac{1}{12}\frac{Mb^3}{1-\sigma^2}\kappa \qquad (3).$$

r to produce spherical curvature, let $\kappa = \lambda$. This gives

in spherical curvature:

$$K = \Lambda = \frac{1}{12}\frac{Mb^3}{1-\sigma}\kappa \qquad (4).$$

r lastly, to produce anticlastic curvature, equal in the two irections, let $\kappa = -\lambda$. This gives

in anticlastic curvature.

$$K = -\Lambda = \frac{1}{12}\frac{Mb^3}{1+\sigma}\kappa \qquad (5).$$

lence, comparing with § 641 (10) and § 642 (16), we have, for the cylindrical rigidity, and for $\mathfrak{h}$ and $\mathfrak{k}$ the synclastic and nticlastic rigidities of a uniform plate of isotropic material,

Flexural rigidities of a plate: (*A*) cylindrical, (𝔥) synclastic, (𝔨) anticlastic

$$A=\tfrac{1}{12}\frac{Mb^3}{1-\sigma^2},$$

$$\mathfrak{h}=\tfrac{1}{12}\frac{Mb^3}{1-\sigma},\quad \mathfrak{k}=\tfrac{1}{12}\frac{Mb^3}{1+\sigma},$$

or [§ 694 (6) and § 698 (5)]

$$\mathfrak{h}=\frac{3nkb^3}{2(3k+4n)}=\frac{n(3m-n)b^3}{6(m+n)},\quad \mathfrak{k}=\tfrac{1}{6}nb^3$$

The coefficient A which appears in the equation of equi... through a plate, urged by any forces [§ 644 (6) and §§ 649... and c, which appears in its boundary conditions, are [§ 64... given in terms of $\mathfrak{h}$ and $\mathfrak{k}$ thus simply:—

$$A=\tfrac{1}{2}(\mathfrak{h}+\mathfrak{k}),\quad c=\tfrac{1}{2}(\mathfrak{h}-\mathfrak{k})$$

Same result for anticlastic flexure of a plate arrived at also by transition from simple torsion of rectangular prism.

721. It is interesting and instructive to investigat... anticlastic flexure of a plate by viewing it as an extrem... of torsion. Consider first a flat bar of rectangular s... uniformly twisted by the proper application of tangential ... tions [§ 706 (10)] on its ends. Let now its breadth be ... parable with its length; equal, for instance, to its length. W... thus have a square plate twisted by opposing couples ap... in the planes of two opposite edges, and so distributed ... these areas as to cause uniform action in all sections para... to them when the other two edges are left quite free. If, la... we suppose the thickness, b, infinitely small in comparison w... the breadth, a, in (46) of § 707, we have

$$N=\tfrac{1}{3}n\tau ab^3$$

The twist τ per unit of length gives $a\tau$ in the length a, w... [§ 640 (4)] is equivalent to an anticlastic curvature ϖ (accor... to the notation of § 639) $=\tau$. And the balancing coup... applied in only one pair of opposite sides of the square i... we see by § 656, equivalent to an anticlastic stress (accor... to the notation of § 637) $\Pi=\tfrac{1}{2}N\div a$. Hence, for the a... clastic rigidity, according to § 642 (13), we have

$$\mathfrak{k}=\Pi\div\varpi=\frac{N}{2\tau a}=\frac{n}{6}b^3$$

which agrees with the value (6) otherwise found in § 72... the composition of flexures.

It is most important to remark—(1.) That one-half of th... $\tfrac{1}{3}n\tau ab^3$ in the value of N given by the formula (46) of § 70... derived from α and β as given by (8) of § 706, and the term —...

of γ by (45);—and (2.) That if we denote by γ' the transcendental series completing the expression (45) for γ, it is the term $n\iint x\frac{d\gamma'}{dy}dxdy$ of § 706 (17), that makes up the other half of the part of N in question, and that it does so as follows, according to the process of integrating by parts, in which it is to be remembered that to change the sign of either x or y, simply changes the sign of γ':— **Analysis of traction in normal section of twisted rectangular prism.**

$$\int_{-\frac{1}{2}a}^{\frac{1}{2}a}\int_{-\frac{1}{2}b}^{\frac{1}{2}b} x\frac{d\gamma'}{dy}dydx = \int_{-\frac{1}{2}a}^{\frac{1}{2}a} xGdx = a\int_0^{\frac{1}{2}a} Gdx - 2\int_0^{\frac{1}{2}a}dx\int_0^x Gdx \quad (10),$$

where

$$= n\int_{-\frac{1}{2}b}^{\frac{1}{2}b}\frac{d\gamma'}{dy}dy = 2n\gamma'_{y=\frac{1}{2}b} = 2n\tau\left(\frac{2}{\pi}\right)^3 b^2\Sigma\frac{1}{(2i+1)^3}\,\frac{\varepsilon^{(2i+1)\frac{\pi x}{b}} - \varepsilon^{-(2i+1)\frac{\pi x}{b}}}{\varepsilon^{(2i+1)\frac{\pi a}{2b}} + \varepsilon^{-(2i+1)\frac{\pi a}{2b}}} \quad (11).$$

Thus in N we have a term

$$a\int_0^{\frac{1}{2}a} Gdx = n\tau a\left(\frac{2}{\pi}\right)^4 b^3\Sigma\frac{1}{(2i+1)^4}\left\{1 - \frac{2}{\varepsilon^{(2i+1)\frac{\pi a}{2b}} + \varepsilon^{-(2i+1)\frac{\pi a}{2b}}}\right\},$$

or, because [as we see, by integrating (40) with reference to y, and putting $y=\frac{1}{2}b$],

$$1 + \frac{1}{3^4} + \frac{1}{5^4} + \text{etc.} = \frac{1}{6}\left(\frac{\pi}{2}\right)^4,$$

$$\int_0^{\frac{1}{2}a} Gdx = \tfrac{1}{6}n\tau ab^3 - n\tau\left(\frac{2}{\pi}\right)^4 ab^3\Sigma\frac{2}{(2i+1)^4\left[\varepsilon^{(2i+1)\frac{\pi a}{2b}} + \varepsilon^{-(2i+1)\frac{\pi a}{2b}}\right]} \quad (12).$$

The transcendental series constituting the second term of this, together with

$$-2\int_0^{\frac{1}{2}a}dx\int_0^x Gdx - n\iint y\frac{d\gamma'}{dx}dxdy$$

makes up the transcendental series which appears in the expression (46) for N. This, when $a \div b$ is infinite, vanishes in comparison with the first term of (46), as we have seen above § 721 (8). But in examining, as now, the composition of the expression, it is to be remarked that, when $a \div b$ is infinite, γ' vanishes except for values of x differing infinitely little from $\pm\frac{1}{2}a$, and therefore we see at once that in this case,

$$n\int_{-\frac{1}{2}a}^{\frac{1}{2}a}dx\int_{-\frac{1}{2}b}^{\frac{1}{2}b}dy\left(x\frac{d\gamma'}{dy} - y\frac{d\gamma'}{dx}\right) = na\int_0^{\frac{1}{2}a}dx\int_{-\frac{1}{2}b}^{\frac{1}{2}b}\frac{d\gamma'}{dy}dy = a\int_0^{\frac{1}{2}a}Gdx,$$

by which, in connexion with what precedes, we see that

722. One half of the couple on each of the edges, by which these conditions are fulfilled, consists of two tangential tractions

Composition of action in normal section of a long rectangular lamina under torsion.

distributed over areas of the edge infinitely near its ends a... perpendicularly to the plate towards opposite parts. The ot... half consists of forces parallel to the length of the edges, u... formly distributed through the length, and varying across i... simple proportion to distance, positive or negative, from middle line.

Uniform distribution of couple applied to its edges to render the stress uniform from the edges inwards.

723. If now we remove the former half, and apply inste... over the edges (BB', AA') hitherto free, a uniform distribu... of couple equal and similar to the latter half, and in the pr... directions to keep up ... same twist through ... plate, we have the pr... edge tractions to f... Poisson's three bounda... conditions (§ 645) for ... case in question; that ... to say, we have such ... distribution of tracti...

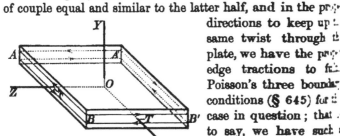

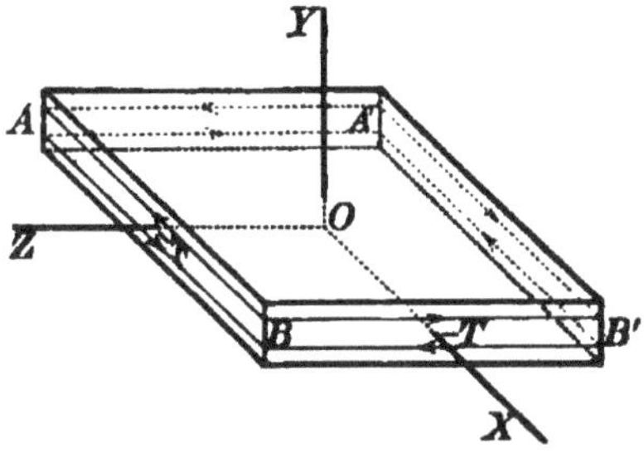

on the four edges of a square plate as produces anticlas... stress (§ 638) uniform not only through all of the plate ... distances from the edges great in comparison with the thick... ness, but throughout the plate up to the very edges. The sta... of strain and stress through the plate is represented by th... following formulæ [as we may gather from §§ 706 and 707 ... (45), (9), (10), (17), and § 722, or, as we see directly, by th... verification which the operations now indicated present]:—

Algebraic solution expressing displacement, strain, and stress, through a plate bent to uniform anticlastic curvature.

$$\left.\begin{array}{l}\alpha=-\tau yz,\ \beta=\tau xz,\ \gamma=-\tau xy\\ \mathfrak{e}=\mathfrak{f}=\mathfrak{g}=0,\ \mathfrak{a}=0,\ \mathfrak{b}=-2\tau y,\ \mathfrak{c}=0\\ P=Q=R=0,\ S=0,\ T=-2n\tau y,\ U=0\\ -L=N=-\int_{-\frac{1}{2}a}^{\frac{1}{2}a}\int_{-\frac{1}{2}b}^{\frac{1}{2}b} Tydydx=\frac{n\tau}{6}ab^3\end{array}\right\}\quad (13$$

where L and N denote the moments (with signs reckoned a... in § 551) of the whole amounts of couple, applied to the tw... edges perpendicular to OX and OZ respectively, in the plane... of these edges.

By turning the axes OX, OZ through 45° in their own plane... we fall back on the formulæ of flexure as in § 719, for the par... ticular case of equal flexures in the two opposite directions.

724. If, on the other hand, we superimpose on the state of strain investigated in § 721, another produced by applying on the pair of edges which it leaves free, precisely the same entire distribution of couple as that described in § 722, but in the direction opposite to the twist which the former gave to the plate (so that now it is not $-L$, but L that is equal to N), we have the square plate precisely in the condition described in § 647, except infinitely near its corners. To find the expressions for components of displacement, strain, and stress, in this case, we must add to the expressions for α, β, γ in (8) of § 706, and (45) of § 707, values obtained by changing the sign of each of these expressions, and interchanging x for z, and α for γ. The consequent values of $\mathfrak{e}$, $\mathfrak{f}$, $\mathfrak{g}$, $\mathfrak{a}$, $\mathfrak{b}$, $\mathfrak{c}$, P, Q, R, S, T, U, are of course obtained in the same way, but need not be written down, as they can be seen in a moment from α, β, γ. Lastly, the strain thus superimposed would, if existing alone, leave the edges parallel to x free from traction, just as the first supposed strain [§ 706 (8)] leaves the edges parallel to z free; and thus, without fresh integration, we see that N has still the value (46), and is the result of the distribution of tractions described in § 722. The parts of the component displacements represented by products of coordinates disappear, and only transcendental series, as follows, remain:—

This rectangular plate subjected to the edge-traction of § 647.

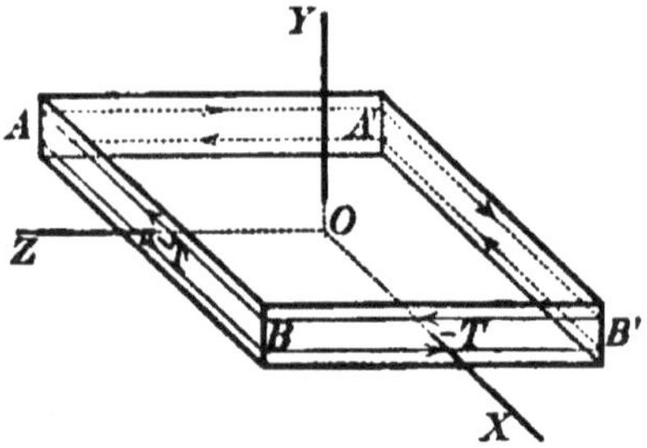

$$\left.\begin{aligned}\alpha &= -\tau\left(\frac{2}{\pi}\right)^3 b^2 \Sigma \frac{(-1)^i}{(2i+1)^3} \frac{\epsilon^{+\frac{(2i+1)\pi z}{b}} - \epsilon^{-\frac{(2i+1)\pi z}{b}}}{\epsilon^{+\frac{(2i+1)\pi a}{2b}} + \epsilon^{-\frac{(2i+1)\pi a}{2b}}} \sin\frac{(2i+1)\pi y}{b} \\ \gamma &= +\tau\left(\frac{2}{\pi}\right)^3 b^2 \Sigma \frac{(-1)^i}{(2i+1)^3} \frac{\epsilon^{+\frac{(2i+1)\pi x}{b}} - \epsilon^{-\frac{(2i+1)\pi x}{b}}}{\epsilon^{+\frac{(2i+1)\pi a}{2b}} + \epsilon^{-\frac{(2i+1)\pi a}{2b}}} \sin\frac{(2i+1)\pi y}{b}\end{aligned}\right\} \quad (14).$$

725. When $a \div b$ is infinite, $\epsilon^{+\frac{(2i+1)\pi a}{2b}}$ becomes infinitely great, and $\epsilon^{-\frac{(2i+1)\pi a}{2b}}$ infinitely small. If then we put $\frac{1}{2}a - z = z'$, and $\frac{1}{2}a - x = x'$, the preceding expressions become

Thin rectangular plate subjected to the edge-traction of § 647.

$$\alpha = -\tau\left(\frac{2}{\pi}\right)^3 b^2 \Sigma \frac{(-1)^i}{(2i+1)^3} \epsilon^{-\frac{(2i+1)\pi x'}{b}} \sin\frac{(2i+1)\pi y}{b}$$

for points not infinitely near the edge $A'B'$;

$$\cdot\gamma = +\tau\left(\frac{2}{\pi}\right)^3 b^2 \Sigma \frac{(-1)^i}{(2i+1)^3} \epsilon^{-\frac{(2i+1)\pi x'}{b}} \sin\frac{(2i+1)\pi y}{b}$$

for points not infinitely near the edge AA';

$\alpha=0$, $\gamma=0$, for all points not infinitely near an edge; (15).

and $\beta=0$ throughout.

Lastly, $L=N=\frac{1}{3}n\tau ab^3$,

of each of which one-half is constituted by tractions uniformly distributed along the corresponding edge, and proportional to distances from the middle line; and the other by tractions infinitely near the corners and perpendicular to the plate.

Transition to plate without corners subjected to edge-traction of § 647.

726. It is clear that if the corners were rounded off, or a plate were of any shape without corners, that is to say, with part of its edge where the radius of curvature is not very gr[eat] in comparison with the thickness, the effect of applying a dis[-]tribution of couple all round its edge in the manner defined in § 647 would be expressed by either of these last formulæ for α and γ. Thus the whole displacement of the substance will be parallel to the edge for all points infinitely near it; will vanish for all other points of the plate; and will be equal to the pre[-]ceding expression (15) for γ if x' denote simply distance from the nearest point of the edge of the plate, and y, as in all the formulæ, distance from the middle surface.

727. We may conclude that if a uniform plate, bounded by an edge everywhere perpendicular to its sides, and of thickness a small fraction of the smallest radius of curvature of the edge at any point, be subjected to the action described in § 647 with the more particular condition that the distribution of tan[-]gential traction is [as asserted in § 634 (3.) for any normal section remote from the boundary of a bent plate] in simple proportion to distance, positive or negative, from the middle line of the edge; the interior strain and stress will be specified by the following statement and formulæ:—

Origin shifted from middle plane to one side of plate.

Let O be any point in one corner of the edge: and let OX perpendicular to the edge inwards, and OY perpendicular to plane of the plate. The displacement of any particle P, (x, y), at any distance from O not a considerable multiple of the thickness, b, will be perpendicular to the plane YOX, and (denoted by γ) will be given by the formula—

Displacement of substance produced by edge-traction of § 647.

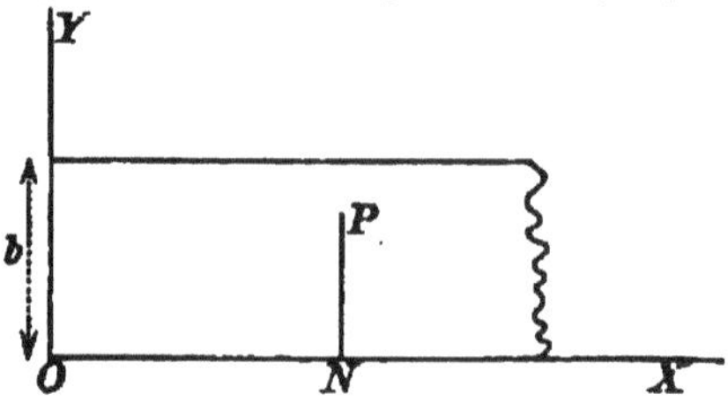

$$\gamma=6\frac{\Omega}{nb}\left(\frac{2}{\pi}\right)^2\left(\epsilon^{-\frac{\pi x}{b}}\cos\frac{\pi y}{b}+\frac{1}{3^2}\epsilon^{-\frac{3\pi x}{b}}\cos\frac{3\pi y}{b}+\frac{1}{5^2}\epsilon^{-\frac{5\pi x}{b}}\cos\frac{5\pi y}{b}+\text{etc.}\right) \quad (16)$$

where Ω denotes the amount of the couple per unit length of the edge, and n the rigidity (§ 680) of the substance. But the simplest and easiest way of arriving at this result is to solve directly by Fourier's analytical method the following problem, a case of one of the general problems of § 696 :—

728. A uniform plane plate of thickness b, extending to infinity in all directions on one side of a straight edge (or plane perpendicular to its sides) being given,—

Case of § 647 independently investigated.

It is required to find the displacement, strain, and stress, produced by tangential traction applied uniformly along the edge, according to a given arbitrary function $[\phi(y)]$ of position on its breadth.

Taking co-ordinates as in § 727, we have to solve equations (2) of § 697, with $X=0$, $Y=0$, $Z=0$, for all points of space for which x is positive, and y between 0 and b, subject to the boundary conditions,

See § 661, or § 662 (1); also § 693 (5), and § 670 (6).

$$\left.\begin{cases} P=0,\ Q=0,\ R=0,\ S=0,\ T=0,\ U=0, \text{ when } y=0 \text{ or } b: \\ P=0,\ Q=0,\ R=0,\ S=0,\ U=0;\ T=\phi(y), \text{ when } x=0: \\ \text{and } \alpha=0,\ \beta=0,\ \gamma=0, \text{ when } x=\infty. \end{cases}\right\} \quad (17).$$

From these, inasmuch as α, β, γ must each be independent of z, we find

$$\left.\begin{aligned} &(a)\quad \frac{d^2\gamma}{dx^2}+\frac{d^2\gamma}{dy^2}=0, \text{ throughout the solid;} \\ &(b)\quad \gamma=0 \text{ when } x=\infty; \\ &(c)\quad n\frac{d\gamma}{dy}=0 \text{ when } y=0 \text{ or } b; \\ \text{and } &(d)\quad n\frac{d\gamma}{dx}=\phi(y) \text{ when } x=0; \end{aligned}\right\} \quad (18);$$

Case of § 647 independently investigated.

and all the equations, both thorough and superficial, involving α and β are satisfied by $\alpha=0$, $\beta=0$, and therefore (App. C) require $\alpha=0$, $\beta=0$. The Fourier solution, of course, is easily seen, because of (a), (b), and (c), to be of the form

$$\gamma=\Sigma A_i \epsilon^{-\frac{i\pi x}{b}}\cos\frac{i\pi y}{b} \qquad (19)$$

and, because of (d), the coefficients A_i are to be found so as to make

$$-\frac{n\pi}{b}\Sigma i A_i \cos\frac{i\pi y}{b}=\phi(y) \qquad (20)$$

They are therefore [as we see by taking in § 77, (13) and (14) ϕ such that $\phi(p-\xi)=\phi\xi$, and putting $p=2b$] as follows:—

$$A_i=-\frac{b}{n\pi}\cdot\frac{1}{i}\cdot\frac{2}{b}\int_0^b\phi(y)\cos\frac{i\pi y}{b}dy \qquad (21)$$

If (for the particular case in question) we take

$$\phi(y)=\frac{12\Omega}{b^2}(y-\tfrac{1}{2}b) \qquad (22)$$

we find $\quad A_{2i}=0$, and $A_{2i+1}=6\frac{\Omega}{nb}\left(\frac{2}{\pi}\right)^2\frac{1}{(2i+1)^2}$ $\qquad (23)$

and so arrive at the result (16).

Rapid decrease of disturbance from edge inwards.

729. It is remarkable how very rapidly the whole disturbance represented by this result diminishes inwards from the edge where the disturbing traction is applied (compare § 58?); also how very much more rapidly the second term diminishes than the first; and so on.

Thus as $\epsilon=2{\cdot}71828$, $\epsilon^{\frac{1}{2}\pi}=4{\cdot}801$, $\epsilon^{2{\cdot}303}=10$, $\epsilon^{\pi}=23{\cdot}141$, $\epsilon^{2\pi}=535{\cdot}5$, we have

$$\text{for } x=\frac{1}{3{\cdot}1416}b,\quad \gamma=6\frac{\Omega}{nb}\left(\frac{2}{\pi}\right)^2\left(\frac{\cos\frac{\pi y}{b}}{2{\cdot}718}-\frac{\cos\frac{3\pi y}{b}}{3^2 2{\cdot}718^3}+\frac{\cos\frac{5\pi y}{b}}{5^2 2{\cdot}718^5}-\text{etc.}\right)$$

$$x=\tfrac{1}{2}b,\quad \gamma=6\frac{\Omega}{nb}\left(\frac{2}{\pi}\right)^2\left(\frac{\cos\frac{\pi y}{b}}{4{\cdot}801}-\frac{\cos\frac{3\pi y}{b}}{3^2 4{\cdot}801^3}+\frac{\cos\frac{5\pi y}{b}}{5^2 4{\cdot}801^5}-\text{etc.}\right)$$

$$x=\frac{2{\cdot}303}{\pi}b,\quad \gamma=6\frac{\Omega}{nb}\left(\frac{2}{\pi}\right)^2\left(\frac{\cos\frac{\pi y}{b}}{10}-\frac{\cos\frac{3\pi y}{b}}{3^2 10^3}+\frac{\cos\frac{5\pi y}{b}}{5^2 10^5}-\text{etc.}\right)$$

$$x=b,\quad \gamma=6\frac{\Omega}{nb}\left(\frac{2}{\pi}\right)^2\left(\frac{\cos\frac{\pi y}{b}}{23{\cdot}14}-\frac{\cos\frac{3\pi y}{b}}{3^2 23{\cdot}14^3}+\frac{\cos\frac{5\pi y}{b}}{5^2 23{\cdot}14^5}-\text{etc.}\right)$$

$$x=2b,\quad \gamma=6\frac{\Omega}{nb}\left(\frac{2}{\pi}\right)^2\left(\frac{\cos\frac{\pi y}{b}}{535{\cdot}5}-\frac{\cos\frac{3\pi y}{b}}{3^2 535{\cdot}5^3}+\frac{\cos\frac{5\pi y}{b}}{5^2 535{\cdot}5^5}-\text{etc.}\right)$$

which proves most strikingly the concluding statement of § 647.

730. We regret that limits of space compel us to leave uninvestigated the torsion-flexure rigidities of a prism and the flexural rigidities of a plate of eolotropic substance: and to still confine ourselves to isotropic substance when, in conclusion, we proceed to find the complete integrals of the equation [§ 697 (2)] of internal equilibrium for an infinite solid under the influence of any given forces, and the harmonic solutions suitable for problems regarding spheres and spherical shells, and solid and hollow circular cylinders (§ 738) under plane strain. The problem to be solved for the infinite solid is this: Problems to be solved.

Let, in (6) *of* § 698, *X, Y, Z be any arbitrary functions whatever of* (*x, y, z*), *either discontinuous and vanishing in all points outside some finite closed surface, or continuous and vanishing at all infinitely distant points with sufficient convergency to make RD converge to* 0 *as D increases to* ∞, *if R be the resultant of X, Y, Z for any point at distance D from origin. It is required to find α, β, γ satisfying those equations* [(6) *of* § 698], *subject to the condition of each vanishing for infinitely distant points* (*that is, for infinite values of x, y, or z*). General problem of infinite solid:

(*a*) Taking $\frac{d}{dx}$ of the first of these equations, $\frac{d}{dy}$ of the second, and $\frac{d}{dz}$ of the third, and adding, we have solved for isotropic substance.

$$(m+n)\nabla^2\delta+\frac{dX}{dx}+\frac{dY}{dy}+\frac{dZ}{dz}=0 \qquad (1).$$

(*b*) This shows that if we imagine a mass distributed through space, with density ρ given by

$$\rho=\frac{1}{4\pi(m+n)}\left(\frac{dX}{dx}+\frac{dY}{dy}+\frac{dZ}{dz}\right) \qquad (2),$$

δ must be equal to its potential at (x, y, z). For [§ 491 (*c*)] if V be this potential we have

$$\nabla^2 V+4\pi\rho=0.$$

Subtracting this from (1) divided by $(m+n)$, we have

$$\nabla^2(\delta-V)=0 \qquad (3),$$

for all values of (x, y, z). Now the convergency of XD, YD, ZD to zero when D is infinite, clearly makes $V=0$ for all infinitely distant points. Hence if S be any closed surface round the origin of co-ordinates, everywhere infinitely distant from it, the function $(\delta-V)$ is zero for all points of it, and satisfies (3) for all points within it. Hence [App. A. (*e*)] we must have $\delta=V$. In other words, the fact that (1) holds for all points of space gives determinately

General equations for infinite isotropic solid integrated.

$$\delta=\frac{1}{4\pi(m+n)}\int_{-\infty}^{\infty}\int_{-\infty}^{\infty}\int_{-\infty}^{\infty}\frac{\left(\frac{dX'}{dx'}+\frac{dY'}{dy'}+\frac{dZ'}{dz'}\right)dx'dy'dz'}{\sqrt{[(x-x')^2+(y-y')^2+(z-z')^2]}} \quad (4),$$

where X', Y', Z' denote the values of X, Y, Z for any point (x', y', z').

(*c*) Modifying by integration by parts, and attending to the prescribed condition of convergences, according to which,

$$\text{when } x'=\infty,\ \int_{-\infty}^{\infty}\int_{-\infty}^{\infty}\frac{X'dy'dz'}{\sqrt{[(x-x')^2+(y-y')^2+(z-z')^2]}}=0 \quad (5),$$

we have

$$\delta=\frac{-1}{4\pi(m+n)}\int_{-\infty}^{\infty}\int_{-\infty}^{\infty}\int_{-\infty}^{\infty}\frac{X'(x-x')+Y'(y-y')+Z'(z-z')}{[(x-x')^2+(y-y')^2+(z-z')^2]^{\frac{3}{2}}}dx'dy'dz' \quad (6),$$

which for most purposes is more convenient than (4).

(*d*) On precisely the same plan as (*b*) we now integrate each of the three equations (6) of § 698 separately for α, β, γ respectively, and find

$$\alpha=u+U,\ \beta=v+V,\ \gamma=w+W \quad (7),$$

where u, v, w, U, V, W denote the potentials at (x, y, z) of distributions of matter through all space of densities respectively

$$\frac{m}{4\pi n}\frac{d\delta}{dx},\ \frac{m}{4\pi n}\frac{d\delta}{dy},\ \frac{m}{4\pi n}\frac{d\delta}{dz},\ \frac{X}{4\pi n},\ \frac{Y}{4\pi n},\ \frac{Z}{4\pi n};$$

in other words, such functions that

$$\nabla^2u+\frac{m}{n}\frac{d\delta}{dx}=0,\ \text{etc., and } \nabla^2U+\frac{X}{n}=0,\ \text{etc.} \quad (8),$$

each through all space. Thus if δ'', X'', Y'', Z'' denote the values of δ, X, Y, Z for a point (x'', y'', z''), we find, for α,

$$\alpha=\frac{1}{4\pi n}\int_{-\infty}^{\infty}\int_{-\infty}^{\infty}\int_{-\infty}^{\infty}\frac{\left(m\frac{d\delta''}{dx''}+X''\right)dx''dy''dz''}{[(x-x'')^2+(y-y'')^2+(z-z'')^2]^{\frac{1}{2}}} \quad (9);$$

if in this we substitute for δ'' its value by (6) we have α expressed by the sum of a sextuple integral and a triple integral, the latter being the U of (7); and similarly for β and γ. These expressions may, however, be greatly simplified, since we shall see presently that each of the sextuple integrals may be reduced to a triple integral.

Force applied uniformly to spherical portion of infinite homogeneous solid.

(*e*) As a particular case, let X, Y, Z be each constant throughout a spherical space having its centre at the origin and radius a, and zero everywhere else. This by (6) will make $-\delta$ the sum of the products of X, Y, Z respectively into the corresponding component attractions of a uniform distribution of matter of density $\frac{1}{4\pi(m+n)}$ through this space. Hence [§ 491 (*b*)]

Dilatation produced by it.

$$\left.\begin{aligned}\delta&=\frac{-a^3}{3(m+n)}\frac{Xx+Yy+Zz}{r^3}\text{ for points outside the spherical space,}\\ \text{and }\delta&=\frac{-1}{3(m+n)}(Xx+Yy+Zz)\text{ for points within the spherical space.}\end{aligned}\right\}\quad(10).$$

Investigation of displacement.

Now we may divide u of (8) into two parts, u' and u'', depending on the values of $\frac{d\delta}{dx}$ within and without the spherical space respectively; so that we have,

$$\left.\begin{aligned}&\text{for } r<a, && \nabla^2 u'=\frac{mX}{3n(m+n)}\text{, a constant,}\\ &\text{for } r>a, && \nabla^2 u'=0;\end{aligned}\right\}\quad(11);$$

for $r<a$, $\nabla^2 u''=0$,

for $r>a$, $\nabla^2 u''=-\frac{m}{n}\frac{d\delta}{dx}$, which is a solid spherical harmonic of degree -3, because δ is given by the first of equations (10). (12).

The solution of (11), being simply the potential due to a uniform sphere of density $-\frac{1}{4\pi}\frac{mX}{3n(m+n)}$, is of course

$$\left.\begin{aligned}u'&=\frac{-mX}{18n(m+n)}(3a^2-r^2)\text{ for } r<a,\\ u'&=\frac{-mX}{9n(m+n)}\frac{a^3}{r}\text{ for } r>a.\end{aligned}\right\}\quad(13).$$

Again, if in (12) of App. B. we put $m=2$, $n=-3$, and $V_{-3}=\frac{d\delta}{dx}$, we have

$$\nabla^2\left(r^2\frac{d\delta}{dx}\right)=-6\frac{d\delta}{dx}\text{ for } r>a\quad(14),$$

since, for $r>a$, $\frac{d\delta}{dx}$ is a spherical harmonic of order -3. And $r^5\frac{d\delta}{dx}$ is [App. B. (13)] a solid harmonic of degree 2: hence if $\left[\frac{d\delta}{dx}\right]$ denote, for any point within the spherical space, the same algebraic expression as $\frac{d\delta}{dx}$ by (10) for the external space, $\frac{r^5}{a^3}\left[\frac{d\delta}{dx}\right]$ is a function which, for all the interior space, satisfies the equation $\nabla^2 u=0$, and is equal to $r^2\frac{d\delta}{dx}$ for points infinitely near the surface, outside and inside respectively. Hence $\frac{r^5}{a^3}\left[\frac{d\delta}{dx}\right]$

Force applied uniformly to spherical portion of infinite homogeneous solid.

for interior space, and $r^3\frac{d\delta}{dx}$ for exterior space, constit[…] potential of a distribution of matter of density $\frac{1}{4\pi}\cdot 6\frac{d\delta}{dx}$ [...] the spherical space and zero within, and, so far as yet [...] any layer of matter whatever distributed over the sepa[...] spherical surface. To find the surface density of this lay[...] first, for an exterior point infinitely near the surface, take $(x\frac{d}{dx}+y\frac{d}{dy}+z\frac{d}{dz})(r^3\frac{d\delta}{dx})$, which we may denote by $-\{rR$ and, for an interior point infinitely near the surface,

$$(x\frac{d}{dx}+y\frac{d}{dy}+z\frac{d}{dz})\,(\frac{r^4}{a^3}[\frac{d\delta}{dx}]), \qquad \ldots -[rR$$

Then, remembering that $x\frac{d}{dx}+y\frac{d}{dy}+z\frac{d}{dz}$ is the same as [...] according to the notation of App. A. (a); we find [by App. B.

$$\{R\}=r\frac{d\delta}{dx}, \text{ and } [R]=-2\frac{r^4}{a^3}[\frac{d\delta}{dx}].$$

Therefore, as $r^3\frac{d\delta}{dx}$ for external space is independent of r, [...] as r differs infinitely little from a for each of the two points,

$$\{R\}-[R]=\frac{3}{a^2}\cdot r^3\frac{d\delta}{dx}$$

But $\{R\}$ and $[R]$ being the radial components of the force at points infinitely near one another outside and inside, correspond-ing to the supposed distribution of potential, it follows from § 47[...] that to produce this distribution there must be a layer of matter on the separating surface, having

$$\frac{1}{4\pi}(\{R\}-[R])$$

for surface density. But, inasmuch as $\{R\}-[R]$ is a surfac[e] harmonic of the second order, the potential due to that surfac[e] distribution alone is [§ 536 (4)]

$$\tfrac{1}{5}(\{R\}-[R])\frac{r^2}{a} \text{ through the inner space,}$$

and $\qquad \tfrac{1}{5}(\{R\}-[R])\frac{a^4}{r^3}$ through the outer space ;

or, according to the value found above for $\{R\}-[R]$,

$$\frac{3}{5a^3}r^4[\frac{d\delta}{dx}] \text{ through the inner space,}$$

and $\qquad \frac{3a^3}{5}\frac{d\delta}{dx}$ through the outer space.

Subtracting now this distribution of potential from the whole distribution formerly supposed, we find

Force applied uniformly to spherical portion of infinite homogeneous solid.

$\frac{2}{5a^3}r^5[\frac{d\delta}{dx}]$ for the inner space, and $(r^2-\frac{3}{5}a^2)\frac{d\delta}{dx}$ for the outer, as the distribution of potential due simply to an external distribution of matter, of density $\frac{1}{4\pi}\cdot 6r^2\frac{d\delta}{dx}$, with no surface layer. Hence, and by (14), we see that the solution of (12) is

$$\left.\begin{aligned} u''&=\tfrac{1}{6}\cdot\tfrac{2}{5}\frac{m}{n}\frac{r^5}{a^3}[\frac{d\delta}{dx}] \text{ for } r<a, \\ \text{and}\qquad u''&=\tfrac{1}{6}\frac{m}{n}(r^2-\tfrac{3}{5}a^2)\frac{d\delta}{dx} \text{ for } r>a. \end{aligned}\right\} \quad (15).$$

And [(8) showing that U is the potential of a distribution of matter of density equal to $\frac{X}{4\pi n}$] as X is constant through the spherical space and zero everywhere outside it, we have

$$\left.\begin{aligned} U&=\frac{X}{6n}(3a^2-r^2) \text{ for } r<a, \\ U&=\frac{X}{3n}\frac{a^3}{r} \text{ for } r>a. \end{aligned}\right\} \quad (16).$$

This, with (13), (15), and (10), gives by (7)

Displacement produced by it.

for $r<a$,

$$\alpha=\frac{1}{18n(m+n)}\{(2m+3n)X(3a^2-r^2)-\tfrac{2}{5}mr^5\frac{d}{dx}\frac{Xx+Yy+Zz}{r^3}\}$$

and for $r>a$,

$$\alpha=\frac{a^3}{18n(m+n)}\{2(2m+3n)\frac{X}{r}-m(r^2-\tfrac{3}{5}a^2)\frac{d}{dx}\frac{Xx+Yy+Zz}{r^3}\} \quad (17),$$

with symmetrical expressions for β and γ.

731. A detailed examination of this result, with graphic illustrations of the displacements, strains, and stresses concerned, is of extreme interest in the theory of the transmission of force through solids; but we reluctantly confine ourselves to the solution of the general problem of § 730.

To deduce which, we have now only to remark that if a become infinitely small, X, Y, Z remaining finite, the expressions for α, β, γ become infinitely small, even within the space of application of the force, and at distances outside it great in comparison with a, they become

Displacement produced by a force applied to an infinitely small part of an infinite elastic solid.

$$\left.\begin{aligned} \alpha&=\frac{V}{24\pi n(m+n)}\{2(2m+3n)\frac{X}{r}-mr^2\frac{d}{dx}\frac{Xx+Yy+Zz}{r^3}\} \\ \beta&=\text{etc.},\ \gamma=\text{etc.} \end{aligned}\right\} \quad (18),$$

where V denotes the volume of the sphere. As these dep- simply on the whole amount of the force (its components b- XV, YV, ZV), and when it is given are independent of the radius of the sphere, the same formulæ express the effect of the same whole amount of force distributed through an infinit- small space of any form not extending in any direction to m- than an infinitely small distance from the origin of co-ordin- Hence, recurring to the notation of § 730 (*b*), we have for the required general solution

Displacement produced by any distribution of force through an infinite elastic solid.

$$\alpha=\frac{1}{24\pi n(m+n)}\iiint dx'dy'dz'\left\{2(2m+3n)\frac{X'}{D}-mD^2\frac{d}{dx}\frac{X'(x-x')+Y'(y-y')+Z'(z-z')}{D^3}\right.$$

$$\beta=\frac{1}{24\pi n(m+n)}\iiint dx'dy'dz'\left\{2(2m+3n)\frac{Y'}{D}-mD^2\frac{d}{dy}\frac{X'(x-x')+Y'(y-y')+Z'(z-z')}{D^3}\right.$$

$$\gamma=\frac{1}{24\pi n(m+n)}\iiint dx'dy'dz'\left\{2(2m+3n)\frac{Z'}{D}-mD^2\frac{d}{dz}\frac{X'(x-x')+Y'(y-y')+Z'(z-z')}{D^3}\right.$$

where $$D=\sqrt{\{(x-x')^2+(y-y')^2+(z-z')^2\}},$$

$\iiint$ denotes integration through all space, and X', Y', Z' th- arbitrary functions of x', y', z' restricted only by the converg- condition of § 730.

This solution was first given, though in a somewhat differ- form, in the *Cambridge and Dublin Mathematical Journal*, 18- *On the Equations of Equilibrium of an Elastic Solid.*

Comparing it with (9), we now see the promised reduction of the sextuple integral involved in that expression to a triple integr-

The process (*e*) by which it is effected consists virtually - the evaluation of a certain triple integral by the proper soluti- of the partial differential equation $\nabla^2 V+4\pi\rho=0$ [like that form- worked out (§ 649) for the much simpler case of ρ merely a function of r]. Proof of the result by direct integration is a good exercise in the integral calculus.

Application to problem of § 696.

732. In §§ 730, 731 the imagined subject has been a hom- geneous elastic solid filling all space, and experiencing the effect of a given distribution of force acting *bodily* on its substance. The solution, besides the interesting applicati- indicated in § 731, is useful for simplifying the practical pro- blem of § 696, by reducing it immediately to the case in wh- no force acts on the interior substance of the body, thus :—

The equations to be satisfied being (6) of § 698, through - the portion of space occupied by the body, and certain equati- for all points of its boundary expressing that the surface displa- ments or tractions fulfil the prescribed conditions; let α, β, - be functions of (x, y, z), which satisfy the equations

$$+m\frac{d`\delta}{dx}+X=0,\quad n\nabla^{2}`\beta+m\frac{d`\delta}{dy}+Y=0,\quad n\nabla^{2}`\gamma+m\frac{d`\delta}{dz}+Z=0, \quad (1),$$

here, for brevity, $`\delta=\frac{d`\alpha}{dx}+\frac{d`\beta}{dy}+\frac{d`\gamma}{dz}$,

through the space occupied by the body. Then, if we put

General problem of § 696 reduced to case of no bodily force.

$$\alpha=`\alpha+\alpha_{\prime},\quad \beta=`\beta+\beta_{\prime},\quad \gamma=`\gamma+\gamma_{\prime} \quad (2),$$

we see that to complete the solution we have only to find $\alpha_{\prime}$, $\beta_{\prime}$, $\gamma_{\prime}$, as determined by the equations

$$\left.\begin{array}{c} n\nabla^{2}\alpha_{\prime}+m\frac{d\delta_{\prime}}{dx}=0,\quad n\nabla^{2}\beta_{\prime}+m\frac{d\delta_{\prime}}{dy}=0,\quad n\nabla^{2}\gamma_{\prime}+m\frac{d\delta_{\prime}}{dz}=0, \\ \delta_{\prime}=\frac{d\alpha_{\prime}}{dx}+\frac{d\beta_{\prime}}{dy}+\frac{d\gamma}{dz}, \end{array}\right\} \quad (3)$$

to be fulfilled throughout the space occupied by the body, and certain equations for all points of its boundary, found by subtracting from the prescribed values of the surface displacement or traction, as the case may be, components of displacement or traction calculated from $`\alpha, `\beta, `\gamma$.

Values for $`\alpha, `\beta, `\gamma$ may always be found according to §§ 730, 731, by supposing equations (1) § 732 to hold through all space, and X, Y, Z to be discontinuous functions, having the given values for all points of the body, and being each zero for all points of space not belonging to it. But all that is necessary is that (1) be satisfied through the space actually occupied by the body; and in some of the most important practical cases this condition may be more easily fulfilled otherwise than by determining $`\alpha, `\beta, `\gamma$ in that way with its superadded condition for the rest of space.

733. Thus, for example, let us suppose the forces to be such that $Xdx+Ydy+Zdz$[1] is the differential of a function, W,

Important class of cases.

[1] Let m be the mass of any small part of the body, x, y, z its co-ordinates at any time, and Pm, Qm, Rm the components of the force acting on it. If the system be conservative, $Pdx+Qdy+Rdz$ must be the differential of a function of x, y, z. Let, for instance, the forces on all parts of the body be due to attractions or repulsions from fixed matter; and let the particle considered be the matter of the body within an infinitely small volume $\delta x\delta y\delta z$. Then we have $Pm=X\delta x\delta y\delta z$, etc.; and therefore, if ρ be the density of the matter of m, so that $\rho\delta x\delta y\delta z=m$, we have, in the notation of the text, $P\rho=X$, $Q\rho=Y$, $R\rho=Z$; and therefore $Xdx+Ydy+Zdz$ is or is not a complete differential according as ρ is or is not a function of the potential; that is to say, according as the density of the body is or is not uniform over the equipotential surfaces for the distribution of force to which (P, Q, R) belongs. Thus the condition of the text, if the system of force is conservative, is satisfied when the body is homogeneous. But it is satisfied whether the system be conservative or not if the density is so distributed, that, were the body to lose its rigidity, and become an incompressible liquid held in a closed rigid vessel, it would (§ 755) be in equilibrium.

Important class of cases,

of x, y, z considered as independent variables. This ass-tion includes some of the most important and inter- practical applications, among which are—

(1.) A homogeneous isotropic body acted on by gravit- sensibly uniform and in parallel lines, as in the case of a b of moderate dimensions under the influence of terrestrial gra-

(2.) A homogeneous isotropic body acted on by any distr- tion of gravitating matter, and either equilibrated at rest the aid of surface-tractions if the attracting forces do n- themselves balance on it; or fulfilling the conditions of ternal equilibrium by the balancing, according to D'Alem- principle (§ 264) of the reactions against acceleration of parts of its mass and the forces of attraction to which it subjected, when the circumstances are such that no acc- ration of rotation has to be taken into account. To this belongs the problem, solved below, of finding the tidal defor- tion of the solid Earth, supposed of uniform specific gravity rigidity throughout, produced by the tide-generating influ- of the Moon and Sun.

(3.) A uniform body strained by centrifugal force due uniform rotation round a fixed axis.

But it does not include a solid with any arbitrary non- uniform distribution of specific gravity subjected to any those influences; nor generally a piece of magnetized st- subjected to magnetic attraction; nor even a uniform b- fulfilling the conditions of internal equilibrium under the fluence of reactions against acceleration round a fixed ax- produced by forces applied to its surface.

reduced to case of no bodily force.

We have, according to the present assumption,

$$\frac{dW}{dx}=X,\ \frac{dW}{dy}=Y,\ \frac{dW}{dz}=Z$$

which give
$$\frac{dX}{dx}+\frac{dY}{dy}+\frac{dZ}{dz}=\nabla^2 W.$$

Hence, for $'\delta$ as in § 730 (a) for δ,

$$(m+n)\nabla^2\,'\delta+\nabla^2 W=0,$$

which is satisfied by the assumption

$$'\delta=-\frac{W}{m+n}$$

Next, introducing these assumptions in (1) of § 732, we see th-

these equations are finally satisfied by values for $`\alpha, `\beta, `\gamma$, assumed as follows:— Important class of cases reduced to case of no bodily force.

$$\left.\begin{array}{l} `\alpha=\frac{1}{m+n}\frac{d\vartheta}{dx},\quad `\beta=\frac{1}{m+n}\frac{d\vartheta}{dy},\quad `\gamma=\frac{1}{m+n}\frac{d\vartheta}{dz} \\ \text{where } \vartheta \text{ is any function satisfying } \nabla^2\vartheta=-W \end{array}\right\} \quad (6).$$

Further, we may remark that if W be a spherical harmonic [App. B. (a)], a supposition including, as we shall see later, the most important applications to natural problems, we have at once, from App. B. (12), an integral of the equation for ϑ, as follows:—

$$\vartheta=\frac{r^2}{2(2i+3)}W_i \quad (7);$$

where the suffix is applied to W to denote that its degree is i.

734. The general problem of § 696 being now reduced to [the] case in which no force acts on the interior substance, it [bec]omes this, in mathematical language:—To find α, β, γ, three [fun]ctions of (x, y, z) which satisfy the equations Problem of § 696 with no force except over surface:

$$\left.\begin{array}{l} n\left(\frac{d^2\alpha}{dx^2}+\frac{d^2\alpha}{dy^2}+\frac{d^2\alpha}{dz^2}\right)+m\frac{d}{dx}\left(\frac{d\alpha}{dx}+\frac{d\beta}{dy}+\frac{d\gamma}{dz}\right)=0 \\ n\left(\frac{d^2\beta}{dx^2}+\frac{d^2\beta}{dy^2}+\frac{d^2\beta}{dz^2}\right)+m\frac{d}{dy}\left(\frac{d\alpha}{dx}+\frac{d\beta}{dy}+\frac{d\gamma}{dz}\right)=0 \\ n\left(\frac{d^2\gamma}{dx^2}+\frac{d^2\gamma}{dy^2}+\frac{d^2\gamma}{dz^2}\right)+m\frac{d}{dz}\left(\frac{d\alpha}{dx}+\frac{d\beta}{dy}+\frac{d\gamma}{dz}\right)=0 \end{array}\right\} \quad (1)$$

[fo]r all points of space occupied by the body, and the proper [eq]uations for all points of the boundary to express one or other [or] any sufficient combination of the two surface conditions [in]dicated in § 696. When these conditions are that the [su]rface displacements are given, the equations expressing them [a]re of course merely the assignment of arbitrary values to [α], β, γ for every point of the bounding surface. On the other [h]and, when force is arbitrarily applied in a fully specified [m]anner over the whole surface, subject only to the conditions [o]f equilibrium of forces on the body supposed rigid (§ 564), in [i]ts actual strained state, and the problem is to find how the [b]ody yields both at its surface and through its interior, the [c]onditions are as follows:—Let $d\Omega$ denote an infinitesimal [e]lement of the surface; and F, G, H functions of position on [t]he surface, expressing the components of the applied traction. [T]hese functions are quite arbitrary, subject only to the follow[i]ng conditions, being the equations [§ 551 (a), (b)] of equili[b]rium of a rigid body:—

Equations of equilibrium to which the surface-tractions are subject.

$$\left.\begin{array}{c}\iint F d\Omega=0,\ \iint G d\Omega=0,\ \iint H d\Omega=0, \\ \iint (Hy-Gz) d\Omega=0,\ \iint (Fz-Hx) d\Omega=0,\ \iint (Gx-Fy) d\Omega=0\end{array}\right\}$$

and the strain experienced by the body must be such satisfy for every point of the surface the following equations

Equations of surface-condition, when tractions are given.

$$\left.\begin{array}{l}\left\{(m+n)\frac{d\alpha}{dx}+(m-n)\left(\frac{d\beta}{dy}+\frac{d\gamma}{dz}\right)\right\}f+n\left(\frac{d\alpha}{dy}+\frac{d\beta}{dx}\right)g+n\left(\frac{d\gamma}{dx}+\frac{d\alpha}{dz}\right)h=F \\ \left\{(m+n)\frac{d\beta}{dy}+(m-n)\left(\frac{d\gamma}{dz}+\frac{d\alpha}{dx}\right)\right\}g+n\left(\frac{d\beta}{dz}+\frac{d\gamma}{dy}\right)h+n\left(\frac{d\alpha}{dy}+\frac{d\beta}{dx}\right)f=G \\ \left\{(m+n)\frac{d\gamma}{dz}+(m-n)\left(\frac{d\alpha}{dx}+\frac{d\beta}{dy}\right)\right\}h+n\left(\frac{d\gamma}{dx}+\frac{d\alpha}{dz}\right)f+n\left(\frac{d\beta}{dz}+\frac{d\gamma}{dy}\right)g=H\end{array}\right\}$$

which we find by (1) of § 662, with (6) of § 670, with (5) of § and (5) of § 698; f, g, h being now taken to denote the tion-cosines of the normal to the bounding surface at

Problem of § 696 solved for spherical shell.

735. The solution of this problem for the spherical (§ 696), found by aid of Laplace's spherical harmonic was first given by Lamé in a paper published in *Journal* for 1854. It becomes much simplified[1] by the we follow of adhering to algebraic notation and symm formulæ [App. B. (1)-(24)], until convenient practical sions of the harmonic functions, whether in algebraic or tr metrical forms, are sought [App. B. (25)-(41), (56)-(66)].

(*a*) Using for brevity the same notation δ and ∇^2 as hit [§ 698 (8) (9)], we find, from (1) of § 734, by the process § 730, $\nabla^2\delta=0$.

Dilatation proved expressible in convergent series of spherical harmonics.

(*b*) Now let the actual values of δ over any two conc spherical surfaces of radii a and a' be expanded, by App. B., in series of surface harmonics, S_0, S_1, S_2, etc., S'_0, S'_1, S'_2, etc.; so that when

$$\left.\begin{array}{ll} & r=a,\ \delta=S_0+S_1+S_2+\ldots S_i+\ldots \\ \text{and} & r=a',\ \delta=S'_0+S'_1+S'_2+\ldots S'_i+\ldots\end{array}\right\}$$

Then, throughout the intermediate space, we must have

$$\delta=\sum_0^\infty \frac{(a^{i+1}S_i-a'^{i+1}S'_i)r^i-(aa')^{i+1}(a'^iS_i-a^iS'_i)r^{-i-1}}{a^{2i+1}-a'^{2i+1}}$$

For (1.) this series converges for all values of r interm between a and a', as we see by supposing a' to be the less of two, and writing it thus:—

$$\delta=\sum_0^\infty\delta_i+\sum_0^\infty\delta_{-i-1}$$

[1] "Dynamical Problems regarding Elastic Spheroidal Shells, and Sp Incompressible Liquid." W. Thomson. *Phil. Trans.*, 1862.

where δ_i, δ_{-i-1} are solid harmonics of degrees i and $-i-1$ given by the following:—

$$\delta_i = \frac{S_i - (\frac{a'}{a})^{i+1} S'_i}{1-(\frac{a'}{a})^{2i+1}} (\frac{r}{a})^i, \text{ and } \delta_{-i-1} = -\frac{(\frac{a'}{a})^i S_i - S'_i}{1-(\frac{a'}{a})^{2i+1}} (\frac{a'}{r})^{i+1}$$

For very great values of i these become sensibly

$$\delta_i = S_i (\frac{r}{a})^i, \text{ and } \delta_{-i-1} = S'_i (\frac{a'}{r})^{i+1},$$

and therefore, as each of the series (4) is necessarily convergent, the two series into which in (6) the expansion (5) is divided, ultimately converge more rapidly than the "geometrical" series

$$(\frac{r}{a})^i, (\frac{r}{a})^{i+1}, (\frac{r}{a})^{i+2}, \ldots, \text{ and } (\frac{a'}{r})^{i+1}, (\frac{a'}{r})^{i+2}, (\frac{a'}{r})^{i+3}, \ldots,$$

respectively.

Again (2.) the expression (5) agrees with (4) at the boundary of the space referred to (the two concentric spherical surfaces).

And (3.) it satisfies $\nabla^2\delta = 0$ throughout the space.

Hence (4.) no function differing in value from that given by (5), for any point of the space between the spherical surfaces, can [App. A. (*e*)] satisfy the conditions (3) and (4) to which δ is subject.

In words, this conclusion is that

736. Any function, δ, of x, y, z, which satisfies the equation $\nabla^2\delta = 0$ for every point of the space between two concentric spherical surfaces, may be expanded into the sum of two series of complete spherical harmonics [App. B. (*c*)] of positive and of negative degrees respectively, which converge for all points of that space.

General theorem regarding expansibility in solid harmonics.

(*c*) We may now write (6), for brevity, thus—

$$\delta = \sum_{-\infty}^{\infty} \delta_i \qquad (7),$$

where δ_i, a complete harmonic of any positive or negative degree, i, is to be determined ultimately to fulfil the actual conditions of the problem. But first supposing it known, we find α, β, γ as in § 730 (*d*), except that now we take advantage of the formulæ appropriate for spherical harmonics instead of proceeding by triple integration. Thus, by (1) and (7), we have

Displacement determined on temporary supposition that dilatation is known.

$$\nabla^2 \alpha = -\frac{m}{n}\Sigma\frac{d\delta_i}{dx};$$

Displacement determined on temporary supposition that dilatation is known.

and therefore, as $\frac{d\delta_i}{dx}$ is a harmonic of degree $i-1$, by tak... App. B. (12), $n=i-1$ and $m=2$, we see that the co... solution of this equation, regarded as an equation for α, is

$$\alpha=u-\frac{mr^2}{2n}\Sigma\frac{1}{2i+1}\frac{d\delta_i}{dx},$$

where u denotes any solution whatever for the equation $\nabla^2 u=$ ... Similarly, if v and w denote any functions such that $\nabla^2 v=$ ... $\nabla^2 w=0$, we have

$$\beta=v-\frac{mr^2}{2n}\Sigma\frac{1}{2i+1}\frac{d\delta_i}{dy}, \text{ and } \gamma=w-\frac{mr^2}{2n}\Sigma\frac{1}{2i+1}\frac{d\delta_i}{dz}.$$

Complete harmonic solution of equations of interior equilibrium.

(*d*) Now, in order that (1) may be satisfied, δ_i must b... related to u, v, w that

$$\frac{d\alpha}{dx}+\frac{d\beta}{dy}+\frac{d\gamma}{dz}=\delta=\Sigma\delta_i.$$

Hence, by differentiating the expressions just found for α, ... and attending to the formula

$$\frac{d}{dx}(r^2\frac{d\phi_i}{dx})+\frac{d}{dy}(r^2\frac{d\phi_i}{dy})+\frac{d}{dz}(r^2\frac{d\phi_i}{dz})=2(x\frac{d}{dx}+y\frac{d}{dy}+z\frac{d}{dz})\phi_i+r^2\nabla^2\phi_i$$
$$=2i\phi_i+r^2\nabla^2\phi_i$$

ϕ_i being any homogeneous function of degree i, we find

$$\Sigma\delta_i=\frac{du}{dx}+\frac{dv}{dy}+\frac{dw}{dz}-\frac{m}{n}\Sigma\frac{i}{2i+1}\delta_i.$$

This gives $\frac{du}{dx}+\frac{dv}{dy}+\frac{dw}{dz}=\Sigma\frac{(2i+1)n+im}{(2i+1)n}\delta_i$

If, therefore, Σu_i, Σv_i, Σw_i be the harmonic expansions (§ ... of u, v, w, we must have

$$\delta_i=\frac{(2i+1)n}{(2i+1)n+im}(\frac{du_{i+1}}{dx}+\frac{dv_{i+1}}{dy}+\frac{dw_{i+1}}{dz})$$

Using this, with i changed into $i-1$, in the preceding expr... for α, β, γ, we have finally, as the spherical harmonic solu... (1), § 734,

$$\alpha=\sum_{i=-\infty}^{i=\infty}\{u_i-\tfrac{1}{2}\frac{mr^2}{(2i-1)n+(i-1)m}\frac{d}{dx}(\frac{du_i}{dx}+\frac{dv_i}{dy}+\frac{dw_i}{dz})\}$$
$$\beta=\sum_{i=-\infty}^{i=\infty}\{v_i-\tfrac{1}{2}\frac{mr^2}{(2i-1)n+(i-1)m}\frac{d}{dy}(\frac{du_i}{dx}+\frac{dv_i}{dy}+\frac{dw_i}{dz})\}$$
$$\gamma=\sum_{i=-\infty}^{i=\infty}\{w_i-\tfrac{1}{2}\frac{mr^2}{(2i-1)n+(i-1)m}\frac{d}{dz}(\frac{du_i}{dx}+\frac{dv_i}{dy}+\frac{dw_i}{dz})\}$$

where u_i, v_i, w_i denote any spherical harmonics of degree i

For the analytical investigations that follow, it is convenient to introduce the following abbreviations:— *Complete harmonic solution of equations of interior equilibrium.*

$$M_i = \tfrac{1}{2}\frac{m}{(2i-1)n+(i-1)m} \qquad (12),$$

and

$$\psi_{i-1} = \frac{du_i}{dx} + \frac{dv_i}{dy} + \frac{dw_i}{dz} \qquad (13),$$

so that (11) becomes

$$\left.\begin{aligned} \alpha &= \sum_{i=-\infty}^{i=\infty} \left(u_i - M_i r^2 \frac{d\psi_{i-1}}{dx}\right) \\ \beta &= \sum_{i=-\infty}^{i=\infty} \left(v_i - M_i r^2 \frac{d\psi_{i-1}}{dy}\right) \\ \gamma &= \sum_{i=-\infty}^{i=\infty} \left(w_i - M_i r^2 \frac{d\psi_{i-1}}{dz}\right) \end{aligned}\right\} \qquad (14).$$

(*e*) It is important to remark that the addition to u, v, w respectively of terms $\frac{d\phi}{dx}$, $\frac{d\phi}{dy}$, $\frac{d\phi}{dz}$ (ϕ being any function satisfying $\nabla^2\phi=0$), does not alter the equation (10). This allows us at once to write down as follows the solution of the problem for the solid sphere with surface displacement given.

Let a be the radius of the sphere, and let the arbitrarily given values of the three components of displacement for every point of the surface be expressed [App. B. (52)] by series of surface harmonics, ΣA_i, ΣB_i, ΣC_i, respectively. The solution is *Solid sphere with surface displacements given.*

$$\left.\begin{aligned} \alpha &= \sum_{i=0}^{i=\infty} \left\{A_i\left(\frac{r}{a}\right)^i + \frac{m(a^2-r^2)}{2a^i[(2i-1)n+(i-1)m]}\frac{d\Theta_{i-1}}{dx}\right\} \\ \beta &= \sum_{i=0}^{i=\infty} \left\{B_i\left(\frac{r}{a}\right)^i + \frac{m(a^2-r^2)}{2a^i[(2i-1)n+(i-1)m]}\frac{d\Theta_{i-1}}{dy}\right\} \\ \gamma &= \sum_{i=0}^{i=\infty} \left\{C_i\left(\frac{r}{a}\right)^i + \frac{m(a^2-r^2)}{2a^i[(2i-1)n+(i-1)m]}\frac{d\Theta_{i-1}}{dz}\right\} \\ \text{where}\quad \Theta_{i-1} &= \frac{d(A_i r^i)}{dx} + \frac{d(B_i r^i)}{dy} + \frac{d(C_i r^i)}{dz} \end{aligned}\right\} \qquad (15).$$

For this is what (11) becomes if we take

$$u_i = A_i\left(\frac{r}{a}\right)^i + \frac{m}{2a^i[(2i+3)n+(i+1)m]}\frac{d\Theta_{i+1}}{dx}, \; v_i = \text{etc. etc.},$$

and it makes

$$\alpha=\Sigma A_i,\ \beta=\Sigma B_i,\ \gamma=\Sigma C_i,\ \text{when } r=a \qquad (16).$$

This result might have been obtained, of course, by a purely analytical process; and we shall fall on it again as a particular case of the following :—

Shell with given displacements of its outer and inner surfaces.

(*f*) The problem for a shell with displacements given arbitrar for all points of each of its concentric spherical bounding faces is much more complicated, and we shall find a pur analytical process the most convenient for getting to its solu Let a and a' be the radii of the outer and inner spherical surfaces, and let ΣA_i, etc., $\Sigma A'_i$, etc., be the series of surface harmo expressing [App. B. (52)] the arbitrarily given components displacement over them; so that our surface conditions are

$$\left.\begin{array}{l}\alpha=\Sigma A_i\\ \beta=\Sigma B_i\\ \gamma=\Sigma C_i\end{array}\right\}\text{when } r=a;\ \text{and}\ \left.\begin{array}{l}\alpha=\Sigma A'_i\\ \beta=\Sigma B'_i\\ \gamma=\Sigma C'_i\end{array}\right\}\text{when } r=a' \Bigg\}\quad (17)$$

Using the abbreviated notation (12) and (13), selecting from all terms of α which become surface harmonics of order i f constant value of r, and equating to the proper harmonic te of (17), we have

$$u_i+u_{-i-1}-r^2\left(M_{i+2}\frac{d\psi_{i+1}}{dx}+M_{-i+1}\frac{d\psi_{-i}}{dx}\right)\left\{\begin{array}{ll}=A_i \text{ when } & r=a\\ =A'_i \quad " & r=a'\end{array}\right\}\ (1$$

Remarking that $r^{-i}u_i$, $r^{i+1}u_{-i-1}$, $r^{-i}\frac{d\psi_{i+1}}{dx}$, and $r^{i+1}\frac{d\psi_{-i}}{dx}$ ar each of them independent of r, we have immediately from (1 the following two equations towards determining these fu functions:—

$$a^i(r^{-i}u_i)+a^{-i-1}(r^{i+1}u_{-i-1})-a^2\left[M_{i+2}a^i\left(r^{-i}\frac{d\psi_{i+1}}{dx}\right)+M_{-i+1}a^{-i-1}\left(r^{i+1}\frac{d\psi_{-i}}{dx}\right)\right]=A_i$$

and

$$a'^i(r^{-i}u_i)+a'^{-i-1}(r^{i+1}u_{-i-1})-a'^2\left[M_{i+2}a'^i\left(r^{-i}\frac{d\psi_{i+1}}{dx}\right)+M_{-i+1}a'^{-i-1}\left(r^{i+1}\frac{d\psi_{-i}}{dx}\right)\right]=A'_i$$

These, and the symmetrical equations relative to y and z, suffice with (13), for the determination of u_i, v_i, w_i for every valu positive and negative, of i. The most convenient order of pro cedure is first to find equations for the determination of the functions by the elimination of the u, v, w, thus:—From (1 we have

$$u_i=\frac{(a^{2i+3}-a'^{2i+3})M_{i+2}\frac{d\psi_{i+1}}{dx}+(a^2-a'^2)M_{-i+1}r^{2i+1}\frac{d\psi_{-i}}{dx}+(a^{i+1}A_i-a'^{i+1}A'_i)r^i}{a^{2i+1}-a'^{2i+1}}$$

$$u_{-i-1}=\frac{-(aa')^{2i+1}(a^2-a'^2)M_{i+2}r^{-2i-1}\frac{d\psi_{i+1}}{dx}+(aa')^2(a^{2i-1}-a'^{2i-1})M_{-i+1}\frac{d\psi_{-i}}{dx}+(aa')^{i+1}(a'^iA\ldots}{a^{2i+1}-a'^{2i+1}}$$

and symmetrical equations for v and w. Or if, for brevity, w put

$$\mathfrak{A}_i=\frac{a^{i+1}A_i-a'^{i+1}A'_i}{a^{2i+1}-a'^{2i+1}},\quad \mathfrak{A}'_i=\frac{(aa')^{i+1}(a^iA'_i-a'^iA_i)}{a^{2i+1}-a'^{2i+1}} \qquad (21),$$

Shell with given displacements of its outer and inner surfaces.

and

$$\mathfrak{M}_{i+2}=\frac{a^{2i+3}-a'^{2i+3}}{a^{2i+1}-a'^{2i+1}}M_{i+2},\quad \mathfrak{N}_{i+2}=\frac{(aa')^{2i+1}(a^2-a'^2)}{a^{2i+1}-a'^{2i+1}}M_{i+2} \qquad (22),$$

$$\left.\begin{aligned} u_i&=\mathfrak{M}_{i+2}\frac{d\psi_{i+1}}{dx}-\mathfrak{N}_{-i+1}r^{2i+1}\frac{d\psi_{-i}}{dx}+\mathfrak{A}_ir^i\\ u_{-i-1}&=-\mathfrak{N}_{i+2}r^{-2i-1}\frac{d\psi_{i+1}}{dx}+\mathfrak{M}_{-i+1}\frac{d\psi_{-i}}{dx}+\mathfrak{A}'_ir^{-i-1}\\ v_i&=\text{etc.},\ v_{-i-1}=\text{etc.},\ w_i=\text{etc.},\ w_{-i-1}=\text{etc.}\end{aligned}\right\} \qquad (23).$$

Performing the proper differentiations and summations to eliminate the u, v, w, functions between these (23) and (13), and taking advantage of the properties of the ψ functions, that

$$\Delta^2\psi_{i+1}=0,\ \Delta^2\psi_{-i}=0,\ x\frac{d\psi_{i+1}}{dx}+y\frac{d\psi_{i+1}}{dy}+z\frac{d\psi_{i+1}}{dz}=(i+1)\psi_{i+1}$$

$$x\frac{d\psi_{-i}}{dx}+y\frac{d\psi_{-i}}{dy}+z\frac{d\psi_{-i}}{dz}=-i\psi_{-i},$$

we find

$$\psi_{i-1}=(2i+1)i\,\mathfrak{N}_{-i+1}r^{2i-1}\psi_{-i}+\frac{d(\mathfrak{A}_ir^i)}{dx}+\frac{d(\mathfrak{B}_ir^i)}{dy}+\frac{d(\mathfrak{C}_ir^i)}{dz}$$

and

$$=(2i+1)(i+1)\mathfrak{N}_{i+2}r^{-2i-3}\psi_{i+1}+\frac{d(\mathfrak{A}'_ir^{-i-1})}{dx}+\frac{d(\mathfrak{B}'_ir^{-i-1})}{dy}+\frac{d(\mathfrak{C}'_ir^{-i-1})}{dz} \qquad (24).$$

Changing i into $i+1$ in the first of these, and into $i-1$ in the second, we have two equations for the two unknown quantities ψ_i and ψ_{-i-1}; which give

$$\left.\begin{aligned}\psi_i&=\frac{\Theta_i+(2i+3)(i+1)\mathfrak{N}_{-i}\Theta'_{-i-1}r^{2i+1}}{1-(2i+3)(2i-1)(i+1)i\mathfrak{N}_{-i}\mathfrak{N}_{i+1}}\\ \psi_{-i-1}&=\frac{(2i-1)i\mathfrak{N}_{i+1}\Theta_ir^{-2i-1}+\Theta'_{-i-1}}{1-(2i+3)(2i-1)(i+1)i\mathfrak{N}_{-i}\mathfrak{N}_{i+1}}\end{aligned}\right\} \qquad (25),$$

where, for brevity

$$\Theta_i=\frac{d(\mathfrak{A}_{i+1}r^{i+1})}{dx}+\frac{d(\mathfrak{B}_{i+1}r^{i+1})}{dy}+\frac{d(\mathfrak{C}_{i+1}r^{i+1})}{dz}$$

and

$$\Theta'_{-i-1}=\frac{d(\mathfrak{A}'_{-i}r^{-i})}{dx}+\frac{d(\mathfrak{B}'_{-i}r^{-i})}{dy}+\frac{d(\mathfrak{C}'_{-i}r^{-i})}{dz} \qquad (26).$$

The functions ψ_i and ψ_{-i-1} for every value of i being thus given, (23) and (14) complete the solution of the problem.

(*g*) The composition of this solution ought to be carefully studied. Thus separating for simplicity the part due to the terms A_i, etc., A'_i, etc., of the single order i, in the surface data, we see that were there no such terms of other orders, all the functions would vanish except ψ_{i-1}, ψ_{i+1}, ψ_{-i}, ψ_{-i-2}. This would give u_{i-2}, u_i, u_{i+2}, u_{-i+1}, u_{-i-1}, and u_{-i-3}; with symmetrical expressions for the v and w functions; of which the composition will be best studied by first writing them out in full, explicitly in terms of $\mathfrak{A}_i$, $\mathfrak{B}_i$, $\mathfrak{C}_i$, $\mathfrak{A}'_i$, $\mathfrak{B}'_i$, $\mathfrak{C}'_i$, and the derived solid harmonics Θ_{i-1} and Θ'_{-i-2}.

Surface-tractions given.

737. When, instead of surface displacements, the force applied over the surface is given, the problem whether for the solid sphere or the shell, is longer because of the preliminary process (*h*), required to express the components of traction on any spherical surface concentric with the given sphere or shell in proper harmonic forms; and its solution is more complicated because of the new solid harmonic function ϕ_{i+1} [(32) below] which, besides the function ψ_{i-1} employed above, we are obliged to introduce in this preliminary process.

(*h*) Taking F, G, H to denote the components of the traction on the spherical surface of any radius r, having its centre at the origin of co-ordinates, instead of merely for the boundary of the body as supposed formerly in § 734 (3), we have still the same formulæ: but in them we have now to put $f=\frac{x}{r}$, $g=\frac{y}{r}$, $h=\frac{z}{r}$. By grouping their terms conveniently, we may, with the notation (28), put them into the following abbreviated forms:—

Component-tractions on any spherical surface concentric with origin.

$$\left.\begin{aligned}
Fr&=(m-n)\delta.x+n\{(r\frac{\partial}{dr}-1)\alpha+\frac{d\zeta}{dx}\}\\
Gr&=(m-n)\delta.y+n\{(r\frac{\partial}{dr}-1)\beta+\frac{d\zeta}{dy}\}\\
Hr&=(m-n)\delta.z+n\{(r\frac{\partial}{dr}-1)\gamma+\frac{d\zeta}{dz}\}
\end{aligned}\right\}\quad(27)$$

where
$$\left.\begin{aligned}
\zeta&=\alpha x+\beta y+\gamma z\\
\text{and}\quad r\frac{\partial}{dr}&=x\frac{d}{dx}+y\frac{d}{dy}+z\frac{d}{dz}
\end{aligned}\right\}\quad(28)$$

so that $\frac{\zeta}{r}$ is the radial component of the displacement at any point, and $\frac{\partial}{dr}$ prefixed to any function of x, y, z denotes the rate of its variation per unit of length in the radial direction.

(*k*) To reduce these expressions to surface harmonics, let us consider homogeneous terms of degree i of the complete solution:

(14), which we shall denote[1] by α_i, β_i, γ_i, and let δ_{i-1}, ζ_{i+1} denote the corresponding terms of the other functions. Thus we have

Component-tractions on any spherical surface concentric with origin,

$$\left.\begin{aligned} Fr&=\Sigma\{(m-n)\delta_{i-1}x+n(i-1)\alpha_i+n\frac{d\zeta_{i+1}}{dx}\},\\ Gr&=\Sigma\{(m-n)\delta_{i-1}y+n(i-1)\beta_i+n\frac{d\zeta_{i+1}}{dy}\},\\ Hr&=\Sigma\{(m-n)\delta_{i-1}z+n(i-1)\gamma_i+n\frac{d\zeta_{i+1}}{dz}\}.\end{aligned}\right\} \quad (29).$$

(*l*) The second of the three terms of order *i* in these equations, when the general solution of § 13 is used, become at the boundary each explicitly the sum of two surface harmonics of orders *i* and $i-2$ respectively. To bring the other parts of the expressions to similar forms, it is convenient that we should first express ζ_{i+1} in terms of the general solution (12) of § 13, by selecting the terms of algebraic degree *i*. Thus we have

$$\alpha_i=u_i-\frac{mr^2}{2[(2i-1)n+(i-1)m]}\frac{d\psi_{i-1}}{dx} \quad (30),$$

and symmetrical expressions for β_i and γ_i, from which we find

$$\alpha_ix+\beta_iy+\gamma_iz=\zeta_{i+1}=u_ix+v_iy+w_iz-\frac{(i-1)mr^2\psi_{i-1}}{2[(2i-1)n+(i-1)m]}.$$

Hence, by the proper formulæ [see (36) below] for reduction to harmonics,

$$\zeta_{i+1}=-\frac{1}{2i+1}\left\{\frac{(2i-1)[(i-1)m-2n]}{2[(2i-1)n+(i-1)m]}r^2\psi_{i-1}+\phi_{i+1}\right\} \quad (31),$$

where

$$\phi_{i+1}=r^{2i+3}\left\{\frac{d(u_ir^{-2i-1})}{dx}+\frac{d(v_ir^{-2i-1})}{dy}+\frac{d(w_ir^{-2i-1})}{dz}\right\} \quad (32),$$

and (as before assumed in § 12)

$$\psi_{i-1}=\frac{du_i}{dx}+\frac{dv_i}{dy}+\frac{dw_i}{dz} \quad (33).$$

Also, by (10) of § 736, or directly from (30) by differentiation, we have

$$\delta_{i-1}=\frac{n(2i-1)}{(2i-1)n+(i-1)m}\cdot\psi_{i-1} \quad (34).$$

Substituting these expressions for δ_{i-1}, α_i, and ζ_{i+1} in (29), we find

$$Fr=\Sigma\left\{n(i-1)u_i+\frac{n(2i-1)[(i+2)m-(2i-1)n]}{(2i+1)[(2i-1)n+(i-1)m]}x\psi_{i-1}-\right.$$
$$\left.\frac{n[2i(i-1)m-(2i-1)n]}{(2i+1)[(2i-1)n+(i-1)m]}r^2\frac{d\psi_{i-1}}{dx}-\frac{n}{2i+1}\frac{d\phi_{i+1}}{dx}\right\} \quad (35).$$

harmonically expressed.

[1] The suffixes now introduced have reference solely to the algebraic degree, positive or negative, of the functions, whether harmonic or not, of the symbols to which they are applied.

Component-tractions on any spherical surface concentric with origin, expressed in terms of r and harmonics.

This is reduced to the required harmonic form by the obviously proper formula

$$x\psi_{i-1}=\frac{1}{2i-1}\left\{r^2\frac{d\psi_{i-1}}{dx}-r^{2i+1}\frac{d(\psi_{i-1}r^{-2i+1})}{dx}\right\} \quad (3\ldots)$$

Thus, and dealing similarly with the expressions for Gr and Hr, we have, finally,

$$Fr=n\Sigma\left\{(i-1)u_i-2(i-2)M_i r^2\frac{d\psi_{i-1}}{dx}-E_i r^{2i+1}\frac{d(\psi_{i-1}r^{-2i+1})}{dx}-\frac{1}{2i+1}\frac{d\phi_{i+1}}{dx}\right\}$$

$$Gr=n\Sigma\left\{(i-1)v_i-2(i-2)M_i r^2\frac{d\psi_{i-1}}{dy}-E_i r^{2i+1}\frac{d(\psi_{i-1}r^{-2i+1})}{dx}-\frac{1}{2i+1}\frac{d\phi_{i+1}}{dy}\right\}$$

$$Hr=n\Sigma\left\{(i-1)w_i-2(i-2)M_i r^2\frac{d\psi_{i-1}}{dz}-E_i r^{2i+1}\frac{d(\psi_{i-1}r^{-2i+1})}{dx}-\frac{1}{2i+1}\frac{d\phi_{i+1}}{dz}\right\}$$

where [as above (12)],

$$M_i=\tfrac{1}{2}\frac{m}{(2i-1)n+(i-1)m}$$

and now, further,

$$E_i=\frac{(i+2)m-(2i-1)n}{(2i+1)[(2i-1)n+(i-1)m]} \quad (3\ldots)$$

Prescribed surface conditions put into harmonics.

(m) To express the surface conditions by harmonic equations for the shell bounded by the concentric spherical surfaces $r=a$, $r=a'$, let us suppose the superficial values of F, G, H to be given as follows:—

$$\left.\begin{array}{l}F=\Sigma A_i\\ G=\Sigma B_i\\ H=\Sigma C_i\end{array}\right\}\text{when } r=a$$

and

$$\left.\begin{array}{l}F=\Sigma A'_i\\ G=\Sigma B'_i\\ H=\Sigma C'_i\end{array}\right\}\text{when } r=a' \quad (3\ldots)$$

where A_i, B_i, C_i, A'_i, B'_i, C'_i denote surface harmonics of order i.

To apply to this harmonic development the conditions § 734 (2) to which the surface traction is subject, let $a^2d\varpi$ and $a'^2d\varpi$ be elements of the outer and inner spherical surfaces subtending at the centre (§ 468) a common infinitesimal solid angle $d\varpi$: and let $\iint d\varpi$ denote integration over the whole spherical surface of unit radius. Equations (2) become

Equations of equilibrium to which the surface-tractions are subject.

$$\iint d\varpi\,\Sigma(a^2A_i-a'^2A'_i)=0,\text{ etc.; and }\iint d\varpi\,[y\Sigma(a^2C_i-a'^2C'_i)-z\Sigma(a^2B_i-a'^2B'_i)]=0,$$

Now App. B. (16) shows that, of the first three of these, all terms except the first (those in which $i=0$) vanishes; and that of the second three all the terms except the second (those for which

$i=1$) vanish because x, y, z are harmonics of order 1. Thus the first three become

Limitations imposed on the otherwise arbitrary harmonic data of surface-tractions, for their equilibrium.

$$\iint d\varpi(a^2A_0-a'^2A'_0), \text{ etc.};$$

which, as A_0, A'_0, etc., are constants, require simply that

$$a^2A_0=a'^2A'_0,\quad a^2B_0=a'^2B'_0,\quad a^2C_0=a'^2C'_0 \qquad (41).$$

The second three are equivalent to

$$r(a^2A_1-a'^2A'_1)=\frac{dH_2}{dx},\; r(a^2B_1-a'^2B'_1)=\frac{dH_2}{dy},\; r(a^2C_1-a'^2C'_1)=\frac{dH_2}{dz} \quad (42),$$

where H_2 is a homogeneous function of x, y, z of the second degree. For [App. B. (a)] rA_1, rA'_1, etc., are linear functions of x, y, z. If therefore (A, x), $(A, y) \ldots (B, x) \ldots$ denote nine constants, we have

$$r(a^2A_1-a'^2A'_1)=(A,x)x+(A,y)y+(A,z)z$$
$$r(a^2B_1-a'^2B'_1)=(B,x)x+(B,y)y+(B,z)z$$
$$r(a^2C_1-a'^2C'_1)=(C,x)x+(C,y)y+(C,z)z.$$

Using these in the second three of (40) of which, as remarked above, all terms except those for which $i=1$ disappear, and remarking that yz, zx, xy are harmonics, and therefore (App. B. (16)] $\iint yz d\varpi=0, \iint zx d\varpi=0, \iint xy d\varpi=0,$

we have $(C,y)\iint y^2 d\varpi-(B,z)\iint z^2 d\varpi=0$: etc.

From these, because $\iint x^2 d\varpi=\iint y^2 d\varpi=\iint z^2 d\varpi$,

it follows that

$$(C,y)=(B,z),\quad (A,z)=(C,x),\quad (B,x)=(A,y),$$

which prove (42).

(n) The terms of algebraic degree i, exhibited in the preceding expressions (37) for Fr, Gr, Hr, become, at either of the concentric spherical surfaces, sums of surface harmonics of orders i and $i-2$, when i is positive, and of orders $-i-1$ and $-i-3$ when i is negative. Hence, selecting all the terms which lead to surface harmonics of order i, and equating to the proper terms of the data (39), we have

Surface conditions expressed in harmonic equations.

$$\left.\begin{array}{l}\left.\begin{array}{l}\ldots 1)u_i-(i+2)u_{-i-1}-2iM_{i+2}r^2\frac{d\psi_{i+1}}{dx}+2(i+1)M_{-i+1}r^2\frac{d\psi_{-i}}{dx}\\ \ldots E_i r^{2i+1}\frac{d(\psi_{i-1}r^{-2i+1})}{dx}-E_{-i-1}r^{-2i-1}\frac{d(\psi_{-i-2}r^{2i+3})}{dx}-\frac{1}{2i+1}\left(\frac{d\phi_{i+1}}{dx}-\frac{d\phi_{-i}}{dx}\right)\end{array}\right\}\\ \qquad =\left\{\begin{array}{l}A_i \text{ when } r=a\\ A'_i \text{ when } r=a'\end{array}\right\}\end{array}\right\} (43),$$

and symmetrical equations relative to y and z.

(o) These equations are to be treated precisely on the same plan as formerly were (18). Thus after finding u_i and u_{-i-1}; we perform on u_i, v_i, w_i the operations of (33), and on u_{-i-1}, v_{-i-1},

Surface-tractions given: general solution; for spherical shell;

w_{-i-1} those of (32), and so arrive at two equations which involve of unknown quantities only ψ_{i-1}, ψ_{-i}, and ϕ_{-i}, and taking the corresponding expressions for u_{i-2}, u_{-i+1}, and applying (32) to u_{i-2}, v_{i-2}, w_{i-2}, and (33) to u_{-i+1}, v_{-i+1}, w_{-i+1} we similarly obtain two equations between ϕ_{i-1}, ψ_{i-1}, and ψ_{-i}. Thus we have in all four simple algebraic equations between ψ_{i-1}, ψ_{-i}, ϕ_{i-1}, ϕ_{-i}, by which we find these four unknown functions: and the u, v, w functions having been already explicitly expressed in terms of them, we thus have, in terms of the data of the problem, every unknown function that appears in (14) its solution.

for solid sphere.

(p) The case of the solid sphere is of course fallen on from the more general problem of the shell, by putting $a'=0$. But if we begin with only contemplating it, we need not introduce any solid harmonics of negative degree (since every harmonic of negative degree becomes infinite at the centre, and therefore is inadmissible in the expression of effects produced throughout a solid sphere by action at its surface); and (43), and all the formulæ described as deducible from it, become much shortened when we thus confine ourselves to this case. Thus, instead of (43), we now have simply

$$\frac{n}{r}\left\{(i-1)u_i-2iM_{i+2}r^2\frac{d\psi_{i+1}}{dx}-E_ir^{2i+1}\frac{d(\psi_{i-1}r^{-2i+1})}{dx}-\frac{1}{2i+1}\frac{d\phi_{i+1}}{dx}\right\}=A_i,$$

when $r=a$

Hence, attending [as formerly in (f)] to the property of a homogeneous function H_j, of any order j, that $r^{-j}H_j$ is independent of r, and depends only on the ratios $\frac{x}{r}$, $\frac{y}{r}$, $\frac{z}{r}$; we have for all values of x, y, z,

$$(i-1)u_i-2iM_{i+2}a^2\frac{d\psi_{i+1}}{dx}-E_ir^{2i+1}\frac{d(\psi_{i-1}r^{-2i+1})}{dx}-\frac{1}{2i+1}\frac{d\phi_{i+1}}{dx}=\frac{A_ir^i}{na^{i-1}}$$

From this and the symmetrical equations for v and w, we have by (33),

$$[i-1+(2i+1)iE_i]\psi_{i-1}=\frac{1}{na^{i-1}}\left\{\frac{d(A_ir^i)}{dx}+\frac{d(B_ir^i)}{dy}+\frac{d(C_ir^i)}{dz}\right\} \quad (4$$

and by (32)

$$2i\phi_{i+1}+2i(i+1)(2i+1)M_{i+2}a^2\psi_{i+1}=\frac{r^{2i+3}}{na^{i-1}}\left\{\frac{d(A_ir^{-i-1})}{dx}+\frac{d(B_ir^{-i-1})}{dy}+\frac{d(C_ir^{-}}{d}\right.$$

Eliminating, by this, ϕ_{i+1} from (45), and introducing the abbreviated notation, Φ_{i+1} [(50) below], we find

$$(i-1)u_i=(i-1)M_{i+2}a^2\frac{d\psi_{i+1}}{dx}+E_ir^{2i+1}\frac{d(\psi_{i-1}r^{-2i+1})}{dx}+\frac{1}{na^{i-1}}\left[A_ir^i+\frac{1}{2i(2i+1)}\right.$$

and (43) gives

Surface-tractions given: general solution; for solid sphere.

$$\psi_{i-1}=\frac{\Psi_{i-1}}{[(i-1)+(2i+1)iE_i]na^{i-1}}=\frac{[(i-1)m+(2i-1)n]\Psi_{i-1}}{[(2i^2+1)m-(2i-1)n]na^{i-1}} \quad (49),$$

where

$$\left.\begin{aligned}\Psi_{i-1}&=\frac{d(A_ir^i)}{dx}+\frac{d(B_ir^i)}{dy}+\frac{d(C_ir^i)}{dz}\\ \text{and }\Phi_{i+1}&=r^{2i+3}\left\{\frac{d(A_ir^{-i-1})}{dx}+\frac{d(B_ir^{-i-1})}{dy}+\frac{d(C_ir^{-i-1})}{dz}\right\}\end{aligned}\right\} \quad (50).$$

With these expressions for ψ_i and u_i, (14) is the complete solution of the problem.

(*q*) The composition and character of this solution is made manifest by writing out in full the terms in it which depend on harmonics of a single order, i, in the surface data. Thus if the components of the surface traction are simply A_i, B_i, C_i, all the Ψ functions except Ψ_{i-1}, and all the Φ functions except Φ_{i+1} vanish. Hence (48) shows that all the u functions except u_{i-2}, and u_i vanish: and for these it gives

$$\left.\begin{aligned}u_{i-2}&=M_ia^2\frac{d\psi_{i-1}}{dx}\\ u_i&=\frac{1}{i-1}\left\{E_ir^{2i+1}\frac{d(\psi_{i-1}r^{-i+1})}{dx}+\frac{1}{na^{i-1}}\left[A_ir^i+\frac{1}{2i(2i+1)}\frac{d\Phi_{i+1}}{dx}\right]\right\}\end{aligned}\right\} \quad (51).$$

Using this in (14) and for E_i and M_i substituting their values by (38), we have, explicitly expressed in terms of the data, and the solid harmonics Ψ_{i-1}, Φ_{i+1}, derived from the data according to the formulæ (50), the final solution of the problem as follows:—

$$\ldots\left\{\tfrac{1}{2}\frac{m(a^2-r^2)}{(2i^2+1)m-(2i-1)n}\frac{d\Psi_{i-1}}{dx}+\frac{1}{i-1}\left[\frac{(i+2)m-(2i-1)n}{(2i^2+1)m-(2i-1)n}\frac{r^{2i+1}d(\Psi_{i-1}r^{-2i+1})}{(2i+1)dx}+\frac{1}{2i(2i+1)}\frac{d\Phi_{i+1}}{dx}+A_ir^i\right]\right\} \quad (52),$$

with symmetrical expressions for β and γ.

Case of homogeneous strain.

(*r*) The case of $i=1$ is interesting, inasmuch as it seems at first sight to make the second part of the expression (52) for α infinite because of the divisor $i-1$. But the terms within the brackets [ ] vanish for $i=1$, owing to the relations (42) proved above, which, for the solid sphere, become

$$rA_1=\frac{dH_2}{dx},\quad rB_1=\frac{dH_2}{dy},\quad rC_1=\frac{dH_2}{dz} \quad (53),$$

Indeterminate rotations without strain, necessarily included in general solution for displacement, when the data are merely of force.

H_2 denoting any homogeneous function of x, y, z of the second degree. The verification of this presents no difficulty, and we leave it as an exercise to the student. The true interpretation of the $\frac{0}{0}$ appearing thus in the expressions for α, β, γ is clearly that they are indeterminate: and that they ought to be so, we see by remarking that an infinitesimal rotation round any diameter

Indeterminate rotations without strain, necessarily included in general solution for displacement, when the data are merely of force.

without strain may be superimposed on any solution without violating the conditions of the problem; in other words (§§ 89, 95),

$$\omega_2 z - \omega_3 y,\ \omega_3 x - \omega_1 z,\ \omega_1 y - \omega_2 x$$

may be added to the expressions for α, β, γ in any solution, and the result will still be a solution.

But though α, β, γ are indeterminate, (50) gives ψ_0 and ϕ_2 determinately. The student will find it a good and simple exercise to verify that the determination of ψ_0 and ϕ_2 determines the state of strain [homogeneous (§ 155) of course in this case] actually produced by the given surface traction.

Plane strain defined.

738. A solid is said (§ 730) to experience a plane strain, or to be strained in two dimensions, when it is strained in a manner subject to the condition that the displacements are all in a set of parallel planes, and are equal and parallel for all points in any line perpendicular to these planes: and any one of these planes may be called the plane of the strain. Thus in plane strain, all cylindrical surfaces perpendicular to the plane of the strain remain cylindrical surfaces perpendicular to the same plane, and nowhere experience stretching along their generating lines.

The condition of plane strain expressed analytically, if we take XOY for the plane, is that γ must vanish, and that α and β must be functions of x and y, without z. Thus we see that

Only two independent variables enter into the analytical expression of plane strain; and thus this case presents a class of problems of peculiar simplicity. For instance, if an infinitely long solid or hollow circular cylinder is the "given solid" of § 696, and if the bodily force (if any) and the surface action consist of forces and tractions everywhere perpendicular to the axis, and equal and parallel at all points of any line parallel to its axis, we have, whether surface displacement or surface traction be given, problems precisely analogous to those of §§ 735, 736, but much simpler, and obviously of very great practical importance in the engineering of long straight tubes under strain.

Problem for cylinders under plane strain,

739. It is interesting to remark, that in these cylindrical problems, instead of surface harmonics of successive orders 1, 2, 3, etc., which are [App. B. (*b*)] functions of spherical surface co-ordinates (as, for instance, latitude and longitude on a globe), we have simple harmonic functions (§§ 54, 75) of the

solved in terms of "plane harmonics."

Plane harmonic functions defined.

me degrees, of the angle between two planes through the .is, and of its successive multiples: and instead of solid ırmonic functions [App. B. (*a*) and (*b*)], we have what we ɔy call *plane harmonic functions*, being the algebraic functions two variables (x, y), which we find by expanding $\cos i\theta$ and n $i\theta$ in powers of sines or cosines of θ, taking

$$\cos\theta = \frac{x}{\sqrt{(x^2+y^2)}}, \text{ and } \sin\theta = \frac{y}{\sqrt{(x^2+y^2)}},$$

ıd multiplying the result by $(x^2+y^2)^{\frac{i}{2}}$.

A plane harmonic function is of course the particular case of a solid harmonic [App. B. (*a*) and (*b*)] in which z does not appear; that is to say, it is any homogeneous function, V, of x and y, which satisfies the equation

$$\frac{d^2V}{dx^2}+\frac{d^2V}{dy^2}=0, \text{ or, as we may write it for brevity, } \nabla^2 V=0.$$

And, as we have seen [§ 707 (23)], the most general expression for a plane harmonic of degree ι (positive or negative, integral or fractional) is

$$\left.\begin{array}{l}\tfrac{1}{2}\{(x+y\sqrt{-1})^{\iota}+(x-y\sqrt{-1})^{\iota}\}-\tfrac{1}{2}B\sqrt{-1}\{(x+y\sqrt{-1})^{\iota}-(x-y\sqrt{-1})^{\iota}\} \\ \text{or in polar co-ordinates, } A\cos\iota\theta+B\sin\iota\theta\end{array}\right\}(1),$$

equations of internal equilibrium [§ 698 (6)] with no bodily force (that is, $X=0$ and $Y=0$) become, for the case of plane strain,

$$\left.\begin{array}{l} n\left(\dfrac{d^2\alpha}{dx^2}+\dfrac{d^2\alpha}{dy^2}\right)+m\dfrac{d}{dx}\left(\dfrac{d\alpha}{dx}+\dfrac{d\beta}{dy}\right)=0 \\ n\left(\dfrac{d^2\beta}{dx^2}+\dfrac{d^2\beta}{dy^2}\right)+m\dfrac{d}{dy}\left(\dfrac{d\alpha}{dx}+\dfrac{d\beta}{dy}\right)=0 \end{array}\right\} \quad (2).$$

Problem for cylinders under plane strain solved in terms of plane harmonics.

The plane harmonic solution of these, found by precisely the same process as § 735 (*a*)...(*e*), but for only two variables instead of three, is

$$\left.\begin{array}{l} \alpha=\Sigma\left\{u_i-\dfrac{m}{2(i-1)(2n+m)}r^2\dfrac{d\psi_{i-1}}{dx}\right\} \\ \beta=\Sigma\left\{v_i-\dfrac{m}{2(i-1)(2n+m)}r^2\dfrac{d\psi_{i-1}}{dy}\right\} \\ \text{where} \qquad \psi_{i-1}=\dfrac{du_i}{dx}+\dfrac{dv_i}{dy} \end{array}\right\} \quad (3),$$

and u_i, v_i denote any two plane harmonics of degree i, so that ψ_{i-1} is a plane harmonic of degree $i-1$. Of course i may be positive or negative, integral or fractional.

This solution may be reduced to polar co-ordinates with advantage for many applications, by putting

Problem for cylinders under plane strain solved in terms of plane harmonics.

$$\left.\begin{array}{l} x=r\cos\theta,\ y=r\sin\theta, \\ u_i=r^i(A_i\cos i\theta+A_i'\sin i\theta) \\ v_i=r^i(B_i\cos i\theta+B_i'\sin i\theta) \end{array}\right\}$$

and taking

which give

$$\psi_{i-1}=ir^{i-1}\{(A_i+B_i')\cos(i-1)\theta+(A_i'-B_i)\sin(i-1)\theta\}$$

and

$$\alpha=\Sigma r^i\{A_i\cos i\theta+A_i'\sin i\theta-\frac{im}{2(2n+m)}[(A_i+B_i')\cos(i-2)\theta-(A_i'-B_i)\sin(i-2)\theta$$

$$\beta=\Sigma r^i\{B_i\cos i\theta+B_i'\sin i\theta-\frac{im}{2(2n+m)}[-(A_i+B_i')\sin(i-2)\theta+(A_i'-B_i)\cos(i-2)$$

The student will find it a good exercise to work out in f... explicit expressions for the displacement of any point of the s... in the cylindrical problems corresponding to the spherical p... blems of § 735 (f), and of § 736 (h)...(r). The process ... the latter may be worked through in the symmetrical algebr... form, as an illustration of the plan we have followed in deal... with spherical harmonics; but the result corresponding to ... of § 737 may be obtained more readily, and in a simpler for... by immediately putting (29) of § 737 into polar co-ordinates, ... (4), (5), (6) of § 739. We intend to use, and to illustrate, the... solutions under "Properties of Matter."

740. In our sections on hydrostatics, the problem of fin... the deformation produced in a spheroid of incompressible li... by a given disturbing force will be solved; and then we sh... consider the application of the preceding result [§ 736 ... for an elastic solid sphere to the theory of the tides and t... rigidity of the earth. This proposed application, howev... reminds us of a general remark of great practical importa... with which we shall leave elastic solids for the pre... Considering different elastic solids of similar substance ... similar shapes, we see that if by forces applied to them in a... way they are similarly strained, the surface tractions in ... across similarly situated elements of surface, whether of th... boundaries or of surfaces imagined as cutting through t... substances, must be equal, reckoned as usual per unit of a... Hence; the force across, or in, any such surface, being res... into components parallel to any directions; the whole am... of each such component for similar surfaces of the diffe... bodies are in proportion to the *squares* of their lineal di... sions. Hence, if equilibrated similarly under the actio...

Small bodies stronger than large ones in proportion to their weights.

avity, or of their kinetic reactions (§ 264) against equal celerations (§ 28), the greater body would be more strained an the less; as the amounts of gravity or of kinetic reaction similar portions of them are as the *cubes* of their linear mensions. Definitively, the strains at similarly situated ints of the bodies will be in simple proportion to their linear mensions, and the displacements will be as the squares of ese lines, provided that there is no strain in any part of any them too great to allow the principle of superposition to hold th sufficient exactness, and that no part is turned through ore than a very small angle relatively to any other part.

Small bodies stronger than large ones in proportion to their weights.

illustrate by a single example, let us consider a uniform ng, thin, round rod held horizontally by its middle. Let its bstance be homogeneous, of density ρ, and Young's modulus, ; and let its length, l, be p times its diameter. Then (as e moment of inertia of a circular area of radius r round a ameter is $\frac{1}{4}\pi r^4$) the flexural rigidity of the rod will (§ 715) $\frac{M}{4}\pi\left(\frac{l}{2p}\right)^4$, which is equal to $\frac{B}{g}$ in the notation of § 610, B is there reckoned in kinetic or absolute measure (§ 223) stead of the gravitation measure in which we now, according engineers' usage (§ 220) reckon M. Also $w=\rho\pi\left(\frac{l}{2p}\right)^2$, and erefore, for § 617,

Example: a straight rod held horizontally by its middle.

$$\frac{gw}{B}=\frac{16p^2\rho}{Ml^2}.$$

is, used in § 617 (10), gives us; for the curvature at the iddle of the rod; the elongation and contraction where eatest, that is, at the highest and lowest points of the normal ction through the middle point; and the droop of the ends; e following expressions

$$\frac{2p^2\rho}{M};\ \frac{pl\rho}{M};\ \text{and}\ \frac{p^2l^2\rho}{8M}.$$

hus, for a rod whose length is 200 times its diameter, if its bstance be iron or steel, for which $\rho=7{\cdot}75$, and $M=194\times10^7$ ammes per square centimetre, the maximum elongation and ntraction (being at the top and bottom of the middle section here it is held), are each equal to $\cdot8\times10^{-6}\times l$, and the droop its ends $2\times10^{-5}\times l^2$. Thus a steel or iron wire, ten centi-

Stiffness of uniform or steel rods of different dimensions.

metres long, and half a millimetre in diameter, held horizontally by its middle, would experience only ·00006 of maximum elongation and contraction, and only ·002 of a centimetre of droop in its ends: a round steel rod, of half a centimetre diameter, and one metre long, would experience ·00008 of maximum elongation and contraction, and ·2 of a centimetre of droop: a round steel rod, of ten centimetres diameter, and twenty metres long, must be of remarkable temper (see Vol. II., Properties of Matter) to bear being held by the middle without taking a very sensible permanent set: and it is probable that no temper of steel is high enough in a round shaft forty metres long, if only two decimetres in diameter, to allow it to be held by its middle without either bending to some great angle, and beyond all appearance of elasticity, or breaking it.

Transition to hydrodynamics.

741. In passing from the dynamics of perfectly elastic solids to abstract hydrodynamics, or the dynamics of perfect fluids, it is convenient and instructive to anticipate slightly some of the views as to intermediate properties observed in real solids and fluids, which, according to the general plan proposed (§ 449) for our work, will be examined with more detail under *Properties of Matter.*

Imperfectness of elasticity in solids.

By induction from a great variety of observed phenomena, we are compelled to conclude that no change of volume or shape can be produced in any kind of matter without dissipation of energy (§ 275); so that if in any case there is return to the primitive configuration, some amount (however small) of work is always required to compensate the energy dissipated away, and restore the body to the same physical and the same palpably kinetic condition as that in which it was given. We have seen (§ 672), by anticipating something of thermodynamic principles, how such dissipation is inevitable even in dealing with the *absolutely perfect* elasticity of volume presented by every fluid, and possibly by some solids, as, for instance, homogeneous crystals. But in metals, glass, porcelain, natural stones, wood, India-rubber, homogeneous jelly, silk fibre, ivory, etc., a distinct *frictional resistance*[1] against every

[1] See Proceedings of the Royal Society, May 1865, "On the Viscosity and Elasticity of Metals" (W. Thomson).

hange of shape, is, as we shall see in Vol. II., under Pro- erties of Matter, demonstrated by many experiments, and is)und to depend on the speed with which the change of hape is made. A very remarkable and obvious proof of rictional resistance to change of shape in ordinary solids, ; afforded by the gradual, more or less rapid, subsidence of ibrations of elastic solids; marvellously rapid in India-rubber, nd even in homogeneous jelly; less rapid in glass and metal prings, but still demonstrably, much more rapid than can be ccounted for by the resistance of the air. This molecular riction in elastic solids may be properly called *viscosity of olids*, because, as being an internal resistance to change of hape depending on the rapidity of the change, it must be lassed with fluid molecular friction, which by general con- ent is called *viscosity of fluids*. But, at the same time, we eel bound to remark that the word viscosity, as used hitherto y the best writers, when solids or heterogeneous semisolid- emifluid masses are referred to, has not been distinctly applied o molecular friction, especially not to the molecular friction of highly elastic solid within its limits of high elasticity, but as rather been employed to designate a property of slow, con- inual yielding through very great, or altogether unlimited, xtent of change of shape, under the action of continued stress. It is in this sense that Forbes, for instance, has used the word in stating that "Viscous Theory of Glacial Motion" which he lemonstrated by his grand observations on glaciers. As, how- ever, he, and many other writers after him, have used the words plasticity and plastic, both with reference to homogeneous solids (such as wax or pitch, even though also brittle; soft metals; etc.), and to heterogeneous semisolid-semifluid masses (as mud, moist earth, mortar, glacial ice, etc.), to designate the property,[1] common to all those cases, of experiencing, under continued stress either quite continued and unlimited change

Viscosity of solids.

Viscosity of fluids.

Forbes' "Viscous Theory of Glacial Motion."

[1] Some confusion of ideas might have been avoided on the part of writers who have professedly objected to Forbes' theory while really objecting only (and we believe groundlessly) to his usage of the word viscosity, if they had paused to consider that no one physical explanation can hold for those several cases; and that Forbes' theory is merely the proof by observation that glaciers have the property that mud (heterogeneous), mortar (heterogeneous), pitch (homogeneous), water (homogeneous), all have of changing shape indefinitely and continuously under the action of continued stress.

Plasticity of solids.

of shape, or gradually very great change at a diminishing (asymptotic) rate through infinite time; and as the use of the term *plasticity* implies no more than does *viscosity*, any physical theory or explanation of the property, the word viscosity is without inconvenience left available for the definition we have given of it above.

Perfect and unlimited plasticity unopposed by internal friction, the characteristic of the ideal perfect fluid, of abstract hydrodynamics.

742. A *perfect fluid*, or (as we shall call it) a fluid, is an unrealizable conception, like a rigid, or a smooth, body: it is defined as a body incapable of resisting a change of shape: and therefore incapable of experiencing distorting or tangential stress (§ 669). Hence its pressure on any surface, whether of a solid or of a contiguous portion of the fluid, is at every point perpendicular to the surface. In equilibrium, all common liquids and gaseous fluids fulfil the definition. But there is finite resistance, of the nature of friction, opposing change of shape at a finite rate; and, therefore, while a fluid is changing shape, it exerts tangential force on every surface other than normal planes of the stress (§ 664) required to keep this change of shape going on. Hence; although the hydrostatical results to which we immediately proceed, are verified in practice; in treating of hydrokinetics, in a subsequent chapter, we shall be obliged to introduce the consideration of fluid friction, except in cases where the circumstances are such as to render its effects insensible.

Fluid pressure.

743. With reference to a fluid the *pressure at any point in any direction* is an expression used to denote the average pressure per unit of area on a plane surface imagined as containing the point, and perpendicular to the direction in question, when the area of that surface is indefinitely diminished.

744. At any point in a fluid at rest the pressure is the same in all directions: and, if no external forces act, the pressure is the same at every point. For the proof of these and most of the following propositions, we imagine, according to § 564, a definite portion of the fluid to become solid, without changing its mass, form, or dimensions.

Suppose the fluid to be contained in a closed vessel, the pressure within depending on the pressure exerted on it by the vessel, and not on any external force such as gravity.

745. The resultant of the fluid pressures on the elements f any portion of a spherical surface must, like each of its components, pass through the centre of the sphere. Hence, if we uppose (§ 564) a portion of the fluid in the form of a plano-onvex lens to be solidified, the resultant pressure on the plane ide must pass through the centre of the sphere; and, therefore, eing perpendicular to the plane, must pass through the centre f the circular area. From this it is obvious that the pressure s the same at all points of any plane in the fluid. Hence, y § 561, the resultant pressure on any plane surface passes hrough its centre of inertia. **Fluid pressure proved equal in all directions.**

Next, imagine a triangular prism of the fluid, with ends erpendicular to its faces, to be solidified. The resultant ressures on its ends act in the line joining the centres of nertia of their areas, and are equal (§ 551) since the resultant ressures on the sides are in directions perpendicular to this ine. Hence the pressure is the same in all parallel planes.

But the centres of inertia of the three faces, and the resultant ressures applied there, lie in a triangular section parallel to he ends. The pressures act at the middle points of the sides f this triangle, and perpendicularly to them, so that their lirections meet in a point. And, as they are in equilibrium, hey must be, by § 557, proportional to the respective sides of he triangle; that is, to the breadths, or areas, of the faces of he prism. Thus the resultant pressures on the faces must be roportional to the areas of the faces, and therefore the pressure s equal in any two planes which meet.

Collecting our results, we see that the pressure is the same t all points, and in all directions, throughout the fluid mass.

746. One immediate application of this result gives us a imple though indirect proof of the second theorem in § 557, or we have only to suppose the polyhedron to be a solidified ortion of a mass of fluid in equilibrium under pressures only. The resultant pressure on each side will then be proportional o its area, and, by § 561, will act at its centre of inertia; which, n this case, is the *Centre of Pressure*. **Application to statics of solids.** **Centre of pressure.**

747. Another proof of the equality of pressure throughout mass of fluid, uninfluenced by other external force than the ressure of the containing vessel, is easily furnished by the nergy criterion of equilibrium, § 289; but, to avoid complica- **Application of the principle of energy**

Proof by energy of the equality of fluid pressure in all directions.

tion, we will consider the fluid to be incompressible. Suppose a number of pistons fitted into cylinders inserted in the sides of the closed vessel containing the fluid. Then, if A be the area of one of these pistons, p the average pressure on it, x the distance through which it is pressed, in or out; the energy criterion is that no work shall be done on the whole, *i.e.*, that

$$A_1p_1x_1+A_2p_2x_2+\ldots=\Sigma(Apx)=0,$$

as much work being restored by the pistons which are forced out, as is done by those forced in. Also, since the fluid is incompressible, it must have gained as much space by forcing out some of the pistons as it lost by the intrusion of the others. This gives

$$A_1x_1+A_2x_2+\ldots=\Sigma(Ax)=0.$$

The last is the only condition to which x_1, x_2, etc., in the first equation, are subject; and therefore the first can only be satisfied if

$$p_1=p_2=p_3=\text{etc.},$$

that is, if the pressure be the same on each piston. Upon this property depends the action of Brahmah's *Hydrostatic Press*.

If the fluid be compressible, the work expended in compressing it from volume V to $V-\delta V$, at mean pressure p, is $p\delta V$.

If in this case we *assume* the pressure to be the same throughout, we obtain a result consistent with the energy criterion.

The work done on the fluid is $\Sigma(Apx)$, that is, in consequence of the assumption, $p\Sigma(Ax)$.

But this is equal to $p\delta V$,
for, evidently, $\Sigma(Ax)=\delta V$.

Fluid pressure depending on external forces.

748. When forces, such as gravity, act from external matter upon the substance of the fluid, either in proportion to the density of its own substance in its different parts, or in proportion to the density of electricity, or of magnetic polarity, or of any other conceivable accidental property of it, the pressure will still be the same in all directions at any one point, but will now vary continuously from point to point. For the preceding demonstration (§ 745) may still be applied by simply taking the dimensions of the prism small enough; since the pressures are as the squares of its linear dimensions, and the effects of the applied forces such as gravity, as the cubes.

749. When forces act on the whole fluid, surfaces of equal ressure, if they exist, must be at every point perpendicular › the direction of the resultant force. For, any prism of the uid so situated that the whole pressures on its ends are equal ıust (§ 551) experience from the applied forces no component ı the direction of its length; and, therefore, if the prism be so mall that from point to point of it the direction of the reultant of the applied forces does not vary sensibly, this direcion must be perpendicular to the length of the prism. From his it follows that whatever be the physical origin, and the ıw, of the system of forces acting on the fluid, and whether it e conservative or non-conservative, the fluid cannot be in quilibrium unless the lines of force possess the geometrical roperty of being at right angles to a series of surfaces.

Surfaces of equal pressure are perpendicular to the lines of force.

750. Again, considering two surfaces of equal pressure ininitely near one another, let the fluid between them be divided nto columns of equal transverse section, and having their engths perpendicular to the surfaces. The difference of presures on the two ends being the same for each column, the esultant applied forces on the fluid masses composing them nust be equal. Comparing this with § 488, we see that if the pplied forces constitute a conservative system, the density of matter, or electricity, or whatever property of the substance hey depend on, must be equal throughout the layer under onsideration. This is the celebrated hydrostatic proposition hat *in a fluid at rest, surfaces of equal pressure are also surfaces f equal density and of equal potential.*

And are surfaces of equal density and of equal potential when the system of force is conservative.

751. Hence when gravity is the only external force considered, surfaces of equal pressure and equal density are (when of moderate extent) horizontal planes. On this depends the action of levels, syphons, barometers, etc.; also the separation of liquids of different densities (which do not mix or combine chemically) into horizontal strata, etc. etc. The free surface of a liquid is exposed to the pressure of the atmosphere simply; and therefore, when in equilibrium, must be a surface of equal pressure, and consequently level. In extensive sheets of water, such as the American lakes, differences of atmospheric pressure, even in moderately calm weather, often produce considerable deviations from a truly level surface.

Gravity the only external force.

Rate of increase of pressure.

752. The rate of increase of pressure per unit of length in the direction of the resultant force, is equal to the intensity of the force reckoned per unit of volume of the fluid. Let F be the resultant force per unit of volume in one of the columns § 750; p and p' the pressures at the ends of the column, l its length, S its section. We have, for the equilibrium of the column,

$$(p'-p)S=SlF.$$

Hence the rate of increase of pressure per unit of length is F.

If the applied forces belong to a conservative system, for which V and V' are the values of the potential at the ends of the column, we have (§ 486)

$$V'-V=-lF\rho,$$

where ρ is the density of the fluid. This gives

$$p'-p=-\rho(V'-V)$$

or

$$dp=-\rho dV.$$

Hence in the case of gravity as the only impressed force the rate of increase of pressure per unit of depth in the fluid is ρ, in gravitation measure (usually employed in hydrostatics). In kinetic or absolute measure (§ 224) it is $g\rho$.

If the fluid be a gas, such as air, and be kept at a constant temperature, we have $\rho=cp$, where c denotes a constant, the reciprocal of H, the "height of the homogeneous atmosphere," defined (§ 753) below. Hence, in a calm atmosphere of uniform temperature we have

$$\frac{dp}{p}=-cdV;$$

and from this, by integration,

$$p=p_0\epsilon^{-cV}$$

where p_0 is the pressure at any particular level (the sea-level, for instance) where we choose to reckon the potential as zero.

When the differences of level considered are infinitely small in comparison with the earth's radius, as we may practically regard them, in measuring the heights of mountains, or of a balloon, by the barometer, the force of gravity is constant, and therefore differences of potential (force being reckoned in units of weight) are simply equal to differences of level. Hence if x denote height of the level of pressure p above that of p_0, we have, in the preceding formulæ, $V=x$, and therefore

$$p=p_0\epsilon^{-cx};\text{ that is,}$$

Pressure in a calm atmosphere of uniform temperature.

753. If the air be at a constant temperature, the pressure iminishes in geometrical progression as the height increases ı arithmetical progression. This theorem is due to Halley. ithout formal mathematics we see the truth of it by remark-ıg that differences of pressure are (§ 752) equal to differences f level multiplied by the density of the fluid, or by the proper ıean density when the density differs sensibly between the two tations. But the density, when the temperature is constant, aries in simple proportion to the pressure, according to Boyle's nd Mariotte's law. Hence differences of pressure between pairs f stations differing equally in level are proportional to the proper nean values of the whole pressure, which is the well-known ompound interest law. The rate of diminution of pressure er unit of length upwards in proportion to the whole pressure t any point, is of course equal to the reciprocal of the height ıbove that point that the atmosphere must have, if of constant lensity, to give that pressure by its weight. The height thus lefined is commonly called "the height of the homogeneous atmosphere," a very convenient conventional expression. It is equal to the product of the volume occupied by the unit mass of the gas at any pressure into the value of that pressure reckoned per unit of area, in terms of the weight of the unit of mass. If we denote it by H, the exponential expression of the law is

Height of the homogeneous atmosphere.

$$p = p_0 \epsilon^{-\frac{z}{H}},$$

which agrees with the final formula of § 752.

The value of H for dry atmospheric air, at the freezing temperature, according to Regnault, is, in the latitude of Paris, 799,020 centimetres, or 26,215 feet. Being inversely as the force of gravity in different latitudes (§ 222), it is 798,533 centimetres, or 26,199 feet, in the latitude of Edinburgh and Glasgow.

Analytical investigation of the preceding theorems.

Let X, Y, Z be the components, parallel to three rectangular axes, of the force acting on the fluid at (x, y, z), reckoned per unit of its mass. Then, inasmuch as the difference of pressures on the two faces $\delta y \delta z$ of a rectangular parallelepiped of the fluid is $\delta y \delta z \frac{dp}{dx} \delta x$, the equilibrium of this portion of the fluid, regarded for a moment (§ 564) as rigid, requires that

$$\delta y \delta z \frac{dp}{dx} \delta x - X \rho \delta x \delta y \delta z = 0.$$

Analytical investigation of the preceding theorems.

From this and the symmetrical equations relative to y and z we have

$$\frac{dp}{dx}=X\rho,\quad \frac{dp}{dy}=Y\rho,\quad \frac{dp}{dz}=Z\rho \qquad (1)$$

which are the conditions necessary and sufficient for the equilibrium of any fluid mass.

From these we have

$$dp=\frac{dp}{dx}dx+\frac{dp}{dy}dy+\frac{dp}{dz}dz=\rho(Xdx+Ydy+Zdz) \qquad (2)$$

This shows that the expression

$$Xdx+Ydy+Zdz$$

must be the complete differential of a function of three independent variables, or capable of being made so by a factor; that is to say, that a series of surfaces exists which cuts the lines of force at right angles; a conclusion also proved above (§ 749).

When the forces belong to a conservative system no factor is required to make the complete differential; and we have

$$Xdx+Ydy+Zdz=-dV$$

if V denote (§ 485) their potential at (x, y, z): so that (2) becomes

$$dp=-\rho dV \qquad (3)$$

This shows that p is constant over equipotential surfaces (or is a function of V); and it gives

$$\rho=-\frac{dp}{dV} \qquad (4)$$

showing that ρ also is a function of V; conclusions of which we have had a more elementary proof in § 752. As (4) is an analytical expression equivalent to the three equations (1), for the case of a conservative system of forces, we conclude that

Conditions of equilibrium of fluid completely filling a closed vessel.

754. It is both necessary and sufficient for the equilibrium of an incompressible fluid completely filling a rigid closed vessel, and influenced only by a conservative system of forces, that its density be uniform over every equipotential surface, that is to say, every surface cutting the lines of force at right angles. If, however, the boundary, or any part of the boundary, of the fluid mass considered, be not rigid; whether it be of flexible solid matter (as a membrane, or a thin sheet of elastic solid), or whether it be a mere geometrical boundary, on the other side of which there is another fluid, or *nothing* [a case which, without believing in vacuum as a reality, we may admit in abstract dynamics (§ 438)], a farther condition is necessary to secure that the pressure from without shall fulfil (4) at every point of the boundary. In the case of a bounding membrane, this condition must be fulfilled either through pressure artificially applied from without, or through the interior elastic forces of the matter of the membrane. In the case of another fluid of different density touching it on the other side of the boundary, all round or over some part of it, with no separating membrane, the condition of equilibrium of a heterogeneous fluid is to be fulfilled relatively to the whole fluid mass made up of the two; which shows that at the boundary the pressure must be constant and equal to that of the fluid on the other side. Thus water, oil, mercury, or any other liquid, in an open vessel, with its free surface exposed to the air, requires for equilibrium simply that this surface be level.

Conditions of equilibrium of fluid completely filling a closed vessel.

Free surface in open vessel is level.

755. Recurring to the consideration of a finite mass of fluid completely filling a rigid closed vessel, we see, from what precedes, that, if homogeneous and incompressible, it cannot be disturbed from equilibrium by any conservative system of forces; but we do not require the analytical investigation to prove this, as we should have "the perpetual motion" if it were denied, which would violate the hypothesis that the system of forces is conservative. On the other hand, a non-conservative system of forces cannot, under any circumstances, equilibrate a fluid which is either uniform in density throughout, or of homogeneous substance, rendered heterogeneous in density only through difference of pressure. But if the forces, though not conservative, be such that through every point of the space occupied by the fluid a surface may be drawn which shall cut at right angles all the lines of force it meets, a heterogeneous fluid will rest in equilibrium under their influence, provided (§ 750) its density, from point to point of every one of these orthogonal surfaces, varies inversely as the product of the resultant force into the thickness of the infinitely thin layer of space between that surface and another of the orthogonal surfaces infinitely near it on either side. (Compare § 488.)

Fluid, in closed vessel, under a non-conservative system of forces.

The same conclusion is proved as a matter of course from (1), since that equation is merely the analytical expression that the force at every point (x, y, z) is along the normal to that surface of the series given by different values of C in $p = C$, which

Fluid under any system of forces.

passes through (x, y, z); and that the magnitude of the resultant force is

$$\frac{\sqrt{\left(\frac{dp^2}{dx^2}+\frac{dp^2}{dy^2}+\frac{dp^2}{dz^2}\right)}}{\rho},$$

of which the numerator is equal to $\frac{\delta C}{\tau}$, if τ be the thickness at (x, y, z) of the shell of space between two surfaces

$$p=C \text{ and } p=C+\delta C,$$

infinitely near one another on two sides of (x, y, z).

The analytical expression of the condition which X, Y, Z must fulfil in order that (1) may be possible is found thus; first, by

$$\frac{d}{dz}\frac{dp}{dy}=\frac{d}{dy}\frac{dp}{dz}, \text{ etc.,}$$

we have

$$\left.\begin{aligned}\frac{d}{dz}(\rho Y)&=\frac{d}{dy}(\rho Z)\\ \frac{d}{dx}(\rho Z)&=\frac{d}{dz}(\rho X)\\ \frac{d}{dy}(\rho X)&=\frac{d}{dx}(\rho Y)\end{aligned}\right\} \quad (5).$$

Performing the differentiations, and multiplying the first of the resulting equations by X, the second by Y, and the third by Z, we have

$$X\left(\frac{dZ}{dy}-\frac{dY}{dz}\right)+Y\left(\frac{dX}{dz}-\frac{dZ}{dx}\right)+Z\left(\frac{dY}{dx}-\frac{dX}{dy}\right)=0 \quad (6);$$

which is merely the well-known condition that the expression

$$Xdx+Ydy+Zdz$$

may be capable of being rendered by a factor the complete differential of a function of three independent variables.

Or if we multiply the first of (5) by $\frac{d\rho}{dx}$, the second by $\frac{d\rho}{dy}$, and the third by $\frac{d\rho}{dz}$, and add, we have

$$\frac{d\rho}{dx}\left(\frac{dZ}{dy}-\frac{dY}{dz}\right)+\frac{d\rho}{dy}\left(\frac{dX}{dz}-\frac{dZ}{dx}\right)+\frac{d\rho}{dz}\left(\frac{dY}{dx}-\frac{dX}{dy}\right)=0 \quad (7).$$

This shows that the line whose direction-cosines are proportional to

$$\frac{dZ}{dy}-\frac{dY}{dz},\ \frac{dX}{dz}-\frac{dZ}{dx},\ \frac{dY}{dx}-\frac{dX}{dy}$$

is perpendicular to the surface of equal density through (x, y, z), and (6) shows that the same line is perpendicular to the resultant force. It is therefore tangential to both the surface of equal density and to that of equal pressure, and therefore to their curve of intersection. The differential equations of this curve are therefore

Fluid under any system of forces.

$$\frac{dx}{\frac{dZ}{dy}-\frac{dY}{dz}}=\frac{dy}{\frac{dX}{dz}-\frac{dZ}{dx}}=\frac{dz}{\frac{dY}{dx}-\frac{dX}{dy}} \qquad (8).$$

756. If we imagine all the fluid to become rigid except an infinitely thin closed tubular portion lying in a surface of equal density, and if the fluid in this tubular circuit be moved any length along the tube and left at rest, it will remain in equilibrium in the new position, all positions of it in the tube being indifferent because of its homogeneousness. Hence the work (positive or negative) done by the force (X, Y, Z) on any portion of the fluid in any displacement along the tube is balanced by the work (negative or positive) done on the remainder of the fluid in the tube. Hence a single particle, acted on only by X, Y, Z, while moving round the circuit, that is moving along any closed curve on a surface of equal density, has, at the end of one complete circuit, done just as much work against the force in some parts of its course, as the forces have done on it in the remainder of the circuit.

Equilibrium condition.

An interesting application of (j) § 190 may be made to prove this result analytically. Thus, if we take for α, β, γ our present force components X, Y, Z; and for the surface there referred to, a surface of equal density in our heterogeneous fluid; the expression

$$\iint dS\left\{l\left(\frac{dZ}{dy}-\frac{dY}{dz}\right)+m\left(\frac{dX}{dz}-\frac{dZ}{dx}\right)+n\left(\frac{dY}{dx}-\frac{dX}{dy}\right)\right\}$$

vanishes because of (7), and we conclude that

$$\int(Xdx+Ydy+Zdz)=0,$$

for any closed circuit on a surface of equal density.

757. The following imaginary example, and its realization in a subsequent section (§ 759), show a curiously interesting practical application of the theory of fluid equilibrium under extraordinary circumstances, generally regarded as a merely abstract analytical theory, practically useless and quite unnatural, "because forces in nature follow the conservative law."

Imaginary example of equilibrium under non-conservative forces.

Imaginary example of equilibrium under non-conservative forces.

758. Let the lines of force be circles, with their centres in one line, and their planes perpendicular to it. They are cut at right angles by planes through this axis; and therefore a fluid may be in equilibrium under such a system of force. The system will not be conservative if the intensity of the force be according to any other law than inverse proportionality to distance from this axial line; and the fluid, to be in equilibrium, must be heterogeneous, and be so distributed as to vary in density from point to point of every plane through the axis, inversely as the product of the force into the distance from the axis. But from one such plane to another it may be either uniform in density, or may vary arbitrarily. To particularize farther, we may suppose the force to be in direct simple proportion to the distance from the axis. Then the fluid will be in equilibrium if its density varies from point to point of every plane through the axis, inversely as the square of that distance. If we still farther particularize by making the force uniform all round each circular line of force, the distribution of force becomes precisely that of the kinetic reactions of the parts of a rigid body against accelerated rotation. The fluid pressure will (§ 749) be equal over each plane through the axis. And in one such plane, which we may imagine carried round the axis in the direction of the force, the fluid pressure will increase in simple proportion to the angle at a rate per unit angle (§ 41) equal to the product of the density at unit distance into the force at unit distance. Hence it must be remarked, that if any closed line (or circuit) can be drawn round the axis, without leaving the fluid, there cannot be equilibrium without a firm partition cutting every such circuit, and maintaining the difference of pressures on the two sides of it, corresponding to the angle 2π. Thus, if the axis pass through the fluid in any part, there must be a partition extending from this part of the axis continuously to the outer bounding surface of the fluid. Or if the bounding surface of the whole fluid be annular (like a hollow anchor-ring, or of any irregular shape), in other words, if the fluid fills a tubular circuit; and

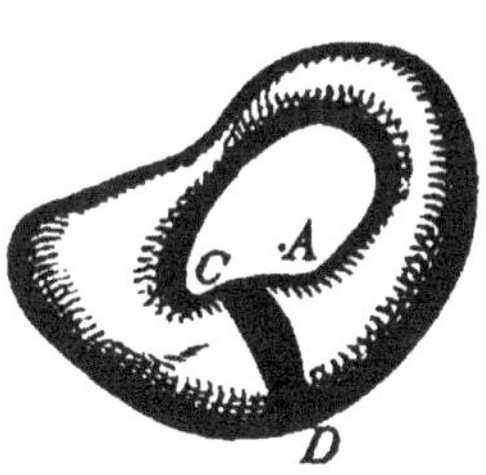

e axis (A) pass through the aperture of the ring (without assing into the fluid); there must be a firm partition (CD) tending somewhere continuously across the channel, or assage, or tube, to stop the circulation of the fluid round it; therwise there could not be equilibrium with the supposed rces in action. If we further suppose the density of the fluid o be uniform round each of the circular lines of force in the ystem we have so far considered (so that the density shall be qual over every circular cylinder having the line of their entres for its axis, and shall vary from one such cylindrical urface to another, inversely as the squares of their radii), we nay, without disturbing the equilibrium, impose any conservaive system of force in lines perpendicular to the axis; that is § 488), any system of force in this direction, with intensity varying as some function of the distance. If this function be the simple distance, the superimposed system of force agrees precisely with the reactions against curvature, that is to say, the centrifugal forces, of the parts of a rotating rigid body.

Imaginary example of equilibrium under non-conservative forces.

759. Thus we arrive at the remarkable conclusion, that if a rigid closed box be completely filled with incompressible heterogeneous fluid, of density varying inversely as the square of the distance from a certain line, and if the box be moveable round this line as a fixed axis, and be urged in any way by forces applied to its outside, the fluid will remain in equilibrium relatively to the box; that is to say, will move round with the box as if the whole were one rigid body, and will come to rest with the box if the box be brought again to rest: provided always the preceding condition as to partitions be fulfilled if the axis pass through the fluid, or be surrounded by continuous lines of fluid. For, in starting from rest, *if* the fluid moves like a rigid solid, we have reactions against acceleration, tangential to the circles of motion, and equal in amount to $\dot{\omega}r$ per unit of mass of the fluid at distance r from the axis, $\dot{\omega}$ being the rate of acceleration (§ 42) of the angular velocity; and (see Vol. II.) we have, in the direction perpendicular to the axis outwards, reaction against curvature of path, that is to say, "centrifugal force," equal to $\omega^2 r$ per unit of mass of the fluid. Hence the equilibrium which we have demonstrated in the preceding section, for the fluid supposed at rest, and

Actual case.

Actual case of fluid equilibrium under non-conservative forces.

arbitrarily influenced by two systems of force (the circ... non-conservative and the radial conservative system) agre... in law with these forces of kinetic reaction, proves for us ... the D'Alembert (§ 264) equilibrium condition for the mot... of the whole fluid as of a rigid body experiencing accelera... rotation; that is to say, shows that this kind of motion ful... for the actual circumstances the laws of motion, and, theref... that it is *the* motion actually taken by the fluid.

Relation between density and potential of applied forces.

760. If the fluid is of homogeneous substance and unif... temperature throughout, but compressible, as all real fluids are, it can be heterogeneous in density, only because of differe... of pressure in different parts; the surfaces of equal dens... must be also surfaces of equal pressure; and, as we have ... above (§ 753), there can be no equilibrium unless the sys... of forces be conservative. The function which the density ... of the pressure must be supposed known (§ 448), as it depen... on physical properties of the fluid. Compare § 752.

Let $$\rho = f(p) \qquad (9)$$

We have, by § 753 (3), integrated,

$$\int \frac{dp}{f(p)} = C - V \qquad (10)$$

or, if F denote such a function that

$$F\{\int \frac{dp}{f(p)}\} = p \qquad (11)$$

$$p = F(C - V),$$

and, by (9), $$\rho = f\{F(C - V)\} \qquad (12)$$

Resultant pressure on a plane area.

761. In § 746 we considered the resultant pressure on a plane surface, when the pressure is uniform. We may now consider (briefly) the resultant pressure on a plane area when the pressure varies from point to point, confining our attention to a case of great importance;—that in which gravity is the only applied force, and the fluid is a nearly incompressible liquid such as water. In this case the determination of the position of the Centre of Pressure is very simple; and the whole pressure is the same as if the plane area were turned about its centre of inertia into a horizontal position.

Kinetic measure $p = g\rho z + p_0$.

The pressure at any point at a depth z in the liquid may be expressed by $$p = \rho z + p_0$$ where ρ is the (constant) density of the liquid, and p_0 the (atm...

spheric) pressure at the free surface, reckoned in units of weight per unit of area. Resultant pressure on a plane area.

Let the axis of x be taken as the intersection of the plane of the immersed plate with the free surface of the liquid, and that of y perpendicular to it and in the plane of the plate. Let a be the inclination of the plate to the vertical. Let also A be the area of the portion of the plate considered, and $\bar{x}$, $\bar{y}$, the co-ordinates of its centre of inertia.

Then the whole pressure is

$$\iint p\,dx\,dy = \iint (p_0 + \rho y \cos a)\,dx\,dy$$
$$= Ap_0 + A\rho\bar{y}\cos a.$$

The moment of the pressure about the axis of x is

$$\iint py\,dx\,dy = Ap_0\bar{y} + Ak^2\rho\cos a,$$

k being the radius of gyration of the plane area about the axis of x.

For the moment about y we have

$$\iint px\,dx\,dy = Ap_0\bar{x} + \rho\cos a\iint xy\,dx\,dy.$$

The first terms of these three expressions merely give us again the results of § 746; we may therefore omit them. This will be equivalent to introducing a stratum of additional liquid above the free surface such as to produce an equivalent to the atmospheric pressure. If the origin be now shifted to the upper surface of this stratum we have

$$\text{Pressure} = A\rho\bar{y}\cos a,$$

$$\text{Moment about } Ox = Ak^2\rho\cos a,$$

$$\text{Distance of centre of pressure from axis of } x = \frac{k^2}{\bar{y}}.$$

But if k_1 be the radius of gyration of the plane area about a horizontal axis in its plane, and passing through its centre of inertia, we have, by § 283,

$$k^2 = k_1^2 + \bar{y}^2.$$

Hence the distance, measured parallel to the axis of y, of the centre of pressure from the centre of inertia is

$$\frac{k_1^2}{\bar{y}};$$

and, as we might expect, diminishes as the plane area is more and more submerged. If the plane area be turned about the line through its centre of inertia parallel to the axis of x, this distance varies as the cosine of its inclination to the vertical; supposing, of course, that by the rotation neither more nor less of the plane area is submerged.

Loss of apparent weight by immersion in a fluid.

762. A body, wholly or partially immersed in any fl. influenced by gravity, loses, through fluid pressure, in appar weight an amount equal to the weight of the fluid displa For if the body were removed, and its place filled with flu homogeneous with the surrounding fluid, there would be equilibrium, even if this fluid be supposed to become rigid. An the resultant of the fluid pressure upon it is therefore a sin force equal to its weight, and in the vertical line through it centre of gravity. But the fluid pressure on the originally immersed body was the same all over as on the solidified port of fluid by which for a moment we have imagined it replaced and therefore must have the same resultant. This proposition is of great use in Hydrometry, the determination of speci gravity, etc. etc.

Analytically, the following demonstration is of interest, especially in its analogies to some preceding theorems, and others which occur in electricity and magnetism.

If V be the potential of the impressed forces, $-\frac{dV}{dx}$ is the force parallel to the axis of x on unit of matter at xyz, and $\rho dxdydz$ is the mass of an element of the fluid, and therefore the whole force parallel to the axis of x on a mass of fluid substituted for the immersed body, is represented by the triple integral

$$-\iiint\rho\frac{dV}{dx}dxdydz$$

taken through the whole space enclosed by the surface. But by § 752,

$$\frac{dp}{dx}=-\rho\frac{dV}{dx}.$$

Hence the triple integral becomes

$$\iiint\frac{dp}{dx}dxdydz=\iint pdydz$$

extended over the whole surface.

Let dS be an element of any surface at x, y, z; λ, μ, ν the direction-cosines of the normal to the element; p the pressure in the fluid in contact with it. The whole resolved pressure parallel to the axis of x is

$$P_x=\iint\lambda pdS$$
$$=\iint pdydz,$$

the same expression as above.

The couple about the axis of z, due to the applied forces on any fluid mass, is (§ 559)

Loss of apparent weight by immersion in a fluid

$$\Sigma dm(Xy - Yx),$$

dm representing the mass of an element of fluid.

This may be written in the form

$$-\iiint \rho dxdydz\left(y\frac{dV}{dx} - x\frac{dV}{dy}\right),$$

the integral being taken throughout the mass.

This is evidently equal to

$$\iiint\left(y\frac{dp}{dx} - x\frac{dp}{dy}\right)dxdydz$$

$$= \iint pydydz - \iint pxdzdx$$

$$= \iint p(\lambda y - \mu x)dS,$$

which is the couple due to surface-pressure alone.

763. The following lemma, while in itself interesting, is of great use in enabling us to simplify the succeeding investigations regarding the stability of equilibrium of floating bodies:—

Lemma.

Let a homogeneous solid, the weight of unit of volume of which we suppose to be unity, be cut by a horizontal plane in $XYX'Y'$. Let O be the centre of inertia, and let XX', YY' be the principal axes, of this area.

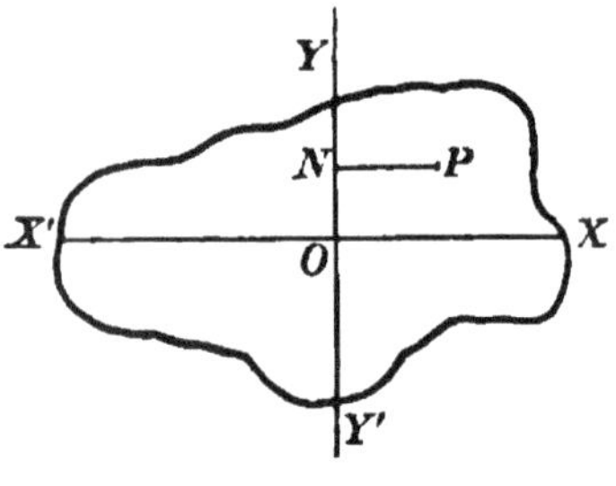

Let there be a second plane section of the solid, through YY', inclined to the first at an infinitely small angle, θ. Then (1.) the volumes of the two wedges cut from the solid by these sections are equal; (2.) their centres of inertia lie in one plane perpendicular to YY'; and (3.) the moment of the weight of each of these, round YY', is equal to the moment of inertia about it of the corresponding portion of the area, multiplied by θ.

Take OX, OY as axes, and let θ be the angle of the wedge: the thickness of the wedge at any point $P(x, y)$ is θx, and the volume of a right prismatic portion whose base is the elementary area $dxdy$ at P is $\theta xdxdy$.

Now let [ ] and ( ) be employed to distinguish integrations extended over the portions of area to the right and left of the axis of y respectively, while integrals over the whole area have no

Lemma.

such distinguishing mark. Let α and α' be these areas, v and v' the volumes of the wedges; $(\bar{x}, \bar{y})$, $(\bar{x}', \bar{y}')$ the co-ordinates of their centres of inertia. Then

$$v = \theta\left[\iint x dx dy\right] = \alpha\bar{x}$$

$$-v' = \theta\left(\iint x dx dy\right) = \alpha'\bar{x}',$$

whence $v - v' = \theta\iint x dx dy = 0$ since O is the centre of inertia

Hence $v = v'$, which is (1.)

Again, taking moments about XX',

$$v\bar{y} = \theta\left[\iint xy dx dy\right],$$

and $$-v'\bar{y}' = \theta\left(\iint xy dx dy\right).$$

Hence $$v\bar{y} - v'\bar{y}' = \theta\iint xy dx dy.$$

But for a principal axis (§ 281) $\Sigma xydm$ vanishes. Hence $v\bar{y} - v'\bar{y}' = 0$, whence, since $v = v'$, we have

$$\bar{y} = \bar{y}', \text{ which proves (2.)}$$

And (3.) is merely a statement in words of the obvious equation

$$\left[\iint x.x\theta dx dy\right] = \theta\left[\iint x^2.dx dy\right].$$

Stability of equilibrium of a floating body.

764. If a positive amount of work is required to produce any possible infinitely small displacement of a body from a position of equilibrium, the equilibrium in this position is stable (§ 291). To apply this test to the case of a floating body, we may remark, first, that any possible infinitely small displacement may (§§ 26, 95) be conveniently regarded as compounded of two horizontal displacements in lines at right angles to one another, one vertical displacement, and three rotations round rectangular axes through any chosen point. If one of these axes be vertical, then three of the component displacements, viz., the two horizontal displacements and the rotation about the vertical axis, require no work (positive or negative), and therefore, so far as they are concerned, the equilibrium is essentially neutral. But so far as the other three modes of displacement are concerned, the equilibrium may be positively stable, or may be unstable, or may be neutral, according to the fulfilment of conditions which we now proceed to investigate.

Vertical displacements.

765. If, first, a simple vertical displacement, downwards let us suppose, be made, the work is done against an increasing resultant of upward fluid pressure, and is of course equal to the mean increase of this force multiplied by the whole space. If this space be denoted by z, the area of the plane of otation by A, and the weight of unit bulk of the liquid by w, ıe increased bulk of immersion is clearly Az, and therefore the ıcrease of the resultant of fluid pressure is wAz, and is in a line ertically upward through the centre of gravity of A. The ıean force against which the work is done is therefore $\frac{1}{2}wAz$, s this is a case in which work is done against a force increasıg from zero in simple proportion to the space. Hence the ork done is $\frac{1}{2}wAz^2$. We see, therefore, that so far as vertical isplacements alone are concerned, the equilibrium is necesarily stable, unless the body is wholly immersed, when the rea of the plane of flotation vanishes, and the equilibrium is eutral.

Work done in vertical displacement.

766. The lemma of § 763 suggests that we should take, as he two horizontal axes of rotation, the principal axes of the lane of flotation. Considering then rotation through an inınitely small angle θ round one of these, let G and E be the

Displacement by rotation about an axis in the plane of flotation.

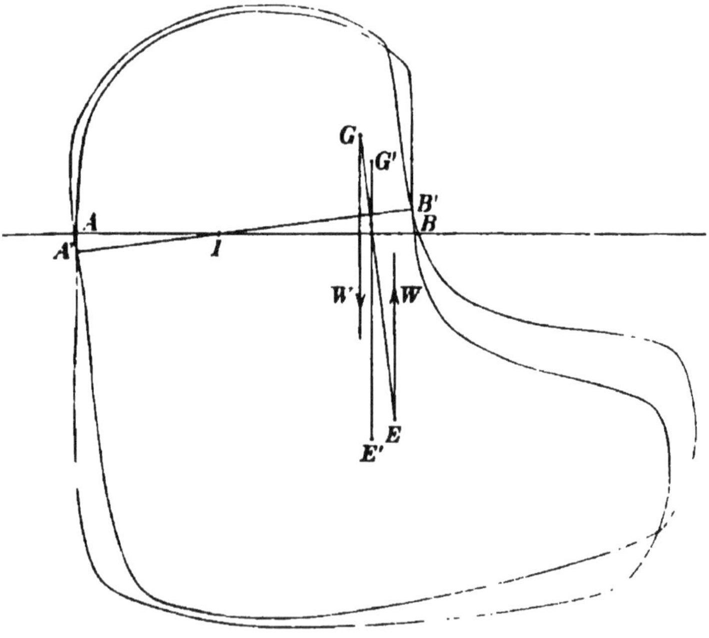

lisplaced centres of gravity of the solid, and of the portion of its volume which was immersed when it was floating in

Displacement by rotation about an axis in the plane of flotation.

equilibrium, and G', E' the positions which they then b... all projected on the plane of the diagram which we supp... be through I the centre of inertia of the plane of flot... The resultant action of gravity on the displaced body is W... weight, acting downwards through G; and that of the f... pressure on it is W upwards through E corrected by the am... (upwards) due to the additional immersion of the wedge AI... and the amount (downwards) due to the extruded wedge BI... Hence the whole action of gravity and fluid pressure on th... displaced body is the couple of forces up and down in vertic... through G and E, and the correction due to the wedges. Th... correction consists of a force vertically upwards through th... centre of gravity of $A'IA$, and downwards through that of BIB... These forces are equal [§ 763 (1)], and therefore constitute a couple which [§ 763 (2)] has the axis of the displacement for its axis, and which [§ 763 (3)] has its moment equal to θw... if A be the area of the plane of flotation, and k its radius of gyration (§ 281) round the principal axis in question. B... since GE, which was vertical ($G'E'$) in the position of equili... brium, is inclined at the infinitely small angle θ to the vertic... in the displaced body, the couple of forces W in the vertica... through G and E has for moment $Wh\theta$, if h denote GE; an... is in a plane perpendicular to the axis, and in the directi... tending to increase the displacement, when G is above E... Hence the resultant action of gravity and fluid pressure on th... displaced body is a couple whose moment is

$$(wAk^2 - Wh)\theta, \text{ or } w(Ak^2 - Vh)\theta,$$

if V be the volume immersed. It follows that when $Ak^2 > V$... the equilibrium is stable, so far as this displacement alone is concerned.

Work done in this displacement.

Also, since the couple worked against in producing the dis... placement increases from zero in simple proportion to the angle of displacement, its mean value is half the above; and therefore the whole amount of work done is equal to

$$\tfrac{1}{2}w(Ak^2 - Vh)\theta^2.$$

General displacement.

767. If now we consider a displacement compounded of a vertical (downwards) displacement z, and rotations through infinitely small angles θ, θ' round the two horizontal principal axes of the plane of flotation, we see (§§ 765, 766) that the work required to produce it is equal to

General displacement. Work required.

$$\tfrac{1}{2}w[Az^2+(Ak^2-Vh)\theta^2+(Ak'^2-Vh)\theta'^2],$$

and we conclude that, for complete stability with reference to all possible displacements of this kind, it is necessary and sufficient that

Conditions of stability.

$$h<\frac{Ak^2}{V}, \text{ and } <\frac{Ak'^2}{V}.$$

768. When the displacement is about any axis through the centre of inertia of the plane of flotation, the resultant of fluid pressures is equal to the weight of the body; but it is only when the axis is a principal axis of the plane of flotation that this resultant is in the plane of displacement. In such a case the point of intersection of the resultant with the line originally vertical, and through the centre of gravity of the body, is called the *Metacentre*. And it is obvious, from the above investigation, that for either of these planes of displacement the condition of stable equilibrium is that the metacentre shall be *above* the centre of gravity.

The metacentre. Condition of its existence.

769. The spheroidal analysis with which we propose to conclude this volume is proper, or practically successful, for hydrodynamic problems only when the deviations from spherical symmetry are infinitely small; or, practically, small enough to allow us to neglect the squares of ellipticities (§ 801); or, which is the same thing, to admit thoroughly the principle of the superposition of disturbing forces, and the deviations produced by them. But we shall first consider a case which admits of very simple synthetical solution, without any restriction to approximate sphericity; and for which the following remarkable theorem was discovered by Newton and Maclaurin:—

770. An oblate ellipsoid of revolution, of any given eccentricity, is a figure of equilibrium of a mass of homogeneous incompressible fluid, rotating about an axis with determinate angular velocity, and subject to no forces but those of gravitation among its parts.

A homogeneous ellipsoid is a figure of equilibrium of a rotating liquid mass.

The angular velocity for a given eccentricity is independent of the bulk of the fluid, and proportional to the square root of its density.

771. The proof of this proposition is easily obtained from

A homogeneous ellipsoid is a figure of equilibrium of a rotating liquid mass.

the results already deduced with respect to the attraction of an ellipsoid and the properties of the free surface of a fluid.

We know, § 522, that if APB be a meridian section of a homogeneous oblate spheroid, AC the polar axis, CB an equatorial radius, and P any point on the surface, the attraction of the spheroid may be resolved into two parts; one, Pp, perpendicular to the polar axis, and varying as the ordinate PM; the other, Ps, parallel to the polar axis, and varying as PN. These components are not equal when MP and PN are equal, else the resultant attraction at all points in the surface would pass through C; whereas we know that it is in some such direction as Pf, cutting the radius BC *between* B and C, but at a point nearer to C than n the foot of the normal at P.

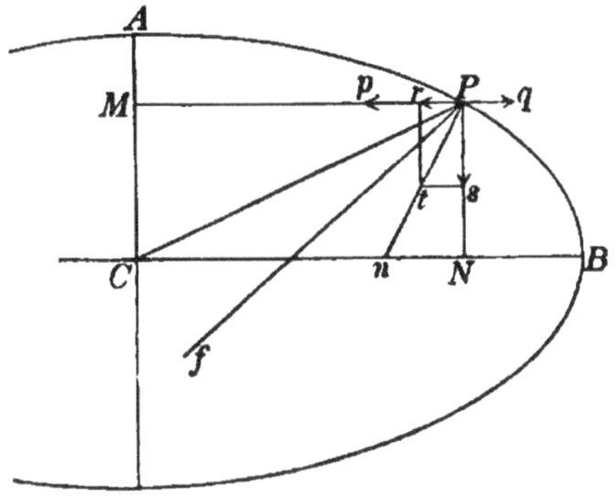

Let then

$$Pp = \alpha . PM,$$

and $$Ps = \beta . PN,$$

where α and β are known constants, depending merely on the density, (ρ), and eccentricity, (e), of the spheroid.

Also, we know by geometry that $Nn = (1 - e^2) CN$.

Hence; to find the magnitude of a force Pq perpendicular to the axis of the spheroid, which, when compounded with the attraction, will bring the resultant force into the normal Pn, make $pr = Pq$, and we must have

$$\frac{Pr}{Ps} = \frac{Nn}{PN} = (1-e^2)\frac{CN}{PN} = (1-e^2)\frac{\beta . Pp}{\alpha . Ps}.$$

Hence $$Pr = (1-e^2)\frac{\beta}{\alpha} Pp$$

$$Pp - Pq = (1-e^2)\frac{\beta}{\alpha} Pp,$$

or $$Pq = \left(1 - (1-e^2)\frac{\beta}{\alpha}\right) Pp$$

$$= \left(\alpha - (1-e^2)\beta\right) PM.$$

Now if the spheroid were to rotate with angular velocity ω about AC, the centrifugal force, §§ 32, 35*a*, 259, would be in the direction Pq, and would amount to

A homogeneous ellipsoid is a figure of equilibrium of a rotating liquid mass.

$$\omega^2 PM.$$

Hence, if we make

$$\omega^2 = \alpha - (1-e^2)\beta,$$

the whole force on P, that is, the resultant of the attraction and centrifugal force, will be in the direction of the normal to the surface, which is the condition for the free surface of a mass of fluid in equilibrium.

Now, § 522, $$\alpha = 2\pi\rho\left(\frac{\sqrt{1-e^2}}{e^3}\sin^{-1}e - \frac{1-e^2}{e^2}\right)$$

$$\beta = 4\pi\rho\left(\frac{1}{e^2} - \frac{\sqrt{1-e^2}}{e^3}\sin^{-1}e\right).$$

Hence $$\omega^2 = 2\pi\rho\left\{\frac{(3-2e^2)\sqrt{1-e^2}}{e^3}\sin^{-1}e - 3\frac{1-e^2}{e^2}\right\}. \qquad (1).$$

This determines the angular velocity, and proves it to be proportional to $\sqrt{\rho}$.

The square of the requisite angular velocity is as the density of the liquid.

772. If, after Laplace, we introduce instead of e a quantity ε defined by the equation

$$\left.\begin{aligned} 1-e^2 &= \frac{1}{1+\varepsilon^2}, \\ \text{or} \quad \varepsilon &= \frac{e}{\sqrt{1-e^2}} = \tan(\sin^{-1}e), \end{aligned}\right\} \qquad (2),$$

the expression (1) for ω^2 is much simplified, and

$$\frac{\omega^2}{2\pi\rho} = \frac{3+\varepsilon^2}{\varepsilon^3}\tan^{-1}\varepsilon - \frac{3}{\varepsilon^2}. \qquad (3).$$

When e, and therefore also ε, is small, this formula is most easily calculated from

$$\frac{\omega^2}{2\pi\rho} = \tfrac{4}{15}\varepsilon^2 - \tfrac{8}{35}\varepsilon^4 + \text{etc.} \qquad (4),$$

of which the first term is sufficient when we deal with spheroids so little oblate as the earth.

The following table has been calculated by means of these simplified formulæ. The last figure in each of the four last

Table of corresponding values of ellipticities and angular velocities.

columns is given to the nearest unit. The two last columns will be explained a few sections later :—

i.	ii.	iii.	iv.	v.
e.	$\frac{1}{e}$.	$\frac{\omega^2}{2\pi\rho}$.	$\frac{2\pi}{\omega}$ when $\rho = 3{\cdot}68 \times 10^{-7}$.	$(1+e^2)^{\frac{1}{3}}\frac{f}{2\pi r}$.
0·1	9·950	0·0027	79,966	0·0027
·2	4·899	·0107	39,397	·0110
·3	3·180	·0243	26,495	·0258
·4	2·291	·0436	19,780	·0490
·5	1·732	·0690	15,730	·0836
·6	1·333	·1007	13,022	·1356
·7	1·020	·1387	11,096	·2172
·8	0·750	·1816	9,697	·3588
·9	·4843	·2203	8,804	·6665
·91	·4556	·2225	8,759	·7198
·92	·4260	·2241	8,729	·7813
·93	·3952	·2247	8,718	·8533
·94	·3629	·2239	8,732	·9393
·95	·3287	·2213	8,783	1·045
·96	·2917	·2160	8,891	1·179
·97	·2506	·2063	9,098	1·359
·98	·2030	·1890	9,504	1·627
·99	·1425	·1551	10,490	2·113
1·00	0·0000	0·0000	∞	∞

From this we see that the value of $\frac{\omega^2}{2\pi\rho}$ increases gradually from zero to a maximum as the eccentricity e rises from zero to about 0·93, and then (more quickly) falls to zero as the eccentricity rises from 0·93 to unity. The values of the other quantities corresponding to this maximum are given in the table.

773. If the angular velocity exceed the value calculated from

$$\frac{\omega^2}{2\pi\rho} = 0{\cdot}2247,$$

when for ρ is substituted the density of the liquid, equilibrium is impossible in the form of an ellipsoid of revolution. If the angular velocity fall short of this limit there are always two ellipsoids of revolution which satisfy the conditions of equilibrium. In one of these the eccentricity is greater than 0·93, in the other less.

774. It may be useful, for special applications, to indicate briefly how ρ is measured in these formulæ. In the definitions

§§ 459, 460, on which the attraction formulæ are based, unit mass is defined as exerting unit force on unit mass at unit distance; and unit volume-density is that of a body which has unit mass in unit volume. Hence, with the foot as our linear unit, we have for the earth's attraction on a particle of unit mass at its surface

Mean density of the earth expressed in attraction units.

$$\frac{\frac{4}{3}\pi\sigma R^3}{R^2}=\frac{4}{3}\pi\sigma R=32\cdot2\,;$$

where R is the radius of the earth (supposed spherical) in feet; and σ its mean density, expressed in terms of the unit just defined.

Taking 20,900,000 feet as the value of R, we have

$$\sigma=0\cdot000000368=3\cdot68\times10^{-7}. \qquad (6).$$

As the mean density of the earth is somewhere about 5·5 times that of water, § 479, the density of water in terms of our present unit is

$$\frac{3\cdot68}{5\cdot5}10^{-7}=6\cdot7\times10^{-8}.$$

775. The fourth column of the table above gives the time of rotation in seconds, corresponding to each value of the eccentricity, ρ being assumed equal to the mean density of the earth. For a mass of water these numbers must be multiplied by $\sqrt{5\cdot5}$, as the time of rotation to give the same figure is inversely as the square root of the density.

Time of rotation for spheroid of given eccentricity.

For a homogeneous liquid mass, of the earth's mean density, rotating in $23^h\,56^m\,4^s$, we find $e=0\cdot093$, which corresponds to an ellipticity of about $\frac{1}{230}$.

776. An interesting form of this problem, also discussed by Laplace, is that in which the moment of momentum and the mass of the fluid are given, not the angular velocity; and it is required to find what is the eccentricity of the corresponding ellipsoid of revolution, the result proving that there can be but one.

Mass and moment of momentum of fluid given.

It is evident that a mass of any ordinary liquid (not a *perfect fluid*, § 742), if left to itself in any state of motion, must preserve unchanged its moment of momentum, § 235. But the viscosity, or internal friction, § 742, will, if the mass remain continuous, ultimately destroy all relative motion among its parts; so that it will ultimately rotate as a rigid

Mass and moment of momentum of fluid given.

solid. If the final form be an ellipsoid of revolution, we can easily show that there is a single definite value of its eccentricity. But, as it has not yet been discovered whether there is any other form consistent with *stable* equilibrium, we do not know that the mass will necessarily assume the form of this particular ellipsoid. Nor in fact do we know whether even the ellipsoid of rotation may not become an *unstable* form if the moment of momentum exceed some limit depending on the mass of the fluid. We shall return to this subject in Vol. II. as it affords an excellent example of that difficult and delicate question *Kinetic Stability*, § 346.

If we call a the equatorial semi-axis of the ellipsoid, e its eccentricity, and ω its angular velocity of rotation, the given quantities are the mass

$$M=\tfrac{4}{3}\pi\rho a^3\sqrt{1-e^2},$$

and the moment of momentum

$$A=\tfrac{8}{15}\pi\rho\omega a^5\sqrt{1-e^2}.$$

These equations, along with (2), determine the three quantities a, e, and ω.

Eliminating a between the two just written, and expressing e as before in terms of ε, we have

$$\frac{A^2}{M^{\frac{10}{3}}}=\tfrac{3}{25}\left(\tfrac{3}{4}\right)^{\frac{1}{3}}\frac{\omega^2(1+\varepsilon^2)^{\frac{2}{3}}}{(\pi\rho)^{\frac{4}{3}}}.$$

This gives

$$\frac{\omega^2}{2\pi\rho}=\frac{k}{(1+\varepsilon^2)^{\frac{2}{3}}},$$

where k is a *given* multiple of $\rho^{\frac{1}{3}}$. Substituting in 772 (2) we have

$$k=(1+\varepsilon^2)^{\frac{2}{3}}\left(\frac{3+\varepsilon^2}{\varepsilon^3}\tan^{-1}\varepsilon-\frac{3}{\varepsilon^2}\right).$$

Now the last column of the table in § 772 shows that the value of this function of ε (which vanishes with ε) continually increases with ε, and becomes infinite when ε is infinite. Hence there is always one, and one only, value of ε, and therefore of e, which satisfies the conditions of the problem.

777. All the above results might without much difficulty have been obtained analytically, by the discussion of the equations; but we have preferred, for once, to show by an actual case that numerical calculation may sometimes be of very great use.

778. No one seems yet to have attempted to solve the neral problem of finding all the forms of equilibrium which a ass of homogeneous incompressible fluid rotating with uniform gular velocity may assume. Unless the velocity be so small at the figure differs but little from a sphere (a case which ill be carefully treated later), the problem presents difficulties an exceedingly formidable nature. It is therefore of some nportance to show by a synthetical process that another form, esides that of the ellipsoid of revolution, may be compatible ith equilibrium; *viz.*, an ellipsoid with three unequal axes, of hich the least is the axis of rotation. This curious theorem as discovered by Jacobi in 1834, and seems, simple as it is, o have been enunciated by him as a challenge to the French athematicians.[1] The proof which follows is virtually the ame as that given by Archibald Smith.[2] Equilibrium ellipsoid of three unequal axes.

By § 522, the components of the attraction of a homogeneous ellipsoid, whose semi-axes are a, b, c, on a point ξ, η, ζ at its surface are

$$\tfrac{3}{2}M\xi\int_0^\infty \frac{d\psi}{(a^2+\psi)\sqrt{(a^2+\psi)(b^2+\psi)(c^2+\psi)}}, \text{ etc.,}$$

which, for the present, we may call $A\xi$, $B\eta$, $C\zeta$.

If the ellipsoid revolve, with angular velocity ω, about the axis of ζ, the components of the centrifugal force are

$$\omega^2\xi,\ \omega^2\eta,\ 0.$$

Hence the components of the whole resultant of gravity and centrifugal force on the particle at ξ, η, ζ are

$$(A-\omega^2)\xi,\ (B-\omega^2)\eta,\ C\xi.$$

But the direction-cosines of the normal to the surface of the ellipsoid at ξ, η, ζ, are proportional to

$$\frac{\xi}{a^2},\ \frac{\eta}{b^2},\ \frac{\zeta}{c^2};$$

and, for equilibrium, the resultant force must be perpendicular to the free surface. Hence

$$a^2(A-\omega^2)=b^2(B-\omega^2)=c^2C \qquad (1).$$

These equations give

$$\omega^2=\frac{a^2A-c^2C}{a^2}=\frac{b^2B-c^2C}{b^2} \qquad (2).$$

We must show, *first*, that for any given values of a and b a value

[1] See a Paper by Liouville, *Journal de l'École Polytechnique*, cahier XXIII., foot-note to p. 290.

[2] *Cambridge Math. Journal*, Feb. 1838.

Equilibrium ellipsoid of three unequal axes.

of c can be assigned which will make these values of ω^2 equal. Then we must show that the value of ω^2 thus found is positive, giving a real value of ω. If we put, as in § 522,

$$\Phi=\int_0^\infty \frac{d\psi}{\sqrt{(a^2+\psi)(b^2+\psi)(c^2+\psi)}}, \qquad (3)$$

we have

$$A=-\tfrac{3}{2}M\frac{d\Phi}{d(a^2)},\quad B=-\tfrac{3}{2}M\frac{d\Phi}{d(b^2)},\quad C=-\tfrac{3}{2}M\frac{d\Phi}{d(c^2)} \qquad (4)$$

Substituting these values in (2), we have

$$\frac{d\Phi}{d(a^2)}-\frac{d\Phi}{d(b^2)}=c^2\left(\frac{1}{a^2}-\frac{1}{b^2}\right)\frac{d\Phi}{d(c^2)},\ \text{or}$$

$$0=(a^2-b^2)\int_0^\infty \frac{d\psi}{\sqrt{(a^2+\psi)(b^2+\psi)(c^2+\psi)}}\left\{-\frac{1}{(a^2+\psi)(b^2+\psi)}+\frac{c^2}{a^2b^2}\frac{1}{c^2+\psi}\right\}. \qquad (5)$$

$a^2=b^2$ gives the ellipsoid of revolution already treated. But the equation may be satisfied without assuming $a^2=b^2$, since the factor in brackets, in the integral, may be written in the form

$$\frac{(c^2a^2+c^2b^2-a^2b^2)\psi+c^2\psi^2}{a^2b^2(a^2+\psi)(b^2+\psi)(c^2+\psi)},$$

of which the numerator alone can change sign. Now if c be greater than the greatest of a and b, the integral is positive; if c be very small, it is evidently negative. Hence for any finite values of a and b whatever the integral may be made to vanish by properly assigning c. With this value of c the integral contains an equal amount of positive and negative elements. But it can have *no* negative elements unless $c^2a^2+c^2b^2-a^2b^2$ is negative, *i.e.*, unless c is less than the least of a and b.

Lastly, by (2) and (4)

$$\omega^2=\tfrac{3}{2}\frac{M}{a^2}\left\{c^2\frac{d\Phi}{d(c^2)}-a^2\frac{d\Phi}{d(a^2)}\right\}$$

$$=\tfrac{3}{2}\frac{M}{a^2}\int_0^\infty \frac{d\psi}{\sqrt{(a^2+\psi)\,(b^2+\psi)\,(c^2+\psi)}}\,\frac{(a^2-c^2)\psi}{(a^2+\psi)\,(c^2+\psi)}.$$

which is positive, as we have shown that c is less than a; and gives the required angular velocity when c has been found from (5).

Digression on spherical harmonics.

779. A few words of explanation, and some graphic illustrations, of the character of spherical surface harmonics may promote the clear understanding not only of the potential and hydrostatic applications of Laplace's analysis, which will occupy us presently, but of much more important applications to be made in Vol. II., when waves and vibrations in spherical fluid or elastic solid masses will be treated. To avoid circumltions, we shall designate by the term ***harmonic spheroid***, or ***herical harmonic undulation***, a surface whose radius to any int differs from that of a sphere by an infinitely small length rying as the value of a surface harmonic function of the sition of this point on the spherical surface. The definitions spherical Solid and Surface Harmonics [App. B. (*a*), (*b*), (*c*)] ow that the harmonic spheroid of the second order is a surface the second degree subject only to the condition of being proximately spherical: that is to say, it may be any elliptic heroid (or ellipsoid with approximately equal axes). Generally a harmonic spheroid of any order i exceeding 2 is a urface of algebraic degree i more restricted than merely to eing approximately spherical. Digression on spherical harmonics. Harmonic spheroid.

Let S_i be a surface harmonic of order i with coefficient of leading term so chosen as to make the greatest maximum value of the function unity. Then if a be the radius of the mean sphere, and c the greatest deviation from it, the polar equation of a harmonic spheroid of order i will be

$$r = a + cS_i \qquad (1)$$

if S_i is regarded as a function of polar angular co-ordinates, θ, ϕ. Considering that $\frac{c}{a}$ is infinitely small, we may reduce this to an equation in rectangular co-ordinates of degree i, thus:—Squaring each member of (1); and putting $\frac{cr^i}{a^{i+1}}$ for $\frac{c}{a}$, from which it differs by an infinitely small quantity of the second order, we have

$$r^2 = a^2 + \frac{2c}{a^i} \cdot r^i S_i \qquad (2).$$

This, reduced to rectangular co-ordinates, is of algebraic degree i.

780. The line of no deviation from the mean spherical surface is called the ***nodal line***, or the ***nodes*** of the harmonic spheroid. It is the line in which the spherical surface is cut by the ***harmonic nodal cone***; a certain cone with vertex at the centre of the sphere, and of algebraic degree equal to the order of the harmonic. An important property of the harmonic nodal line, indicated by an interesting hydrodynamic theorem due to Rankine,[1] is that when self-cutting at any point or points, the different branches make equal angles with one another round each point of section. Harmonic nodal cone and line.

[1] "Summary of the Properties of certain Stream-Lines." *Phil. Mag.*, Oct. 1864.

Digression on spherical harmonics.

Denoting $r^i S_i$ of § 779 by V_i, we have

$$V_i = 0$$

Theorem regarding nodal cone.

for the equation of the harmonic nodal cone. As V_i is [App. B. (a)] a homogeneous function of degree i, we may write

$$V_i = H_0 z^i + H_1 z^{i-1} + H_2 z^{i-2} + H_3 z^{i-3} + \text{etc.} \quad (4).$$

where H_0 is a constant, and H_1, H_2, H_3, etc., denote integral homogeneous functions of x, y of degrees 1, 2, 3, etc.; and then the condition $\nabla^2 V_i = 0$ [App. B. (a)] gives

$$\left.\begin{array}{c} \nabla^2 H_2 + i(i-1)H_0 = 0, \quad \nabla^2 H_3 + (i-1)(i-2)H_1 = 0, \\ \nabla^2 H_s + (i-s+2)(i-s+1)H_{i-s+2} = 0 \end{array}\right\} \quad (5),$$

which express all the conditions binding on H_0, H_1, H_2, etc.

Now suppose the nodal cone to be autotomic, and, for brevity and simplicity, take OZ along a line of intersection. Then $z = c$ makes (3) the equation in x, y, of a curve lying in the tangent plane to the spherical surface at a double or multiple point of the nodal line, and touching both or all its branches in this point. The condition that the curve in the tangent plane may have a double or multiple point at the origin of its co-ordinates is, when (4) is put for V_i,

$$H_0 = 0; \text{ and, for all values of } x, y; \ H_1 = 0.$$

Hence (5) gives $\qquad \nabla^2 H_2 = 0,$

so that, if $\qquad H_2 = Ax^2 + By^2 + 2Cxy,$

we have $A + B = 0$. This shows that the two branches cut one another at right angles.

If the origin be a triple, or n-multiple point, we must have

$$H_0 = 0, \ H_1 = 0 \ldots H_{n-1} = 0,$$

and (5) gives $\qquad \nabla^2 H_n = 0.$

Hence [§ 707 (23)]

$$H_n = A\{(x+y\sqrt{-1})^n + (x-y\sqrt{-1})^n\} + B\sqrt{-1}\{(x+y\sqrt{-1})^n - (x-y\sqrt{-1})^n\},$$

or, if $x = \rho\cos\phi$, $y = \rho\sin\phi$,

$$H_n = 2\rho^n (A\cos n\phi + B\sin n\phi)$$

which shows that the n branches cut one another at equal angles round the origin.

Cases of solid harmonics resolvable into factors.

781. The harmonic nodal cone may, in a great variety of cases [V_i resolvable into factors], be composed of others of lower degrees. Thus (the only class of cases yet worked out) each of

Polar harmonics.

the $2i + 1$ elementary polar harmonics [as we may conveniently call those expressed by (36) or (37) of App. B., with any one alone of the $2i + 1$ coefficients A_s, B_s] has for its nodes circles f the spherical surface. These circles, for each such harmonic lement, are either (1.) all in parallel planes (as circles of latiude on a globe), and cut the spherical surface into zones, in which case the harmonic is called zonal; or (2.) they are all in planes through one diameter (as meridians on a globe), and cut the surface into equal sectors, in which case the harmonic is called sectorial; or (3.) some of them are in parallel planes, and the others in planes through the diameter perpendicular to those planes, so that they divide the surface into rectangular quadrilaterals, and (next the poles) triangular segments, as areas on a globe bounded by parallels of latitude, and meridians at equal successive differences of longitude.

Digression on spherical harmonics.

Zonal and sectorial harmonics defined.

With a given diameter as axis of symmetry there are, for complete harmonics [App. B. (c), (d)], just one zonal harmonic of each order and two sectorial. The zonal harmonic is a function of latitude alone ($\frac{\pi}{2} - \theta$, according to the notation of App. B.); being the $\Theta_i^{(0)}$ given by putting $s = 0$ in App. B. (38). The sectorial harmonics of order i, being given by the same with $s = i$, are

$$\sin^i\theta \cos i\phi, \text{ and } \sin^i\theta \sin i\phi \quad (1).$$

The general polar harmonic element of order i, being the $\Theta_i^{(s)}\cos s\phi$ and $\Theta_i^{(s)}\sin s\phi$ of B. (38), with any value of s from 0 to i, has for its nodes $i-s$ circles in parallel planes, and s great circles intersecting one another at equal angles round their poles; and the variation from maximum to minimum along the equator, or any parallel circle, is according to the simple harmonic law. It is easily proved (as the mathematical student may find for himself) that the law of variation is *approximately* simple harmonic along lengths of each meridian cutting but a small number of the nodal circles of latitude, and not too near either pole, for any polar harmonic element of high order having a large number of such nodes (that is, any one for which $i-s$ is a large number). The law of variation along a meridian in the neighbourhood of either pole, for polar harmonic elements of high orders, will be carefully examined and illustrated in Vol. II., when we shall be occupied with vibrations and waves of water in a circular vessel, and of a circular stretched membrane.

Tesseral division of surface by nodes of a polar harmonic.

Digression on spherical harmonics.

782. The following simple and beautiful investigation the zonal harmonic due to Murphy[1] may be acceptable to analytical student; but (§ 453) we give it as leading to a r- ful formula, with expansions deduced from it, differing fr any of those investigated above in App. B.:—

"Prop. I.

Murphy's analytical invention of the zonal harmonics.

"To find a rational and entire function of given dimension "with respect to any variable, such that when multiplied "*any* rational and entire function of lower dimensions, th "integral of the product taken between the limits 0 and "shall always vanish.

"Let $f(t)$ be the required function of n dimensions with respe "to the variable t; then the proposed condition will evidently re- "quire the following equations to be separately true; namely.

"$(a)\ldots\ldots\int f(t)dt=0,\ \int f(t).tdt=0,\ \int f(t).t^2dt=0,\ldots\ldots\int f(t).t^{n-1}dt=$

"each integral being taken between the given limits.

"Let the indefinite integral of $f(t)$, commencing when $t=0$, be "represented by $f_1(t)$; the indefinite integral of $f_1(t)$, commencing "also when $t=0$, by $f_2(t)$; and so on, until we arrive at the "function $f_n(t)$, which is evidently of $2n$ dimensions. Then the "method of integrating by parts will give, generally,

"$\int f(t).t^x dt=t^x f_1(t)-xt^{x-1}f_2(t)+x.(x-1).t^{x-2}.f_3(t)-\text{etc.}$

"Let us now put $t=1$, and substitute for x the values 1, 2, 3 "......$(i-1)$ successively; then in virtue of the equations (... "we get,

"$(b)\ldots\ldots\ldots f_1(t)=0,\ f_2(t)=0,\ f_3(t)=0,\ldots\ldots\ldots f_i(t)=0.$

"Hence, the function $f_i(t)$ and its $(i-1)$ successive differentia "coefficients vanish, both when $t=0$, and when $t=1$; therefore "t^i and $(1-t)^i$ are each factors of $f_i(t)$; and since this function is "of $2i$ dimensions, it admits of no other factor but a constant c.

"Putting $1-t=t'$, we thus obtain

$$f_i(t)=c.(tt')^i;$$

"and therefore
$$f(t)=c\frac{d^i(tt')^i}{dt^i}.$$

"*Corollary.*—If we suppose the first term of $f(t)$, when arranged "according to the powers of t, to be unity, we evidently have "$c=\dfrac{1}{1.2.3\ldots\ldots i}$; on this supposition we shall denote the abo "quantity by Q_i.

[1] *Treatise on Electricity.* Cambridge, 1833.

" PROP. II.

" The function Q_i which has been investigated in the preceding proposition, is the same as the coefficient of e^i in the expansion of the quantity Digression on spherical harmonics. Murphy's analysis.

$$\{1-2e.(1-2t)+e^2\}^{-\frac{1}{2}}.$$

" Let u be a quantity which satisfies the equation

$$(c)\ldots\ldots\ldots\ldots u=t+e.u(1-u);$$

" that is, $u=-\frac{1-e}{2e}+\frac{1}{2e}\cdot\{1-2e.(1-2t)+e^2\}^{\frac{1}{2}};$

" therefore $\frac{du}{dt}=\{1-2e(1-2t)+e^2\}^{-\frac{1}{2}}.$

" But if, as before, we write t' for $1-t$, we have, by Lagrange's " theorem, applied to the equation (c)

$$u=t+e.tt'+\frac{e^2}{1.2}\cdot\frac{d(tt')^2}{dt}+\frac{e^3}{1.2.3}\cdot\frac{d^2.(tt')^3}{dt^2}+\text{etc.}$$

" If we differentiate, and put for $\frac{d^i(tt')^i}{dt^i}$ its value $1.2.3\ldots iQ_i$ given " by the former proposition, we get

$$\frac{du}{dt}=1+Q_1e+Q_2e^2+Q_3e^3+\text{etc.}$$

" Comparing this with the above value of $\frac{du}{dt}$ the proposition is " manifest.

" PROP. V.

" To develop the function Q_i. Expansions of zonal harmonic.

" *First Expansion.*—By Prop. I., we have

$$Q_i=\frac{1}{1.2.3\ldots i}\cdot\frac{d^i.(tt')^i}{dt^i}.$$

" Hence $Q_i=\frac{1}{1.2.3\ldots i}\cdot\frac{d^i}{dt^i}\{t^i-it^{i+1}+\frac{i.(i-1)}{1.2}\cdot t^{i+2}-\text{etc.}\}$

$$\text{" }(e)\ldots\ldots=1-\frac{i}{1}\,\frac{i+1}{1}\cdot t+\frac{i.(i-1)}{1.2}\cdot\frac{(i+1)\,(i+2)}{1.2}\cdot t^2-\text{etc.}$$

" *Second Expansion.*—If u and v are functions of any variable t, " then the theorem of Leibnitz gives the identity

$$\frac{d^i(uv)}{dt^i}=v\frac{d^iu}{dt^i}+i\frac{dv}{dt}\cdot\frac{d^{i-1}u}{dt^{i-1}}+\frac{i(i-1)}{1.2}\cdot\frac{d^2v}{dt^2}\cdot\frac{d^{i-2}v}{dt^{i-2}}+\text{etc.}$$

" Put $u=t^i$ and $v=t'^i$, and dividing by $1.2.3\ldots i$, we have

Digression on spherical harmonics.

Formulæ for zonal,

$$\text{“}(f)\ldots\ldots Q_i = t'^i - \left(\frac{i}{1}\right)^2 t'^{i-1}t$$

$$+\left\{\frac{i(i-1)}{1.2}\right\}^2 t'^{i-2}t^2 - \left\{\frac{i(i-1)(i-2)}{1.2.3}\right\}^2 t'^{i-3}t^3 + \text{\&c.”}$$

“*Third Expansion.*—Put $1-2t=\mu$, and therefore $tt'=\frac{1-\mu^2}{4}$

$$\text{“hence } Q_i = \frac{1}{2^i}\cdot\frac{1}{1.2.3\ldots i}\cdot\frac{d^i(\mu^2-1)^i}{d\mu^i}$$

$$=\frac{1}{2.4.6\ldots 2i}\cdot\frac{d^i}{d\mu^i}\cdot\{\mu^{2i} - i\mu^{2i-2} + \frac{i.(i-1)}{1.2}\cdot\mu^{2i-4} - \text{etc.}\}$$

$$\text{“}(g)\ldots\ldots\ldots = \frac{1.3.5\ldots(2i-1)}{1.2.3\ldots i}\cdot\{\mu^i - \frac{i(i-1)}{2(2i-1)}\cdot\mu^{i-2}$$

$$+\frac{i(i-1)(i-2)(i-3)}{2.4.(2i-1)(2i-3)}\cdot\mu^{i-4} - \text{etc.}\}\text{”}$$

The t, t' and μ of Murphy's notation are related to the θ we have used, thus:—

$$\left.\begin{array}{c} t=(2\sin\tfrac{1}{2}\theta)^2, \quad t'=(2\cos\tfrac{1}{2}\theta)^2 \\ \mu=\cos\theta \end{array}\right\} \qquad (2).$$

Also it is convenient to recall from App. B. (v'), (38), (40), and (42), that the value of Q_i [or $\vartheta_i^{(0)}$ of App. B. (61)], when $\theta=0$ is unity, and that it is related to the $\Theta_i^{(s)}$, of our notation for polar harmonic elements, thus:—

$$\vartheta_i^{(0)} = Q_i = \cdot\frac{1.3.5\ldots(2i-1)}{1.2.3\ldots i}\Theta_i^{(0)} \qquad (3).$$

and tesseral harmonics.

as is proved also by comparing (g) with App. B. (38). We add the following formula, manifest from (38), which shows a derivation of $\Theta_i^{(s)}$ from $\Theta_i^{(0)}$, valuable if only as proving that the $i-s$ roots of $\Theta_i^{(s)}=0$ are all real and unequal, inasmuch as App. B. (p) proves that the i roots of $\Theta_i^{(0)}=0$ are all real and unequal:—

$$\frac{\Theta_i^{(s)}}{\sin\theta} = \frac{1}{i-s+1}\,\frac{d}{d\mu}\left[\frac{\Theta_i^{(s-1)}}{\sin^{s-1}\theta}\right] \qquad (4).$$

From this and (3) we find

$$\Theta_i^{(s)} = \frac{1.2.3\ldots(i-s)}{1.3.5\ldots(2i-1)}\sin^s\theta\,\frac{d^sQ_i}{d\mu^s} \qquad (5).$$

And lastly, referring to App. B. (w); let

$$Q'_i \text{ and } Q_i[\cos\theta\cos\theta' + \sin\theta\sin\theta'\cos(\phi-\phi')]$$

denote respectively what Q_i becomes when $\cos\theta$ is replaced by $\cos\theta'$, and again by $\cos\theta\cos\theta'+\sin\theta\sin\theta'\cos(\phi-\phi')$: and let μ denote $\cos\theta$; and μ', $\cos\theta'$. By what precedes, we may put (61)

* See Errata.

of App. B. into the following much more convenient form, agreeing with that given by Murphy (*Electricity*, p. 24):—

Biaxal harmonic expanded.

$$Q_i[\cos\theta\cos\theta'+\sin\theta\sin\theta'\cos(\phi-\phi')]$$

$$Q_i'+2\left\{\frac{\cos(\phi-\phi')}{i(i+1)}\sin\theta\sin\theta'\frac{dQ_i}{d\mu}\frac{dQ_i'}{d\mu}+\frac{\cos 2(\phi-\phi')}{(i-1)i(i+1)(i+2)}\sin^2\theta\sin^2\theta'\frac{d^2Q_i}{d\mu^2}\frac{d^2Q_i'}{d\mu^2}+\text{etc.}\right\} \quad (6).$$

783. Elementary polar harmonics become, in an extreme case of spherical harmonic analysis, the proper harmonics for the treatment, by either polar or rectilineal rectangular co-ordinates, of problems in which we have a plane, or two parallel planes, instead of a spherical surface, or two concentric spherical surfaces, thus:—

Physical problems relative to plane rectangular and circular plates.

First, let S_i be any surface harmonic of order i, and V_i and V_{-i-1} the solid harmonics [App. B. (b)] equal to it on the spherical surface of radius a: so that

$$V_i=\left(\frac{r}{a}\right)^i S_i, \text{ and } V_{-i-1}=\left(\frac{a}{r}\right)^{i+1} S_i.$$

Now [compare § 655]

$$\left(\frac{r}{a}\right)^i=\varepsilon^{i\log\frac{r}{a}};$$

and, therefore, if a be infinite, and $r-a$ a finite quantity denoted by x, which makes

$$\log\frac{r}{a}=\frac{x}{a},$$

and if i be infinite, and $\frac{a}{i}=p$, we have

$$\left(\frac{r}{a}\right)^i=\varepsilon^{\frac{i}{a}x}=\varepsilon^{\frac{x}{p}}, \text{ and similarly } \left(\frac{a}{r}\right)^{i+1}=\varepsilon^{-\frac{i+1}{a}x}=\varepsilon^{-\frac{x}{p}};$$

the solid harmonics then become

$$\varepsilon^{\frac{x}{p}}S_i \text{ and } \varepsilon^{-\frac{x}{p}}S_i.$$

Supposing now S_i to be a polar harmonic element, and considering, as Green did in his celebrated Essay on Electricity, an area sensibly plane round either pole, or considering any sensibly plane portion far removed from each pole, it is interesting and instructive to examine how the formulæ [App. B. (36)...(40), (61), (65); and § 782, (e), (f), (g)] wear down to the proper plane polar or rectangular formulæ. This we may safely leave to the analytical student. In Vol. II. the plane polar solution will be fully examined. At present we merely remark that, in rectangular surface co-ordinates (y, x) in the spherical surface become plane, S_i may be any function whatever fulfilling the equation

Physical problems relative to plane rectangular and circular plates.

$$\frac{d^2S_i}{dy^2}+\frac{d^2S_i}{dz^2}+\frac{S_i}{p^2}=0,$$

and that the rectangular solution into which the elementary spherical harmonic wears down, for sensibly plane portions of a spherical surface far removed from the poles, is

$$S_i=\cos\frac{y}{q}\cos\frac{z}{q'},$$

where q and q' are two constants such that

$$q^2+q'^2=p^2.$$

Examples of polar harmonics.

784. The following tables and graphic representations of all the polar harmonic elements of the 6th and 7th orders may be useful in promoting an intelligent comprehension of the subject.

Sixth order: Zonal,

$$Q_6=\frac{1}{16}(231\mu^6-315\mu^4+105\mu^2-5) \qquad =\frac{231}{16}\Theta_6^{(0)}.$$

$$\frac{1}{21}\cdot\frac{dQ_6}{d\mu}=\frac{1}{8}(33\mu^4-30\mu^2+5)\mu \qquad =\frac{33}{8}\Theta_6^{(1)}(1-\mu^2)^{-\frac{1}{2}}$$

Tesseral,

$$\frac{1}{210}\cdot\frac{d^2Q_6}{d\mu^2}=\frac{1}{16}(33\mu^4-18\mu^2+1) \qquad =\frac{33}{16}\Theta_6^{(2)}(1-\mu^2)^{-1}$$

$$\frac{1}{1260}\cdot\frac{d^3Q_6}{d\mu^3}=\frac{1}{8}(11\mu^2-3)\mu \qquad =\frac{11}{8}\Theta_6^{(3)}(1-\mu^2)^{-\frac{3}{2}}$$

$$\frac{1}{4725}\cdot\frac{d^4Q_6}{d\mu^4}=\frac{1}{10}(11\mu^2-1) \qquad =\frac{11}{10}\Theta_6^{(4)}(1-\mu^2)^{-2}$$

$$\frac{1}{10395}\cdot\frac{d^5Q_6}{d\mu^5}=\mu \qquad =\Theta_6^{(5)}(1-\mu^2)^{-\frac{5}{2}}$$

Sectorial.

$$\frac{1}{10395}\cdot\frac{d^6Q_6}{d\mu^6}=1 \qquad =\Theta_6^{(6)}(1-\mu^2)^{-3}.$$

Seventh order: Zonal,

$$Q_7=\frac{1}{16}(429\mu^6-693\mu^4+315\mu^2-35)\mu=\frac{429}{16}\Theta_7^{(0)}.$$

Tesseral,

$$\frac{1}{28}\cdot\frac{dQ_7}{d\mu}=\frac{1}{64}(429\mu^6-495\mu^4+135\mu^2-5) \qquad =\frac{429}{64}\Theta_7^{(1)}(1-\mu^2)^{-\frac{1}{2}}$$

$$\frac{1}{378}\cdot\frac{d^2Q_7}{d\mu^2}=\frac{1}{48}(143\mu^4-110\mu^2+15)\mu \qquad =\frac{143}{48}\Theta_7^{(2)}(1-\mu^2)^{-1}$$

$$\frac{1}{3150}\cdot\frac{d^3Q_7}{d\mu^3}=\frac{1}{80}(143\mu^4-66\mu^2+3) \qquad =\frac{143}{80}\Theta_7^{(3)}(1-\mu^2)^{-\frac{3}{2}}$$

$$\frac{1}{17325}\cdot\frac{d^4Q_7}{d\mu^4}=\frac{1}{10}(13\mu^2-3)\mu \qquad =\frac{13}{10}\Theta_7^{(4)}(1-\mu^2)^{-2}$$

$$\frac{1}{62370}\cdot\frac{d^5Q_7}{d\mu^5}=\frac{1}{12}(13\mu^2-1) \qquad =\frac{13}{12}\Theta_7^{(5)}(1-\mu^2)^{-\frac{5}{2}}$$

$$\frac{1}{135135}\cdot\frac{d^6Q_7}{d\mu^6}=\mu \qquad =\Theta_7^{(6)}(1-\mu^2)^{-3}$$

Sectorial.

$$\frac{1}{135135}\cdot\frac{d^7Q_7}{d\mu^7}=1 \qquad =\Theta_7^{(7)}(1-\mu^2)^{-\frac{7}{2}}$$

Polar harmonics of sixth order.

μ.	Q_6.	$\frac{1}{21}\frac{dQ_6}{d\mu}$.	$\frac{33}{8}\Theta_6^{(1)}$.	$\frac{1}{210}\frac{d^2Q_6}{d\mu^2}$.	$\frac{33}{16}\Theta_6^{(2)}$.
·0	— ·3125	·0000	·0000	+ ·0625	+ ·0625
·01	— ·3118	......	......	......	......
·05	— ·2961	+ ·0308	+ ·0307	+ ·0597	+ ·0595
·08	— ·2738	......	......	......	......
·10	— ·2488	+ ·0588	+ ·0585	+ ·0515	+ ·0510
·13	— ·2072	......	......	......	......
·15	— ·1746	+ ·0814	+ ·0805	+ ·0382	+ ·0378
·17	— ·1390	......	......	......	......
·2	— ·0806	+ ·0963	+ ·0944	+ ·0208	+ ·0200
·24	+ ·0029	......	......	......	......
·25	+ ·0248	+ ·1017	+ ·0984	+ ·0002	+ ·0002
·2506	......	......	......	·0000	·0000
·3	+ ·1293	+ ·0966	+ ·0921	— ·0221	— ·0201
·34	+ ·2058	......	......	......	......
·35	+ ·2225	+ ·0796	+ ·0745	— ·0441	— ·0387
·36	+ ·2388	......	......	......	......
·4	+ ·2926	+ ·0522	+ ·0479	— ·0647	— ·0544
·43	+ ·3191	......	......	......	......
·45	......	+ ·0157	+ ·0140	— ·0807	— ·0644
·46	+ ·3314	......	......	......	......
·4688	......	·0000	·0000	......	......
·469	+ ·3321	......	......	......	......
·5	+ ·3233	— ·0273	— ·0237	— ·0898	— ·0674
·54	+ ·2844	......	......	......	......
·55	......	— ·0726	— ·0606	— ·0891	— ·0622
·56	+ ·2546	......	......	......	......
·6	+ ·1721	— ·1142	— ·0914	— ·0752	— ·0481
·63	+ ·0935	......	......	......	......
·65	......	— ·1450	— ·1102	— ·0446	— ·0258
·66	+ ·0038	......	......	......	......
·7	— ·1253	— ·1555	— ·1110	+ ·0064	+ ·0033
·74	— ·2517	......	......	......	......
·75	— ·2808	— ·1344	— ·0889	+ ·0823	+ ·0360
·76	— ·3087	......	......	......	......
·8	— ·3918	— ·0683	— ·0410	+ ·1873	+ ·0674
·82	— ·4119	......	......	......	......
·8302	— ·4147	·0000	·0000	......	......
·84	— ·4119	......	......	......	......
·85	— ·4030	+ ·0586	+ ·0308	+ ·3263	+ ·0905
·87	— ·3638	......	......	......	......
·90	— ·2412	+ ·2645	+ ·1153	+ ·5044	+ ·0958
·92	— ·1084	+ ·1764	+ ·1464	......	......
·93	......	+ ·4346	+ ·1597	......	......
·9325	·0000	......	......	......	......
94	+ ·0751	+ ·5002	+ ·1706	......	......
·95	......	+ ·5704	+ ·1778	+ ·7271	+ ·0709
·96	+ ·3150	......	......	......	......
·97	......	+ ·7260	+ ·1764	......	......
·98	+ ·6203	+ ·8117	+ ·1615	+ ·8844	+ ·0350
·99	+ ·8003	+ ·9029	+ ·1274	+ ·9411	+ ·0187
1·00	+ 1·0000	+ 1·0000	·0000	+ 1·0000	+ ·0000

Polar harmonics of sixth order.

μ.	$\frac{1}{1260}\frac{d^3Q_6}{d\mu^3}$.	$\frac{11}{8}\Theta_6^{(3)}$.	$\frac{1}{4725}\frac{d^4Q_6}{d\mu^4}$.	$\frac{11}{10}\Theta_6^{(4)}$.	$\Theta_6^{(5)}$.
·0	·0000	·0000	— ·1000	— ·1000	·0000
·05	— ·0186	— ·0185	— ·0975	— ·0970	+ ·0497
·1	— ·0361	— ·0356	— ·0890	— ·0886	+ ·0975
·15	— ·0516	— ·0499	— ·0753	— ·0720	+ ·1417
·2	— ·0640	— ·0602	— ·0560	— ·0516	+ ·1806
·25	— ·0723	— ·0656	— ·0313	— ·0275	+ ·2127
·3	— ·0754	— ·0655	— ·0010	— ·0008	+ ·2370
·35	— ·0767	— ·0630	+ ·0348	+ ·0268	+ ·2524
·4	— ·0620	— ·0477	+ ·0760	+ ·0536	+ ·2586
·45	— ·0435	— ·0310	+ ·1227	+ ·0773	+ ·2555
·5	— ·0156	— ·0101	+ ·1750	+ ·0984	+ ·2436
·55	+ ·0225	+ ·0131	+ ·2327	+ ·1132	+ ·2234
·6	+ ·0720	+ ·0369	+ ·2960	+ ·1211	+ ·1966
·63	......	......	+ ·3366	+ ·1224	......
·65	+ ·1338	+ ·0587	+ ·3647	+ ·1204	+ ·1647
·7	+ ·2091	+ ·0750	+ ·4390	+ ·1139	+ ·1300
·75	+ ·2988	+ ·0865	+ ·5188	+ ·0991	+ ·0949
·8	+ ·4040	+ ·0873	+ ·6040	+ ·0783	+ ·0622
·83	......	......	+ ·6578	+ ·0637	......
·85	+ ·5257	+ ·0768	+ ·6947	+ ·0535	+ ·0344
·87	......	......	+ ·7326	+ ·0433	......
·89	......	......	+ ·7718	+ ·0333	......
·9	+ ·6649	+ ·0551	+ ·7910	+ ·0285	+ ·0150
·92	......	......	......	......	+ ·0085
·93	+ ·7572	+ ·0376	+ ·8514	+ ·0155	......
·95	+ ·8226	+ ·0249	+ ·8928	+ ·0084	+ ·0028
·96	+ ·8565	......	+ ·9138	......	......
·97	+ ·8911	+ ·0128	+ ·9350	+ ·0032	......
·98	+ ·9216	+ ·0078	+ ·9564	+ ·0015	......
·99	+ ·9629	......	+ ·9781	+ ·0004	......
1·00	+ 1·0000	·0000	+ 1·0000	·0000	·0000

Graphic representation of polar harmonics sixth order.

Diag. No. 1.

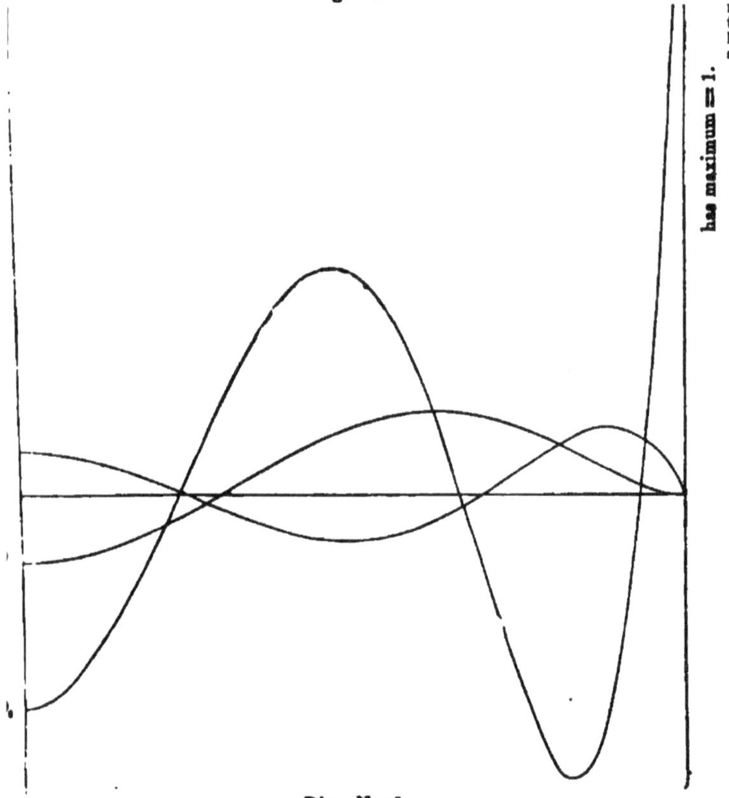

Diag. No. 2.

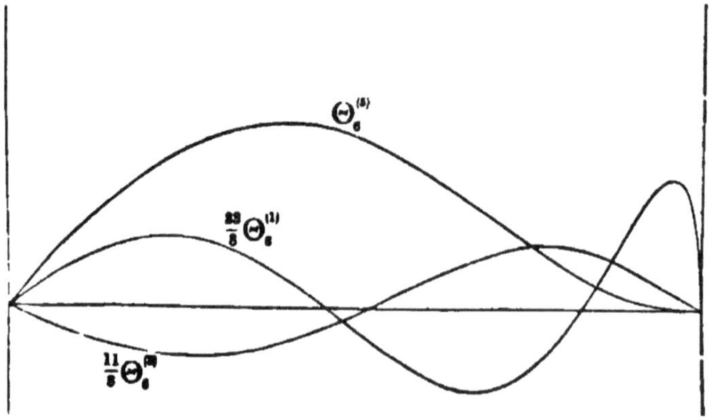

Graphic representation of polar harmonics of seventh order.

Diag. No. 3.

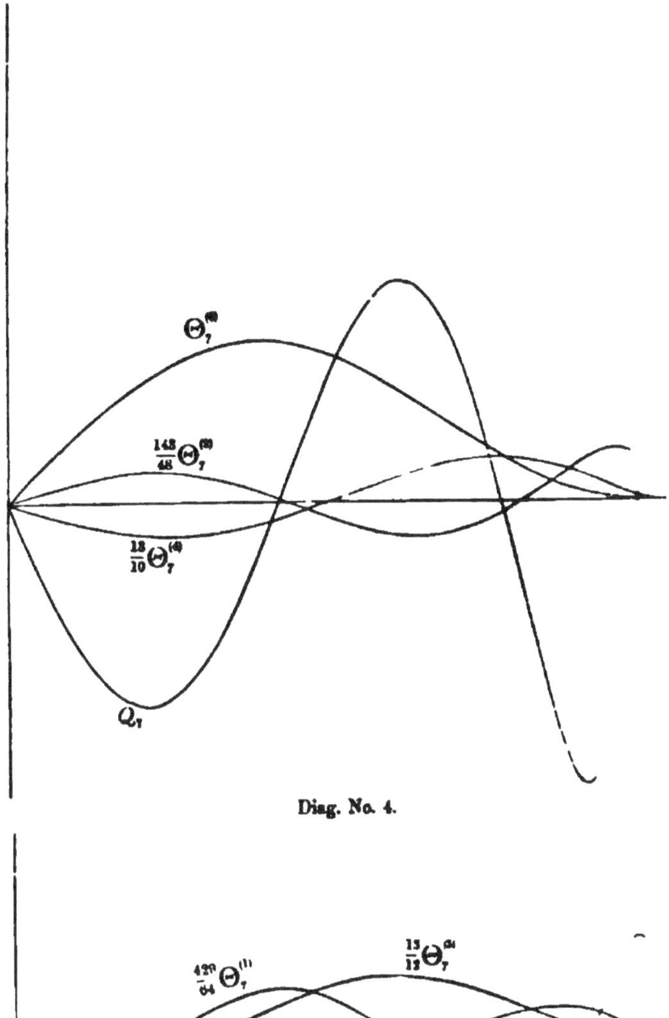

Diag. No. 4.

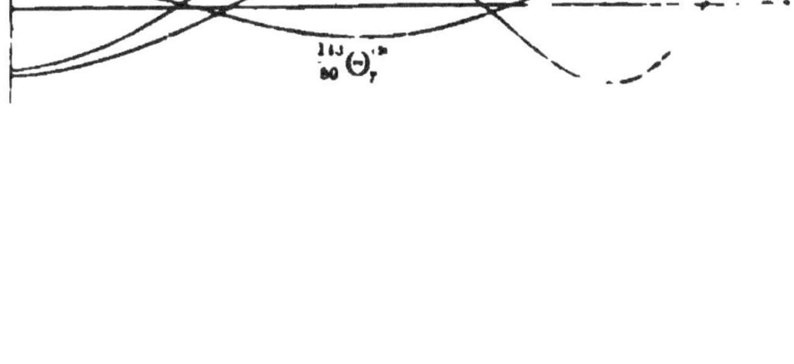

Polar harmonics of seventh order.

μ.	Q_7.	$\frac{1}{28}\frac{dQ_7}{d\mu}$.	$\frac{429}{64}\Theta_7^{(1)}$.	$\frac{1}{278}\frac{d^2Q_7}{d\mu^2}$.	$\frac{143}{48}\Theta_7^{(2)}$.
·0	·0000	— ·0781	— ·0781	·0000	·0000
·05	— ·1069	— ·0720	— ·0719	+ ·0153	+ ·0153
·1	— ·1995	— ·0578	— ·0522	+ ·0290	+ 0287
·15	— ·2649	— ·0345	— ·0341	+ ·0394	+ ·0385
·18	— ·2873	......	......	......	......
·2	— ·2935	— ·0057	— ·0056	+ ·0451	+ ·0433
·2098	......	·0000	·0000	......	......
·2261	......	......	......	+ 0459	......
·23	— ·2905	......	......	......	......
·24	......	+ ·0190	......	......	......
·25	— ·2799	+ ·0251	+ ·0248	+ ·0452	+ ·0424
·3	— ·2240	+ ·0540	+ ·0515	+ ·0391	+ ·0356
·35	— ·1318	+ ·0765	+ ·0717	+ ·0268	+ ·0235
·38	— ·0635	......	......	......	......
·4	— ·0365	+ ·0888	+ ·0814	+ ·0084	+ ·0074
·42	+ ·0356	......	......	......	......
·4209	......	+ ·0901	......	·0000	·0000
·45	+ ·1106	+ ·0875	+ ·0782	— ·0132	— ·0105
·5	+ ·2231	+ ·0706	+ ·0611	— ·0371	— ·0278
·53	......	......	......	......	— ·0366
·55	+ ·3007	+ ·0378	+ ·0315	......	— ·0415
·57	+ ·3207	......	......	......	......
·58	......	......	......	......	— ·0415
·5917	+ ·3236	·0000	·0000	......	......
·6	+ ·3226	— ·0115	— ·0092	— ·0758	— ·0485
·62	+ ·3121	......	......	......	......
·6406	......	......	......	— ·0809	......
·65	+ ·2737	— ·0619	— ·0471	— ·0806	— ·0465
·7	+ ·1502	— ·1129	— ·0806	— ·0666	— ·0340
·7415	·0000	......	......	......	......
·75	— ·0342	— ·1458	— ·0964	— ·0254	— ·0111
·7694	......	— ·1490	......	......	......
·7695	......	......	......	·0000	·0000
·8	— ·2397	— ·1390	— ·0834	+ ·0529	+ ·0190
·82	— ·3134	......	......	......	......
·85	— ·3918	— ·0634	— ·0334	+ ·1801	+ ·0500
·86	— ·4054	......	......	......	......
·8717	— ·4117	·0000	·0000	......	......
·88	— ·4082	......	......	......	......
·9	— ·3678	+ ·1183	+ ·0515	+ ·3698	+ ·0723
·92	— ·2713	......	......	......	......
·93	......	......	......	+ ·5276	+ 0712
·9491	·0000	......	......	......	......
·95	+ ·0112	+ ·4533	+ ·1413	+ ·6378	+ ·0621
·97	+ ·3165	+ ·6421	+ ·1563	+ ·7699	+ ·0455
·98	+ ·5115	+ ·7517	+ ·1458	......	......
·99	+ ·7384	+ ·8706	+ ·1230	+ ·9190	+ ·0184
1·00	+ 1·0000	+ 1·0000	·0000	+ 1·0000	·0000

Polar harmonics of seventh order.

μ.	$\frac{1}{3150}\frac{d^3Q_7}{d\mu^3}$.	$\frac{143}{80}\Theta_7^{(3)}$.	$\frac{1}{17325}\frac{d^4Q_7}{d\mu^4}$	$\frac{13}{10}\Theta_7^{(4)}$;	$\frac{1}{62370}\frac{d^5Q_7}{d\mu^5}$	$\frac{13}{12}\Theta_7^{(5)}$.	Θ.
·0	+ ·0375	+ ·0375	·0000	·0000	— ·0833	— ·0833	[illegible]
·05	+ ·0355	·0353	— ·0148	— ·0147	— ·0806	— ·0801	+ [illegible]
·1	+ ·0294	·0290	— ·0287	— ·0281	— ·0725	— ·0707	+ [illegible]
·15	+ ·0198	·0192	— ·0406	— ·0387	— ·0590	— ·0557	+ [illegible]
·2	+ ·0074	·0068	— ·0496	— ·0457	— ·0400	— ·0361	+ [illegible]
·2261	·0000	·0000	......	......	......	......	......
·25	— ·0071	— ·0064	— ·0544	— ·0478	— ·0156	— ·0133	+ [illegible]
·2773	......	......	— ·0555	......	·0000	·0000	......
·3	— ·0225	— ·0195	— ·0549	— ·0454	+ ·0142	+ ·0112	+ [illegible]
·35	— ·0367	— ·0302	— ·0493	— ·0378	+ ·0494	+ ·0356	+ ·2364
·4	— ·0487	— ·0375	— ·0368	— ·0260	+ ·0900	+ ·0582	+ [illegible]
·45	— ·0563	— ·0400	— ·0165	— ·0104	+ ·1361	+ ·0773	+ [illegible]
·4804	— ·0577	......	·0000	·0000	......	......	......
·5	— ·0570	— ·0370	+ ·0125	+ ·0070	+ ·1875	+ ·0913	+ ·211
·55	— ·0485	— ·0282	+ ·0513	+ ·0248	+ ·2564	+ ·1041	+ [illegible]
·6	— ·0278	— ·0142	+ ·0708	+ ·0412	+ ·3067	+ ·1004	+ ·1573
·6406	·0000	·0000	......	......	......	......	......
·65	+ ·0080	+ ·0035	+ ·1620	+ ·0540	+ ·3744	+ ·0948	+ ·125
·7	+ ·0624	+ ·0227	+ ·2359	+ ·0613	+ ·4475	+ ·0831	+ [illegible]
·75	+ ·1390	+ ·0401	+ ·3234	+ ·0619	+ ·5260	+ ·0665	+ [illegible]
·8	+ ·2417	+ ·0521	+ ·4256	+ ·0551	+ ·6100	+ ·0474	+ ·03
·85	+ ·3745	+ ·0546	+ ·5434	+ ·0418	+ ·6994	+ ·0283	+ [illegible]
·9	+ ·5420	+ ·0448	+ ·6777	+ ·0244	+ ·7942	+ ·0132	+ [illegible]
·92	+ ·6197	+ ·0373	+ ·7363	+ ·0170	......	......	+ [illegible]
·95	+ ·7489	+ ·0227	+ ·8732	+ ·0083	+ ·8944	+ ·0026	+ [illegible]
·97	+ ·8062	+ ·0116	+ ·9230	+ ·0032	......	......	+ [illegible]
·98	+ ·9564	+ ·0076	......	......	......	......	......
·99	......	......	......	......	......	......	......
1·00	1·0000	·0000	+ 1·0000	·0000	+ 1·0000	·0000	·0000

785. A short digression here on the theory of the potential, and particularly on equipotential surfaces differing little from concentric spheres, will simplify the hydrostatic examples which follow. First we shall take a few cases of purely synthetical investigation, in which, distributions of matter being given, resulting forces and level surfaces (§ 487) are found; and then certain problems of Green's and Gauss's analysis, in which, from data regarding amounts of force or values of potential over individual surfaces, or shapes of individual level surfaces, the distribution of force through continuous void space is to be determined. As it is chiefly for their application to physical geography that we admit these questions at present, we shall occasionally avoid circumlocutions by referring at once to the Earth, when any attracting mass with external equipotential surfaces approximately spherical would answer as well. We shall also sometimes speak of "*the sea level*" (§§ 750, 754) merely as a "level surface," or "surface of equilibrium" (§ 487) just enclosing the solid, or enclosing it with the exception of comparatively small projections, as our dry land. Such a surface will of course be an equipotential surface for mere gravitation, when there is neither rotation nor disturbance due to attractions of other bodies, as the moon or sun, and "change of motion" produced by these forces on the Earth; but it may be always called an equipotential surface, as we shall see (§ 793) that both centrifugal force and the other disturbances referred to may be represented by potentials.

Digression on theory of potential.

Sea level.

Level surface relatively to gravity and centrifugal force.

786. To estimate how the sea level is influenced, and how much the force of gravity in the neighbourhood is increased or diminished by the existence within a limited volume underground of rocks of density greater or less than average, let us imagine a mass equal to a very small fraction, $\frac{1}{n}$, of the earth's whole mass to be concentrated in a point somewhere at a depth below the sea level which we shall presently suppose to be small in comparison with the radius, but great in comparison with $\frac{1}{\sqrt{n}}$ of the radius. Immediately over the centre of disturbance, the sea level will be raised in virtue of the disturbing attraction, by a height equal to the same fraction of the radius

Disturbance of sea level by denser than average matter underground.

that the distance of the disturbing point from the chief centre is of n times its depth below the sea level as thus disturbed. The augmentation of gravity at this point of the sea level will be the same fraction of the whole force of gravity that n times the square of the depth of the attracting point is of the square of the radius. This fraction, as we desire to limit ourselves to natural circumstances, we must suppose to be very small. The disturbance of *direction* of gravity will, for the sea level, be a maximum at points of a circle described from A as centre, with $\frac{D}{\sqrt{2}}$ as radius; D being the depth of the centre of disturbance. The amount of this maximum deflection will be $\frac{2}{3\sqrt{3}}\frac{a^2}{nD^2}$ of the unit angle of $57°\cdot296$ (§ 41), a denoting the earth's radius.

Intensity and direction of gravity altered by underground local excess above average density.

Let C be the centre of the chief attracting mass $(1-\frac{1}{n})$, and

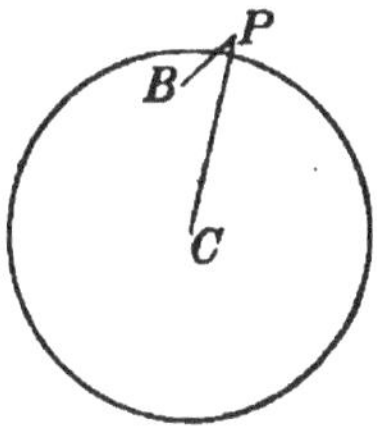

B that of the disturbing mass $(\frac{1}{n})$, the two parts being supposed to act as if collected at these points. Let P be any point on the equipotential surface for which the potential is the same as what it would be over a spherical surface of radius a, and centre C if the whole were collected in C. Then (§ 491),

$$(1-\frac{1}{n})\cdot\frac{1}{CP}+\frac{1}{n}\cdot\frac{1}{BP}=\frac{1}{a},$$

which is the equation of the equipotential surface in question. It gives

$$CP-a=\frac{a}{nBP}(CP-BP).$$

This expresses rigorously the positive or negative elevation of the disturbed equipotential at any point above the undisturbed surface of the same potential. For the point A, over the centre of disturbance, it gives

$$CA-a=\frac{a}{n.BA}\cdot CB,$$

which agrees exactly with the preceding statement: and it proves the approximate truth of that statement as applied to the sea level when we consider that when BP is many times BA, $CP-a$

is many times smaller than its value at A. We leave the proof of the remaining statements of this and the following sections (§§ 787...792) as an exercise for the student.

787. If ρ be the general density of the upper crust, and σ earth's mean density, and if the disturbance of § 786 be e to there being matter of a different density, ρ', throughout pherical portion of radius b, with its centre at a depth D low the sea level, the value of n will be $\frac{\sigma a^3}{(\rho'-\rho)b^3}$; and the vation of the sea level, and the proportionate augmentation gravity at the point right over it, will be respectively

Effects of local excess above average density on sea level, and on direction and intensity of gravity:

$$\frac{(\rho'-\rho)b^3}{\sigma a D}\cdot b, \text{ and } \frac{(\rho'-\rho)b^3}{\sigma a D^2}.$$

he actual value of σ is about double that of ρ. And let us ppose, for example, that $D = b = 1000$ feet, or $\frac{1}{21000}$ of the rth's radius, and ρ' to be either equal to 2ρ or to zero. The revious results become

example.

$$\pm \tfrac{1}{42} \text{ of a foot, and } \pm \tfrac{1}{42000} \text{ of gravity,}$$

hich are therefore the elevation or depression of sea level, and he augmentation or diminution of gravity, due to there being natter of double or zero density through a spherical space 2000 eet in diameter, with its centre 1000 feet below the surface. The greatest deviation of the plummet is at points of the circle f 707 feet radius round the point; and it amounts to $\frac{1}{100000}$ f the unit angle, or nearly 2″.

788. It is worthy of remark that, to set off against the increase in the amount of gravity due to the attraction of the disturbing mass, which we have calculated for points of the sea level in its neighbourhood, there is but an insensible deduction on account of the diminution of the attraction of the chief mass, owing to increase of the distance of the sea level from its centre, produced by the disturbing influence. The same remark obviously holds for disturbances in gravity due to isolated mountains, or islands of small dimensions, and it will be proved (§ 794) to hold also for deviations of figure represented by harmonics of high orders. But we shall see (§ 789) that it is otherwise with harmonic deviations of low orders, and consequently with wide-spread disturbances, such as are produced by great tracts of elevated land or of deep sea. We intend to return to the subject in Vol. II., under Properties of Matter when we shall have occasion to examine the phenomenal and experimental foundations of our knowledge of gravity; and we shall then apply §§ 477 (*b*) (*c*) (*d*), 478, 479, and solutions of other allied problems, to investigate the effects on magnitude and direction of gravity, and on the level surfaces, produced by isolated hills, mountain-chains, large table lands, and by corresponding depressions, as lakes or circumscribed deep places in the sea, great valleys or clefts, large tracts of deep ocean.

Harmonic spheroidal levels.

789. All the level surfaces relative to a harmonic spheroid (§ 779) of homogeneous matter are harmonic spheroids of the same order and type. That one of them, which lies as much inside the solid as outside it, cuts the boundary of the solid in a line (or group of lines)—the mean level line of the surface of the solid. This line lies on the mean spherical surface, and therefore (§ 780) it constitutes the nodes of each of the two harmonic spheroidal surfaces which cut one another in it. If i be the order of the harmonic, the deviation of the level spheroid is (§§ 545, 815) just $\frac{3}{2i+1}$ of the deviation of the bounding spheroid, each reckoned from the mean spherical surface.

Thus if $i=1$, the level coincides with the boundary of the solid: the reason of which is apparent when it is considered that any spherical harmonic deviation of the first order from a given spherical surface constitutes an equal spherical surface round a centre at some infinitely small distance from the centre of the given surface.

If $i=2$, the level surface deviates from the mean sphere by $\frac{3}{5}$ of the deviation of the bounding surface. This is the case of an ellipsoidal boundary differing infinitely little from spherical figure. It may be remarked that, as is proved readily from § 522, those of the equipotential surfaces relative to a homogeneous ellipsoid which lie wholly within it are exact ellipsoids, but not so those which cut its boundary or lie wholly without it; these being approximately ellipsoidal only when the deviation from spherical figure is very small.

790. The circumstances for very high orders are sufficiently illustrated if we confine our attention to sectorial harmonics

781). The figure of the line in which a sectorial harmonic ıeroid is cut by any plane perpendicular to its polar axis is 781 (1)], as it were, a harmonic curve (§ 62) traced from a cular instead of a straight line of abscissas. Its *wave length* · double length along the line of abscissas from one zero or dal point to the next in order) will be $\frac{1}{i}$ of the circumference the circle. And when i is very great, the factor $\sin^i\theta$ makes e sectorial harmonic very small, except for values of θ differ-g little from a right angle, and therefore a sectorial harmonic ıheroid of very high order consists of a set of parallel ridges ıd valleys perpendicular to a great circle of the globe, of ɜarly simple harmonic form in the section by the plane of ıis circle (or equator), and diminishing in elevation and ɜpression symmetrically on the two sides of it, so as to be ısensible at any considerable angular distance (or "latitude") ·om it on either side. The level surface due to the attraction f a homogeneous solid of this figure is a figure of the same ind, but of much smaller degree of elevations and depressions, hat is, as we have seen, only $\frac{3}{2i+1}$ of those of the figure: or ıpproximately three times the same fraction of the inequalities ıf figure that the half-wave length is of the circumference of he globe. It is easily seen that when i is very large the level surface at any place will not be sensibly affected by the in-ɜqualities in the distant parts of the figure.

Harmonic spheroidal levels of high orders.

791. Thus we conclude that, if the substance of the earth were homogeneous, a set of several parallel mountain-chains and valleys would produce an approximately corresponding un-dulation of the level surface in the middle district: the height to which it is raised, under each mountain-crest, or drawn down below the undisturbed level, over the middle of a valley, being three times the same fraction of the height of mountain above or depth of valley below mean level, that the breadth of the mountain or of the valley is of the earth's circumference.

Undulation of level due to parallel mountain-ridges and valleys.

792. If the globe be not homogeneous, the disturbance in magnitude and direction of gravity, due to any inequality in the figure of its bounding surface, will (§ 787) be $\frac{\rho}{\sigma}$ of what it would be if the substance were homogeneous; and further, it

Practical conclusions as to disturbances of sea level, and amount and direction of gravity.

may be remarked that, as the disturbances are supposed to be small, we may superimpose such as we have now described, on any other small disturbances, as, for instance, on the general oblateness of the earth's figure, with which we shall be occupied presently.

Practically, then, as the density of the upper crust is somewhere about $\frac{1}{2}$ the earth's mean density, we may say that the effect on the level surface, due to a set of parallel mountain-chains and valleys, is, of the general character explained in § 791, but of half the amounts there stated. Thus, for instance a set of several broad mountain-chains and valleys twenty nautical miles from crest to crest, or hollow to hollow, and of several times twenty miles extent along the crests and hollows, and 7,200 feet vertical height from hollow to crest, would raise and lower the level by $2\frac{1}{2}$ feet above and below what it would be were the surface levelled by removing the elevated matter and filling the valleys with it.

Determinateness of potential through space from its value over every point of a surface.

793. Green's theorem [App. A. (e)][1] and Gauss's theorem (§ 497) show that if the potential of any distribution of matter, attracting according to the Newtonian law, be given for every point of a surface completely enclosing this matter, the potential, and therefore also the force, is determined throughout all space external to the bounding surface of the matter, whether this surface consist of any number of isolated closed surfaces each simply continuous, or of a single one. It need scarcely be said that no general solution of the problem has been obtained. But further, even in cases in which the potential has been fully determined for the space outside the surface over which it is given, mathematical analysis has hitherto failed to determine it through the whole space between this surface and the attracting mass within it. We hope to return, in later volumes, to the grand problem suggested by Gauss's theorem of § 497. Meantime, we restrict ourselves to questions practically useful for physical geography.

Example (1.)—Let the enclosing surface be spherical, of radius a; and let $F(\theta, \phi)$ be the given potential at any point of it.

[1] First apply Green's theorem to the surface over which the potential is given. Then Gauss's theorem shows that there cannot be two distributions of potential agreeing through all space external to this surface, but differing for any part of the space between it and the bounding surface of the matter.

specified in the usual manner by the polar co-ordinates θ, ϕ. Green's solution [§ 499 (3) and App. B. (46)] of his problem for the spherical surface is immediately applicable to part of our present problem, and gives

Determination of potential from its value over a spherical surface enclosing the mass.

$$= \frac{1}{4\pi a}\int_0^{2\pi}\int_0^{\pi} \frac{(r^2-a^2)F(\theta',\phi')r^2\sin\theta' d\theta' d\phi'}{\{r^2-2ar[\cos\theta\cos\theta'+\sin\theta\sin\theta'\cos(\phi-\phi')]+a^2\}^{\frac{3}{2}}} \quad (3)$$

for the potential at any point (r, θ, ϕ) external to the spherical surface. But inasmuch as Laplace's equation $\nabla^2 u=0$ is satisfied through the whole internal space as well as the whole external space by the expression (46) of App. B., and in our present problem $\nabla^2 V=0$ is only satisfied [§ 491 (c)] for that part of the internal space which is not occupied by matter, the expression (3) gives the solution for the exterior space only. When $F(\theta, \phi)$ is such that an expression can be found for the definite integral in finite terms, this expression is necessarily the solution of our problem through all space exterior to the actual attracting body. Or when $F(\theta, \phi)$ is such that the definite integral, (3), can be transformed into some definite integral which varies continuously across the whole or across some part of the spherical surface, this other integral will carry the solution through some part of the interior space: that is, through as much of it as can be reached without discontinuity (infinite elements) of the integral, and without meeting any part of the actual attracting mass. To this subject we hope to return later in connexion with Gauss's theorem (§ 497); but for our present purpose it is convenient to expand (3) in ascending powers of $\frac{a}{r}$, as before in App. B. (*s*). The result [App. B. (51)] is

$$V=\frac{a}{r}F_0(\theta,\phi)+\left(\frac{a}{r}\right)^2 F_1(\theta,\phi)+\left(\frac{a}{r}\right)^3 F_2(\theta,\phi)+\text{etc.} \quad (3 \text{ bis})$$

where $F_0(\theta,\phi)$, $F_1(\theta,\phi)$, etc., are the successive terms of the expansion [App. B. (52)] of $F(\theta,\phi)$ in spherical surface harmonics; the general term being given by the formula

$$F_i(\theta,\phi)=\frac{2i+1}{4\pi}\int_0^{2\pi}\int_0^{\pi} Q_i F(\theta',\phi')d\theta' d\phi' \quad (4),$$

where Q_i is the function of (θ,ϕ) (θ',ϕ') expressed by App. B. (61).

In any case in which the actual attracting matter lies all within an interior concentric spherical surface of radius a', the harmonic expansion of $F(\theta,\phi)$ must be at least as convergent as the geometrical series

$$\frac{a'}{a}+\left(\frac{a'}{a}\right)^2+\left(\frac{a'}{a}\right)^3+\text{etc.};$$

and therefore (3 bis) will be convergent for every value of r exceeding a', and will consequently continue the solution in the interior at least as far as this second spherical surface.

Determination of potential from the form of an approximately spherical equipotential surface round the mass.

Example (2.)—Let the attracting mass be approximately centrobaric (§ 526), and let one equipotential surface completely enclosing it be given. It is required to find the distribution of force and potential through all space external to the smallest spherical surface that can be drawn round it from its centre of gravity as centre. Let a be an approximate or mean radius, and, taking the origin of co-ordinates exactly coincident with the centre of inertia (§ 230), let

$$r=a[1+F(\theta,\phi)] \qquad (5)$$

be the polar equation of the surface; F being for all values of θ and ϕ so small that we may neglect its square and higher powers. Consider now two proximate points (r, θ, ϕ) (a, θ, ϕ). The distance between them is $aF(\theta, \phi)$ and is in the direction through O, the origin of co-ordinates. And if M be the whole mass, the resultant force at any point of this line is approximately equal to $\frac{M}{a^2}$ and is along this line. Hence the difference of potentials (§ 486) between them is $\frac{MF(\theta,\phi)}{a}$. And if a be the proper mean radius, the constant value of the potential at the given surface (5) will be precisely $\frac{M}{a}$. Hence, to a degree of approximation consistent with neglecting squares of $F(\theta, \phi)$, the potential at the point (a, θ, ϕ) will be

$$\frac{M}{a}+\frac{M}{a}F(\theta,\phi) \qquad (6)$$

Hence the problem is reduced to that of the previous example and remarking that the part of its solution depending on the term $\frac{M}{a}$ of (6) is of course simply $\frac{M}{r}$, we have, by (3 bis), for the potential now required,

$$U=M\left\{\frac{1}{r}+\frac{a}{r^2}F_1(\theta,\phi)+\frac{a^2}{r^3}F_2(\theta,\phi)+\text{etc.}\right\} \qquad (7)$$

where F_i is given by (4). F_0 is zero in virtue of a being the "proper" mean radius; the equation expressing this condition being $\iint F(\theta,\phi)\sin\theta d\theta d\phi=0$ (8).

If further O be chosen in a proper mean position, that is to say, such that $\iint Q_1 F(\theta, \phi) \sin\theta d\theta d\phi = 0$ (9) F_1 vanishes and [§ 539 (12)] O is the centre of gravity of the attracting mass; and the harmonic expansion of $F(\theta, \phi)$ becomes

Determination of potential from the form of an approximately spherical equipotential surface round the mass.

$$F(\theta, \phi) = F_2(\theta, \phi) + F_3(\theta, \phi) + F_4(\theta, \phi) + \text{etc.} \qquad (10).$$

If a' be the radius of the smallest spherical surface having O for centre and enclosing the whole of the actual mass, the series (7) necessarily converges for all values of θ and ϕ, at least as rapidly as the geometrical series

$$1 + \frac{a'}{r} + \left(\frac{a'}{r}\right)^2 + \left(\frac{a'}{r}\right)^3 + \text{etc.} \qquad (11)$$

for every value of r exceeding a'. Hence (7) expresses the solution of our present particular problem. It may carry it even further inwards; as the given surface (6) may be such that the harmonic expansion (10) converges more rapidly than the series

$$1 + \frac{a'}{a} + \left(\frac{a'}{a}\right)^2 + \left(\frac{a'}{a}\right)^3 + \text{etc.}$$

Resultant force.

The direction and magnitude of the resultant force are of course [§§ 486, 491] deducible immediately from (7) throughout the space through which this expression is applicable, that is all space through which it converges that can be reached from the given surface without passing through any part of the actual attracting mass. It is important to remark that as the resultant force deviates from the radial direction by angles of the same order of small quantities as $F(\theta, \phi)$, its magnitude will differ from the radial component by small quantities of the same order as the square of this: and therefore, consistently with our degree of approximation, if R denote the magnitude of the resultant force

$$R = -\frac{dU}{dr} = \frac{M}{r^2}\left\{1 + 3\left(\frac{a}{r}\right)^2 F_2(\theta,\phi) + 4\left(\frac{a}{r}\right)^3 F_3(\theta,\phi) + \text{etc.}\right\} \quad (12).$$

For the resultant force at any point of the spherical surface agreeing most nearly with the given surface we put in this formula $r = a$, and find

$$\frac{M}{a^2}\{1 + 3F_2(\theta, \phi) + 4F_3(\theta, \phi) + \text{etc.}\} \qquad (13).$$

And at the point (r, θ, ϕ) of the given surface we have $r = a$ nearly enough for our approximation, in all terms except the first, of the series (12): but in the first term, $\frac{M}{r^2}$, we must put $r = a\{1 + F(\theta, \phi)\}$; so that it becomes

$$\frac{M}{r^2} = \frac{M}{a^2\{1 + F(\theta, \phi)\}^2} = \frac{M}{a^2}\{1 - 2F(\theta, \phi)\} = \frac{M}{a^2}\{1 - 2[F_2(\theta,\phi) + \text{etc.}]\} \ (14),$$

Resultant force at any point of approximately spherical level surface, for gravity alone.

and we find for the normal resultant force at the point (θ, ϕ) of the given approximately spherical equipotential surface

$$\frac{M}{a^2}\{1+F_2(\theta, \phi)+2F_3(\theta, \phi)+3F_4(\theta, \phi)+\ldots\} \qquad (15).$$

Taking for simplicity one term, F_i, alone, in the expansion of F, and considering, by aid of App. B. (38), (40), (p), and §§ 779...784, the character of spherical surface harmonics, we see that the maximum deviation of the normal to the surface

$$r=a\{1+F_i(\theta, \phi)\} \qquad (16)$$

from the radial direction is, in circular measure (§ 404), just i times the half range from minimum to maximum in the values of $F_i(\theta, \phi)$ for all harmonics of the second order (case $i=2$), and for all sectorial harmonics (§ 781) of every order; and that it is approximately so for the equatorial regions of all zonal harmonics of very high degree. Also, for harmonics of high degree contiguous maxima and minima are approximately equal. We conclude that

794. If a level surface (§ 487), enclosing a mass attracting according to the Newtonian law, deviate from an approximately spherical figure by a pure harmonic undulation (§ 779) of order i; the amount of the force of gravity at any point of it will exceed the mean amount by $i-1$ times the very small fraction by which the distance of that point of it from the centre exceeds the mean radius. The maximum inclination of the resultant force to the true radial direction, reckoned in fraction of the unit angle 57°·3 (§ 404) is, for harmonic deviations of the second order, equal to the ratio which the whole range from minimum to maximum bears to the mean magnitude. For the class described above under the designation of *sectorial harmonics*, of whatever degree, i, the maximum deviation in direction bears to the proportionate deviation in magnitude from the mean magnitude, exactly the ratio $i\div(i-1)$; and approximately the ratio of equality for *zonal harmonics* of high degrees.

Example (3.)—The attracting mass being still approximately centrobaric, let it rotate with angular velocity ω round OZ, and let one of the level surfaces (§ 487) completely enclosing it be expressed by (5), § 793. The potential of centrifugal force (§§ 800, 813), will be $\frac{1}{2}\omega^2(x^2+y^2)$, or, in solid spherical harmonics, $\frac{1}{3}\omega^2 r^2+\frac{1}{6}\omega^2(x^2+y^2-2z^2)$.

This for any point of the given surface (5) to the degree of approximation to which we are bound, is equal to

Resultant force at any point of approximately spherical level surface for gravity and centrifugal force.

$$\tfrac{1}{3}\omega^2 a^2 + \tfrac{1}{2}\omega^2 a^2(\tfrac{1}{3} - \cos^2\theta);$$

which, added to the gravitation potential at each point of this surface, must make up a constant sum. Hence the gravitation potential at (θ, ϕ) of the given surface (5) is equal to

$$\frac{M}{a} - \tfrac{1}{2}\omega^2 a^2(\tfrac{1}{3} - \cos^2\theta);$$

and therefore, all other circumstances and notation being as in Example 2 (§ 793), we now have instead of (6) for gravitation potential at (a, θ, ϕ), the following:

$$\frac{M}{a} + \frac{M}{a}F(\theta, \phi) - \tfrac{1}{2}\omega^2 a^2(\tfrac{1}{3} - \cos^2\theta) \qquad (16).$$

Hence, choosing the position of O, and the magnitude of a, according to (9) and (8), we now have, instead of (7), for potential of pure gravitation, at any point (r, θ, ϕ),

$$V = M\{\frac{1}{r} + \frac{a^2}{r^3}[F_2(\theta,\phi) - \tfrac{1}{2}m(\tfrac{1}{3} - \cos^2\theta)] + \frac{a^3}{r^4}F_3(\theta,\phi) + \frac{a^4}{r^5}F_4(\theta,\phi) + \ldots\} \quad (17).$$

where m denotes $\omega^2 a \div \frac{M}{a^2}$, or the ratio of centrifugal force at the equator, to pure gravity at the mean distance, a. The force of pure gravity at the point (θ, ϕ) of the given surface (5) is consequently expressed by the following formula instead of (15):—

$$\frac{M}{a^2}\{1 + F_2(\theta,\phi) - 3.\tfrac{1}{2}m(\tfrac{1}{3} - \cos^2\theta) + 2F_3(\theta,\phi) + 3F_4(\theta,\phi) + \ldots\} \quad (18).$$

From this must be subtracted the radial component of the centrifugal force, which is (in harmonics)

$$\tfrac{2}{3}\omega^2 a + \omega^2 a(\tfrac{1}{3} - \cos^2\theta),$$

to find the whole amount of the resultant force, g (apparent gravity), normal to the given surface: and therefore

$$g = \frac{M}{a^2}\{1 - \tfrac{2}{3}m + F_2(\theta,\phi) - \tfrac{5}{2}m(\tfrac{1}{3} - \cos^2\theta) + 2F_3(\theta,\phi) + 3F_4(\theta,\phi) + \ldots\} \quad (19).$$

If in a particular case we have

$F_i(\theta, \phi) = 0$, except for $i = 2$; and $F_2(\theta, \phi) = e(\tfrac{1}{3} - \cos^2\theta)$:

this becomes

$$g = \frac{M}{a^2}\{1 - \tfrac{2}{3}m - (\tfrac{5}{2}m - e)(\tfrac{1}{3} - \cos^2\theta)\} \qquad (20).$$

795. Hence if outside a rotating solid the lines of resultant force of gravitation and centrifugal force are cut at right angles

Resultant force at any point of approximately spherical level surface for gravity and centrifugal force:

by an elliptic spheroid[1] symmetrical round the axis of rotation the amount of the resultant differs from point to point of the surface as the square of the sine of the latitude: and the excess of the polar resultant above the equatorial bears to the whole amount of either a ratio which added to the ellipticity of the figure is equal to two and a half times the ratio of equatorial centrifugal force to gravity.

Clairaut's theorems.

For the case of a rotating fluid mass, or solid with density distributed as if fluid, these conclusions, of which the second is now generally known as Clairaut's theorem, were first discovered by Clairaut, and published in 1743 in his celebrated treatise *La Figure de la Terre.* Laplace extended them by proving the formula (19) of § 794 for any solid consisting of approximately spherical layers of equal density. Ultimately Stokes[2] pointed out that, only provided the surfaces of equilibrium relative to gravitation alone, and relative to the resultant of gravitation and centrifugal force, are approximately spherical; whether the surfaces of equal density are approximately spherical or not, the same expression (19) holds. A most important practical deduction from this conclusion is that irrespectively of any supposition regarding the distribution of the earth's density, the true figure of the sea level can be determined from pendulum observations alone, without any hypothesis as to the interior condition of the solid.

Figure of the sea level determinable from measurements of gravity.

Let, for brevity,

$$g\{1+\tfrac{3}{2}m(\tfrac{1}{3}-\cos^2\theta)\}=f(\theta,\phi) \tag{21}$$

where m (§ 801) is $\frac{1}{289}$, and g is known by observation in different localities, with reduction to the sea level according to the square of the distance from the earth's centre (not according to Young's rule). Let the expansion of this in spherical surface harmonics be

$$f(\theta,\phi)=f_0+f_1(\theta,\phi)+f_2(\theta,\phi)+\text{etc.} \tag{22}$$

We have, by (19),

$$F_i(\theta,\phi)=\frac{1}{i}\,\frac{f_i(\theta,\phi)}{f_0} \tag{23}$$

[1] Following the best French writers, we use the term spheroid to designate any surface differing very little from spherical figure. The commoner English usage of confining it to an ellipsoid symmetrical round an axis; and extending it to such figures though not approximately spherical; is bad.

[2] "On the Variation of Gravity at the surface of the Earth."—*Trans. of the Camb. Phil. Soc.*, 1849.

and therefore the equation (5) of the level surface becomes

$$r = a\left\{1 + \frac{1}{f_0}\left[\tfrac{1}{2}f_2(\theta,\phi) + \tfrac{1}{3}f_3(\theta,\phi) + \text{etc.}\right]\right\} \quad (24).$$

Figure of the sea level determinable from measurements of gravity;

Confining our attention for a moment to the first two terms we have for f_2, by App. B. (38), explicitly

$$f_2(\theta,\phi) = A_0(\cos^2\theta - \tfrac{1}{3}) + (A_1\cos\phi + B_1\sin\phi)\sin\theta\cos\theta + (A_2\cos 2\phi + B_2\sin 2\phi)\sin^2\theta \quad (25).$$

Substituting in (24) squared, putting

$$\cos\theta = \frac{z}{r}, \quad \sin\theta\cos\phi = \frac{x}{r}, \quad \sin\theta\sin\phi = \frac{y}{r},$$

and reducing to a convenient form, we find

$$(f_0 + \tfrac{1}{3}A_0 - A_2)x^2 + (f_0 + \tfrac{1}{3}A_0 + A_2)y^2 + (f_0 - \tfrac{2}{3}A_0)z^2 - B_1 yz - A_1 zx - 2B_2 xy = f_0 a^2 \quad (26).$$

Now from §§ 539, 534, we see that, if OX, OY, OZ are principal axes of inertia, the terms of f_2 which, expressed in rectangular co-ordinates, involve the products yz, zx, xy must disappear: that is to say, we must have $B_1=0$, $A_1=0$, $B_2=0$. But whether B_2 vanishes or not, if OZ is a principal axis we must have both $A_1=0$ and $B_1=0$; which therefore is the case, to a very minute accuracy, if we choose for OZ the average axis of the earth's rotation, as will be proved in Vol. II., on the assumption rendered probable by the reasons adduced below, that the earth experiences little or no sensible disturbance in its motion from want of perfect rigidity. Hence the expansion (22) is reduced to

if ellipsoid with three unequal axes must have one of them coincident with axis of revolution.

$$f(\theta,\phi) = f_0 + A_0(\cos^2\theta - \tfrac{1}{3}) + (A_2\cos 2\phi + B_2\sin 2\phi)\sin^2\theta + f_3(\theta,\phi) + \text{etc.} \quad (27).$$

If $f_3(\theta,\phi)$ and higher terms are neglected the sea level is an ellipsoid, of which one axis must coincide with the axis of the earth's rotation. And, denoting by e the mean ellipticity of meridional sections, e' the ellipticity of the equatorial section, and I the inclination of one of its axes to OX, we have

$$e = \tfrac{1}{2}\frac{A_0}{f_0}, \quad e' = \frac{\sqrt{(A_2^2 + B_2^2)}}{f_0}, \quad I = \tfrac{1}{2}\tan^{-1}\frac{B_2}{A_2}.$$

In general, the constants of the expansion (22); f_0 (being the mean force of gravity), A_0, A_2, B_2, the seven coefficients in $f_3(\theta,\phi)$, the nine in $f_4(\theta,\phi)$, and so on; are to be determined from sufficiently numerous and wide-spread observations of the amount of gravity.

796. A first approximate result thus derived from pendulum observations and confirmed by direct geodetic measurements is that the figure of the sea level approximates to an oblate spheroid of revolution of ellipticity about $\frac{1}{297}$. Both

Figure of the sea level determinable from measurements of gravity:

methods are largely affected by local irregularities of the solid surface and underground density, to the elimination of which a vast amount of labour and mathematical ability have been applied with as yet but partial success. Considering the general disposition of the great tracts of land and ocean, we can scarcely doubt that a careful reduction of the numerous accurate pendulum observations that have been made in localities widely spread over the earth[1] will lead to the determination of an ellipsoid with three unequal axes coinciding more nearly on the whole with the true figure of the sea level than does any spheroid of revolution. Until this has been either accomplished or proved impracticable it would be vain to speculate as to the possibility of obtaining, from attainable data, a yet closer approximation by introducing a harmonic of the third order [$f_3(\theta, \phi)$ in (27)]. But there is little probability that harmonics of the fourth or higher orders will ever be found useful: and local quadratures, after the example first set by Maskelyne in his investigation of the disturbance produced by Schehallien, must be resorted to to interpret irregularities in particular districts; whether of the amount of gravity shown by the pendulum; or of its direction, by geodetic observation. We would only remark here, that the problems presented by such local quadratures with reference to the *amount* of gravity seem about as much easier and simpler than those with reference to its direction as pendulum observations are than geodetic measurements: and that we expect much more knowledge regarding the true figure of the sea level from the former than from the latter, although it is to the reduction of the latter that the most laborious efforts have been hitherto applied. We intend to return to this subject in Vol. II. in explaining, under Properties of Matter, the practical foundation of our knowledge of gravity.

rendered difficult by local irregularities.

797. During the last seven years geodetic work of extreme importance has been in progress, through the co-operation of

[1] In 1672, a pendulum conveyed by Richer from Paris to Cayenne first proved variation of gravity. Captain Kater and Dr. Thomas Young, *Trans. R. S.*, 1819. Biot, Arago, Mathieu, Bouvard, and Chaix; *Base du Système Métrique*, Vol. III., Paris 1821. Captain Edward Sabine, R.E., "Experiments to determine the Figure of the Earth by means of the Pendulum;" published for the Board of Longitude, London, 1825. Stokes "On the Variation of Gravity at the Surface of the Earth."—*Camb. Phil. Trans.*, 1849.

Results of geodesy.

the Governments of Prussia, Russia, Belgium, France, and England, in connecting the triangulation of France, Belgium, Russia, and Prussia, which were sufficiently advanced for the purpose in 1860, with the principal triangulation of Great Britain and Ireland, which had been finished in 1851. With reference to this work, Sir Henry James makes the following remarks:—"Before the connexion of the triangulation of the "several countries into one great network of triangles extend-"ing across the entire breadth of Europe, and before the dis-"covery of the Electric Telegraph, and its extension from "Valentia (Ireland) to the Ural mountains, it was not possible "to execute so vast an undertaking as that which is now in "progress. It is, in fact, a work which could not possibly "have been executed at any earlier period in the history "of the world. The exact determination of the Figure and "Dimensions of the Earth has been one great aim of astrono-"mers for upwards of two thousand years; and it is fortunate "that we live in a time when men are so enlightened as to "combine their labours to effect an object which is desired by all, "and at the first moment when it was possible to execute it."

For a short time longer, however, we must be contented with the results derived from the recent British Triangulation, with the separate measurements of arcs of meridians in Peru, France, Prussia, Russia, Cape of Good Hope, and India. The investigation of the ellipsoid of revolution agreeing most nearly with the sea level for the whole Earth, has been carried out with remarkable skill by Captain A. R. Clarke, R.E., and published in 1858, by order of the Master General and Board of Ordnance, in a volume (of 780 pages, quarto, almost every page of which is a record of a vast amount of skilled labour) drawn up by Captain Clarke, under the direction of Lieutenant-Colonel (now Sir Henry) James, R.E. The following account of conclusions subsequently worked out regarding the ellipsoid of three unequal axes most nearly agreeing with the sea level, is extracted from the preface to another volume recently published as one item of the great work of comparison with the recent triangulations of other countries:[1]—

[1] "Comparisons of the Standards of Length of England, France, Belgium, Prussia, Russia, India, Australia, made at the Ordnance Survey Office, Southampton, by Captain A. R. Clarke, R.E., under the direction of Colonel Sir Henry James, R.E., F.R.S." Published by order of the Secretary of State for War, 1866.

Results of geodesy.

"In computing the figures of the meridians and of the "equator for the several measured arcs of meridian, it is found "that the equator is slightly elliptical, having the longer "diameter of the ellipse in 15° 34′ east longitude. In the "eastern hemisphere the meridian of 15° 34′ passes through "Spitzbergen, a little to the west of Vienna, through the Straits "of Messina, through Lake Chad in North Africa, and along "the west Coast of South Africa, nearly corresponding to the "meridian which passes over the greatest quantity of land in "that hemisphere. In the western hemisphere this meridian "passes through Behring's Straits and through the centre of the "Pacific Ocean, nearly corresponding to the meridian which "passes over the greatest quantity of water of that hemi-"sphere.

"The meridian of 105° 34′ passes near North-East Cape, in "the Arctic Sea, through Tonquin and the Straits of Sunda, and "corresponds nearly to the meridian which passes over the "greatest quantity of land in Asia; and in the western hemi-"sphere it passes through Smith's Sound in Behring's Straits, "near Montreal, near New York, between Cuba and St. Do-"mingo, and close along the western coast of South America, "corresponding nearly to the meridian passing over the greatest "amount of land in the western hemisphere.

"These meridians, therefore, correspond with the most re-"markable physical features of the globe.

	Feet.
"The longest semi-diameter of the equatorial ellipse is	20926350
"And the shortest	20919972
"Giving an ellipticity of the equator equal to .	$\frac{1}{3269\cdot5}$
"The polar semi-diameter is equal to . .	20853429
"The maximum and minimum polar compressions "are	$\frac{1}{285\cdot97}$ and $\frac{1}{313\cdot38}$
"Or a mean compression of very closely . .	$\frac{1}{300}$."

Captain Clarke had previously found ("Account of Principal Triangulation," 1858) for the spheroid of revolution most nearly representing the same set of observations, the following :—

Equatorial semi-axis $= a = 20926062$ feet,
Polar semi-axis $= b = 20855121$ feet;

$$\text{whence } \frac{b}{a} = \frac{293 \cdot 98}{294 \cdot 98}; \text{ and ellipticity} = \frac{a-b}{a} = \frac{1}{294 \cdot 98}.$$

"In this figure, however, the sum of the squares of the "latitude corrections is 153·9939 against 138·3020 in the figure "of three unequal axes."[1]

798. As an instructive example of the elementary principles of fluid equilibrium, useful also because it includes the celebrated hydrostatic theories of the Tides and of the Figure of the Earth, let us suppose a finite mass of heterogeneous incompressible fluid resting on a rigid spherical shell or solid sphere, under the influence of mutual gravitation between its parts, and of the attraction of the core supposed symmetrical; to be slightly disturbed by any attracting masses fixed either in the core or outside the fluid; or by force fulfilling any imaginable law, subject only to the condition of being a conservative system; or by centrifugal force. Hydrostatic examples resumed.

First we may remark that were there no such disturbance the fluid would come to rest in concentric spherical layers of equal density, the denser towards the centre, this last characteristic being essential for stability, which clearly requires also that the mean density of the nucleus shall be not less than that of the layer of fluid next it; otherwise the nucleus would, as it were, float up from the centre, and either protrude from the fluid at one side, or (if the gradation of density in the fluid permits) rest in an eccentric position completely covered; fulfilling in either case the condition (§ 762) for the equilibrium of floating bodies.

799. The effect of the disturbing force could be at once found without analysis if there were no mutual attraction between parts of the fluid, so that the influence tending to maintain the spherical figure would be simply the symmetrical attraction of the fixed core. For the equipotential surfaces would then be known (as directly implied by the data), and the fluid would (§ 750) arrange itself in layers of equal density defined by these surfaces. No mutual force between portions of the liquid.

800. *Examples of* § 799.—(1.) Let the nucleus act according

[1] From p. 287 of "Comparison of Standards of Length" (1866).

No mutual force between portions of the liquid: Example (1.)

to the Newtonian law, and be either symmetrical round a point or (§ 526) of any other centrobaric arrangement; and let the disturbing influence be centrifugal force. In Vol. II. it will appear, as an immediate consequence from the elementary dynamics of circular motion, that kinetic equilibrium under centrifugal force in any case will be the same as the static equilibrium of the imaginary case in which the same material system is at rest, but influenced by repulsion from the axis in simple proportion to distance.

If z be the axis of rotation, and ω the angular velocity, the components of centrifugal force (§§ 32, 35*a*, 259) are $\omega^2 x$ and $\omega^2 y$. Hence the potential of centrifugal force is

$$\tfrac{1}{2}\omega^2(x^2+y^2),$$

reckoned from zero at the axis, and increasing in the direction of the force, to suit the convention (§ 485) adopted for gravitation potentials. The expression for the latter (§§ 491, 528) is

$$\frac{E}{\sqrt{(x^2+y^2+z^2)}}$$

where E denotes the mass of the nucleus, and the co-ordinates are reckoned from its centre of gravity (§ 526) as origin. Hence the "level surfaces" (§ 487) external to the nucleus are given by assigning different values to C in the equation

$$\frac{E}{\sqrt{(x^2+y^2+z^2)}}+\tfrac{1}{2}\omega^2(x^2+y^2)=C \qquad (1),$$

and the fluid when in equilibrium has its layers of equal density and its outer boundary in these surfaces. If ρ be the density and p the pressure of the fluid at any point of one of these surfaces, regarded as functions of C, we have (§ 760)

$$p=\int\rho dC \qquad (2).$$

Unless the fluid be held in by pressure applied to its bounding surface, the potential must increase from this surface inwards (or the resultant of gravity and centrifugal force, perpendicular as it is to the surface, must be directed inwards), as negative pressure is practically inadmissible. The student will find it an interesting exercise to examine the circumstances under which this condition is satisfied; which may be best done by tracing the meridional curves of the series of surfaces of revolution given by equation (1).

Let a and $a(1-e)$ be the equatorial and polar semidiameters of one of these surfaces. We have

$$\frac{E}{a}+\tfrac{1}{2}\omega^2a^2=\frac{E}{a(1-e)},$$

No mutual force between portions of the liquid: Example (1.)

whence
$$e=\frac{\tfrac{1}{2}\omega^2a}{\frac{E}{a^2}+\tfrac{1}{2}\omega^2a}=\frac{m}{2+m} \quad (3),$$

if m denote the ratio of centrifugal force at its equator to pure gravity at the same place. (Contrast approximately agreeing definition of m, § 794.) From this, and the form of (1), we infer that

801. In the case of but small deviation from the spherical figure, which alone is interesting with reference to the theory of the Earth's figure and internal constitution, the bounding surface and the surfaces of equal density and pressure are very approximately oblate ellipsoids of revolution;[1] the ellipticity[2] of each amounting to half the ratio of centrifugal force in its largest circle (or its equator, as we may call this) to gravity at any part of it; and therefore increasing from surface to surface outwards as the cubes of the radii. The earth's equatorial radius is 20926000 feet, and its period (the sidereal day) is 86164 mean solar seconds. Hence in British absolute measure (§ 225) the equatorial centrifugal force is $(\frac{2\pi}{86164})^2\times 20926000$, or ·11127. This is $\frac{1}{289}$ of 32·158; or very approximately the same fraction of the mean value, 32·14, of apparent gravity over the whole sea level, as determined by pendulum observations. It is therefore [§ 794 (20)] $\frac{1}{289\cdot 66}$, or approximately $\frac{1}{290}$, of the mean value of true gravitation. Hence, if the solid earth attracted merely as a point of matter collected at its centre, and there were no mutual attraction between the different parts of the sea, the sea level would be a spheroid of ellipticity $\frac{1}{580}$. In reality, we find by observation that the ellipticity of the spheroid of revolution which most nearly coincides with the sea level is about $\frac{1}{295}$. The difference between these, or $\frac{1}{600}$, must therefore be due to deviation of true gravity from spherical

[1] Airy has estimated 24 feet as the greatest deviation of the bounding surface from a true ellipsoid.

[2] A term used by writers on the figure of the Earth to denote the ratio which the difference between the two axes of an ellipse bears to the greater. Thus if e be the ellipticity, and ϵ the eccentricity of an ellipse, we have $\epsilon^2=2e+e^2$. Hence, when the eccentricity is small, the ellipticity is a small quantity of the same order as its square; and the former is equal approximately to the square root of twice the latter.

No mutual force between portions of the liquid: Example (1.)

symmetry. Thus the whole ellipticity of the actual sea level $\frac{1}{295}$, may be regarded as made up of two nearly equal parts: of which the greater, $\frac{1}{580}$, is due directly to centrifugal force, and the less, $\frac{1}{590}$, to deviation of solid and fluid attracting mass from any truly centrobaric arrangement (§ 526). A little later (§§ 820, 821) we shall return to this subject.

802. The amount of the resultant force perpendicular to the free surface of the fluid is to be found by compounding the force of gravity towards the centre with the centrifugal force from the axis; and it will be approximately equal to the former diminished by the component of the latter along it, when the deviation from spherical figure is small. And as the former component varies inversely as the square of the distance from the centre, it will be less at the equator than at either pole by an amount which bears to either a ratio equal to twice the ellipticity, and which is therefore (§ 801) equal to the centrifugal force at the equator. Thus in the present case half the difference of apparent gravity between poles and equator is due to centrifugal force, and half to difference of distance from the centre. The gradual increase of apparent gravity in going from the equator towards either pole is readily proved to be as the square of the sine of the latitude; and this not only for the result of the two combined causes of variation, but for each separately. These conclusions needed, however, no fresh proof, as they constitute merely the applications to the present case, of Clairaut's general theorems demonstrated above (§ 795).

Analytically, for the present case, we have

$$g=-\frac{\partial V}{dr}$$

if g denote the magnitude of the resultant of true gravity and centrifugal force; $\frac{\partial}{dr}$ [as in App. B. (g)] rate of variation per unit of length along the direction of r; and V the first member of (1) § 800. Hence taking $z^2=r^2\cos^2\theta$, and $x^2+y^2=r^2\sin^2\theta$ we find

$$g=\frac{E}{r^2}-\omega^2 r\sin^2\theta \qquad (4).$$

On the hypothesis of infinitely small deviation from spherical figure this becomes

$$g=\frac{E}{a^2}(1-2u)-\omega^2 a\sin^2\theta \qquad (5),$$

if in the small term we put a, a constant, for r, and in the other $r=a(1+u)$. By (1) we see that $\frac{E}{C}$ is an approximate value for r, and if we take it for a, that equation gives

$$u=\tfrac{1}{2}\frac{\omega^2 a^3}{E}\sin^2\theta \qquad (6);$$

and using this in (5) we have

$$g=\frac{E}{a^2}(1-2\frac{\omega^2 a^3}{E}\sin^2\theta)=\frac{E}{a^2}(1-2m\sin^2\theta) \qquad (7),$$

where, as before, m denotes the ratio of equatorial centrifugal force to gravity.

803. *Examples of* § 799 *continued.*—(2.) The nucleus being held fixed, let the fluid on its surface be disturbed by the attraction of a very distant fixed body attracting according to the Newtonian law. No mutual force between portions of the liquid: Example (2)

Let r, θ be polar co-ordinates referred to the centre of gravity of the nucleus as origin, and line from it to the disturbing body as axis; let, as before, E be the mass of the nucleus; lastly, let M be the mass of the disturbing body, and D its distance from the centre of the nucleus. The equipotentials have for their equation

$$\frac{E}{r}+\frac{M}{\sqrt{(D^2-2rD\cos\theta+r^2)}}=\text{const.} \qquad (8),$$

which, for very small values of $\frac{r}{D}$, becomes approximately

$$\frac{E}{r}+\frac{M}{D}(1+\frac{r}{D}\cos\theta)=\text{const.} \qquad (9).$$

And if, as in corresponding cases, we put $r=a(1+u)$ where a is a proper mean value of r, and u is an infinitely small numerical quantity, a function of θ, we have finally

$$u=\frac{M.a^3}{E.D^3}\cos\theta \qquad (10).$$

This is a spherical surface harmonic of the first order, and (§ 789) we conclude that

The fluid will not be disturbed from its spherical figure, but it will be drawn towards the disturbing body, so that its centre will deviate from the centre of the nucleus by a distance amounting to the same fraction of its radius that the attraction of the disturbing body is of the attraction of the nucleus, on a

No mutual force between portions of the liquid: Example (2.)

point of the fluid surface. This fraction is about $\frac{1}{3000000}$ (being $\frac{1}{83\times 60\times 60}$) for the earth and moon, as the moon's distance is 60 times the earth's radius, and her mass about $\frac{1}{83}$ of the earth's. Hence if the earth's and moon's centres were both held fixed, there would be a rise of level at the point nearest to the moon, and fall of level at the point farthest from it, each equal to $\frac{1}{3000000}$ of the earth's radius, or about 70 feet. Or if we consider the sun's influence under similar unreal circumstances, we should have a tide of 12,500 feet rise on the side next the sun, and the same fall on the remote side [being (§ 812) $\frac{1}{38{\cdot}7\times 10^6}$ of the sun's distance].

Example for tides:

804. *Examples of* § 799 *continued.*—(3.) With other conditions, the same as in Example (2.) (§ 803), let one-half of the disturbing body be removed and fixed at an equal distance on the other side.

The equation of the equipotentials, instead of (8), is now

$$\frac{E}{r}+\tfrac{1}{2}M\left[\frac{1}{\sqrt{(D^2-2rD\cos\theta+r^2)}}+\frac{1}{\sqrt{(D^2+2rD\cos\theta+r^2)}}\right]=\text{const.} \quad (11)$$

and as the first approximation for $\frac{r}{D}$ very small, instead of (9), we now have

$$\frac{E}{r}+\frac{M}{D}\left[1+\tfrac{1}{2}\frac{r^2}{D^2}(3\cos^2\theta-1)\right]=\text{const.} \quad (12);$$

whence finally, instead of (10), with corresponding notation;

$$u=\tfrac{1}{2}\frac{M.a^2}{E.D^3}(3\cos^2\theta-1) \quad (13).$$

This is a spherical surface harmonic of the second order, and $\frac{M.a^3}{E.D^3}$ is one-quarter of the ratio that the difference between the moon's attraction on the nearest and farthest parts of the earth bears to terrestrial gravity. Hence

result agrees with ordinary equilibrium theory.

The fluid will be disturbed into a prolate ellipsoidal figure, with its long axis in the line joining the two disturbing bodies, and with ellipticity (§ 801) equal to $\frac{3}{4}$ of the ratio which the difference of attractions of one of the disturbing bodies on the nearest and farthest points of the fluid surface bears to the surface value of the attraction of the nucleus. If, for instance, we suppose the moon to be divided into two halves, and these to be fixed on opposite sides of the earth at distances each equal to the true moon's mean distance; the ellipticity of the disturbed terrestrial level would be $\frac{3}{2\times 60\times 300000}$, or $\frac{1}{12.000.000}$; and the whole difference of levels from highest to lowest would be about $1\frac{3}{4}$ feet. We shall have much occasion to use this hypothesis in Vol. II., in investigating the kinetic theory of the tides. We shall see that it (or some equivalent hypothesis) is essential to Laplace's evanescent diurnal tide on a solid spheroid covered with an ocean of equal depth all over; but, on the other hand, we find presently (§ 814) that it agrees very closely with the actual circumstances so far as the foundation of the equilibrium theory is concerned.

The tides: results of ordinary equilibrium theory.

805. The rise and fall of water at any point of the earth's surface we may now imagine to be produced by making these two disturbing bodies (moon and anti-moon, as we may call them for brevity) revolve round the earth's axis once in the lunar twenty-four hours, with the line joining them always inclined to the earth's equator at an angle equal to the moon's declination. If we assume that at each moment the condition of hydrostatic equilibrium is fulfilled; that is, that the free liquid surface is perpendicular to the resultant force, we have what is called the "equilibrium theory of the tides."

Correction of ordinary equilibrium theory.

806. But even on this equilibrium theory, the rise and fall at any place would be most falsely estimated if we were to take it, as we believe it is generally taken, as the rise and fall of the spheroidal surface that would bound the water were there no dry land (uncovered solid). To illustrate this statement, let us imagine the ocean to consist of two circular lakes *A* and *B*, with their centres 90° asunder, on the equator, communicating with one another by a narrow channel. In the course of the lunar twelve hours the level of lake *A* would rise and fall, and that of lake *B* would simultaneously fall and rise to maximum deviations from the mean level. If the areas of the two lakes were equal, their tides would be equal, and would amount in each to about one foot above and below the mean level; but not so if the areas were unequal. Thus, if the diameter of the greater be but a small part of the earth's quadrant, not more, let us say, than 20°, the amounts of the rise and fall in the two lakes will be inversely as their areas to a close degree of approximation. For instance, if the dia-

The tides: correction of ordinary equilibrium theory.

meter of B be only $\frac{1}{10}$ of the diameter of A, the rise and fall at A will be scarcely sensible; while the level of B will rise and fall by about two feet above and below its mean; just as the rise and fall of level in the open cistern of an ordinary barometer is but small in comparison with fall and rise in the tube. Or, if there be two large lakes A, A' at opposite extremities of an equatorial diameter, two small ones B, B' at two ends of the equatorial diameter perpendicular to that one, and two small lakes C, C' at two ends of the polar axis, the largest of these being, however, still supposed to extend over only a small portion of the earth's curvature, and all the six lakes communicate with one another freely by canals, or underground tunnels: there will be no sensible tides in the lakes A and A'; in B and B' there will be high water of two feet above mean level when the moon or anti-moon is in the zenith, and low water of two feet below mean when the moon is rising or setting; and at C and C' there will be tides rising and falling one foot above and below the mean, the time of low water being when the moon or anti-moon is in the meridian of A, and of high water when they are on the horizon of A. The simplest way of viewing the case for the extreme circumstances we have now supposed is, first, to consider the spheroidal surface that would bound the water at any moment if there were no dry land, and then to imagine this whole surface lowered or elevated all round by the amount required to keep the height at A and A' invariable. Or, if there be a large lake A in any part of the earth, communicating by canals with small lakes over various parts of the surface, having in all but a small area of water in comparison with that of A, the tides in any of these will be found by drawing a spheroidal surface of two feet difference between greatest and least radius, and, without disturbing its centre, adding or subtracting from each radius such a length, the same for all, as shall do away with rise or fall at A.

807. It is, however, only on the extreme supposition we have made, of one water area much larger than all the others taken together, but yet itself covering only a small part of the earth's curvature, that the rise and fall can be done away with nearly altogether in one place, and doubled in another place. Taking e actual figure of the earth's sea-surface, we must subtract a ertain positive or negative quantity α from the radius of the pheroid that would bound the water were there no land, α eing determined according to the moon's position, to fulfil ne condition that the volume of the water remains unchanged, nd being the same for all points of the sea, at the same time. Iany writers on the tides have overlooked this obvious and ssential principle; indeed we know of only one sentence[1] itherto published in which any consciousness of it has been ndicated.

The tides, mutual attraction of the waters neglected: correction of the ordinary equilibrium theory.

808. The quantity α is a spherical harmonic function of the econd order of the moon's declination, and hour-angle from he meridian of Greenwich, of which the five constant co-fficients depend merely on the configuration of land and water, nd may be easily estimated by necessarily very laborious uadratures, with data derived from the inspection of good naps.

Let as above

$$r=a(1+u) \tag{14}$$

be the spheroidal level that would bound the water were the whole solid covered; u being given by (13) of § 804. Thus, if $\iint d\sigma$ denote surface integration over the whole surface of the sea,

$$a\iint u d\sigma$$

expresses the addition (positive or negative as the case may be) to the volume required to let the water stand to this level everywhere. To do away with this change of volume we must suppose the whole surface lowered equally all over by such an amount α (positive or negative) as shall equalize it. Hence if Ω be the whole area of sea, we have

$$\alpha=\frac{a\iint u d\sigma}{\Omega} \tag{15}.$$

And

$$\mathfrak{r}=r-\alpha=a\left\{1+u-\frac{\iint u d\sigma}{\Omega}\right\} \tag{16},$$

is the corrected equation of the level spheroidal surface of the sea. Hence

$$h=a\left\{u-\frac{\iint u d\sigma}{\Omega}\right\} \tag{17},$$

where h denotes the height of the surface of the sea at any

[1] "Rigidity of the Earth," § 17, *Phil. Trans.*, 1862.

2 T

The tides, mutual attraction of the waters neglected: corrected equilibrium theory.

place, above the level which it would take if the moon were removed.

To work out (15), put first, for brevity,

$$e=\tfrac{3}{2}\frac{M.a^3}{E.D^3} \qquad (18)$$

and (13) becomes

$$u=e(\cos^2\theta-\tfrac{1}{3}) \qquad (19)$$

Now let l and λ be the geographical latitude and west longitude of the place, to which u corresponds; and ψ and δ the moon's hour-angle from the meridian of Greenwich, and her declination. As θ is the moon's zenith distance at the place (corrected for parallax), we have by spherical trigonometry

$$\cos\theta=\cos l\cos\delta\cos(\lambda-\psi)+\sin l\sin\delta;$$

which gives

$$3\cos^2\theta-1=\tfrac{3}{2}\cos^2 l\cos^2\delta\cos 2(\lambda-\psi)+6\sin l\cos l\sin\delta\cos\delta\cos(\lambda-\psi)+\tfrac{1}{2}(3\sin^2\delta-1)(3\sin^2 l-1) \qquad (20)$$

Hence if we take $\mathfrak{A}$, $\mathfrak{B}$, $\mathfrak{C}$, $\mathfrak{D}$, $\mathfrak{E}$ to denote five integrals depending solely on the distribution of land and water, expressed as follows:—

$$\left.\begin{aligned}
&\mathfrak{A}=\frac{1}{\Omega}\iint\cos^2 l\cos 2\lambda d\sigma, \quad \mathfrak{B}=\frac{1}{\Omega}\iint\cos^2 l\sin 2\lambda,\\
&\mathfrak{C}=\frac{1}{\Omega}\iint\sin l\cos l\cos\lambda d\sigma, \quad \mathfrak{D}=\frac{1}{\Omega}\iint\sin l\cos l\sin\lambda d\sigma,\\
&\mathfrak{E}=\frac{1}{\Omega}\iint(3\sin^2 l-1)d\sigma,
\end{aligned}\right\} \qquad (21)$$

where of course $d\sigma=\cos l dl d\lambda$

we have

$$\mathfrak{a}=\frac{a}{\Omega}\iint u d\sigma=\tfrac{1}{3}ae\{\tfrac{3}{2}\cos^2\delta(\mathfrak{A}\cos 2\psi+\mathfrak{B}\sin 2\psi)+6\sin\delta\cos\delta(\mathfrak{C}\cos\psi+\mathfrak{D}\sin\psi)+\tfrac{1}{2}\mathfrak{E}(3\sin^2\delta-1)\}$$

This, used with (19) and (20) in (17), gives for the full conclusion of the equilibrium theory,

Lunar or solar semi-diurnal tide.
Lunar or solar diurnal tide.
Lunar fortnightly tide or solar semi-annual tide.

$$\left.\begin{aligned}
h=\;&\frac{ae}{2}[(\cos^2 l\cos 2\lambda-\mathfrak{A})\cos 2\psi+(\cos^2 l\sin 2\lambda-\mathfrak{B})\sin 2\psi]\cos^2\delta\\
&+2ae[(\sin l\cos l\cos\lambda-\mathfrak{C})\cos\psi+(\sin l\cos l\sin\lambda-\mathfrak{D})\sin\psi]\sin\delta\cos\delta\\
&+\tfrac{1}{6}ae(3\sin^2 l-1-\mathfrak{E})(3\sin^2\delta-1)
\end{aligned}\right\} \qquad (22)$$

in which the value of e may be taken from (18) for either the moon or the sun: and δ and ψ denote the declination and Greenwich hour-angle of one body or the other, as the case may be. In this expression we may of course reduce the semi-diurnal terms to the form $A\cos(2\psi-\epsilon)$, and the diurnal terms to $A'\cos(\psi-\epsilon')$. Interpreting it we have the following conclusions:—

809. In the equilibrium theory, the whole deviation of level at any point of the sea, due to sun and moon acting jointly, is expressed by the sum of six terms, three for each body.

The tides, mutual attraction of the waters neglected: corrected equilibrium theory.

(1.) The lunar or solar semi-diurnal tide rises and falls in proportion to a simple harmonic function of the hour-angle from the meridian of Greenwich, having for period 180° of this angle (or in time, half the period of revolution relatively to the earth), with amplitude varying in simple proportion to the square of the cosine of the declination of the Sun or Moon, as the case may be, and therefore varying but slowly, and through but a small entire range.

(2.) The lunar or solar diurnal tide varies as a simple harmonic function of the hour-angle of period 360°, or twenty-four hours, with an amplitude varying always in simple proportion to the sine of twice the declination of the disturbing body, and therefore changing from positive maximum to negative, and back to positive maximum again, in the tropical[1] period of either body in its orbit.

(3.) The lunar fortnightly or solar semi-annual tide is a variation on the average height of water for the twenty-four lunar or the twenty-four solar hours, according to which there is on the whole higher water all round the equator and lower water at the poles, when the declination of the disturbing body is zero, than when it has any other value, whether north or south; and maximum height of water at the poles and lowest at the equator, when the declination has a maximum, whether north or south. Gauss's way of stating the circumstances on which "secular" variations in the elements of the solar system depend is convenient for explaining this component of the tides. Let the two parallel circles of the north and south declination of the moon and anti-moon at any time be drawn on a geocentric spherical surface of radius equal to the moon's distance, and let the moon's mass be divided into two halves and distributed over them. As these circles of matter gradually vary each fortnight from the equator to maximum declina-

Explanation of the lunar fortnightly and solar semi-annual tides.

[1] The tropical period differs from the sidereal period in being reckoned from the first point of Aries instead of from a line fixed in space; the difference being only one day in 26,000 years.

The tides, mutual attraction of the waters neglected: corrections on ordinary equilibrium results.

tion and back, the tide produced will be solely and exactly the "fortnightly tide."

810. In the equilibrium theory as ordinarily stated there is, at any place, high water of the semi-diurnal tide, *precisely* when the disturbing body, or its opposite, crosses the meridian of the place; and its amount is the same for all places in the same latitude; being as the square of the cosine of the latitude, and therefore, for instance, zero at each pole. In the corrected equilibrium theory, high water of the semi-diurnal tides may be either before or after the disturbing body crosses the meridian, and its amount is very different at different places in the same latitude, and is certainly not zero at the poles. In the ordinarily stated equilibrium theory, there is, *precisely* at the time of transit, high water or low water of diurnal tides in the northern hemisphere, according as the declination of the body is north or south; and the amount of the rise and fall is in simple proportion to the sine of twice the latitude, and therefore vanishes both at the equator and at the poles. In the corrected equilibrium theory, the time of high water may be considerably either before or after the time of transit, and its amount is very different for different places in the same latitude, and certainly not zero at either equator or poles. In the ordinary statement there is no lunar fortnightly or semi-annual diurnal tide in the latitude 35° 16′ (being $\sin^{-1}\frac{1}{\sqrt{3}}$), and its amount in other latitudes is in proportion to the deviations of the squares of their sines from the value $\frac{1}{3}$. In the corrected equilibrium theory each of these tides is still the same in the same latitude, and vanishes at a certain latitude, and in any other latitudes is in simple proportion to the deviation of the squares of their sines from the square of the sine of that latitude. But the latitude where there is no tide of this class is not $\sin^{-1}\frac{1}{\sqrt{3}}$, but $\sin^{-1}(\sqrt{\frac{1+\mathfrak{E}}{3}})$, where $\mathfrak{E}$ is the mean value of $3\sin^2 l - 1$, for the whole covered portion of the earth's surface, a quantity easily estimated by a laborious quadrature, from sufficiently complete geographical data of the coast lines for the whole earth.

As the fortnightly and semi-annual tides most probably

›llow in reality very nearly the equilibrium law, it becomes a ıatter of great importance to evaluate this quantity; but we egret that hitherto we have not been able to undertake the rork. Conversely it is probable that careful determination of he fortnightly and semi-annual tides at various places, by ›roper reductions of tidal observations, may contribute to ;eographical knowledge as to the amount of water surface in he hitherto unexplored districts of the arctic and antarctic egions.

The tides, mutual attraction of the waters neglected: practical importance of correction for equilibrium fortnightly and semi-annual tides.

811. The superposition of the solar semi-diurnal on the lunar semi-diurnal tide has been investigated above as an example of the composition of simple harmonic motions; and the well-known phenomena of the "spring-tides" and "neap-tides," and of the "priming" and "lagging" have been explained (§ 60). We have now only to add that observation proves the proportionate difference between the heights of spring-tides and neap-tides, and the amount of the priming and lagging to be much less in nearly all places than estimated in § 60 on the equilibrium hypothesis; and to be very different in different places, as we shall see in Vol. II. is to be expected from the kinetic theory.

Spring and neap tides: "priming" and "lagging."

Discrepance from observed results, due to inertia of water.

812. The potential expressions used in the preceding investigation are immediately convenient (§§ 802, 804) for the hydrostatic problem. But it is interesting, in connexion with this problem, to know the amount of the disturbing influence on apparent terrestrial gravity at any point of the earth's surface, produced by the lunar or solar influence. We shall therefore —still using the convenient static hypothesis of § 804—determine convenient rectangular components for the resultant of the two approximately equal and approximately opposed disturbing forces assumed in that hypothesis. First, we may remark that these two forces are approximately equivalent to a force equal to their difference in a line parallel to that of the centres of the earth and moon, compounded with another perpendicular to this and equal to twice either, multiplied into the cosine of half the obtuse angle between them.

Lunar and solar influence on apparent terrestrial gravity.

Resolving each of these components along and perpendicular to the earth's radius through the place, we obtain, by a process, the details of which we leave to the student for an exercise, the following results, which are stated in gravitation measure:—

Lunar and solar influence on apparent terrestrial gravity.

Vertical component $=\frac{M}{E}\left(\frac{a}{D}\right)^3(2\cos^2\theta-\sin^2\theta)$, upwards

Horizontal component $=3\frac{M}{E}\left(\frac{a}{D}\right)^3\sin\theta\cos\theta$, towards point of horizon under moon or anti-moon.

Here, as before, E and M denote the masses of the earth and moon, D the distance between their centres, a the earth's radius, and θ the moon's zenith distance.

Or from the potential expression (12), by taking $\frac{d}{dr}$ and $\frac{d}{rd\theta}$ we find the same expressions.

The vertical component is a maximum upwards, amounting to

$$2\frac{M}{E}\left(\frac{a}{D}\right)^3,$$

when the moon or anti-moon is overhead; and a maximum downwards of half this amount when the moon is on the horizon. The horizontal component has its maximum value amounting to

$$\frac{3}{2}\frac{M}{E}\left(\frac{a}{D}\right)^3,$$

when the moon or anti-moon is 45° above the horizon. Similar statements, of course, apply to the disturbing influence of the sun. For the moon $\frac{M}{E}\left(\frac{a}{D}\right)^3$ is probably equal to about $\frac{1}{83\times(60\cdot3)^3}$, or $\frac{1}{18\cdot2\times10^6}$: and the corresponding measure of the sun's influence is very approximately $\left(1+\frac{1}{83}\right)\left(\frac{27\cdot3}{365}\right)^2\frac{1}{(60\cdot3)^3}$, or $\frac{1}{39\cdot1\times10^6}$. Hence, as the moon or anti-moon rises from the horizon to the zenith of any place on the earth's surface, the intensity of apparent gravity is diminished by about one six-millionth part: and the plummet is deflected towards the point of the horizon under either moon or anti-moon, by an amount which reaches its maximum value, $\frac{1}{12\times10^6}$ of the unit angle (57°·3) when the altitude is 45°. The corresponding effects of solar influence are of nearly half these amounts.

813. *Examples of* § 799 *continued.*—(4.) All other circumstances as in Example (2.), let the two bodies be not fixed, but let them revolve in circles round their common centre of inertia with angular velocity such as to give centrifugal force to each just equal to the force of attraction it experiences from the other.

Real tide-generating influence explained by method of centrifugal force.

Let the centre of the earth be origin of rectangular co-ordinates, and OZ perpendicular to the plane of the circular orbits, and let OX revolve so as always to pass through the disturbing body. Then, dealing with centrifugal force by the potential method, as in § 794; for the equation of a series of surfaces cutting everywhere at right angles the resultant of gravity and centrifugal force, we find

$$\frac{E}{\sqrt{(x^2+y^2+z^2)}}+\frac{M}{\sqrt{[(D-x)^2+y^2+z^2]}}+\tfrac{1}{2}\omega^2[(b-x)^2+y^2]=\text{const. } (24),$$

where ω denotes the angular velocity of revolution of the two bodies round their centre of inertia, and b the distance of this point from the earth's centre:—so that

$$M(D-b)\omega^2=Eb\omega^2=\frac{ME}{D^2} \qquad (25).$$

Hence $$\frac{Mx}{D^2}-\omega^2 bx=0.$$

Using this in (24), expanded and dealt with generally as (12) in Example (3.), we see that the first power of x disappears; and, omitting terms of third and higher orders, we have

$$\frac{E}{r}+\frac{M}{D}\left(1+\tfrac{1}{2}\frac{3x^2-r^2}{D^2}\right)+\tfrac{1}{2}\omega^2(x^2+y^2)=\text{const.} \qquad (26).$$

To reduce to spherical harmonics we have

$$x^2+y^2=\tfrac{2}{3}r^2-\tfrac{1}{3}(3z^2-r^2),$$

and therefore, as according to our approximation we may take ω^2a^2 for ω^2r^2, we find [with the notation $r=a(1+u)$ as above]

$$u=\tfrac{1}{2}\frac{Ma}{ED}\cdot\frac{3x^2-r^2}{D^2}-\tfrac{1}{6}\omega^2\frac{a}{E}\cdot(3z^2-r^2),$$

or in polar co-ordinates (27).

$$u=\tfrac{1}{2}\frac{Ma^3}{ED^3}(3\sin^2\theta\cos^2\phi-1)-\tfrac{1}{6}\frac{\omega^2a^3}{E}(3\cos^2\theta-1).$$

This interpreted is as follows:—

The surface of the fluid will be a harmonic spheroid of the second order [that is (§ 779), an ellipsoid differing infinitely little from a sphere], which we may regard as the result of superimposing on the deviation from spherical figure investigated in § 804, another consisting of the oblateness due to rotation with angular velocity ω round the diameter of the

Real tide-generating influence explained by method of centrifugal force.

earth perpendicular to the plane of the disturbing body's orbit. We may prove this conclusion with less analysis by supposing the purely static system of Example (3.) to rotate, first with any angular velocity ω, about any diameter of the earth perpendicular to the straight line through its centre in which the disturbing bodies are placed; and then supposing this angular velocity to be just such as to balance the earth's attraction on the two disturbing bodies, so that the holdfasts by which they were prevented from falling together may be removed. Then it is easy to prove analytically that the effect of carrying either disturbing body to the other side, and uniting the two, will be a small disturbance in the figure of the fluid amounting to some such fraction of the deviation investigated in Example (3.) as the earth's radius is of the distance of the disturber.

814. The purely static system of Example (3.) gives the simplest and most symmetrical foundation for the equilibrium theory of the tides. The kinetic system of Example (4.) is indeed not less purely static in relation to the earth, and is equivalent to an absolutely static imaginary system in which repulsion from a fixed line, on parts of a non-rotating system, is substituted for the centrifugal force of the rotating system. But it is complicated by the oblateness of the fluid surface produced by the centrifugal force or repulsion. This oblateness, as we see from § 801, would amount to as much as $\frac{1}{(27{\cdot}4)^2}\times\frac{1}{580}$, or $\frac{1}{435{,}000}$, being about 27·8 times the ellipticity of the lunar tide level for the case of the earth and moon; although only to $\frac{1}{366^2}\times\frac{1}{580}$, or $\frac{1}{77{,}700{,}000}$ for the case of the earth and sun.

Augmentation of result by mutual gravitation of the disturbed water.

815. When the attraction of the fluid on itself is sensible, the disturbance in its distribution gives rise to a counter disturbing force, which increases the deviation of the equipotential surfaces from the spherical figure. The general hydrostatic condition (§ 750), that the surfaces of equal density must still coincide with the equipotential surfaces, thus presents an exquisite problem for analysis. It has called forth from Legendre and Laplace an entirely new method in mathematics, commonly referred to by English writers as "Laplace's coefficients" or "Laplace's Functions." The principles have been

Augmentation of result by mutual gravitation of the disturbed water.

cetched in the second Appendix to our first Chapter; from hich, and the supplementary investigations of §§ 778...784, re have immediately the solution for the case in which the uid is homogeneous, and the nucleus (being a solid of any hape, and with any internal distribution of density, subject nly to the condition that its external equipotential surfaces re approximately spherical) is wholly covered by the fluid. The conclusion may be expressed thus:—Let ρ be the density of the fluid, and let σ be the mean density of the whole mass, fluid and solid. Let the disturbing influence, whether of external disturbing masses, or of deviation from accurate centrobaric (§ 526) quality in the nucleus, or of centrifugal force due to rotation, be such as to render the level surfaces harmonic spheroids of order i, when the liquid is kept spherical by a rigid envelope in contact with it all round. The tendency of the liquid surface would be to take the figure of that one of these level surfaces which encloses the proper volume. But in changing its figure, if permitted, it would *increase* the deviation of this level surface. The result is, that if the constraint be removed, the level surface of the liquid in equilibrium will be a harmonic spheroid of the same type, but of deviation from sphericity augmented in the ratio of 1 to $1-\frac{3\rho}{(2i+1)\sigma}$.

Let the potential at or infinitely near the bounding surface be

$$\frac{4\pi\sigma a^3}{3r}+S_i \qquad (1)$$

when the liquid is held fixed in shape by a spherical envelope, of radius a. In these circumstances

$$r=a\left(1+\frac{3S_i}{4\pi\sigma a^2}\right) \qquad (2)$$

is the equipotential surface of mean radius a. If now the bounding surface of the liquid be changed into the harmonic spheroid

$$r=a(1+cS_i) \qquad (3),$$

the potential (§ 543) becomes changed from (1) to

$$\frac{4\pi\sigma a^3}{3r}+\left(1+\frac{4\pi\rho c a^2}{2i+1}\right)S_i \qquad (4),$$

and the equipotential surface becomes, instead of (2),

$$r=a\left\{1+\left(1+\frac{4\pi\rho c a^2}{2i+1}\right)\frac{3S_i}{4\pi\sigma a^2}\right\} \qquad (5).$$

Augmentation of result by mutual gravitation of the disturbed water.

Hence that the boundary (3) of the liquid may be an equipotential surface,

$$c=\left(1+\frac{4\pi\rho c a^3}{2i+1}\right)\cdot\frac{3}{4\pi\sigma a^3},$$

which gives
$$4\pi c a^3=\frac{1}{\frac{\sigma}{3}-\frac{\rho}{2i+1}},$$

whence
$$1+\frac{4\pi\rho c a^3}{2i+1}=\frac{1}{1-\frac{3\rho}{(2i+1)\sigma}} \quad (6).$$

Using this in (5), and comparing with (2), we prove the proposition.

Stability of the ocean.

816. The instability of the equilibrium in the case in which the density of the liquid is greater than the mean density of the nucleus, already remarked as obvious, is curiously illustrated by the present result, which makes the deviation infinite when $i=1$ and $\sigma=\rho$. But it is to be remarked that it is only when the nucleus is completely covered that the equilibrium would be unstable. However dense the liquid may be, there would be a position of stable equilibrium with the nucleus protruding on one side; and if the bulk of the liquid is either very small or very large in comparison with that of the nucleus, the figure of its surface in stable equilibrium would clearly be approximately spherical. Excluding the case of a very small nucleus of lighter specific gravity (which would become merely a small floating body, not sensibly disturbing the general liquid globe), we have, in the apparently simple question of finding the distribution of a small quantity of liquid on a symmetrical spherical nucleus of less specific gravity, a problem which utterly transcends mathematical skill as hitherto developed.

Augmentations of results by mutual gravitation of water calculated for examples of § 799.

817. The cases of $i=1$ and $i=2$ give the solutions of the several examples of § 799 when the attraction of the liquid on itself is taken into account, provided always that the solid is wholly covered. Thus [§ 799, Example (2.)] if the earth and moon were stopped, and each held fixed, the moon's attraction would still not disturb the figure of the liquid surface from true sphericity, but would render it eccentric to a greater degree than that previously estimated, in the ratio of 1 to

$1 - \frac{\rho}{\sigma}$. For the earth and sea, $\frac{\rho}{\sigma}$ is about $\frac{2}{11}$, and therefore the spherical liquid surface would be drawn towards the moon by 86 feet, being $1\frac{2}{9}$ times the amount of 70 feet found above (§ 803). And the tidal and rotational ellipticities estimated in §§ 800, 814, 813 would, on the supposition now made, be augmented each in the ratio of 1 to $1 - \frac{3}{5}\frac{\sigma}{\rho}$; or 55 to 49 for the case of earth and sea. The true correction for the attraction of the sea, as altered by tidal disturbance, in the equilibrium theory of the tides, must be less than this, as the liquid does not cover more than about $\frac{2}{3}$ of the surface of the solid. To find the true amount of the correction for the attraction of the water on itself when the whole solid is not covered, even if the arrangement of dry land and sea were quite symmetrical and simple (as, for instance, one circular continent and the rest ocean), belongs to the transcendental problem already referred to (§ 816). It can be practically solved, if necessary, by laborious methods of approximation; but the irregular boundaries of land and sea on the real earth, and the true kinetic circumstances of the tides, are such as to render nugatory any labours of this kind. Happily the error committed in neglecting altogether the correction in question may be safely estimated as less than 10 per cent. ($\frac{6}{49}$ being 12·3 per cent.), and may be neglected in our present uncertainty as to absolute values of causes and effects in the theory of the tides.

Augmentations of results by mutual gravitation of water calculated for examples of § 799.

818. But although the influence on the tides produced by the attraction of the water itself as it rises and falls is not considerable even in any one place; it is a manifest, though not an uncommon, error to suppose that the moon's disturbing influence on terrestrial gravity is everywhere insensible. It was pointed out long ago by Robison[1] that the great tides of the Bay of Fundy should produce a very sensible deflection on the plummet in the neighbourhood, and that observation of this effect might be turned to account for determining the earth's mean density. But even ordinary tides must produce, at places close to the sea shore, deviations in the plummet considerably exceeding the greatest direct

Local influence of high water on direction of gravity.

[1] *Mechanical Philosophy*, 1804. See also Forbes, *Proc. R. S. E.*, April 1849.

Attraction of high water on a plummet at the sea-side.

effect of the moon, which, as we have seen (§ 812), amounts to $\frac{1}{12.000.000}$ of the unit angle (57°·3). Thus, at a point on or not many feet above the mean sea level, and 100 yards from low-water mark, a deflection, amounting to more than $\frac{1}{8.000.000}$ of the unit angle on each side of the mean vertical, will be produced by tides of five feet rise and fall on each side of the mean, if the line of coast does not deviate very much from one average direction for 50 miles on either side, and if the rise and fall is approximately simultaneous and equal for 50 miles out to sea. For, a point placed as O in the

Vertical section through O.

sketch will, as the water rises from low tide to high tide, experience the attraction of a plate of water indicated in section by $HKK'L'L$. If we neglect the small part of the whole effect due to the long bar (extending along the coast) shown in section by HKL, we have only to find the attraction of the rectangular plate of water by hypothesis of 50 miles' breadth from KL, 100 miles' length parallel to the coast, and 10 feet thickness (KL). This will not be sensibly altered if O is precisely in the continuation of the middle plane EE' (instead of a few feet above it, as would generally be the case in a convenient sea-side gravitation observatory), and the whole matter of the plate were condensed into its middle plane. But

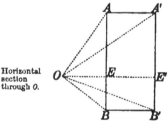

Horizontal section through O.

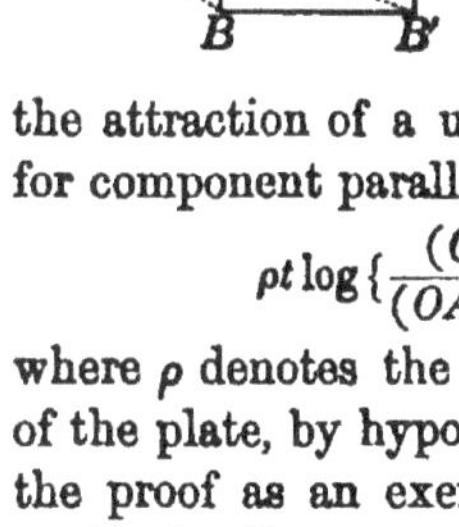

the attraction of a uniform rectangular plate on a point O has, for component parallel to AB,

$$\rho t \log \left\{ \frac{(OA+AE).(OB+BE).OE'^2}{(OA'+A'E').(OB'+B'E').OE^2} \right\} \tag{7}$$

where ρ denotes the density of the water, and t the thickness of the plate, by hypothesis a small fraction of OE. (We leave the proof as an exercise to the student.) Now, taking the nautical mile as 2000 yards, we have, according to the assumed data, very approximately

$$\frac{AE}{OE}=\frac{OA}{OE}=\frac{OE'}{OE}=1000, \text{ and } \frac{OA'}{OE}=1000\sqrt{2},$$

nd B, B' are at the same distances on one side of OE' as AA' n the other. Hence the preceding expression becomes

Attraction of high water on a plummet at the sea-side.

$$2\rho t \log \frac{2100}{1+\sqrt{2}}, \text{ which is equal to } 13{\cdot}5 \times \rho t.$$

The ratio of this to $\frac{4\pi}{3}\sigma r$, the earth's whole attraction on O, is $\frac{3 \times 13{\cdot}4}{4\pi} \frac{\rho t}{\sigma r}$: which (as $\frac{t}{r}$ is $\frac{1}{2{,}100{,}000}$ by hypothesis, and $\frac{\rho}{\sigma}$ is about $\frac{2}{11}$) amounts to $\frac{1}{3{,}580{,}000}$. The plummet will therefore, at high tide, be disturbed from the position it had at low tide, by a horizontal force of somewhat more than $\frac{1}{4{,}000{,}000}$ of the vertical force; and its deviation will of course be this fraction of $57^{\circ}{\cdot}3$, the unit angle.

819. Recurring to the case of $\rho = \sigma$, we learn from § 817 that a homogeneous liquid in equilibrium under the influence of centrifugal force, or of tide-generating action, has $2\frac{1}{2}$ times as much ellipticity as it would have if mutual attraction between the parts of the fluid were done away with (§ 800), and gravity were towards a fixed interior centre of force. For a homogeneous liquid of the same mean density as the earth, rotating in a time equal to the sidereal day, the ellipticity is therefore $\frac{1}{232}$, being $2\frac{1}{2}$ times the result, $\frac{1}{580}$, which we found in § 801. This agrees with the conclusion for the case of approximate sphericity, which we derived (§ 775) from the theorem of § 771, regarding the equilibrium of a homogeneous rotating liquid. But even for this case Laplace's spherical harmonic analysis is most important, as proving that the solution is *unique*, when the figure is approximately spherical; so that neither an ellipsoid with three unequal axes, nor any other figure than the oblate elliptic spheroid of revolution, can satisfy the hydrostatic conditions, when the restriction to approximate sphericity is imposed. Our readers will readily appreciate this item of the debt we owe to the great French naturalist, when we tell them that one of us had actually for a time speculated on three unequal axes as a possible figure of terrestrial equilibrium.

Application of § 817 to theory of the earth's figure.

820. As another example of the result of § 817 for the case $i = 2$, let us imagine the earth, rotating with the actual angular

Application of § 817 to theory of the earth's figure.

velocity, to consist of a solid centrobaric nucleus covered with a thin liquid layer of density equal to the true density of the upper crust, that is, we may say, half the mean density of the nucleus. The ellipticity of the free surface would be

$$\frac{1}{580}\times\frac{1}{1-\frac{3}{5}\times\frac{1}{2}}, \text{ or } \frac{1}{386}.$$

Or, lastly, let it be required to find the density of a superficial liquid layer on a centrobaric nucleus which, with the actual angular velocity of rotation, would assume a spheroidal figure with ellipticity equal to $\frac{1}{295}$, the actual ellipticity of the sea level. We should have

$$\frac{1}{1-\frac{3}{5}\frac{\rho}{\sigma}}=\frac{580}{295},$$

which gives $\rho = \cdot 819 \times \sigma$.

821. Bringing together the several results of §§ 801, 817, 819 for a centrobaric nucleus revolving with the earth's angular velocity, and covered with a thin layer of liquid of density ρ, the mean density of the whole being σ, we have—

$$\text{(1.) for } \frac{\rho}{\sigma}=0, \quad e=\tfrac{1}{580},$$

$$\text{(2.) ,, } \frac{\rho}{\sigma}=\tfrac{2}{11}, \quad e=\tfrac{1}{517},$$

$$\text{(3.) ,, } \frac{\rho}{\sigma}=\tfrac{1}{2}, \quad e=\tfrac{1}{386}.$$

$$\text{(4.) ,, } \frac{\rho}{\sigma}=\cdot 819, \quad e=\tfrac{1}{295},$$

$$\text{(5.) ,, } \frac{\rho}{\sigma}=1, \quad e=\tfrac{1}{232},$$

where e denotes the ellipticity of the free bounding surface of the liquid. The density of the earth's upper crust may be roughly estimated as $\frac{1}{2}$ the mean density of the entire mass, and is certainly in every part less than ·812 of this mean density. The ellipticity of the sea level does not differ from $\frac{1}{297}$ by more than 2 or 3 per cent., and is therefore decidedly too great to be accounted for by centrifugal force, and ellipticity in the upper crust alone, on the hypothesis that there is a rigid centrobaric nucleus, covered by only a thin upper crust with surface on the whole agreeing in ellipticity with the free liquid surface. It is therefore quite certain that there must be on the

Observation shows so great an ellipticity of sea level that there must be oblateness of the solid not only in its bounding surface, but also in interior layers of equal density.

hole some degree of oblateness in the lower strata, in the me direction as that which centrifugal force would produce the mass were fluid. There is, as we shall see in later lumes, a great variety of convincing evidence in support of e common geological hypothesis that the upper crust was t one time all melted by heat. This would account for the eneral agreement of the boundary of the solid with that of uid equilibrium, though largely disturbed by upheavals, and hrinkings, in the process of solidification which (App. D.) has robably been going on for a few million years, but is not yet uite complete (witness lava flowing from still active volcanoes). The oblateness of the deeper layers of equal density which we now infer from the figure of the sea level, the observed density of the upper crust, and Cavendish's weighing of the earth as a whole, renders it highly probable that the earth has been at one time melted not merely all round its surface, but either throughout, or to a great depth all round.

Equilibrium of rotating spheroid of heterogeneous liquid, investigated.

822. We therefore, as our last hydrostatic example, proceed to investigate the conditions of a heterogeneous liquid resting on a rigid spherical centrobaric core or nucleus, and slightly disturbed, as explained in § 815, by attracting masses fixed either externally or in the core (among which, of course, fall to be included deviations, if any, from a rigorously centrobaric distribution in the matter of the core).

For any point (r, θ, ϕ) in space let

N be the potential due to the core,
V „ „ undisturbed fluid,
Q „ „ disturbing force,
U „ „ disturbance in the distribution of the fluid.

Thus the whole potential at the point in question is $N+V$ when the fluid is undisturbed, and $N+Q+V+U$ when the disturbing force is introduced and equilibrium supervenes. Let also ρ be the density of the undisturbed fluid at (r, θ, ϕ) (which of course would vanish if the point in question were situated in any other part of space than that occupied by the fluid); and let $\rho+\varpi$ be the altered density at the same point (r, θ, ϕ) when the fluid rests under the disturbing influence. It is to be noticed that N, V, ρ are functions of r alone; while Q, U, ϖ are functions of r, θ, ϕ.

Equilibrium of rotating spheroid of heterogeneous liquid, investigated.

Let now δr be an infinitely small variation of r. The density of the liquid at the point $(r+\delta r, \theta, \phi)$ will be $\rho+\varpi+\frac{d(\rho+\varpi)}{dr}\delta r$. or simply

$$\rho+\varpi+\frac{d\rho}{dr}\delta r,$$

as ϖ is infinitely small by hypothesis. If we equate this to ρ we have

$$\varpi+\frac{d\rho}{dr}\delta r=0,$$

Spheroidal surface of equal density.

and deduce

$$\delta r=-\frac{\varpi}{\frac{d\rho}{dr}} \qquad (1)$$

for the equation expressing the deviation from the spherical surface of radius r, of the spheroidal surface over which the density in the disturbed liquid is ρ. The liquid being incompressible, the volume enclosed by this spheroidal surface must be equal to that enclosed by the spherical surface, and therefore, if $d\sigma$ denote an element of the spherical surface, and $\iint$ integration over the whole of it,

$$\iint \delta r d\sigma=0 \qquad (2).$$

Hence, by (1), as $\frac{d\rho}{dr}$ is independent of θ, ϕ,

Expression of incompressibility.

$$\iint \varpi d\sigma=0 \qquad (3).$$

Now, as before for density, we have for the disturbed potential at $(r+\delta r, \theta, \phi)$

$$N+Q+V+U+\frac{d(N+Q+V+U)}{dr}\delta r,$$

or, because $Q+U$ is infinitely small,

$$N+Q+V+U+\frac{d(N+V)}{dr}\delta r.$$

And, therefore, to express that the spheroidal surface corresponding to (1), with r constant, is an equipotential surface in the disturbed liquid, we have

Hydrostatic equation.

$$Q+U-\frac{\frac{d}{dr}(N+V)}{\frac{d\rho}{dr}}\varpi+N+V=F(r) \qquad (4),$$

which (§ 750) is the equation of hydrostatic equilibrium. In this equation we must suppose N and ρ to be functions of r, and Q a function of r, θ, ϕ; all given explicitly: and from ρ we have, by putting $i=0$, in (15) and (16) of § 542,

$$V=4\pi\left(\int_r^{\mathrm{r}} r'\rho' dr' + \frac{1}{r}\int_a^r r'^2\rho' dr'\right) \qquad (5).$$

where ρ' is the value of ρ at distance r' from the centre, and $\mathfrak{r}$ the radius of the outer bounding surface of the undisturbed fluid, and a that of the fixed spherical surface of the core on which it rests. To find $V+U$, following strictly the directions of § 545, we add the potential of a distribution of matter with density $\rho+\varpi$ through the space between the spherical surfaces of radii a and $\mathfrak{r}$ to that of the shell B of positive and negative matter there defined. Let the thickness of the latter at the point $(\mathfrak{r}, \theta, \phi)$ be called h, being the value of δr at the surface; and let q denote its density, being the surface value of ρ. Then, subtracting the undisturbed potential V, we have

Equilibrium of rotating spheroid of heterogeneous liquid.

$$U=\iiint\frac{\varpi' r'^2 d\sigma' dr'}{D}+\left[\iint\frac{q'h'd\sigma'}{D}\right] \qquad (6),$$

Part of the potential, due to oblateness:

if as usual D denote the distance between the points (r, θ, ϕ), (r', θ', ϕ'), and the accented letters denote the values of the corresponding elements in the latter; and if [ ] denotes surface values and integration.

Let us now suppose the required deviation of the surfaces of equal pressure density and potential to be expressed as follows in surface harmonics, of which the term R_0 disappears because of (2):—

$$\left.\begin{array}{l}\text{for the interior of the fluid, } \delta r=R_1+R_2+R_3+\text{etc.,}\\ \text{and for the outer bounding surface, } h=\mathfrak{R}_1+\mathfrak{R}_2+\mathfrak{R}_3+\text{etc.}\end{array}\right\} \quad (7).$$

Hence by (1) $$\varpi=\frac{-d\rho}{dr}(R_1+R_2+R_3+\text{etc.}) \qquad (8).$$

Using this in (6) according to §§ 544, 542, 536, we have

developed in harmonics.

$$4\pi\sum_1^\infty\frac{1}{2i+1}\left\{r^i\int_r^{\mathfrak{r}} r'^{-i+1}\frac{-d\rho'}{dr'}R_i'dr'+r^{-i-1}\int_a^r r'^{i+2}\frac{-d\rho'}{dr'}R_i'dr'+q\mathfrak{R}_i\frac{\mathfrak{r}^i}{r^{i-1}}\right\} \quad (9),$$

where R_i' denotes the value of R_i for the point (r', θ, ϕ) instead of (r, θ, ϕ).

To complete the expansion of the hydrostatic equation (4) we may suppose the harmonic expression for Q to be either directly given, or to be found immediately by Appendix B. (52), or by (8) of § 539, according to the form in which the data are presented. Thus let us have

$$Q=\sum_{i=0}^{i=\infty}\sum_{s=0}^{s=i}(A_i^{(s)}\cos s\phi+B_i^{(s)}\sin s\phi)\Theta_i^{(s)} \qquad (10),$$

Harmonic development of disturbing potential.

according to the notation of App. B. (37) and (38), $A_i^{(s)}$, $B_i^{(s)}$

Equilibrium of rotating spheroid of heterogeneous liquid.

denoting known functions of r. Using now this and (8) in (4), we have

$$\sum_{i=1}^{i=\infty}\left\{\sum_{s=0}^{s=i}(A_i^{(s)}\cos s\phi+B_i^{(s)}\sin s\phi)\Theta_i^{(s)}-\frac{-d(N+V)}{dr}R_i\right.$$

$$+\frac{1}{2i+1}\left(r^i\int_r^{\mathfrak{r}} r'^{-i+1}\frac{-d\rho'}{dr'}R_i'dr'+r^{-i-1}\int_a^r r'^{i+2}\frac{-d\rho'}{dr'}R_i'dr'+\mathfrak{R}_i\frac{r^i}{\mathfrak{r}^{i-1}}\right)$$

$$+A_0^{(0)}+N+V=F(r) \quad (11)$$

where $\left[\frac{-d\rho}{dr}\right]$ denotes the value of $\frac{-d\rho}{dr}$ for $r=\mathfrak{r}$. Hence: first, for the terms of zero order,

$$A_0^{(0)}+N+V=F(r) \quad (12)$$

which merely shows the value of $F(r)$, introduced temporarily in (4) and not wanted again: and, by terms of order i,

Equation of equilibrium for general harmonic term:

$$\frac{-d(N+V)}{dr}R_i-\frac{1}{2i+1}\left\{r^i\int_r^{\mathfrak{r}} r'^{-i+1}\frac{-d\rho'}{dr'}R_i'dr'+r^{-i-1}\int_a^r r'^{i+2}\frac{-d\rho'}{dr'}R_i'dr'+\mathfrak{g}\right\}$$

$$=\sum_{s=0}^{s=i}(A_i^{(s)}\cos s\phi+B_i^{(s)}\sin s\phi)\Theta_i^{(s)} \quad (13)$$

Lastly, expanding R_i (as above for the i term of Q) by App. B. (37), let us have

$$R_i=\sum_{s=0}^{s=i}(u_i^{(s)}\cos s\phi+v_i^{(s)}\sin s\phi)\Theta_i^{(s)} \quad (14)$$

where $u_i^{(s)}$, $v_i^{(s)}$ are functions of r, to the determination of which the problem is reduced. Hence equating separately the coefficients of $\cos s\phi\,\Theta_i^{(s)}$, etc., on the two sides, and using u_i to denote any one of the required functions $u_i^{(s)}$, $v_i^{(s)}$, and A_i any of the given functions $A_i^{(s)}$, $B_i^{(s)}$, and u_i', $\mathfrak{u}_i$ the values of u_i for $r=r'$ and $r=\mathfrak{r}$ respectively, we have

leading to equation for general coefficient, u_i, regarded as a function of r.

$$\frac{-d(N+V)}{dr}u_i-\frac{4\pi}{2i+1}\left\{r^i\int_r^{\mathfrak{r}} r'^{-i+1}\frac{-d\rho'}{dr'}u_i'dr'+r^{-i-1}\int_a^r r'^{i+2}\frac{-d\rho'}{dr'}u_i'dr'+q\frac{\mathfrak{u}_i r^i}{\mathfrak{r}^{i-1}}\right\}=A_i$$

or, as it will be convenient sometimes to write it, for brevity, $\sigma_i(u_i)=A_i$

where σ_i denotes a determinate operation, involving definite integrations, and such that $\sigma_i(u)$ for u any function of r, continuous or discontinuous, necessarily disappears through any range of values of r, for all of which $u=0$. To reduce this to a differential equation, divide by r^i, differentiate, multiply by r^{2i+2}, and differentiate again. If, for brevity, we put

$$\frac{-d(N+V)}{dr}=r\psi \quad (16)$$

the result is

$$\frac{d}{dr}\{r^{2i+2}\frac{(r^{-i+1}\psi u_i}{dr}\}+4\pi r^{i+2}\frac{-d\rho}{dr}u_i=\frac{d}{dr}\{r^{2i+2}\frac{d(r^{-i}A_i)}{dr}\} \quad (17),$$

Equilibrium of rotating spheroid of heterogeneous liquid.

a linear differential equation, of the second order, for u_i, with coefficients and independent term, known functions of r. The general solution, as is known, is of the form

Differential equation to be integrated.

$$u_i=CP+C'P'+\alpha \quad (18),$$

where α is a function of r satisfying the equation

$$\sigma_i(\alpha)=A_i \quad (19),$$

C and C' are two arbitrary constants, and P and P' two distinct functions of r.

Equation (15) requires that $C=0$ and $C'=0$; in other words, u_i, if satisfying it, is fully determinate. This is best shown by remarking that if, instead of (15), we have

$$\sigma_i(u)=A_i+Kr^i+K'r^{-i-1} \quad (20)$$

Determination of constants to complete the required solution.

where K, K' are any two constants, these constants disappear in the differentiations, and we have still the same differential equation, (17); and that the two arbitrary constants C and C' of the general solution (18) of this are determined by (20) when any two values are given for K and K'. In fact, the expression (18), used for u_i, reduces (20) to

$$C\sigma_i(P)+C'\sigma_i(P')=Kr^i+K'r^{-i-1} \quad (21),$$

which shows that $\sigma_i(P)$ and $\sigma_i(P')$ must be distinct linear functions of r^i and r^{-i-1}, and determines C and C'.

Thus we see that whatever be A_i we have, in the integration of the differential equation (19), and the determination of the arbitrary constants (14), the complete solution of our problem.

Introduction of the Newtonian law of force

Unless it is desired, as a matter of analytical curiosity, or for some better reason, to admit the supposition that N is any arbitrary function of r, it is unnecessary to retain both ψ and ρ as two distinct given functions. For the external force of the nucleus, or that part of it of which N is the potential, being by hypothesis symmetrical, relatively to the centre, it must in nature vary inversely as the square of the distance from this point; that is to say,

$$\frac{-dN}{dr}=\frac{\mu}{r^2} \quad (22),$$

μ being a constant, measuring in the usual unit (§ 459) the mass of the nucleus. And by (5)

$$\frac{-dV}{dr}=\frac{4\pi}{r^2}\int_a^r \rho' r'^2 dr' \quad (23).$$

From this, with (22) and (17), we have

Equilibrium of rotating spheroid of heterogeneous liquid.

$$\psi=\frac{4\pi}{r^3}\int_a^r \rho' r'^2 dr' + \frac{\mu}{r^3} \qquad (24).$$

which gives $4\pi\rho=\frac{d(\psi r^3)}{r^2 dr}$ and $4\pi\frac{d\rho}{dr}=r\frac{d^2\psi}{dr^2}+4\frac{d\psi}{dr}$. (25).

Simplification by introducing the Newtonian law of force.

Using this last in (17), and reducing by differentiation, we have

$$\frac{d^2u_i}{dr^2}+2\left(\frac{d\log\psi}{dr}+\frac{2}{r}\right)\frac{du_i}{dr}-\frac{(i-1)(i+2)}{r^2}u_i=\frac{1}{r^{i+2}}\frac{d}{dr}\left\{r^{2i+2}\frac{d(r^{-i}A_i)}{dr}\right\} \qquad (26)$$

Another form, convenient for cases in which the disturbing force is due to *external* attracting matter, or to centrifugal force of the fluid itself, if rotating, is got by putting, in (17),

$$r^{-i+1}u_i=e_i \qquad (27),$$

and reducing by differentiation. Thus

Differential equation for proportionate deviation from sphericity.

$$\frac{d^2e_i}{dr^2}+2\left(\frac{d\log\psi}{dr}+\frac{i+1}{r}\right)\frac{de_i}{dr}+\frac{2(i-1)}{r}\frac{d\log\psi}{dr}e_i=\frac{1}{r^{2i+2}}\frac{d}{dr}\left\{r^{2i+2}\frac{d(r^{-i}A_i)}{dr}\right\} \qquad (28)$$

With this notation the intermediate integral, obtained from (15) by the first step of the process of differentiating executed in the order specified, gives

$$\frac{de_i}{dr}+\frac{d\log\psi}{dr}e_i-r^{-2i-2}\int_a^r\left(r\frac{d^2\psi}{dr}+4\frac{d\psi}{dr}\right)r^{2i+1}e_i=\frac{d(r^{-i}A_i)}{dr} \qquad (29).$$

Important conclusions, readily drawn from these forms, are that if Q is a solid harmonic function (as it is when the disturbance is due either to disturbing bodies in the core, or in the space external to the fluid, or to centrifugal force of the fluid rotating as a solid about an axis); then (1.) e_i, regarded as positive, and as a function of r, can have no maximum value, although it might have a minimum; and (2.) if the disturbance is due to disturbing masses outside, or to any other cause (as centrifugal force) which gives for potential a solid harmonic of order i with only the r^i term, and no term r^{-i-1}, e_i can have no minimum except at the centre, and must increase outwards throughout the fluid.

Layers of greatest and least proportionate deviation from sphericity.

To prove these, we must first remark that ψ necessarily diminishes outwards. To prove this, let n denote the excess of the mass of the nucleus above that of an equal solid sphere of density s equal to that of the fluid next the nucleus. Then we may put (24) under the form

$$\psi=4\pi s-\frac{4\pi}{r^3}\int_a^{r'}(s-\rho')r'^2dr'+\frac{n}{r^3} \qquad (30).$$

For stability it is necessary that n and $s-\rho'$ be each positive; and therefore the last term of the second member is positive, and diminishes as r increases, while the second term of the same is negative, and in absolute magnitude increases, and the first term is constant. Hence ψ diminishes as r increases. Again, when the force is of the kind specified, we must [App. B. (58)] have

Equilibrium of rotating spheroid of heterogeneous liquid.

$$A_i = Kr^i + K'r^{-i-1} \qquad (31),$$

and therefore the second member of (28) vanishes. Hence if, for any value of r,

$$\frac{de_i}{dr} = 0,$$

for the same,

$$\frac{d^2e_i}{dr^2} = -\frac{2(i-1)}{r}\frac{d\log\psi}{d\psi}e_i,$$

and is therefore positive, which proves (1.) Lastly, when the force is such as specified in (2.), we have $A_i = Kr^i$ simply, and therefore the second member of (29) vanishes. This equation then gives, for values of r exceeding a by infinitely little,

$$\frac{de_i}{dr} = -\frac{d\log\psi}{dr}e_i,$$

Proportionate deviation for case of centrifugal force, or of force from without.

which is positive. Hence e_i commences from the nucleus increasing. But it cannot have a minimum (1.), and therefore it increases throughout, outwards.

823. When the disturbance is that due to rotation of the liquid, the potential of the disturbing force is

Case of centrifugal force.

$$\frac{\omega^2}{2}(x^2+y^2),$$

which is equal to a solid harmonic of the second degree with a constant added. From this it follows [§§ 822, 779] that the surfaces of equal density are concentric oblate ellipsoids of revolution, with a common axis, and with ellipticities diminishing from the surface inwards.

We have, in (10) of last section,

$$Q = \frac{\omega^2}{2}(x^2+y^2) = \frac{r^2\omega^2}{6}(\Theta_0^{(0)} + \Theta_2^{(0)}).$$

This gives at once, by (7) and (14),

$$\delta r = u_2\Theta_2^{(0)}.$$

Hence

$$r + \delta r = r\left(1 + \frac{u_2}{r}\Theta_2^{(0)}\right)$$

$$= r\left[1 + \frac{u_2}{r}\left(\tfrac{1}{3} - \cos^2\theta\right)\right]$$

$$= r\left(1 - \frac{2u_2}{3r}\right)\left(1 + \frac{u_2}{r}\sin^2\theta\right) \qquad (1),$$

neglecting terms of the second order because ω, and therefore also $\frac{u_2}{r}$, are very small.

quilibrium rotating pheroid heterogeneous quid. ase of entrifugal rce.

Thus the sphere, whose radius was r, has become an oblate ellipsoid of revolution whose ellipticity [§ 822 (27)] is

$$e_2 = \frac{u_2}{r} \quad (2).$$

Its polar diameter is diminished by the fraction $\frac{2u_2}{3r}$ or $\frac{2e_2}{3}$, and its equatorial diameter is increased by $\frac{e_2}{3}$; the volume remaining unaltered.

In order to find the value of u_2, we must have data or assumptions which will enable us to integrate equation (15). These may be given in many forms; but one alone, to which we proceed, has been worked out to practical conclusions.

aplace's ypothetical w of density within he earth.

824. To apply the results of the preceding investigation to the determination of the law of ellipticity of the layers of equal density within the earth, on the hypothesis of its original fluidity, it is absolutely essential that we commence with some assumption (in default of information) as to the law which connects the density with the distance from the earth's centre. For we have seen (§ 821) how widely different are the results obtained when we take two extreme suppositions, viz., that the mass is homogeneous; and that the density is infinitely great at the centre. In few measurements hitherto made of the Compressibility of Liquids (see Vol. II., *Properties of Matter*) has the pressure applied been great enough to produce condensation to the extent of one-half per cent. The small condensations thus experimented on have been found, as might be expected, to be very approximately in simple proportion to the pressures in each case; but experiment has not hitherto given any indication of the law of compressibility for any liquid under pressures sufficient to produce considerable condensations. In default of knowledge, Laplace assumed as a hypothesis the law of compressibility of the matter of which, before its solidification, the earth consisted,

Assumed relation between density and pressure.

to be that the *increase of the square of the density is proportional to the increase of pressure.* This leads, by the ordinary equation of hydrostatic equilibrium, to a very simple expression for the law of density, which is still further simplified if we assume that the density is everywhere finite.

Neglecting the disturbing forces, we have (§§ 822, 752)

$$dp = \rho d(V+N) \qquad (1).$$

Laplace's hypothetical law of density within the earth.

But, by the hypothesis of Laplace, as above stated, k being some constant,

$$dp = k\rho d\rho \qquad (2).$$

Hence

$$k\rho + C = V + N$$

or, by § 822 (5),

$$= 4\pi\int_r^{\mathfrak{r}} r'\rho' dr' + \frac{4\pi}{r}\int_a^r r'^2\rho' dr' + \frac{\mu}{r}.$$

Multiplying by r, and differentiating, we get

$$k\frac{d}{dr}(r\rho) + C = 4\pi\int_r^{\mathfrak{r}} r'\rho' dr'$$

and

$$\frac{d^2}{dr^2}(r\rho) = -\frac{4\pi}{k}r\rho.$$

If we write

$$\frac{4\pi}{k} = \frac{1}{\kappa^2},$$

the integral may be thus expressed—

$$r\rho = F\sin\left(\frac{r}{\kappa} + G\right).$$

If we suppose the whole mass to be liquid, *i.e.*, if there be no solid core, or, at all events, the same law of density to hold from surface to centre, G must vanish, else the density at the centre would be infinite. Hence, in what follows, we shall take

$$\rho = \frac{F}{r}\sin\frac{r}{\kappa} \qquad (3).$$

Law of density.

With this value of ρ it is easy to see that

$$\int_0^r r'^2\rho' dr' = -\kappa^2 r^2\frac{d\rho}{dr}, \qquad (4),$$

the common value of these quantities being

$$F\kappa^2\left(\sin\frac{r}{\kappa} - \frac{r}{\kappa}\cos\frac{r}{\kappa}\right).$$

Determination of ellipticities of surfaces of equal density.

We are now prepared to find the value of u_2 in § 823, upon which depends the ellipticity of the strata. For (15) of § 822 becomes, by (23) of that section and the late equation (4),

$$\left(\frac{\mu-\mu'}{r^3} - 4\pi\kappa^2\frac{d\rho}{dr}\right)u_2 + \frac{4\pi}{5}\left[r^2\int_r^{\mathfrak{r}}\frac{u_2'}{r'}\frac{d\rho'}{dr'}dr' + r^{-3}\int_a^r r'^4 u_2'\frac{d\rho'}{dr'}dr'\right] - \frac{4\pi}{5}q\frac{\mathfrak{u}_2\mathfrak{r}^3}{\mathfrak{r}} = \frac{\omega^2 r^2}{2} \qquad (5)$$

where μ' is the mass of fluid, following the density law (3), which is displaced by the core μ. In the terrestrial problem we may assume $\mu' = \mu$, and of course $a = 0$. For simplicity put

$$r\frac{d\rho}{dr}u_2 = v, \qquad (6),$$

then divide by r^2 and differentiate, and we have

Laplace's hypothetical law of density within the earth.

Consequences as regards ellipticities of surfaces of equal density.

$$\frac{d}{dr}\left(\frac{v}{r^3}\right)+\frac{1}{\kappa^2 r^4}\int^r r'^2 v' dr' = 0.$$

Multiply by r^4, and again differentiate; the result is

$$\frac{d^2 v}{dr^2}+\left(\frac{1}{\kappa^2}-\frac{6}{r^2}\right)v=0 \qquad (7).$$

The integral of this equation is known to be

$$v=C\left[\left(\frac{3}{r^2}-\frac{1}{\kappa^2}\right)\sin\left(\frac{r}{\kappa}+C'\right)-\frac{3}{\kappa r}\cos\left(\frac{r}{\kappa}+C'\right)\right] \qquad (8),$$

so that u_2 is known from (6). Now we have already proved that u increases from the centre outwards, so that we must have

$$C'=0,$$

for otherwise u_2 would be infinite at the centre. Thus we have finally

$$e_2=\frac{u_2}{r}=-\frac{C}{F}\,\frac{\left(\frac{3}{r^2}-\frac{1}{\kappa^2}\right)\tan\frac{r}{\kappa}-\frac{3}{\kappa r}}{\tan\frac{r}{\kappa}-\frac{r}{\kappa}} \qquad (9).$$

The constants are, of course, to be determined by the known values of the ellipticity of the surface and of the angular velocity of the mass.

Now (5) becomes, at the surface,

$$\frac{4\pi}{\mathfrak{r}^2}\mathfrak{u}_2\int_0^{\mathfrak{r}}\rho r^2 dr+\frac{4\pi}{5\mathfrak{r}^3}\int_0^{\mathfrak{r}} r^4\frac{d\rho}{dr}u_2 dr=\frac{\mathfrak{r}^2\omega^2}{2}+\frac{4\pi}{5}q\mathfrak{u}_2\mathfrak{r} \qquad (10).$$

We may next eliminate ρ, $\frac{d\rho}{dr}$, and q by means of (3), (4), (6), and (8), and substitute everywhere re_2 for u_2. Also, if m be the ratio $(\frac{1}{289})$ of centrifugal force to gravity at the equator, ω is to be eliminated by means of the equation

$$m=\frac{\mathfrak{r}\omega^2}{\frac{4\pi}{\mathfrak{r}^2}\int_0^{\mathfrak{r}}\rho r^2 dr},$$

from which ρ is to be removed by (3). By the help of these substitutions (10) becomes transformed as follows:—

$$\frac{4\pi F\mathfrak{e}}{\mathfrak{r}}\int_0^{\mathfrak{r}} r\sin\frac{r}{\kappa}dr+\frac{4\pi C}{5\mathfrak{r}^3}\int_0^{\mathfrak{r}} r^3\left[\left(\frac{3}{r^2}-\frac{1}{\kappa^2}\right)\sin\frac{r}{\kappa}-\frac{3}{\kappa r}\cos\frac{r}{\kappa}\right]dr$$

$$=\frac{4\pi mF}{2\mathfrak{r}}\int_0^{\mathfrak{r}} r\sin\frac{r}{\kappa}dr+\frac{4\pi F}{5}\mathfrak{r}\mathfrak{e}\sin\frac{\mathfrak{r}}{\kappa}.$$

If we put $\tan\frac{\mathfrak{r}}{\kappa}=t$, and $\frac{\mathfrak{r}}{\kappa}=\theta$, the integrated expression, divided by $\frac{4\pi F\mathfrak{t}}{5\mathfrak{r}}$, becomes

Laplace's hypothetical law of density within the earth.

Ratio of ellipticity of surface to the fraction expressing centrifugal force at equator in terms of surface gravity.

$$5(t-\theta)-\frac{t-\theta}{(3-\theta^2)t-3\theta}[15(t-\theta)+\theta^3-6t\theta^2]=\frac{5m}{2\epsilon}(t-\theta)+\theta^2t.$$

Hence at once

$$\frac{5m}{2\epsilon}=\frac{\theta^4+\theta^3t+\theta^4t^2-2t^2\theta^3}{(t-\theta)[(3-\theta^2)t-3\theta]}. \qquad (11).$$

If we put $1-z$ for $\frac{\theta}{t}$, *i.e.*, for $\frac{\frac{\mathfrak{r}}{\kappa}}{\tan\frac{\mathfrak{r}}{\kappa}}$, this becomes somewhat simpler, and may be written

$$\frac{5m}{2\epsilon}=\frac{\theta^4-3z\theta^2+z^2\theta^2}{z(3z-\theta^2)}. \qquad (12).$$

The mean density is, of course,

$$\frac{\int_0^{\mathfrak{r}}\rho r^2dr}{\int_0^{\mathfrak{r}}r^2dr}=\frac{F\kappa^2(\sin\frac{\mathfrak{r}}{\kappa}-\frac{\mathfrak{r}}{\kappa}\cos\frac{\mathfrak{r}}{\kappa})}{\frac{\mathfrak{r}^3}{3}}=\frac{3F}{\mathfrak{r}}\frac{\sin\frac{\mathfrak{r}}{\kappa}-\frac{\mathfrak{r}}{\kappa}\cos\frac{\mathfrak{r}}{\kappa}}{(\frac{\mathfrak{r}}{\kappa})^2}.$$

Let q_0 be the mean density, and q, as before, the surface-density,

Mean density

$$q_0=\frac{3F}{\mathfrak{r}}\frac{\sin\theta-\theta\cos\theta}{\theta^2},$$

$$q=\frac{F}{\mathfrak{r}}\sin\theta.$$

Hence $$\frac{q_0}{q}=3\frac{t-\theta}{t\theta^2}=\frac{3z}{\theta^2}. \qquad (13).$$

If we put f for this ratio of the mean density to the surface-density, a quantity which may be determined by experiment, (13) gives

Ratio of mean, to surface, density.

$$f=3\frac{t-\theta}{t\theta^2}.$$

From this equation θ may be found by approximation, and then (12) gives ϵ in terms of known quantities. In fact, it becomes

$$\frac{5m}{2\epsilon}=3\frac{1-f+\frac{f^2\theta^2}{9}}{f(f-1)}=\frac{f\theta^2}{3(f-1)}-\frac{3}{f} \qquad (14).$$

From (13) and (14) the numbers in columns iv. and v. of the following table are easily calculated. Column vii. shows the ratio of the moment of inertia about a mean diameter, on the assumed law of density, to what it would be if the earth were homogeneous :—

aplace's ypothetical ıw of density within ıe earth.

i.	ii.	iii.	iv.	v.	vi.	vii.
$1-\frac{\theta}{t}$	$\frac{\theta}{\pi}180°$.	θ.	f.	ϵ.	$\frac{C-A}{C}$.	$\frac{C}{\frac{2}{3}Mr^2}$.
3·91	140°	2·444	1·966	$\frac{1}{292}$	·00335	·843
4·24	142°·5	2·487	2·057	$\frac{1}{295}$	·00330	·835
4·61	145°	2·531	2·161	$\frac{1}{299}$	·00325	·826
5·04	147°·5	2·574	2·282	$\frac{1}{302·5}$	·00321	·818
5·53	150°	2·618	2·423	$\frac{1}{306·5}$	·00315	·810
6·11	152°·5	2·662	2·589	$\frac{1}{311}$	·00309	·801
6·80	155°	2·705	2·788	$\frac{1}{315}$	·00304	·792

ynamical rigin of recession nd Nutation.

825. The phenomena of Precession and Nutation result from the Earth's being not centrobaric (§ 526), and therefore attracting the Sun and Moon, and experiencing reactions from them, in lines which do not pass precisely through the Earth's centre of inertia, except when they are in the plane of its equator. The attraction of either body transferred (§ 555) from its actual line to a parallel line through the Earth's centre of inertia, gives therefore a couple which, if we first assume, for simplicity, gravity to be symmetrical round the polar axis, tends to turn the Earth round a diameter of its equator, in the direction bringing the plane of the equator towards the disturbing body. The moment of this couple is [§ 539 (14)] equal to

$$S\frac{3(C-A)\sin\delta\cos\delta}{D^3} \qquad (14),$$

where S denotes the mass of the disturbing body, D its distance, and δ its declination; and C and A the Earth's moments of inertia round polar and equatorial diameters respectively. In all probability (§§ 796, 797) there is a very sensible difference between the moments of inertia round the two principal axes in the plane (§ 795) of the equator: but it is obvious, and

Dynamical origin of Precession and Nutation.

will be proved in Vol. II., that Precession and Nutation are the same as they would be if the earth were symmetrical about an axis, and had for moment of inertia round equatorial diameters, the arithmetical mean between the real greatest and least values. From (12) of § 539 we see that in general the *differences* of the moments of inertia round principal axes, or, in the case of symmetry round an axis, the value of $C-A$, may be determined solely from a knowledge of surface or external gravity, or [§§ 794, 795] from the figure of the sea level, without any data regarding the internal distribution of density.

Equating § 539 (12) to § 794 (17), in which, when the sea level is supposed symmetrical, $F_2(\theta, \phi)$ becomes simply $\epsilon(\frac{1}{3}-\cos^2\theta)$, we find

$$\frac{M\mathrm{r}^2}{r^3}(\epsilon-\tfrac{1}{2}m)(\tfrac{1}{3}-\cos^2\theta)=\tfrac{3}{2}\frac{C-A}{r^3}(\tfrac{1}{3}-\cos^2\theta),$$

whence $$C-A=\tfrac{2}{3}M\mathrm{r}^2(\epsilon-\tfrac{1}{2}m) \qquad (15).$$

Similarly we may prove the same formula to hold for the real case, in which the sea level is an ellipsoid of three unequal axes, one of which coincides with the axis of rotation; provided e denotes the mean of the ellipticities of the two principal sections of this ellipsoid through the axis of rotation, and A the mean of the moments of inertia round the two principal axes in the plane of the equator.

Precession gives information as to the distribution of the earth's mass, while surface-gravity does not.

826. The angular accelerations produced by the disturbing couples are (§ 281) directly as the moments of the couples, and inversely as the earth's moment of inertia round an equatorial diameter. But (Vol. II.) the integral results, observed in Precession and Nutation, would, if the earth's condition varied, vary directly as $C-A$, and inversely as C. We have seen (§ 794) that if the interior distribution of density were varied in any way subject to the condition of leaving the surface [and consequently (§ 793) the exterior] gravity unchanged, $C-A$ remains unchanged. But it is not so with C, which will be the less or the greater, according as the mass is more condensed in the central parts, or more nearly homogeneous to within a small distance of the surface: and thus it is that a comparison between dynamical theory and observation of Precession and Nutation gives us information as to the interior distribution of the earth's density (just as from the rate of

Precession gives information as to the distribution of the earth's mass.

acceleration of balls or cylinders rolling down an inclined plane we can distinguish between solid brass gilt, and hollow gold, shells of equal weight and equal surface dimensions: while no such information can be had from the figure of the sea level, the surface distribution of gravity, or the disturbance of the moon's motion, without hypothesis as to primitive fluidity or present agreement of surfaces of equal density with the surfaces which would be of equal pressure were the whole deprived of rigidity.

The constant of Precession deduced from Laplace's law.

827. But we shall first find what the magnitude of the terrestrial constant $\frac{C-A}{C}$ of *Precession* and *Nutation* would be, if Laplace's were the true law of density in the interior of the earth; and if the layers of equal density were level for the present angular velocity of rotation. Every moment of inertia involving the latter part of this assumption will be denoted by a Saxon capital.

The moment of inertia about the polar axis is, by § 281,

$$\mathfrak{C}=2\int_0^{\mathrm{r}}\int_0^{+\frac{\pi}{2}}\int_0^{2\pi}\rho \mathrm{r}^2\sin\theta d\mathrm{r}d\theta d\phi.\mathrm{r}^2\sin^2\theta,$$

the first factor under the integral sign being an element of the mass, the second the square of its distance from the axis.

For the moment of inertia about another principal axis (which may be any equatorial radius, but is here taken as that lying in the plane from which ϕ is measured), we have

$$\mathfrak{A}=2\int_0^{\mathrm{r}}\int_0^{+\frac{\pi}{2}}\int_0^{2\pi}\rho \mathrm{r}^2\sin\theta d\mathrm{r}d\theta d\phi.\mathrm{r}^2(1-\sin^2\theta\sin^2\phi).$$

Now, by § 823, we have

$$\mathrm{r}=r[1+e_2(\tfrac{1}{3}-\cos^2\theta)],$$

where r denotes the mean radius of the surface of equal density passing through r, θ, ϕ; whence

$$\mathrm{r}^4d\mathrm{r}=\tfrac{1}{5}\frac{d\mathrm{r}^5}{dr}dr=r^4dr+\frac{d.r^5e_2}{dr}(\tfrac{1}{3}-\cos^2\theta)dr.$$

$$\left.\begin{array}{ll}\text{Let} & \int_0^{\mathrm{r}}\rho r^4dr=K,\\ \text{and} & \int_0^{\mathrm{r}}\rho\frac{d.r^5e_2}{dr}dr=K_1\end{array}\right\} \quad (16).$$

Then $\mathfrak{C}=2\int_0^{+\frac{\pi}{2}}\int_0^{2\pi}\sin^3\theta d\theta d\phi[K+K_1(\tfrac{1}{3}-\cos^2\theta)]$

or $$C=\frac{8\pi}{3}K \text{ nearly.} \qquad (17).$$ *The constant of Precession deduced from Laplace's law.*

$$\mathfrak{C}-\mathfrak{A}=2\int_0^{+\frac{\pi}{2}}\int_0^{2\pi}\sin\theta d\theta d\phi[K+K_1(\tfrac{1}{3}-\cos^2\theta)](\sin^2\theta-1+\sin^2\theta\sin^2\phi)$$
$$=\frac{8\pi}{15}K_1 \qquad (18).$$

Now we have
$$K=\int_0^{\mathfrak{r}}\rho r^4 dr=F\int_0^{\mathfrak{r}} r^3\sin\frac{r}{\kappa}dr,$$
or, if we put as before $\theta=\frac{\mathfrak{r}}{\kappa}$,
$$K=F\kappa^4\cos\theta(-\theta^3+3\theta^2 t+6\theta-6t).$$
Again $$K_1=\int_0^{\mathfrak{r}}\rho\frac{d}{dr}(r^5 e_2)dr=\mathfrak{r}^5\epsilon q-\int_0^{\mathfrak{r}} r^5 e_2\frac{d\rho}{dr}dr,$$
and this, by (10) of last section, becomes
$$K_1=5\mathfrak{r}^2\epsilon\int_0^{\mathfrak{r}}\rho r^2 dr-\frac{5\mathfrak{r}^5\omega^2}{8\pi} \qquad (19).$$
$$=5(\epsilon-\frac{m}{2})F\kappa^4\theta^2(t-\theta)\cos\theta.$$
Thus, finally,
$$\frac{\mathfrak{C}-\mathfrak{A}}{C}=\frac{1}{5}\frac{K_1}{K}=(\epsilon-\frac{m}{2})\frac{\theta^2(t-\theta)}{-\theta^3+3\theta^2 t+6\theta-6t}$$
$$=(\epsilon-\frac{m}{2})\frac{z}{2+(1-\frac{6}{\theta^2})z} \qquad (20).$$

From this formula the numbers in Column vi. of the table in § 824 were calculated. By (18) and (19) we see that
$$\mathfrak{C}-\mathfrak{A}=\frac{8\pi}{3}(\mathfrak{r}^2\epsilon\int_0^{\mathfrak{r}}\rho r^2 dr-\frac{\mathfrak{r}^5\omega^2}{8\pi})$$
$$=\frac{2}{3}M\mathfrak{r}^2(\epsilon-\frac{m}{2})$$
which agrees, as it ought to do, with (15) of § 825.

828. From the elaborate investigations of Precession and Nutation made by Le Verrier and Serret, it appears that the true value of $\frac{C-A}{C}$ is, very approximately, ·00327.[1] This, according to the table of § 824, agrees with $\frac{\mathfrak{C}-\mathfrak{A}}{C}$ for $f=2{\cdot}1$, which gives $\epsilon=\frac{1}{297}$. These are (§§ 792, 796, 797) about the most probable values which we can assign to these elements *Comparison of Laplace's hypothesis with observation.*

[1] *Annales de l'Observatoire Impérial de Paris*, 1859, p. 324.

Comparison of Laplace's hypothesis with observation; The compressibility involved in the hypothesis.

by observation. Thus, so far as we have the means of testing it, Laplace's hypothesis is verified.

829. But, as a further check upon Laplace's assumption, it is necessary to inquire whether the results involve anything inconsistent with experimental knowledge of the compressibility of matter under such pressures as we can employ in the laboratory. For this purpose the first column has been added to the preceding table. From it may be deduced the compressibility of the upper stratum of liquid matter, which composes the crust of the earth, required by the assumed law of density, for the respective values of θ. In fact, the numbers in Col. I. are those by which the earth's radius must be divided to find the lengths of the modulus of compression (§ 688) of the uppermost layer of fluid, according to the surface value of gravity.

We have, by § 824 (3),

$$q=\frac{F}{r}\sin\frac{r}{\kappa},\quad \frac{dq}{dr}=-\frac{F}{r}\left(\frac{\sin\frac{r}{\kappa}}{r}-\frac{\cos\frac{r}{\kappa}}{\kappa}\right),$$

whence, at the surface, $\left[-\frac{1}{q}\frac{dq}{dr}\right]=\frac{1}{\mathfrak{r}}\left(1-\frac{\theta}{t}\right).$

The corresponding numbers for several different liquid and solid substances are as follows:—

Compressibility of lava required by Laplace's hypothesis, compared with experimental data.

Alcohol, ...	37
Water, ...	27
Mercury, ...	27
Glass, ...	5·0
Copper, ...	8·1
Iron, ...	4·1
Melted Lava, by Laplace's law, with $f=2·1$,	4·42

This comparison may be considered as decidedly not adverse to Laplace's law, but actual experiments on the compressibility of melted rock are still a desideratum.

Argument from ellipticity of earth and tidal friction.

830. In § 276 it was proved that the tides must tend to diminish the angular velocity of the earth's rotation; and it will be proved in a later volume that this tendency is not counterbalanced to more than a very minute degree by the tendency to acceleration which results from secular cooling and shrinking. Since the printing of § 276, results of physical astronomy have become known to us which overthrow the conclusion quoted in § 405, and so allow a practical importance to § 276 which that conclusion denied to it. The conclusion quoted in § 405 was drawn by Laplace from the agreement

Argument from ellipticity of earth and tidal friction, to determine the date of consolidation.

between observation and his dynamics of the moon's mean motion. In 1853 Adams pointed out an error in Laplace's work, which had till then escaped the notice of physical astronomers; and showed that only about half of the observed acceleration of the moon's mean motion relative to the angular velocity of the earth's rotation, was accountable for by Laplace's theory. In 1859 he communicated to Delaunay his final result:—that at the end of a century the moon is 5″·7 before the position she would have, relatively to a meridian of the earth, according to the angular velocities of the two motions, at the beginning of the century, and the acceleration of the moon's motion truly calculated from the various disturbing causes then recognised. Delaunay soon after verified this result: and about the beginning of 1866 suggested that the true explanation may be a retardation of the earth's rotation by tidal friction. Using this hypothesis, and allowing for the consequent retardation of the moon's mean motion by tidal reaction (§ 276), Adams, in an estimate which he has communicated to us, founded on the rough assumption that the parts of the earth's retardation due to solar and lunar tides are as the squares of the respective tide-generating forces, finds 22^s as the error by which the earth would in a century get behind a perfect clock rated at the beginning of the century. If the retardation of rate giving this integral effect in a century were uniform (§ 35, *b*), the earth, as a timekeeper, would be going slower by ·22 of a second per year in the middle, or ·44 of a second per year at the end, of the century. The latter is $\frac{1}{70000000}$ of the present angular velocity; and if the rate of retardation had been uniform since ten million centuries back, the earth must have been then rotating faster by $\frac{1}{7}$ than at present, and the centrifugal force greater in the proportion of 64 to 49. If the consolidation took place then or earlier, the ellipticity of the upper layers of equal density must have been $\frac{1}{230}$ instead of about $\frac{1}{300}$, as it certainly is. It is impossible to escape the conclusion that the date of the consolidation is considerably more recent than a thousand million years ago. In Appendix D., it is shown from the theory of the conduction of heat that the date of consolidation may be about a hundred million years ago, but cannot possibly have been so remote as five hundred million years.

Abrupt changes of interior density, not improbable.

831. From the known facts regarding compressibilities of terrestrial substances, referred to above (§ 829), it is most probable that even in chemically homogeneous substance there is a continuous increase of density downwards at some rate comparable with that involved in Laplace's law. But it is not improbable that there may be abrupt changes in the quality of the substance, as, for instance, if a large portion of the interior of the earth had at one time consisted of melted metals, now consolidated. We therefore append a solution of the problem of determining the ellipticities of the surfaces of a rotating mass consisting of two non-mixing fluids of different densities, each, however, being supposed incompressible.

Two non-mixing liquids of different densities, each homogeneous.

Let the densities of the two liquids be ρ and $\rho+\rho'$, the latter forming the spheroid

$$r=a'[1+\epsilon'(\tfrac{1}{3}-\cos^2\theta)] \qquad (1),$$

and the former filling the space between this spheroid and the exterior concentric and coaxal surface

$$r=a[1+\epsilon(\tfrac{1}{3}-\cos^2\theta)] \qquad (2).$$

Also let the whole revolve with uniform angular velocity ω. The conditions of equilibrium are that the surface of each spheroid must be an equipotential surface.

Now the potential at a point r, θ, in the outer fluid is

$$\left.\begin{aligned}&\frac{4\pi\rho}{3}\Big[\frac{3a^2-r^2}{2}+\tfrac{2}{5}r^2\epsilon(\tfrac{1}{3}-\cos^2\theta)\Big]\\ &+\frac{4\pi\rho'}{3}\Big[\frac{a'^3}{r}+\tfrac{3}{5}\frac{a'^5}{r^3}\epsilon'(\tfrac{1}{3}-\cos^2\theta)\Big]\\ &+\tfrac{1}{3}\omega^2r^2+\tfrac{1}{2}\omega^2r^2(\tfrac{1}{3}-\cos^2\theta).\end{aligned}\right\} \qquad (3).$$

The first line is the potential due to a liquid of density ρ filling the larger spheroid, the second that due to a liquid of density ρ' filling the inner spheroid, the third is the potential $(\frac{1}{2}\omega^2r^2\sin^2\theta)$ of centrifugal force arranged in solid harmonics.

Substituting in (3) the values of r from (1) and (2) successively, neglecting squares, etc., of the ellipticities, and equating to zero the sum of the coefficients of $(\frac{1}{3}-\cos^2\theta)$; we have two equations from which we find

$$\epsilon=\frac{\rho+\frac{1}{5}\rho'(2+3\frac{a'^5}{a^5})}{(\rho+\frac{3}{5}\rho')(\frac{2}{5}\rho+\frac{a'^5}{a^5}\rho')-\frac{9}{25}\rho\rho'\frac{a'^5}{a^5}}\cdot\frac{3\omega^2}{8\pi} \qquad (4).$$

The corresponding value of ϵ' is to be found from the equation

$$\epsilon(\rho+\frac{a'^5}{a^5}\rho')=\epsilon'\{\rho+\tfrac{1}{5}\rho'(2+3\frac{a'^5}{a^5})\}.$$

Expressing ω^2 in terms of the known quantity m we have

Two non-mixing fluids of different densities.

$$\frac{3\omega^2}{8\pi}=\frac{m}{2}\left(\rho+\frac{a'^3}{a^3}\rho'\right) \qquad (5).$$

Also, to a sufficient approximation, we have

$$\left.\begin{aligned} C&=\frac{8\pi}{15}a^5\left(\rho+\frac{a'^5}{a^5}\rho'\right)\\ M&=\frac{4\pi}{3}a^3\left(\rho+\frac{a'^3}{a^3}\rho'\right)\end{aligned}\right\} \qquad (6),$$

and the mean density is obviously

$$\rho+\frac{a'^3}{a^3}\rho' \qquad (7).$$

The numerical values of the expressions (4) and (7) are approximately known from observation and experiment, so that if we assume a value of $\frac{a'}{a}$ we can at once find ρ and ρ', and, from them, the value of $\frac{C-A}{C}$.

From the formulæ just given it is easy to show that results closely agreeing with observation as regards precession, ratio of surface density to mean density, and ellipticity of sea level, may be obtained without making any inadmissible hypotheses as to the relative volumes and densities of the two assumed liquids. But this must be left as an exercise for the student.

832. These estimates, and all dynamical investigations (whether static or kinetic) of tidal phenomena, and of precession and nutation, hitherto published, with the exception referred to below, have assumed that the outer surface of the solid earth is absolutely unyielding. A few years ago,[1] for the first time, the question was raised: Does the Earth retain its figure with practically perfect rigidity, or does it yield sensibly to the deforming tendency of the Moon's and Sun's attractions on its upper strata and interior mass? It must yield to *some* extent, as no substance is infinitely rigid: but whether these solid tides are sufficient to be discoverable by any kind of observation, direct or indirect, has not yet been ascertained. The negative result of attempts to trace their influence on ocean and lake tides, as hitherto observed, and on precession and nutation, suffices, as we shall see, to disprove

Rigidity of the earth

[1] "On the Rigidity of the Earth." W. Thomson. *Trans. R. S.*, May 1862.

Rigidity of the earth: great enough to negative the geological hypothesis of a thin solid crust.

the hypothesis, hitherto so prevalent, that we live on a mere thin shell of solid substance, enclosing a fluid mass of melted rocks or metals, and proves, on the contrary, that the Earth as a whole is much more rigid than any of the rocks that constitute its upper crust.

833. The character of the deforming influence will be understood readily by considering that if the whole Earth were perfectly fluid, its bounding surface would coincide with an equipotential surface relatively to the attraction of its own mass, the centrifugal force of its rotation, and the tide-generating resultant (§ 804) of the Moon's and Sun's forces, and their kinetic reactions.[1] Thus (§§ 819, 824) there would be the full equilibrium lunar and solar tides; of $2\frac{1}{2}$ times the amount of the disturbing deviation of level if the fluid were homogeneous, or of nearly twice this amount if it were heterogeneous with Laplace's hypothetical law of increasing density. If now a very thin layer of lighter liquid were added, this layer would rest covering the previous bounding surface to very nearly equal depth all round, and would simply rise and fall with that surface, showing only infinitesimal variations in its own depth under tidal influences. Hence had the solid part of the earth so little rigidity as to allow it to yield in its own figure very nearly as much as if it were fluid, there would be very nearly nothing of what we call tides—that is to say, rise and fall of the sea relatively to the land; but sea and land together would rise and fall a few feet every twelve lunar hours. This would, as we shall see, be the case if the geological hypothesis of a thin crust were true. The actual phenomena of tides, therefore, give a secure contradiction to that hypothesis. We shall see indeed, presently, that even a continuous solid globe, of the same mass and diameter as the earth, would, if homogeneous

[1] It will be proved in Vol. II. that the "equilibrium theory" of the tides for an ocean, whether of uniform density or denser in the lower parts, completely covering a solid nucleus, requires correction, on account of the diurnal rotation, but less and less correction the smaller this nucleus is; and that it agrees perfectly with the "kinetic theory" when there is no nucleus, always provided the angular velocity is not too great for the ordinary approximations (§§ 794, 801, 802, 815) which require that there be not, on any account, more than an infinitely small disturbance from the spherical figure. It is interesting to remark that this proposition does not require the tidal deformations to be small in comparison with the 70,000 feet deviation due to centrifugal force of rotation.

and of the same rigidity (§ 680) as glass or as steel, yield in its shape to the tidal influences three-fifths as much, or one-third as much, as a perfectly fluid globe; and further, it will be proved that the effect of such yielding in the solid, according as its supposed rigidity is that of glass or that of steel, would be to reduce the tides to about $\frac{2}{5}$ or $\frac{2}{3}$ of what they would be if the rigidity were infinite. Rigidity of the earth.

834. To prove this, and to illustrate this question of elastic tides in the solid earth, we shall work out explicitly the solution of the general problem of § 696, for the case of a homogeneous elastic solid sphere exposed to no surface traction; but deformed infinitesimally by an equilibrating system of forces acting *bodily* through the interior, which we shall ultimately make to agree with the tide-generating influence of the Moon or Sun. In the first place, however, we only limit the deforming force by the final assumption of § 733. Elastic solid tides

Homogeneous elastic solid globe free at surface; deformed by bodily harmonic force.

Following the directions of § 732, we are to find, for the two constituents $(`\alpha, `\beta, `\gamma)$ and $(\alpha_{,}, \beta_{,}, \gamma_{,})$ for the complete solution; of which the first, given by (6) and (7) of § 733, is as follows:—

$$`\alpha = -\frac{1}{2(2i+5)(m+n)}\frac{d(r^2 W_{i+1})}{dx}\;\frac{1}{m+n}\left\{-\frac{r^2}{2(2i+3)}\frac{dW_{i+1}}{dx} + \frac{r^{2i+5}}{(2i+3)(2i+5)}\frac{d(W_{i+1}r^{-2i-3})}{dx}\right\} \quad (1),$$

with symmetrical formulæ for $`\beta$ and $`\gamma$; which [§ 733 (6)], give

$$\left.\begin{aligned} `\delta &= -\frac{W_i}{m+n}, \\ \text{and, [§ 737 (28)],}\quad `\zeta &= -\frac{(i-3)r^2 W_{i+1}}{2(2i+5)(m+n)} \end{aligned}\right\} \quad (2).$$

These, used in (29) of § 737 with $i+2$ for i, give

$$-`F.r = \frac{1}{m+n}\left\{(m-n)W_{i+1}x + \frac{(i+2)n}{2i+5}\frac{d(r^2 W_{i+1})}{dx}\right\} \quad (3);$$

which, reduced to harmonics by the proper formula [§ 737 (36)], becomes

$$`F_r = \frac{-1}{(2i+3)(m+n)}\left\{[m+(i+1)n]r^2\frac{dW_{i+1}}{dx} - \frac{(2i+5)m-n}{2i+3}r^{2i+5}\frac{d(W_{i+1}r^{-2i-3})}{dx}\right\} \quad (4).$$

This and the symmetrical formulæ for $`Gr$ and $`Hr$, with r taken equal to a, express the components of the force per unit area which would have to be balanced by the application from without of surface traction to the bounding surface of the globe, if the strain through the interior were exactly that expressed by (1).

Homogeneous elastic solid globe free at surface; deformed by bodily harmonic force.

Hence, still according to the directions of § 732, we must now find $(\alpha_{,}, \beta_{,}, \gamma_{,})$ the state of interior strain which with no force from without acting bodily through the interior, would result from surface traction equal and opposite to that (4). Of this part of the problem we have the solution in § 737 (52), the particular data being now

$$\frac{A_i}{a^{i+1}}=\frac{m+(i+1)n}{(2i+3)(m+n)}r^{-i}\frac{dW_{i+1}}{dx}\,;\ \frac{A_{i+2}}{a^{i+1}}=-\frac{(2i+5)m-n}{(2i+3)(2i+5)(m+n)}r^{i+3}\frac{d(W_{i+1}r^{-2i-3})}{dx}$$

with symmetrical terms for B_i, C_i, and B_{i+2}, C_{i+2}; but none of other orders than i and $i+2$. Hence for the auxiliary functions of § 737 (50)

$$\left.\begin{aligned}\Psi_{i-1}=0,\ \Phi_{i+1}&=-\frac{(i+1)(2i+1)[m+(i+1)n]a^{i+1}}{(2i+3)(m+n)}W_{i+1}\\ \Psi_{i+1}&=\frac{(i+2)[(2i+5)m-n]a^{i+1}}{(2i+3)(m+n)}W_{i+1},\ \text{and}\ \Phi_{i+2}=0\end{aligned}\right\}\quad(6).$$

Now (52), with the proper terms for $i+2$ instead of i added, is to be used to give us $\alpha_{,}$; and through the vanishing of Ψ_{i-1} and Φ_{i+2}, it becomes

$$\alpha_{,}=\frac{1}{n}\Big\{\frac{1}{i-1}\Big[\frac{a^{-i+1}}{2i(2i+1)}\frac{d\Phi_{i+1}}{dx}+\frac{A_i r^i}{a^{i-1}}\Big]+\tfrac{1}{2}\frac{m(a^2-r^2)a^{-i-1}}{[2(i+2)^2+1]m-(2i+3)n}\frac{d\Psi_{i-1}}{dx}$$
$$+\frac{a^{-i-1}}{i+1}\Big[\frac{[(i+4)m-(2i+3)n]}{[2(i+2)^2+1]m-(2i+3)n}\frac{r^{2i+5}d(\Psi_{i+1}r^{-2i-3})}{(2i+5)dx}+A_{i+2}r^{i+2}\Big]\Big\}\quad(7)$$

To this we must add $`\alpha$, given by (1), to obtain, according to § 732, the explicit solution, α, of our problem. Thus, after somewhat tedious algebraic reductions in which $m+n$, appearing as a factor in the numerator and denominator of each fraction, is removed, we find a remarkably simple expression for α. This, and the symmetrical formulæ for β and γ, are as follows:—

$$\left.\begin{aligned}\alpha&=(\mathfrak{E}_{i+1}a^2-\mathfrak{F}_{i+1}r^2)\frac{dW_{i+1}}{dx}-\mathfrak{G}_{i+1}r^{2i+5}\frac{d(W_{i+1}r^{-2i-3})}{dx}\\ \beta&=(\mathfrak{E}_{i+1}a^2-\mathfrak{F}_{i+1}r^2)\frac{dW_{i+1}}{dy}-\mathfrak{G}_{i+1}r^{2i+5}\frac{d(W_{i+1}r^{-2i-3})}{dy}\\ \gamma&=(\mathfrak{E}_{i+1}a^2-\mathfrak{F}_{i+1}r^2)\frac{dW_{i+1}}{dz}-\mathfrak{G}_{i+1}r^{2i+5}\frac{d(W_{i+1}r^{-2i-3})}{dz}\end{aligned}\right\}\quad(8),$$

where

$$\left.\begin{aligned}\mathfrak{E}_{i+1}&=\frac{(i+1)[(i+3)m-n]}{2in\{[2(i+2)^2+1]m-(2i+3)n\}}\\ \mathfrak{F}_{i+1}&=\frac{(i+2)(2i+5)m-(2i+3)n}{2(2i+3)n\{[2(i+2)^2+1]m-(2i+3)n\}}\\ \mathfrak{G}_{i+1}&=\frac{(i+1)m}{(2i+3)n\{[2(i+2)^2+1]m-(2i+3)n\}}\end{aligned}\right\}\quad(9).$$

The infinitely great value of $\mathfrak{E}_{i+1}$ for the case $i=0$ depends on the circumstance that the bodily force for this case, being uniform and in parallel lines through the whole mass, is not self equilibrating, and therefore surface stress would be required for equilibrium.

Homogeneous elastic solid globe free at surface; deformed by bodily harmonic force.

The case of $i=1$ is that with which we are concerned in the tidal problem. In it the formulæ (9) for the numerical coefficients become

$$\mathfrak{E}_2=\frac{4m-n}{n(19m-5n)},\quad \mathfrak{F}_2=\frac{21m-5n}{10n(19m-5n)},\quad \mathfrak{G}_2=\frac{2m}{5n(19m-5n)} \quad (10).$$

To prepare for terrestrial applications we may conveniently reduce to polar co-ordinates (distance from the centre, r; latitude, l; longitude, λ) such that

$$x=r\cos l\cos\lambda,\ y=r\cos l\sin\lambda,\ z=r\sin l \quad (11);$$

and denote by ρ, μ, ν, the corresponding components of displacement. The expressions for these will be precisely the same as those for α, β, γ, except that instead of $\frac{d}{dx}$, as it appears in the expression for α, we have $\frac{d}{dr}$ in the expression for ρ; $\frac{d}{rdl}$ in that for μ, and $\frac{d}{r\cos l d\lambda}$ in that for ν. Thus if we put

$$W_{i+1}=S_{i+1}r^{i+1} \quad (12),$$

so that S_{i+1} may denote the surface harmonic, or the harmonic function of directional angular co-ordinates l, λ, corresponding to W_{i+1}, we have

$$\left.\begin{aligned}\rho&=\{(i+1)\mathfrak{E}_{i+1}a^2-[(i+1)\mathfrak{F}_{i+1}-(i+2)\mathfrak{G}_{i+1}]r^2\}r^iS_{i+1}\\ \mu&=\{\mathfrak{E}_{i+1}a^2-(\mathfrak{F}_{i+1}-\mathfrak{G}_{i+1})r^2\}r^i\frac{dS_{i+1}}{dl}\\ \nu&=\{\mathfrak{E}_{i+1}a^2-(\mathfrak{F}_{i+1}-\mathfrak{G}_{i+1})r^2\}r^i\frac{dS_{i+1}}{\cos l d\lambda}\end{aligned}\right\} \quad (13),$$

whence, for ρ, finally, by (9),

$$\rho=\frac{(i+1)\{(i+1)[(i+3)m-n]a^2-i[(i+2)m-n]r^2\}}{2in\{[2(i+2)^2+1]m-(2i+3)n\}}r^iS_{i+1} \quad (14).$$

The expressions for μ and ν are more conveniently left as exhibited in (12) and (9). From (13) we have

$$\frac{d\rho}{dr}=\frac{(i+1)^2\{ma^2+[(i+2)m-n](a^2-r^2)\}}{2n\{[2(i+2)^2+1]m-(2i+3)n\}}r^{i-1}S_{i+1} \quad (15),$$

which is always positive when $r<a$; as i is at least equal to 1, and [§ 698 (1)] m necessarily (§ 694) exceeds $\frac{1}{3}n$. But although ρ therefore always increases for the successive concentric spheroids outwards, we easily see, by writing down the expression for

Homogeneous elastic solid globe free at surface; deformed by bodily harmonic force.

$\frac{d\frac{\rho}{r}}{dr}$, that $\frac{\rho}{r}$, when $i>1$ increases first from the centre outwards, comes to a maximum at a certain distance, and diminishes thence to the surface. When $i=1$, we have

$$\rho=\frac{2(4m-n)a^2-(3m-n)r^2}{n(19m-5n)}rS_2 \qquad (16):$$

and therefore in this case $\frac{\rho}{r}$ diminishes from the centre outwards to the surface; and its extreme values are

$$\left.\begin{array}{ll}\text{at the centre} & \frac{\rho}{r}=\frac{2(4m-n)a^2}{n(19m-5n)}S_2=\frac{8a^2}{19n}\Big(1+\frac{\frac{1}{76}\frac{n}{m}}{1-\frac{5}{19}\frac{n}{m}}\Big)S_2 \\ \text{at the free surface} & \frac{\rho}{r}=\frac{(5m-n)a^2}{n(19m-5n)}S_2=\frac{5a^2}{19n}\Big(1+\frac{\frac{6}{95}\frac{n}{m}}{1-\frac{5}{19}\frac{n}{m}}\Big)S_2\end{array}\right\} \qquad (17):$$

Cases:— centrifugal force:—

When the disturbing action is the centrifugal force of uniform rotation with angular velocity ω, we have as found above (§ 794) for the whole potential

$$W=w\{\tfrac{1}{3}\omega^2r^2+\tfrac{1}{2}\omega^2r^2(\tfrac{1}{3}-\cos^2\theta)\} \qquad (18),$$

where w denotes the mass of the solid per unit volume. The effect of the term $\frac{1}{3}w\omega^2r^2$ is merely a drawing outwards of the solid from the centre symmetrically all round; which we may consider in detail later in illustrating properties of matter in our second volume. The remainder of the expression gives us according to our present notation

$$W_2=\tfrac{1}{3}\tau(x^2+y^2-2z^2);\ \text{or}\ S_2=w\tau(\tfrac{1}{3}-\cos^2\theta) \qquad (19),$$

where

$$\tau=\tfrac{1}{2}\omega^2 \qquad (20).$$

tide-generating force.

For tide-generating force the same formulæ (15) and (16) hold if (§§ 804, 808, 813) we take

$$\tau=\tfrac{3}{2}\frac{M}{D^3} \qquad (21)$$

and alter signs so as to make the strain-spheroids prolate instead of oblate. The deformed figure of each of the concentric spherical surfaces of the sphere is of course an ellipsoid of revolution, and from (15) and (17) we find for the extremes:—

$$\left.\begin{array}{ll}\text{ellipticity of central strain spheroids} & =\frac{8a^2}{19n}\Big(1+\frac{\frac{1}{76}\frac{n}{m}}{1-\frac{5}{19}\frac{n}{m}}\Big)\cdot w\tau \\ \quad\text{,,}\quad \text{of free surface} & =\frac{5a^2}{19n}\Big(1+\frac{\frac{6}{95}\frac{n}{m}}{1-\frac{5}{19}\frac{n}{m}}\Big)\cdot w\tau\end{array}\right\} \qquad (22).$$

From these results, (8) to (22), we conclude that

Homogeneous elastic solid globe free at surface; deformed by bodily harmonic force.

835. The bounding surface and concentric interior spherical surfaces of a homogeneous elastic solid sphere strained slightly by balancing attractions from without, become deformed into harmonic spheroids of the same order and type as the solid harmonic expressing the potential function of these forces, when they are so expressible: and the direction of the component displacement perpendicular to the radius at any point is the same as that of the component of the attracting force perpendicular to the radius. These concentric harmonic spheroids although of the same type are not similar. When they are of the second degree (that is when the force potential is a solid harmonic of the second degree), the proportions of the ellipticities in the three normal sections of each of them are the same in all: but in any one section the ellipticities of the concentric ellipsoids increase from the outermost one inwards to the centre, in the ratio of

Case of second degree gives elliptic deformation, diminishing from centre outwards:—

$$1 : \frac{5m-n}{2(4m-n)}.$$

For harmonic disturbances of higher orders the amount of deviation from sphericity, reckoned of course, in proportion to the radius, increases from the surface inwards to a certain distance, and then decreases to the centre. The explanation of this remarkable conclusion is easily given without analysis, but we shall confine ourselves to doing so for the case of ellipsoidal disturbances.

higher degrees give greatest proportionate deviation from sphericity neither at centre nor surface.

836. Let the bodily disturbing force cease to act, and let the surface be held to the same ellipsoidal shape by such a distribution of surface traction (§§ 693, 662) as shall maintain a homogeneous strain throughout the interior. The interior ellipsoidal surfaces of deformation will now become similar concentric ellipsoids: and the inner ones must clearly be less elliptic than they were when the same figure of outer boundary was maintained by forces acting throughout all the interior; and, therefore, they must have been greater for the inner surface. And we may reason similarly for the portion of the whole solid within any one of the ellipsoids of deformation, by supposing all cohesive and tangential force between it and the

Synthetic proof of maximum ellipticity at centre, for deformation of second order.

Synthetic proof of maximum ellipticity at centre, for deformation of second order.

solid surrounding it to be dissolved; and its ellipsoidal figure to be maintained by proper surface traction to give homogeneous strain throughout the interior when the bodily force ceases to act. We conclude that throughout the solid from surface to centre, when disturbed by bodily force without surface traction, the ellipticities of the concentric ellipsoids increase inwards.

837. When the disturbing action is centrifugal force, or tide-generating force (as that of the Sun or Moon on the Earth), the potential is, as we have seen, a harmonic of the second degree, symmetrical round an axis. In one case the spheroids of deformation are concentric oblate ellipsoids of revolution; in the other case prolate. In each case the ellipticity increases from the surface inwards, according to the same law [§ 834 (16)] which is, of course, independent of the radius of the sphere.

Oblateness induced in homogeneous elastic solid globe, by rotation.

For spheres of different dimensions and similar substances the ellipticities produced by equal angular velocities of rotation are as the squares of the radii. Or, if the equatorial surface velocity (V) be the same in rotating elastic spheres of different dimensions but similar substance, the ellipticities are equal. The values of the surface and central ellipticities are respectively

$$\frac{3}{11}\frac{V^2 w}{2n} \text{ and } \frac{14}{33}\frac{V^2 w}{2n}$$

for solids fulfilling Poisson's hypothesis (§ 685), according to which $m = 2n$.

Numerical results for iron and glass.

For steel or iron the values of n and m are respectively 780×10^6 and about 1600×10^6 grammes weight per square centimetre, or 770×10^9 and about 1600×10^9 gramme-centimetre-seconds, absolute units (§ 223), and the specific gravity (w) is about 7·8. Hence a ball of steel of any radius rotating with an equatorial velocity of 10,000 centimetres per second will be flattened to an ellipticity (§ 801) of $\frac{1}{7240}$. For a specimen of flint glass of specific gravity 2·94 Everett finds $n = 244 \times 10^6$ grammes weight per square centimetre and very approximately $m = 2n$. Hence for this substance $\frac{n}{w} = 83 \times 10^6$ [being the length of the modulus of rigidity (§ 687) in centimetres]. But the numbers used above for steel give $\frac{n}{w} = 100 \times 10^6$ centimetres;

and therefore (§ 838) the flattening of a glass globe is $\frac{1}{\cdot 83}$, or $1\frac{1}{3}$ times that of a steel globe with equal velocities.

838. For rotating or tidally deformed globes of glass or metals, the amount of deformation is but little influenced by compressibility, as we see in a moment from (22) of § 834; because for such substances (§§ 684, 694), the value of m is either about equal to $2n$, or still greater. Thus for any substance for which $m \gtreqless 2n$ the surface ellipticity is diminished by three per cent. or by less than three per cent., and the centre ellipticity by $\frac{3}{4}$ per cent., or less than $\frac{3}{4}$ per cent. if we suppose the rigidity to remain in any case unchanged, but the substance to become absolutely incompressible. For the surface ellipticity, § 834 (22) gives on this supposition

Rotational or tidal ellipticities but little influenced by compressibility, in globes of metallic, vitreous, or gelatinous elastic solid.

$$e = \frac{5a^2 w}{19n}\tau \qquad (23),$$

or with $n = 770 \times 10^6$ as for steel (§ 837),
$a = 640 \times 10^6$, the earth's radius in centimetres,
and $w = 5{\cdot}5$, " " mean density,

we have, in anticipation of § 839,

$$e = 77 \times 10^4 . \tau \qquad (24).$$

839. If now we consider a globe as large as the Earth, and of incompressible homogeneous material, of density equal to the Earth's mean density, but of the same rigidity as steel or glass; and if, in the first place, we suppose the matter of such a globe to be deprived of the property of mutual gravitation between its parts: the ellipticities induced by rotation, or by tide generating force, will be those given by the preceding formulæ [§ 834 (22)], with the same values of n as before; with $\frac{n}{m} = 0$; with 640×10^6 for a, the earth's radius in centimetres; and with 5·5 for w instead of the actual specific gravities of glass and steel.

Value of surface ellipticities for globe same size and mass as Earth, of non-gravitating material, homogeneous, incompressible, and same rigidity as steel.

But without rigidity at all, and mutual gravitation between the parts alone opposing deviation from the spherical figure, we found before (§ 819), for the ellipticity

$$e = \frac{5}{2}\frac{a}{g}\tau = 162 \times 10^4 . \tau \qquad (25).$$

840. Hence of these two influences which we have con-

Comparison between spheroidal-maintaining powers of gravitation and rigidity, for large homogeneous solid globes.

sidered separately :—on the one hand, elasticity of figure, even with so great a rigidity as that of steel; and, on the other hand, mutual gravitation between the parts: the latter is considerably more powerful than the former, in a globe of such dimensions as the Earth. When, as in nature, the two resistances against change of form act jointly, the actual ellipticity of form will be the reciprocal of the sum of the reciprocals of the ellipticities that would be produced in the separate cases of one or other of the resistances acting alone. For we may imagine the disturbing influence divided into two parts: one of which alone would maintain the actual ellipticity of the solid, without mutual gravitation; and the other alone the same ellipticity if the substance had no rigidity but experienced mutual gravitation between its parts. Let τ be the disturbing influence as measured by § 834 (20), (21); and let $\frac{\tau}{\mathfrak{r}}$ and $\frac{\tau}{\mathfrak{g}}$ be the ellipticities of the spheroidal figure into which the globe becomes altered on the two suppositions of rigidity without gravity and gravity without rigidity, respectively. Let e be the actual ellipticity and let τ be divided into τ' and τ'' proportional to the two parts into which we imagine the disturbing influence to be divided in maintaining that ellipticity. We have

$\mathfrak{g}$, $\mathfrak{r}$ denote resistances to deformation due respectively to gravity and to rigidity.

$$\tau=\tau'+\tau''$$

and
$$e=\frac{\tau'}{\mathfrak{r}}=\frac{\tau''}{\mathfrak{g}}.$$

Whence
$$\frac{\tau}{e}=\mathfrak{r}+\mathfrak{g},$$

or
$$\frac{1}{e}=\frac{\mathfrak{r}}{\tau}+\frac{\mathfrak{g}}{\tau},$$

which proves the proposition. It gives

$$e=\frac{\tau}{\mathfrak{r}+\mathfrak{g}}=\frac{\frac{\tau}{\mathfrak{g}}}{\frac{\mathfrak{r}}{\mathfrak{g}}+1} \qquad (26).$$

By §§ 838, 839 we have

$$\mathfrak{r}=\frac{19n}{5a^2w}, \text{ and } \mathfrak{g}=\frac{2}{5}\frac{g}{a} \qquad (27).$$

and
$$\frac{\mathfrak{r}}{\mathfrak{g}}=\frac{19n}{2gaw}=\frac{19n\div g}{2aw} \qquad (28).$$

where $n+g$ is the rigidity in grammes weight per square centimetre. For steel and glass as above (§§ 837, 839) the values of $\frac{\mathfrak{r}}{\mathfrak{g}}$ are respectively 2·1 and ·66.

Rigidity of the earth:

841. Hence it appears that if the rigidity of the earth, on the whole, were only as much as that of steel or iron, the earth as a whole would yield about one-third as much to the tide-generating influences of the sun and moon as it would if it had no rigidity at all; and it would yield by about three-fifths of the fluid yielding, if its rigidity were no more than that of glass.

unless greater than that of steel would be very imperfectly effective in maintaining figure against tide-generating force.

842. To find the effect of the earth's elastic yielding on the tides, we must recollect (§ 819) that the ellipticity of level due to the disturbing force, and to the gravitation of the undisturbed globe, which [§§ 804, 808, (18), (19)] is $\frac{a}{g}\tau$, will be augmented by $\frac{3}{5}e$ on account of the alteration of the globe into a spheroid of ellipticity e: so that if (§ 799) we neglect the mutual attraction of the waters, we have for the disturbed ellipticity of the sea level (§ 785)

Influence of elastic yielding of the solid earth on the surface-liquid tides.

$$\frac{a}{g}\tau+\tfrac{3}{5}e \qquad (29).$$

The rise and fall of the water relatively to the solid earth will depend on the excess of this above the ellipticity of the solid. Denoting this excess, or the ellipticity of relative tides, by ϵ, we have

$$\epsilon=\frac{a}{g}\tau-\tfrac{2}{5}e \qquad (30),$$

or by (26) and (27)

$$\epsilon=\frac{a}{g}\tau\frac{\mathfrak{r}}{\mathfrak{r}+\mathfrak{g}} \qquad (31).$$

Hence the rise and fall of the tides is less than it would be were the earth perfectly rigid, in the proportion that the resistance against tidal deformation of the solid due to its rigidity bears to sum of the resistances due to rigidity of the solid and to mutual gravitation of its parts. By the numbers at the end of § 840 we conclude that if the rigidity were as great as that of steel, the relative rise and fall of the water would be reduced by elastic yielding of the solid to $\frac{2}{3}$, or if the rigidity were only

Rigidity of the earth:

that of glass, the relative rise and fall would be actually reduced to $\frac{2}{5}$, of what it would be were the rigidity perfect.

probably greater on the whole than that of a solid glass globe.

843. Imperfect as the comparisons between theory and observation as to the actual height of the tides has been hitherto, it is scarcely possible to believe that the height is in reality only two-fifths of what it would be if, as has been universally assumed in tidal theories, the earth were perfectly rigid. It seems, therefore, nearly certain, with no other evidence than is afforded by the tides, that the tidal effective rigidity of the Earth must be greater than that of glass.

Dynamic theory of tides too imperfect to give any estimate of absolute values for main phenomena:

but not so for the fortnightly and semi-annual tides.

844. The actual distribution of land and water, and of depth where there is water, over the globe is so irregular, that we need not expect of even the most powerful mathematical analysis any approach to a direct dynamical estimate of what the ordinary semi-diurnal tides in any one place ought to be if the earth were perfectly rigid. In water 10,000 feet deep (which is considerably less than the general depth of the Atlantic, as demonstrated by the many soundings taken within the last few years, especially those along the whole line of the Atlantic Telegraph Cable, from Valencia to Newfoundland), the velocity of long free waves, as will be proved in Vol. II., is 567 feet per second.[1] At this rate the time of advancing through 57° (or a distance equal to the earth's radius) would be only ten hours. Hence it may be presumed that, at least at all islands of the Atlantic, any tidal disturbance, whose period amounts to several days or more, ought to give very nearly the true equilibrium tide, not modified sensibly, or little modified, by the inertia of the fluid. Now such tidal disturbances (§ 808) exist in virtue of the moon's and sun's changes of declination, having for their periods the periods of these changes.

Amounts of fortnightly tide estimated on various suppositions as to rigidity.

845. The sum of the rise from lowest to highest at Teneriffe, and simultaneous fall from highest to lowest at Iceland, in the lunar fortnightly tide, would amount to 4·5 inches if the earth were perfectly rigid, or 3 inches if the tidal effective rigidity were only that of steel, or 1·8 inches if the tidal effective rigidity were only that of glass. The amounts of the semi-annual tide, whatever be the actual rigidity of the earth, would of course be about half that of the fortnightly tide. The amount

[1] Airy, *Tides and Waves*, § 170.

either in any one place would be discoverable with certainty to a small fraction of an inch by a proper application of the method of least squares, such as has hitherto not been made, to the indications of an accurate self registering tide-gauge. For our present object, the semi-annual tide, though it may have the advantage of being more certainly not appreciably different from the true equilibrium amount, may be sensibly affected by the melting of ice from the arctic and antarctic polar regions, and by the fall of rain and drainage of land elsewhere, which will probably be found to give measurable disturbances in the sea level, exhibiting, on the average of many years, an annual and semi-annual harmonic variation. This disturbance will, however, be eliminated for any one fortnight or half-year, by combining observations at well-chosen stations in different latitudes. It seems probable, therefore, that a somewhat accurate determination of the true amount of the Earth's elastic yielding to the tide-generating forces of the Moon and Sun may be deduced from good self-registering tide-gauges maintained for several years at such stations as Iceland, Teneriffe, Cape Verde Islands, Ascension Island, and St. Helena. It is probable also that the ratio of the Moon's mass to that of the Sun may be determined from such observations more accurately than it has yet been. It is to be hoped that these objects may induce the British Government, which has done so much for physical geography in many ways, to establish tide-gauges at proper stations for determining with all possible accuracy the fortnightly and semi-annual tides, and the variations of sea level due to the melting of ice in the polar regions, and the fall of rain and drainage of land over the rest of the world.

Rigidity of the earth: probably to be best learned from observations giving amounts of fortnightly tides.

Tide-gauges to be established at ocean stations.

Reduction of results by method of least squares.

846. More observation, and more perfect reduction of observations already made, are wanted to give any decided answer to the questions, how much the fortnightly tide and the semi-annual tide really are. "In the *Philosophical Transactions*, "1839, p. 157, Mr. Whewell shows that the observations of "high and low water at Plymouth give a mean height of water "increasing as the moon's declination increases, and amounting "to 3 inches when the moon's declination is 25°. This is the "same direction as that corresponding in the expression above

Scantiness of information regarding fortnightly tides, hitherto drawn from observation.

Scantiness of information regarding fortnightly tides, hitherto drawn from observation.

"to a high latitude. The effect of the sun's declination is not "investigated from the observations. In the *Philosophical* "*Transactions*, p. 163, Mr. Whewell has given the observations "of some most extraordinary tides at Petropaulofsk, in Kams "chatka, and at Novo-Arkhangelsk, in the Island of Sitkhi, on "the west coast of North America.

"From the curves in the *Philosophical Transactions*, as well "as from the remaining curves relating to the same places "(which, by Mr. Whewell's kindness, we have inspected), there "appears to be no doubt that the mean level of the water at "Petropaulofsk and Arkhangelsk rises as the moon's declina- "tion increases. We have no further information on this "point."—(Airy's *Tides and Waves*, § 533.)

Effect of elastic yielding on precession and nutation:

847. We intend, in our second volume, to give a dynamical investigation of precession and nutation, in which it will be proved that the earth's elastic yielding influences these phenomena to the same proportionate degree as it influences the tides. We have seen already (§§ 825, 826, 796, 797) that the only datum wanted for a comparison between their observed amounts and their theoretical amounts on the hypothesis of perfect rigidity, to an accuracy of within one per cent., is a knowledge of the earth's moment of inertia about any diameter within one per cent. We have seen (§ 828) that the best theoretical estimates of precession hitherto made are in remarkable accordance with the observed amount. But it is not at all improbable that better-founded estimates of the earth's

not yet discovered by observation:

moment of inertia (§ 826), and more accurate knowledge than we yet have from observation, of the harmonic of the second degree in the expression of external gravity (§§ 825, 795), may show that the true amount of precession (which is known at present with extreme accuracy) is somewhat smaller than it

but might probably be so if whole moment of inertia round any diameter were known.

would be if the rigidity were infinite. Such a discrepancy, if genuine, could only be explained by some small amount of deformation experienced by the solid parts of the earth under lunar and solar influence. The agreement between theory, on the hypothesis of perfect rigidity, and observation as to precession and nutation are, however, on the whole so close as to allow us to infer that the earth's elastic yielding to the disturbing influence of the sun and moon is very small—much smaller or instance, than it would be if its effective rigidity were no more than the rigidity of steel.

848. It is interesting to remark, that the popular geological hypothesis that the earth is a thin shell of solid material, having a hollow space within it filled with liquid, involves two effects of deviation from perfect rigidity, which would influence in opposite ways the amount of precession. The comparatively easy yielding of the shell must, as we shall see in our second volume, render the effective moving couple, due to Sun and Moon, much smaller than it would be if the whole interior were solid, and, on this account, must tend to diminish the amount of precession and nutation. But the effective moment of inertia of a thin solid shell containing fluid, whether homogeneous or heterogeneous, in its interior, would be much less than that of the whole mass if solid throughout; and the tendency would be to much greater amounts of precession and nutation on this account. It seems excessively improbable that the defect of moment of inertia due to fluid in the Earth's interior should bear at all approximately the same ratio to the whole moment of inertia, that the actual elastic yielding bears to the perfectly easy yielding which would take place if the Earth were quite fluid. But we must either admit this supposition, improbable as it seems, or conclude (from the close agreement of precession and nutation, with what they would be if the Earth were perfectly rigid) that the defect of moment of inertia, owing to fluid in the interior, is small in comparison with the whole moment of inertia of the Earth, about any diameter; and that the deformation experienced by the Earth, from lunar and solar influence, is small in comparison with what it would be if the Earth were perfectly fluid. It is, however, certain that there is some fluid matter in the interior of the Earth: witness eruptions of lava from volcanoes. But this is probably quite local, as has been urged by Hopkins, who first adduced the phenomena of precession and nutation to disprove the hypothesis that the solid part of the Earth's mass is merely a thin shell.

Geological hypothesis of a thin shell filled with liquid would make moment of inertia very small, and on this account give very great precession if the shell were perfectly rigid.

Probable conclusion that there is little of fluid in the earth, and that the solid parts are rendered very rigid by great pressure.

849. A curiously similar remark is applicable to the tides; but only in virtue of the greater density in the deeper parts of the hypothetical fluid. The oblateness induced by the tide-generating influence in the layers of equal density would (§ 815) augment the surface tidal ellipticity of level; and therefore, if owing to perfect rigidity of the containing shell the bounding surface were absolutely constant in figure, the tides would be greater than they would be if the earth were perfectly rigid throughout.

APPENDIX TO CHAPTER VII.

Appendix C.

(C.)—Equations of Equilibrium of an Elastic Solid deduced from the Principle of Energy.

(*a*.) Let a solid composed of matter fulfilling no condition of isotropy in any part, and not homogeneous from part to part, be given of any shape, unstrained, and let every point of its surface be altered in position to a given distance in a given direction. It is required to find the displacement of every point of its substance, in equilibrium. Let x, y, z be the co-ordinates of any particle, P, of the substance in its undisturbed position, and $x+\alpha$, $y+\beta$, $z+\gamma$ its co-ordinates when displaced in the manner specified: that is to say, let α, β, γ be the components of the required displacement. Then, if for brevity we put

Strain, not necessarily small, specified by six elements.

$$\left.\begin{aligned}
A&=\left(\frac{d\alpha}{dx}+1\right)^2+\left(\frac{d\beta}{dx}\right)^2+\left(\frac{d\gamma}{dx}\right)^2\\
B&=\left(\frac{d\alpha}{dy}\right)^2+\left(\frac{d\beta}{dy}+1\right)^2+\left(\frac{d\gamma}{dy}\right)^2\\
C&=\left(\frac{d\alpha}{dz}\right)^2+\left(\frac{d\beta}{dz}\right)^2+\left(\frac{d\gamma}{dz}+1\right)^2\\
a&=\frac{d\alpha}{dy}\frac{d\alpha}{dz}+\left(\frac{d\beta}{dy}+1\right)\frac{d\beta}{dz}+\frac{d\gamma}{dy}\left(\frac{d\gamma}{dz}+1\right)\\
b&=\frac{d\alpha}{dz}\left(\frac{d\alpha}{dx}+1\right)+\frac{d\beta}{dz}\frac{d\beta}{dx}+\left(\frac{d\gamma}{dz}+1\right)\frac{d\gamma}{dx}\\
c&=\left(\frac{d\alpha}{dx}+1\right)\frac{d\alpha}{dy}+\frac{d\beta}{dx}\left(\frac{d\beta}{dy}+1\right)+\frac{d\gamma}{dx}\frac{d\gamma}{dy}
\end{aligned}\right\}\quad(1);$$

these six quantities A, B, C, a, b, c are proved [§ 190 (*e*) and § 181 (5)] to thoroughly determine the strain experienced by the substance infinitely near the particle P (irrespectively of any rotation it may experience), in the following manner:—

(*b*.) Let ξ, η, ζ be the undisturbed co-ordinates of a particle infinitely near P, relatively to axes through P parallel to those of x, y, z respectively; and let $\xi_,$, $\eta_,$, $\zeta_,$ be the co-ordinates relative still to axes through P, when the solid is in its strained condition. Then

$$\xi_,^2+\eta_,^2+\zeta_,^2=A\xi^2+B\eta^2+C\zeta^2+2a\eta\zeta+2b\zeta\xi+2c\xi\eta \quad (2);$$

and therefore all particles which in the strained state lie on spherical surface

$$\xi_,^2+\eta_,^2+\zeta_,^2=r_,^2,$$

Appendix C.

are in the unstrained state, on the ellipsoidal surface,

$$A\xi^2+B\eta^2+C\zeta^2+2a\eta\zeta+2b\zeta\xi+2c\xi\eta=r_1^{\,2}.$$

This (§§ 155-165) completely defines the homogeneous strain of the matter in the neighbourhood of P.

Anticipatory application of the Carnot and Clausius thermodynamic law:

(*c.*) Hence, the thermodynamic principles by which, in a paper on the Thermo-elastic Properties of Matter, in the first number of the *Quarterly Mathematical Journal* (April 1855), of which an account will be given in Vol. III., Green's dynamical theory of elastic solids was demonstrated as part of the modern dynamical theory of heat, show that if $wdxdydz$ denote the work required to alter an infinitely small undisturbed volume, $dxdydz$, of the solid, into its disturbed condition, when its temperature is kept constant, we must have

Its combination with Joule's law expressed analytically for elastic solid.

$$w=f(A, B, C, a, b, c) \qquad (3)$$

where f denotes a positive function of the six elements, which vanishes when $A-1$, $B-1$, $C-1$, a, b, c each vanish. And if W denote the whole work required to produce the change actually experienced by the whole solid, we have

Potential energy of deformation;

$$W=\iiint wdxdydz \qquad (4)$$

where the triple integral is extended through the space occupied by the undisturbed solid.

(*d.*) The position assumed by every particle in the interior of the solid will be such as to make this a minimum subject to the condition that every particle of the surface takes the position given to it; this being the elementary condition of stable equilibrium. Hence, by the method of variations

a minimum for stable equilibrium.

$$\delta W=\iiint \delta w dxdydz=0 \qquad (5).$$

But, exhibiting only terms depending on $\delta\alpha$, we have

$$\delta w=\left\{2\frac{dw}{dA}\left(\frac{d\alpha}{dx}+1\right)+\frac{dw}{db}\frac{d\alpha}{dz}+\frac{dw}{dc}\frac{d\alpha}{dy}\right\}\frac{d\delta\alpha}{dx}$$
$$+\left\{2\frac{dw}{dB}\frac{d\alpha}{dy}+\frac{dw}{da}\frac{d\alpha}{dz}+\frac{dw}{dc}\left(\frac{d\alpha}{dx}+1\right)\right\}\frac{d\delta\alpha}{dy}$$
$$+\left\{2\frac{dw}{dC}\frac{d\alpha}{dz}+\frac{dw}{da}\frac{d\alpha}{dy}+\frac{dw}{db}\left(\frac{d\alpha}{dx}+1\right)\right\}\frac{d\delta\alpha}{dz}$$
$$+\text{etc.}$$

Hence, integrating by parts, and observing that $\delta\alpha$, $\delta\beta$, $\delta\gamma$ vanish at the limiting surface, we have

$$\delta W=-\iiint dxdydz\left\{\left(\frac{dP}{dx}+\frac{dQ}{dy}+\frac{dR}{dz}\right)\delta\alpha+\text{etc.}\right\} \qquad (6)$$

where for brevity P, Q, R denote the multipliers of $\frac{d\delta\alpha}{dx}$, $\frac{d\delta\alpha}{dy}$, $\frac{d\delta\alpha}{dz}$

respectively, in the preceding expression. In order that δW may vanish, the multipliers of $\delta\alpha$, $\delta\beta$, $\delta\gamma$ in the expression now found for it must each vanish, and hence we have, as the equations of equilibrium

Appendix C.

Equations of internal equilibrium of an elastic solid experiencing no bodily force.

$$\left.\begin{array}{l}\frac{d}{dx}\{2\frac{dw}{dA}(\frac{d\alpha}{dx}+1)+\frac{dw}{db}\frac{d\alpha}{dz}+\frac{dw}{dc}\frac{d\alpha}{dy}\}\\ +\frac{d}{dy}\{2\frac{dw}{dB}\frac{d\alpha}{dy}+\frac{dw}{da}\frac{d\alpha}{dz}+\frac{dw}{dc}(\frac{d\alpha}{dx}+1)\}\\ +\frac{d}{dz}\{2\frac{dw}{dC}\frac{d\alpha}{dz}+\frac{dw}{da}\frac{d\alpha}{dy}+\frac{dw}{db}(\frac{d\alpha}{dx}+1)\}=0.\\ \qquad\text{etc. etc.}\end{array}\right\} \quad (7),$$

of which the second and third, not exhibited, may be written down merely by attending to the symmetry.

(*e.*) From the property of w that it is necessarily positive when there is any strain, it follows that there must be some distribution of strain through the interior which shall make $\iiint w\,dx\,dy\,dz$ *the least possible* subject to the prescribed surface condition; and therefore that the solution of equations (7) subject to this condition, is possible. If, whatever be the nature of the solid as to difference of elasticity in different directions, in any part, and as to heterogeneousness from part to part, and whatever be the extent of the change of form and dimensions to which it is subjected, there cannot be any internal configuration of unstable equilibrium, nor consequently any but one of stable equilibrium, with the prescribed surface displacement, and no disturbing force on the interior; then, besides being always positive, w must be such a function of A, B, etc., that there can be only one solution of the equations. This is obviously the case when the unstrained solid is homogeneous.

Their solution proved possible and unique when surface displacement is given, unless there can be unstable equilibrium: hence necessarily unique for a homogeneous solid.

(*f.*) It is easy to include, in a general investigation similar to the preceding, the effects of any force on the interior substance, such as we have considered particularly for a spherical shell, of homogeneous isotropic matter, in §§ 730...737 above. It is also easy to adapt the general investigation to superficial data of *force*, instead of displacement.

Extension of the analysis to include bodily force, and data of surface force, easy.

(*g.*) Whatever be the general form of the function f for any part of the substance, since it is always positive it cannot change in sign when $A-1$, $B-1$, $C-1$, a, b, c, have their signs changed; and therefore for infinitely small values of these quantities it must be a homogeneous quadratic function of them with constant coefficients. (And it may be useful to observe that for all values of

Appendix C.

Case of infinitely small strains:— Green's theory:

the variables A, B, etc., it must therefore be expressible in the same form, with varying coefficients, each of which is always finite, for all values of the variables.) Thus, for infinitely small strains we have Green's theory of elastic solids, founded on a homogeneous quadratic function of the components of strain, expressing the work required to produce it. Thus, putting

$$A-1=2e,\ B-1=2f,\ C-1=2g \quad (8)$$

and denoting by $\frac{1}{2}(e,e)$, $\frac{1}{2}(f,f)$,...(e,f),...(e,a),... the coefficients, we have, as above (§ 673),

$$\left.\begin{aligned} w=\tfrac{1}{2}\{(e,e)e^2+(f,f)f^2+(g,g)g^2+(a,a)a^2+(b,b)b^2+(c,c)c^2\} \\ +(e,f)ef+(e,g)eg+(e,a)ea+(e,b)eb+(e,c)ec \\ +(f,g)fg+(f,a)fa+(f,b)fb+(f,c)fc \\ +(g,a)ga+(g,b)gb+(g,c)gc \\ +(a,b)ab+(a,c)ac \\ +(b,c)bc \end{aligned}\right\} \quad (9).$$

(*h*.) When the strains are infinitely small the products $\frac{dw}{dA}\frac{da}{dx}$, $\frac{dw}{db}\frac{da}{dz}$, etc., are each infinitely small, of the second order. We therefore omit them; and then attending to (8), we reduce (7) to

dynamic equations of internal equilibrium:

$$\left.\begin{aligned} \frac{d}{dx}\frac{dw}{de}+\frac{d}{dy}\frac{dw}{dc}+\frac{d}{dz}\frac{dw}{db}=0 \\ \frac{d}{dx}\frac{dw}{dc}+\frac{d}{dy}\frac{dw}{df}+\frac{d}{dz}\frac{dw}{da}=0 \\ \frac{d}{dx}\frac{dw}{db}+\frac{d}{dy}\frac{dw}{da}+\frac{d}{dz}\frac{dw}{dg}=0 \end{aligned}\right\} \quad (10),$$

which are the equations of interior equilibrium. Attending to (9) we see that $\frac{dw}{de}\cdots\frac{dw}{da}\cdots$ are linear functions of e, f, g, a, b, c the components of strain. Writing out one of them as an example we have

$$\frac{dw}{de}=(e,e)e+(e,f)f+(e,g)g+(e,a)a+(e,b)b+(e,c)c \quad (11).$$

And, α, β, γ denoting, as before, the component displacements of any interior particle, P, from its undisturbed position (x, y, z) we have, by (8) and (1)

and relative kinematic equations.

$$\left.\begin{aligned} e=\frac{d\alpha}{dx},\quad f=\frac{d\beta}{dy},\quad g=\frac{d\gamma}{dz} \\ a=\frac{d\beta}{dz}+\frac{d\gamma}{dy},\quad b=\frac{d\gamma}{dx}+\frac{d\alpha}{dz},\quad c=\frac{d\alpha}{dy}+\frac{d\beta}{dx} \end{aligned}\right\} \quad (12).$$

It is to be observed that the coefficients (e, e) (e, f), etc., will be in general functions of (x, y, z), but will be each constant when the unstrained solid is homogeneous. Appendix C. Case of infinitely small strains:—

(*i*.) It is now easy to prove directly, for the case of infinitely small strains, that the solution of the equations of interior equilibrium, whether for a heterogeneous or a homogeneous solid, subject to the prescribed surface condition, is unique. For, let α, β, γ be components of displacement fulfilling the equations, and let α', β', γ' denote any other functions of x, y, z, having the same surface values as α, β, γ, and let $e', f', \ldots, w'$ denote functions depending on them in the same way as $e, f, \ldots, w$ depend on α, β, γ. Thus by Taylor's theorem, solution proved unique whether the body is homogeneous or heterogeneous, when surface displacement is given:

$$w'-w=\frac{dw}{de}(e'-e)+\frac{dw}{df}(f'-f)+\frac{dw}{dg}(g'-g)+\frac{dw}{da}(a'-a)+\frac{dw}{db}(b'-b)+\frac{dw}{dc}(c'-c)+H,$$

where H denotes the same homogeneous quadratic function of $e'-e$, etc., that w is of e, etc. If for $e'-e$, etc., we substitute their values by (12), this becomes

$$w'-w=\frac{dw}{de}\frac{d(\alpha'-\alpha)}{dx}+\frac{dw}{db}\frac{d(\alpha'-\alpha)}{dz}+\frac{dw}{dc}\frac{d(\alpha'-\alpha)}{dy}+\text{etc.}+H.$$

Multiplying by $dxdydz$, integrating by parts, observing that $\alpha'-\alpha$, $\beta'-\beta$, $\gamma'-\gamma$ vanish at the bounding surface, and taking account of (10), we find simply

$$\iiint(w'-w)dxdydz=\iiint Hdxdydz \qquad (13).$$

But H is essentially positive. Therefore every other interior condition than that specified by α, β, γ, provided only it has the same bounding surface, requires a greater amount of work than w to produce it: and the excess is equal to the work that would be required to produce, from a state of no displacement, such a displacement as superimposed on α, β, γ, would produce the other. And inasmuch as α, β, γ, fulfil only the conditions of satisfying (11) and having the given surface values, it follows that no other than one solution can fulfil these conditions.

(*j*.) But (as has been pointed out to us by Stokes) when the surface data are of force, not of displacement, or when force acts from without, on the interior substance of the body, the solution is not in general unique, and there may be configurations of unstable equilibrium even with infinitely small displacement. For instance, let part of the body be composed of a steel-bar magnet; and let a magnet be held outside in the same line, and with a pole of the same name in its end nearest to one end of the inner magnet. The equilibrium will be unstable, and there will but not necessarily so when the surface data are of force.

Appendix C.

be positions of stable equilibrium with the inner bar slightly inclined to the line of the outer bar, unless the rigidity of the rest of the body exceed a certain limit.

Condition that substance may be isotropic, without limitation to infinitely small strains:

(*k*.) Recurring to the general problem, in which the strains are not supposed infinitely small; we see that if the solid is isotropic in every part, the function of A, B, C, a, b, c which expresses w, must be merely a function of the roots of the equation [§ 181 (11)]

$$(A-\zeta^2)(B-\zeta^2)(C-\zeta^2)-a^2(A-\zeta^2)-b^2(B-\zeta^2)-c^2(C-\zeta^2)+2abc=0 \quad (14)$$

which (that is the positive values of ζ) are the ratios of elongation along the principal axes of the strain-ellipsoid. It is unnecessary here to enter on the analytical expression of this condition. For the case of $A-1$, $B-1$, $C-1$, a, b, c each infinitely small, it obviously requires that

$$\left.\begin{array}{l}(e,e)=(f,f)=(g,g);\ (f,g)=(g,e)=(e,f);\ (a,a)=(b,b)=(c,c);\\ (e,a)=(f,b)=(g,c)=0;\ (b,c)=(c,a)=(a,b)=0;\ \text{and}\\ (e,b)=(e,c)=(f,c)=(f,a)=(g,a)=(g,b)=0.\end{array}\right\} \quad (15).$$

Thus the 21 coefficients are reduced to three—

(e, e) which we may denote by the single letter $\mathfrak{A}$,
(f, g) " " " " $\mathfrak{B}$,
(a, a) " " " " n.

expressed in equations among the moduli of elasticity for case of infinitely small strains.

It is clear that this is necessary and sufficient for insuring *cubic isotropy*; that is to say, perfect equality of elastic properties with reference to the three rectangular directions OX, OY, OZ. But for *spherical isotropy*, or complete isotropy with reference to all directions through the substance, it is further necessary that

$$\mathfrak{A}-\mathfrak{B}=2n \quad (16);$$

as is easily proved analytically by turning two of the axes of co-ordinates in their own plane through 45°; or geometrically by examining the nature of the strain represented by any one of the elements a, b, c (a "simple shear") and comparing it with the resultant of c, and $f=-e$ (which is also a simple shear). It is convenient now to put

$$\mathfrak{A}+\mathfrak{B}=2m;\ \text{so that}\ \mathfrak{A}=m+n,\ \mathfrak{B}=m-n \quad (17);$$

and thus the expression for the potential energy per unit of volume becomes

Potential energy of infinitely small strain in isotropic solid.

$$2w=m(e+f+g)^2+n(e^2+f^2+g^2-2fg-2ge-2ef+a^2+b^2+c^2) \quad (18).$$

Using this in (9), and substituting for e, f, g, a, b, c their values by (12), we find immediately for the equations of internal equilibrium, equations the same as (6) of § 698.

(*l*.) To find the mutual force exerted across any surface within the solid, as expressed by (1) of § 662, we have clearly, by considering the work done respectively by P, Q, R, S, T, U (§ 662) on any infinitely small change of figure or dimensions in the solid, Appendix C.

$$P=\frac{dw}{de}, \quad Q=\frac{dw}{df}, \quad R=\frac{dw}{dg}, \quad S=\frac{dw}{da}, \quad T=\frac{dw}{db}, \quad U=\frac{dw}{dc} \quad (19).$$

Components of stress required for infinitely small strain.

Hence, for an isotropic solid, (18) gives the expressions which we have used above, (12) of § 673.

(m.) To interpret the coefficients m and n in connexion with elementary ideas as to the elasticity of the solid; first let $a=b=c=0$, and $e=f=g=\frac{1}{3}\delta$: in other words, let the substance experience a uniform dilatation, in all directions, producing an expansion of volume from 1 to $1+\delta$. In this case (18) becomes

Moduli of resistance to compression and of rigidity.

$$w=\tfrac{1}{2}(m-\tfrac{1}{3}n)\delta^2;$$

and we have

$$\frac{dw}{d\delta}=(m-\tfrac{1}{3}n)\delta.$$

Hence $(m-\frac{1}{3}n)\delta$ is the normal force per unit area of its surface required to keep any portion of the solid expanded to the amount specified by δ. Thus $m-\frac{1}{3}n$ measures the elastic force called out by, or the elastic resistance against, change of volume: and viewed as a *modulus of elasticity*, it may be called the *elasticity of volume*. [Compare §§ 692, 693, 694, 688, 682, and 680.] What is commonly called the "compressibility" is measured by $\frac{1}{m-\frac{1}{3}n}$.

And let next $e=f=g=b=c=0$; which gives

$$w=\tfrac{1}{2}na^2; \text{ and, by (19), } S=na.$$

This shows that the tangential force per unit area required to produce an infinitely small shear (§ 171), amounting to a, is na. Hence n measures the innate power of the body to resist change of shape, and return to its original shape when force has been applied to change it: that is to say, it measures *the rigidity* of the substance.

(D.)—On the Secular Cooling of the Earth.[1]

(a.) For eighteen years it has pressed on my mind, that essential principles of Thermo-dynamics have been overlooked by those geologists who uncompromisingly oppose all paroxysmal hypotheses, and maintain not only that we have examples now

Dissipation of energy, disregarded by many followers of Hutton.

[1] *Transactions of the Royal Society of Edinburgh*, 1862 (W. Thomson).

Appendix D.

before us, on the earth, of all the different actions by which its crust has been modified in geological history, but that these actions have never, or have not on the whole, been more violent in past time than they are at present.

Dissipation of energy from the solar system.

(*b*.) It is quite certain the solar system cannot have gone on, even as at present, for a few hundred thousand or a few million years, without the irrevocable loss (by dissipation, not by *annihilation*) of a very considerable proportion of the entire energy initially in store for sun heat, and for Plutonic action. It is quite certain that the whole store of energy in the solar system has been greater in all past time than at present; but it is conceivable that the rate at which it has been drawn upon and dissipated, whether by solar radiation, or by volcanic action in the earth or other dark bodies of the system, may have been nearly equable, or may even have been less rapid, in certain periods of the past. But it is far more probable that the secular rate of dissipation has been in some direct proportion to the total amount of energy in store, at any time after the commencement of the present order of things, and has been therefore very slowly diminishing from age to age.

Terrestrial climate influenced by the probably hotter sun of a few million years ago.

(*c*.) I have endeavoured to prove this for the sun's heat, in an article recently published in *Macmillan's Magazine* (March 1862), where I have shown that most probably the sun was sensibly hotter a million years ago than he is now. Hence, geological speculations assuming somewhat greater extremes of heat, more violent storms and floods, more luxuriant vegetation, and hardier and coarser grained plants and animals, in remote antiquity, are more probable than those of the extreme quietist, or "uniformitarian" school. A "middle path," not generally safest in scientific speculation, seems to be so in this case. It is probable that hypotheses of grand catastrophes destroying all life from the earth, and ruining its whole surface at once, are greatly in error; it is impossible that hypotheses assuming an equability of sun and storms for 1,000,000 years, can be wholly true.

(*d*.) Fourier's mathematical theory of the conduction of heat is a beautiful working out of a particular case belonging to the general doctrine of the "Dissipation of Energy."[1] A characteristic of the practical solutions it presents is, that in each case a distribution of temperature, becoming gradually equalized through

[1] *Proceedings of Royal Soc. Edin.*, Feb. 1852. "On a Universal Tendency in Nature to the Dissipation of Mechanical Energy." Also, "On the Restoration of Energy in an Unequally Heated Space," *Phil. Mag.*, 1853, first half-year.

an unlimited future, is expressed as a function of the time, which is infinitely divergent for all times longer past than a definite determinable epoch. The distribution of heat at such an epoch is essentially *initial*—that is to say, it cannot result from any previous condition of matter by natural processes. It is, then, well called an "*arbitrary* initial distribution of heat," in Fourier's great mathematical poem, because it could only be realized by action of a power able to modify the laws of dead matter. In an article published about nineteen years ago in the *Cambridge Mathematical Journal*,[1] I gave the mathematical criterion for an essentially initial distribution; and in an inaugural essay, "De Motu Caloris per Terræ Corpus," read before the Faculty of the University of Glasgow in 1846, I suggested, as an application of these principles, that a perfectly complete geothermic survey would give us data for determining an initial epoch in the problem of terrestrial conduction. At the meeting of the British Association in Glasgow in 1855, I urged that special geothermic surveys should be made for the purpose of estimating absolute dates in geology, and I pointed out some cases, especially that of the salt-spring borings at Creuznach, in Rhenish Prussia, in which eruptions of basaltic rock seem to leave traces of their igneous origin in residual heat.[2] I hope this suggestion may yet be taken up, and may prove to some extent useful; but the disturbing influences affecting underground temperature, as Professor Phillips has well shown in a recent inaugural address to the Geological Society, are too great to allow us to expect any very precise or satisfactory results.

Mathematicians' use of word "arbitrary" metaphysically significant.

Criterion of an essentially "initial" distribution of heat in a solid:

now applied to estimate date of earth's consolidation, from data of present underground temperature.

Value of local geothermic surveys, for estimation of absolute dates in geology.

(*e.*) The chief object of the present communication is to estimate from the known general increase of temperature in the earth downwards, the date of the first establishment of that *consistentior status*, which, according to Leibnitz's theory, is the initial date of all geological history.

(*f.*) In all parts of the world in which the earth's crust has been examined, at sufficiently great depths to escape large influence of the irregular and of the annual variations of the superficial temperature, a gradually increasing temperature has been found in going deeper. The rate of augmentation (estimated at only $\frac{1}{110}$th of a degree, Fahr., in some localities, and as much as $\frac{1}{15}$th of a degree in other, per foot of descent) has not been observed in a sufficient number of places to establish any fair

Increase of temperature downwards in earth's crust; but very imperfectly observed hitherto.

[1] Feb. 1844.—"Note on Certain Points in the Theory of Heat."

[2] See British Association Report of 1855 (Glasgow) Meeting.

Appendix D.

average estimate for the upper crust of the whole earth. But $\frac{1}{50}$th is commonly accepted as a rough mean; or, in other words, it is assumed as a result of observation, that there is, on the whole, about 1° Fahr. of elevation of temperature per 50 British feet of descent.

Secular loss of heat out of the earth *demonstrated:*

(*g.*) The fact that the temperature increases with the depth implies a continual loss of heat from the interior, by conduction outwards through or into the upper crust. Hence, since the upper crust does not become hotter from year to year, there must be a secular loss of heat from the whole earth. It is possible that no cooling may result from this loss of heat, but only an exhaustion of potential energy, which in this case could scarcely be other than chemical affinity between substances forming part of the earth's mass. But it is certain that either the earth is becoming on the whole cooler from age to age, or the heat conducted out is generated in the interior by temporary dynamical (that is, in this case, chemical) action. To suppose, as Lyell, adopting the chemical hypothesis, has done,[1] that the substances, combining together, may be again separated electrolytically by thermo-electric currents, due to the heat generated by their combination, and thus the chemical action and its heat continued in an endless cycle, violates the principles of natural philosophy in exactly the same manner, and to the same degree, as to believe that a clock constructed with a self-winding movement may fulfil the expectations of its ingenious inventor by going for ever.

but not so any present or past secular cooling, however probable.

Fallacy of a thermo-electric perpetual motion.

(*h.*) It must indeed be admitted that many geological writers of the "Uniformitarian" school, who in other respects have taken a profoundly philosophical view of their subject, have argued in a most fallacious manner against hypotheses of violent action in past ages. If they had contented themselves with showing that many existing appearances, although suggestive of extreme violence and sudden change, may have been brought about by long-continued action, or by paroxysms not more intense than some of which we have experience within the periods of human history, their position might have been unassailable; and certainly could not have been assailed except by a detailed discussion of their facts. It would be a very wonderful, but not an absolutely incredible result, that volcanic action has never been more violent on the whole than during the last two or three

Exception to the soundness of arguments adduced in the promulgation and prosecution of the Huttonian reform.

[1] *Principles of Geology*, chap. xxxi. ed. 1853.

centuries; but it is as certain that there is now less volcanic energy in the whole earth than there was a thousand years ago, as it is that there is less gunpowder in a "Monitor" after she has been seen to discharge shot and shell, whether at a nearly equable rate or not, for five hours without receiving fresh supplies, than there was at the beginning of the action. Yet this truth has been ignored or denied by many of the leading geologists of the present day, because they believe that the facts within their province do not demonstrate greater violence in ancient changes of the earth's surface, or do demonstrate a nearly equable action in all periods.

Appendix D.

Secular diminution of whole amount of volcanic energy quite certain: but not hitherto admitted by some of the chief geologists.

(*i.*) The chemical hypothesis to account for underground heat might be regarded as not improbable, if it was only in isolated localities that the temperature was found to increase with the depth; and, indeed, it can scarcely be doubted that chemical action exercises an appreciable influence (possibly negative, however) on the action of volcanoes; but that there is slow uniform "combustion," *eremacausis*, or chemical combination of any kind going on, at some great unknown depth under the surface everywhere, and creeping inwards gradually as the chemical affinities in layer after layer are successively saturated, seems extremely improbable, although it cannot be pronounced to be absolutely impossible, or contrary to all analogies in nature. The less hypothetical view, however, that the earth is merely a warm chemically inert body cooling, is clearly to be preferred in the present state of science.

Chemical hypothesis to account for ordinary underground heat not impossible, but very improbable.

(*j.*) Poisson's celebrated hypothesis, that the present underground heat is due to a passage, at some former period, of the solar system through hotter stellar regions, cannot provide the circumstances required for a palæontology continuous through that epoch of external heat. For from a mean of values of the conductivity, in terms of the thermal capacity of unit volume, of the earth's crust, in three different localities near Edinburgh, deduced from the observations on underground temperature instituted by Principal Forbes there, I find that if the supposed transit through a hotter region of space took place between 1250 and 5000 years ago, the temperature of that supposed region must have been from 25° to 50° Fahr. above the present mean temperature of the earth's surface, to account for the present general rate of underground increase of temperature, taken as 1° Fahr. in 50 feet downwards. Human history negatives this supposition. Again, geologists and astronomers will,

Poisson's proved impossible without destruction of life.

Appendix D.

Poisson's hypothesis to account for ordinary underground heat disproved as any acceptable mitigation of Leibnitz's theory.

I presume, admit that the earth cannot, 20,000 years ago, have been in a region of space 100° Fahr. warmer than its present surface. But if the transition from a hot region to a cool region supposed by Poisson took place more than 20,000 years ago, the excess of temperature must have been more than 100° Fahr., and must therefore have destroyed animal and vegetable life. Hence, the further back and the hotter we can suppose Poisson's hot region, the better for the geologists who require the longest periods; but the best for their view is Leibnitz's theory, which simply supposes the earth to have been at one time an incandescent liquid, without explaining how it got into that state. If we suppose the temperature of melting rock to be about 10,000° Fahr. (an extremely high estimate), the consolidation may have taken place 200,000,000 years ago. Or, if we suppose the temperature of melting rock to be 7000° Fahr. (which is more nearly what it is generally assumed to be), we may suppose the consolidation to have taken place 98,000,000 years ago.

Probable limits of uncertainty as to thermal conductivities and capacities of surface rocks.

(*k.*) These estimates are founded on the Fourier solution demonstrated below. The greatest variation we have to make on them, to take into account the differences in the ratios of conductivities to specific heats of the three Edinburgh rocks, is to reduce them to nearly half, or to increase them by rather more than half. A reduction of the Greenwich underground observations recently communicated to me by Professor Everett of Windsor, Nova Scotia, gives for the Greenwich rocks a quality intermediate between those of the Edinburgh rocks. But we are very ignorant as to the effects of high temperatures in altering the conductivities and specific heats of rocks, and as to their latent heat of fusion. We must, therefore, allow very wide limits in such an estimate as I have attempted to make; but I think we may with much probability say that the consolidation cannot have taken place less than 20,000,000 years ago, or we should have more underground heat than we actually have, nor more than 400,000,000 years ago, or we should not have so much as the least observed underground increment of temperature. That is to say, I conclude that Leibnitz's epoch of "emergence" of the *consistentior status* was probably between those dates.

Extreme admissible limits of date of earth's consolidation.

(*l.*) The mathematical theory on which these estimates are founded is very simple, being, in fact, merely an application of one of Fourier's elementary solutions to the problem of finding at any time the rate of variation of temperature from point to point, and the actual temperature at any point, in a solid extending to infinity in all directions, on the supposition that at an initial epoch the temperature has had two different constant values on the two sides of a certain infinite plane. The solution for the two required elements is as follows:— Appendix D.

Mathematical expression for interior temperature near the surface of a hot solid commencing to cool;

$$\frac{dv}{dx}=\frac{V}{\sqrt{\pi\kappa t}}\epsilon^{-\frac{x^2}{4\kappa t}}$$

$$v=v_0+\frac{2V}{\sqrt{\pi}}\int_0^{\frac{x}{2\sqrt{\kappa t}}}dz\epsilon^{-z^2}$$

where κ denotes the conductivity of the solid, measured in terms of the thermal capacity of the unit of bulk;

V, half the difference of the two initial temperatures;

v_0, their arithmetical mean;

t, the time;

x, the distance of any point from the middle plane;

v, the temperature of the point x at time t;

and, consequently (according to the notation of the differential calculus), $\frac{dv}{dx}$ the rate of variation of the temperature per unit of length perpendicular to the isothermal planes.

(*m.*) To demonstrate this solution, it is sufficient to verify—(1.) That the expression for v fulfils the partial differential equation,

$$\frac{dv}{dt}=\kappa\frac{d^2v}{dx^2},$$

Fourier's equation for the "linear conduction of heat;" (2.) That when $t=0$, the expression for v becomes v_0+V for all positive, and v_0-V for all negative values of x; and (3.) That the expression for $\frac{dv}{dx}$ is the differential coefficient with reference to x, of the expression for v. The propositions (1.) and (3.) are proved directly by differentiation. To prove (2.) we have, when $t=0$, and x positive, proved.

$$v=v_0+\frac{2V}{\sqrt{\pi}}\int_0^{\infty}dz\epsilon^{-z^2}$$

or according to the known value, $\frac{1}{2}\sqrt{\pi}$, of the definite integral $\int_0^{\infty}dz\epsilon^{-z^2}$,

$$v=v_0+V;$$

and for all values of t, the second term has equal positive and negative values for equal positive and negative values of x, so that when $t=0$ and x negative,

$$v=v_0-V.$$

Appendix D.

The admirable analysis by which Fourier arrived at solutions including this, forms a most interesting and important mathematical study. It is to be found in his *Théorie Analytique de la Chaleur*. Paris, 1822.

Expression for interior temperature near surface of a hot body commencing to cool:

(*n*.) The accompanying diagram (page 719) represents, by two curves, the preceding expressions for $\frac{dv}{dx}$ and v respectively.

(*o*.) The solution thus expressed and illustrated applies, for a certain time, without sensible error, to the case of a solid sphere, primitively heated to a uniform temperature, and suddenly exposed to any superficial action, which for ever after keeps the surface at some other constant temperature. If, for instance, the case considered is that of a globe 8000 miles diameter of solid rock, the solution will apply with scarcely sensible error for more than 1000 millions of years. For, if the rock be of a certain average quality as to conductivity and specific heat, the value of κ, as found in a previous communication to the Royal Society,[1] will be 400, for unit of length a British foot and unit of time a year; and the equation expressing the solution becomes

proved to be practically approximate for the earth for 1000 million years.

$$\frac{dv}{dx}=\frac{V}{35{\cdot}4}\cdot\frac{1}{t^{\frac{1}{2}}}\cdot\epsilon^{-\frac{x^2}{1600t}};$$

and if we give t the value 1,000,000,000, or anything less, the exponential factor becomes less than $\epsilon^{-5{\cdot}6}$ (which being equal to about $\frac{1}{270}$, may be regarded as insensible), when x exceeds 3,000,000 feet, or 568 miles. That is to say, during the first 1000 million years the variation of temperature does not become sensible at depths exceeding 568 miles, and is therefore confined to so thin a crust, that the influence of curvature may be neglected.

(*p*.) If, now, we suppose the time to be 100 million years from the commencement of the variation, the equation becomes

$$\frac{dv}{dx}=\frac{V}{354000}\epsilon^{-\frac{x^2}{160000000000}}.$$

Distribution of temperature 100 million years after commencement of cooling of a great enough mass of average rock.

The diagram, therefore, shows the variation of temperature which would now exist in the earth if, its whole mass being first solid and at one temperature 100 million years ago, the temperature of its surface had been everywhere suddenly lowered by V degrees, and kept permanently at this lower temperature: the scales used being as follows:—

[1] "On the Periodical Variations of Underground Temperature." *Trans. Roy Soc. Edin.*, March 1860.

INCREASE OF TEMPERATURE DOWNWARDS IN THE EARTH.

Appendix D. Distribution of temperature 100 million years after commencement of cooling of a great enough mass of average rock:

$$ON = x.$$

$$NP' = b\epsilon^{-\frac{x^2}{a^2}} = y'.$$

$$NP = area\ ONP'A \div a = \frac{1}{a}\int_0^x y'dx.$$

$$a = 2\sqrt{\kappa t}.$$

$$\frac{dv}{dx} = \frac{V}{a} \cdot \frac{NP}{b\frac{1}{2}\sqrt{\pi}}.$$

$$v - v = V. \frac{NP}{b.\frac{1}{2}\sqrt{\pi}}.$$

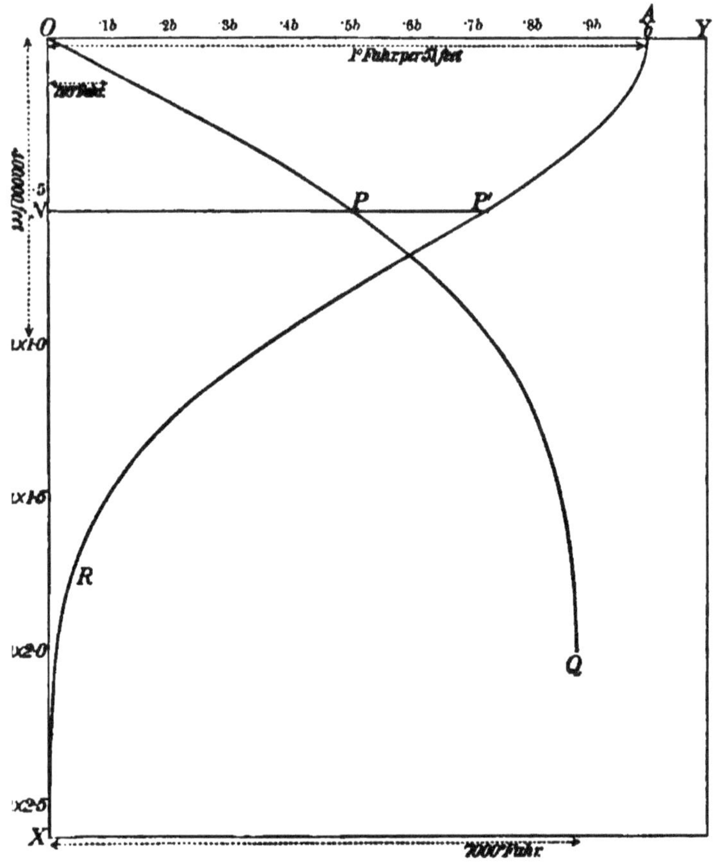

graphically represented.

OPQ curve showing excess of temperature above that of the surface.
AP'R curve showing rate of augmentation of temperature downwards.

Appendix D.

Distribution of temperature in a great enough mass of average rock, 100 million years after commencement of cooling from a temperature of 7000°.

(1.) For depth below the surface,—scale along OX, length a. represents 400,000 feet.

(2.) For rate of increase of temperature per foot of depth,—scale of ordinates parallel to OY, length b, represents $\frac{1}{354000}$ of V per foot. If, for example, $V=7000°$, this scale will be such that b represents $\frac{1}{50 \cdot 6}$ of a degree per foot.

(3.) For excess of temperature,—scale of ordinates parallel to OY, length b, represents $\dfrac{V}{\frac{1}{2}\sqrt{\pi}}$, or 7900°, if $V=7000°$.

Thus the rate of increase of temperature from the surface downwards would be sensibly $\frac{1}{51}$ of a degree per foot for the first 100,000 feet or so. Below that depth the rate of increase per foot would begin to diminish sensibly. At 400,000 feet it would have diminished to about $\frac{1}{141}$ of a degree per foot. At 800,000 feet it would have diminished to less than $\frac{1}{50}$ of its initial value,—that is to say, to less than $\frac{1}{2550}$ of a degree per foot; and so on, rapidly diminishing, as shown in the curve. Such is, on the whole, the most probable representation of the earth's present temperature, at depths of from 100 feet, where the annual variations cease to be sensible, to 100 miles; below which the whole mass, or all except a nucleus cool from the beginning, is (whether liquid or solid), probably at, or very nearly at, the proper melting temperature for the pressure at each depth.

Terrestrial climate not sensibly influenced by underground heat.

(q.) The theory indicated above throws light on the question so often discussed—Can terrestrial heat have influenced climate through long geological periods? and allows us to answer it very decidedly in the negative. There would be an increment of temperature at the rate of 2° Fahr. per foot downwards near the surface, 10,000 years after the beginning of the cooling, in the case we have supposed. The radiation from earth and atmosphere into space (of which we have yet no satisfactory absolute measurement) would almost certainly be so rapid in the earth's actual circumstances, as not to allow a rate of increase of 2° Fahr. per foot underground to augment the temperature of the surface by much more than about 1°; and hence I infer that the general climate cannot be sensibly affected by conducted heat at any time more than 10,000 years after the commencement of superficial solidification. No doubt, however, in particular places there might be an elevation of temperature by thermal springs, or by eruptions of melted lava, and everywhere vegetation would, for the first three or four million years, if it existed so soon after the epoch of consolidation, be influenced by the sensibly higher temperature met with by roots extending a foot or more below the surface.

Appendix D.

Rates of increase of temperature inwards in a great enough mass of average rock, at various times after commencement of cooling from a primitive temperature of 7000°.

(*r*.) Whatever the amount of such effects is at any one time, it would go on diminishing according to the inverse proportion of the square roots of the times from the initial epoch. Thus, if at 10,000 years we have 2° per foot of increment below ground,

At	40,000	years we	should have	1°	per foot.
,,	160,000	,,	,,	$\frac{1}{2}$°	,,
,,	4,000,000	,,	,,	$\frac{1}{10}$°	,,
,,	100,000,000	,,	,,	$\frac{1}{50}$°	,,

It is therefore probable that for the last 96,000,000 years the rate of increase of temperature under ground has gradually diminished from about $\frac{1}{10}$th to about $\frac{1}{50}$th of a degree Fahrenheit per foot, and that the thickness of the crust through which any stated degree of cooling has been experienced has gradually increased in that period from $\frac{1}{5}$th of what it is now to what it is. Is not this, on the whole, in harmony with geological evidence, rightly interpreted? Do not the vast masses of basalt, the general appearances of mountain-ranges, the violent distortions and fractures of strata, *the great prevalence of metamorphic action* (which must have taken place at depths of not many miles, if so much), all agree in demonstrating that the rate of increase of temperature downwards must have been much more rapid, and in rendering it probable that volcanic energy, earthquake shocks, and every kind of so-called plutonic action, have been, on the whole, more abundantly and violently operative in geological antiquity than in the present age?

Objections to terrestrial application raised and removed.

(*s*.) But it may be objected to this application of mathematical theory—(1), That the earth was once all melted, or at least melted all round its surface, and cannot possibly, or rather cannot with any probability, be supposed to have been ever a uniformly heated solid, 7000° warmer than our present surface temperature, as assumed in the mathematical problem; and (2), No natural action could possibly produce at one instant, and maintain for ever after, a seven thousand degrees' lowering of the surface temperature. Taking the second objection first, I answer it by saying, what I think cannot be denied, that a large mass of melted rock, exposed freely to our air and sky, will, after it once becomes crusted over, present in a few hours, or a few days, or at the most a few weeks, a surface so cool that it can be

Appendix D.

Objections to terrestrial application raised and removed.

walked over with impunity. Hence, after 10,000 years, or, indeed, I may say after a single year, its condition will be sensibly the same as if the actual lowering of temperature experienced by the surface had been produced in an instant, and maintained constant ever after. I answer the first objection by saying, that if experimenters will find the latent heat of fusion, and the variations of conductivity and specific heat of the earth's crust up to its melting point, it will be easy to modify the solution given above, so as to make it applicable to the case of a liquid globe gradually solidifying from without inwards, in consequence of heat conducted through the solid crust to a cold external medium. In the meantime, we can see that this modification will not make any considerable change in the resulting temperature of any point in the crust, unless the latent heat parted with on solidification proves, contrary to what we may expect from analogy, to be considerable in comparison with the heat that an equal mass of the solid yields in cooling from the temperature of solidification to the superficial temperature. But, what is more to the purpose, it is to be remarked that the objection, plausible as it appears, is altogether fallacious, and that the problem solved above corresponds much more closely, in all probability, with the actual history of the earth, than does the modified problem suggested by the objection. The earth, although once all melted, or melted all round its surface, did, in all probability, really become a solid at its melting temperature all through, or all through the outer layer, which had been melted; and not until the solidification was thus complete, or nearly so, did the surface begin to cool. That this is the true view can scarcely be doubted, when the following arguments are considered.

Present underground temperature probably due to heat generated in the original building of the earth.

(1.) In the first place, we shall assume that at one time the earth consisted of a solid nucleus, covered all round with a very deep ocean of melted rocks, and left to cool by radiation into space. This is the condition that would supervene, on a cold body much smaller than the present earth meeting a great number of cool bodies still smaller than itself, and is therefore in accordance with what we may regard as a probable hypothesis regarding the earth's antecedents. It includes, as a particular case, the commoner supposition, that the earth was once melted throughout, a condition which might result from the collision of two nearly equal masses. But the evidence which has convinced most geologists that the earth had a fiery beginning, goes but a very small depth below the surface, and affords us absolutely no means of

Appendix D.

distinguishing between the actual phenomena, and those which would have resulted from either an entire globe of liquid rock, or a cool solid nucleus covered with liquid to any depth exceeding 50 or 100 miles. Hence, irrespectively of any hypothesis as to antecedents from which the earth's initial fiery condition may have followed by natural causes, and simply assuming, as rendered probable by geological evidence, that there was at one time melted rock all over the surface, we need not assume the depth of this lava ocean to have been more than 50 or 100 miles; although we need not exclude the supposition of any greater depth, or of an entire globe of liquid.

Primitive heating may have been throughout, or merely through a superficial layer of no greater depth than $\frac{1}{40}$ of the radius.

(*u.*) In the process of refrigeration, the fluid must [as I have remarked regarding the sun, in a recent article in *Macmillan's Magazine* (March 1862), and regarding the earth's atmosphere, in a communication to the Literary and Philosophical Society of Manchester[1]] be brought by convection, to fulfil a definite law of distribution of temperature which I have called "convective equilibrium of temperature." That is to say, the temperatures at different parts in the interior must differ according to the different pressures by the difference of temperatures which any one portion of the liquid would present, if given at the temperature and pressure of any part, and then subjected to variation of pressure, but prevented from losing or gaining heat. The reason for this is the extreme slowness of true thermal conduction; and the consequently preponderating influence of great currents throughout a continuous fluid mass, in determining the distribution of temperature through the whole.

"Convective equilibrium of temperature" defined: must have been approximately fulfilled until solidification commenced.

(*v.*) The thermo-dynamic law connecting temperature and pressure in a fluid mass, not allowed to lose or gain heat, investigated theoretically, and experimentally verified in the cases of air and water, by Dr. Joule and myself,[2] shows, therefore, that the temperature in the liquid will increase from the surface downwards, if, as is most probably the case, the liquid contracts

[1] *Proceedings*, Jan. 1862. "On the Convective Equilibrium of Temperature in the Atmosphere."

[2] Joule, "On the Changes of Temperature produced by the Rarefaction and Condensation of Air," *Phil. Mag.* 1845. Thomson, "On a Method for Determining Experimentally the Heat evolved by the Compression of Air; Dynamical Theory of Heat, Part IV.," *Trans. R. S. E.*, Session 1850-51; and reprinted *Phil. Mag.* Joule and Thomson, "On the Thermal Effects of Fluids in Motion," *Trans. R. S. Lond.*, June 1853 and June 1854. Joule and Thomson, "On the Alterations of Temperature accompanying Changes of Pressure in Fluids," *Proceedings R. S. Lond.*, June 1857.

Appendix D.

Alternative cases as to distribution of temperature before solidification.

in cooling. On the other hand, if the liquid, like water near its freezing-point, expanded in cooling, the temperature, according to the convective and thermo-dynamic laws just stated (§§ *u*, *v*), would actually be lower at great depths than near the surface, even although the liquid is cooling from the surface; but there would be a very thin superficial layer of lighter and cooler liquid, losing heat by true conduction, until solidification at the surface would commence.

Effect of pressure on the temperature of solidification.

(*w*.) Again, according to the thermo-dynamic law of freezing, investigated by my brother,[1] Professor James Thomson, and verified by myself experimentally for water,[2] the temperature of solidification will, at great depths, because of the great pressure, be higher there than at the surface if the fluid contracts, or lower than at the surface if it expands, in becoming solid.

(*x*.) How the temperature of solidification, for any pressure, may be related to the corresponding temperature of fluid convective equilibrium, it is impossible to say, without knowledge, which we do not yet possess, regarding the expansion with heat, and the specific heat of the fluid, and the change of volume, and the latent heat developed in the transition from fluid to solid.

Question whether solidification commenced at surface or centre or bottom.

(*y*.) For instance, supposing, as is most probably true, both that the liquid contracts in cooling towards its freezing-point, and that it contracts in freezing, we cannot tell, without definite numerical data regarding those elements, whether the elevation of the temperature of solidification, or of the actual temperature of a portion of the fluid given just above its freezing-point, produced by a given application of pressure, is the greater. If the former is greater than the latter, solidification would commence at the bottom, or at the centre, if there is no solid nucleus to begin with, and would proceed outwards; and there could be no complete permanent incrustation all round the surface till the whole globe is solid, with, possibly, the exception of irregular, comparatively small spaces of liquid.

(*z*.) If, on the contrary, the elevation of temperature, produced by an application of pressure to a given portion of the fluid, is greater than the elevation of the freezing temperature produced by the same amount of pressure, the superficial layer of the fluid would be the first to reach its freezing-point, and the first actually to freeze.

[1] "Theoretical Considerations regarding the Effect of Pressure in lowering the Freezing-point of Water," *Trans. R. S. E.*, Jan. 1849.

[2] *Proceedings R. S. E.*, Session 1849-50.

(*aa.*) But if, according to the second supposition of § *v*, the liquid expanded in cooling near its freezing-point, the solid would probably likewise be of less specific gravity than the liquid at its freezing-point. Hence the surface would crust over permanently with a crust of solid, constantly increasing inwards by the freezing of the interior fluid in consequence of heat conducted out through the crust. The condition most commonly assumed by geologists would thus be produced. Appendix D.

(*bb.*) But Bischof's experiments, upon the validity of which, so far as I am aware, no doubt has ever been thrown, show that melted granite, slate, and trachyte, all contract by something about 20 per cent. in freezing. We ought, indeed, to have more experiments on this most important point, both to verify Bischof's results on rocks, and to learn how the case is with iron and other unoxydised metals. In the meantime we must consider it as probable that the melted substance of the earth did really contract by a very considerable amount in becoming solid.

Importance of experimental investigation of contraction or expansion of melted rocks in solidification.

(*cc.*) Hence if, according to any relations whatever among the complicated physical circumstances concerned, freezing did really commence at the surface, either all round or in any part, before the whole globe had become solid, the solidified superficial layer must have broken up and sunk to the bottom, or to the centre, before it could have attained a sufficient thickness to rest stably on the lighter liquid below. It is quite clear, indeed, that if at any time the earth were in the condition of a thin solid shell of, let us suppose 50 feet or 100 feet thick of granite, enclosing a continuous melted mass of 20 per cent. less specific gravity in its upper parts, where the pressure is small, this condition cannot have lasted many minutes. The rigidity of a solid shell of superficial extent so vast in comparison with its thickness, must be as nothing, and the slightest disturbance would cause some part to bend down, crack, and allow the liquid to run out over the whole solid. The crust itself would in consequence become shattered into fragments, which must all sink to the bottom, or to meet in the centre and form a nucleus there if there is none to begin with.

Bischof's experiments proving contraction make it probable that the surface was never allowed to cool till solidification was very nearly complete through the interior.

(*dd.*) It is, however, scarcely possible, that any such continuous crust can ever have formed all over the melted surface at one time, and afterwards have fallen in. The mode of solidification conjectured in § *y*, seems on the whole the most consistent with what we know of the physical properties of the matter concerned. So far as regards the result, it agrees, I believe, with the view

Appendix D.

adopted as the most probable by Mr. Hopkins.[1] But whether from the condition being rather that described in § z, which seems also possible, for the whole or for some parts of the heterogeneous substance of the earth, or from the viscidity as of mortar, which necessarily supervenes in a melted fluid, composed of ingredients becoming, as the whole cools, separated by crystallizing at different temperatures before the solidification is perfect, and which we actually see in lava from modern volcanoes; it is probable that when the whole globe, or some very thick superficial layer of it, still liquid or viscid, has cooled down to near its temperature of perfect solidification, incrustation at the surface must commence.

(ee.) It is probable that crust may thus form over wide extents of surface, and may be temporarily buoyed up by the vesicular character it may have retained from the ebullition of the liquid in some places, or, at all events, it may be held up by the viscidity of the liquid; until it has acquired some considerable thickness sufficient to allow gravity to manifest its claim, and sink the heavier solid below the lighter liquid. This process must go on until the sunk portions of crust build up from the bottom a sufficiently close ribbed solid skeleton or frame, to allow fresh incrustations to remain bridging across the now small areas of lava pools or lakes.

Probable cause of volcanoes and earthquakes.

(*ff.*) In the honey-combed solid and liquid mass thus formed, there must be a continual tendency for the liquid, in consequence of its less specific gravity, to work its way up; whether by masses of solid falling from the roofs of vesicles or tunnels, and causing earthquake shocks, or by the roof breaking quite through when very thin, so as to cause two such hollows to unite, or the liquid of any of them to flow out freely over the outer surface of the earth; or by gradual subsidence of the solid, owing to the thermodynamic melting, which portions of it, under intense stress, must experience, according to views recently published by Professor James Thomson.[2] The results which must follow from this tendency seem sufficiently great and various to account for all that we see at present, and all that we learn from geological investigation, of earthquakes, of upheavals, and subsidences of solid, and of eruptions of melted rock.

[1] See his Report on "Earthquakes and Volcanic Action." British Association Report for 1847.

[2] *Proceedings of the Royal Society of London*, 1861, "On Crystallization and Liquefaction as influenced by Stresses tending to Change of Form in Crystals."

(*gg.*) These conclusions, drawn solely from a consideration of the necessary order of cooling and consolidation, according to Bischof's result as to the relative specific gravities of solid and of melted rock, are in perfect accordance with §§ 832...849, regarding the present condition of the earth's interior,—that it is not, as commonly supposed, all liquid within a thin solid crust of from 30 to 100 miles thick, but that it is on the whole more rigid certainly than a continuous solid globe of glass of the same diameter, and probably than one of steel. Appendix D.

END OF VOL. I.

EDINBURGH : T. CONSTABLE,
PRINTER TO THE QUEEN, AND TO THE UNIVERSITY.

www.ingramcontent.com/pod-product-compliance
Lightning Source LLC
LaVergne TN
LVHW012209040326
832903LV00003B/208